D1333241

Fiber-Optic Systems for Telecommunications

Fiber-Optic Systems for Telecommunications

Roger L. Freeman

A JOHN WILEY & SONS, INC., PUBLICATION

This book is printed on acid-free paper. ♾

Published simultaneously in Canada.

For ordering and customer service, call 1-800-CALL-WILEY

Library of Congress Cataloging-in-Publication Data is available.

ISBN 0-471-41477-8

Printed in the United States of America

10 9 8 7 6 5 4 3 2 1

CONTENTS

PREFACE

Fiber optics has come a long way in the last 20 years. It is the transmission medium of choice for terrestrial systems. In my telecommunication seminars I tell the attendees that optical fiber has an infinite bandwidth for all intents and purposes. Of course it does not, but for those who are entering the field from the wireless/radio arena, it surely seems so. In fact the entire usable radio-frequency spectrum can be placed on a single fiber with a lot of spectral space to spare.

The worldwide demand for bit rate capacity seems to double every three years. Only optical fiber can satisfy the transport of such information capacity requirements. The amount of optical fiber cable that is going into the ground and under the sea is truly breathtaking. Telecommunication systems engineers have to design and install this myriad of fiber-optic systems that have gone in and that are being planned for now and years to come. Once installed, these systems have to be maintained like any other.

This book addresses the needs of these network architects, engineers, and craftsmen who plan, install, and maintain fiber-optic systems. In early chapters, enough of fiber-optic theory is instilled into the reader to allow the remainder of the chapters to concentrate on practice. A good fourth-year high-school mathematics course certainly will provide enough bases for the math level employed throughout the text. No more than algebra, plane trigonometry, and logarithms are required. A semester course in the physics of optics would be helpful, but certainly not mandatory.

The book is well-documented. I identify my sources. Many of these sources are very familiar such as Telcordia with their excellent design documents, EIA/TIA standards and test procedures, and ITU-T documentation, and, in particular the G.600 and G.900 series of recommendations. I have also reached into the industry for assistance. Among the list of supporters are familiar names such as Alcatel Raleigh, Times Fiber and Cable, Corning Cable Systems, Belden, and AMP.

There is a tutorial flavor throughout the text. A fiber link is either loss-limited or dispersion-limited. I address these two issues to show how best to reduce link loss or how to overcome dispersion of a digital signal. The causes of loss and dispersion are examined. This paves the way to an appreciation of how best to design links where we can minimize the effects of these two overriding impairments.

I have had several consulting projects on large optical fiber systems. On each of these projects I learned a lot. System availability tops the list. Thus, I stress this issue and come at it from a number of angles. Should force majeure events[1] count against us? We lie to ourselves and to our customers if we do not count such occurrences.

Also, during these practical exercises of calculating availability from both theoretical and practical aspects, I found out the importance of maintaining a continuous supply of prime power, both 48 VDC and 120/240 VAC. Because of its importance for effective availability, I decided to dedicate a chapter to the subject. Of course, the heart of the prime power chapter is no-break power. A discussion of remote power sources where no commercial power is available is also offered. We suggest that gas turbine generation might be one of the most cost-effective and efficient alternatives for this application.

A complete chapter is devoted to DWDM (dense wave-division multiplexing), the great capacity multiplier. An overrriding question is minimum separation of light carriers. During the period of preparation of this book, we stood at 160 light carriers on one fiber and counting.

Discussion of light transmission impairments pervades the text. Yet this grouping of topics is so important that a whole chapter is dedicated to this family of critical issues. Topics such as Brillioun scattering, Raman scattering, four-wave mixing, and other nonlinear effects are among the 14 impairments to light transmission covered.

A major market for optical fiber is the CATV industry with hybrid fiber-coaxial cable (HFC) systems. These systems have brought cable television from a simple entertainment medium to a formidable player in the two-way broadband digital market, especially the internet. Fiber technology has its shining day with this technology application of fiber-optic transmission. The chapter contains a short introduction to cable television and then proceeds to introduce digital transmission and the application of fiber optics nearly to the last mile of residential and small/medium-size business users.

My goal in this book is to address a wide audience. I explain difficult concepts in simple terms. What should a potential user look for in performance parameters in a well-designed fiber-optic system? That is the primary objective of the designer. Then what are the most cost-effective means to reach that goal? The answers to these two primary questions are the mainstay of this text.

ROGER L. FREEMAN

[1]Force majeure events are "acts of God" such as earthquake, avalanche, flood, and so on.

ACKNOWLEDGMENTS

I was not alone in assembling this text. I had a whole army of helpers. It all started in Alaska on our fiber line project where we helped to troubleshoot the design and improvement plan, even recommended fiber burial depth for an operational system nearly 1000 miles long, much of it across tundra. Besides my two partners on the fiber line project, Ronald Brown and John Lawlor, both independent consultants in telecommunications, there were many others that gave me invaluable help in the preparation of the manuscript. Perhaps topping that list is David Heimke of Palmer, AK, who is another independent consultant. A close second was Karl Ebhardt of Alcatel (Raleigh-Durham) who provided me system handbooks of their gear. David Charlton of Corning critiqued my outline and offered many fine suggestions. Chris Currie of IEEE-USA helped open some doors for me. John Bellamy, who wrote *Digital Telephony* (Wiley), reviewed the book outline and caught me in some missing topics. Joe Golden of Telica Systems, Marlborough, MA made some cogent suggestions, as did my son, Bob Freeman. Stephen Sattler gave us an excellent tour of the existing fiber line facility above North Pole (AK). Steve is now with Qwest.

Shelly Schmitt, Chairman of the IEEE Alaska Section, invited me to give a paper on *bandwidth*, one of my favorite subjects. If one is interested in the supply of bandwidth, fiber optic transmission must be your keynote. The preparation of the paper made me look behind some doors I would never have thought of. Thank you, Shelly, for giving me that little push. Shelly is a "60-Hzer" but we will forgive her for that. She does keep that IEEE section on its toes.

There were others from Corning who facilitated some of the wealth of knowledge available in the upstate New York town. Among these are Judy AuPont and Mike Banach, specifically in clarifying this issue of Raman gain and how it is measured. Dr. Alan Evans who kindly provided several specific figures dealing with Raman amplifiers.

Dr. Bob Shine of Wavesplitter Technologies, Inc. of Fremont, CA, opened up the concept of de-interleavers for me. I acknowledge the gracious support of Dr. John Panagiotopoulos, the technical director of Iolon, Inc., in the arena of Bragg gratings and Bragg reflector lasers. Eric Pineau and Pierre Talbot of INO in Quebec City supplied a neat comparison of EDFA operation at 1536 and 1550 nm. My thanks to Luc Ceuppens of Calient Networks for figures dealing with multiprotocol lambda switching and to Dr. Yiquin Hu for the rapid turnaround on my request for information on M × N switching by wavelength selection. I am also indebted to Zeke Kruglic of OMM Inc. for his ideas on the MEMS approach to light wave switching.

Marshall Call of Agilent Technologies supplied excellent reproducible eye pattern diagrams. I am indebted to Prof. Mark Sceats of Australian Photonics PLC for numerous drawings of passive light wave components and to Kevin Green, COO of ExceLight Communications (Sumimoto Electric Co.) for data provided on operational characteristics of fiber Bragg filters. Chuck Grothaus of ADC Communications gave me a gold mine of information on HFC networks. I also wish to acknowledge the kind cooperation from Guy Paul Allard of EXFO and the fine little book they produced, "Guide to WDM Technology Testing." Also, may I say *thank* you to Manny Schrecter of IBM for information on fiber connectors. Susan Hoyler of TIA was there for me when I needed her support. To those many people from the light wave industry that I may have overlooked naming, please accept my apologies. The response was tremendous. Many thanks to you all.

My son-in-law, Peter Sills, introduced me to fiber installation activities in the Phoenix area to see how fiber worked out in a MAN environment. This brings up the fact that hybrid fiber-coax (HFC) has become the heart of a CATV transmission installation. I wish to thank Mike Schwartz of Cable Labs for answering some crucial questions on such CATV installations. Dr. Ted Woo of the Society of Cable Telecommunication Engineers kept me informed on two-way broadband transmission on CATV and supplied me with DOCSIS specification.

1

INTRODUCTION TO FIBER-OPTIC TRANSMISSION

1.1 DEMAND FOR BANDWIDTH

Optical fiber is the transmission medium of choice for terrestrial telecommunications. It can transport huge quantities of information. If we equate spectral bandwidth with bit rate capacity, assuming one bit per hertz of occupancy, its capacity seems to be nearly infinite. In fact, the entire usable radio-frequency spectrum[1] can be transported on a single fiber strand.

Fiber lends itself well to digital transmission. For example, coaxial cable and wire-pair transmission require far more repeaters per unit length than does optical fiber, from 20:1 to 100:1. As a result, the accumulation of jitter is far less with fiber-optic transmission than with transmission systems based on copper. This is because the buildup of systematic jitter is a function of the number of repeaters in tandem.

With current technology, fiber capacity can be as high as 10 Gbps per bit stream.As many as 80 bit streams using wave division multiplex can be accommodated on a single glass strand of fiber. By simple multiplication, a single fiber can carry 800 Gbps. During the early period of the lifetime of this book, expect 40 Gbps on a bit stream. Thus with 80 bit streams on a fiber, we are looking at 3200 Gbps on a glass strand. Suppose a fiber-optic cable has 24 strands where we allow 4 spares, leaving 20 strands. For full duplex symmetric operation, there would be 10 transmit strands in one direction and 10 in the other. We have 3200 Gbps per strand, times 10, making the total cable capacity of 32,000 Gbps. This should satisfy the demand for bit rate capacity for some time to come.

With some sophisticated bit packing techniques, an 18-GHz radio/wireless carrier 40 MHz wide can carry 655 Mbps with present technology. If we allow 10 radio-frequency (RF) carriers in one direction and 10 in the other, the total bit rate transport capacity of the system is 6 Gbps, or 1/500th of its fiber counterpart. Of course on the fiber-optic system, with current transmission methods, no bit-packing techniques are employed.

[1] Assume this spectrum lies between 3 kHz and 200 GHz.

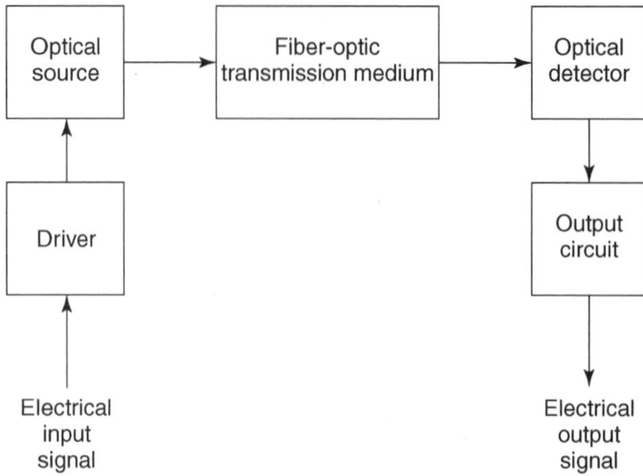

Figure 1.1. A simplified fiber-optic communication link model.

1.2 FIBER-OPTIC SYSTEM MODEL

Figure 1.1 gives a simple fiber-optic system model. We will refer back to this model repeatedly throughout the text. With a little imagination we can see that a fiber-optic link is analogous to a radio system or wire system.

Table 1.1 highlights the analogy comparison. The table sets the stage for the reader when progressing through the text. It shows that in many respects, fiber-optic transmission is not all that different from copper line and radio transmission.

Operation may be analog or digital. Many cable television systems use an analog format; more and more are turning to a digital format. Another analog application is to transport RF signal(s) in their native waveform format without frequency translation.

Turning back to Figure 1.1, we briefly review the function of each block in the diagram. Moving from left to right, the driver converts the digital waveform to either NRZ, RZ, or Manchester-coded. It also sets the required DC level of the incoming pulses. Waveform and coding are covered in Chapter 4.

TABLE 1.1 Analogy Comparison

Fiber-Optic Link	Radio/Wireless or Wireline Link	Comments
Driver	Modulator or signal conditioner	All three cases require some form of waveform manipulation such as AMI to NRZ
Source	Signal source in transmitter or modem	Signal source output commonly low level, except SatCom
Fiber-optic strand transmission medium	RF signal through atmosphere or RF or audio signal over a copper medium	
Optical detector	Receiver or modem demodulator	Receiver threshold in all three cases determines performance
Output circuit	Receiver or modem output and waveform conditioner	

The principal thrust of Chapter 4 is the fiber-optic light source. Some texts may call the source a transmitter. There are two distinct types of light sources in common use today: LED (light-emitting diode) and LD (laser diode). Both are comparatively low level devices with outputs in the range of −10 dBm to +6 dBm. They use intensity modulation, which, for our initial discussion here, we'll call on–off modulation.

The source is connected to a light detector at the distant end by means of a strand of optical fiber that is contained in a cable with other fiber strands enclosed by various means of protection. Fiber-optic cable is discussed in detail in Chapter 2. The optical glass strands inside the cable may be of the monomode or multimode varieties. A strand's physical dimensions determine which mode it is designed for. There are economic and performance trade-offs that will determine which type of fiber to employ on a certain project.

Fiber-optic cable comes on reels of cable sections that are 1 km, 2 km, 5 km, and 10 km in length. Connectors are used at the bitter ends of a cable length to connect the cable to the source and to the detector. For longer links, several reels of cable will probably be required. These cable lengths are connected one to another by splices. Splices and connectors are covered in Chapter 3. Two important parameters must be considered with splices and connectors. These are insertion loss and return loss. Insertion loss due to a splice should be less than 0.1 dB, and for a connector it should be less than 1 dB. Return loss (reflection loss), giving a measure of the impedance match between the splice and the cable, should be greater than 30 dB or higher.

The receiver or light detector at the far end of a fiber-optic link is a photon counter. On most fiber-optic systems, there are two types of receivers in common use today: PIN (*p*-intrinsic-*n*) diode and APD (avalanche photodiode). The PIN diode is simpler and less sensitive to the environment. It has no inherent gain. An APD is more complex; it is more sensitive to the environment and can provide 10- to 20-dB gain. The fiber-optic link designer selects a receiver threshold for the desired link bit error rate (BER). The threshold is an input power level in the negative dBm range which is a function of the receiver type and, to some degree, receiver design, the bit rate, and, of course, the desired BER. In system design, care must be taken that the signal level entering a detector is not excessive. On short links, often a light attenuator is added in series to place the incoming signal level in the desired range (Ref. 1).

Table 1.2 compares impairments and expected digital performance: fiber-optic transmission, radio/wireless transmission, and wireline transmission.

1.2.1 Operational Wavelength Bands for Fiber-Optic Transmission

Radio, wire, and cable transmission systems use *frequency* to describe the operational region of the radio-frequency spectrum to be utilized. Of course, frequency is measured in hertz. It is said that because fiber-optic transmission was researched and developed by physicists, *wavelength* is used to describe the location of operation in the spectrum.

Allow that light is an extension of the radio-frequency spectrum at the high end. This concept of continuous spectrum is illustrated in Figure 1.2. The notation commonly used for wavelength is λ. It is *length*, and its basic measurement unit is

TABLE 1.2 Performance and Impairments Comparison

Performance/Impairment	Radio/Wireless	Wireline	Fiber Optics
Bit error rate	1×10^{-9}	1×10^{-10}	1×10^{-12}
Link loss (dB)	Principal impairment	Principal impairment	Principal impairment
Dispersion	Can be principal impairment at high bit rates	Secondary or tertiary impairment	At high bit rates, may be principal impairment
Fading	Yes	No	No
Jitter accumulation	Medium	High	Low
Vulnerability	Low	Medium	High
Bit rate capacity	Low/medium	Low/medium	Very high
Rainfall absorption	Above 10 GHz, major impairment	No	No
EMC susceptibility	Yes	Yes	No
EMC emanation	Yes	Some	No

Notes

Bit error rate (BER). Fiber-optic links are commonly designed to 1×10^{-12} BER.

Fiber-optic links are either loss-limited or dispersion-limited. High-bit-rate radio/wireless systems may be dispersion-limited. Space diversity and automatic IF equalizers can accommodate the dispersion. The deleterious effects are the same, namely, intersymbol interference resulting in degraded to severely degraded error performance. Wirelines are loss-limited.

Vulnerability. Wireline and fiber-optic systems are vulnerable to accidental or purposeful severing of cable. Both may suffer environmental damage such as water damage, then freezing.

The major disadvantage of a fiber-optic system is its vulnerability. In Chapter 9 we discuss ways to mitigate, but not completely eliminate, the vulnerability problem.

Rainfall absorption. The wider bandwidths on radio/wireless systems are only permitted on frequencies above 10 GHz, resulting in limiting link length because of rainfall absorption of the signal; the higher the frequency, the greater the limitation for a given time availability (i.e., propagation reliability). Of course, time availability is not an issue on wireless or fiber-optic systems.

EMC (eletromagnetic compatibility) has two subsets: susceptibility and emanation. Emanation means that it can be a source of radio-frequency interference (RFI). Susceptibility means that it is susceptible to RFI. For radio systems, both emanation and susceptibility can be, and often are, major issues. Wireline systems certainly are susceptible to RFI. A fiber-optic system neither emanates nor is susceptible (Ref 2.).

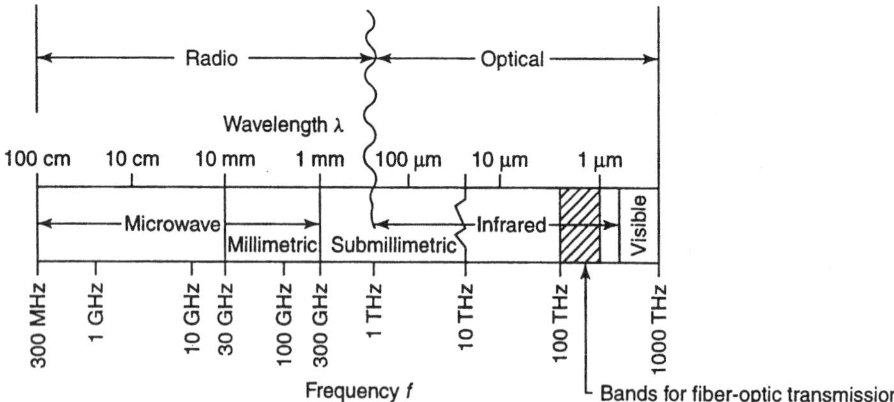

Figure 1.2. The frequency spectrum above 300 MHz showing the general spectrum location of light transmission on fiber optics.

the meter. We can relate frequency (Hz) and wavelength (meters) by the traditional formula

$$F\lambda = 3 \times 10^8 \text{ m/s} \qquad \text{(the velocity of light in a vacuum)} \tag{1.1}$$

Again, F is in hertz and λ is in meters.

Examples. Your favorite FM station for music is at 104 MHz. What is the equivalent wavelength?

$$104 \times 10^6 \lambda = 3 \times 10^8$$

$$\lambda = 3 \times 10^8 / 104 \times 10^6$$

$$= 2.8846 \text{ m}$$

The operational wavelengths in fiber optics are commonly given in nanometers (nm). A nanometer is as follows: 1 nm = 1×10^{-9} m, or 0.000000001 m.

A popular wavelength for fiber-optic operation is 1310 nm. This is where, by its very nature, dispersion is at a minimum. For 1310 nm, what is its equivalent frequency?

$$1310 \times 10^{-9} F = 3 \times 10^8 \text{ m/s}$$

$$F = 3 \times 10^8 / 1310 \times 10^{-9}$$

$$= 2.29 \times 10^{14} \text{ Hz, or } 2.29 \times 10^5 \text{ GHz or 229 THZ}$$

Figure 1.3 shows the three basic windows of operation for optical fiber. These are

- 820–900 nm
- 1280–1350 nm
- 1528–1561 nm

This last window can be extended to 1620 nm. Looking at the extension of this latter band, for example, to impress our radio-frequency friends, consider using equation (1.1). Calculate the frequency at

$$F = 3 \times 10^8 / 1528 \times 10^{-9}$$

$$= 1.96 \times 10^{14} \text{ Hz} = 196 \text{ THz}$$

Calculate the frequency at the other end of the band (i.e., at 1620 nm)

$$F = 3 \times 10^8 / 1620 \times 10^{-9}$$

$$= 185 \text{ THz}$$

By subtraction we see that the useful band of operation at this window is 11 THz, or 11,000 GHz. This is about 110 times the useful radio-frequency spectrum (Ref. 1).

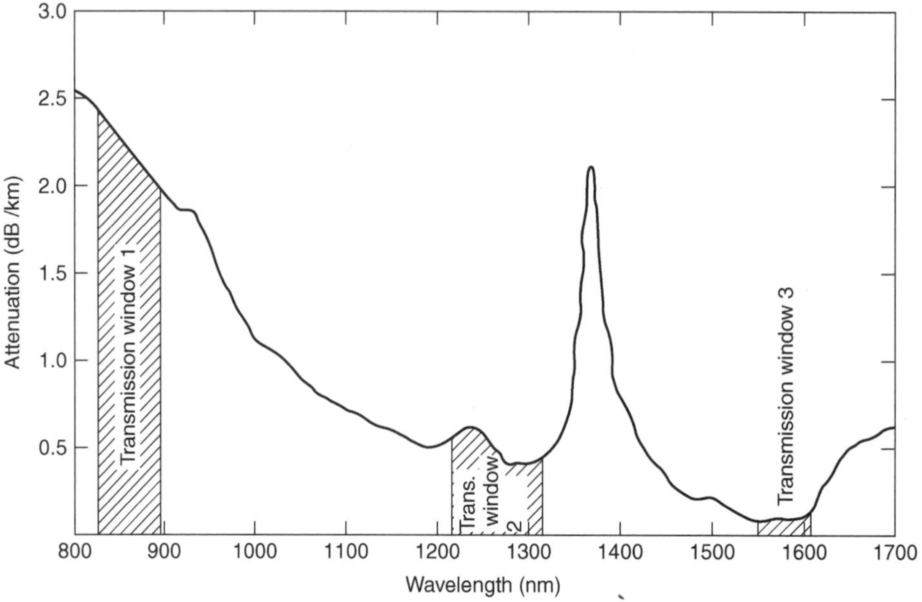

Figure 1.3. Optical fiber attenuation versus wavelength showing the three common transmission windows.

The useful bandwidth of a single light emission is determined by the impulse response function of the fiber in question. A mathematical derivation of the impulse response of a fiber is complex and rigorous and beyond the scope of this book. But in simple terms we will provide some relationships. Let B_o be the bandwidth of the optical fiber and let B_c be the resulting bandwidth of the signal after detection (i.e., the electrical signal). In that the optical bandwidth of the fiber is determined by the impulse response of the fiber, it can be shown that the 3-dB optical bandwidth B_o is related to the full width at half-maximum (FWHM) of the function by

$$B_o = 441/\text{FWHM} \tag{1.2}$$

assuming that the function is Gaussian in shape, where B_o is in megahertz and FWHM is in nanoseconds.

It further can be shown that the rise time, t, is related to the optical bandwidth, B_o, by

$$t = 315/B_o \tag{1.3}$$

Optical bandwidth is defined in a similar manner as the bandwidth of an RF signal, defined by its 3-dB power points. This relates directly to the current, I, in the optical detector. However, the electrical power generated in the detector, as we know, is proportional to I^2; therefore, a 3-dB drop in optical power (indicated by a 50% drop in I) results in a 6-dB drop in electrical power (indicated by a 75% drop in I_2). Thus a 3-dB optical bandwidth is equal to a 6-dB electrical bandwidth.

This is not used and will not be defined further. It follows, then, that the 3-dB electrical bandwidth must be smaller than the 3-dB optical bandwidth. While the mathematics of the function are not simple, if the function is Gaussian in shape, it can be shown that

$$B_c = B_o/\sqrt{2} = 0.707B_o \tag{1.4}$$

In Figure 1.3, the reader should take note of the so-called "water" absorption line at about 1400 nm. Water is the cause of fiber impurity, but it is actually the OH^- radical absorption line to which we refer. There is high attenuation in the region around 1400 nm as a result (Ref. 3).

1.3 FIBER-OPTIC LIGHT-GUIDE AS A TRANSMISSION MEDIUM

1.3.1 Light-Guide Construction

The fiber strand may be called a *light-guide*. One can guess that the term has been robbed from the radio folks and their waveguide. Figure 1.4 illustrates a strand of fiber and its component parts. Of course the figure is exaggerated and idealized to make several points. As we can see, a fiber strand or light-guide basically consists of an inner core surrounded by a cladding. Any additional coverings are protective. The figure shows a plastic covering.

Conventionally, the index of refraction of the core is assigned the notation n_1, and the index of refraction of the cladding is assigned the notation n_2. These are important parameters that lead to our discussion below. When a strand of fiber so constructed such that $n_1 > n_2$, the structure (core and cladding) will act like a waveguide. Silica glass (SiO_2) is the base material for both the core and cladding. Dopants such as boron or germanium are used to adjust the refractive index.

As we remember from physics, the index of refraction equals the speed of light in a vacuum divided by the speed of light in the medium of interest. The refractive index in a vacuum equals 1, by definition.

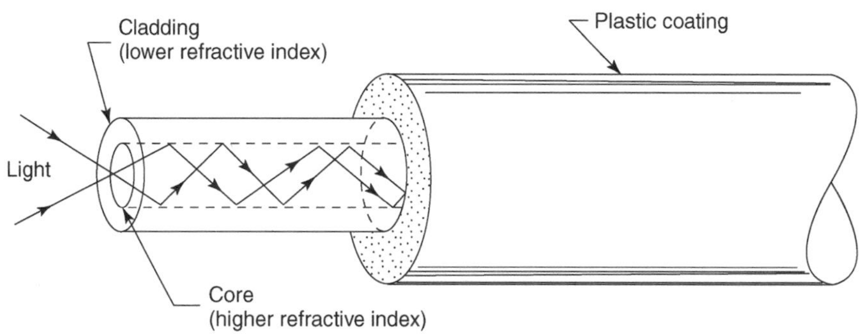

Figure 1.4. Basic composition of optical fiber.

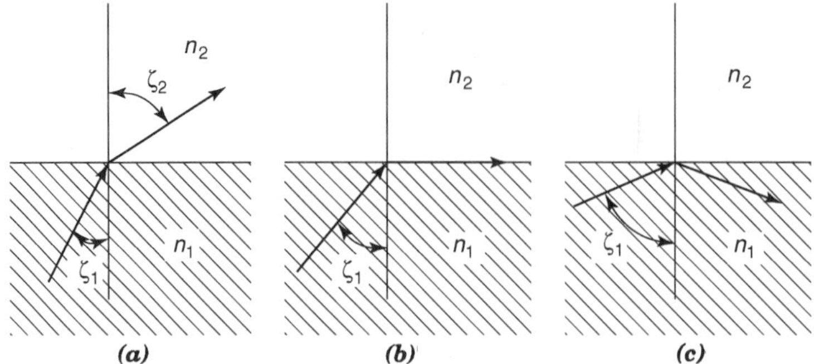

Figure 1.5. Ray paths for several angles of incidence, $n_1 > n_2$, where n_1 and n_2 are indices of refraction of the different media. (From Figure 4, page 15, ref. 2.)

1.3.2 How Light Is Propagated in a Fiber Light-Guide

The practical propagation of light through an optical fiber is best explained using ray theory and Snell's law. Simply stated, we can say that when light passes from a medium of higher refractive index into a medium of lesser refractive index, the refracted ray is bent away from the normal. For instance, a ray traveling in water and passing into an air region is bent away from the normal to the interface between the two regions. As the angle of incidence becomes more oblique, the refracted ray is bent more until finally the refracted ray emerges at an angle of 90° with respect to the normal and just grazes along the surface. Figure 1.5 shows the various incidence angles. Figure 1.5a illustrates an angle of incidence where the ray of light escapes entirely into free space. Figure 1.5b illustrates what is called the critical angle, where the refracted ray just grazes along the surface. Figure 1.5c is an example of total reflection. This is when the angle of incidence exceeds the critical angle. A glass fiber, when utilized for the transmission of light, requires total internal reflection.

Another property of the fiber for a given wavelength λ is the normalized frequency V; then

$$V = \frac{2\pi a}{\lambda}\sqrt{n_1^2 - n_2^2} \approx n_1\sqrt{2\Delta} \tag{1.5}$$

where a = the core radius, n_2 for unclad fiber = 1, and $\Delta = (n_1 - n_2)/n_1$.

The term $\sqrt{n_1^2 - n_2^2}$ in equation (1.5) is called the numerical aperture (NA). In essence, the numerical aperture is used to describe the light-gathering ability of a fiber. In fact, the amount of optical power accepted by a fiber varies as the square of its NA. It is also interesting to note that the numerical aperture is independent of any physical dimension of the fiber.

To better understand numerical aperture (NA), consider Figure 1.6, which illustrates the acceptance cone. As shown below in the figure, $\sin\theta_A = \text{NA}$. The concept of light-gathering capability of a fiber, expressed numerically by the NA, is well illustrated graphically with the acceptance cone.

As shown in Figure 1.1, there are three basic elements in an optical fiber transmission system: the source, the fiber link, and the optical detector. Regarding the fiber link, there are two basic design parameters that limit the length of a link

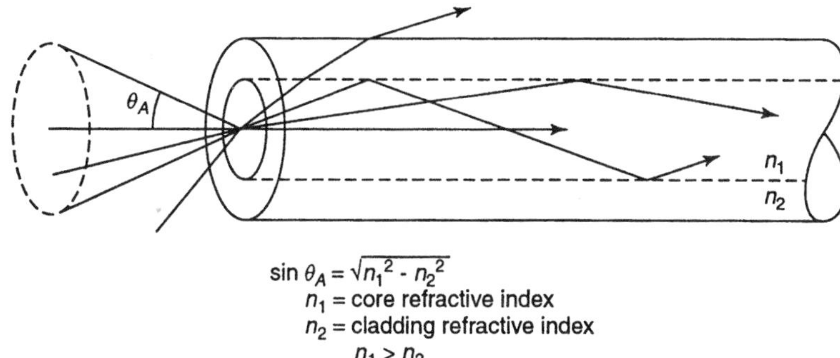

$\sin \theta_A = \sqrt{n_1^2 - n_2^2}$
n_1 = core refractive index
n_2 = cladding refractive index
$n_1 > n_2$

Figure 1.6. Acceptance cone.

without repeaters, or limit the distance between repeaters. These most important parameters are loss, usually expressed in dB/km, and dispersion, which is often expressed as an equivalent bandwidth–distance product in MHz/km. A link may be power limited (loss limited) or it may be dispersion limited.

Dispersion, manifesting itself with intersymbol interference at the far end, is brought about by two factors. One is material dispersion and the other is modal dispersion. Material dispersion is caused by the fact that the refractive index of the material changes with frequency. If the fiber waveguide supports several modes, we have a modal dispersion. Because the different modes have different phase and group velocities, energy in the respective modes arrives at the detector at different times. Consider that most optical sources excite many modes; and if these modes propagate down the fiber waveguide, delay distortion (dispersion) will result. The degree of distortion depends on the amount of energy in the various modes at the detector input.

One way of limiting the number of propagating modes in the fiber is in the design and construction of the fiber waveguide itself. Return again to equation (1.5). The modes propagated can be limited by increasing the radius a and keeping the ratio n_1/n_2 as small as practical, often 1.01 or less.

We can approximate the number of modes N that a fiber can support by applying formula (1.6). If $V = 2.405$, only one mode will propagate (HE_{11}). If V is greater than 2.405, more than one mode will propagate, and when a reasonably large number of modes propagate we obtain

$$N = \left(\tfrac{1}{2}\right)V^2 \tag{1.6}$$

Dispersion is discussed in more detail later in Chapter 7.

REFERENCES

1. Roger L. Freeman, *Telecommunication Transmission Handbook*, 4th ed., John Wiley & Sons, New York, 1998.
2. *Fiber Optics System Design*, MIL-HDBK-415, US Department of Defense, Washington, DC, 1985.
3. Govind P. Agrawal, *Fiber-Optic Communication Systems*, 2nd ed., John Wiley & Sons, New York, 1997.

2

FIBER-OPTIC CABLE

2.1 TYPES OF FIBER

There are three generic types of fiber as distinguished by their mode characteristics and physical properties:

1. Single-mode (monomode) fiber
2. Step-index (multimode) fiber
3. Graded-index (multimode) fiber

2.1.1 Defining Core Diameters

Figure 2.1 is a graphical representation of a multimode fiber (left) and of a single-mode fiber (right). The principal message of the figure is the comparative dimensions of single-mode versus multimode fiber.

Note in Figure 2.1 that the fiber diameter is the same, whether monomode or multimode, nominally 125 μm. However, there is a drastic difference between the core diameters, 50 μm for multimode fiber and between 8.6 and 9.5 μm for monomode (single-mode) fiber. Another common core diameter for multimode fiber is 62.5 μm with 125-μm fiber diameter.

2.1.2 Three Types of Optical Fiber

Figure 2.2 illustrates the fiber construction and refractive index profile for step-index fiber (Figure 2.2a) and graded-index fiber (Figure 2.2b). Step-index fiber is characterized by an abrupt change in refractive index, and graded-index fiber is characterized by a continuous and smooth change in refractive index (i.e., from n_1 to n_2).

Step-index multimode fiber is more economical than graded-index fiber. For step-index fiber, the multimode bandwidth–distance product, the measure of dispersion discussed previously, is on the order of 10–100 MHz/km. With repeater spacings at some 10 km, only a few megahertz of bandwidth is possible.

Graded-index fiber is more expensive than step-index fiber, but it is one alternative for improved bandwidth–distance products. When a laser diode is used as a

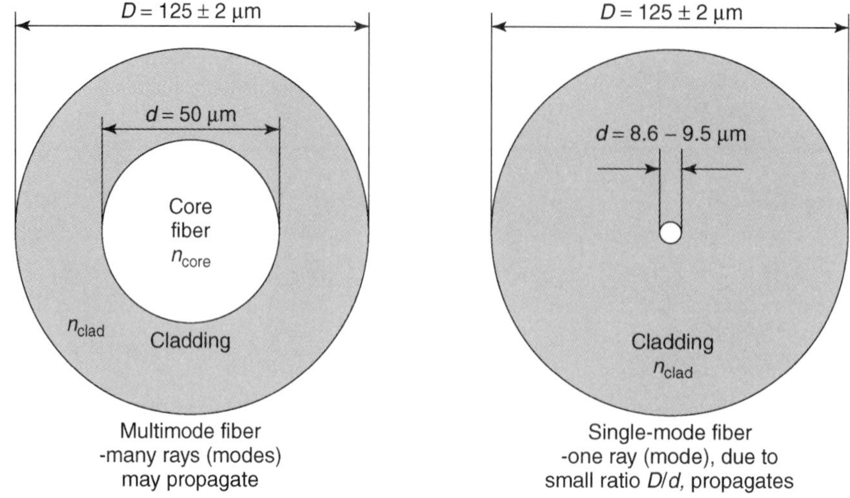

Figure 2.1. Cross section of a multimode fiber (left) and a single-mode (monomode) fiber (right).

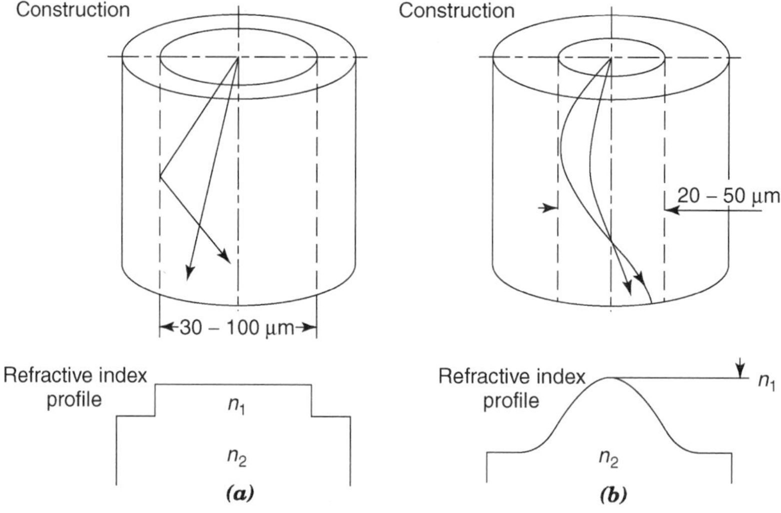

Figure 2.2. Construction and refractive index properties for (a) step-index fiber and (b) graded-index fiber.

light source, values of 400–1000 MHz/km are possible. But if an LED source is used with its much broader emission spectrum (see Chapter 4 for a discussion of emission spectrum), bandwidth–distance products with graded-index fiber can be achieved up to only 300 MHz/km or greater. The principal limiting factor of achievable bandwidth in this case is material dispersion. (See Chapter 6 for a discussion of material dispersion.)

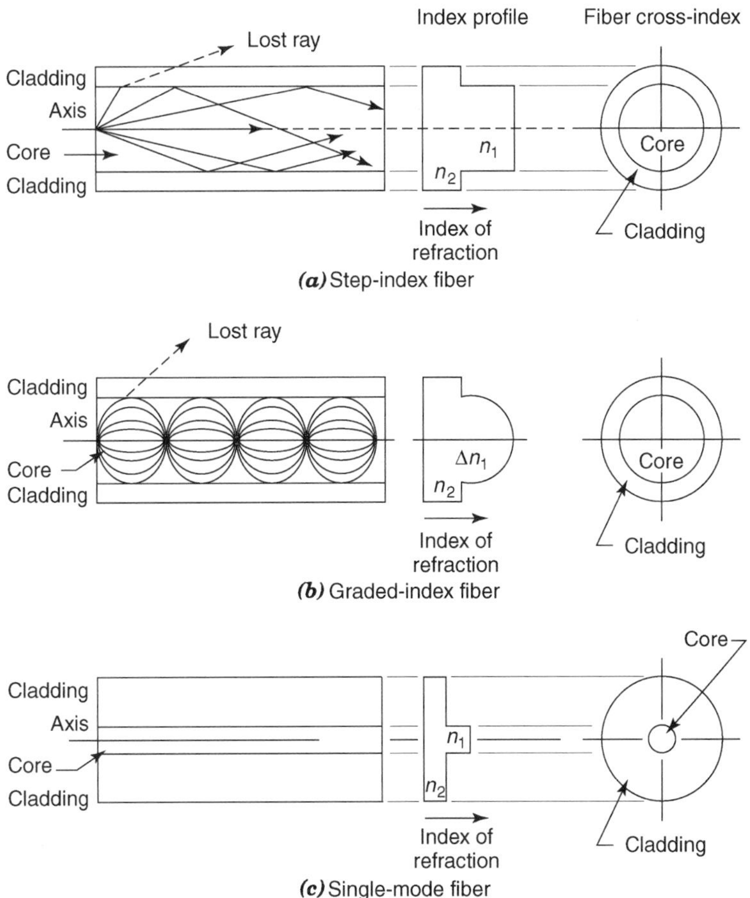

Figure 2.3. Index profiles and modes of propagation for the three types of fiber.

Figure 2.3 illustrates refractive index profiles and the resulting modes of propagation for the three types of common silica-based optical fiber.

Single-mode or monomode fiber is designed such that only one mode is propagated. To do this, $V < 2.405$ [see equation (1.5)]. Such a fiber exhibits no modal dispersion because only one mode propagates. Typically, we might encounter a fiber with indices of refraction on $n_1 = 1.48$ and on $n_2 = 1.46$. If the optical source wavelength were 820 nm, for single-mode operation, the maximum core diameter would be 2.6 μm, a very small diameter indeed. Figure 2.3c gives the refractive index profile for monomode fiber. It is this fiber that displays by far the best bandwidth–distance product of the three fiber types described.

As we discussed in Chapter 1, numerical aperture (NA) gives a measure of the light coupling between a source and the fiber core. By observation in Figure 2.1, the much larger diameter of the core in multimode fiber results in a larger NA, around 0.22, whereas the much narrower diameter of a monomode fiber inhibits light collection from the source. Here the NA runs around 0.11.

2.2 PROPAGATION OF DIFFERENT MODES THROUGH AN OPTICAL FIBER

Multimode fiber with its comparatively broad core allows several or many modes to propagate down the fiber. Some of these modes may only travel a short distance and then disappear; others may propagate the full length of the fiber. Multimode propagation is illustrated in Figure 2.4. A major problem arises when these modes reach the distant end receiver. Consider a pulse traveling down the fiber some distance. The pulse is represented by light energy of several modes. The lowest-order mode arrives at the receiver first, followed by other light energy delayed some because it or they traveled further. The arriving pulse, made up of light energy components arriving somewhat later, tend to broaden the pulse energy of the initial arriving pulse of the lowest-order mode as illustrated in Figure 2.4. The trouble is that these pulses, or lack of pulses, represent 1s and 0s. Let's say that a pulse represents a binary 1, and no pulse represents a binary 0. Suppose now we transmit an initial 1 followed by a 0. Dispersed energy from the first pulse (as shown in Figure 2.4, lower right) spills into the second bit position, which is supposed to be a zero. If enough energy spills into the space for the second bit, there is a good chance that the light detector will interpret the second bit as a binary 1 when it was supposed to be a binary 0. A bit error has occurred. This is a rather simplistic description showing the deleterious result of dispersion causing intersymbol interference.

Under these circumstances, as the bit rate is increased, the pulse width gets shorter, and the effects of dispersion are more and more deleterious and the BER on the link approaches a value that is completely unacceptable.

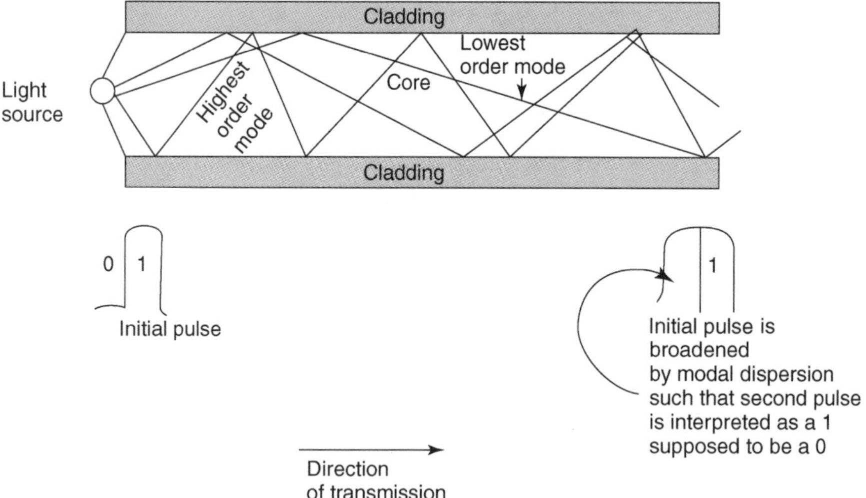

Figure 2.4. An idealized sketch of a light source into an acceptance cone of a section of multimode light-guide (optical fiber) showing three modes of optical light transmission. The lowest-order mode has just two reflections in the figure, and the highest-order mode has nearly seven reflections over the same distance traveled. As a result, the highest-order mode signal energy is delayed compared to the lowest-order mode.

This situation can be mitigated or eliminated by

- Shortening the fiber link (mitigates)
- Reducing the bit rate (mitigates and can possibly eliminate)
- Using monomode fiber (eliminates modal dispersion because only one mode propagates).

Source: Refs. 1 and 2.

2.3 MICROBENDS AND MACROBENDS

A *microbend* is a fiber imperfection. Microbends cause an increase in cable loss. This loss can result in an excessively large loss in excess of 100 dB/km in some cases (Ref. 1). A major cause of this loss occurs during cable manufacture. It is related to axial distortions that invariably occur during cabling when the fiber is pressed against a surface that is not perfectly smooth. Microbend-induced loss is a function of mode-field diameter, cable design, and cable construction. Losses due to microbend-induced attenuation consistently decrease with mode-field diameter.

A *macrobend* refers to the specified minimum bending radius. The cable manufacturer should specify the minimum bending radius. When fiber is on reels, of course it is bent around the reel. When it is installed, particularly in buildings, fiber cable must be bent around corners. The installer must not exceed the specified minimum bending radius by making a still sharper bend than the specification calls for. We might expect a typical bending radius of a fiber-optic cable to be between 10 cm (4 in.) and 30 cm (11 in.), depending to a certain degree on the fiber count in the cable. Bending a fiber-optic cable tighter than the specified bending radius can cause damage, even break the fiber carried in the cable. It can also cause a dramatic increase in fiber attenuation (Refs. 3 and 4).

Figure 2.5 illustrates and differentiates the macrobend and the microbend.

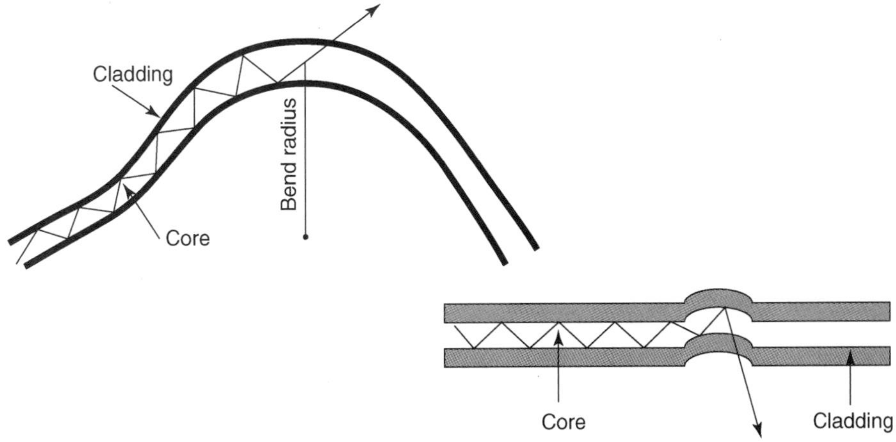

Figure 2.5. Illustrative sketches of macrobend (left) and microbend (right).

2.4 CABLE CONSTRUCTION

2.4.1 Fiber Diameters

Fiber-optic system designers and installers often refer to fiber cable sizes in the following format: "core/cladding." For instance, a fiber is specified as 50/125, and this means that the core has a diameter of 50 μm and that the cladding diameter is 125 μm. We immediately know that this fiber is of the multimode variety due to its core size. If it were monomode, the core diameter would be in the range of 7–10 μm.

The outer surface of the cladding is covered with a coating that has a diameter of 250 or 500 μm. Table 2.1 gives the basic physical parameters of the more common types of optical fiber. The 900-μm value indicates tight buffering, and the 2000- to 3000-μm values represent loose buffering.

Source: Refs. 3 and 7.

2.4.2 Tight Buffering or Loose Buffering

Buffering helps isolate fiber from external forces. Two types of buffering have been developed: loose buffer and tight buffer, as illustrated in Figure 2.6.

In the loose-buffer construction, the fiber is contained in a plastic tube that has an inner diameter considerably larger than the fiber itself, as shown in Table 2.1. The interior of the plastic tube is usually filled with a gel material.

TABLE 2.1 Basic Dimensional Parameters of Optical Fiber
[Dimensions in microns (μm)]

Type	Core	Cladding	Coating	Buffer or Tube
I	7–10	125	250 or 500	900 or 2000–3000
II	50	125	250 or 500	900 or 2000–3000
III	62.5	125	250 or 500	900 or 2000–3000
IV	85	125	250 or 500	900 or 2000–3000
V	100	140	250 or 500	900 or 2000–3000

Notes
Type I fiber, of course, is monomode fiber.
Type II fiber is multimode fiber, as are Types III, IV, and V.
Type IV fiber, 85/125, is more popular in Europe than in North America.
Type V fiber, 100/140, has the largest NA coupling the most light because of its large core diameter. Its potential bandwidth is the lowest and is used on spans of short and intermediate distance. Because of its size, it is the easiest to install, particularly when connectors are employed rather than fusion splices. Expect to find this type of fiber in buildings.

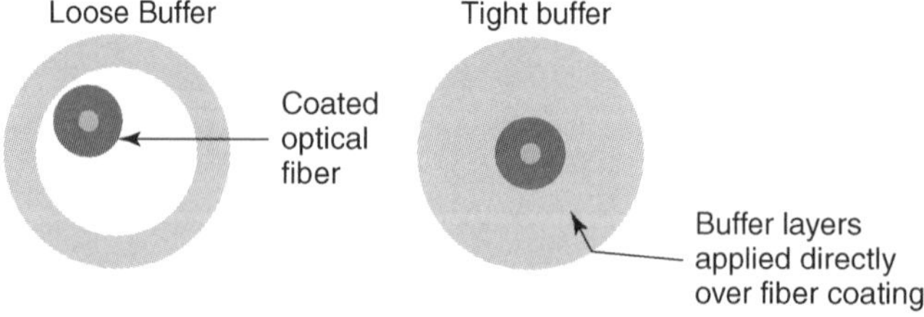

Figure 2.6. Simplified examples of loose- and tight-buffer construction.

Loose-tube construction isolates the fiber from exterior mechanical forces acting on the cable. For cables containing many fibers, a number of these tubes, each containing one or more fibers, are combined with strength members to keep the fibers free of stress.

Tight buffering is achieved during the cable manufacture using a direct extrusion of plastic over the basic fiber coating. The tight-buffer design results in lower isolation for the fiber from stresses by temperature variation. On the other side of the coin, tight-buffer construction can withstand much greater crush and impact forces without damage to the fiber.

Breakout cable is a derivative form of tight buffer construction. In breakout cable, a tightly buffered fiber is surrounded by *aramid yam* and a jacket, typically PVC. These single-fiber subunit elements are then covered by a common sheath for the breakout cable. This "cable" within a cable offers the advantage of direct, simplified connector attachment and installation.

Loose-tube cables are optimized for outside plant applications. The major constituents of an optical cable structure are silica glass and polymeric plastics. For a given temperature change, the rate/magnitude or material expansion and contraction will be different because each material possesses a different coefficient of thermal expansion. The loose-tube cable establishes a strain-free environment for the optical fiber by mitigating the influences of this effect. Loose-tube cable manufacture ensures that the optical fiber to buffer tube length ratio is controlled such that no optical fiber is compressed against the tube wall when the tube expands or contracts with changes in temperature. The strain-free environment established in the loose-tube cable design compensates for movement in the cable structure without inducing mechanical forces on the fiber. This characteristic enhances the operating temperature range of the loose-tube design. On the other hand, tight-buffered cables do not isolate the fibers from external forces to the same extent; therefore, temperature-related expansion and contraction effects applied to any component of a tight-buffered cable can be translated directly to the optical fiber. As a result, tight-buffered cables are typically more sensitive to temperature extremes and mechanical disturbances than are loose-tube cables. The tight-buffered design is well-suited for indoor application, which requires low flame spread or smoke release characteristics, and 900-μm buffered fiber is ideal for direct termination. However, the same design characteristics that make tight-buffered cable optimum for indoor applications limit its performance in an outdoor environment.

Another attribute of loose-tube fiber cable is its ability to withstand ice crush in locations where standing water and freezing temperatures coexist. Water migration inside the cable outer jacket can result in the formation of ice crystals within the optical fiber cable core. This ice will impart stresses in close proximity to the optical fibers and may result in an unacceptable increase in attenuation or even breakage. Therefore, it is essential to prevent the intrusion and uncontrolled movement of water inside the cable.

The design of loose-tube cable affords protection against water penetration and migration using two distinct water intrusion preventative measures: core water-blocking protection and buffer tube filling compound. Waterblocking protection is accomplished by surrounding the cable core with a gel and/or dry water-swellable material to stop the entry and migration of water should the cable's outer jacket be breached. This protective measure is primarily to maintain the mechanical integrity of the cable itself (e.g., prevent ice crush from within the cable, fungus growth, or corrosion of the metallic cable members when present). Filling compound provides a mechanical cushion for the fiber, allowing it to float within the buffer tube, and provides an additional barrier between the optical fiber and water/moisture in the operating environment. Standard tight-buffered cables do not have filling compounds or waterblocking protection, making them susceptible to damage caused by water penetration and migration.

Ultraviolet (UV) light protection is another matter. In aerial installation, optical fiber cable must be able to withstand direct exposure to UV sunlight. Loose-buffered cable incorporates carbon black into its jacket material to provide maximum UV protection. Tight-buffered cables do not contain carbon black in their outer jacket and thus should not be used in installations exposed to UV-rich sunlight.

Source: Ref. 6.

2.4.3 Strength Members

Strength members are an important part of a fiber-optic cable, especially during the "pulling" process used during installation. The levels of stress on the cable during pulling and other installation procedures may cause microbending losses that can result in an attenuation increase and possible fatigue effects. To transfer these stress loads in short-term installation and long-term application, the internal strength members are added to the optical cable structure.

These strength members provide the tensile load properties similar to outside plant telephone and other electronic cable structures. They keep the fiber from stress by minimizing elongation and contraction. It should be kept in mind that optical fiber stretches very little before breaking. Thus, the strength members must have low elongation at the expected tensile loads.

Three different strength member types are commonly used in cable construction: fiberglass epoxy rod, steel, and aramid. The load to break for the first two is 480 lb, and that for aramid is 944 lb. The percentage elongation before break is 3.5 for fiberglass epoxy rod, 0.7 for steel, and 2.4 for aramid. Impact resistance, flexing, and bending are other mechanical factors affecting the choice of strength members.

Source: Ref. 5.

Some typical fiber-optic cable layup cross sections are shown in Figures 2.7a and 2.7b. Figure 2.7a gives four examples of fiber-optic cable for long-haul and CATV services. This is followed by Table 2.2a, which gives the typical physical characteristics of this type of cable. The cables shown are Belden Beloptix fiber cables. All four cable examples are of loose-tube construction. Note that "matched-clad" and "depressed-clad" are discussed in Section 2.6.

Figure 2.7b gives typical characteristics of tight-buffered cables for indoor use in risers. These are Corning Cable Systems Unitized MIC® cables providing from 24 to 144 individual cabled fiber strands. The figure refers to Table 2.2b, which provides their corresponding physical and transmission characteristics.

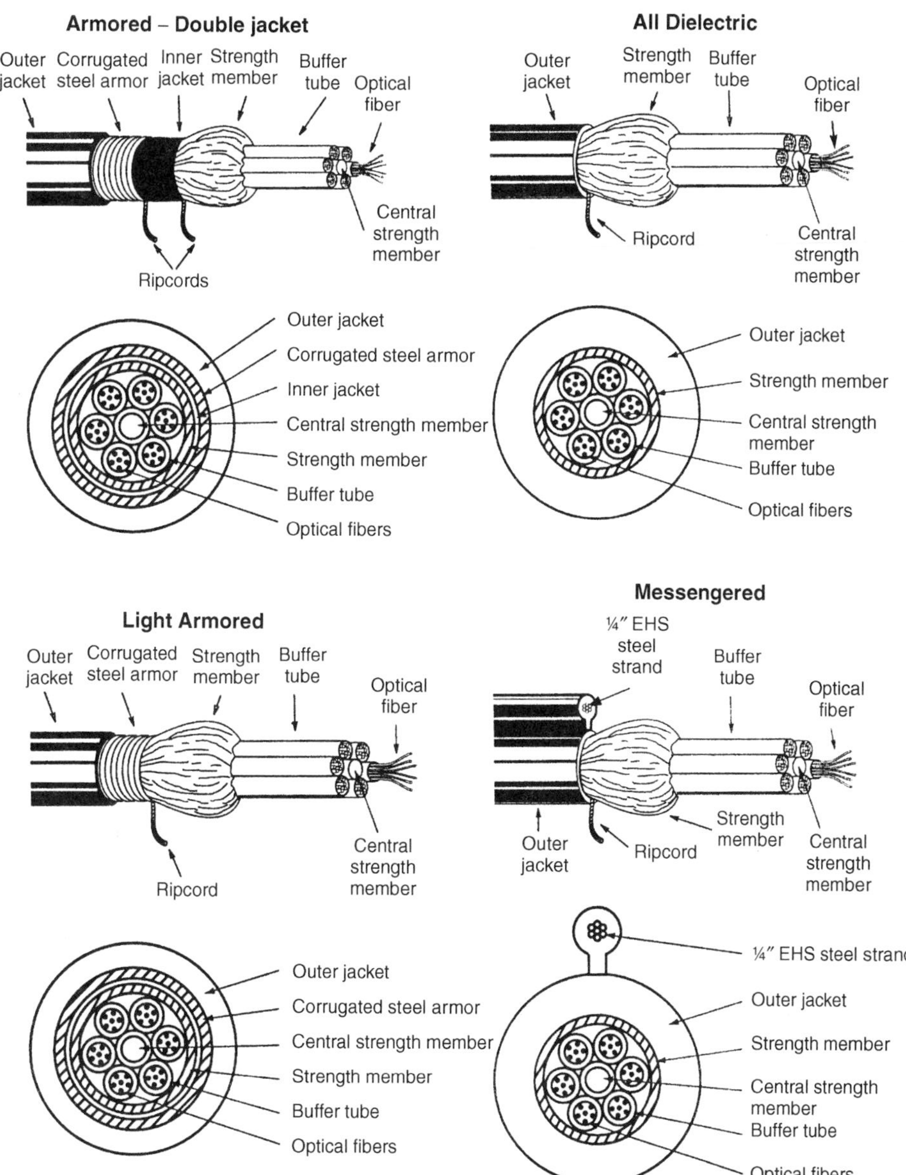

Figure 2.7a. Layup cross section of four Belden loose-buffered cables designed for outside plant application in the long-haul public/private network and CATV HFC plant. (Courtesy of Belden Wire and Cable Inc., Richmond, Indiana [Ref. 5].)

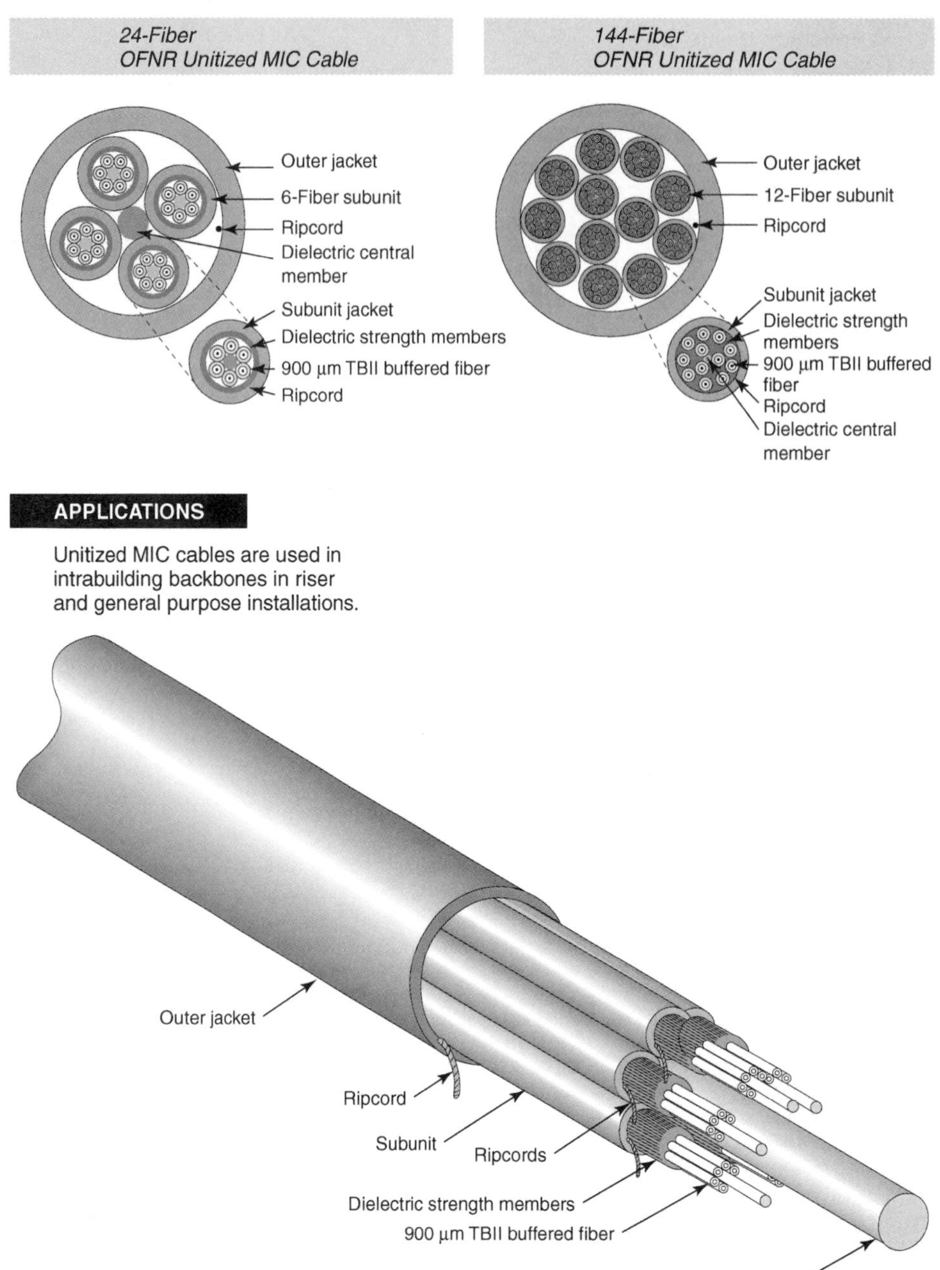

Figure 2.7b. Layup cross section and physical construction of Corning Cable Systems tight-buffered fiber-optic cables for routing inside buildings. The cables are flame-retardant and meet the required NEC codes (unitized MIC® riser cables). (Courtesy of Corning Cable Systems, Ref. 7, LANscape Solutions Catalog, page 1.30.)

TABLE 2.2a Physical Characteristics of Belden Public Network Fiber Cables

Physical Data	Matched Clad	Depressed Clad
Fiber counts	4 thru 288	4 thru 288
Fiber type	Single mode	Single mode
Cladding diameter	125 ± 1 μm	125 ± 1 μm
Coating diameter	245 ± 10 μm, dual layer	245 ± 10 μm
Coating type	UV acrylate	UV acrylate
Jacket	Medium density polyethylene (black)	Medium density polyethylene (black)
Strength member	Aramid and central FGE[a] rod	Aramid and central FGE rod
Operating temperature range	−40°C to +70°C	−40°C to +70°C
Maximum load @ installation	600 lb (2700 N)	600 lb (2700 N)
Maximum load–long term	135 lb (600 N)	135 lb (600 N)
Mode field diameter	9.3 ± 0.5 μm @ 1310 nm 10.5 ± 1.0 μm @ 1550 nm	8.8 ± 0.5 μm @ 1310 nm 9.7 ± 0.6 μm @ 1550 nm

[a]FGE, fiberglass epoxy.
Maximum attenuation at 1310 nm, 0.35 dB/km. Maximum attenuation at 1550 nm, 0.25 dB/km.
Minimum bending radius (inches) at installation: 20 × diameter.
Minimum bend radius (inches) for long-term application: 10 × diameter.
Source: Courtesy of Belden Wire and Cable Company (Ref. 5).

2.5 CHARACTERISTICS OF OPTICAL FIBERS

2.5.1 Optical Characteristics

As we discussed, with single-mode optical fiber, only one mode propagates down the fiber at operating wavelengths. In this category of optical fiber we find single-mode dispersion-unshifted, dispersion-shifted, and nonzero-dispersion fibers. These characteristics depend on the fiber construction. When testing these fibers with a signal source, one should keep in mind that the light source, whether laser diode or LED, is not monochromatic, but has an output that covers a range of wavelengths (or optical frequencies). Pulse spreading will result as the transit times of spectral components differ. The magnitude of the pulse spreading is proportional to the spectral width of the source. Nearly monochromatic (single longitudinal mode or SLM) laser sources, typically the distributed feedback (DFB) lasers, permit the operation of a single-mode fibers at wavelengths farther from the zero-dispersion wavelength than do multilongitudinal mode (MLM) lasers.

The EIA/TIA has classified single-mode fibers according to their dispersion characteristics. There are dispersion-unshifted single-mode fibers, which have a zero-dispersion wavelength in the 1310-nm region. This fiber type is classified as TIA/EIA IVa. There are two types of dispersion-shifted fiber: One has its zero-dispersion wavelength in the 1550-nm region and is classified as TIA/EIA IVb. The other, nonzero-dispersion-shifted fiber, has a nonzero specified dispersion range in the 1550-nm operating window. This fiber type is called Class IVd fiber by TIA/EIA.

In the 1550-nm region, the attenuation can be significantly lower than at 1310 nm for each fiber class. However, the dispersion of the shifted design at 1310 nm is substantially greater than that of the dispersion unshifted design at that wavelength.

TABLE 2.2b Summary of Physical and Transmission Characteristics of Corning Cable Systems Unitized MIC® Fiber-Optic Riser Cables

Storage Temperature	−40°C to +70°C (−40°F to +158°F)
Operating Temperature	−20°C to +70°C (−4°F to 158°F)
NEC/CSA Listing	NEC "OFNR", CSA "FT-4"
Flame Resistance	UL-1666 (for riser and general building applications)

Fiber Count	Unit Count	Nominal Outer Diameter, mm (in.)	Nominal Weight kg/km (lb/1000 ft)	Central Member	Maximum Tensile Load: Short Term, N (lbf)	Maximum Tensile Load: Long Term, N (lbf)	Minimum Bend Radius: Loaded, cm (in.)	Minimum Bend Radius: Installed, cm (in.)
6-Fiber Subunits								
24	4	12.2 (0.5)	120 (81)	G	2,500 (563)	1,000 (225)	18.2 (7.2)	12.2 (4.8)
30	5	13.6 (0.5)	159 (107)	G	3,500 (788)	1,700 (383)	20.4 (8.0)	13.6 (5.4)
36	6	15.2 (0.6)	189 (127)	JG	4,000 (900)	2,000 (450)	22.8 (9.0)	15.2 (6.0)
48	8	17.9 (0.7)	264 (177)	JG	5,000 (1125)	2,500 (563)	26.8 (10.5)	17.9 (7.0)
60	10	21.1 (0.8)	380 (255)	JG	5,500 (1238)	3,000 (675)	31.7 (12.5)	21.1 (8.3)
72	12(9/3)	20.3 (0.8)	301 (203)	G	5,500 (1238)	3,000 (675)	30.5 (12.0)	20.3 (8.0)
12-Fiber Subunits								
72	6	22.3 (0.9)	373 (251)	JG	7,000 (1575)	3,500 (788)	33.5 (13.2)	22.3 (8.8)
84	7	24.5 (1.0)	458 (308)	JG	7,000 (1575)	3,500 (788)	36.8 (14.5)	24.5 (9.6)
96	8	26.6 (1.0)	543 (365)	JG	8,800 (1980)	4,000 (900)	39.9 (15.7)	26.6 (10.5)
108	12 (9/3***)	30.0 (1.2)	492 (331)		10,000 (2250)	4,000 (900)	45.0 (17.7)	30.0 (11.8)
120	12 (9/3**)	30.0 (1.2)	527 (354)		10,000 (2250)	4,000 (900)	45.0 (17.7)	30.0 (11.8)
132	12 (9/3*)	30.0 (1.2)	567 (381)		10,000 (2250)	4,000 (900)	45.0 (17.7)	30.0 (11.8)
144	12 (9/3)	30.0 (1.2)	572 (384)		10,000 (2250)	4,000 (900)	45.0 (17.7)	30.0 (11.8)

TRANSMISSION PERFORMANCE

Performance Option (850/1300 μm	12 62.5/125 μm (850/1,300 nm)	30 62.5/125 μm (850/1,300 nm)	50 62.5/125 μm (850/1,300 nm)	31 50/125 μm (850/1,300 nm)	31 Single-mode (1,310/1,550 nm)
Maximum attenuation (dB/km)	3.5/1.0	3.5/1.0	3.5/1.0	3.5/1.5	1.0/0.75
Typical attenuation (dB/km)	3.0/1.0	3.0/1.0	3.0/1.0	3.0/1.0	0.5/0.4
Minimum LED bandwidth (MHz · km)	200/500	200/500	200/500	500/500	—
Minimum RML bandwidth (MHz · km)	—	220/—	385/—	—	—
Gigabit ethernet distance guarantee (m)	275/550	300/550	500/1,000	600/600	5,000

Note: * = 1 Filler Unit; ** = 2 Filler Units; *** = 3 Filler Units; *, **, and *** all on inner layer. Unit count includes the arrangement when in a dual layer design. Example: 12(9/3**) = 9 outside units around 3 inner units where 2 of the inner units are fillers.

Source: Courtesy of Corning Cable System, Ref. 7, LANscape Catalog, page 1.31.

2.5.2 Mechanical Characteristics

An inherent property of optical fiber is strength. However, during the manufacturing process, microscopic surface flaws appear which notably reduce a fiber's basic strength. There is further strength degradation in the fiber due to cabling, handling, and installation of the fiber. Strength degradation and failure inducing the growth of flaws have three causes: dynamic fatigue, static fatigue, and zero-stress aging. Many fiber installers have a background of installing copper cable, which has decidedly different mechanical characteristics. Dynamic fatigue will occur when high tensile loads are applied over short periods of time. This is typical of the installation scenario where the fiber is "pulled" into place through a duct or along a cable way. Static fatigue is brought about when a constant load is applied to fiber cable over a long period of time. Zero-stress aging is a type of strength degradation that may happen to a fiber-optic cable without loads but in environments of high temperature and high humidity.

2.5.3 Fiber-Optic Units

The function of a fiber-optic unit is to organize fibers so as to ease fiber identification and handleability so that fiber maintains its structure both when inside a fiber cable and once the sheath has been removed. A fiber-optic unit may be a fiber-optic bundle, fiber ribbons, or loose-buffer tubes.

A fiber-optic unit typically consists of 6 or 12 fibers bound together loosely with helical binders. These binders must be such that they maintain their position in the cable and facilitate fiber unit identification once the sheath of the cable has been removed.

One method of mounting fiber strands is in a ribbon structure. These fiber-optic ribbons may house several thousand fiber strands in a cable. Typically, fiber ribbons accommodate 4, 6, 8, 12, or 24 fiber strands in a linear array to form a unit.

Fiber ribbons are mounted in a matrix for ease of identification, as well as for strength and protection. The matrix material is optimized for performance and reliability. The material should also be compatible with the matrix coating when a technician uses heat-stripping tools to promote coating and matrix material removal when splicing, connectorizing, and fulfilling other installation requirements. The matrix material should be transparent such that individual fiber strands can be identified.

Fibers can also be placed inside a buffer tube to isolate the fibers from outside stresses. Typically, 6 or 12 fibers are placed in a tube. A tube should facilitate fiber unit identification once the cable sheath has been removed.

2.6 MATCHED-CLAD VERSUS DEPRESSED-CLAD FIBER

Matched-clad is the simplest single-mode fiber design. Matched-clad fiber results in a constant refractive index profile throughout the cladding, or from the edge of the core. This produces a fiber consistency that is especially important when different fibers need to be joined together in an existing network. Matched-clad fiber typically results in slightly lower fiber attenuation and a larger mode-field diameter[1] than its depressed-clad fiber counterpart.

Depressed-clad fiber is usually produced from the inside vapor deposition (IVD) or modified chemical vapor deposition (MCVD) process. It is called *depressed* because the index of refraction of the inner cladding region is lower than the rest of the cladding. A depression occurs where the two indices of refraction do not match. That is where the glass tube and the deposited glass meet. This

[1]*Mode-field diameter.* The region occupied by light in a single-mode fiber. Typically, it is larger than the core diameter.

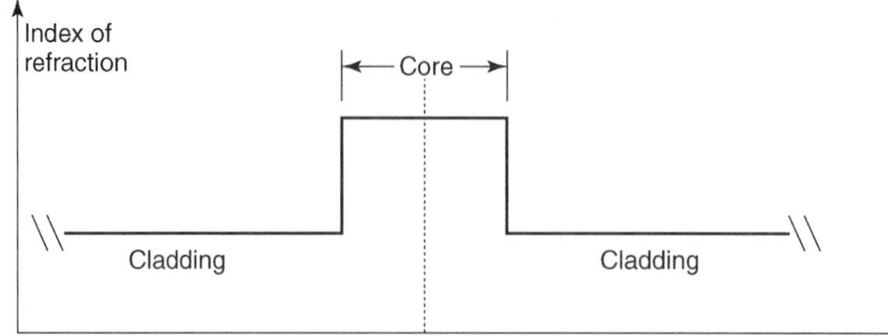

Figure 2.8a. Index of refraction profile for matched-clad fiber. (Courtesy of *Corning Fiber News & Views*, Ref. 8.)

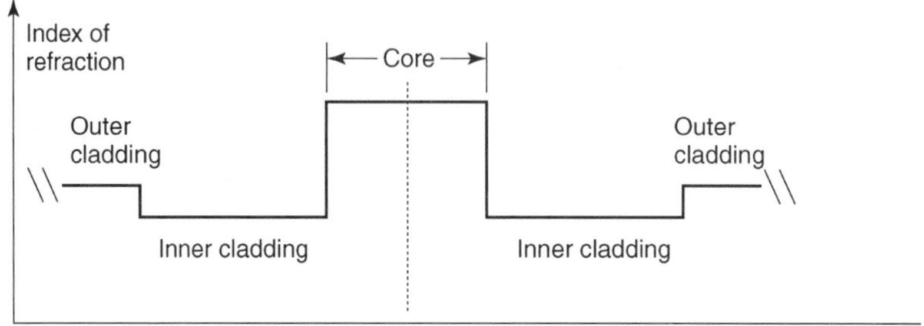

Figure 2.8b. Index of refraction profile for depressed-clad fiber. (Courtesy of *Corning Fiber News & Views*, Ref. 8.)

depression is caused by the addition of dopant chemicals to the cladding just prior to depositing the core material.

The type of cladding used affects bend losses. As we mentioned above, the smaller the mode-field diameter, the smaller the bending loss. This is true for both matched-clad and depressed-clad fibers.

However, for the same mode-field diameter (MFD), induced losses are greater in depressed-clad fiber for bend diameters greater than 50 mm. These types of bends are usually encountered in splice enclosures and in cables. Thus, one would expect to find that standard production depressed-clad fibers have a somewhat smaller mode-field diameter to achieve satisfactory cable-fiber performance. The smaller MFD is designed into a depressed clad fiber.

Figure 2.8a illustrates conceptually an index of refraction profile for matched-clad fiber, and Figure 2.8b illustrates that for depressed-clad fiber.

Source: Ref. 8.

2.7 TYPICAL CHARACTERISTICS OF AN OPTICAL FIBER OF HIGH-GRADE DESIGN

Corning's large effective area fiber (LEAF®) is a proprietary design of Corning Fiber Systems. It is ideal for extended bandwidth 1550-nm operation using DWDM. Table 2.3 gives an overview of the fiber's optical characteristics, Table 2.4 provides

TABLE 2.3 Optical Specifications of Corning's Leaf® Fiber

Parameter/Item	Value	Comments
Attenuation at 1550 nm	$\leq$ 0.25 dB/km	
at 1625 nm	$\leq$ 0.25 dB/km	
Point discontinuity	None greater than 0.10 dB at 1550 nm	
Attenuation at 1383 $\pm$ 3 nm	$\leq$ 1.0 dB/km	
Attenuation vs. wavelength 1525–1575 nm	Maximum increase at 0.05 dB/km	Ref. λ 1550 nm
Attenuation with bending	Induced attenuation: $\leq$ 0.05 dB	Mandrel diameter 32 mm, 1 turn Ref. λ 1550 & 1625 nm
Attenuation with bending	Induced attenuation: $\leq$ 0.50 dB	Mandrel diameter 75 mm, 100 turns, Ref. λ 1550 & 1625 nm
Mode-field diameter	9.20–10.00 μm at 1550 nm	
Dispersion, total	2.0–6.0 ps/(nm.km)	Over range of 1530–1565 nm
Dispersion: Polarization mode dispersion (PMD)	4.5–11.2 ps/(nm.km)	Over range of 1565–1625 nm
PMD link value	$\leq$ 0.08 ps $(\text{km})^{1/2}$ (see Note)	
Maximum individual fiber	$\leq$ 0.2 ps $(\text{km})^{1/2}$	

Note: The PMD link value is a term used to describe the PMD of concatenated lengths of fiber (also known as the link quadrature average). The value is used to determine a statistical upper limit for PMD system performance.

TABLE 2.4 Dimensional Specifications

Item/Parameter	Value	Comments
Standard Length	4.4–25.2 km/reel	
Fiber curl	$\geq$ 4.0 m radius of curvature	
Cladding diameter	125.0 $\pm$ 1.0 μm	
Core/clad concentricity	$\leq$ 0.5 μm	
Cladding noncircularity	$\leq$ 1.0% (see Note)	
Coating diameter	245 $\pm$ 5 μm	
Coating/cladding concentricity	< 12 μm	

Note: Defined as

$$[1 - (\text{minimum cladding diameter})/(\text{max. cladding diameter})] \times 100$$

TABLE 2.5 Performance Characterizations (typical values)

Item/Parameter	Value	Comments
Proof test: The entire length of fiber is subjected to tensile proof stress	$\geq$ 100 kpsi (0.7 GN/m^2)[a]	
Effective area (A_{eff})	72 μkm^2	
Effective group index of refraction (N_{eff})	1.469 at 1550 nm	
Fatigue resistance parameter (n_d)	20	
Coating strip force, dry	0.6 lb (3.0 N)	
Coating strip force, wet, 14 days room temperature	0.6 lb (3.0 N)	

[a]Higher proof test available at a premium.

an overview of its dimensional specifications, and Table 2.5 gives its performance characterizations.

REFERENCES

1. Govind P. Agrawal, *Fiber-Optic Communication Systems*, 2nd ed., John Wiley & Sons, New York, 1997.
2. *Fiber Optics System Design*, MIL-HDBK-415, US Department of Defense, Washington, DC, 1985.
3. Donald J. Sterling, Jr., *A Technician's Guide to Fiber Optics*, 3rd ed., Delmar Publishers, Albany, NY, 2000.
4. *Introduction of Fiber Optics*, Corning Cable Systems, Hickory, NC, 2000.
5. *Fiber Optic Catalog*, Belden Wire and Cable Company, Richmond, IN, 1998.
6. *Loose Tube vs. Tight Buffered Cable in Outdoor Applications*, AEN 26 Rev. 4, Corning Cable Systems, Hickory, NC, 2000.
7. *Corning Cable Systems LANscape Catalog*, Premises Fiber Optic Products Catalog, 7th ed., Corning Cable Systems, Hickory, NC, January 2000.
8. The Facts: Matched-Clad vs. Depressed-Clad Fiber, *Corning News & Views*, Corning, NY, April 1999.
9. *Generic Requirements for Optical Fiber and Optical Fiber Cable*, Telcordia (Bellcore) GR-20-CORE Issue 2, Piscataway, NJ, July 1998.
10. Product flyer, "Corning LEAF® Non-Zero Dispersion-Shifted Single-Mode Fiber," Corning, NY, 1999.

3

CONNECTORS, SPLICES, AND OTHER PASSIVE DEVICES

3.1 CHAPTER OBJECTIVE

The purpose of this chapter is to describe and characterize the various passive devices used in a fiber-optic network. There are five generic types of fiber-optic networks:

1. Long-haul public or private networks.
2. In-building and enterprise network applications, also called premises networks.
3. Local distribution networks including cable television, especially hybrid fiber-coax (HFC) networks.
4. Metropolitan area networks (MANs) favoring data connectivity as found in an enterprise network. They provide local delivery of bulk data around a metropolitan area.
5. Specialty applications, usually very short-haul signal transport.

Each network type will require a number of general passive device types and some specialty devices. For example, cable television networks will widely employ signal splitters.

WDM/DWDM uses a wide array of specialty passive devices. We leave the description of these in our chapter on WDM (see Chapter 8).

Our first concern in this chapter will be fiber connectors and cable splicing. We also cover the following passive devices:

- Optical couplers, signal splitters, branching components
- Optical isolators
- Fiber-optic filters
- Light attenuators
- Optical switches
- Passive (chromatic) dispersion compensators
- Terminations

Each device, when employed in a circuit carrying a light signal, will incur an insertion loss. It will also display *reflectance*, sometimes called reflection loss or return loss. This is conventionally measured in decibels. With the exception of attenuators, we wish the insertion loss to be very low and the return loss to be very high. For example, we would like to see splices with an insertion loss less than 0.1 dB and with a return loss of over 40 dB.

3.2 SELECTED DEFINITIONS (From ITU-T Rec. G.671 [Ref. 8].)

3.2.1 Fiber-Optic Branching Component (Wavelength Nonselective)

This is a passive component (wavelength nonselective) possessing three or more ports that shares power among its ports in a predetermined fashion, without any amplification, switching or other active modulation.

3.2.2 Coupler (Splitter–Combiner)

This is a term that is used as a synonym for a branching device. The term is also used to define a structure for transferring optical power between two fibers or between an active device and a fiber.

3.2.3 Attenuator

This is a passive component that produces a controlled signal attenuation in an optical fiber transmission line.

3.2.4 Fiber-Optic Filter

This is a passive component used to modify the optical radiation passing through it, generally by altering the spectral distribution. In particular, fiber-optic filters are usually employed to reject or absorb optical radiation in particular ranges of wavelength, while passing optical radiation in other ranges of wavelength.

3.2.5 Fiber-Optic Isolator

This is a nonreciprocal optical device intended to suppress backward reflection along an optical fiber transmission line while having minimum insertion loss in the forward direction.

3.2.6 Fiber-Optic Termination

This is a component used to terminate a fiber (connectorized or not) in order to suppress reflections.

3.2.7 Fiber-Optic Switch

This is a passive component possessing one or more ports that selectively transmits, redirects, or blocks optical power in an optical fiber transmission line.

3.2.8 Passive (Chromatic) Dispersion Compensator

This is a passive component used to compensate the chromatic dispersion of an optical path.

3.2.9 Fiber-Optic Connector

This is a component normally attached to an optical cable or piece of apparatus for the purpose of providing frequent optical interconnection/disconnection of optical fibers or cables.

3.2.10 Fiber-Optic Splice

This is a permanent or semipermanent joint whose purpose is to couple optical power between two optical fibers.

- *Fusion splice:* A splice in which the fiber ends are joined in a permanent manner by means of fusion.
- *Mechanical splice:* A splice in which the fiber ends are joined in a permanent or separable manner by means other than fusion.

3.3 FUNCTIONAL PARAMETER DEFINITIONS (From ITU-T Rec. G.671, Section 3.2.)

3.3.1 Insertion Loss (IL)

This is the reduction in optical power between an input and output port of a passive component in decibels. It is defined as

$$IL = -10\log(P_1/P_0) \tag{3.1}$$

where P_0 is the optical power launched into the input port and P_1 is the optical power received from the output port.

Note 1. For a (fiber-optic) branching component, it is an element a_{ij} (where $i \neq j$) of the logarithmic transfer matrix (1.3.7 of IEC 875-1).

Note 2. For a WDM device, it is an element a_{ij} (where $i \neq j$) of the logarithmic transfer matrix, and it shall be specified at each operating wavelength range.

Note 3. For a (fiber-optic) switch, it is an element a_{ij} (where $i \neq j$) of the logarithmic transfer matrix. It depends on the state of the switch (1.3.9 of IEC 876-1).

Note 4. For a (fiber-optic) filter, it shall be specified in each operating wavelength range.

Insertion loss is the ratio of the optical power launched at the input port of an appropriate optical device to the optical power from any single output port, expressed in decibels. Insertion loss includes such parameters as splitting loss in the case of couplers, and with other branching devices we use excess loss. These are the most useful parameters for system design. The maximum and minimum insertion loss is the upper and lower limit, respectively, of the insertion loss of the device of interest and applies over the entire wavelength range specified in the bandpass. Typical insertion loss is the expected value of the insertion loss measured at the specified center wavelength. The IEEE (Ref. 10) defines insertion loss as "the total optical power loss caused by the insertion of an optical component such as a connector, splice or coupler."

3.3.2 Return Loss (RL)

This is the fraction of input power that is returned from the input port of a passive component. It is defined as

$$RL = -10\log(P_r/P_i) \tag{3.2}$$

where P_i is the optical power launched into the input port and P_r is the optical power received back from the same port.

> **Note**. For clarity, return loss values for fiber-optic devices do not include the return loss contributions of connectors. Return loss contributions from connectors will be considered separately.

Also see "Notes" under Section 3.3.1.

3.3.3 Reflectance

This is the ratio R of reflected power P_r to incident power P_i at a given port of a passive component, for given conditions of spectral composition, polarization, and geometrical distribution, generally expressed in decibels as

$$R = -10\log(P_r/P_i) \tag{3.3}$$

When referring to reflected power from an individual component, reflectance is the term preferred to return loss. For clarity, reflectance values for fiber-optic devices do not include the reflectance contributions of connectors. Reflectance contributions from connectors are considered separately.

3.3.4 Directivity, Return Loss, and Reflectance—Introductory Discussion

Directivity is the ratio of the optical power launched into an input port to the optical power returning from any other port. Directivity has been referred to as near-end isolation or near-end crosstalk. Return loss is the ratio of optical power launched into an input port to the optical power returning from the same input port. Both directivity and return loss are expressed as positive decibels and are measured with all output ports optically terminated. Reflectance is the negative of return loss. In many cases, reflectance and return loss are used synonymously. Minimum directivity and return loss are the lower limits that apply over the entire wavelength range specified in the bandpass.

3.3.5 Operating Wavelength Range

This is the specified range of wavelengths from $\lambda_{i\,\mathrm{min}}$ to $\lambda_{i\,\mathrm{max}}$ about a nominal operating wavelength λJ, within which a passive component is designed to operate with a specified performance.

Note 1. For a fiber-optic branching component with more than one operating wavelength, the corresponding wavelength ranges are not necessarily equal.

Note 2. The component including attenuators, terminations, connectors, and splices may operate with a specified performance or acceptable performance even outside the specified range of applications.

3.3.6 Polarization-Dependent Loss (PDL)

This is maximum variation of insertion loss due to a variation of the state of polarization over all states of polarization.

3.3.7 Polarization-Dependent Reflectance

This is maximum variation of reflectance due to a variation of the state of polarization over all states of polarization.

3.3.8 Backward Loss (Isolation) for Fiber-Optic Isolator

This is a measure of the decrease in optical power (decibels) resulting from the insertion of an isolator in its backward direction. The launching port is the output port and the receiving port is the input port of the isolator. It is given by the following formula:

$$BL = -10\log(P_{\mathrm{ob}}/P_{\mathrm{ib}}) \tag{3.4}$$

where P_{ob} is the optical power measured at the input port of the isolator when P_{ib} is launched into the operating port. In operating conditions, P_{ib} is the optical power reflected at the far-end optical circuit devices in the backward direction into the output port of the isolator being measured.

3.3.9 Directivity

For a fiber-optic branching component, this is the value a_{ij} of the logarithmic transfer matrix between two isolated ports.

3.3.10 Uniformity

The logarithmic transfer matrix of a branching component may contain a specified set of coefficients that are nominally finite and equal. In the case the range of these coefficients, a_{ij} (expressed in decibels) is termed the uniformity of the branching component.

3.3.11 Port

This is an optical fiber or an optical fiber connector attached to a fiber-optic component for the entry and/or exit of optical power.

3.3.12 Transfer Matrix for a Fiber-Optic Branching Device and a WDM Device

The optical properties of a fiber-optic branching device can be defined in terms of an $n \times n$ matrix of coefficients. N is the number of ports, and the coefficients represent the fractional power transferred between designated ports. In general, the transfer matrix T is

$$T = \begin{bmatrix} t_{11} & t_{12} & \cdot & t_{1n} \\ \cdot & \cdot & \cdot & \cdot \\ \cdot & \cdot & t_{ij} & \cdot \\ t_{n1} & \cdot & \cdot & t_{nn} \end{bmatrix}$$

where t_{ij} is the ratio of optical power P_{ij} transferred out of port j with respect to input power P_i into port i, that is,

$$t_{ij} = P_{ij}/P_i \qquad \text{(1.3.4 of IEC 875-1)}$$

Note. Generally, t_{ij} could be wavelength-dependent.

3.3.13 Transfer Coefficient for a Fiber-Optic Branching Device and a WDM Device

This is an element t_{iy} of the transfer matrix.

3.3.14 Logarithmic Transfer Matrix Coefficient for a Fiber-Optic Branching Device and a WDM Device

In general, the logarithmic transfer matrix is

$$A = \begin{bmatrix} a_{11} & a_{12} & \cdot & a_{1n} \\ \cdot & \cdot & \cdot & \cdot \\ \cdot & \cdot & a_{ij} & \cdot \\ a_{n1} & \cdot & \cdot & a_{nn} \end{bmatrix}$$

where a_{ij} is the optical power reduction in decibels out of port j with unit power into port i, that is,

$$a_{ij} = -10 \log t_{ij}$$

where t_{ij} is the transfer matrix coefficient (1.3.6 of IEC 875-1).

3.3.15 Transfer Matrix for a Fiber-Optic Switch

The optical properties of a fiber-optic switch can be defined in an $m \times n$ matrix of coefficient (n is the number of ports). The T matrix represents the on-state paths (worst-case transmission), and the T^o matrix represents the off-state paths (worst-case isolation) (1.3.6 of IEC 876-1).

$$T = \begin{bmatrix} t_{11} & t_{12} & \cdot & t_{1n} \\ t_{21} & & & t_{2n} \\ \cdot & \cdot & \cdot & \cdot \\ \cdot & \cdot & t_{ij} & \cdot \\ t_{n1} & \cdot & \cdot & t_{nn} \end{bmatrix}$$

$$T^o = \begin{bmatrix} t^o_{11} & t^o_{12} & \cdot & t^o_{1n} \\ t^o_{21} & & & t^o_{2n} \\ \cdot & \cdot & \cdot & \cdot \\ \cdot & \cdot & t^o_{ij} & \cdot \\ t^o_{n1} & \cdot & \cdot & t^o_{nn} \end{bmatrix}$$

3.3.16 Transfer Coefficient for a Fiber-Optic Switch

This is an element t_{ij} or t^o_{iy} of the transfer matrix. Each coefficient t_{ij} is the worst-case (minimum) fraction of power transferred from port i to port j for any state with path *ij* switched on. Each coefficient t^o_{iy} is the worst-case (maximum) fraction of power transferred from port i to port j for any state with path *ij* switched off.

3.3.17 Logarithmic Transfer Matrix for a Fiber-Optic Switch

In general a logarithmic transfer matrix is

$$A = \begin{bmatrix} a_{11} & a_{12} & \cdot & a_{1n} \\ \cdot & \cdot & \cdot & \cdot \\ \cdot & \cdot & a_{ij} & \cdot \\ a_{n1} & \cdot & \cdot & a_{nn} \end{bmatrix}$$

where a_{ij} is the optical power reduction in decibels out of port j with unit power into port i, that is,

$$a_{ij} = -10\log(t_{ij})$$

where t_{ij} is the transfer matrix coefficient.

Similarly, for the off state, $a_y^o = -10\log(t_y^o)$.

3.3.18 Excess Loss for a Fiber-Optic Branching Device

This is the total power lost in a branching device when an optical signal is launched into port 1. It is defined as

$$EL_i = -10\log \textstyle\sum_j t_{ij}$$

where the summation is performed only over those values of j for which i and j are conducting ports. For a branching device with N input ports, there will be an array of N values of excess loss, one for each input port i (1.3.12 of IEC 875-1).

3.3.19 Coupling Ratio

For a given input port I, this is the ratio of light at a given output port k to the total light from all output ports. It is defined as

$$CR_{ik} = t_{ik}/\textstyle\sum_j t_{ij}$$

where the j's are the operational output ports (1.3.14 of IEC 875-1).

3.3.20 Operating Wavelength

This is a nominal wavelength λ at which a passive component is designed to operate with the specified performance.

3.3.21 Switching Time Matrix for a Fiber-Optic Switch

$$S = \begin{bmatrix} s_{11} & s_{12} & \cdot & s_{1n} \\ \cdot & \cdot & \cdot & \cdot \\ \cdot & \cdot & s_{ij} & \cdot \\ s_{n1} & \cdot & \cdot & s_{nn} \end{bmatrix}$$

This is a matrix of coefficients in which each coefficient S_{ij} is the longest switching time to turn path ij on or off from any initial state (1.3.21 of IEC 876-1).

3.4 CONNECTORS AND SPLICES

Connectors and splices are used to interconnect fiber sections. Fiber-optic cable is delivered to the telecommunications contractor on reels of from 1 to 25 km long. For longer networks, as distinguished from premises networks, reels consist of fiber cable segments that have to be interconnected to make a working system. Either connectors or splices can be used for this purpose.

Industrial practice and good sense dictates the use of connectors at the bitter ends of a link and splices for the intervening sections. The reasons for this practice are as follows:

- Splices have insertion losses as low as 0.04 dB per splice joint; connectors have a somewhat larger insertion loss. There is a certain permanency to splices.
- It behooves us to use connectors for locations where we expect to have some or many connections/disconnections such as at fiber-optic patch panels or cross-connects. Many connectors are made for easy connect/disconnect operations.

Connectors should be considered where a fiber enters either a passive device or an active device. If we wish to replace the device, a connector is much better suited than a splice for this purpose.

3.4.1 Connectors

There are a large number of proprietary connectors on the market. Fiber connectors are available in two size classes: standard size and miniature connectors. There are connectors that individually connect a number of fibers, and, of course, there are individual fiber connectors.

Fiber-optic connectors may be field-installable, or, under certain circumstances they are factory installed. The factory-installed connector is sometimes called the *pigtail connector*. A pigtail is a short length of fiber connected to a light device such as a light source or light detector. The other end of the pigtail (that short length of fiber) has a connector that has been installed by the manufacturer of that device. Once in the field, the installer must provide the mate to this connector and install the mate on the bitter end of the fiber under consideration.

The remainder of our discussion of connectors will concentrate on field-installable connectors.

3.4.1.1 Construction of a Connector—General Case. The basic connector consists of three parts:

1. Ferrule
2. Coupling receptacle
3. Coupling nut

A typical connector assembly is illustrated in Figure 3.1.

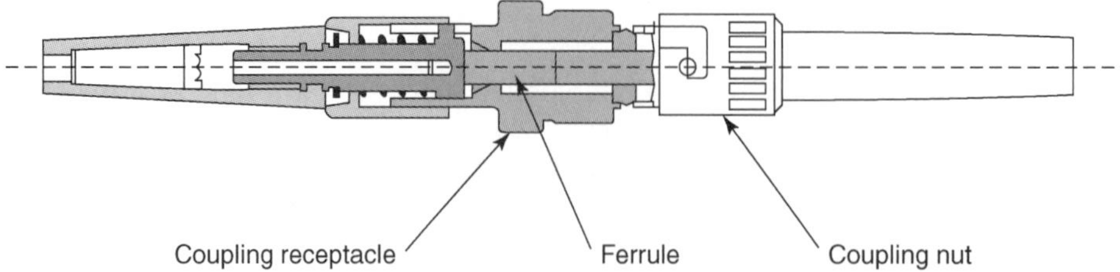

Figure 3.1. Basic connector structure. (Courtesy of Tyco Electronics, Harrisburg, PA, Ref. 1.)

Usually the connector consists of connector shells that carry a concentric ceramic ferrule with a precision concentric bore. The bare fiber is inserted into the ferrule and retained with a suitable resin or hot-melt adhesive. The fiber end is then cleaved and polished to a flush, flat mirror finish. The metal shells align and butt the ceramic ferrule under gentle pressure. The most popular ferrule has a 2.5-mm diameter. There are also small-form-factor connectors that employ a 1.25-mm ferrule.

For single-mode fibers, alignment accuracies greater than 0.1 μm and less than 5° angular misalignment are required. It is advisable to check the insertion loss of the installed connector before it is placed into service. This loss should be measured by a transmission technique, not by an optimal time-domain reflectometer (OTDR). However, it might also be wise to check the return loss of the connector interface. This can be done by an OTDR. A design objective is a return loss of 40 dB or better.

Source: Ref. 2.

There are numerous types of connectors on the market, each with slightly different installation procedures. However, all connectors require two important installation steps that are common among all connectors.

First, the fiber is epoxied into the connector. The epoxy process is important to the long-term reliability of the connector. The epoxy keeps fiber movement over temperature at a minimum, allows polishing without fear of fracturing the fiber, and seals the fiber from effects of the environment. In addition, it allows the fiber to be aggressively cleaned on the endface. Therefore, it is very important that the epoxy be present around the entire length of the bare fiber, around the buffer as the fiber enters the connector, and as a bead surrounding the fiber on the endface of the connector. (See Figure 3.2.)

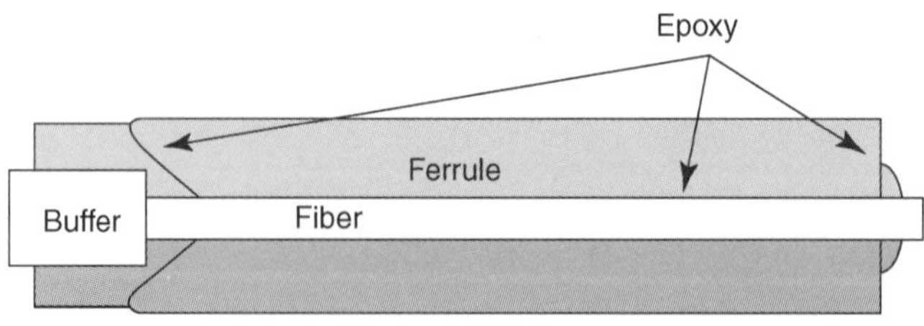

Figure 3.2. Epoxy application. (Reprinted with permission from Corning Cable Systems, Ref. 2, Figure 7.4, page 7.4.)

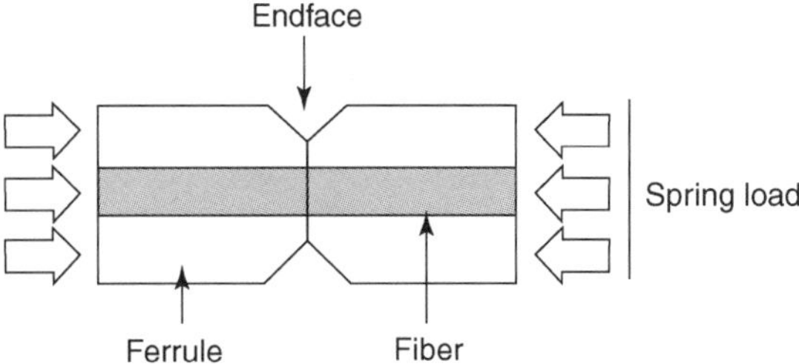

Figure 3.3. Illustrating physical contact (PC) in an optical fiber connector. (Reprinted with permission from Corning Cable Systems, Ref. 2, Figure 7.5, page 7.5.)

Second, the connector endface must be polished. A physical contact (PC) finish is recommended. This means the fiber will be physically touching inside the connector adapter as they are held under compression. Lack of a PC finish results in an airgap between the fibers and increased attenuation. Figure 3.3 illustrates this point.

There are several polishing methods recommended, which are typically dependent on the ferrule material used. In general, if a ferrule material is very hard—ceramic, for example—it is common for the ferrule to be radiused on the endface, which is referred to as preradiused. Softer ferrule materials such as composite thermoplastic or glass-in-ceramic may be polished flat. These materials wear away at approximately the same rate as the fiber and can be polished aggressively and still maintain a PC finish.

There are new approaches to fiber preparation and connector assembly. The butt ends of the fiber to be "connectorized" are now rounded instead of being flat, perpendicular ends. The advantage of this approach is that light no longer reflects directly back toward the source (i.e., the angle of reflection is equal to the angle of incident light). It reflects back at an angle and is usually lost from the fiber. The rounded finish first cuts the amount of light reflected and then redirects what is reflected so that it does not reach the laser diode light source. This approach is called a *physical contact* or PC-end finish. A second advantage to PC-end finish with rounded fiber ends is that the fibers always touch on the high side near the light-carrying fiber core. The user is virtually assured that the fibers touch and that the airgap is eliminated.

Source: Ref. 3.

Return reflection can be lowered even further with angle PC (APC). This angle reflects the light into the cladding rather than back down the core.

Connector return loss should be at least 40 dB. Another performance parameter is *number of matings*. This refers to the number of times a connector can be mated (i.e., connected–disconnected) before its performance starts to deteriorate. The value from our experience runs between 200 and 600 matings.

3.4.1.2 Connector Types

ST-Style Connectors. ST-style connectors use a quick-release bayonet coupling, which requires only a quarter turn to engage or disengage. Built-in keying provides repeatable performance because the connector will always mate with a coupling bushing in the same way. The ST-style connector is being superseded by the SC connector. The insertion loss of the ST-style connector is on the order of 0.5 dB.

SC Connectors. This is a widely used connector for both single-mode and multimode fibers. The term SC derives from "subscriber connector." The term came from popular early applications in the subscriber plant.

The SC connector is a general-purpose connector used in long-haul network applications and for premise cabling. It uses a push–pull mechanism for mating. The basic SC connector consists of a plug assembly containing a ferrule. These plugs mate in a connector housing that aligns the ferrules. One advantage of the SC connector is that several plugs can be joined together to form a multiposition connector. This multifiber capability is especially useful for building duplex (two-position) connectors. In this case, one fiber can carry information in one direction while the other fiber carries information in the other direction. The connectors are keyed to prevent mismating. The insertion loss of the SC connector is under 0.4 dB.

FC Connectors. This connector was originally developed in Japan by its national telephone company, Nippon Telegraph and Telephone Company. This connector is popular with single-mode fiber. It has an insertion loss of about 0.4 dB.

The FC connector has a means of tunable keying such that the key can be adjusted to the point of lowest loss. Insertion loss can be reduced up to several tenths of a decibel by tuning. Once the lowest loss position is found, the key is locked into place. The connector is available in both single-mode and multimode versions.

D4-Style Connectors. This is a popular connector usually used with single-mode fiber. It is similar in most respects to the FC connector but has a smaller ferrule, only 2.00 mm in diameter. The insertion loss for the D4 is about 0.4 dB.

568C Connectors. This is the favored connector for building cabling. Its nomenclature is taken from the EIA/TIA-568 Commercial Building Telecommunications Cabling Standard. It is actually a duplex SC connector. The 568SC has a latching mechanism that allows easier mating than the bayonet-style (ST-compatible) connector. In addition, the connector has an adapter that will accept either a simplex or a duplex connector, and the connector may be used in either duplex or simplex applications. Expect an insertion loss to be around 0.3 dB for the 568SC connector.

FDDI Connectors. The FDDI (fiber distributed data interface) connector is a two-channel device using two ceramic ferrules and a side-latch mechanism. A rigid shroud protects the ferrules from accidental damage, while a floating interface ensures consistent mating without stubbing. Different keying arrangements are provided to key the connector to different FDDI requirements. Expect an insertion loss for the connector of around 0.3 dB for the single-mode application and

TABLE 3.1 Comparison of Four Miniature Connector Features

	LC	MT-RJ	SC-DC	VF-45
Fiber spacing	6.25 mm	0.75 mm	0.75	4.5 mm
Number of ferrules	2	1	1	0
Ferrule material	Ceramic	Plastic	Plastic	None
Alignment	Bore and ferrule	Pin and ferrule	Rail and ferrule	V-groove
Ferrule size	ϕ 1.25 mm	2.5 mm $\times$ 4.4 mm	ϕ 2.5 mm	None
Trx opening	11.1 mm	7.2 mm	11 mm	12.1 mm
(width	5.7 mm	5.7 mm	7.5 mm	8 mm
length)	14.6 mm	14 mm	12.7 mm	21 mm
Fiber cable	Duplex	Duplex or ribbon	Duplex or ribbon	GGP polymer-coated
Field term—plug	Pot and polish	Prepolished stub	prepolished stub	Not available
Field term—socket	Plug + coupler	Plug + coupler and socket	Plug + coupler	Cleave and polish socket
Latch	RJ—top two latches coupled	RJ—top latch	SC push-pull	RJ—top latch

Note that the insertion loss of these connectors varies between 0.3 and 0.6 dB.

Source: Performance Comparison of Small Form Factor Fiber-Optic Connectors. Jean Trewhella, IBM T. J. Watson Research Center, Yorktown Heights, NY, page 1, Ref. 12. Courtesy of IBM Corp.

0.5 dB for the multimode application. FDDI connectors are not restricted to FDDI applications. FDDI is a 100-Mbps (125-megabaud) local area network arrangement governed by an ANSI standard.

Miniature Connectors. Miniature connectors, also called small form-factor connectors, are about half the size of their standard connector counterparts (e.g., SC, FC, ST) with 1.25-mm diameter rather than 2.5-mm diameter, thus allowing denser cable packing typically at patch panels and dense rack mounting installations. Table 3.1 gives an overview of the principal parameters of four of the more popular small form factor connectors.

3.4.2 Splices

A splice permanently joins two fibers. There are two types of splices:

1. Mechanical splice
2. Fusion splice

It is extremely important in a splicing procedure to precisely align the two fiber ends when joining. Good polish of fiber ends and cleanliness are also important during the splicing procedure.

3.4.2.1 Mechanical Splice. A mechanical splice is a small fiber connector some 6 cm long and 1 cm in diameter. The connector precisely aligns the two bare fibers and makes a secure and permanent mechanical connection. The splice is fastened with a snap-type cover or adhesive binding or both may be used. Mechanical splices are available as temporary or permanent splices. The insertion loss of a mechanical splice is somewhat greater than that of a fusion splice and is on the order of 0.1–0.8 dB.

Because single-mode fibers have small optical cores and hence small mode-field diameters, they are less tolerant of misalignment at a fiber joint. Consequently, mechanical splices capable of achieving acceptable performance within a single-mode system loss budget are somewhat more expensive to purchase, are more time-consuming to install, and may require capital equipment outlays equivalent to fusion splicing.

3.4.2.2 Fusion Splice. Single-fiber fusion splicing is the most widely used permanent method of joining optical fibers. Obtaining a good fusion splice is much easier today due to continued improvements to fusion splice equipment, procedures, and practices, in addition to the evolutionary improvements in controlling optical fiber geometries. As a result, insertion losses typically range from 0.04 to 0.10 dB for both single-mode and multimode fibers.

Fusion Splice Quality. Two parameters affect the quality of a fusion splice: splice insertion loss and tensile strength.

In the case of multimode fiber, there are fiber-related factors that include core diameter mismatch, numerical aperture (NA) mismatch, index profile mismatch, and core/cladding concentricity error. Concentricity refers to how well a fiber core is centered in the cladding circle—that is, that there is no offset from circularity. This loss can be reduced, however, by using a splicing technique that aligns the fiber cores at the point of joining. From Figure 3.4 we can estimate the theoretical intrinsic splice loss for its major contributors: core diameter and NA mismatch.

It should be noted that the splice loss is directional with regard to these variables (i.e., loss occurs only when optical propagation is across a joint in which the receiving fiber has the smaller core diameter of NA). Splice loss values are additive, so if two multimode fibers that display mismatches in both core diameter and NA are joined, then their contribution to intrinsic loss is the sum of the two losses.

In the case of single-mode dispersion nonshifted fibers, the dominant fiber-related factor in MFD[1] mismatch. Figure 3.5 may be used to estimate the intrinsic loss contribution due to MFD mismatch.

As illustrated in Figure 3.5, the actual splice loss (bidirectional average) is practically nondirection (i.e., the same fiber-related losses will be seen across the joint regardless of the direction of optical propagation). It should be taken into account that intrinsic loss is relatively low for MFD mismatches expected within typical manufacturer's tolerances. For instance, the loss would be approximately 0.04 dB for the worst-case fiber-related bidirectional loss for fiber having a 9.3 ± 0.5-μm

[1]MFD stands for mode-field diameter. This is a measure of the width of the guided optical power's intensity distribution in a single-mode fiber.

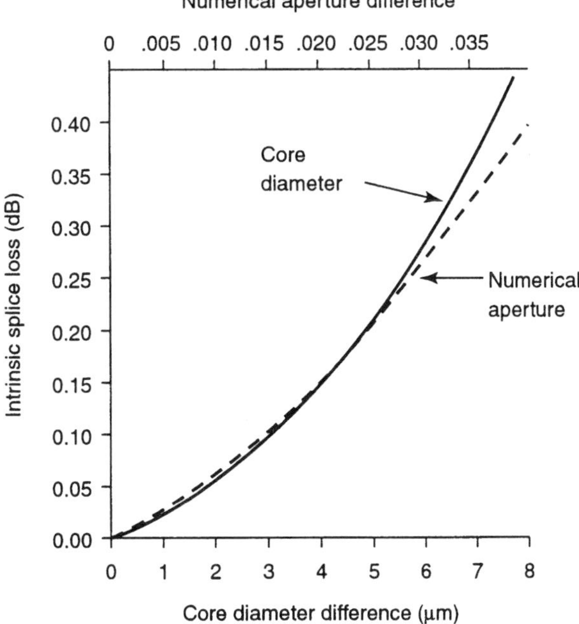

Figure 3.4. Intrinsic splice loss due to core diameter and NA mismatch. (Reprinted with permission from Corning, Inc., AN103, Ref. 4, Figure 1.)

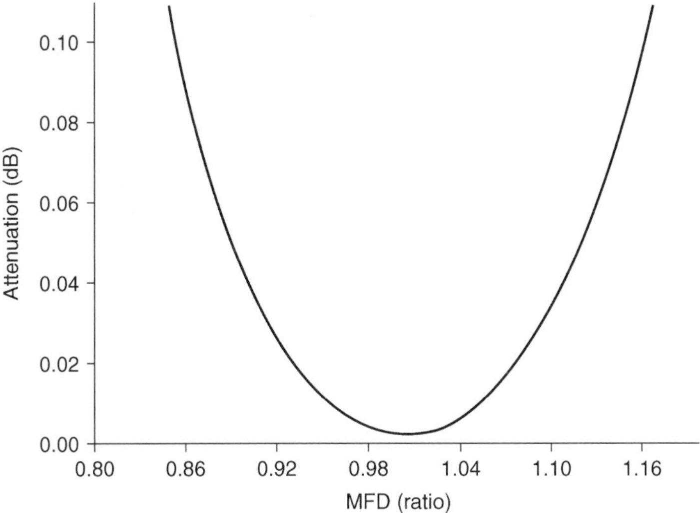

Figure 3.5. Single-mode intrinsic splice loss due to MFD mismatch. (Reprinted with permission from Corning, Inc., AN103, Ref. 4, Figure 2, page 3.)

MFD specification. There are additional splice process factors for mechanical (butt-spliced) joints. These include fiber end separation, fiber end angle, and Fresnel reflection.

There are splice process-related factors as well. These factors are those induced by the splicing methods and procedures. Splice process factors involve lateral and angular misalignment, contamination, and core deformation. The effects can be minimized by using properly trained technicians, automated fiber-alignment equipment, and fusing cycles on the more modern equipment.

Preparation of the fiber-optic cable for splicing involves the following tasks: fiber stripping, surface cleaning, and fiber end angle.

The fiber coating can be removed by a number of techniques such as chemically, by using thermal stripping equipment or by mechanical stripping. For typical acrylate-coated fibers, Corning recommends mechanical stripping because it is safe, fast, and inexpensive and creates a well-defined coating termination.

Surface cleaning is very important. Any acrylate coating residue that remains after stripping should be removed from the bare fiber surface. Handling the bare fiber before the fusion splice is completed should be avoided. This will minimize the chance of contaminating the fibers with dust or body oils, which may contribute to higher splice losses and lower tensile strengths.

The primary attribute affecting single-fusion splicing is *end angle*. Thus proper fiber end angle preparation is a fundamental step in obtaining an acceptable fusion splice. Requirements for fiber end angle vary from user to user and the type of cleaver used. In general, however, fiber end angles less than two degrees usually yield acceptable field fusion splices. However, one should expect end angles around half a degree when well-controlled cleavers are used.

Fiber Alignment. There are manual and automatic fusion splice alignment units. The operator first mounts the clean, cleaved fibers into alignment blocks and/or other holding mechanism of the splicer. The fibers are then visually aligned in the lateral (X–Y) directions. Visual alignment requires maintaining the smallest gap possible between the fibers to reduce the visual errors that may occur when manually aligning the edges of the fibers under magnification.

In the case of automatic alignment, the initial alignment involves nothing more than placing the fibers in the V-groove chucks. The alignment unit automatically aligns the fibers.

There are five equipment alternatives for final fiber core alignment:

1. Power monitoring using a source and detector
2. Use of optical time-domain reflectometer (OTDR) power monitoring
3. Local injection and detection (LID) techniques
4. Profile alignment techniques
5. Passive V-groove alignment

The power monitoring technique bases optimum fiber alignment on the amount of optical power transferred through the splice point. A light source is connected to the input end of one of the fibers to be fusion spliced. The light signal passes through the splice point, and its level is read on a power meter at the output end. The alignment is achieved by moving the fibers in the X and Y lateral directions until a maximum power reading is achieved. Two people are required with this technique. One reads the power meter while the other, some distance away, operates the splicer. This method is an improvement over visual alignment in that it optimally aligns the fiber cores rather than the cladding.

An OTDR may be used instead of a remote power meter as in the method above. It should be noted that OTDR alignment depends on the ability of the OTDR to provide a real-time display of power for alignment optimization.

Many fusion splicers employ a local injection and detection (LID) system. This is basically another power-alignment system that is self-contained at the fusion site. LIDs eliminate the need for remote monitoring of power level. With this device the fibers on either side of the splicing point are bent around cylindrical mandrels that are small enough to allow the injection of light through the fiber coating on the input side and detection on the output side.

Profile alignment systems present an image of the splice point to allow the technician to properly align the two fibers for a fusion splice. Colliminated light is directed through the fibers at right angles to the fiber axis right at the splice point. This produces an image of the fiber that can be brought into alignment. One specialized type of profile alignment creates a computer-generated image of the core centerlines where the computer automatically brings the fibers into alignment prior to fusing.

Another profile alignment system carries out the alignment procedure using the fiber-clad profile. It should be kept in mind that quality alignment depends a lot on the core-clad concentricity. With passive fixed V-groove alignment, the fiber alignment is a result of precision-machined V-grooves and precisely controlled fiber-clad diameter and core/clad concentricity.

The Fusing Procedure. The fusion process involves an electric arc for heating and fusing. Some technicians will apply one or more short bursts of arc current to remove any contaminants from the fiber points before starting the fusing procedures.

Pre-fusion is the next step. This is a heating process to soften the fiber ends to be joined. Pre-fusion is done to ensure that the fiber ends are at an optimum temperature during the final fusion step that allows the fibers to flow together upon physical contact. If the pre-fusion temperature is too high, there will be excessive fiber-end deformation that may result in a change in glass geometry. If the temperature is too low, mechanical deformation of the fiber ends may result. When this happens there will be fiber buckling as the fiber ends are forced together during the fusion step.

Optimum splicer settings include arc current and duration, gap distance, and overlap for the pre-fusion and fusion steps. The settings must be determined on a job-by-job basis.

Splice quality involves two principal parameters as discussed above. These are fiber strength and induced loss at the splice point. Some splicers have a pull test capability. Experienced technicians are known to use a manual pull test such that they can fairly estimate fiber strength.

Splice loss may be checked by remote OTDR or power meter using a method similar to the one described above for fiber alignment. Accurate measurement of splice loss by an OTDR requires averaged bidirectional measurements.

Section 3.4.2.2 is based on the Corning document AN103 (6/99) [Ref. 4].

3.5 FIBER-OPTIC BRANCHING COMPONENTS OR COUPLERS

3.5.1 Introduction

Branching components either divide a light signal into separate paths or take several light signals and combine them into one path. Some of the devices that carry out this function are called couplers. Different types of branching components with a brief description of their function(s) are listed below.

Combiner typically has one output port and two or more input ports. It can be used for unidirectional or bidirectional operation.

Splitter is a device that typically will have one input and several outputs. It can also be used for bidirectional transmission or to distribute a signal to two or more devices or end users.

Tree coupler accepts a single input and distributes the signal to several outputs or vice versa. It is commonly used to distribute signals from a single source to many recipients.

Star coupler is a multiple port device with at least two input ports and two or more output ports. It can distribute or combine signals from multiple input ports to a single output port, or it can accept a single light signal and distribute it to multiple output ports.

Broadband coupler or wideband coupler, also called wavelength-insensitive coupler, operates at both long wavelength windows, 1310 and 1550 nm. All branching components should have this dual-band capability. Another desirable capability is that branching components should be insensitive to variations in wavelength within a single window. In other words, insertion loss should be the same for any wavelength within a window.

Access coupler or tap is a three- or four-port branching component that facilitates an add–drop function, usually with some small fraction of optical power. Its coupling ratio is highly nonuniform. We find this type of coupler used in the HFC (hybrid fiber coax) plant, for signal line-status monitoring, and for add–drop multiplexing.

Wavelength division multiplexers/demultiplexers are nothing more than branching components or couplers. These devices distribute light signals based on their wavelength. A wavelength division multiplexer is used to transmit multiple light signals, each at a different wavelength, on a single fiber. A wavelength division demultiplexer takes an aggregate of light signals on a single fiber and separates them into their different wavelength components where each component wavelength is placed on a different individual fiber.

Branching components or couplers have broad application in local area networks where the transmission medium is optical fiber. The coupler provides a two-way coupling between the backbone fiber and an LAN station. Couplers also have application in a broadband wireless (radio) system called a local multipoint distribution system (LMDS). In this case they couple a fiber trunk to an LMDS node that houses an add–drop multiplexer, access control devices, and radio terminal(s). Couplers are also widely used in optical switches.

3.5.2 Coupler/Branching Component Concepts

Consider Figure 3.6. This shows what can happen if we place two pieces of fiber side by side in a flame and draw them into a *fused biconical tapered coupler*. Within

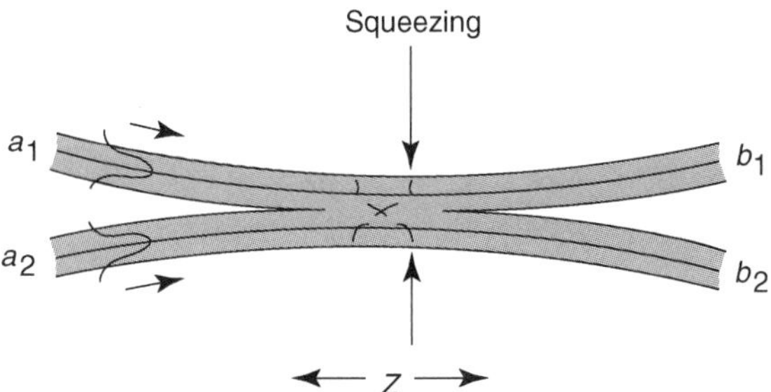

Figure 3.6. Conceptual sketch of a fused biconical coupler. Note the coupling region of length Z and the squeezing of the core field out into the "air cladding."

each fiber, there is a long tapered section, then a uniform section of length Z where they are fused together, and then another positive taper back to the original cross-section configuration of two separate fibers.

The tapers are very gradual, so that a negligible fraction of the energy incident from either of the left-hand ports is reflected back to either left-hand port. It is for this reason that these devices are sometimes called *directional couplers*.

Using this technique, a variety of couplers can be made that exploit the fact that the power coupled from one fiber to the other can be varied by changing the following: Z, the length of the *coupling region* over which the fields from the two fibers interact; a, the core radius in the coupling region; and Δa, the difference in core radii in the coupling region. A number of different couplers based on this concept will be described below.

Lower-order modes remain in the original fiber as the incident angle is still greater than the *critical angle*. The output converts the cladding modes back into *core modes*. The *splitting ratio* is determined by the length of the taper (Z in Figure 3.6) and the thickness of the cladding.

A typical power split for a coupler such as this may be 50 : 50, with half of the output power going to one output port and half going to the other output port. A first attempt at analysis might yield the following: If the input light signal were -10 dBm, the output at each port would seem to be -13 dBm. Unfortunately, this will be true. We did not include insertion loss. This is the loss brought about internally in the coupler, and the optical power dissipated one way or another in the coupler itself. A typical insertion loss in a coupler such as this one would be 0.7 dB. Thus the output signal level at each port would be -13.7 dBm. This coupler (power splitter) is part of the generic class of couplers employing the concept of the fused biconical tapered coupler described above. Many different types of couplers can be made out of this basic coupler such as power splitters as shown in Figure 3.7, combiners, Y-junction, star couplers, directional couplers, and so on.

Figure 3.8 shows a coupler that operates with single-mode fiber and is wavelength-dependent. When two couplers are brought close to each other in a biconical taper arrangement as in Figure 3.8, resonance occurs. The light in fiber A couples to the core of fiber B. The light power level transmitted into fiber B and fiber A depends on *coupling length*. Light from fiber A will couple 100%; that is, it will pass entirely into fiber B after a certain distance called the coupling length or odd multiple of the coupling length. The coupling length varies with the wavelength of the light in the fiber. The *splitting ratio* can be tuned by selecting the right coupling length.

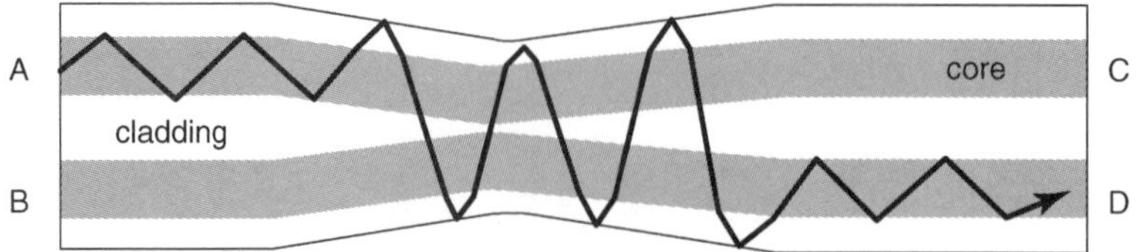

Figure 3.7. Coupler with multimode fibers. (Courtesy of Australian Photonics CRC, from the Internet, Ref. 5.)

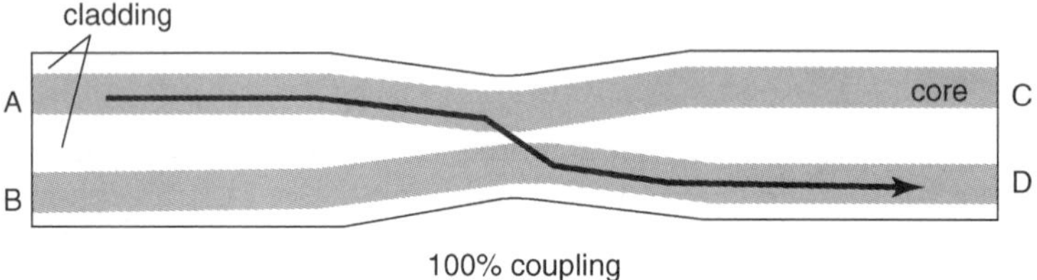

Figure 3.8. An optical coupler operating with single-mode fiber. (Courtesy of Australian Photonics CRC, from the Internet, Ref. 5.)

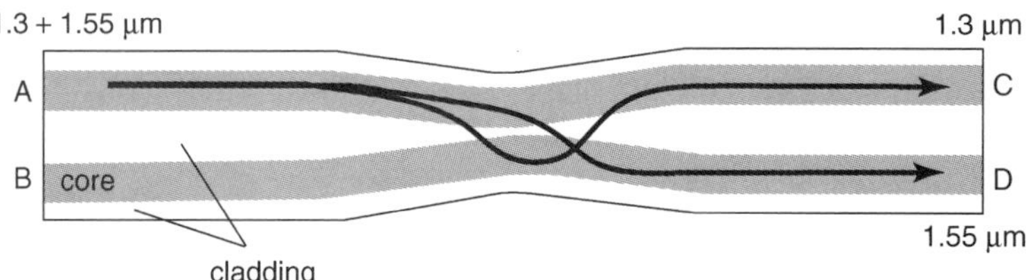

Figure 3.9. A coupler made of single-mode fiber showing wavelength separation. (Courtesy of Australian Photonics CRC, from the Internet, Ref. 5.)

Figure 3.10. A Y-junction or 1×2 coupler. (Courtesy of Australian Photonics CRC, from the Internet, Ref. 5.)

There is an important consequence of the fact that the coupling length depends on the wavelength of light in a single-mode coupler. Let's imagine that we transmit two wavelengths along a fiber such as 1.3 μm and 1.55 μm. The coupling length for 1.55 μm is greater than the coupling length of 1.3 μm. This results in the light at 1.3 μm coupling 100% into the core of fiber B, then back to the core of fiber A. Also, the light at 1.55 μm couples 100% to the core of fiber B. By carefully selecting the coupling length, it is possible to combine or separate two wavelengths. This concept is illustrated in Figure 3.9.

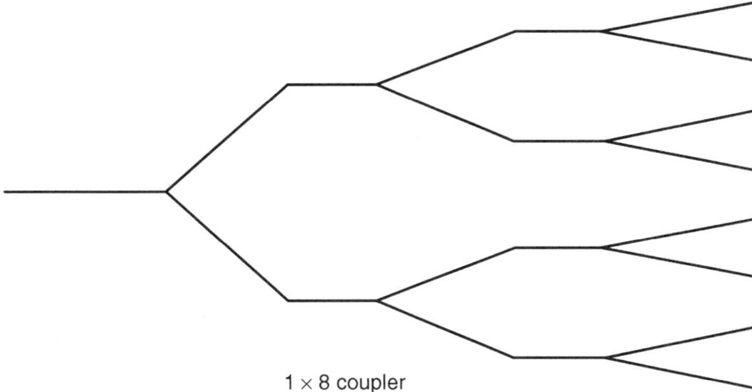

Figure 3.11. Y-junctions can be concatenated. (Courtesy of Australian Photonics CRC, from the Internet, Ref. 5.)

Figure 3.10 shows a coupler operating as a Y-junction or 1 × 2 power splitter. In this case, ideally, the light is split equally between the two arms. Y-junctions are difficult to make joining three fibers and the resulting device is lossy. The more practical approach is to create waveguides within a glass substrate.

Y-junctions can be concatenated to create 1 × 4 or 1 × 8 couplers as shown in Figure 3.11.

3.5.3 Coupler/Branching Components Operational Parameters

Table 3.2 gives functional transmission parameters for fiber-optic coupler or branching components.

3.5.4 Selected Coupler/Branching Component Definitions

Coupling Ratio. Coupling ratio or splitting ratio is defined as the ratio of the optical power from one output port of the coupler to the sum of the total power from all output ports. The coupling ratio is measured at the specified center wavelength. Multimode couplers are measured with an equilibrium mode fill.

TABLE 3.2 Transmission Parameters for Couplers/Branching Components[a]

	All Networks	
Parameter	Max	Min
Insertion loss (dB)	$4.0 \log_2 n$	Not applicable
Optical reflectance (dB)	−40	Not applicable
Operating wavelength range (nm)[b]	$\frac{1580}{1360}$	$\frac{1480}{1260}$
Polarization–dependence loss (ΔdB)	$0.1 (1 + \log_2 n)$	Not applicable
Directivity (dB)	Not applicable	50
Uniformity (dB)	$1.0 \log_2 n$	Not applicable

[a]Note that 2 × n devices for $2 \leq n \leq 32$ are under study.

[b]Assumes operation at either or both passbands; but if a restricted wavelength exists over a passband, then parameter values like loss apply only over that restricted band as well.

Source: Table 6.2, page 11, ITU-T Rec. G.671, Ref. 8.

Center Wavelength and Bandpass. The performance of all couplers varies with wavelength. Couplers are usually specified over a wavelength window or in some cases multiple windows. The center wavelength is the nominal wavelength of operation of the coupler, while the bandpass is the range of wavelengths over which the specifications are guaranteed.

Bandpass criteria recommended by Telcordia (Ref. 9) are the following:

For non-WDM and 1310/1550 nm operation and WDM branching components for all digital application except long-reach SONET, the following bandpass requirements for both long-wavelength operational bands should be met:

1260–1360 nm and 1480–1580 nm

For WDM applications and for long-reach SONET, the following bandpass requirements have been established by Telcordia:

1280–1335 nm and 1525–1575 nm

For DWDM, the recommended bandpass criteria are

1285–1325 nm and 1530–1566 nm

For HFC systems using AM-VSB transmission, the recommended bandpass criteria are

1290–1330 nm and 1530–1570 nm

It should be noted that fiber-optic component developers are now marketing various devices that extend well into the 1600-nm band.

Insertion Loss. For definition and discussion, see Section 3.3.1.

Typical Excess Loss. Excess loss is the ratio of the optical power launched at the input port of the coupler to the total optical power from any single output port, expressed in decibels. Typical excess loss is the expected value of the excess loss measured at the specified center wavelength. Multimode couplers are measured with an equilibrium mode fill.[2]

Excess Insertion Loss. In an optical waveguide coupler, excess insertion loss is the optical loss associated with that portion of the light which does not emerge from the nominally operational ports of the device (Ref. 10).

Uniformity. Uniformity is a measure of how evenly power is distributed between the output ports of the coupler. Uniformity applies to couplers with a nominally equal coupling ratio and is defined as the difference between the highest and lowest insertion loss between all the coupler output ports, expressed in decibels. Uniformity is a typical value across the entire bandpass. For additional thoughts on uniformity, see Section 3.3.9.

[2]*Equilibrium mode fill* is the condition in a multimode optical waveguide in which the relative power distribution (i.e., "fill") among propagating modes is independent of length (Ref. 10). Synonym: steady-state condition.

Telcordia, in Ref. 9, defines uniformity, ΔL, as the maximum variation of insertion loss between one port i and any two output ports j and k, or between input ports j and k and one output port i.

Branching components intended for use in digital systems at bit rates up to 10 Gbps should have a uniformity of

$$\Delta L \leq 0.8 \log_2 N$$

where N = number of ports being coupled.

For AM-VSB systems, uniformity is given by

$$\Delta L \leq 0.5 \log_2 N.$$

Uniformity is particularly important in DWDM and AM-VSB systems.

Directivity, Return Loss, and Reflectance. Directivity is the ratio of the optical power launched into an input port to the optical power returning from any other input port. Directivity has been referred to as near-end isolation or near-end crosstalk. Return loss is the ratio of optical power launched into an input port to the optical power returning from the same input port. Both directivity and return loss are expressed as positive decibels and are measured with all output ports optically terminated. Reflectance is the negative of return loss. In many cases, reflectance and return loss are used synonymously. Minimum directivity and return loss are the lower limits that apply over the entire wavelength range specified in the bandpass. For additional thoughts on reflectance, see Section 3.3.3.

3.5.5 Star Couplers and Directional Couplers—Additional Discussion

Star couplers have more than four ports. There are two types of star couplers: transmission star coupler and reflection star coupler. A transmission star coupler is illustrated in Figure 3.12.

Light arriving at an input port of a transmission star coupler is split evenly among all output ports. For example, in Figure 3.12, light arriving at input port E is split evenly among output ports G, H, I, J, K, and L.

There are tree and branch directional couplers with a splitting ratio of 1 × N: 2 × 2. Directivity is achieved with one main I/O (input–output) port and two branch I/O ports. The main fiber (tree trunk) can transmit optical power in two directions. The branch ports are unidirectional where optical power is routed to/from the main fiber. Figure 3.13 shows a reflection star coupler application.

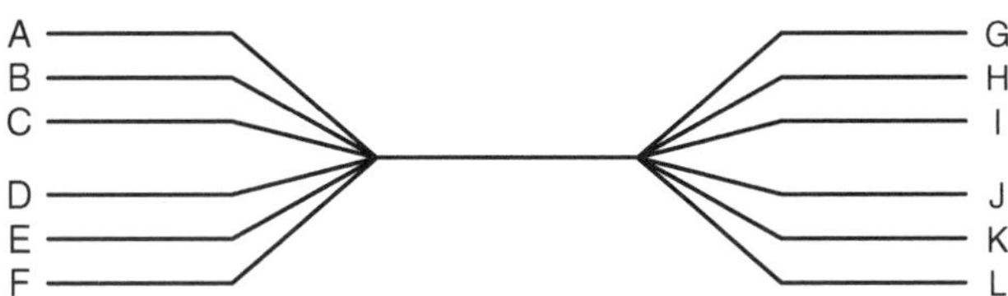

Figure 3.12. A transmission star coupler. (Courtesy of Australian Photonics CRC, from the Internet, Ref. 5.)

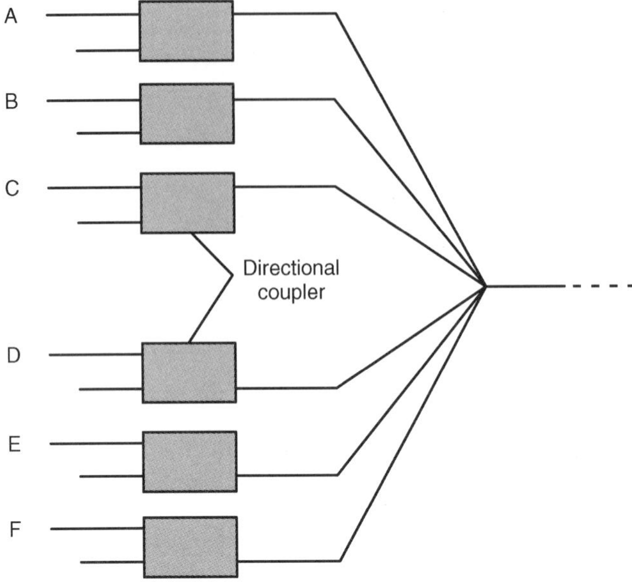

Figure 3.13. Reflection star coupler application. Note passive LAN star architecture. (Courtesy of Australian Photonics CRC, from the Internet, Ref. 5.)

3.6 OPTICAL ATTENUATORS

An attenuator is a device that reduces the intensity of the light signal passing through it. Attenuators are often used in a link circuit after a laser transmitter to tailor the output power to meet the needs of subsequent devices in the circuit such as EDFA (erbium-doped fiber amplifier; see Chapter 7). Care must be taken that these attenuators have excellent return loss properties (i.e., > 40 dB) to ensure that any light reflected back into the transmitter is of a very low level.

Other key parameters of attenuators are

- Stability
- Reliability
- ORL (optical return loss)
- PDL (polarization-dependent loss)
- Accuracy
- Repeatability
- Insertion loss
- PMD (polarization mode dispersion)

One of the most elementary applications is on short fiber-optic links where the light level is so high that it is outside the dynamic range of the light detector (receiver). An attenuator is placed in the circuit to reduce the light level such that it is in the receiver's dynamic range.

Another application of an attenuator is in WDM systems where the power in each channel is tailored so that the power delivered to the first in-line EDFA is spectrally flat. Consequently, flatness over a channel width is a key performance parameter.

Source: Ref. 7.

There are variable optical attenuators (VOAs) and fixed attenuators, sometimes called *pads*.

3.6.1 Selected Operational Parameters for Attenuators

The insertion loss tolerance should be no greater than $\pm 15\%$. For fixed attenuators we should expect loss (insertion loss) values of 3, 5, 10, 15, 20, 25, or 30 dB. The optical reflectance value should be -40 dB maximum. An attenuator's operating wavelength should extend from 1360 nm to 1580 nm maximum and from 1260 nm to 1480 nm minimum. The typical wavelength passband is 1310–1580 nm. Its polarization-dependent loss (PMD) should be no greater than 0.3 ΔdB.

3.7 ISOLATORS

An isolator exhibits a low loss to light passing through in one direction and a high loss in the opposite direction. Isolators are commonly placed in output circuits of devices with a high output light level such as laser diode transmitter and EDFAs. Their function is to reduce the level of reflected light back into the laser diode or EDFA.

The following are critical parameters that determine the performance of isolators:

- Wavelength-dependence, especially for so-called narrowband isolators that are designed to operate in a spectral range narrower than 20 nm. Isolators are described by peak reverse direction attenuation and by the bandwidth for which the isolation is within 3 dB of the peak value.
- Low insertion loss. < 1 dB in the forward direction, in excess of 35 dB (single-stage isolator) or 60 dB (double-stage isolator) in the reverse direction, and low polarization dependence.
- Polarization mode dispersion (PMD). Isolators are constructed using high birefringent elements; and they are very prone to PMD (typically 50–100 fs),[3] especially for single-stage designs. Double-stage isolators can be designed so that the PMD induced by the first stage is largely canceled by the second stage.
- Polarization-dependent loss (PDL). This degrades the performance of an optical isolator.

[3]The abbreviation fs stands for femtosecond (i.e., 10^{-15} s).

Common practice in isolators is to take advantage of the *Faraday effect*. It governs the rotation of the plane of polarization of the optical beam in the presence of a magnetic field. The rotation is in the same direction for light propagating parallel or antiparallel to the magnetic field direction. Optical isolators consist of a rod of Faraday material such as yttrium iron garnet (YIG), whose length is selected to provide 45° rotation. The YIG rod is sandwiched between two polarizers whose axes are tilted by 45° with respect to each other. Light propagating in one direction passes through the second polarizer because of the Faraday rotation. By contrast, light propagating in the opposite direction is blocked by the first polarizer (Ref. 6). The isolation should be greater than 30 dB. The reflectance of an optical isolator should be 40 dB or greater.

3.8 FIBER-OPTIC FILTERS

An optical filter requires a wavelength-selective mechanism and can be classified into two broad categories depending on whether optical interference or diffraction is the underlying physical mechanism. There are fixed wavelength and tunable filters. The desirable properties of a tunable filter include:

- Wide tuning range to maximize the number of channels that can be selected
- Negligible crosstalk to avoid interference from adjacent channels
- Fast tuning speed to minimize access time
- Small insertion loss

Filters carry out an extremely important function in WDM/DWDM equipment on the demultiplex side. ITU-T Rec. G.671 (Ref. 8) recommends the following parameter values:

Insertion loss: maximum, 1.5 dB, passband
Insertion loss: minimum, 40 dB, stopband
Optical reflectance: -40 dB

There will be further discussion of optical filters in Chapter 8.

3.9 CROSS-CONNECTS, PATCH PANELS, AND OPTICAL SWITCHES

Patch panels are straightforward. They are simply a panel that contains coupling adapters to allow cables to be plugged and unplugged. Most patch panels permit plugging on both sides to achieve maximum flexibility in arranging and rearranging the fiber cables. Others provide cable management to keep the routing and placement of cable orderly.

Patch panels are used as circuit access points for testing and trouble-shooting. They are used for rearrangement of equipment. Suppose an equipment or device has failed. A properly placed patch panel allows the craftsperson to route the circuit through standby or other equipment.

Distribution units and enclosures are similar to patch panels, except they provide additional space to organize fiber. It is advisable to leave some extra fiber. This slack is then available if it becomes necessary to reterminate a connection. Distribution units provide an area to store extra fiber in a large loop.

Cross-connects come in several varieties. Some technicians and field engineers use the term "cross-connect" synonymously with a patch panel where all operations are manual. We will define a cross-connect as synonymous with a light-wave switch.

One basic purpose of a switch in the network is to rearrange links. They carry out signal rerouting and are used for configuring a path or restoring a link. These switches operate in the optical domain and are automatic, being configured by some stimulus. Switches are also used in conjunction with optical add–drop multiplexers. They may also be used for optical routing. This routing is by optical wavelengths. The key parameters that determine performance of switches, and thus their suitability for particular network applications, are as follows:

- Insertion and coupling loss
- Return loss
- PDL (polarization-dependent loss)
- Crosstalk and isolation
- Reliability
- Switching time
- Stability
- Complexity

TABLE 3.3 Recommended Transmission Parameters for an Optical Switch

	$1 \times n$ Switches[a]		2×2 Switches	
Parameter	Max	Min	Mean	Standard
Insertion loss (dB)	$2.5 \mid \log_2 n$	Not applicable	Under study	Not applicable
Optical reflectance (dB)	−40	Not applicable	−40	Not applicable
Operating wavelength range (nm)	Under study	Under study	Under study	Under study
Polarization-dependent loss (δdB)	Under study $\mid 0.1(1 + \log_2 n)$	Not applicable	Under study	Not applicable
Switching time	$10 \mid 20$ ms	Not applicable	Under study	Not applicable
Repeatability (dB)	0.25	Not applicable	Under study	Not applicable
Uniformity (dB)	Under study $\mid 0.4 \log_2 n$	Not applicable	Under study	Not applicable
Crosstalk (dB)	Not applicable	Under study[b]	Under study	Not applicable
Directivity (dB)	Not applicable	50	Under study	Not applicable

[a]Dual values (a|b) indicate values for "slow" and "fast" switches respectively.
[b]A value of 25 dB is under consideration, pending an agreed definition of crosstalk.
Source: Paragraph 6.7, ITU-T Rec. G.671, Ref. 8.

Table 3.3 lists values of transmission parameters for an optical switch from ITU-T Rec. G.671.

Optical switching will be covered more extensively in Chapter 17.

REFERENCES

1. Tyco Electronics, Fiber Optic Division, catalog 1307895, Harrisburg, PA, May 2000.
2. *Corning Cable Systems Design Guide*, Release 4, Corning Cable Systems, Hickory, NC, 1999.
3. *An Introduction to Fiber Optic Networking*, AMP (Tyco Electronics), Harrisburg, PA, 1995.
4. *Single Fiber Fusion Splicing*, Corning Application Note AN103, Corning, Inc., Corning, NY, June 1999.
5. Australian Photonics CRC lecture series. From the Internet at http://central.vislab.usyd.edu.au/photonics/devices/networks/coupler, October 2001.
6. Govind P. Agrawal, *Fiber-Optic Communication Systems*, 2nd ed., John Wiley & Sons, New York, 1997.
7. *Guide to WDM Technology*, 2nd ed., EXFO Electrical-Optical Engineering Co., Varnier, Quebec, Canada, 2000.
8. *Transmission Characteristics of Passive Optical Components*, ITU-T Rec. G.671, ITU Geneva, November 1996.
9. *Generic Requirements for Fiber Branching Components*, Telcordia GR-1209-CORE, Issue 3, Piscataway, NJ, March 2001.
10. *The IEEE Standard Dictionary of Electrical and Electronic Terms*, 6th ed., IEEE, New York, 1996.
11. Overview of Selected IEC (International Electrical Commission) References.

    Most of the definitions of functional parameters specified in Section 3.3, for each of the above mentioned passive components, are given in the corresponfing IEC generic specification and are sumarized below:

    IEC Publication 869-1 (1994), *Fibre optic attenuators—Part 1: Generic specifications.*

    IEC Publication 874-1 (1993), *Connectors for optical fibres and cables—Part 1: Generic* specification.

    IEC Publication 875-1 (1992), *Fibre optic branching devices—Part 1: Generic specification.*

    IEC Publication 876-1 (1994), *Fibre optic switches—Part 1: Generic specification.*

    IEC Publication 1073-1 (1994), *Splices for optical fibres and cables—Part 1: Generic* specification—Hardware and accessories.

    IEC Publication 1202-1 (1994), *Fibre optic isolators—Part 1: Generic specification.*

    IEC Publication 1931-1, *Fibre-optic terminolgy.*

4

LIGHT SOURCES

4.1 CHAPTER OBJECTIVE AND SCOPE

Turning to Figure 1.1, we see that a fiber link in the light domain consists of a light source or transmitter (this chapter), combined with a light detector or receiver (Chapter 5) interconnected by sections of fiber-optic cable (Chapter 2) that are jointed by connectors and splices (Chapter 3). Chapters 1–5 set the groundwork for the remainder of the book.

Modern light sources or transmitters contain integrated circuits and laser diodes (LDs) or light-emitting diodes (LEDs). These are modulated by separate integrated circuits, which have now largely replaced early fiber-optic transmitters made up of discrete electrical components and electro-optical devices. Today very large scale integration has arrived on scene to meet the need of ever-higher modulation rates and to improve reliability. Figure 4.1 is a simplified functional block diagram of a fiber-optic transmitter.

There are two generic types of light sources employed in fiber-optic circuits for telecommunications: LEDs and LDs. In this chapter we will consider these devices alone and treat them as separate entities, knowing well that they really will form part of an integrated circuit scheme just described. We will also discuss a new source type, the VCSEL (vertical-cavity surface-emitting laser).

4.2 LIGHT-EMITTING DIODES

The LEDs used in telecommunications emit light in the near-infrared region. They are inexpensive when compared to most lasers. LEDs are primarily used with multimode fiber because they emit light in a broad cone that can only be captured efficiently by the large numerical aperture of multimode fiber.

As described by Govind Agrawal (Ref. 1), an LED in its simplest form is a forward-biased p–n homojunction. Radiative recombination of electron–hole pairs in the depletion region generates light; some of it escapes from the device and can be coupled into an optical fiber. The emitted light is incoherent with a relatively wide spectral width (30–60 nm) and a relatively large angular spread.

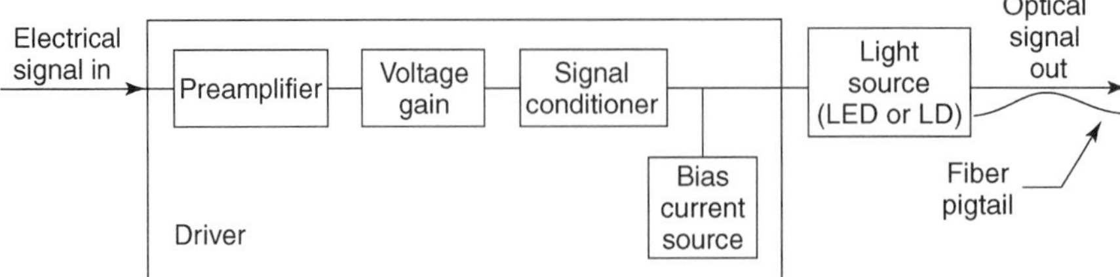

Figure 4.1. A simplified functional block diagram of a fiber-optic transmitter.

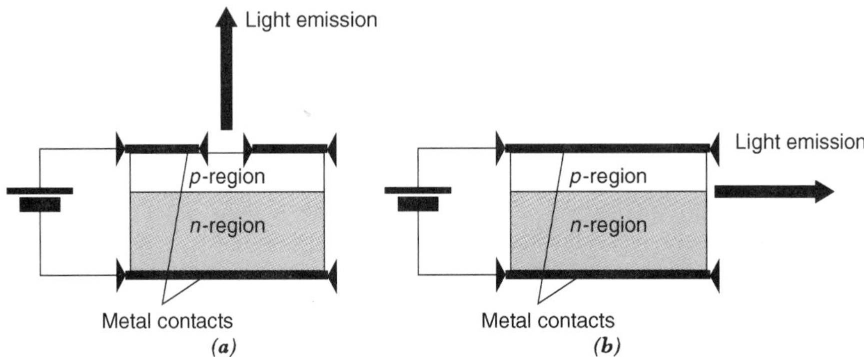

Figure 4.2. Cross-section sketch of (a) a surface-emitting LED and (b) an edge-emitting LED.

An LED structure can be classified as surface-emitting or edge-emitting, depending on whether the LED emits light from a surface that is parallel to the junction plane or from the edge of the junction region. Figure 4.2 shows the design of the two types of LEDs. Both types can be made using either a p–n homojunction or a heterostructure design in which the active region is surrounded by p- and n-type cladding layers. The heterostructure design leads to superior performance. It provides a control over the emissive area and eliminates internal absorption because of the transparent cladding layers. The LED is notoriously inefficient. With proper design, surface-emitting LEDs can couple 1% of the internally generated power into an optical fiber.

4.3 LASER DIODES

When system requirements are less stringent, the LED is the light source of choice. Laser diodes (LDs) are warranted for use on long-distance links and/or with high bit rates (e.g., above 155 Mbps). There are several types of LDs:

- Multilongitudinal mode (MLM) or *Fabry–Perot*
- Single longitudinal mode (SLM)
- Single longitudinal mode with distributed feedback (DFB), more often called a DFB laser

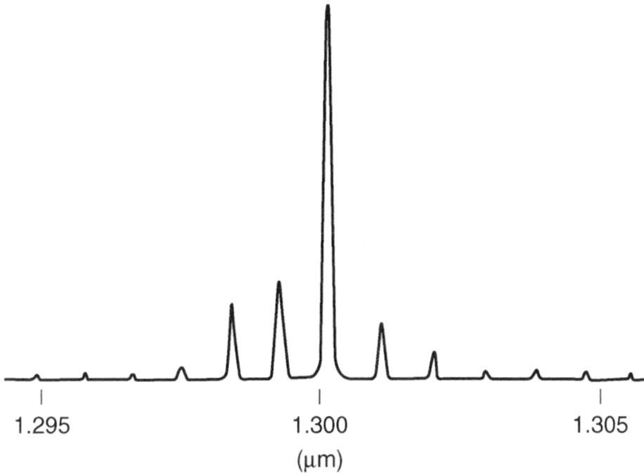

Figure 4.3. CW spectrum for a 1300-nm wavelength MLM (Fabry–Perot) laser.

- DFB laser with external modulator
- Vertical-cavity surface-emitting laser (VCSEL)

The listing of laser types is in the order of presentation; they are also listed in chronological order of development.

4.3.1 The MLM or Fabry–Perot Laser

The MLM or Fabry–Perot laser emits a multimode emission pattern as illustrated in Figure 4.3. What we mean here is that the output of an MLM laser has a dominant spectral line on the desired wavelength and subsidiary lines separated by about 1 nm that have some lesser amplitudes. As the laser is modulated the subsidiary modes are modulated right along with the dominant mode. The full width at half-maximum (FWHM)[1] of the laser, when modulated, is approximately 4 or 5 nm.

A closer examination of the laser spectrum shows that although the total output power is relatively stable, there can be significant variation in the power of each individual mode. This phenomenon, known as *mode partitioning*, has important practical implications. When the laser signal is transmitted over fiber with wavelength-dependent group delay (*chromatic dispersion*), mode partitioning gives rise to noise in the output signal. This introduces a power-independent error rate floor into the system response which cannot be overcome by making allowances in the system power budget. With systems operating at data rates higher than a few hundred Mbps on low-loss fiber, that can constitute a major limit to span length. Moreover, even small reflections back into the laser from external surfaces (connectors) can cause significant change in the mode partitioning behavior and thus the performance of the system.

[1]FWHM (full width at half-maximum) is the spectral width in nanometers (nm) of an optical source at one-half peak optical power level.

It has been found that even modes which, on average, have only a few percent of the total power have a finite probability for having more than half the total power. It may therefore be appropriate to define the effective spectral width of a laser as the spectral range covered by modes that, on average, carry 1% or more of the total power (Ref. 2).

4.3.2 Single Longitudinal Mode (SLM) Laser

SLM lasers are designed such that cavity losses are different for different longitudinal modes of the cavity, in contrast with the MLM whose losses are independent. In the MLM laser the longitudinal mode with the smallest cavity loss reaches threshold first and becomes the dominant mode. The other nearby modes are discriminated against by their high losses, which prevent buildup from spontaneous emission. In this case the power carried by these secondary modes is usually of low level, less than 1% of the total emitted power. When an SLM is operating properly, we can expect the first side mode to be at least 30 dB down from the dominant mode.

4.3.3 Distributed Feedback Semiconductor Laser (DFB)

DFB laser structure has built-in wavelength selectivity by means of a feedback mechanism. The feedback is not localized at the facets but is distributed throughout the cavity length. This type of laser contains a periodic grating between the two layers of the laser structure (typically at the interface n-InP substrate and n-InGaAsP layers) to provide feedback at a fixed wavelength that is determined by the grating pitch. This corresponds to a periodic variation in the mode index.

A DFB laser is very sensitive to light feedback, particularly from the connector where the laser interfaces with the main-line fiber. Even a relatively small amount of feedback—for instance, less than 0.1% (Ref. 1)—can destabilize the laser and affect system performance. For example, when line width can be broadened, mode hopping and RIN[2] enhancement can occur. Several steps can be taken to reduce the intensity of the feedback or to mitigate the effects of the feedback. One step commonly taken is to use antireflective coatings. Feedback can also be reduced by cutting the fiber tip at a slight angle (see Chapter 3) so that the reflected light does not strike the active region of the laser. Another, more drastic step is to install an isolator (see Chapter 3) between the laser and fiber connector interface.

An important parameter of a DFB laser is the *mode suppression ratio* (MSR). The principal objective with this type of semiconductor laser is to attenuate subsidiary longitudinal modes and to obtain maximum power in the dominant mode (seeFigure 4.3). We can expect an MSR > 30 dB for a DFB operating continuously. Our interest here is to transmit a light signal from the laser with a singular narrow spectral line (i.e., the dominant mode). The FWHM we can expect under ideal conditions is less than 0.2 nm. When a DFB structure is combined with multiple quantum well (MQW) structures to improve line width of the produced laser light, the line width can be down to a few hundred kHz (Ref. 3, page 103). As

[2] RIN stands for *relative-intensity noise*, a form of laser-generated noise.

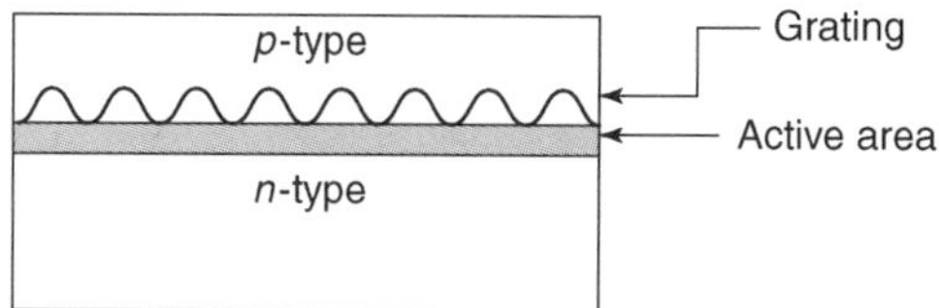

Figure 4.4. Sketch outline of a DFB laser.

linewidths get larger, chromatic dispersion increases (see Chapter 6 for a discussion of chromatic dispersion). This is highly undesirable for bit rate systems above 1 Gbps. The DFB laser has the narrowest line width among practical production-type lasers on the market. It is almost universally employed on long-haul fiber-optic circuits.

A DFB laser is expensive and its operation is vital to a fiber-optic circuit. To ensure optimum operation of a DFB laser and to monitor its vital operation, several components are added to the DFB assembly. For example, a photodiode (PIN receiver; see Chapter 5) monitors its output; a thermoelectric cooler (TEC) or heatpump and heatsink control the junction temperature of the laser chip, and a feedback circuit controls its output and maintains the desired frequency. The ideal temperature for the laser chip is 25°C.

Figure 4.4 is a sketch of a DFB laser.

4.3.4 DFB Laser with External Modulator

Up to this point we have discussed or alluded to optical sources with direct modulation, sometimes called intensity modulation. Conceptually, all we are doing is turning the laser on and off, where on may represent a binary 1 and off may represent a binary 0. In actuality, the laser is never turned completely off. The equivaalent "off" is a point on the laser's operational curve at just above threshold (i.e., there is a very small output level) or just below threshold. This threshold setting is important to reduce *chirp*, which will be discussed below.

Another approach to impinging (modulating) the binary 1s and 0s is to use an optical modulator. The concepts of direct modulation and that of employing an optical modulator are illustrated in Figure 4.5. Note that the optical modulator is positioned between the laser carrier wave (CW) source and the fiber cable output interface. A *CW source* is a light source that is "on" or operational all the time at its rated power output.

Optical modulators are integrated components designed to control the amount of continuous optical power transmitted in an optical waveguide. They act like shutters; the shutter is closed for a binary 0 and open for a binary 1. There are commonly three types of modulators:

1. Mach–Zehnder (M–Z)
2. Electro-refraction
3. Electro-absorption (semiconductor) MQW

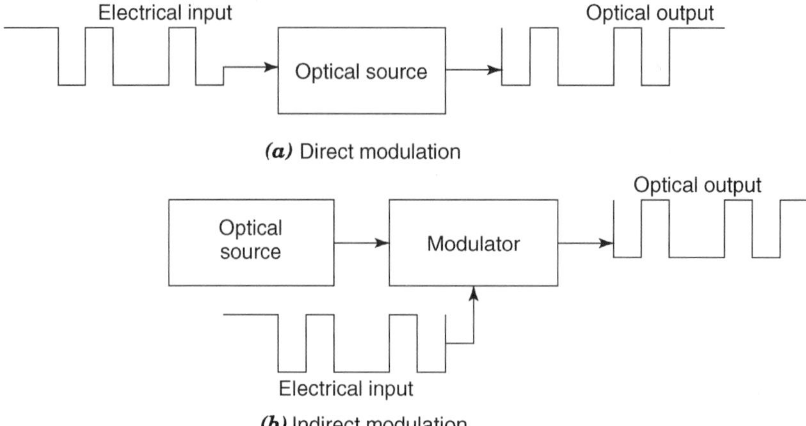

Figure 4.5. Conceptual drawings of (a) a DFB laser transmitter using direct modulation and (b) the same laser diode using an external modulator.

The Mach–Zehnder (M–Z) modulator is an interferometer that makes use of titanium-diffused $LiNbO_3$ waveguides or a directional coupler configuration. The waveguides in an M–Z form a Y-configuration. The refractive index of electro-optic materials such as $LiNbO_3$ can be changed by applying an external voltage. In the absence of the external voltage, the optical fields in the two arms of the M–Z interferometer experience identical phase shifts and interfere constructively. The additional phase shift introduced in one of the arms through voltage-induced index changes destroys the constructive nature of the interference and reduces the transmitter output. In particular, no light is transmitted when the phase difference between the two arms equals π, because of the destructive interference occurring in that case. As a result, the electrical bit stream applied to the modulator produces an optical replica of the bit stream.

The performance of an external modulator is quantified through the on-to-off ratio, more often called the *extinction ratio* and modulation bandwidth. $LiNbO_3$ modulators provide an extinction ratio greater than 20 (13 dB) and can be modulated at bit rates up to 75 Gbps.

Modulators have been fabricated using electro-optical polymers. Here modulation bit rates have been achieved up to 60 Gbps. Such modulators are often integrated monolithically with the driving circuitry.

Another type of modulator is made using semiconductors. These are *electro-absorption* modulators. The technique takes advantage of the Franz–Keldysh effect, according to which the bandgap of the semiconductor decreases when an electric field is applied across it. Thus, a transparent semiconductor layer begins to absorb light when its bandgap is reduced by applying an external voltage. This is when photon energy exceeds bandgap energy. Because the electroabsorption effect is stronger in MQW structures, MQW has become the structure of choice in this type of modulator. An extinction ratio of 15 dB or more for an applied reverse bias of 2 V can be realized at a bit rate of several Gbps. Low chirp-rate transmission has been achieved at 5 Gbps. These types of modulators are used on fiber-optic circuits at bit rates in excess of 20 Gbps and have been demonstrated up to 60 Gbps.

We reiterate that the basic purpose of using a modulator is to reduce pulse broadening caused by chirp. Many of these modulators are monolithically incorporated on the same chip of the transmitter they control.

4.3.5 Vertical-Cavity Surface-Emitting Laser (VCSEL)

The MLM (Fabry–Perot), SLM, and DFB lasers require current on the order of tens of milliamperes to operate. In addition, their output beam to interface the circular optical fiber has an elliptical cross section, typically with an aspect ratio of 3 : 1. Such a beam makes a bad match of the cylindrical beam capture of the fiber. A noncylindrical beam often requires additional optics to enhance the interface to a circular cross-section fiber core. A VCSEL emits the desirable circular beam. A comparison of these beams is illustrated in Figure 4.6.

A VCSEL consists of a vertical sandwich of *p*-type multilayer, an active region, and an *n*-type multilayer. The number of layers depends on the wavelength desired. The multilayers comprise Bragg reflectors that are fabricated with In + Ga + As + (Al or P). For instance, an In + Ga + As + P combination is used for lasers in the wavelength window of 1310–1550 nm. The layers are made with epitaxial growth followed by planar processing. VCSELs operate in a single longitudinal mode by virtue of an extremely small cavity length ($\sim 1\ \mu$m) for which the mode spacing exceeds the gain bandwidth. They emit light in a direction normal to the active-layer plane in a manner analogous to that of a surface-emitting LED. Operation in a single transverse mode can be realized by reducing the VCSEL diameter to 2–3 μm. The output power and bandwidth of VCSELs are typically lower than those of edge-emitting BFB lasers, and VCSELs have application in fiber-optic LANs and local loops. They are comparatively low-cost devices when compared to DFB lasers, for example. Another application is to operate VCSELs

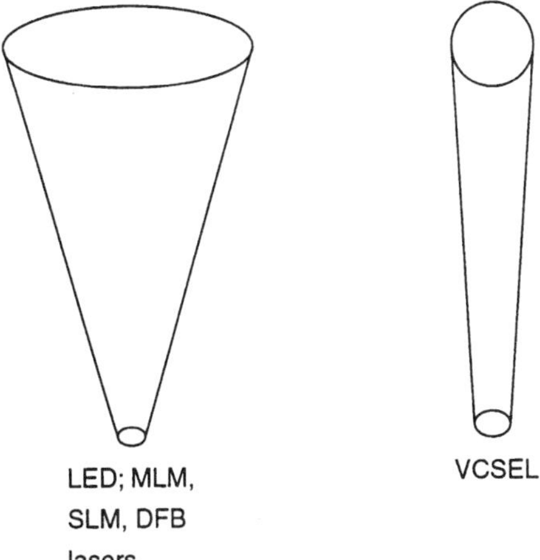

Figure 4.6. Comparison of the elliptical output beam of an LED and of MLM, SLM, and DFB lasers versus circular output beam typical of a VCSEL.

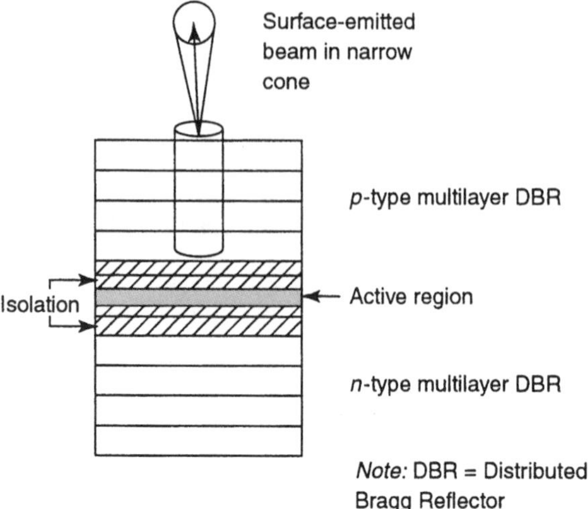

Figure 4.7. A conceptual sketch of a VCSEL.

in arrays where each laser operates at a different wavelength, which makes them ideally suited for WDM lightwave systems.

A conceptual sketch of a VCSEL is shown in Figure 4.7 (Refs. 1 and 3).

4.4 CHIRP

4.4.1 Introduction to Frequency Chirp

Frequency chirp can limit the performance of a 1550-nm lightwave system even with DFB lasers with large values of MSR[3] ($\sim$ 40 dB). As we have discussed, intensity modulation (direct modulation) in semiconductor lasers is invariably accompanied by phase modulation because of the carrier-induced change in the refractive index governed by the linewidth enhancement factor. Optical pulses with a time-dependent phase shift are called "chirped." As a result of the frequency chirp imposed on an optical pulse, its spectrum is considerably broadened. Such a spectral broadening affects the pulse shape at the fiber because of fiber dispersion (see Chapter 6) and will degrade system BER.

Source: Ref. 1.

4.4.2 Detailed Discussion of Chirp

A pulse is said to be chirped if its carrier frequency changes with time. The frequency change is related to the phase derivative. The time-dependent frequency shift is called the *chirp*. In the literature we will read about the parameter C. This parameter governs the *linear frequency chirp* imposed on the pulse. Chirped pulses may broaden or compress. Pulse broadening, of course, is undesirable, taking on

[3]*MSR* stands for mode suppression ratio. See Section 4.3.3.

the characteristics of dispersion. Broadening of a pulse spills pulse energy into the next bit position and, if the broadening is sufficient, will cause at least 1 bit in error if the bit that is supposed to be in that position is a binary 0.

The chirp parameter C can take on negative values as well as positive values. The BL product[4] is reduced dramatically for negative values of C. This is due to enhanced broadening of a laser diode pulse. For directly modulated semiconductor lasers, C is generally negative with a typical value of -6 at 1.55 μm. Because $BL < 100$ Gbps-km under such conditions, fiber dispersion limits the bit rate to about 2 Gbps for $L = 50$ km (i.e., a link 50 km long). This problem can be overcome by using dispersion-shifted fibers or by using dispersion-compensation schemes (see Chapter 6). Frequency chirp is essentially eliminated with the use of an external modulator. The reason for this is that the light source, typically a DFB, is on all the time. However, with direct modulation, chirp becomes very evident and, as we said, can be destructive.

Source: Ref. 1.

4.4.3 Pulse Parameters Showing the Effects of Chirp

Figure 4.8 shows a typical waveform from a directly modulated semiconductor laser. The y axis represents amplitude, and the x axis is time. Important concepts illustrated in this figure are the initial rise of the pulse and its overshoot. This is chirp-related. The fact that the following pulse peak characteristics are different than those of the initial pulse is also caused by chirp. Each pulse top will be different from the next in a random fashion.

Pulse rise time is an extremely important parameter. It sets a limit maximum on system bit rate. Setting of threshold affects system performance. Threshold should be set as low as possible without losing the lasing condition. This will affect the

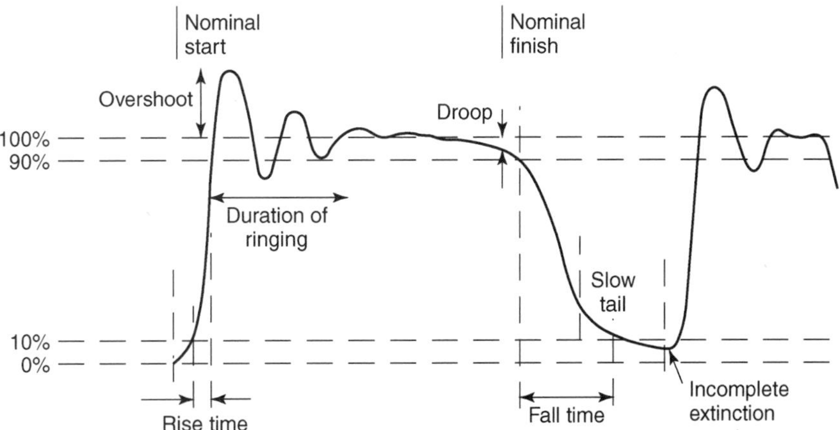

Figure 4.8. Pulse parameters. (From Ref. 2, Figure 3.12.)

[4]BL product stands for bandwidth–distance product (e.g., 10 GHz-km).

extinction ratio, which we want as high as possible. This is the ratio of pulse peak amplitude to the threshold amplitude.

4.5 POWER PENALTIES

Agrawal (Ref. 1) lists five physical phenomena that, in combination with fiber dispersion, degrade the signal-to-noise ratio at the far-end receiver on high-bit-rate fiber-optic systems (i.e., > 500 Mbps). These sources of degradation are

1. Modal noise
2. Dispersion broadening
3. Mode-partition noise
4. Frequency chirp
5. Reflection feedback and noise
6. Extinction ratio

Modal noise is an impairment typical on fiber-optic systems using multimode fiber. Its cause is the interference among the various propagation modes in multimode fiber. It manifests itself by degrading signal-to-noise ratio at the far-end receiver because of the fluctuating amplitude of the received signal. This is why most multimode systems use LEDs rather than laser diodes. This is because of an LED's very wide spectral bandwidth.

Dispersion broadening can limit the bit-rate–distance (BL) product. This pulse broadening affects receiver performance in two ways. The first we have discussed previously. We refer here to the fact that the broadening can spill pulse energy into the subsequent bit slot which will cause intersymbol interference (ISI).

The second factor is that the pulse peak energy is reduced because of the broadening. So we have to increase incoming signal level at the receiver to compensate for this factor. To quantify the power penalty, let us define δ_d:

$$\delta_d = 10 \log f_b$$

where f_b is the pulse broadening factor.

The power penalty can be calculated by

$$\delta_d = -5 \log 10\left[1 - (4BLD\sigma_\lambda)^2\right]$$

where B is the bit rate, L is the length of the link (in km), D is the dispersion factor (in ps/nm-km), and σ_λ is the rms width of the source spectrum, which is assumed to be Gaussian. (A Gaussian pulse is a rounded pulse versus a square-wave pulse.)

Agrawal (Ref. 1) gives some values for δd as follows:

When $BLD\sigma_\lambda = 0.1$, $\delta_d = 0.38$ dB

When $BLD\sigma_\lambda = 0.2$, $\delta_d = 2.2$ dB

When $BLD\sigma_\lambda = 0.25$, $\delta_d =$ infinite

Mode-partition noise (MPN) occurs in multimode fiber systems and is brought about by the semiconductor laser source. It is a phenomenon that occurs due to anticorrelation among pairs of longitudinal modes. There may be many such modes present, and individual modes exhibit notable power fluctuation even though the total power remains relatively constant. These various modes, as they travel down the fiber, become desynchronized because they are traveling at different velocities. This causes fluctuations in the receiver signal current degrading the signal-to-noise ratio. A power penalty must be paid for this.

The calculation of this penalty is complex, involving the mode-partition coefficient k whose value ranges from 0 to 1. δ_{mpn} is the power penalty in dB for MPN. The power penalty can be reduced to a negligible level ($\delta_{mpn} < 0.5$ dB) by designing the optical system such that $BLD\sigma_\lambda < 0.1$. Consider the following example (Ref. 1) based on a 1.3-μm system. We assume that the operating wavelength is matched to the zero-dispersion wavelength within 10 nm, $D \approx 1$ ps(km-nm). A typical value of σ_λ for multimode semiconductor lasers is 2 nm. The MPN-induced power penalty would be negligible if the BL product were below 50 Gbps-km. At $B = 2$ Gbps, the transmission distance is limited to 25 km. The MPN power penalty is quite sensitive to spectral bandwidth of the multimode laser. It can be reduced by reducing the spectral bandwidth.

When we turn to 1.55-μm operation with DFB lasers, MPN can be severe enough for a power penalty. MPN can be reduced in this situation by using a DFB laser with good MSR characteristics (i.e., numerically > 30 dB).

Frequency chirping was discussed in Section 4.3 above. With directly modulated transmitters there is an optimal setting of bias current which will achieve minimum chirp. This is a adjustment to obtain a certain r_{ex} value, where r_{ex} is the extinction ratio or ratio of the on power to the "off" power. We have $r_{ex} = P_0/P_1$ or the laser output power in the binary 1 condition to its output power in the binary 0 condition. Above-threshold bias setting increases the extinction. This decreases the receiver "sensitivity." The total penalty can be reduced below 2 dB by operating the system with an extinction ratio of about 0.1.

Source: Ref. 1.

Reflection feedback provides light reflected back into the laser source. This reflected light, even at low levels, can be a source of system upset or cause major degradation in performance. In fact it can degrade system performance to the extent that the system cannot achieve the required BER despite an infinite increase in receive power. Most reflection in a fiber link originates at the fiber–air interfaces. It should be kept in mind that a considerable fraction of the transmitted signal can be reflected back unless precautions are taken to reduce this optical feedback. In our discussion up to this point, we have mentioned a number of ways of reducing the reflection level or the effects of the reflection(s).

Generally speaking, most lightwave systems operate satisfactorily when the reflection feedback is below -30 dB. In practice, the problem can be nearly eliminated by using an optical isolator in the transmitter assembly fiberguide well prior to using the pigtail connector.

Extinction Ratio. An insufficient extinction ratio can incur a power penalty. A fiber-optic light source has an on state and an off state. We assign a binary 1 to the on state and assign a binary 0 to the off state. The trouble lies in the fact that in the off state the light transmitter is not completely off. The reason for doing this is to greatly reduce the initial rise time of the light transmitter, allowing the transmitter to have a much greater bit rate than it would have if it was completely off in the binary 0-state. If we allow P_0 to be the off-state output power and allow P_1 to be the on-state output power of the light transmitter, then the extinction ratio is defined as

$$r_{ex} = P_0/P_1$$

Stated another way using decibels, the extinction ratio is

$$EX = 10\log(A/B)$$

where A is the average optical power of a logical 1 and B is the average optical power level of a logical 0.

Based on values of r_{ex}, the following are equivalent extinction ratio penalties. These are based on using a PIN detector.

r_{ex}	Penalty (dB)
0.5	3
0.4	2.2
0.3	1.7
0.2	1.0
0.1	0.5
0.07	0.3
0.05	0.2
0.02	0.1

Source: Ref. 8.

4.6 TYPICAL PERFORMANCE PARAMETERS OF COMMERCIAL LIGHT SOURCES

4.6.1 Light-Emitting Diodes

The output power of an LED, depending on the design and manufacturer, is between 0.01 and 0.1 mW (−20 to −10 dBm).

There are two undesirable performance characteristics of LEDs that the system designer must take into consideration:

1. Emitted beamwidth
   a. Surface-emitting LED ~ 120°
   b. Edge-emitting LED ~ 30°
2. Emitted spectral width: 30–80 nm

LEDs are available that operate on all three band windows, namely, 850, 1310, and 1550 nm. However, they are more commonly used in the 850- and 1310-nm bands.

TABLE 4.1 Summary of ELED Characteristics

Characteristic	Value
Output power into single-mode fiber (25°C)	2–50 μW
Numerical aperture (NA)	0.1–0.6
Rise/fall times	3 ns max
Half-power linewidth (25°C)	30–60 nm
Output power temperature coefficient	1.2% /°C typical
Center wavelength variation with temperature	0.5–0.8 nm/°C
Spectral broadening	0.4 nm/°C, typical

Sources: Based on data from Refs. 2, 4, and 5.

LEDs are considerably cheaper than laser diodes; they have a much longer operational life and do not require temperature control if operated in normal ambient. Regarding coupled light, their efficiency is low. However, between the two types the edge-emitting LED is much more efficient than the surface-emitting LED. Their application is generally limited to circuits carrying 155 Mbps or less. The LED is the light source of choice for premises systems. Table 4.1 presents a summary of edge-emitting LED characteristics.

4.6.2 Laser Diodes

The semiconductor laser diode is the light source of choice for high-bit-rate (e.g., > 155 Mbps), long-distance systems. Laser output power from many laser manufacturers is on the order of +3 to +10 dBm (1 to 10 mW). Laser diodes with more output power (e.g., +20 dBm) are beginning to appear on the market. Now place a fiber-optic amplifier, typically an EDFA (see Chapter 7), at the output of a laser diode source, and up to 500 mW or more of power can be developed. One reason that manufacturers have been forced to develop lasers with much greater output power is the use of dense wavelength division multiplexing (DWDM). DWDM in-line components tend to be very lossy. These higher power levels plus the use optical amplifers help overcome the DWDM losses. Lasers operating at these high power levels have brought up a raft of new issues, especially the effects of such power levels on downstream optical components. Isolators are used universally to reduce reflected light energy back into the laser source. Such isolators have return losses greater than 80 dB, greatly reducing reflected power.

We place in tandem a DFB laser diode (typically a Lucent D2500 type), a Mach–Zehnder-based external modulator (typically Lucent Lithium-Niobate [developmental]), and a Lucent EDFA with 28-dB gain. Let the output of the laser diode be 0.0 dBm and let the external modulator have a 6-dB insertion loss. There are 2 dB of other losses such as isolator loss and connector loss. The power output of the combination will be

$$0.0 \text{ dBm} - 6 \text{ dB} - 2 \text{ dB} + 28 \text{ dB} = +20 \text{ dBm}$$

If the recommended (by Lucent) dispersion penalty is applied, the output drops to +18 dBm.

TABLE 4.2 Comparison Summary: ELED Versus Several LDs[a]

Characteristic	Edge-emitting LED	FP Laser (MLM laser)	DFB Laser with External Modulator	VCSEL
Wavelength (nm)	850/1310	1310/1550	1550	850/1310
Coupled power, typical (dBm)	−10 to −15	0.0	0.0	0.0
Spectral width (nm)	30–60	< 3	< 0.1	< 3
Maximum bit rate (Gbps)	< 0.155	> 2	> 10[b]	2
Fiber type, typical	Multimode	Single mode	Single, dispersion shifted	Multimode or single mode
Cost	Economical	Moderately expensive	Expensive	Moderate
MTBF[c] (hours)	10^9	10^8	10^7	10^8

[a]The spectral line width is given in terms of full width at half-maximum (FWHM). This is the spectral width of an emitted light signal from an optical source measured in nanometers at one-half the peak optical power level.

[b]Presently the value is being extended to 40 Gbps with no end in sight.

[c]MTBF (mean time between failure). The values are estimated, and the device is being operated in a normal ambient.

Source: Based on Refs. 1, 2, 4, 6 and 7.

4.6.3 Comparison of LED Performance Characteristics to the Performance Characteristics of Several Types of Laser Diodes

Table 4.2 shows the comparisons of ELED versus laser diodes.

***4.6.3.1 Spectral Width Comparison*—LED Versus LD.** Figure 4.9 is a sketch comparing the spectral line width of an LED output to that of a laser diode.

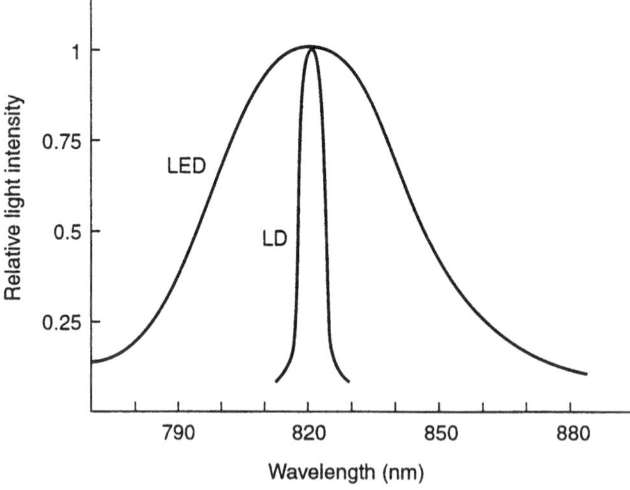

Figure 4.9. Typical spectral linewidths for LED and LD. Note that the peak intensities have been normalized to the same value; actual peak intensity of the LD is much greater than the LED.

TABLE 4.3 Basic Optical Characteristics of a Single Longitudial; Mode (DFB) Laser Transmitter

Parameter	Typical	G.957 Value	Industry	Comments
Average optical power output	0 dBm	+3 dBm	+2 dBm to +6 dBm	With OFA: over +20 DBm [1, 2]
Output power variation	0.5 dB	+3 to −2 dB	0.5 dB	Definition
Dispersion penalty	1.0 dB	Under study	2.0 dB for long reach	Value placed in link budget
Operating wavelength range	1280-1480; 1500-1650	1260-1360 1500-1580	1260-1480 1500-1660	Extending to the 1600 nm band is desirable[3]
Minimum return loss	24-30 dB	24 dB	24dB	
Extinction ratio	10 dB	8,2, 8.3 dB	11 dB	> 11 dB desirable

[1]Power output should be adjustable to accomodate short links.
[2]High output power is required to compensate for WDM component losses.
[3]The range of operation in the 1500 nm band may be limited to the range of operation of the EDFAs employed

Source: ITU-T Rec. G.957, Ref. 11, and Refs. 1, 6, 7, and 8.

TABLE 4.4 Tanable Laser Comparisons

LaserType	Advantages	Disadvantages	Applications
Distributed Feedback (DFB)	Wavelength stability; in production	Comparatively low power output[1]; limited tuning range	Narrow tuning range; widely used on long-haul links
Distributed Bragg reflector (DBR)	Fast switching speed [2]	Broad Line Width; wavelength instability	Access Links, OADM application
Sampled-grating (DBR)	Broad tuning range; fast switching speed[2]	Low output power; broad line width; non-continuous tuning	Access links; metro switching; OADM
Vertical cavity surface-emitting laser (VCSEL)	Narrow line width for O/P[3]; low power consumption circular beam, broad tuning range	Low output power for E/P[3], confined to short wavelengths (850/1300 nm)	Metro network and links
Micro-external cavity lasers	High-Power, narrow line width, low RIN, continuous tuning; broad tuning range	Switching speed shock/vibration sensitivity	Long haul and ultra-long haul links; metro area OADM; switching

Source: Iolin Inc., San Jose, CA. Refs. 12 and 13.

4.6.4 Operational Characteristics of Several Commercial Laser Diodes

4.6.4.1 Lucent Technologies C488-Type 2.5-Gbps Tunable, Long-Haul Laser Transmitters. The C488 transmitter is based on a 1500-nm laser with an electro-absorptive modulator and an integrated stabilizer. Lucent states that the device is capable of 360-km or 640-km transmission at 2.5 Gbps. It may be used in DWDM systems with 50-GHz or less channel spacing. Table 4.3 gives its basic optical characteristics, and Table 4.4 lists dispersion performance.

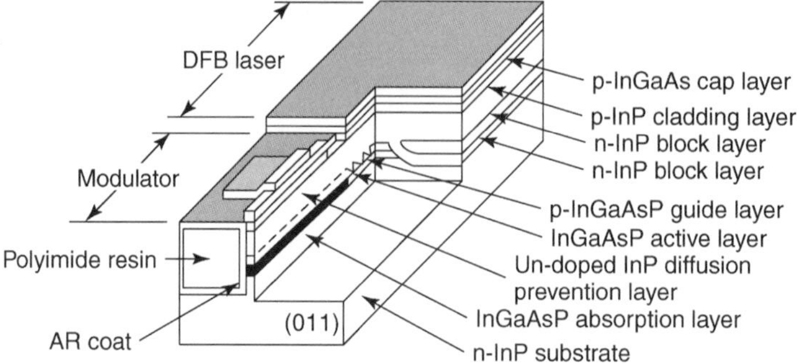

Figure 4.10. Schematic structure of the OKI DFB laser with modulator. (From Ref. 7, Figure 1.)

4.6.4.2 OKI 10-Gbps DFB Laser Transmitter

Overview—Structure. Figure 4.10 illustrates the structure of the OKI DFB laser with modulator. MOVPE (metal organic vapor-phase epitaxy) was used for all crystal growth processes. The DFB laser region is a buried structure that has an ordinary *pn* block laser. The active layer is a five-layer MQW structure where well layers are InGaAs and the barrier layers are InGaAsP. The modulator region has a high mesa ridge structure where an InGaAsP bulk with bandgap wavelength of 1.48 μm is used for the absorption layer. Both ends of the ridge are buried in with polymide resin to decrease electrode capacitance. The front-end surface of the modulator is treated with antireflection coating. The optical coupling between the DFB laser and the modulator is based on the butt-joint method. This laser has a laser region length of 350 μm, and the modulator region length is 200 μm. These regions are separated by a 30-μm-long electrode separation region. The separation resistance of this region is 10–20 kΩ.

The optical output power of the laser is about 4 mW (+6 dBm) at a laser drive current of 100 mA, the light extinction efficiency is 10 dB/V, and the −3-dB frequency bandwidth is 16 GHz. The sideband parameter is 0.3 to 0.4.

Driver IC. The driver IC drives the modulator at 10 Gbps with an output amplitude of 2 V peak-to-peak or more. The output waveform is close to rectangular (the rise and fall times are short with a little jitter). The drive IC consists of a two-stage differential amplification circuit. This amplification is linear, and there is limiting of amplitude in the differential amplification area in the second stage. This makes the amplitude of the output signals constant. The FET used for this IC is an InGaAs/GaAs stained BP-MESFET (buried P-layer MESFET) with a gate length of 0.2 μm.

Figure 4.11 shows an eye diagram of the output waveform of the driver IC at 10-Gbps operation. The test input signals were pseudorandom NRZ signals (see Section 4.7) with an amplitude of 0.8 Vpp. The output amplitude is 2.7 Vpp and the rise and fall time at 20–80% amplitude were 32 ps and 30 ps, respectively. For small-signal-frequency characteristics, the gain was 15 dB, and the −3 dB bandwidth is 9.5 GHz. Power consumption was about 1.8 W. The output is sufficient to drive the modulator at 10-Gbps operation.

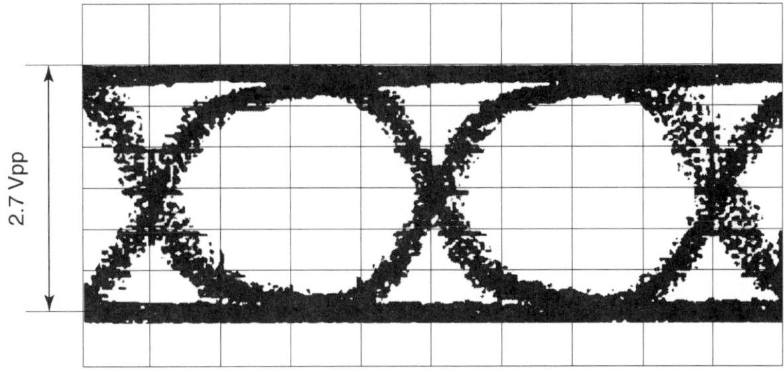

Figure 4.11. Eye diagram of driver IC output waveform at 10 Gbps. (From Ref. 7, Figure 2.)

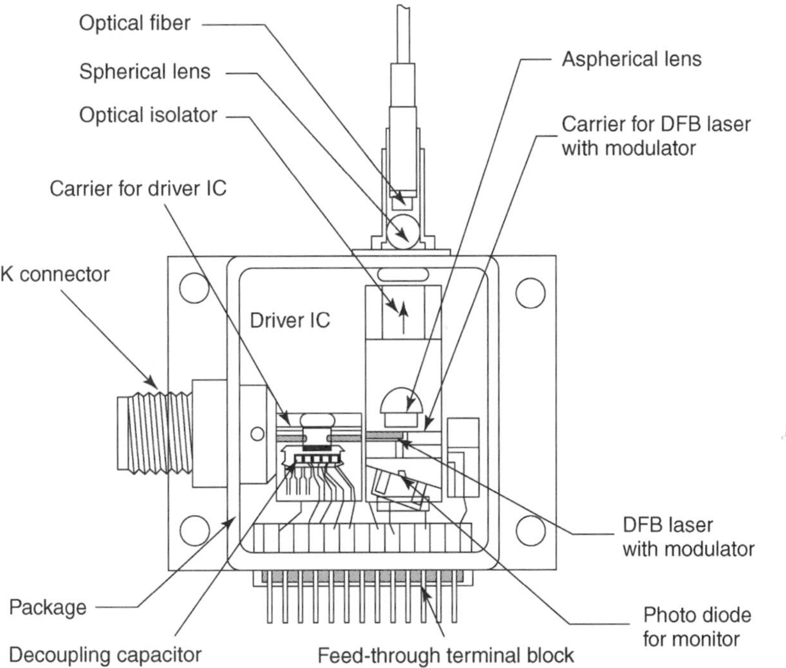

Figure 4.12. Schematic structure of the optical transmitter module (top view). (Courtesy of OKI Electric Co., Ref. 7, Figure 3.)

Module Configuration. Figure 4.12 shows the top-view structure of the optical transmitter module. The module encloses a DFB laser with modulator, the driver IC, a photodiode for a monitor, a thermistor, a thermoelectric cooler to stabilize internal ambient, a terminating resistor, a lens, and an optical isolator.

Characteristics. The optical coupling efficiency of the optical transmitter module is over 60%. The extinction ratio is 12 dB when the input amplitude of the driver IC is 0.6 V. Figure 4.13 shows the small-signal-frequency response characteristics

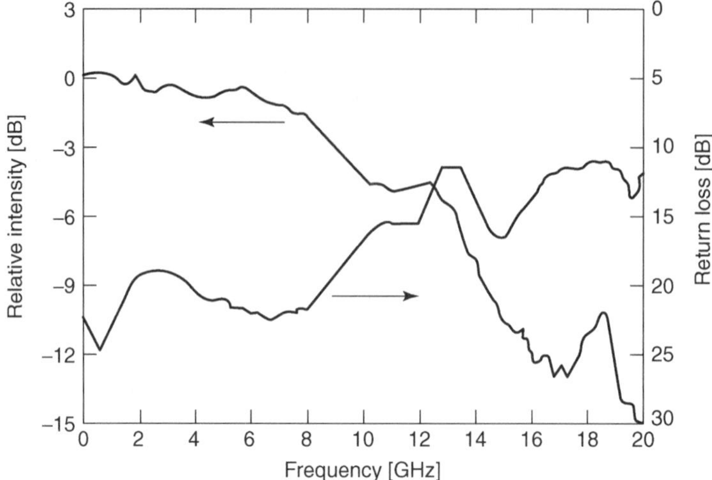

Figure 4.13. Small signal frequency characteristics. (Courtesy of OKI Electric Co., Ref. 7, Figure 4.)

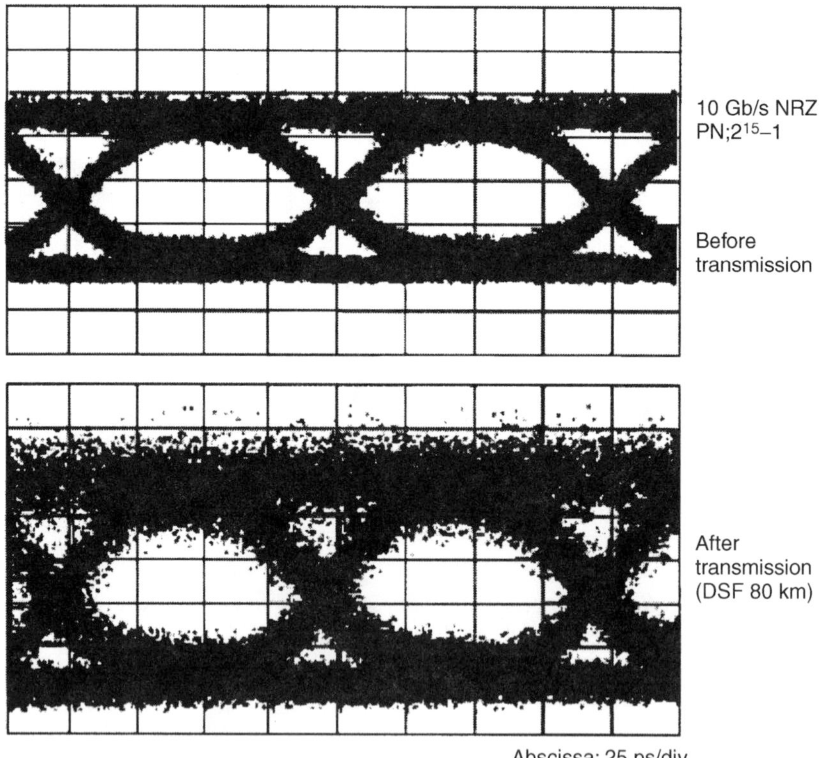

Figure 4.14. Eye diagram before and after transmission. DSF = dispersion-shifted fiber. (Courtesy of OKI Electric Co., Ref. 7, Figure 5.)

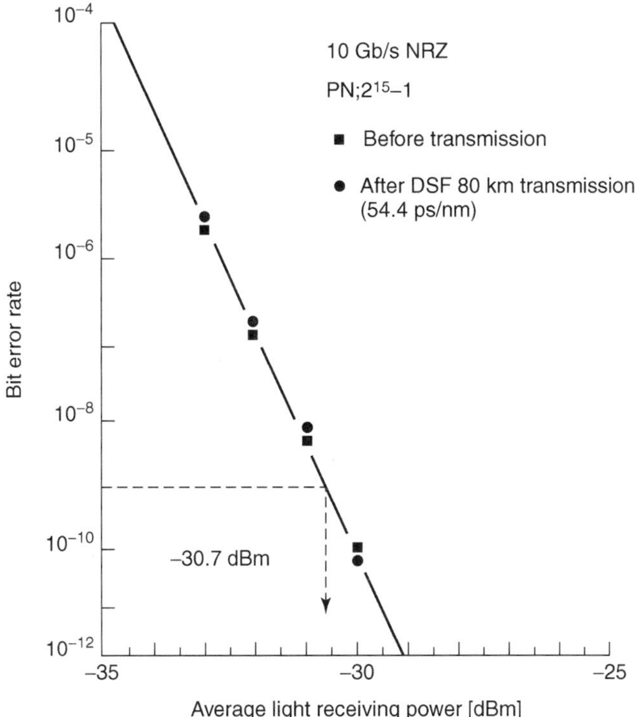

Figure 4.15. BER versus light signal level at receiver. (Courtesy of OKI Electric Co., Ref. 7, Figure 6.)

and return loss values. The −3-dB bandwidth is 9 GHz, and the return loss is 15 dB or more up to 12 GHz.

On a 10-Gbps transmission experiment using the module, the input electrical signals to the module was 10-Gbps NRZ pseudorandom signals. The output from the module was amplified by an EDFA and then was input to an optical fiber line. The link length was 80 km using 1.55-μm dispersion-shifted fiber. The total dispersion was 54.4 ps/nm where the operational wavelength was 1.554 μm. The far-end receiver converted the optical input signals to an electrical equivalent. These electrical signals were passed through a timing extraction circuit and then were passed to a decision circuit.

Figure 4.14 shows the eye diagram before and after transmission. Figure 4.15 shows the resulting BER characteristics versus incoming signal level at the receiver. For a BER of 1×10^{-9}, the receive signal level was −30.7 dBm. There was no power penalty required and no error rate floor was evident, even after transmission.

4.7 TUNABLE LASERS

The tunable laser will be the heart of optical switching in a DWDM optical network. It will also serve an important role in an advanced optical ADM application. The optical network is discussed in Chapter 17.

There are three major performance requirements for DWDM optical networking, namely link length, channel spacing (e.g., 100, 50, or 25 GHz), and transmission rate (e.g., 2.5, 10, and 40 Gbps). Different switching and ADM applications will drive distinct sets of performance requirements. These requirements are dependent on output power, tuning time, and tuning range, all of which are parameters of tunable lasers included in the listing below.

Four distinct applications are identified for tunable lasers:

1. Optical add–drop multiplexing (OADM)
2. Metropolitan area optical networks (MANs)
3. Long-haul links as part of optical network
4. Ultra-long-haul links

The choice of tunable laser technology will be governed by many performance parameters. Among these parameters are:

- Output power
- Line width
- Tuning range
- Relative intensity noise
- Tuning time
- Stability

There are a variety of laser structures that can be selected for the production of tunable lasers. We identify five such structures. Each structure has its own advantages and disadvantages. Several of these devices have been discussed previously in this chapter. Our discussion here is from the viewpoint of a tunable laser rather than a fixed wavelength device. These are the DFB lasers (Sections 4.3.3 and 4.3.4), distributed Bragg reflector (DBR), sampled-grating DBR (SGDBR), vertical-cavity surface-emitting laser (VCSEL) (Section 4.3.5), and external-cavity lasers (ECL).

4.7.1 Tunable Distributed Feedback Laser

As described in Section 4.3.3, the DFB laser is a comparatively simple structure that uses an internal grating to change the wavelength of operation. Ideally the designer would tune these lasers to ITU grid wavelengths (see Chapter 8, Table 8.3). Wavelength—or, if you will, frequency—is changed by changing the temperature of the medium. This can be done through drive current changes or with a temperature-controlled heat sink. The heat sink changes the refractive index of the internal waveguide. Modern thermoelectric coolers can precisely control the temperature to generate a narrow-line-width frequency output.

The tuning range of DFB lasers is limited to about 5 nm; and as the tuning temperature increases, the DFB's efficiency and output power decreases. One way to extend the tuning range is to use an ensemble of these devices integrated side-by-side as an array, typically three lasers on a chip which are coupled into a single output. Only one laser at a time is driven to select a wavelength. It has been found that this approach is not continuously tunable, the resulting chip size leads

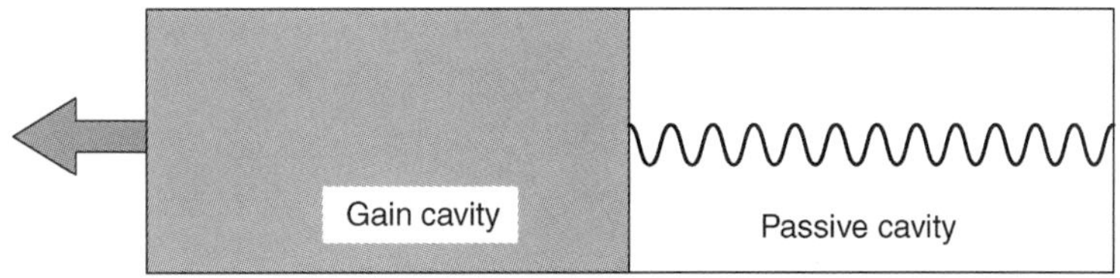

Figure 4.16. Sketch of distributed Bragg reflector laser. (Courtesy of Iolon, Inc., Ref. 11.)

to yield issues, and the combining mechanism is inefficient. It is difficult to achieve mode stability for each of the laser sections.

4.7.2 Distributed Bragg Reflector Laser

Distributed Bragg reflector (DBR) lasers are made up of two or more sections with at least one active region (gain cavity) and one passive region as illustrated in Figure 4.16.

As shown in the figure, the passive region contains a grating. Each end of the laser cavity has a reflective surface. The laser wavelength is tuned by current changes to the passive region to change the refractive index. The difference between a DFB laser and a DBR laser is that the active region and the grating region are separated in the DBR, while with the DFB the regions are combined.

The tuning range of a DBR is about 40 nm; its tuning is very fast. One drawback is that designs can be limited by current saturation. Another drawback is that it is difficult to control the optical path length between the two reflectors at each end of the cavity. This results in a broad line width and instability.

4.7.3 Sampled-Grating DBR

The SGDBR is a tunable laser that uses grating reflectors at either end of the cavity to produce a spectral comb response. Because the back and front sampled gratings have slightly different pitches, the resulting spectral combs have slightly different mode spacing. By changing the current in the two grating sections, alignment of the two combs at the selected wavelength is achieved.

To improve mode stability and decrease noise, an additional contact is required to adjust the phase. This allows an integral number of half-wavelengths to exist. When changing wavelengths, the laser appears to "hop" between wavelengths.

SGDBRs have a wide tuning range. However, they tend to be complex and to suffer from low output power, and the output has a broad line width. The fabrication of the device is very intricate.

4.7.4 VCSELs

The vertical-cavity surface-emitting laser (VCSEL) is described in Section 4.3.5. It offers several advantages as a tunable laser. Its emission line width is narrow, shows low power consumption, and can offer continuous tunability without mode hops. Its principal disadvantage is its limited output power. The low power output is fundamental to the VCSEL design. It is due to the constraint to maintain a single spatial mode of operation with a very small active region.

4.7.5 External-Cavity Laser

The external-cavity laser (ECL) is of straightforward design. With the external-cavity approach, the beam wavelength is altered mechanically by adjusting the laser cavity. Other tuning approaches use current or temperature changes that are applied to the semiconductor material.

Figure 4.17 illustrates a grating-based ECL that is designed in the Littman–Metcalf cavity configuration. The laser is a simple Fabry–Perot laser diode (see Section 4.3.1). The laser consists of a separately fabricated gain medium and an external cavity. This cavity is made up of separately fabricated optical structures. These are a diffraction grating and mirror integrated at an assembly step. Tuning is achieved by applying a voltage to the actuator of the MEMS (microelectromechanical system). This rotates a mirror to allow a particular diffracted wavelength to couple back into the laser diode. The actual wavelength of the laser output is determined by combining the gain bandwidth of the diode, the grating dispersion, and the external-cavity mode structure.

ECLs have many favorable characteristics for a switched optical network and for optical add–drop multiplexers. They have continuous tuning across the bandwidth range of interest. They display narrow line widths with high stability and low noise. An ECL does not display mode hops as DBRs do, for example. An ECL has comparatively high power output. Their disadvantages are their size and cost. They are insensitive to shock and other environmental influences.

The use of MEMS technology has negated many of these disadvantages. For example, with the use of MEMS in optical component designs that readily fit on standard transmitter cards, ECLs have become more cost-competitive.

Reference 9 states that one key breakthrough technology in the development of MEMS-based ECLs is the use of *deep reactive ion etching* (DRIE) techniques to

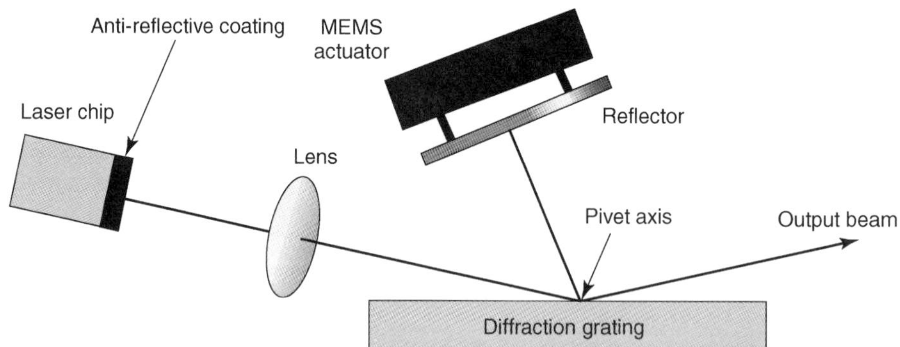

Figure 4.17. Tunable external cavity laser. Note that a mirror rotating on a MEMS actuator is combined with a laser-chip gain medium and a diffraction grating to couple a unique wavelength of light back into the laser chip, forming a tunable external-cavity laser. (Courtesy of Iolon, Inc., Ref. 11.)

TABLE 4.5 Tunable Laser Comparisons

Laser Type	Advantages	Disadvantages	Applications
Distributed feedback (DFB)	Wavelength stability; established fabrication process	Low output power,[a] limited tuning range	Narrowly tunable applications
Distributed Bragg reflector (DBR)	Fast switching speed[b]; established fab process	Broad line width; wavelength instability	Access; switching; optical add–drop multiplexing (OADM)
Sampled-grating DBR	Broad tuning range; fast switching speed[b]	Low output power; broad line width; noncontinuous tuning	Access; metro; switching OADM
Vertical-cavity surface-emitting laser (VCSEL)	Narrow line width (for O/P)[c]; low power consumption; mode stability; circular beam emitted; test at water level	Low output power (for E/P)[c]; traditionally confined to short wave-lengths (850/1300 nm)	Metro; access

[a]Fixed-wavelength DFB lasers with > 20-mW output power, but tunable DFBs have limited power.
[b]When thermal stabilization is not required. Thermal stabilization typically leads to > 25 ms tuning time.
[c]The O/P (optically pumped); VCSEL demonstrates narrow line width but higher output power. The E/P (electrically pumped); VCSEL has a large line width (~ 150 MHz) and low output power.

Source: Courtesy of Iolon, Inc., Ref. 11.

fabricate the MEMS actuators. These DRIE techniques allow cost-effective and reliable production of rigid mechanical drive structures. These provide a suitable force for high-speed and high-precision movement of optical elements over large linear and angular deflections. MEMS actuators are low cost; they are quite accurate and insensitive to shock, vibration, temperature changes, or creep.

These same devices can be repackaged to produce tunable receivers, polarization controllers, optical monitors, variable attenuators, optical switches, and tunable filters. The typical DRIE MEMS ECL has a 10-mW output and a 13-nm tuning range. Units are also available with 20-mW output and tune across a 40-nm range.

Table 4.5 gives comparisons, advantages, and disadvantages of the four of the tunable lasers discussed.

4.8 MODULATION WAVEFORMS

An important digital design consideration is the format of the electrical signal driving the light source. In other words, the manner in which the 1s and 0s are presented to the light modulator or to the laser input itself in the case of direct modulation is important for a number of reasons.

First, amplifiers for fiber-optic receivers are usually AC-coupled. As a result, each light pulse that impinges on the detector produces a linear electrical output response with a low-amplitude negative tail of comparatively long duration. At high bit rates, tails from a sequence of pulses may tend to accumulate, giving rise to a condition known as baseline wander, and such tails cause intersymbol interference. If the number of "on" pulses and "off" pulses can be kept fairly balanced for periods that are short compared to the tail length, the effect of AC coupling is then merely to introduce a constant offset in the linear output of the receiver, which can be compensated for by adjusting the threshold of the regenerator. A line-signaling format can be selected that will provide such a balance. The selection is also important on synchronous systems for self-clocking at the receiver.

Figure 4.18 illustrates five commonly used binary formats. Each is briefly discussed in the following:

1. *NRZ* (*Nonreturn to Zero*). This is the most common signal format in data communications that one is likely to encounter. Care must be taken with signal sense, because it is reversed from what is standardized in, say, EIA-232 or ITU-T Recs. V.10 and V.11. In the context of fiber optics, the binary 1 is the active signal element and binary 0 is the passive element. With the NRZ waveform, a change of

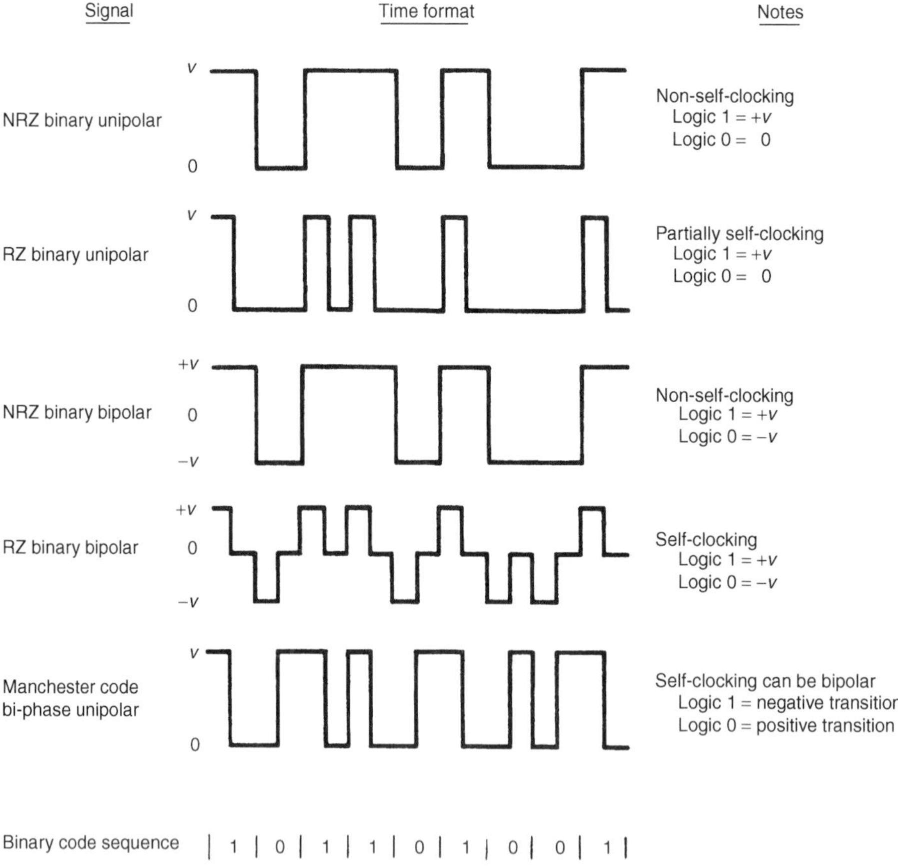

Figure 4.18. Five examples of binary signal formats.

state only occurs when there is a 1-to-0 or 0-to-1 transition. A string of binary 1s is a continuous pulse or "on" condition, and a string of 0s is a continuous "off" condition. In NRZ, information is extracted from transitions or lack of transitions in a synchronous format, and a single pulse completely occupies the designated bit interval (or bit position).

2. *RZ* (*Return to Zero*). In this case there is a transition for every bit transmitted, whether a binary 1 or a binary 0, as shown in Figure 4.18; as a result, a pulse width is less than the bit interval to permit the return-to-zero condition. (*Note:* When we say "return-to-zero," we mean the signal returns to zero volts. Of course the "zero" condition is not exactly zero volts. Its value is determined by the laser or LED threshold setting, often some very small positive voltage.)

3. *Bipolar NRZ*. This is similar to NRZ, except the binary 1s alternate in polarity.

4. *Bipolar RZ*. This is the same as bipolar NRZ, but in this case there is a return-to-zero condition for each signal element (i.e., bit), and again the pulse width is always less than the bit interval.

5. *Manchester Code*. This code format is used on a number of fiber-optic systems. Here the binary information is carried in the transition that occurs at midpulse. By convention a logic 0 is defined as a positive-going transition, and a logic 1 is defined as a negative-going transition. The signal can be either unipolar or bipolar.

The choice of code format is important in fiber-optic communication system design, and there are a number of trade-offs to be considered. For instance, RZ formats assist in reducing baseline wander. To extract timing on synchronous systems, the Manchester code and the RZ bipolar codes are good candidates because of their self-clock capability. However, it will be appreciated from Figure 4.18 that they will require twice the bandwidth as the NRZ unipolar code format. An advantage of the Manchester code is that it can be unipolar, which adapts well to direct intensity modulations of LED and LD sources and also provides at least one transition per unit interval (i.e., the bit) for self-clocking.

With the NRZ format we can attain the highest power per information bit if we wish to tolerate baseline wander.[5] Achieving this power is especially desirable with LED sources. On the other hand, LD sources can be driven to high power levels for short intervals, thereby conserving the life of the laser diode and making the RZ format attractive. Longer life with a shorter duty cycle can be traded off for higher modulation rate, which will result in greater bandwidth requirements. As we have seen, RZ systems require more than twice the bandwidth of NRZ systems for a given bit rate. This is because RZ uses many more transitions per unit of time for the same binary sequence as NRZ.

[5]For a discussion of baseline wander, consult *Telecommunication Transmission Handbook*, 4th ed., (Ref. 8).

REFERENCES

1. Govind P. Agrawal, *Fiber-Optic Communication Systems*, 2nd ed., John Wiley & Sons, NY, 1997.
2. *Optical Fibres Systems Planning Guide*, CCITT, Geneva, 1998.
3. Stamatios K. Kartalopoulos, *Introduction to DWDM Technology*, IEEE Press, New York, 1999.
4. *Fiber Optics System Design*, MIL-HDBK-415, US Department of Defense, Washington, DC, February 1985.
5. *Lightwave* (Magazine), World-wide Directory of Fiber Optic Communications Products and Services, Pennwell Corp., Tulsa, OK, March 25, 2000.
6. Lucent Technologies Advance Data Sheet (from the Web), C488-Type Laser Transmitter, Lucent Technologies, Murray Hill, NJ, February 2000. (Now Agere Technologies.)
7. OKI Technical Review, "Optical Transmitter Module for 10 Gbps Optical Communication Systems," OKI Electric Industry, Inc., No. 158, Vol. 63, Sunnyvale, CA, April 1997.
8. Roger L. Freeman, *Telecommunication Transmission Handbook*, 4th ed., John Wiley & Sons, New York, 1998.
9. Cindana Tukatte, Iolon Inc. "Tunable-laser Technologies vs. Optical-networking Requirements," *Lightwave*, March 2001, page 136.
10. Vince Sykes, K2 Optronics, "External-Cavity Diode Lasers for Ultra-Dense WDM Networks," *Lightwave*, March 2001, page 130.
11. Private communication, Iolon, Inc., March 30, 2002.

5

LIGHT DETECTORS

5.1 INTRODUCTION AND SCOPE

In this chapter we describe the construction, operation, and performance parameters of light receivers, which are more commonly called *light detectors*. With light sources we worked on more familiar ground such as output power in milliwatts or dBm and bandwidth measured in hertz. In this chapter, the language and terms of expression—such as responsivity, dark current, and noise equivalent power—will be new to many of the readers.

A light detector is no more than a photon counter, converting impinging light energy into electric energy. In general, in this chapter we will be dealing with two generic types of light detectors: PIN diodes and avalanche photodiodes (APDs). The term *PIN* derives from the semiconductor construction of the device where an intrinsic (I) material is used between the *p–n* junction of the diode. Before returning to our general discussion of light detectors, we will define a number of terms that will be used throughout this chapter, many of which will be used in the remainder of the book. A great number of these terms derive from solid-state physics.

5.2 DEFINITIONS

Photoconductive Detector. A photon detector that exhibits increased conductivity with incident radiant power.

Photovoltaic Detector. A photon detector with a *p–n* or *p–i–n* junction that converts radiant power directly into electrical current; it is also called a photodiode.

***D*-Star (*D**)**. A relative measure of sensitivity used to compare the detecting capabilities of different detectors. D^* is the signal-to-noise ratio at a specific electrical frequency with a 1-Hz bandwidth when radiant power is incident on the detector active area.

Responsivity. A value indicating signal output from radiation incident on the detector element. The value where the detector has a maximum spectral response is called *peak responsivity*. It is a function of the detector area, wavelength (of the radiant signal), and circuit parameters.

Noise-Equivalent Power (NEP). The amount of required signal radiant power on the detector element area to yield a signal-to-noise ratio of 1. It indicates the minimum detection radiation level; the smaller the NEP value, the better the performance.

Resistivity. The square area resistance of a thin film detector, where L and W are equal. L is the separation between electrodes and W is the length of the detector active area. Resistivity is a function of the detector element temperature and the level of irradiance.

rms Signal Voltage or Current. The element of the electrical output (voltage or current) which is coherent with the monochromatic or blackbody input signal radiant power. It is a function of electrical frequency, signal power, spectral characteristics, operating temperature, and other circuit parameters such as the load resistor and bias voltage.

rms Noise Voltage or Current. The element of the electrical output (voltage or current) which is incoherent with the signal radiant power, usually measured with no signal radiation incident on the detector element and is related to the detector area. It is a function of frequency response, noise equivalent bandwidth, operating temperature, other circuit parameters such as load resistor, and, in some cases, detector solid angle and background temperature.

Dark Resistance. The ratio of the DC voltage across the detector to the DC current through it when no radiation is incident on the detector.

Dark Current. The measured current in a detector circuit when operated with no signal radiation incident on the detector element. For a good photodiode the dark current should be < 10 nA (Ref. 1).

Bias Voltage. The voltage applied to the detector circuit, normally DC volts. It is sometimes called *optimum bias* for values that give optimum signal-to-noise ratios and maximum bias values for values that produce the maximum signal voltage output. It is called *reverse bias* when applied to a p–n junction of solid crystal detectors in a reverse mode to increase speed or response or to increase the long wavelength response.

Background Temperature. The effective temperature of all radiation sources viewed by the detector, excluding signal source.

Spectral Response. Most of the time this is shown as D^* versus wavelength, usually presented as a curve showing relative signal level as a function of wavelength of the incident radiant power.

Load Resistor. A resistance element that is in series with the detector element and bias voltage; typically matched to the detector's dark resistance.

Open-Circuit Voltage. A DC voltage produced by a photovoltaic detector when connected to a high impedance load.

Time Constant. A measurement of a detector's speed of response when the detector is exposed to a square wave pulse of radiation. The rise-time constant is the time required for the signal voltage to reach 0.63 times its asymptotic value. The decay time constant is the time required for the signal voltage to decay to 0.37 of the asymptotic value. This can also be measured by determining the chopping frequency at which the signal response is 0.707 of its maximum value.

Rise Time and Fall Time. Rise time and fall time are the times (in seconds) required for the signal response to rise from 10% to 90% and fall from 90% to 10% of the maximum observed signal value. This happens when detectors are exposed to pulses of signal radiant power.

Cutoff Wavelength. The long wavelength point where the detector responsivity has degraded to a specified percent of the peak responsivity; usually 20% or 50% of the peak responsivity.

All of the above definitions have been taken from 1998 New England Photoconductor (from the Web, at www.netcorp@ici.net, Ref. 2).

Noise Figure (F or f). $f = S/N_{in}/S/N_{out}$ of the device in question where f is a numeric. Noise figure is often given in dB and derived from $F_{dB} = 10\log(f)$.

Quantum Limit. A bound on what is ultimately achievable from a particular type of telecommunication link. It is usually stated in terms of the minimum number of photons per bit that will allow a light detector to achieve a given BER with a specified modulation format and type of receiver.

Johnson noise = thermal noise (Ref. 1)

IMPORTANT CONSTANTS

Planck's constant: (h) 6.626×10^{-34} joule-second (Ref. 3)

Joule: 1 joule = 1 watt-second (Ref. 3)

Boltzmann's constant = 1.38×10^{-22} joule per kelvin (from Ref. 4) or -228.6 dBW or -198.6 dBm (from Ref. 5).

Using Planck's constant we provide the following formula, valid for 1.5-μm band:

$$1 \text{ mW} = 7.5 \times 10^{15} \text{ photon/second}$$

(from Ref. 6, page 270). See equation (5.1).

5.3 INTRODUCTORY RELATIONSHIPS

The most commonly used light detectors for fiber-optic communication systems are either PIN photodiodes, or avalanche photodiodes (APD).

A photodiode can be considered a photon counter. The photon energy E is a function of frequency and is give by

$$E = hv \tag{5.1}$$

where h is Planck's constant (given above in Section 5.2) and v is frequency (Hz). E is measured in watt-seconds or kilowatt-hours.

The receive power in the optical domain can be measured by counting, in quantum steps, the number of photons received by a light detector per second. The power in watts may be derived by multiplying this count by the photon energy, as given in equation (5.1). Also consult the relationship given at the end of Section 5.2, converting milliwatts to photons/second.

Quantum Efficiency. The efficiency of the optical-to-electrical power conversion, expressed as a percentage, is defined by a photodiode *quantum efficiency* η, which is a measure of the average number of electrons released by each incident photon. The sensitivity of a photodiode may also be expressed in practical units of amperes of photodiode current per watt of incident illumination. The quantum efficiency, η, is related to a photodiode's responsivity by the following equation:

$$\eta\ (\%) = 1.24 \times 10^5 R/\lambda\ (\text{nm}) \tag{5.2}$$

where R is the responsivity in amperes per watt (A/W) and λ is the wavelength (in nm) of the light signal.

Operating under ideal conditions of reflectance, crystal structure, and internal resistance, a high-quality silicon photodiode of optimum design would be capable of approaching a quantum efficiency of 80%. A quantum efficiency of 100% is unattainable.

Source: From the Web at www.west.net/~ centro/tech2.htm (Ref. 7).

For the fiber-optic communication system engineer, *responsivity* is a most important parameter when dealing with photodiode detectors. Responsivity is expressed in amperes per watt or volts per watt and is sometimes called sensitivity. Responsivity is the ratio of the rms value of the output current or voltage of a photodetector to the rms value of the amount of electrical power.

In other words, responsivity is a measure of the amount of electrical power we can expect at the output of a photodiode, given a certain incident light power signal input. For a photodiode the responsivity R is related to the wavelength λ of the light flux and to the quantum efficiency η, the fraction of the incident photons that produce a hole–electron pair. Thus

$$R = \frac{\eta\lambda}{1234} \qquad (\text{A/W}) \tag{5.3}$$

with λ measured in nanometers.

Responsivity can also be related to electron charge Q by the following equation:

$$R = \frac{\eta Q}{hv} \tag{5.4}$$

where hv = photon energy [equation (5.1)] and Q = electron charge, 1.6×10^{-19} coulombs (C).

Noise Equivalent Power (NEP). Noise equivalent power is the minimum detectable light (power) of a photodiode. The minimum incident power required on a photodiode to generate a photocurrent equal to the total photodiode noise current is defined as the noise equivalent power. NEP is calculated by the following relationship:

$$\text{NEP} = \text{noise current (A)/responsivity (A/W)} \tag{5.5}$$

NEP is dependent on bandwidth of the measuring system. To remove this dependence, the NEP value is divided by the square root of the bandwidth. This gives the NEP in units of watts/$Hz^{1/2}$. Because the photodiode light power to current conversion depends on the radiation wavelength, the NEP power is quoted at a particular wavelength. The NEP is nonlinear over its wavelength range, as is responsivity.

Noise. Before we enter that all-important realm of noise, let us accept that photocurrent I_p is directly proportional to the incident optical power P_{in}. This is expressed by the following relationship:

$$I_p = RP_{in} \tag{5.6}$$

where R is the responsivity. The responsivity is given in equation (5.3).

There are two fundamental noise mechanisms we must deal with in PIN and APD light detectors. These are:

1. Shot noise
2. Thermal noise (called Johnson noise in some texts)

Equation (5.6) considers the system noise-free. Of course, no circuit is noise-free. In the case of light detectors, noise causes current fluctuations affecting receiver performance. This relationship in (5.6) still holds true, however, if signal current I_p is considered average current.

Reference 1 provides the following relationship to calculate shot noise (σ):

$$\sigma_s^2 = 2q(I_p + I_d)\,\Delta f \tag{5.7}$$

where q is the electron charge (1.6×10^{-19} coulombs), I_d is the dark current noise, and Δf is the bandwidth (of the receiver).

Derived from equation (5.7), the following relationship may be used to calculate shot noise current (I_s):

$$I_s = [2qI_d\,\Delta f]^{1/2} \tag{5.8}$$

where I_d is the dark current leakage (A).

Souce: From the Web at www.west.net/ ~ centro/tech2.htm (Ref. 7).

Thermal noise contributions are provided by the shunt resistance of the device, series resistance, and load resistance. Thermal noise can be calculated from the following equation where I_t is thermal noise:

$$I_t = [4KT\,\Delta f/R_{es}]^{1/2} \tag{5.9}$$

where K is Boltzmann's constant (given above), T is absolute temperature in kelvins, R_{es} is the resistance value in ohms (this is the resistance giving rise to thermal noise), and Δf is the bandwidth.

Source: From the Web at www.west.net/ ~ centro/tech2.htm (Ref. 7).

Reference 1 gives the following equation to calculate thermal noise, using the reference's notation:

$$\sigma_T^2 = (4kT_B/R_L)f_n\,\Delta f \tag{5.10}$$

where σ_T is the thermal noise in watts/Hz, f_n represents the factor by which thermal noise is enhanced by various resistors used in the preamplifiers and main amplifiers, and R_L is the resistance of the load resistor.

Signal-to-Noise Ratio (S/N or SNR). This subsection deals only with the SNR of PIN diodes. SNR for any device is simply

$$\text{SNR} = \text{average signal power/noise power} \tag{5.11}$$

We know that electrical power varies by the square of the current, and from equation (5.6) we obtain $I_p = RP_{in}$. This then becomes the numerator of equation (5.11). Combine equations (5.7) and (5.10) for the denominator and we have

$$\text{SNR} = R^2P_{in}^2/[2q(RP_{in} + I_d)\,\Delta f + 4(k_BT/R_L)f_n\,\Delta f] \tag{5.12}$$

where $R = \eta q/hv$ is the responsivity of a *PIN* diode.

In most cases, thermal noise dominates receiver performance when it has a much larger value than shot noise. Then dropping the shot noise term in the denominator of equation (5.12), SNR based on thermal noise alone is calculated by the following expression:

$$\text{SNR} = R_L R^2 P_{in}^2/4k_BTf_n\,\Delta f \tag{5.13}$$

Response Time. As we know, simplistically, present fiber-optic communication systems transmit information by expressing a binary 1 as the presence of a pulse and a binary 0 as the absence of a pulse.[1] An extremely important parameter in this process is *rise time* (and *fall time*) (see Figure 4.8). Rise time determines the maximum bit rate the device can handle. We can calculate rise time (T_r) as

$$T_r = \ln 9(\tau_{tr} + \tau_{RC}) \tag{5.14}$$

where τ_{tr} is the *transit time* and τ_{RC} is the time constant of the equivalent RC circuit.

Because of the RC relationship, we can see the importance of the device capacitance. The internal capacitance is inversely proportional to the depletion region thickness (Ref. 6). Photodiode designers want to make the depletion region as wide as possible in order to maximize the quantum efficiency, but this increases the transit time of carriers across the region. As long as carriers in the depletion region see an electric field strength of several kilovolts, the velocity of travel is at

[1]The expression *absence of current* is not exactly true. Laser diodes are usually biased so that the binary 0 condition allows a low level of current to flow. Thus the laser diode is "on," producing output current all the time. This, then, is a high level of current for a binary 1 and a low level of current for a binary 0.

its maximum of around 8×10^6 cm/s for electrons and about half that for holes. A typical PIN diode has a depletion region of 20 μm. Its contribution to the time constant for finite carrier mobility is about 0.2 ns (based on Ref. 6).

At bit rates above ~ 1 Gbps, electrical parasitics may be generated, thereby affecting the RC time constant limiting the maximum bit rate sustainable.

The numerical values of τ_{tr} and τ_{RC} depend on detector design and can vary over a wide range. The bandwidth (affecting the bit rate) is related to these two parameters as follows:

$$\Delta f = \left[2\pi(\tau_{tr} + \tau_{RC})\right]^{-1} \tag{5.15}$$

Source: (Ref. 1).

It should be noted that the bias voltage setting influences rise time. The higher the voltage, the faster the rise time. Rise times for a good photodiode should be in the tenth nanosecond region. Depending on design, they vary from < 0.2 ns to 1 ns for an Si photodiode and from 0.04 ns to 0.5 ns for a well-designed InGaAs photodetector.

5.4 PIN PHOTODIODES

There are several photodetectors that may be employed in light receivers for fiber-optic communication systems. There are two, however, that have the most appeal for the fiber-optic system designers. These are the silicon PIN diode and the InGaAs PIN diode. Figure 5.1a shows the responsivity curve versus wavelength for an Si (silicon) photodiode and Figure 5.1b shows the responsivity curve for an

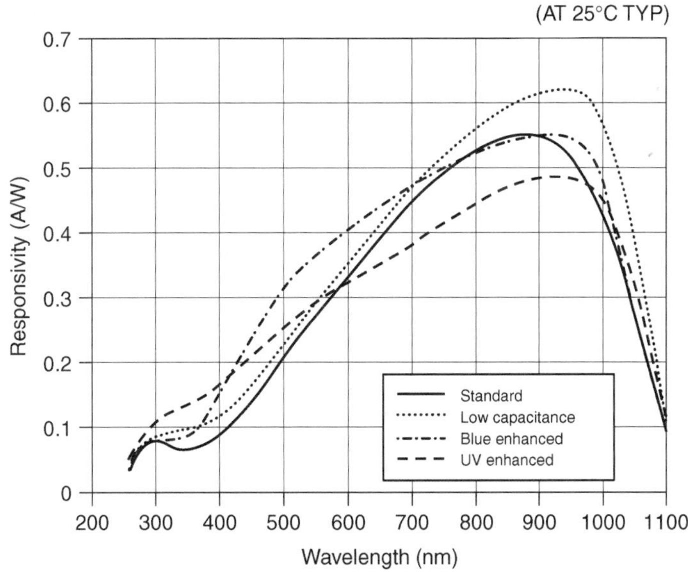

Figure 5.1a. Responsivity curves versus wavelength for a silicon-based photodiode detector. (From Silicon Sensors, from the Web at www.siliconsensors.com, Ref. 9.)

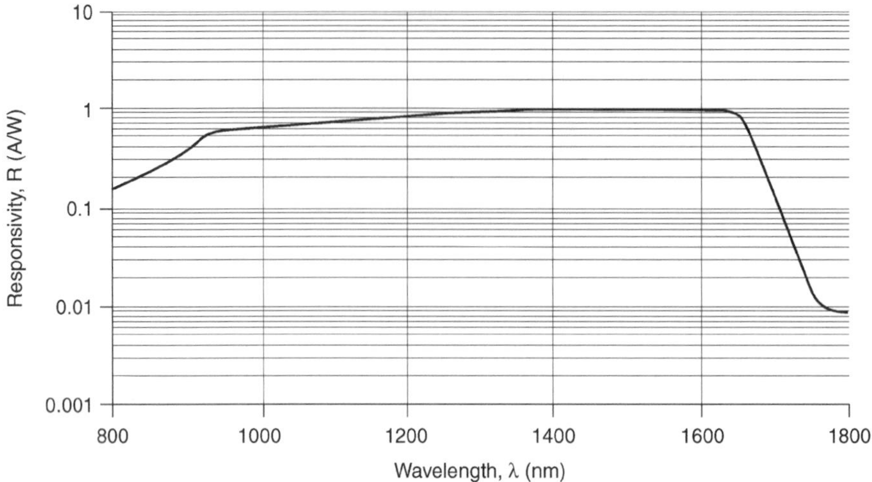

Figure 5.1b. Responsivity versus wavelength curve for a standard InGaAs photodiode detector. (From the Web at www.fermionics.com/R1300.htm, Ref. 8.)

InGaAs photodiode detector. The silicon-based photodiodes would find application on short wavelength (850 nm) applications, and the InGaAs-based photodiode would be used for 1310-nm and 1550-nm circuits.

5.4.1 Construction of an Si Photodiode Detector

Silicon photodiodes are fabricated in a similar manner as integrated circuits (ICs) in that they are constructed from single crystal silicon wafers. However, Si photodiodes require a higher purity of silicon. The purity is directly related to its resistivity.[2] The higher the resistivity, the higher the purity of silicon.

Figure 5.2 shows a cross section of a silicon photodiode. The basic material is n-type silicon. There is a thin p-layer formed on the front surface of the device. The forming is by thermal diffusion or ion implantation of the appropriate doping material. The doping material is usually boron. The p–n junction is the interface between the p-layer and the n-silicon. There are small metal contacts applied to the front surface of the photodiode. The entire back of the photodiode is coated with a contact metal. In familiar terms, the front contact is the anode and the back circuit is the cathode. The active area is coated with either silicon nitride, silicon dioxide, or silicon monoxide to serve as an antireflection coating. The thickness of the coating is optimized for particular wavelength bands.

What makes photodiode junctions unusual compared to the more common p–n junction is that the top "p" layer is very thin. There is a relationship between the thickness of the layer and the operating wavelength to be detected. The silicon becomes depleted of electrical charges near the p–n junction. This region is known as the *depletion region*. By applying a reverse bias voltage across the

[2]*Resistivity* is the reciprocal of volume conductivity (measured in siemens per centimeter), which is a steady-state parameter (Ref. 3).

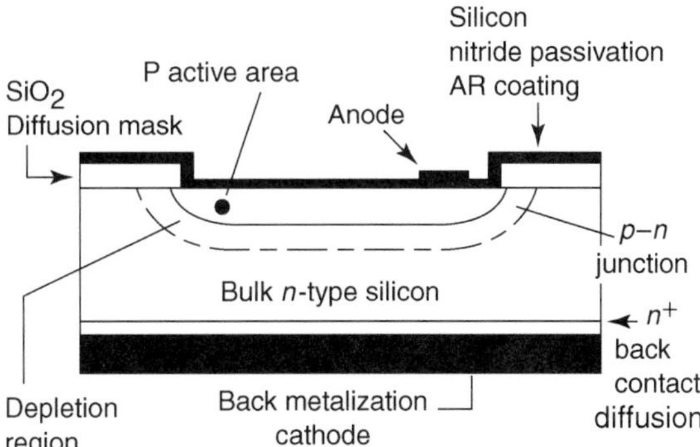

Figure 5.2. Cross section of a silicon photodiode. (From the first figure in *A Primer on Photodiode Technology*, from the Web at www.west.net/~ centro/tech2.htm, Ref. 7.)

junction, the depth of the depletion region can be varied. The diode is said to be *fully depleted* when the depletion region reaches the back of the diode. This region is important to the photodiode's performance because most of the sensitivity to light radiation originates there.

As we said, the capacitance of the *p–n* junction depends on the thickness of the variable depletion region. The bias voltage controls the depth of the region, and with increasing depletion the capacitance is lowered until the fully depleted condition is achieved. Figure 5.3 shows the relationship among capacitance, bias voltage, and area.

An electron–hole is formed when light is absorbed in the active area. The electrons are separated and pass to the *n* region and the holes pass to the *p* region. This results in a current generated by the impinging light. This migration of holes and electrons to their respective regions is called *the photovoltaic effect*.

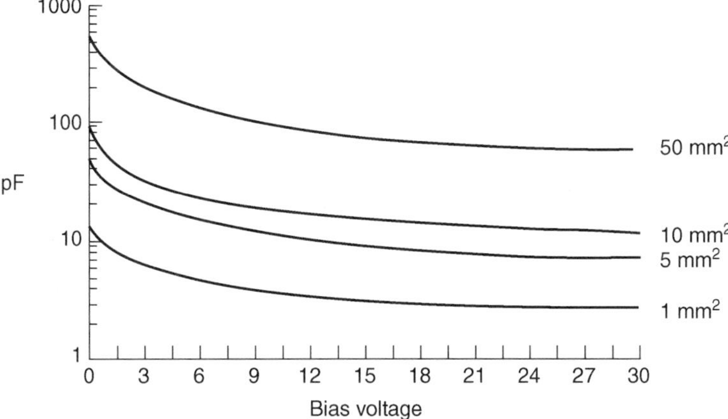

Figure 5.3. Junction capacitance of a photodiode depends on its area and bias voltage. (Si photodiode.) (From the second figure in "A Primer on Photodiode Technology," from the Web at www.west.net/~ centro/tech2.htm, Ref. 7.)

The current generated, usually defined as the short-circuit current, is a linear function of the light radiated on the active area. The current can vary over a very wide range, at least seven orders of magnitude. The notation I_{sc} defines this current's magnitude. The current is only slightly affected by temperature change, varying less than 0.2% per degree centigrade for the visible wavelengths.

The polarity of the voltage of the photodiode's two terminals (cathode and anode) is based on the fact that there is a low forward resistance (anode positive) and a high reverse resistance (anode negative). An Si photodiode usually has negative bias on the active area, which is the anode, or has positive bias on the backside of the device, which is the cathode. In the zero bias and photovoltaic modes the generated current or voltage is in the diode forward direction. Thus the generated polarity is opposite to that required in the biased mode.

5.4.2 Overview of an InGaAs Photodiode Detector

Figure 5.4 is a generalized sketch of an InGaAs PIN photodiode. These types of diodes are used as photodetectors for the longer wavelengths (i.e., 1310- and 1550-nm bands).

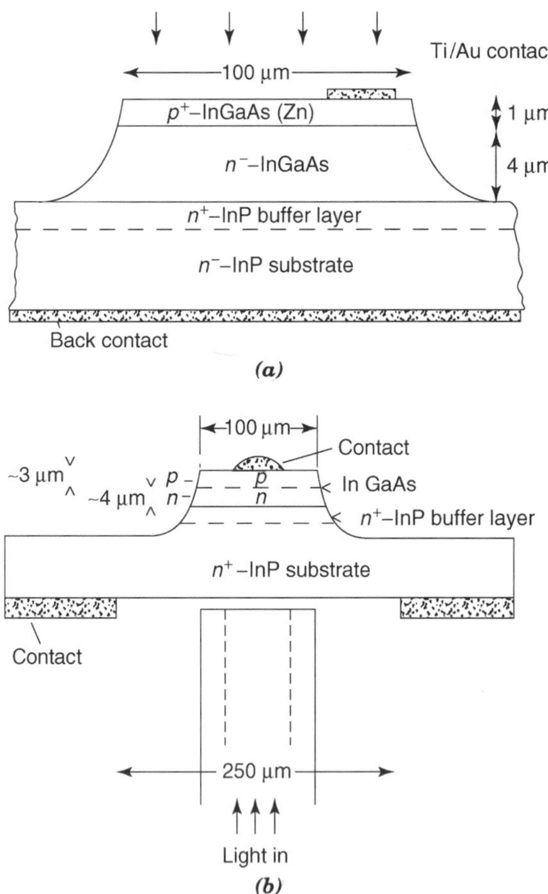

Figure 5.4. A generalized sketch of an InGaAs PIN diode detector. (a) Front entry. (b) Substrate entry. (Reprinted with permission of the ITU, Ref. 10, Figure 3.18, page 52.)

From Figure 5.4 the layers consist of InP material for the p layer, InGaAs material for the i layer, and InP material for the n layer. Because the bandgap for InP is 1.35 eV, InP is transparent for light whose wavelength exceeds 0.92 μm. In contrast, the bandgap for the i layer consisting of special InGaAs material is about 0.75 eV. This value corresponds to a cutoff wavelength of 1650 nm. The middle InGaAs layer thus absorbs strongly in the wavelength region of 1.3–1.6 μm. This is a heterostructure photodiode of the detector that eliminates completely the diffusive component of the detector because photons are absorbed only in the depletion region. These types of PIN diodes have excellent operation characteristics in the two longer wavelength bands. For example, we can expect the responsivity to be between 0.6 A/W and 0.9 A/W and expect quantum efficiency (%) to be between 60 and 70.

5.4.3 Avalanche Photodiode (APD)

An APD is simply a PIN diode detector with gain. Figure 5.5 gives a schematic cross section of a typical structure of an APD. The figure shows an absorption region A and a multiplication region M. Across region A is an electric field E that serves to separate the photogenerated holes and electrons and sweeps one carrier toward the multiplication region. The multiplication region M exhibits a high electric field so as to provide internal photocurrent gain by *impact ionization.* This gain region is broad enough to provide a useful gain, M, of at least 100 (20 dB) for silicon APDs and 10–40 for germanium or InGaAs APDs. In addition, the multiplying electric field profile must enable effective gain to be achieved at a field strength below the breakdown field value for the diode.

5.4.3.1 APD Performance Parameters. Because of its internal photoelectronic signal gain, an APD differs from a PIN photodiode which does not display gain.

The output signal current I_s of an APD is

$$I_s = MR_0(1)P_s \tag{5.16}$$

where $R_0(1)$ is the intrinsic responsivity of the APD at a gain where $M = 1$ and at wavelength 1; M is the gain of the APD and P_s is the incident optical power.

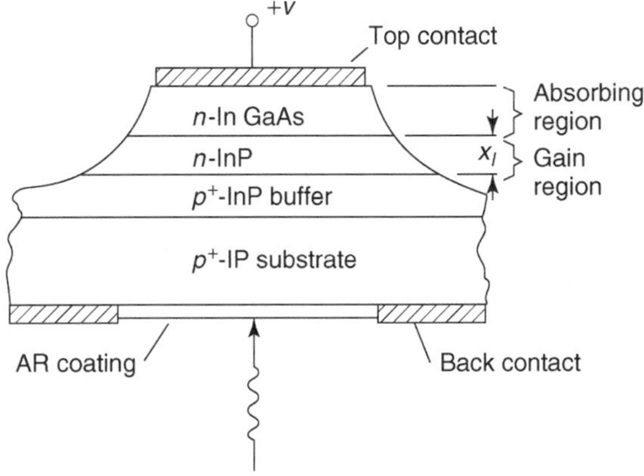

Figure 5.5. (Courtesy of the ITU, Ref. 10.)

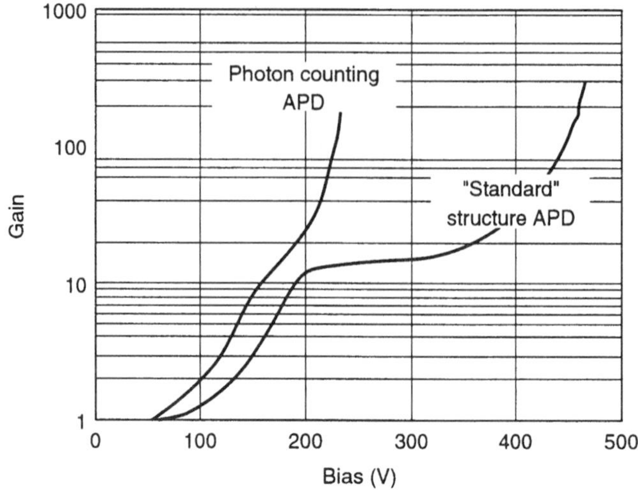

Figure 5.6. Typical gain–voltage curve for Si APDs. (From Figure 2, PerkinElmer, *Avalanche Photodiodes: A Users Guide*, from the Web, Ref. 11.)

The gain is a function of the reverse voltage V_R of the APD and varies with the applied bias voltage. A typical gain–voltage curve for a silicon APD manufactured by PerkinElmer is shown in Figure 5.6.

Spectral noise is a key parameter when selecting an APD. Like other photodetectors, an APD will normally be operating in one of two regimes: (a) detector noise limited at low power levels or (b) photon shot noise limited at higher power levels. An APD operates with reverse bias. Shot noise and APD leakage current limit sensitivity at low light levels. This differs from a PIN diode detector, in that bulk leakage current I_{DB} is multiplied by the gain, M, of the APD. The total leakage current I_{D} is equal to

$$I_{\text{D}} = I_{\text{DS}} + I_{\text{DB}} M \tag{5.17}$$

where I_{DS} is the surface leakage current.

An APD's performance is degraded by an excess noise factor (F) compared to a PIN. Total spectral noise current for an APD under dark conditions is given by

$$i_N = \left[2q\left(I_{\text{DS}} + I_{\text{DB}} M^2 F\right)B\right]^{1/2} \tag{5.18}$$

where q is the electron charge and B is the system bandwidth.

At the higher signal level, there is a transition to the photon shot noise-limited regime where the APD's sensitivity is limited by photon shot noise on the current generated by the optical signal. Total APD noise $[i_{N(\text{total})}]$ equals the quadratic sum of the detector noise plus the signal shot noise with radiant light present. Total APD noise may be expressed as follows:

$$i_{N(\text{total})} = \left[2q\left(I_{\text{DS}} + \left(I_{\text{DB}} M^2 + R_0(I) M^2 P_S\right)F\right)B\right]^{1/2} \tag{5.19}$$

where F is the excess noise factor, B is the system bandwidth, M is the multiplication factor, I_{DS} is the surface leakage current, and I_{DB} is the bulk leakage current.

Reference 11 reports that in the absence of other noise sources, an APD thus can provide a signal-to-noise ratio (SNR) that is $F^{1/2}$ worse than a PIN detector with the same quantum efficiency. In cases where the APD internal gain increases the signal level without dramatically affecting the overall system noise, an APD can produce a better overall system signal-to-noise ratio than a PIN detector.

NEP cannot be used as the only measure of a detector's relative performance, but rather detector SNR at a specific wavelength and bandwidth should be used to determine the optimum detector type for a given application. It should be noted that optimum SNR occurs at a gain M where total detector noise equals the input noise of the amplifier or load resistor. For silicon APDs, M ranges from 100 to 1000; and for germanium and InGaAs APDs, M is between 30 and 40. Optimum gain depends in part on the excess noise factor, F, of the APD.

5.4.3.2 Types of APDs. APDs are available for the wavelength band between 300 and 1700 nm. Silicon APDs can be used for wavelengths between 300 and 1100 nm. Germanium APDs cover the wavelengths between 800 and 1600 nm, and InGaAs APDs cover those from 900 to 1700 nm.

InGaAs APDs are notably more expensive than germanium APDs, and they can have significantly lower current, exhibit extended spectral response to 1700 nm, and provide higher-frequency bandwidth for a given active area.

5.4.3.3 SAM APDs. Conventional APDs have a number of drawbacks. High-level electric fields are required to achieve avalanche multiplication (the factor M). Their narrow bandgap (InGaAs – $\epsilon_g \sim 0.75$ eV) causes large tunneling leakage currents at electric fields lower than those required to exhibit sufficient avalanche multiplication in this material.

A separate *absorption and multiplication* (SAM) APD structure has been adopted to solve this problem. With this approach the functions and absorption and multiplication are carried out in different layers in the device. A new field-control layer is imposed on the structure. It consists of moderately doped InP to maintain a low electric field in the InGaAs absorption layer while supporting a high field in an InP multiplication layer. This is illustrated in Figures 5.5 and 5.7.

This wider bandgap InP ($\epsilon_g \sim 1.35$ eV) provides avalanche multiplication without tunneling. It operates at wavelengths longer than 950 nm.

This SAM structure has been adopted by nearly all commercially available APDs for long wavelength links. However, beyond this generic concept, design details vary widely from manufacturer to manufacturer. Unlike its PIN diode cousin, InGaAs/InP APDs exist in many variations.

Edge breakdown became a problem with these types of APDs. The key to mitigate edge breakdown is to reduce the electric field intensity at the device edges. There have been numerous approaches to solving the problem. Among these are:

- Low doping density at junction edges
- Control of the total charge profile in the field-control layer
- Control of the junction profile

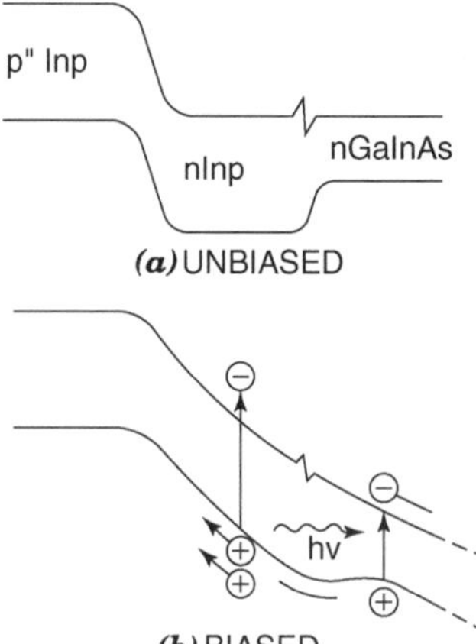

Figure 5.7. (Courtesy of Epitaxx, Inc., Ref. 12.)

Another problem in designing higher bit rate devices (e.g., > 2.5 Gbps) was to attain a sufficiently high bandwidth. For example, for 10-Gbps operation, the bandwidth should be at least 7–8 GHz to achieve yet maintain a high gain work against each other. The maximum sensitivity is generally found for multiplication gains of $M \sim 10$. This dictates that the gain–bandwidth products of 80 GHz are required. The epitaxial layer thickness and device area are designed so that the transit time and RC time constant do not limit the bandwidth to less than 8 GHz. If the APD is not the bandwidth-limiting element in the receiver when the gain is optimized for sensitivity (i.e., $M \sim 10$), the designer can compromise high-gain bandwidth at lower gain and then the receiver's dynamic range will be improved. APD receivers based on this design can achieve a -26-dBm sensitivity at a BER of 1×10^{-10}. This device uses a GaAs pseudomorphic high electron mobility transistor (p-HEMT). This performance represents a 5- to 6-dB sensitivity improvement relative to comparably designed PIN diode receivers. Figure 5.7 is a schematic outline drawing of a SAM APD.

5.4.3.4 APD Performance Parameters

Responsivity and Gain. APD gain varies as a function of the applied reverse voltage, as shown in Figure 5.6. Furthermore, for many APDs it is not possible or practical to obtain an accurate measurement of intrinsic responsivity, $R_0(I)$, at a gain where $M = 1$. Thus one should not expect a manufacturer to state typical gain and diode sensitivity at $M = 1$, as a method for specifying diode responsivity at a given operating voltage. To characterize APD response, the APD responsivity is specified in A/W at a given operating voltage. It should be noted that due to diode variations in the exact gain–voltage curve of each APD, the specific operating voltage for a given responsivity varies from one APD to another. A specification should therefore state a voltage range within which a specific responsivity will be achieved. Diode responsivity should be stated following this example for an InGaAs APD:

$$R_{\text{MIN}}(1300\text{ nm}) = 9.0\text{ A/W}, \qquad V_{\text{OP}} = 50\text{–}90\text{ V}, \qquad M \sim 10$$

Dark Current and Noise Current. Referring to equation (5.19), one can see that the total APD current (and its corresponding spectral noise current) is only meaningful when specified at a given operating gain. Dark current noise and spectral noise current are a function of APD gain; these two parameters should be specified at a stated responsivity level. Using an InGaAs diode as an example, dark current and spectral noise current should be specified as follows:

$$I_D(R = 9.0\text{ A/W}) = 10\text{ nA(max)}, \qquad M \sim 10$$

$$i_N(R = 6.0\text{ A/W, 1 MHz, 1 Hz BW}) = 0.8\text{ pA/Hz}^{1/2}\text{(max)}, \qquad \text{M} > 5$$

5.4.3.5 Excess Noise Factor. Due to the statistical nature of the avalanche process, all APDs generate excess noise. The notation F is the excess noise factor. $F^{1/2}$ is the factor by which the statistical noise on the APD current, which is the sum of the multiplied current photocurrent plus the multiplied APD bulk dark current [see equations (5.15) and (5.16)], exceeds that which would be expected from a noiseless multiplier on the basis of Poissonian statistics (shot noise) alone.

F is a function of the carrier ionization ratio k, where k is usually defined as the ratio of hole-to-electron ionization probabilities (where $k < 1$). F may be calculated using a model developed by McIntyre (Ref. 13) which considers the statistical nature of avalanche multiplication. F may be calculated from the following:

$$F = k_{\text{EFF}} M + (1 - k_{\text{EFF}})\{2 - 1/M\} \tag{5.20}$$

From the equation, the lower we can have the values of k and M, the lower the excess noise factor will be. The variable k_{EFF} is the effective k factor for an APD. It can be measured experimentally by fitting the McIntyre formula to the measured dependence of F on gain. This should be done under radiant light conditions. From the carrier ionization coefficients and electric field profile of the APD structure, F may also be theoretically calculated.

The ionization ratio k is strongly related to the electric field across the APD structure. It takes its lowest value at low electric fields (for Si APDs only). Because the electric field profile is dependent on the doping profile, the k factor is also a function of the doping profile. Both the electric field profile traversed by a photogenerated carrier and subsequent avalanche-ionized carriers may vary according to the photon absorption depth, depending, of course, on the APD structure. The mean absorption depth is a function of wavelength for indirect bandgap semiconductors such as silicon. With these types of semiconductors, the absorption coefficient varies slowly at the longer wavelengths. The value of k_{EFF} and gain, M, for such a silicon APB is therefore a function of wavelength for some doping profiles.

The McIntyre formula can be approximated for $k < 0.1$ and $M > 20$, without much loss of accuracy, as

$$F = 2 + kM \tag{5.21}$$

TABLE 5.1 Typical Values of *k*, *X*, and *F* for Si, Ge, and InGaAs APDs

Detector Type	Ionization Ratio (k)	X-factor	Typical Gain (M)	Excess Noise Factor (F)
Silicon (reach through structure)	0.02	—	150	4.9
Silicon (reach through structure)	0.002	—	500	3.0
Germanium	0.9	0.95	10	9.2
InGaAs	0.45	0.7–0.75	10	5.5

Source: Avalanche Photodiodes: A User's Guide from the Web at perkinelmer.com/library, (Ref. 11).

Some vendors of APDs employ an empirical formula to calculate F, as follows:

$$F = MX \tag{5.22}$$

where X is derived as a log-normal linear fit of measured F values for given values of M (gain). The approximation provides sufficient accuracy for many applications, especially for APDs with a high k factor such as InGaAs and germanium APDs.

Table 5.1 provides typical values of k, X, and F for silicon, germanium, and InGaAs APDs.

Section 5.4.3 is based on *Avalanche Photodiodes: A User's Guide*, from the Web, perkinelmer.com/library at (Ref. 11).

APD Signal-to-Noise Ratio. In practical APD receivers, thermal noise dominates. Thus where thermal noise is much greater than shot noise in an APD, SNR is

$$\text{SNR} = \left(R_L R^2/4k_B T f_n \, \Delta f\right) M^2 P_{\text{in}}^2 \tag{5.23}$$

The SNR is improved by a factor M^2 when compared to its PIN counterpart.

5.4.4 APD Application Notes

For low bit-rate systems (i.e., < 655 Mbps), the use of an APD provides no significant benefit when compared to a PIN diode receiver. At higher bit rates such as 2.5 and 10 Gbps, however, the sensitivity enhancement of an APD-based receiver can be considerable to a comparably designed PIN-based receiver.

For InGaAs/InP APDs appropriate to long-wavelength fiber-optic circuits, APD receivers designed for 2.5-Gbps operation typically provide at least a 7-dB improvement relative to PIN diode receiver. For 10-Gbps circuits, this improvement is currently 5–6 dB.

5.5 AN OPTICAL RECEIVER

Figure 5.8 is a simplified block diagram of a fiber-optic receiver. A PIN diode or an APD resides at the far left of the figure. The greater portion of the receiver is devoted to electrical circuitry.

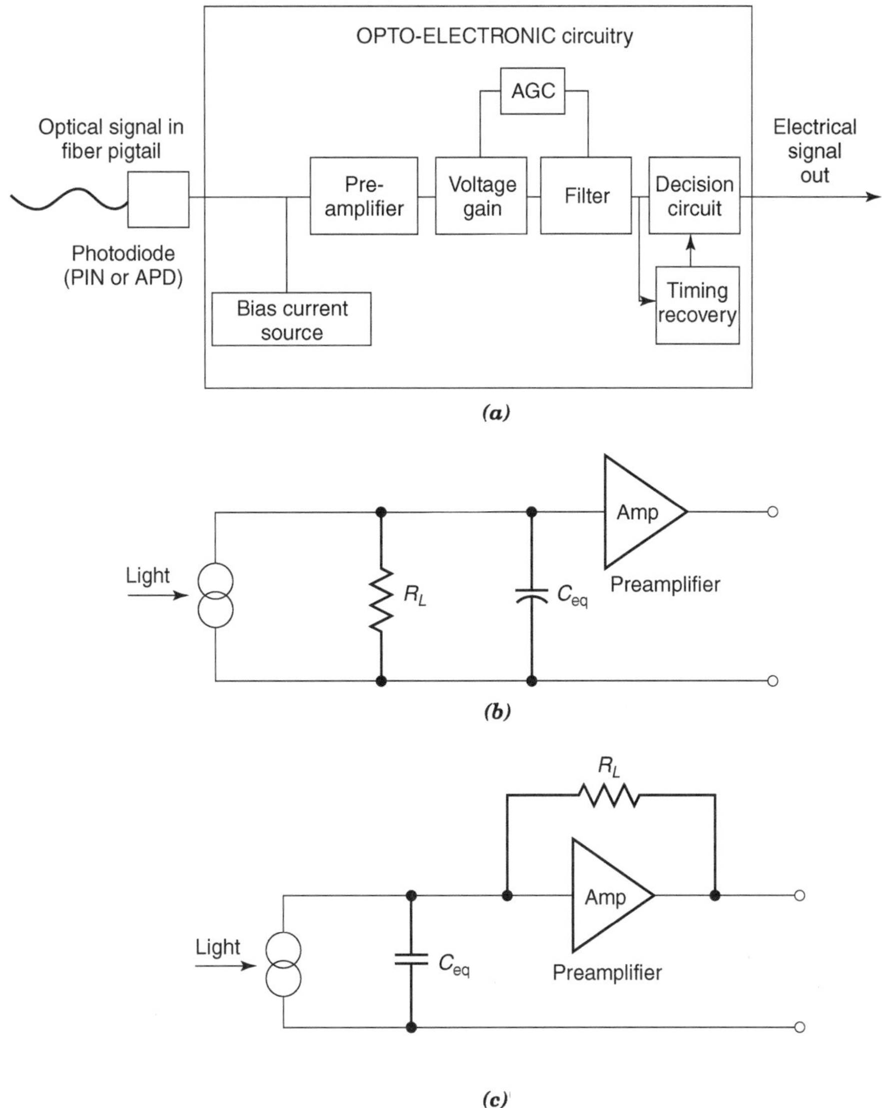

Figure 5.8. (a) Simplified functional block diagram of a photodiode receiver. (b) Simplified electrical model, high-impedance front end. (c) Transimpedance front end.

5.5.1 Electrical Amplifier, Receiver Back End

The various components of a fiber-optic receiver are shown in Figure 5.8, where the photodiode, whether PIN or APD, is just a part. The preamplifier is the other key component that determines receiver performance. The detected signal output of the photodiode is where the signal is at its lowest level. This signal is input to the preamplifier. Here is where the signal is most prone to contamination by noise. The role of the preamplifier is to amplify the electrical signal for further processing.

There is a trade-off in the design of the receiver front end between high bit rate and sensitivity. The input voltage to the preamplifier can be increased by using a large load resistor R_L. Thus a *high-impedance front end* (Figure 5.8b) is often used. A large R_L will reduce thermal noise and improve receiver sensitivity. The drawback to this approach is its low bandwidth. The receiver bandwidth is limited by its slowest component. If the bandwidth of a high-impedance front end (Figure 5.8b) is considerably less than the bit rate, this type of amplifier cannot be used. To overcome this situation, an equalizer is sometimes incorporated to increase the bandwidth. The filter attenuates the low-frequency components of the signal more than the high-frequency components, which effectively increases the bandwidth. Where sensitivity is not important, R_L can be decreased to increase bandwidth. These front ends are called *low-impedance front ends*.

Figure 5.8c shows a *transimpedance front end*. This approach provides a configuration that has a large bandwidth and high sensitivity. Here R_L is connected as a feedback resistor around an inverting amplifier. In this case, R_L can be large because the negative feedback reduces the effective input impedance by a factor, G, the gain of the amplifier. The bandwidth is increased by the factor G compared to high-impedance front ends. Many types of optical receivers employ transimpedance front ends because of the high bandwidth and excellent sensitivity. However, there is a major design issue with these types of receivers related to the stability of the feedback loop.

Subsequent components of the receiver are a high-gain voltage amplifier and a low-pass filter. The amplifier gain is controlled automatically (the AGC circuit) to limit the average voltage to a fixed level regardless of the incident average optical power at the receiver. The low-pass filter shapes the voltage pulse. The purpose of this filter is to reduce noise without introducing intersymbol interference (ISI). It is this filter that determines the receiver bandwidth. Its bandwidth is less than the equivalent bit rate, whereas other receiver components are designed for bandwidths greater than the equivalent bit rate.

The last component in Figure 5.8a is the decision circuit. The clock recovery provides synchronization and bit-slot timing. The decision circuit compares the output of the voltage amplifier filter circuit to a threshold level and determines, at each bit slot, whether the signal element is a binary 1 or a binary 0. The bit slot duration for an NRZ waveform is $1/B$, where B is the bit rate. For example, a 1-Mbps NRZ signal will have a bit-slot 1-μs duration. A 1-Gbps will have a bit-slot duration of 1.0×10^{-9} s, or 1.0 ns, or 0.001 μs. A 10-Gbps serial bit stream in NRZ format will have a bit duration of 0.1 ns, or 100 ps.

The *dynamic range* of a photodiode receiver is another important characteristic. Say that a receiver is designed for optimum sensitivity and bandwidth. One such receiver we will discuss below, operating at 10 Gbps, has a threshold for a BER of 1×10^{-10} of -34.0 dBm. Its dynamic range is 26 dB. Thus any received signal greater in level than -8 dBm will overload the receiver. The system designer may run into this situation on short links. In this case an attenuator must be used to keep the received signal comfortably within the dynamic range of the receiver.

5.5.2 Eye Pattern

Using an *eye pattern* is a quick and dirty method of getting a good estimate of signal quality. Figure 5.9a is an idealized eye pattern with no signal degradation

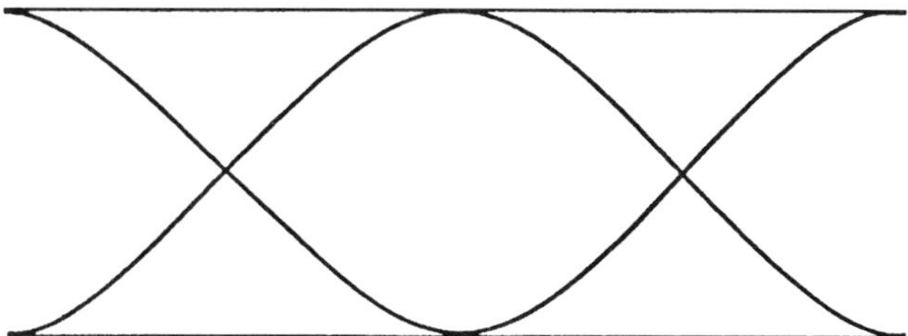

Figure 5.9a. An idealized eye diagram. No degradation to the signal.

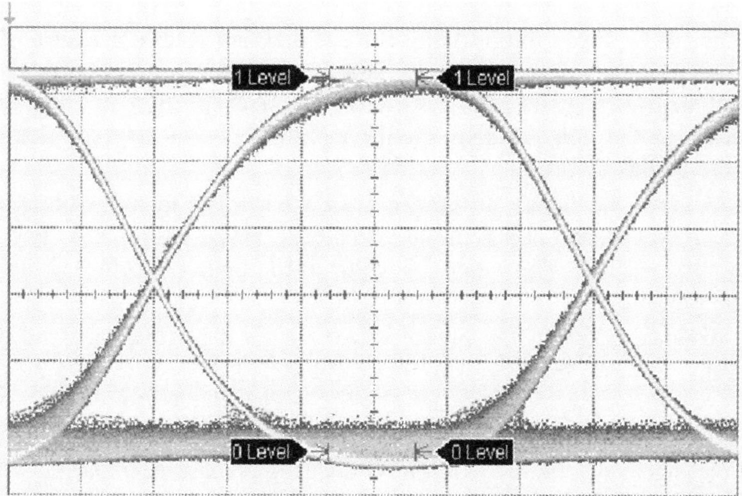

Figure 5.9b. Eye diagram, 622 Mbps, NRZ, fiber transmitter and receiver back-to-back with suitable attenuator. (Courtesy of Agilant Technologies, Inc.)

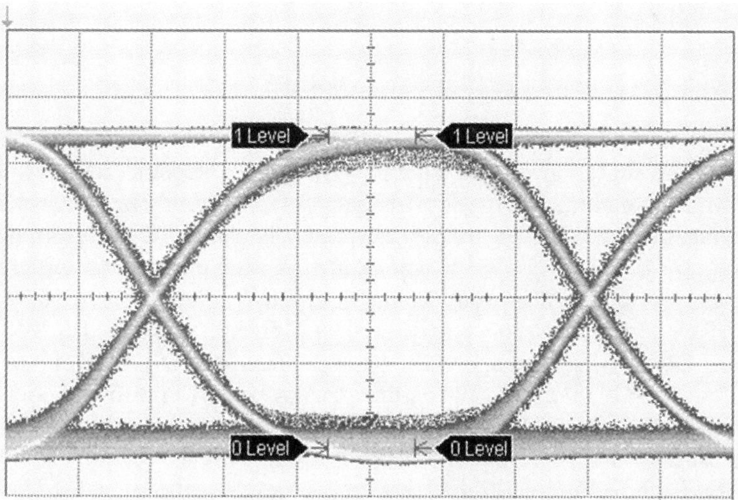

Figure 5.9c. A somewhat degraded signal, 622 Mbps, NRZ waveform over a 15-Em link. Time is shown on the *x* axis, and amplitude is shown on the *y* axis. The space between the two crossovers represent sa bit period. NRZ waveform assumed. (Courtesy of Agilant Technologies, Inc.)

whatsoever. Figure 5.9b shows a perfect eye pattern with no signal degradation. In this case the setup is in the laboratory with the fiber transmitter and receiver connected back-to-back with a suitable attenuator. An eye pattern is an oscilloscope display of two or more bit sequences, one on top of the other. If the receiver decision circuit gate exactly corresponds to the bit period of the developed bit stream, we should get an eye pattern approaching what we see in Figure 5.9b. This should show the maximum opening of the "eye." As the eye begins to close, we see ever-increasing signal degradation such as in Figure 5.9c.

The following comments will be helpful in interpreting an eye pattern or eye diagram:

- The height from top to bottom of the eye diagram gives a measure of noise in the signal. As the lines get thicker and fuzzier, the circuit is noisy and we can expect degraded BER. The height of the clear portion of the eye gives a measure of the noise margin. As the circuit begins to degrade due to noise, the eye begins to close.
- The width of the signals at the center of the eye is a measure of accumulated jitter. If the lines are thin, as in Figure 5.9a, there is a low level of accumulated jitter. The wider the width of the lines the center of the eye, the worse the jitter.
- The distance between the line crossovers gives a relative measure of the bit period.

Sometimes a mask is superimposed on the oscilloscope display (as in Figure 5.9b). If the signal display lines remain outside the mask boundaries, a circuit is acceptable. The display with mask gives qualitative values for noise, jitter, rise times, fall times, and pulse (bit) duration. An eye pattern does not provide precise quantitative quality values.

The reader should consult TIA/EIA-526-4A (Ref. 18), which gives useful guidance on fiber-optic-derived eye patterns.

5.5.3 Level of Received Signal Versus BER

In the design of a fiber-optic link, one of the first steps we take is to set the receive signal level threshold, given the characteristics of a specific receiver. For each receiver type the manufacturer will provide a curve or family of curves where BER is plotted against signal level, usually expressed in dBm. The thresholds will vary between 1×10^{-9} to 1×10^{-12}, depending on the entity that either owns or operates the system. Sprint sets its threshold for a link at 1×10^{-12}. Govind Agrawal (Ref. 1) uses 1×10^{-9}; MIL-HDBK-415 (Ref. 14) uses 1×10^{-9}, ITU-T Rec. G.957 (Ref. 15) requires 1×10^{-10}, and Telcordia TSGR (Ref. 16) specifies 2×10^{-10} at a DSX interface. We assume this to be an "end-to-end" performance requirement. Thus the underlying lightwave network carrying these DSX configurations must have a considerably better performance. If we consider that the number of ADM (add–drop multiplex) and regenerator sections in tandem was 100 to cross the U.S. continent, each link would require a bit error rate of 2×10^{-12} to achieve 1×10^{-10} at the receive ends assuming random errors.

Figure 5.10 shows a waterfall curve with no real values. The curve looks like water flowing over a waterfall. BER values are on the y axis, and signal level values (usually in rate) are 10 Gbps. A rough rule of thumb to extrapolate other values is as follows: Increase signal level 1 dB for the steeper part of the traditional waterfall curve, and the BER will improve 2 orders of magnitude. Figure 5.11 shows input power to a typical receiver operating at 10 Gbps. The y-axis is the BER value for a particular input level in dBm.

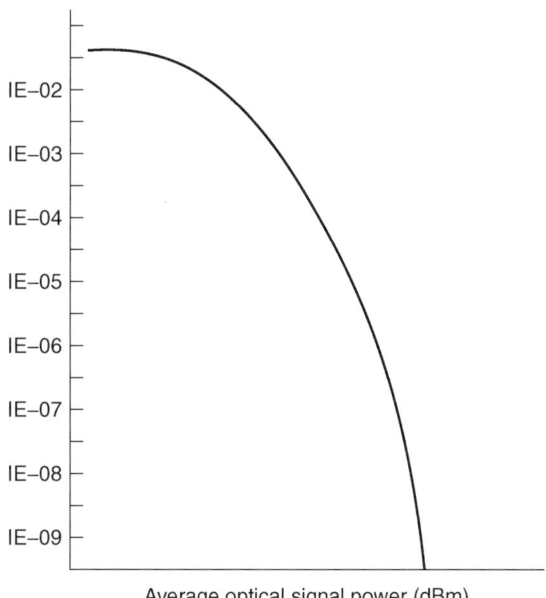

Figure 5.10. The waterfall curve, plotting received power versus BER.

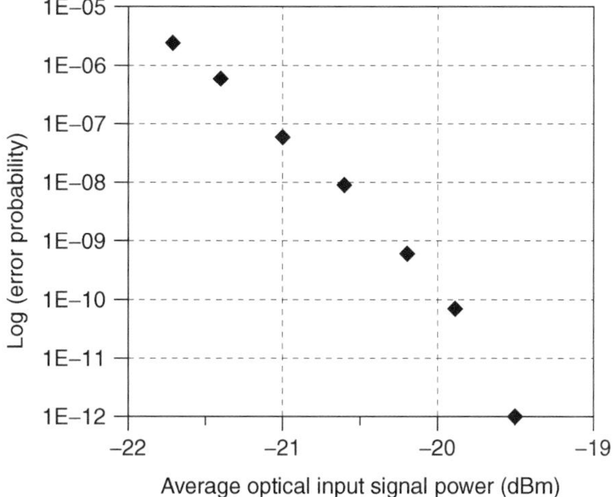

Figure 5.11. 10-Gbps optical receiver, responsivity w/350 V/W, BER versus input power (dBm). (Courtesy of Discovery Semiconductors, Ref. 17.)

TABLE 5.2 Receive Levels, BER Values, and Bit Rates

Bit Rate	BER	Level (dBm)	Comments
155 Mbps	1×10^{-10}	−33	Alcatel, PIN
2.5 Gbps	1×10^{-10}	−26	Alcatel, APD
622 Mbps	1×10^{-10}	−27	Alcatel, PIN
155 Mbps	1×10^{-10}	−35	Alcatel, PIN
622 Mbps	1×10^{-10}	−28	ITU-T G.957
2.5 Gbps	1×10^{-10}	−23	Lucent, PIN
2.5 Gbps	1×10^{-10}	−32	Lucent, APD
155 Mbps	1×10^{-10}	−38	Fujitsu, PIN
10 Gbps	1×10^{-10}	−16.3	Discovery PIN
10 Gbps	1×10^{-10}	−26	Epitaxx APD

Table 5.2 provides critical operational circuit characteristics for PIN diodes and APDs.

5.6 APPLICATION NOTES

The link budget, discussed in detail in Chapter 10, governs the optical system design. The cornerstone of the link budget is the optical receiver. The system engineer establishes a BER for a link. From the BER, which probably will be between 1×10^{-10} and 1×10^{-12}, the engineer will determine an equivalent input light power for the receiver. Normally this value will be given in dBm. Consider two conditions: (1) The link is "short" and (2) the link is "long." In the case of a short link, the optics system engineer raises a red flag and asks himself/herself if there will be an excess of light power placing the PIN diode out of range of the light signal. The manufacturer's data sheet should give that range expressed between a minimum signal level and maximum signal level. In this case our interest is the maximum signal level. We are exceeding the maximum. This simply calls for an attenuator in the link of such dB value to place the light signal inside the detector range.

In the case of condition 2, the "long" link, the opposite situation exists. There is insufficient signal power to meet the BER requirement. Here the link designer may be forced to shorten the link (thereby reducing link loss), use an optical preamplifier, use fiber with less loss per unit length, use a more sensitive detector such as a more sensitive PIN detector, or replace the PIN with an APD.

System availability may come into play here, where the PIN diode may have a much greater MTBF than an APD. Most experienced system designers will opt for the PIN diode. It is cheaper than its APD counterpart, much less complex, much less temperature-sensitive, and more technician-friendly. It may have 10 times the MTBF. If we are dealing with DWDM aggregate signals, the situation becomes more complex.

These matters are discussed in greater detail in Chapter 10.

REFERENCES

1. Govind P. Agrawal, *Fiber-Optic Communication Systems*, 2nd ed., John Wiley & Sons, New York, 1997.
2. 1998 New England Photoconductor, from the Web at www.netcorp.net.ici.

3. *The IEEE Standard Dictionary of Electrical and Electronic Terms*, 6th ed., IEEE New York, 1996.
4. *Reference Data for Radio Engineers*, 5th ed., ITT Howard W. Sams, Indianapolis, IN, 1968.
5. Roger L. Freeman, *Reference Manual for Telecommunication Engineers*, 2nd ed., John Wiley & Sons, New York, 1994.
6. Paul E. Green, *Fiber Optic Networks*, Prentice-Hall, Englewood Cliffs, NJ, 1993.
7. *A Primer on Photodiode Technology*, Centro Vision, Newbury Park, CA, 2000, from the Web at www.west.net/ ~ centro/tech2.htm.
8. *Responsivity of Standard InGaAs Photodiodes*, Fermionics, Inc., Simi Valley, CA, from the Web at www.fermionics.com/R1300.htm.
9. *Photodiode Basics*, Silicon Sensors, Inc., Dodgeville, WI, 2000, from the Web at www.siliconsensors.com.
10. *Optical Fibres Systems Planning Guide*, CCITT, ITU Geneva, 1989.
11. *Avalanche Photodiodes: A Users' Guide*, PerkinElmer Optoelectronics, Santa Clara, CA, 2000, from the Web at www.perkinelmer.com.
12. Private communication, Epitaxx Inc., Mark Itzler on AADs for 10 Gbps applications.
13. R. J. McIntyre, Multiplication Noise in Uniform Avalanche Photodiodes, *IEEE Transactions on Electron Devices*, **ED-13**, pages 164–168, 1966.
14. *Fiber Optic System Design*, MIL-HDBK-415, US Department of Defense, Washington, DC, February 1985.
15. *Optical Interfaces for Equipment and Systems Relating to the Synchronous Digital Hierarchy*, ITU-T Rec. G.957, ITU Geneva, July 1999.
16. *Transport Systems Generic Requirements (TSGR): Common Requirements*, Bellcore (Telcordia) GR-499-CORE, Issue 2, December 1998, Bellcore (Telcordia), Piscataway, NJ.
17. Abhay M. Joshi, *DC to 65 GHz Wide Bandwidth InGaAs Photodiodes and Photoreceivers*, Fiber Optics Forum, Discovery Semiconductors, Inc., Princeton, NJ, March 2000, from the Web at www.chipsat.com.
18. *Optical Eye Pattern Measurement Procedure*, EIA/TIA-526-4 Rev. A, EIA/TIA, Washington, DC, 1997.

6

IMPAIRMENTS TO LIGHT TRANSMISSION

6.1 INTRODUCTION AND SCOPE

At the outset in Chapter 1, we stated that a fiber-optic link was either loss (attenuation)-limited or dispersion-limited. In this chapter we will discuss in detail the causes of loss and dispersion and the steps that can be taken to mitigate these impairments. In the 1980s and early 1990s, these two basic effects had to be accommodated in fiber-optic link design. Thus, the process of designing a link was fairly simple and straightforward.

With the introduction of optical fiber amplifiers and then WDM followed by DWDM, other impairments began to degrade performance. They have been there all the time or were of such low level that the link designer was able to neglect them entirely. Still others appeared because of the use of new techniques. *Four-wave mixing* (FWM) is an example. Once a second channel is multiplexed on the fiber, FWM will emerge, probably at a very low level.

When I started to implement optical fiber links, signals universally operated at very low levels, around −2 dBm to +3 dBm. One reason was to extend the life of the laser diode. Once WDM, and especially DWDM, entered the picture, signal levels went up 100-fold. By combining the output of a laser transmitter with a fiber amplifier, signal levels around +20 dBm are generated to compensate for the loss of the WDM passive components. These high levels exacerbated many of the impairments, thereby corrupting performance. The increase in bit rates was still another factor. Bit streams of 40 Gbps will become commonplace.

The chapter begins with the causes of loss in optical fiber. There are ways to mitigate fiber loss such as selecting the right type of fiber and wavelength or the job. Of course there will be a cost–benefit trade-off here. Dispersion comes in a number of garden varieties. We will begin with modal and ordinary material dispersion. We then proceed with chromatic dispersion, then to polarization-mode dispersion and differential delay. Next, we cover nonlinear effects, then polarization-dependent effects, and finally other specific impairments such as self-filtering and optical surge generation.

6.2 LOSS OR ATTENUATION IN OPTICAL FIBER

There are four causes of loss in optical fiber:

1. Intrinsic loss
2. Impurity loss (sometimes called extrinsic loss)
3. Rayleigh scattering
4. Loss due to fiber imperfections

6.2.1 Intrinsic Loss

Intrinsic material absorption is the loss caused by pure silica, whereas extrinsic loss is that loss caused by impurities in the fiber (discussed in Section 6.2.2). Given a specific material, due to its molecular makeup, there will be signal absorption at certain wavelengths. In the case of silica (SiO_2), there are electronic resonances in the ultraviolet region where $\lambda < 0.4$ μm. There are vibrational resonances in the infrared region where $\lambda > 7$ μm. Fused silica, which is the material of an optical waveguide, is amorphous in nature. Therefore these resonances are in the form of absorption bands with tails extending into the visible light region. In the longer wavelength band (1310 and 1550 nm), this type of absorption amounts to no more than 0.03 dB/km (Ref. 1). There is nothing the manufacturer can do about intrinsic absorption unless some other material is selected to transport a light signal.

6.2.2 Impurity Losses (Extrinsic Absorption Losses)

Extrinsic absorption losses are brought about by impurities in the fiber. Modern manufacturing techniques have reduced these losses to a very low value. Included in this group are the following impurities: iron, copper, nickel, magnesium, and chromium, which are strong sources of absorption in our wavelength bands of interest. In modern manufacturing processes their content has been reduced below one part in a billion and thus they add little to the total extrinsic loss. However, the loss due to the residual hydroxyl ion (OH) produces an absorption line at 2730 nm, and its harmonics and tonal combinations at 1390, 1240, and 950 nm contribute to extrinsic loss. These losses are due to the water left in the fiber during the fabrication process. The OH ion content of the silica fiber should be reduced to less than one part in 100 million to maintain the fiber loss objectives. Even a concentration of one in 10^6 of water can cause a 50-dB loss at the 1.39-μm "water line."

6.2.3 Rayleigh Scattering

This imperfection is intrinsically due to minute density fluctuations and molecular concentration variations by bulk imperfections in the fiber such as bubbles, inhomogeneities, and cracks or by waveguide imperfections due to core and cladding interfacial irregularities. There is a point in the spectrum at about 1550 nm where the infrared and ultraviolet absorption are a minimum. Around this point the Rayleigh scattering is the dominant loss. Rayleigh scattering is an inverse function of wavelength. As wavelength increases, Rayleigh scattering decreases. At just above 1.6 μm, infrared absorption becomes dominant.

6.2.4 Fiber Imperfections

Fiber imperfections are another source of loss. These include macro- and microbend losses and were discussed in Section 2.3. Fiber geometry is still another important consideration in the area of fiber imperfections.

6.2.4.1 Glass Geometry. Glass geometry describes the on-end dimensional characteristics of an optical fiber. It has long been recognized as a primary factor in determining splice loss and splice yield. A major goal of fiber manufacturers is for tighter fiber geometry. Fiber that displays tightly controlled geometry tolerances is easier and faster to splice and ensures predictable, high-quality splice performance.

Three parameters that have proven to have most impact on splicing performance are core/clad concentricity, cladding diameter, and fiber curl.

Core/clad concentricity represents how well a fiber's core is centered in the cladding glass region. Improving this characteristic in the fiber manufacture reduces the chance of core misalignment; therefore better lower-loss splices occur.

Outer cladding diameter determines the size of the fiber. The tighter the specification of cladding diameter, the less chance that the fibers will be of a different size. Cladding diameter is especially important when using sizing ferrules or mating field-installable connectors. These devices rely on cladding diameter to align the fibers for connection.

Fiber curl refers to the amount of curvature along a length of fiber. Large amounts of fiber curl can result in too much offset in a mass fusion or V-groove alignment splicer, which can cause high loss splices and poor yields.

6.3 DISPERSION

The effect of dispersion is the broadening of a light pulse as it travels down a length of optical fiber. There are four types of dispersion that relate to its several causes:

1. Intermodal dispersion
2. Material dispersion
3. Chromatic dispersion (in many texts, item 2 is included in item 3 and vice versa)
4. Polarization mode dispersion (PMD)

6.3.1 Intermodal Dispersion

In Section 2.2 we saw that light propagated down multimode fiber had multiple ray trajectories, each taking a different path inside the fiber core. Turning back to

Chapter 1, equation (1.6) gives the number of modes we can expect to propagate down the fiber or

$$M = V^2/2 \tag{6.1}$$

where V is the normalized frequency defined in equation (1.5). If $V = 2.405$ or less, only one mode will propagate. If V is large, many modes will propagate. Figure 2.4 illustrates an example of multimode transmission of light down a fiber light-guide. It also shows the broadening effects on a light pulse after traveling down the fiber some distance, and it shows the destructive effects on a sequential train of NRZ pulses. This type of distortion can be eliminated by the use of monomode fiber where $V < 2.405$.

We used the term *distortion* rather than dispersion following the IEEE (Ref. 2) definition: "In addition, the signal suffers degradation from multimode *distortion*, which is often (erroneously referred to as multimode *dispersion*.)"

At the receive end of the fiber, energy from the different modes arrives somewhat delayed in time in relation to the principal mode (HE_{11}). This causes a smearing of the received pulse. It is destructive because some of that smeared energy falls into the slot of pulse position two. If sufficient smeared energy disperses into slot two, there is a 50% chance that pulse two will be in error.

6.3.2 Material Dispersion

Material dispersion (D_M) is caused by the fact that different wavelengths passing through a certain material travel at different velocities. We are familiar with the relationship that defines the index of refraction (n):

$$n = c/v \tag{6.2}$$

where c is the velocity of light in a vacuum and v is the velocity of the same wavelength in the material in question. Of course the material of interest here is silica glass (SiO_2). Our problem arises because each wavelength propagated through the material travels at a slightly different velocity.

The IEEE (Ref. 2) defines *material dispersion* as "that dispersion attributable to the wavelength dependence of the refractive index of the material used to form the waveguide."

As we proceed in the discussion on dispersion, the insightful reader will pick up on the fact that *chromatic dispersion* should be included under the umbrella of material dispersion. The reader is right. However, for reasons that will become evident, chromatic dispersion is covered under a separate heading.

In Chapter 4, an LED emitted a broad spectrum of wavelengths anywhere in the range of 30 nm to over 100 nm in width, whereas a DFB laser emitted a spectral line width of < 0.1 nm up to 1.0 nm. Obviously, if dispersion were a concern on a certain link, we would opt for a DFB laser rather than an LED and use single-mode fiber.

Regarding velocities of propagation through "material," there is an interesting phenomenon. In the 850-nm band, the longer wavelengths propagate faster than the shorter wavelengths (e.g., an emission at 865 nm will travel faster through a silica glass light-guide than will a similar emission at 835 nm).

Quite to the contrary, in the 1550-nm band, the shorter wavelengths propagate at a greater velocity than longer wavelengths (e.g., 1535-nm wave travels faster than a 1560-nm wave).

An interesting occurrence happens in the 1310-nm band. There is a wavelength, λ_{ZD}, above which D_M is positive and below which D_M is negative. This is called the *zero-dispersion wavelength* which for pure silica is at 1276 nm. It can vary in the range of 1270–1290 nm for optical fibers whose core and cladding are doped to vary the refractive index. The zero-dispersion wavelength for optical fibers also depends on the core radius and the index step (Δ) through the waveguide contribution to the total dispersion (Ref. 1).

It should be pointed out that *waveguide dispersion* will shift the zero-dispersion wavelength by an amount 30–40 nm so that the total dispersion is zero near 1310 nm for a production type of optical fiber.

Material dispersion is a major issue with single-mode fiber systems. With multimode fiber systems, material dispersion contribution to total dispersion is virtually insignificant. Modal dispersion is by far the major contributor.

When wearing my consultancy hat, I advise my clients that if they used a DFB laser source with monomode fiber, a link would be power-limited up to some 1.2 Gbps. Then 2.5-Gbps links came onto the scene. If a client contemplated using this greater bit rate, I advised "Watch it!"

Consider dispersion's effect on a received bit stream. As the bit rate speeds up, the bit's time-slot width gets smaller. The waveform is NRZ. The time-slot width is equal to the bit *period*. Thus we can say

$$\text{Bit period (seconds)} = 1/(\text{bit rate}) \tag{6.3}$$

Here are some simple examples:

For a bit stream at 1 Mbps, the period of a bit is 1 μs.

For a bit stream at 10 Mbps, the bit period is 100 ns.

For a bit stream at 1 Gbps, the bit period is 1 ns.

For a bit period of 10 Gbps, the bit period is 100 ps.

The time slot gets smaller and smaller. The smaller it gets, the more vulnerable it is to dispersion!

As fiber-optic telecommunication systems evolved over time, operation near the zero-dispersion wavelength was very attractive. Lower-bit-rate systems operated in the 1550-nm band where loss per kilometer was lowest. Wouldn't it be neat if we could get those zero-dispersion characteristics into the 1550-nm band?

6.3.3 Chromatic Dispersion

Chromatic dispersion is an extension of material dispersion. When we deal with dispersion affecting the higher bit rates (e.g., > 1 Gbps), the bit period and thus the bit time slot is so small that even with very narrow DFB laser line widths, there was this form of material dispersion.

Reference 4 comes at it a little differently. All glass, including that used to fabricate fiber, exhibits material dispersion because its index of refraction varies with wavelength (as we discussed above). In addition, when a single-mode fiber is drawn from glass, the geometric form and refractive index profile contribute significantly to the wavelength dependence of the velocity of propagation of pulses riding on this fiber—that is, the *waveguide dispersion*. Together, the material dispersion (D_M) and waveguide dispersion (D_W) yield what is termed *chromatic dispersion*.

The corrupting effect of chromatic dispersion is the same as with other forms of dispersion, namely the broadening of the received pulse. Some texts call chromatic dispersion *group velocity dispersion* (GVD) due to the wavelength dependence of the group velocity of light in fiber. Chromatic dispersion, D, is measured in units of picoseconds per nanometer-kilometer (ps/km-nm). That is the amount of broadening in picoseconds that occurs in a pulse with a bandwidth of 1 nm while propagating through 1 km of optical fiber. For example, in the 1550-nm minimum loss window we might expect dispersion in conventional single-mode fiber typically around 17 ps/km-nm. We may want to talk about *dispersion slope*. This describes how the dispersion for a certain fiber value varies with wavelength, or more accurately we could say that it is the rate of change of dispersion with wavelength.

Our interest is the dispersion parameter D as expressed in units of ps/(km-nm) and

$$D = D_M + D_W \tag{6.4}$$

We will use the criterion $B\,\Delta T < 1$ to determine the effect of dispersion on bit rate by using equation (6.5), where ΔT is a time interval that should be shorter than the bit time-slot (bit period). Remember that for an NRZ waveform, the bit period is 1/bit rate (units are seconds).

Building on the criterion expressed above, then

$$BLD\,\Delta\lambda < 1 \tag{6.5}$$

(only valid for the real value of D, which is the dispersion parameter defined above). Equation (6.5) (Ref. 1) gives an order of magnitude estimate of the effect of dispersion on bit rate.

ΔT may be derived as follows:

$$\Delta T = DL\,\Delta\lambda \tag{6.6}$$

where D is the dispersion parameter defined above, L is the length of the link, and $\Delta\lambda$ is equivalent to the spectral width of the pulse.

The BL product[1] from equation (6.6) for single-mode fibers can exceed 1 Tbps/km when SLM DFB lasers are used where $\Delta\lambda$ is below 1 nm. The idea here is to make the laser spectral line width just as narrow as possible. Even in this situation, chromatic dispersion will dominate.

[1] BL product is the bit-rate–distance product.

A successful effort has been made to shift the zero-dispersion wavelength λ_{ZD} to the low-loss 1550-nm band. This is called *dispersion-shifted fiber*. It is described in ITU-T Rec. G.653 (Ref. 3). This dispersion shifting is based on manipulating waveguide dispersion parameters because D_W depends on such fiber parameters as the core radius a and the index (of refraction) difference Δ. It is also possible to tailor the waveguide contribution such that the total dispersion D is relatively small over a wide wavelength range extending from 1.3 to 1.6 μm. This fiber type is called *dispersion-flattened fiber*. It is described in ITU-T Rec. 655, where the chromatic dispersion is 6 ps/km-nm or less between 1530 and 1565 nm, the popular WDM band at the time of this writing.

Chromatic dispersion of a fiber link is cumulative with distance, and it is stated as the change in group delay per unit wavelength (in ps/nm). Chromatic dispersion on an optical fiber link is sensitive to

- An increase in the number of links in tandem and the link lengths.
- An increase in the bit rate. (*Note:* Increasing the bit rate increases the modulation rate of the laser thus increasing the width of its sidebands.)

In WDM systems, chromatic dispersion is not significantly influenced by

- A decrease in channel spacing.
- An increase in the number of channels.

The effects of chromatic dispersion decrease with

- A decrease in the absolute value of the fiber's chromatic dispersion (a decrease in the value of D).
- Dispersion compensation.

Control of chromatic dispersion is especially critical in WDM systems.

6.3.4 Polarization Mode Dispersion (PMD)

In single-mode fiber, only one mode is present, the HE_{11} mode. However, from the viewpoint of polarization, two modes are launched into a single-mode fiber. At the launch we imagine these two polarization modes to be orthogonal, one to the other, and that the polarization is linear. One of the modes is dominant and is supported along the x axis, and the other mode is supported on the y axis. Some references identify the axes as the fast axis and the slow axis. This perfect situation would be on ideal fiber with precise geometry. Reference 4 points out that these axes do not necessarily correspond to a linear state of polarization.

In the real world, when the fibers are cabled and installed in the field, it is difficult to achieve the ideal fiber. There are stresses placed upon the fiber during the manufacturing process. The fiber core and cladding are installed in place during the fiber drawing process, causing an unpredictable *birefringence*[2] to the

[2]*Birefringence* leads to a periodic power exchange between the two polarization components (Ref. 1). The property whereby the effective propagation speed of a light wave in a medium depends on the orientation of the electric field (state of polarization) of the light (Ref. 4).

fiber. The mechanical action of winding fiber on mandrels causes asymmetrical strains. When the cable is installed, further stresses are induced. These actions deform the fiber distorting the fiber's circularity or concentricity of the core within the cladding. There may be fiber elongation and bending.

After the fiber is installed, having undergone all of these stresses, the orientation of these axes and the relative difference in propagation speed corresponding to each axis (directly related to the magnitude of the local birefringence) changes along the optical path. One might imagine in some idealized situation that different sections of the fiber would have different orientations of these local birefringence axes. In each fiber segment a time delay will be introduced between that portion of the light aligned with the local fast versus the local slow birefringence axes. Because the relative orientation of these axes in adjoining segments is different, the pulse will experience a statistical spreading over time. Thus we have polarization mode dispersion.

Source: Ref. 4.

PMD is measured in picoseconds for a particular span of installed fiber. Flaws in the fiber either add or counteract PMD, but there is a gradual increase in PMD as a light pulse proceeds along the fiber and then span to span. The appropriate units for the coefficient that characterizes the fiber itself are $\text{ps}/\text{km}^{1/2}$. For a fiber connectivity made of some several spans, a root mean square summation of PMD for each span must be used.

Chromatic dispersion effects can be mitigated by *dispersion compensators*. However, as of now, there is no way of mitigating PMD.

PMD will affect a fiber-optic system with:

- An increase in channel bit rate (the bit period gets smaller)
- An increase in link length (between regenerator sections)
- An increase in the number of channels (denser WDM)

PMD can be reduced through tight quality control during the fiber fabrication stage.

Source: Ref. 4.

6.3.5 Dispersion Compensation

There are two different devices that can be used to compensate for chromatic dispersion. The first is *dispersion compensation fiber* (DCF) and the second type uses chirped, in-fiber *Bragg grating*.

6.3.5.1 Dispersion Compensation Fiber (DCF). Chromatic dispersion is essentially cumulative linearly along the length of a fiber. This is fortunate. If we then add to our operational fiber link a highly dispersive fiber, where D, the dispersion coefficient, has an opposite sign and with approximately equal magnitude of that of the operational fiber link, then the dispersion of the operational fiber is canceled out. The DCF can have a dispersion coefficient (D) as high as -200 ps/km-nm.

Source: Ref. 1.

The length of the DCF should be as short as possible. The length of the DCF (L_2) can be calculated by the following formula:

$$L_2 = -(D_1/D_2)L_1 \tag{6.7}$$

where L_1 is the length of the operational fiber, L_2 is the length of the DCF, D_1 is the dispersion coefficient of the operational fiber, and D_2 is the dispersion coefficient of the DCF.

Several problems arise from the use of a DCF. First, the special fiber employed in a DCF is considerably more lossy than operational fiber. The loss per unit length ranges from 0.4 to 1.0 dB/km. This adds to the total link loss. Second, it takes about 1 km of DCF to compensate for 10–12 km of operational fiber. Third, as described by Agrawal (Ref. 1), because of a relatively small-mode diameter, the optical intensity is larger at a given input power, resulting in enhanced nonlinear effects.

For dispersion compensating techniques to be effective, it is necessary to measure the total dispersion of the fiber link being installed as well as the dispersion coefficient of the DCF.

6.3.5.2 Bragg Grating Compensators. Another way to compensate for chromatic dispersion is by use of Bragg grating techniques. Here advantage is taken of the changes in a chirped pattern of index of refraction creating wavelength-selective mirrors. A fiber Bragg grating acts as an optical filter because of the existence of a "stop band." The stop band is centered at the *Bragg wavelength.* The Bragg wavelength is a function of the grating period and the mode index.

Source: Ref. 1

Bragg gratings have a relatively narrow stop band. With present-day fiber-optic systems, broad-band gratings are required. A solution is provided by the *chirped fiber grating*. In this case the optical period varies linearly over its length. Because the Bragg wavelength also varies along the grating length, different frequency components of an incident pulse are reflected at different points, depending on where the Bragg condition is satisfied locally.

Key to understanding the chirped Bragg grating dispersion compensation is the velocity of propagation of light. Using standard fiber in the 1550-nm band, the higher-frequency components of an optical pulse propagate faster than the low-frequency components. Considering the Bragg wavelength increases along the grating length, the low-frequency components travel further into the grating before being reflected. They experience more delay induced by the grating than the high frequency components. It is the relative delay induced by the grating, which is just the opposite of the fiber, that compensates for the fiber dispersion.

Source: Refs. 1 and 4.

6.4 NONLINEAR EFFECTS

In this section, seven phenomena will be briefly covered:

1. Stimulated Brillouin scattering (SBS)
2. Stimulated Raman scattering (SRS)
3. Self-phase modulation (SPM)
4. Four-wave mixing (FWM)
5. Modulation instability (MI)
6. Soliton formation
7. Cross-phase modulation (CPM)

As capacity of fiber-optic systems increases, the tendency to increase signal power is evident. However, signal level may not be increased indefinitely because optical nonlinearities will act to modify the system performance as the power level is raised. This will tend to set a maximum limit on optical power, which, in itself, will set a limit on S/N, resulting in a maximum capacity limit. Other factors setting limits to maximum capacity will be covered as we move along in our discussion.

Why then can we not put these nonlinearities to our advantage? Such considerations gave rise to nonlinear dispersion compensation and soliton transmission systems.

6.4.1 Introduction to Optical Nonlinearities

Nonlinear interactions between the optical signal and the silica fiber transmission medium begin to appear as optical signal powers are increased. The signal powers are increased to compensate for the high insertion losses of WDM equipment as bit rates are increased (e.g., 2.5 to 10 to 40 Gbps) and to achieve longer span lengths. As a consequence, nonlinear fiber behavior has emerged as an important consideration along unregenerated routes and with the introduction of WDM and dense WDM. These nonlinearities can be classified into two general areas: scattering effects (typically stimulated Brillouin scattering and stimulated Raman scattering) or effects related to the Kerr effect. The *Kerr effect* is the change in refractive index of a material under the influence of an electric field. It brings about an intensity dependence of the refractive index. In this group of nonlinear effects we will find self-phase modulation, cross-phase modulation, modulation instability, soliton formation, and four-wave mixing. The severity of these effects are influenced, more or less, by the following parameters of the fiber and signal(s) on the fiber: fiber dispersion characteristics, the effective core area of the fiber, the number and spacing of optical channels in multichannel systems, and overall unregenerated system length, as well as signal intensity and source line width.

6.4.2 Stimulated Brillouin Scattering

Stimulated scattering comes about when an incident signal is scattered. The scattering may be either forward or backward and is brought about by one of several possible mechanisms. In every case, light is shifted to longer wavelengths. For example, at 1550 nm the scattered light is downshifted or Brillouin-shifted by approximately 11 GHz.

Of the nonlinearities described herein, stimulated Brillouin scattering (SBS) has the lowest threshold power. It has been shown that the SBS threshold can vary between the different types of fiber and among individual fibers. It is typically on the order of 5–10 nW for extremely modulated, narrow linewidth sources. On directly modulated lasers, its power may be on the order of 20–30 mW. On G.653 fiber the SBS threshold is slightly lower than that for a system on G.652 fiber. This is due to the smaller effective area of the G.653 fiber. It can also be said that this will be true for all of the nonlinear effects that we will discuss. The SBS threshold is sensitive to the source linewidth and level of power. However, it is independent of the number of WDM channels.

6.4.2.1 *Impairment Effects on Transmission.*

Effectively, SBS limits the amount of light that can be transmitted down a fiber path. Figure 6.1 shows this effect for a narrow-band source, where all of the signal power falls within the Brillouin bandwidth. The transmitted power becomes saturated, and the backscattered power rapidly increases. The input power level to the fiber at which this rapid increase occurs is defined as the SBS threshold. In the general case the SBS threshold is expressed as

$$P_{th} = 21 \frac{KA_{eff}}{gL_{eff}} \cdot \frac{\Delta\upsilon_p - \Delta\upsilon_B}{\Delta\upsilon_B}$$

where g denotes the Brillouin gain coefficient, and A_{eff} is the effective core area. K is a constant determined by the degree of freedom of the polarization state (in Recommendation G.652 fibers, $K = 2$). The variables $\Delta\upsilon_B$ and $\Delta\upsilon_p$ represent the

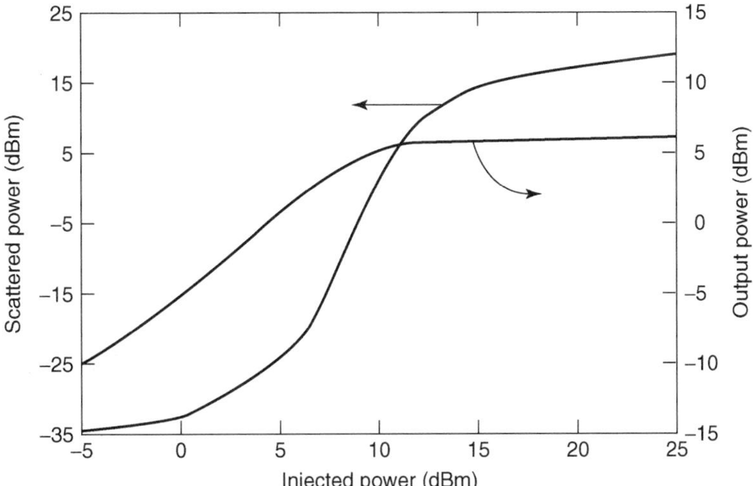

Figure 6.1. Stimulated Brillouin scattering threshold for narrow-band source. (From Figure II.3-1/G.663, page 11, ITU-T Rec. G.663, Ref. 10.)

Brillouin bandwidth and a linewidth of a pump light, respectively. L_{eff} denotes the effective length defined as

$$L_{\text{eff}} = \frac{1 - \exp(-\alpha L)}{\alpha}$$

where α is the fiber attenuation coefficient and L is the fiber length.

The SBS threshold, P_{th} depends on the linewidth of pump light, $\Delta\upsilon_p$. When the linewidth of the pump light is small compared to the Brillouin bandwidth, the SBS threshold power can be estimated using the following relation:

$$P_{\text{th}} = 21\frac{KA_{\text{eff}}}{gL_{\text{eff}}}$$

6.4.2.2 Mitigation of the Impairment. SBS impairments will not arise in systems where the source linewidth significantly exceeds the Brillouin bandwidth or where the signal power is below the threshold power.

6.4.3 Stimulated Raman Scattering

Stimulated Raman scattering (SRS) only becomes an impairment when light transmission levels are high. Its effects are somewhat similar to Brillouin scattering, but the light is shifted to much lower frequencies, between 10 and 15 THz for the 1550-nm window. Also, the downshifted frequency has a much wider bandwidth than its Brillouin scattering counterpart, some 7 THz. In WDM systems, the effect is to transfer power from the shorter-wavelength channels to the longer-wavelength ones. In this case, the fiber acts as a Raman amplifier and the long-wavelength channels are amplified by the short-wavelength channels as long as the wavelength difference is within the bandwidth of the Raman gain. This phenomenon can occur with silica fiber where amplification can result for spacing as great as 200 nm.

Reference 1 reports that the shortest-wavelength channel is most depleted in a WDM configuration, because it can pump many channels simultaneously. Such an energy transfer among channels can be determined for system performance because it depends on bit pattern. Amplification occurs only when binary 1 bits are present in both channels simultaneously. The signal-dependent amplification leads to increased power fluctuation, which adds to receiver noise and degrades receiver performance. *Raman crosstalk* can be avoided if channel powers are made so small that Raman amplification is negligible over the fiber length. Particular care should be taken with SRS when there are fiber amplifiers in tandem along the length of a link. These amplifiers add noise that experiences less Raman loss than the desired signal. The result is degradation in signal-to-noise ratio at the far-end receiver.

Source: Ref. 4.

6.4.3.1 Impairment Effects on Transmission. SRS can occur in both single- and multiple-channel systems. Signal powers on the order of +30 dBm or more are needed to experience SRS with only a single channel, without light amplifiers. However, shorter-wavelength signals in WDM systems with widely spaced channels can suffer degraded S/N performance when a portion of their power is transferred to longer wavelength channels through the SRS phenomenon. This results in total system capacity limitations based on the total number of WDM channels, their spacing, overall system length, and average input power. Reference 10 comments that, in particular, the threshold for observation of a 1-dB penalty in a multichannel system due to Raman gain in dispersion unshifted fiber can be estimated to be

$$P_{\text{tot}} \cdot \Delta\lambda \cdot L_{\text{eff}} < 40 \text{ mW} \cdot \text{nm} \cdot \text{Mm} \tag{6.8}$$

where P_{tot} is the combined power of all the WDM channels, $\Delta\lambda$ is the optical spectrum over which the channels are distributed, and L_{eff} is the effective length [in units of 10^6 meters (Mm)]. The SRS threshold for a system using G.653 fiber is slightly lower than that for a system deploying G.652 fiber, due to the smaller effective area of G.653 fiber. SRS does not practically degrade single-channel systems. However, it may limit the capability of WDM systems.

6.4.3.2 Mitigation of Impairment. With single-channel systems the unwanted spectrum can be removed with filters. However, no practical techniques to eliminate the effects of SRS in WDM systems have as yet been reported. The effects of SRS can be reduced by decreasing the input optical power. Reference 10 reports that SRS does not appear to present a practical limitation to the deployment of WDM systems.

6.4.4 Self-Phase Modulation

When the coupled output level from a light source becomes too high, the signal can modulate its own phase. As the name implies, this is *self-phase modulation* (SPM). As illustrated in Figure 6.2, this results in a broadening of the transmitted

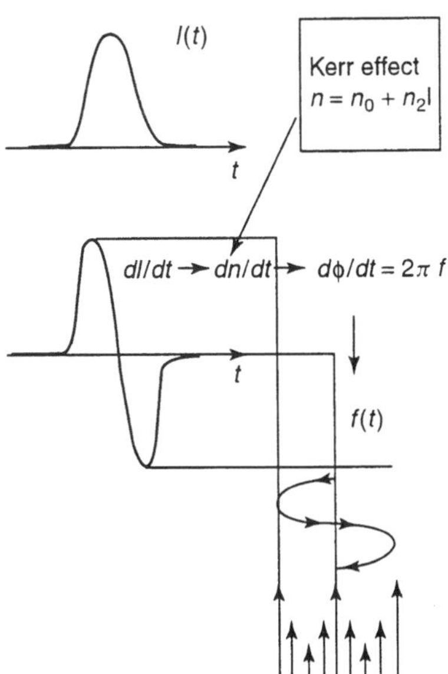

Figure 6.2. Spectral broadening mechanism due to self-phase modulation. (From Figure II.3-6/G.663, ITU-T Rec. G.663, page 17, Ref. 10.)

pulse and temporarily broadens or compresses the signal. Whether it does one or the other depends on the sign (positive or negative) of the chromatic dispersion present. The result is a shift to long wavelengths at the leading edge of the pulse and a shift to short wavelengths at the trailing edge.

6.4.4.1 *Impairment Effects on Transmission.* As one would expect, self-phase modulation will increase with an increase in transmit power. It is more destructive as the channel bit rate increases and pulse rise times are tighter. It will also increase with negative chromatic dispersion.

SPM is not significantly influenced by a decrease in channel spacing in WDM systems or by an increase in the number of channels. Self-phase modulation decreases when there is a zero or small value of chromatic dispersion or an increase in fiber effective area.

Generally, the effects of SPM are significant only in systems with high cumulative dispersion or on very long systems. Fiber-optic systems that are dispersion-limited may not tolerate the effects due to SPM. In WDM systems with very closely spaced channels, the spectral broadening induced by SPM may also create interference between adjacent channels.

On G.652 fibers, SPM on low-chirp intensity-modulated signals leads to pulse compression; and on G.655 fibers with anomalous dispersion as a function of transmit power. The pulse compression counteracts the chromatic dispersion and offers some dispersion accommodation. However, limits of maximum dispersion and related transmission length exist.

6.4.4.2 *Mitigation of Impairment Effects.* Selecting an operational wavelength for G.653 fiber near the zero dispersion wavelength tends to reduce the impact of SPM. On shorter fiber connectivities, of, say, less than 1000 km, SPM can be controlled through the use of dispersion compensation devices at indicated intervals along the length of a G.652 fiber. As we mentioned previously, the effects of many of these impairments may be mitigated by reducing signal power or by operating above the zero-dispersion wavelength of G.655 fiber.

6.4.5 Four-Wave Mixing

For those of us from the radio or wireless world, four-wave mixing (FWM) reminds us of third-order products. It can play havoc with WDM systems. It appears when the intensity of the laser signal reaches a critical level. FWM manifests itself with ghost signals, some of which may fall on top of desired channels. Whenever three or more light signals propagate down a fiber, we can expect four-wave mixing. These three light signals, ω_i, ω_j, and ω_k, will generate a fourth signal ω_{ijk} obeying the relationship

$$\omega_{ijk} = \omega_i + \omega_j - \omega_k \tag{6.9}$$

FWM can even be generated on single-channel systems, between the desired wavelength and OFA ASE and between main mode and side modes. In the case of two signals, the intensity modulation at their beat frequency modulates the fiber refractive index and produces a phase modulation at a difference frequency. The

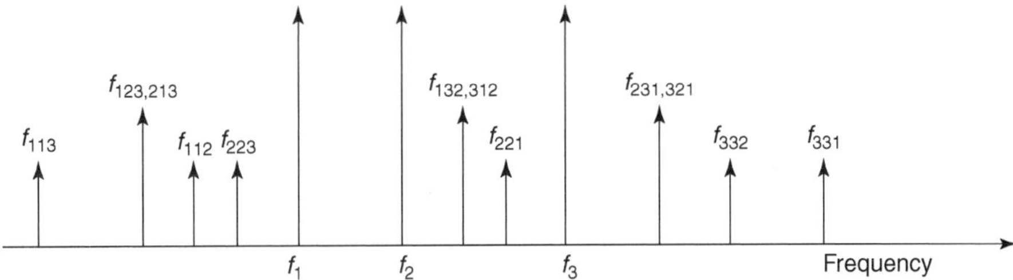

Figure 6.3. Mixing products generated due to four-wave mixing of three optical signals. (From Figure II.3-2/G.663, page 12, ITU-T Rec. G.663, Ref. 10.)

phase modulation develops two sidebands at frequencies given by this difference. In the case of three signals, more and stronger mixing products are produced (see Figure 6.3) which will fall directly on adjacent signal channels when the channel spacings are equal in frequency. Two optical waves propagating along a fiber produce FWM with high efficiency if the phase matching condition is achieved between sidebands and initial signals. For relatively low optical powers, this means

$$2\pi\Delta v^2 D\lambda_t^2 L/c \ll 1 \qquad (6.10)$$

where Δv is the frequency spacing among the wavelengths, D is the chromatic dispersion of the fiber, λ_t is the central wavelength, L is the fiber length, and c is the speed of light in a vacuum. [Losses can be taken into account by slightly modifying equation (6.10).] The efficiency of FWM is also sensitive to overall optical power in the fiber. Consider the following: For two signals with optical P_1 and P_2, the maximum parametric gain coefficient for the sidebands, $g_{\max}$, can be estimated as

$$g_{\max} \approx 2g(P_1 + P_2) \qquad (6.11)$$

where g is the nonlinear Kerr coefficient.

6.4.5.1 Impairment Effects on Transmission. With WDM, and especially DWDM, FWM can be very destructive. In a DWDM system have N channels, the derived four-wave mixing channels can be calculated by

$$N^2(N-1)/2 \qquad (6.12)$$

For example, a four-channel WDM system would develop 24 ghost channels. An eight-channel system would potentially develop 224 ghost channels, and so on.

FWM is a particularly serious issue in systems using G.653 (Ref. 3) dispersion-shifted fibers. Conversely, the placement of an optical signal channel directly at or near the dispersion zero can result in a very significant buildup of FWM products over a relatively short fiber length (i.e., tens of kilometers). It is less important with G.655 (Ref. 6) nonzero dispersion-shifted fibers, especially those with large effective areas. It is less important for dispersion-unshifted fiber, typically Rec. G.652 (Ref. 5) fiber, because the dispersion is relatively flat.

FWM is sensitive to the following system characteristics:

- An increase in channel power
- An increase in the number of channels
- Closer channel spacing

With WDM based on 200-GHz spacing, FWM is drastically reduced when compared to WDM based on 100-GHz spacing.

Four-wave mixing decreases with the absolute value of chromatic dispersion.

The generation of FWM sidebands can result in significant depletion of signal power. In addition, when mixing products fall directly on signal channels, they cause parametric interference that manifests as amplitude gain or loss in the signal pulse, depending on the phase interaction of the signal and sideband.

Parametric loss causes closure of the eye pattern at the receiver output, thereby degrading BER. Frequency spacing and chromatic dispersion reduce the efficiency of the FWM process by destroying the phase matching between the interacting waves. Systems deployed over G.652 fiber experience less FWM impairment compared to systems deployed over G.653 fiber. Conversely, the placement of a signal channel directly at or near the dispersion-zero can result in a very serious buildup of FWM products over a relatively short fiber length (e.g., tens of kilometers). FWM is also sensitive to channel separation.

FWM can create a serious system impairment in WDM systems on G.653 fiber, because the channel signals experience only a small value of chromatic dispersion. The impact of dispersion on achievable system capacity for a four-channel system over three amplifier spans is shown in Figure 6.4. This illustrates what can happen

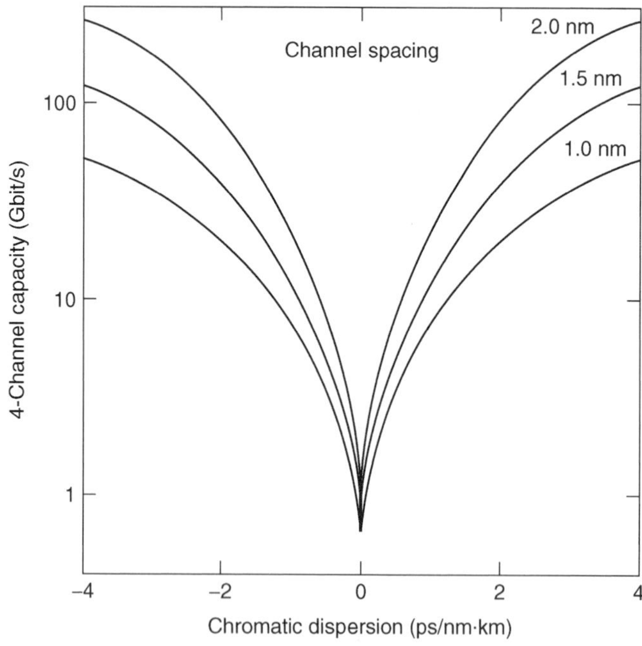

Figure 6.4. Impact of dispersion on system capacity in an FWM environment. (From Figure II.3-3/G.663, page 13, ITU-T Rec. G.663, Ref. 10.)

at high power levels when conditions promote the generation of mixing products. The capacity limitation is based on a worst-case calculation of mixing products generated by the FWM process with four +8-dBm signal channels centered around the dispersion value shown. This system develops intolerable levels of distortion due to FWM as the dispersion experienced by the signal channels approaches zero.

In single-channel systems, the FWM interaction can occur between the OFA ASE noise and the transmission channel, as well as between the main mode and the side modes of the optical transmitter. Phase noise is added to the signal carrier due to the accumulated ASE via the fiber Kerr effect, thus broadening the signal spectral tails.

6.4.5.2 Mitigation of Impairment Effects. As we noted above, dispersion may be used to suppress the generation of FWM sidebands. Typically, this may be the dispersion found in G.655 fibers. Nonuniform channel spacings may also be used to mitigate the severity of the FWM impairment. Reduction of the input power levels in G.653 fiber systems could permit multiple channel operation, but might compromise the economic advantages of optical amplification.

In order to adequately suppress the generation of mixing products, use of a fiber with a minimum permitted (i.e., nonzero) dispersion within the region of the OFA amplification band has been proposed by the industry. Alternating spans of such nonzero-dispersion fiber with opposite dispersion characteristics has also been considered as a potential option, because the resultant cable would maintain a net chromatic dispersion of approximately zero. However, Ref. 10 notes that this alternative may present difficulties in the areas of installation, operations, and maintenance by introducing a second fiber type into the outside plant environment. Similar approaches using long spans of fiber with small finite dispersion and short lengths of opposite and higher dispersion fiber to provide compensation have also been demonstrated. In particular, in links with periodic amplification, a short piece of compensating fiber can be located inside the box in which the optical amplifier is located.

Uneven channel spacing and larger channel spacing have been proposed as a means to mitigate the effects of nonlinearities and allow deployment of DWDM systems on G.653 fiber. Uneven channel spacing ensures that mixing products generated by three or more channels do not fall directly on other channels wavelengths. However, the transfer of power from the signals into the mixing products (i.e., signal depletion) remains unaffected by making the channel spacing uneven and may still cause significant eye closure. Increased channel spacing also reduces the effects of FWM. Use of these mitigation techniques may be constrained by channel narrowing due to the concatenation of optical amplifiers, which reduces the width of the usable amplification spectrum.

6.4.6 Modulation Instability

Modulation instability (MI) breaks a CW signal or pulse into a modulated structure. Reference 10 reports that it can be observed in the anomalous dispersion regime (i.e., above the zero-dispersion wavelength), where a quasi-monochromatic signal spontaneously tends to generate two symmetric spectral sidebands, as shown in Figure 6.5. Frequency separation and gain of the sidebands are determined by

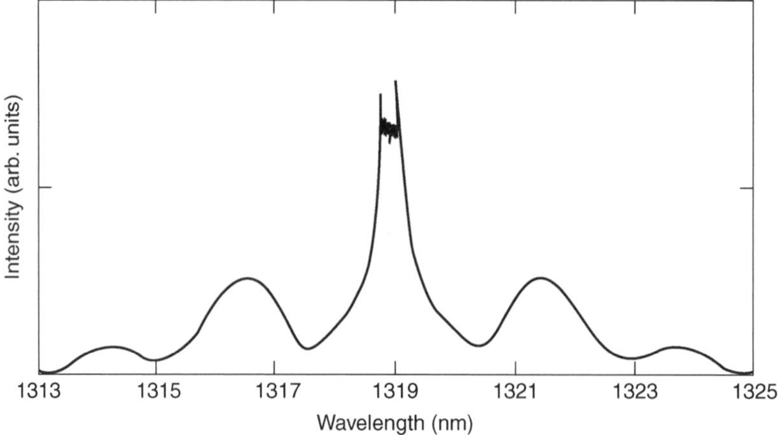

Figure 6.5. Power spectrum of a pulse after propagation in 1-km-long fiber. Input pulse width is 100 ps, and peak power is 7 W. The spectral sidelobes appeared due to MI. (From Figure II.3-4/G.663, page 15, ITU-T Rec. G.663, Ref. 10.)

the intensity of the wave and by dispersion and nonlinear coefficients of the fiber. The maximum conversion efficiency occurs at a frequency separation given by

$$\Omega_{\max} = \pm\left[\frac{8\pi^2 c n_2 P_0}{\lambda^3 A_{\text{eff}} D(\lambda)}\right]^{1/2}$$

where n_2 is the silica nonlinear coefficient, A_{eff} is the effective area of the fiber, P_0 is the launched power, $D(\lambda)$ is the chromatic dispersion coefficient, and λ is the operation wavelength; sidebands located at $\pm\Omega_{\max}$ from the carrier experience a gain per unit length: $g_{\max} = 4\pi P_0/(\lambda \mathrm{A}_{\text{eff}})$. Fiber loss can be taken into account by slightly modifying the equations above. Dependence of MI gain on the frequency deviation with respect to the signal is given in Figure 6.6 in the presence of fiber loss and for various values of fiber dispersion.

The MI can be viewed a a particular case of FWM where two photons of the intense incoming signal are converted into two photons at different frequencies.

6.4.6.1 ***Impairment Effects on Transmission.*** Modulation instability (MI) may decrease the signal-to-noise ratio due to the generation of sidebands either spontaneously or seeded by the amplified spontaneous emission (ASE). Because the maximum degradation of the signal is expected for high values of $g_{\max}$ and for Ω close to the bandwidth of the signal, MI may be critical when using very powerful boosters in dispersion-shifted fiber links with directly modulated lasers. On long-haul unrepeated systems, MI can be observed at lower power levels and may cause excess amplification of the spontaneous emission noise of the cascaded OFAs. This Kerr-effect-induced broadening at signal spectral tails may cause signal-carrier depletion, and the tails may be attenuated by the narrow-band ASE filters or by the self-filtering effect in very long systems.

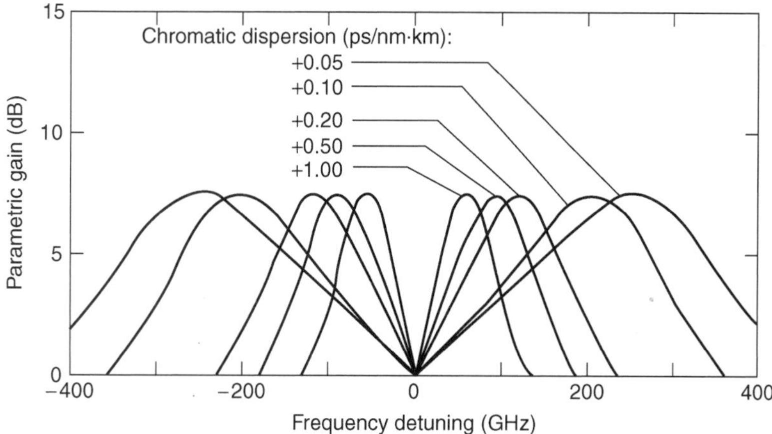

Figure 6.6. Calculated MI gain versus frequency detuning, from 30-km-long fibers (0.24-dB/km loss) with five different dispersion coefficient values, for +16-dBm launched power. (From Figure II.3-5/G.663, page 16, ITU-T Rec. G.663, Ref. 10.)

6.4.6.2 Methods to Mitigate Impairment Effects. The effect of MI can be minimized either by decreasing the power level or by operating at wavelengths below the zero-dispersion wavelength of the link. Dispersion management is another possibility to reduce the formation of MI sidebands. Otherwise, the received signal should be electrically filtered to lower the level of the spurious amplified noise. The impact of MI can be decreased considerably by external modulation of lasers giving narrower spectra.

6.4.7 Soliton Formation

6.4.7.1 Taking Advantage of the Soliton. If somehow we could rid ourselves of the dispersion in an optical fiber line and operate around the point of minimum attenuation in the 1550-nm window while transmitting at 10 Gbps, we could extend the distance between regenerative repeaters to some 1000 km. What a boon this would be for undersea fiber-optic systems!

The effect of dispersion on a transmitted pulse is a pulse broadening as the pulse traverses down the length of a fiber line. A *soliton* is a pulse that does not change its shape as it traverses long fiber links. It does not broaden due to the effects of dispersion and fiber nonlinearities.

A soliton represents a balance between nonlinearity and dispersion. The fiber nonlinearity counteracts the accumulating dispersion as a pulse propagates along a fiber-optic line.

Being more specific and abstracting from Ref. 1: Fiber solitons are a result of a balance between *group velocity*[3] *dispersion* (GVD) and *self-phase modulation* (SPM), described in Section 6.4.2. Taken individually, each of these limit the performance of a fiber-optic system. GVD broadens optical pulses during their propagation

[3]*Group velocity* (fiber optics): "For a particular mode, the reciprocal of the rate of change of the phase constant with respect to angular frequency" (Ref. 2).

inside the fiber when the pulse is initially chirped in the right manner. More specifically, a chirped pulse can be compressed during the early stage of propagation imparting the GVD parameter β_2 and the chirp parameter C with opposite signs, so that $\beta_2 C$ is negative. SPM, resulting from the intensity dependence of the refractive index, imposes a chirp on the optical pulse such that $C > 0$. Because $\beta_2 < 0$ in the 1550-nm region, the condition $\beta_2 C < 0$ is satisfied. Moreover, because the SPM-induced chirp is power-dependent, it is not difficult to imagine that under certain conditions, SPM and GVD may cooperate in such a way that the SPM-induced chirp is just right to cancel the GVD-induced broadening of the pulse. The optical pulse would then propagate undistorted in the form of a soliton.

Solitons must maintain a certain separation one from the other when transporting a serial bit stream. This prevents soliton interaction that can be destructive. To achieve the necessary separation of two adjacent soliton pulses, the system designer turns to the use of the RZ (return-to-zero) format rather than the NRZ format, which is more commonly employed. With the RZ format the soliton occupies a small fraction of the bit slot (bit period or bit duration) to ensure that neighboring bits are sufficiently separated.

What limits the length or supportable bit rate on a soliton fiber link?

- *Loss-Induced Soliton Broadening.* A soliton pulse must maintain sufficient amplitude. If it does not, the soliton pulse begins to broaden. The reduced peak power weakens the nonlinear effect necessary to counteract the GVD. The use of fiber-optic amplifiers (e.g., EDFAs) will return the soliton required peak energy.
- *Amplifier Noise.* The amplifiers needed to restore the soliton energy also add noise originating from *amplified spontaneous emission* (ASE) (see Chapter 7).
- *Timing Jitter Deriving from the In-Line Optical Amplifiers.* Timing jitter is a mechanism that induces deviations in the soliton position from its original location at the bit center. Ideal operation is when all the solitons arrive at the far-end receiver at the center of their assigned bit slot. Deviations cause soliton interaction and can degrade BER.

6.4.7.2 Impairment Effects on Transmission. Effects due to soliton formation may be relevant in the following fiber-type systems: Recs. 652, 653, and 655. As we have shown, fundamental soliton formation can be useful; however, other solitons generally give rise to a strong degradation of the transmitted signal. Thus, higher-order soliton formation sets a limit to the maximum power that can be launched into a fiber.

6.4.7.3 Mitigation of Impairment Effects. Soliton formation can be avoided by operating the fiber system below the zero-dispersion wavelength of the link. However, in this regime, soliton transmission is not supported and both dispersion and nonlinearity contribute to pulse broadening. Signal degradation can be minimized by proper management of dispersion along the link.

6.4.8 Cross-Phase Modulation

In WDM systems, particularly DWDM systems, cross-phase modulation (XPM) will gradually broaden the signal spectrum when changes in optical intensity result in changes due to interactions between adjacent channels. The amount of spectral broadening introduced by XPM is related to the channel separation, because the dispersion-induced differential group velocities will cause the interacting pulses to separate as they propagate down the fiber. Once spectral broadening is introduced by XPM, the signal experiences a greater temporal broadening as it propagates along the length of the fiber due to the effects of chromatic dispersion.

6.4.8.1 Impairment Effects on Transmission. Impairments from XPM are more significant in G.652 fiber systems, relative to G.653 and G.655 systems. The

TABLE 6.1 A Review of Nonlinear Optical Effects in Glass Fibers

Nonlinear Optical Effect	Cause	Characteristics	Critical Light Power in SMF	Impact
Self-phase modulation (SPM) and cross-phase modulation (XPM)	Optical Kerr effect: intensity-dependent refractive index	Phase shift Self-induced (SPM) Adjacent channel (XPM) Spectral frequency broadening	$P_c > \sim 10$ mW	Spectral broadening increases effect of dispersion Power/dispersion limited high bit-rate transmission Pulse compression Positive dispersion Pulse propagation (solitons) Limitations in PSK systems by AM/PM conversion
Stimulated Raman scattering (SRS)	Interaction of photon-optical phonons	Raman lines $f - n\Delta f$(Stokes) $\Delta f = 12$ THz $\Delta\lambda = 70$ nm (1310 nm) $\Delta\lambda = 102$ nm (1550 nm)	$P_c > \sim 1$ W (for single channel) $P_c > \sim 1$ mW for Raman amplification in a WDM system with critical channel spacing, $\Delta\lambda$	Optical loss in fiber Optical crosstalk in WDM system Signal power depletion
Stimulated Brillouin scattering (SBS)	Interaction: photon-acoustic phonons	Brillouin lines in backward direction $f \pm \Delta f$ $\Delta f = 13.2$ GHz (1310 nm) $\Delta f = 11.1$ GHz (1550 nm)	$P_c > \sim 5$ mW (for narrow linewidth optical source) P_c increases with signal line width	Signal instability Optical loss in fiber Optical crosstalk in bidirectional coherent multi-channel systems
Four-photon mixing or four-wave mixing (FWM)	Multiphoton interaction	Mixing products generated $f_4 = f_1 \pm f_2 \pm f_3$	$P_c > \sim 10$ mW (for Recommendation G.653 fibers) depends on specific parameters (e.g., channel spacing and closeness to λ_0)	Optical crosstalk in WDM systems Signal power depletion

Source: Table II.1/G.663, page 20, ITU-T Rec. G. 663, Ref. 10.

broadening due to XPM may result in interference between adjacent channels in WDM systems.

6.4.8.2 Mitigation of Impairment Effects. XPM can be controlled by the proper selection of channel spacing in WDM/DWDM systems. Studies (Ref. 10) have shown that only adjacent channels contribute significantly to XPM-induced signal distortion in WDM systems. The S/N of the center channel of a three-channel system will approach that of a single-channel system as channel separation is increased. As a result, the effect of XPM can be rendered negligible with adequate spacing between optical channels. Channel separations of 100 GHz were shown to be sufficient to reduce XPM effects in a simulation of a system with 5 mW of power per channel. Dispersion penalties due to XPM may also be controlled by the implementation of dispersion compensation at appropriate intervals along the length of the system.

A review of nonlinear optical effects in glass fibers is shown in Table 6.1.

6.5 POLARIZATION PROPERTIES

6.5.1 Polarization Mode Dispersion (PMD)

See Section 6.3.4 for an introductory discussion of PMD.

6.5.1.1 Impairment Effects on Transmission. In a digital system, the principal effect of PMD is to cause intersymbol interference (ISI). As an approximate rule of thumb, a 1-dB penalty occurs for a total dispersion equal to 0.4 T, where T is the bit period. According to Ref. 10, this is the accepted value for the maximum tolerable system power penalty. Although still unresolved, according to Ref. 10, current studies indicate that optical fibers and cables will be specified according to the mean level of polarization mode dispersion, a view reflected by studies of single-channel and multichannel systems where the mean level will also be specified. This corresponds to a mean differential group delay equal to one-tenth the bit period, 0.1 T. Computer simulations have predicted that if the polarization mode dispersion has a Maxwellian distribution with a mean value not to exceed 0.1 T, there is less than a 10^{-9} probability that the system power penalty will exceed 1 dB.

Furthermore, Ref. 10 states, that in long-haul amplifier systems employing polarization scrambles (devices that deliberately modulate the polarization state of a signal laser so that it appears to be unpolarized), the polarization mode dispersion causes an increase in the degree of polarization of the signal. This degrades system performance through interactions with polarization-dependent loss and polarization hole burning (see below). In an analog system, the interaction loss and mode dispersion with laser chirp leads to a second-order distortion proportional to the modulation frequency. A further second-order penalty, independent of modulation frequency, is incurred when additional polarization-dependent loss is present in the system.

It has also been shown, and mentioned briefly above, that a second-order effect can cause a coupling between polarization mode dispersion and chromatic dispersion. This is caused by the wavelength dependence of differential group delay. This leads to a statistical contribution to the chromatic dispersion. This is an area that is not well understood and that is presently under study.

6.5.1.2 *Mitigation of Impairment Effects.* Given that PMD arises from induced birefringence, much of the effort in reducing the effects of polarization mode dispersion have been concerned with minimizing the birefringence introduced by fiber or cable manufacture. Care is taken to optimize fiber production to ensure concentricity of the fiber core. Optical cables are manufactured using materials and processes that minimize the residual strain in the cable structure across the fiber core. Elaborate cable structures can also be used which introduce a circular component to the induced birefringence. By careful design, such an effect can counteract linear birefringence to produce a cable with a resultant zero polarization mode dispersion. Typically, the mean polarization mode dispersion of fibers and cables lies in the range of

$$0 < \langle \Delta\tau \rangle < 0.5\ \text{ps}/\sqrt{\text{km}}$$

Another method makes use of the concept of principal states that was introduced earlier. In this scheme, a polarization controller is inserted at the input and output ends of the system. A polarization beam splitter follows the output polarization controller and is used to generate an error signal. The output polarizer searches for the error signal, and the input polarizer is adjusted to minimize this error signal. At the point of no error signal, the input polarization state is one of the principal states for the system. Using such a technique up to 1-bit period of delay has been compensated for in a 5-Gbps system. A similar technique has been applied to coherent frequency-division multiplexing (FDM) systems.

6.5.2 Polarization-Dependent Loss

Polarization-dependent loss (PDL) arises from dichroism of the passive optical components, such as isolators, couplers, and so on, in the signal path. When the signal passes through the dichroic element, the component of its electric field parallel to the lossy axis is attenuated. As in the case of polarization mode dispersion, the axes that define the PDL are oriented randomly with respect to each other.

6.5.2.1 *Impairment Effects on Transmission.* Let us examine a typical system configuration to identify and then control PDL. In amplified systems, one mode of amplifier control is to operate at a constant signal power. Both the signal and the noise are affected by PDL. However, because the noise is unpolarized, the signal and the noise are affected differently. The noise can be resolved into a component parallel to the signal and a component orthogonal to the signal. It can be shown that the combined effect of PDL and optical amplification is always to increase the component of the noise orthogonal to the signal. Furthermore, the magnitude of the orthogonal noise component changes with time as the signal polarization changes due to polarization mode dispersion. This leads to a reduction in signal-to-noise ratio and the Q value at the receiver. In addition, the fluctuations time lead to fading of the signal-to-noise and Q value at the receiver, both of which lead to an impairment in system performance.

In analog systems, the PDL can interact with laser chirp and polarization mode dispersion to reduce the system performance in terms of composite second order distortion. As would be expected, this impairment is time-varying and leads to fluctuations in the system composite second order with time.

6.5.2.2 Mitigation of Impairment Effects. First, it should be noted that the impact of PDL on system performance increases as the number of amplifiers increases. In long-haul submarine systems, for example, the requirements are extremely tight, because the number of amplifiers in tandem can be in the hundreds. In short-haul terrestrial systems, where there are only a few amplifiers in tandem, the impact of PDL on system performance is still under study by the ITU-T Organization.

6.5.3 Polarization Hole Burning

Polarization hole burning (PHB) results from an anisotropic saturation created by a polarized saturating signal launched into an erbium fiber. This results in a selective depopulation of excited states aligned with the polarized field. Consequently, the available gain in the orthogonal direction is higher. Although the erbium ions are distributed randomly within the glass matrix, on a microscopic level the dipole associated with the erbium ion is anisotropic. The PHB effect is maximum where the linearly polarized saturating signal is aligned with the major axis of the dipole and is reduced where the polarization state of the saturating signal is elliptical or circular. Both the signal laser and the pump laser contribute to the total effect, the total differential gain being the vector sum of the two contributions. The degree of hole burning is proportional to the degree of polarization of the saturating signal. For an unpolarized saturating signal there is no hole burning. In principle, this is similar to the case of a circularly polarized signal.

6.5.3.1 Impairment Effects on Transmission. Polarization hole burning (PHB) impacts the system performance by causing the noise buildup along the amplifier chain to be greater than would be predicted from simple linear theory. That is, the signal-to-noise ratio is reduced by PHB; as for cases of polarization mode dispersion and polarization-dependent loss (PDL), the measured Q fluctuates in time. Because there are two contributions to PHB, there are two ways in which the system performance is affected. The total effect is proportional to the gain saturation, increasing with an increased degree of saturation.

First of all, we treat the effect of the polarized pump laser. Let us consider for this discussion that the pump polarization is fixed and invariant. The pump causes a differential gain in the direction orthogonal to its polarization axis. Noise aligned orthogonally to the pump experiences a higher gain than noise aligned with the pump. However, the polarization axes of the pump lasers in each amplifier along a chain are uncorrelated with each other. Therefore, the cumulative effect is similar to a random walk, and the pump-induced PHB can be considered as a contribution to the PDL of the amplifier. Thus, averaged over a number of amplifiers, the noise buildup should be linear as expected from simple theory.

The signal-laser-induced PHB is slightly different. As the laser signal propagates along the system, the noise polarized along parallel to the laser signal will see the same gain as the signal. However, noise polarized orthogonal to the laser signal will always experience higher gain because it will always be orthogonal to the signal polarization axis. Therefore, the total noise will increase in a nonlinear way along the chain of amplifiers.

The total differential gain due to PHB varies as the polarization state of the signal changes (due to polarization mode dispersion) along the amplifier chain. It varies because the signal hole burning effect is correlated with the pump effect. As with the relative polarization states of the signal and pump lasers, the magnitude of the differential gain changes. Therefore, although the total noise increases nonlinearly along the chain, it does so in such a way that the total noise fluctuates in time. Consequently, as we explained above, the signal-to-noise ratio is reduced and fluctuates in time. The system Q is, therefore, reduced and fluctuates with time.

6.5.3.2 Mitigation of Impairment Effects. One way to mitigate the effect of PHB is to operate amplifiers at lower signal levels. However, often this is not possible or not desirable. Reference 10 suggests that the simplest approach is to use depolarized signals. A depolarized signal can be generated in many ways, but it is most commonly generated by polarization scrambling. Using a phase modulator, the polarization state is varied between two orthogonal states in time. The signal then appears to be depolarized.

Practice has shown that it is optimal to impose the polarization modulation at twice the bit rate. This is because the PDL in the amplifier converts the polarization modulation to amplitude modulation. By polarization modulating at twice the bit rate, the amplitude fluctuations are at a rate higher than the detector bandwidth and thus are not seen by the receiver. Using such techniques, the performance of very long-haul systems has been improved to the point where the predicted performance is met with a high degree of confidence. Polarization modulation is now a standard practice on transoceanic amplified systems.

However, in long amplified systems, polarization mode dispersion causes a repolarization of the signal, thus allowing PHB again to degrade the system performance. Such an effect illustrates the complex nature of the interaction of polarization phenomena in amplified links.

6.6 OTHER SYSTEM DEGRADATIONS

6.6.1 OFA-Related Noise Accumulation

With cascaded OFAs (see Section 7.4.2) along an optical fiber connectivity, the ASE noise generated at an OFA repeats a cycle of attenuation and amplification in a manner identical to the desired optical signal. Because the incoming ASE noise is amplified at each OFA and added to the ASE noise generated at that OFA, the total ASE noise power increases almost proportionally with the number of OFAs, and the signal power accordingly decreases. Reference 10 points out that the noise power can exceed the signal power.

The ASE noise spectral profile also evolved along the system length. When ASE noise from the first OFA is input to a second OFA, the gain profile of the second OFA changes due to the ASE noise power via the gain saturation effect. Similarly, the effective gain profile of the third OFA is then modified by the output-power spectrum of the second OFA. Such an effect is repeated all the way down to the last OFA. The ASE noise accumulates even if narrow-band filters are used at each OFA because the noise exists over frequency ranges that include the signal frequency.

6.6.1.1 *Impairment Effects on Transmission.* ASE noise accumulation affects system SNR because the degradation in the received signal SNR is due predominantly to ASE-related beat noise. Such beat noise increases linearly with the number of OFAs. Thus, the error rate degrades as the number of OFAs increase. Furthermore, noise accumulates exponentially with the magnitude of the amplifier gain.

As a result of the gain spectrum of the OFA, the ASE noise spectrum after passing through many OFAs tends to have a peak at a certain wavelength due to the self-filtering effect (see Section 6.6.2). If a closed optical ring network is considered, expect ASE noise accumulation as if an infinite number of OFAs were cascaded. Although the accumulation of ASE noise in filtered systems is considerably reduced by the filters, in-band ASE still increases with the number of cascaded OFAs. As a result, the SNR degrades as the number of OFAs increases.

6.6.1.2 *Mitigation of Impairment Effects.* The ASE noise accumulation can be reduced by decreasing the OFA spacing while at the same time maintaining the total gain equal to the total loss of the transmission path because ASE noise accumulates exponentially with the magnitude of the amplifier gain. There are two filtering techniques that can reduce ASE noise still further:

1. ASE noise filters
2. Self-filtering effect or self-filtering method

The self-filtering method is used when there are tens of OFAs or more in tandem. With this method, the system designer aligns the system wavelength with the self-filtering wavelength so that there is reduction of ASE noise at the detector input. It is just as if a narrow-band filter had been used. Reference 10 reports that this is very effective when used on shortened OFA spans and with low-gain OFAs to reduce the initial ASE noise.

It is not advisable to use the self-filtering method on closed-ring WDM networks employing OFAs. The principal reason is that there is a resulting peak in the overall OFA gain spectrum which may strongly affect the system performance. For cases such as these, it is advisable to use ASE filters minimizing ASE noise. This is achieved by filtering the WDM channels not dedicated to the network node before switching them out of the node.

When there are only a few OFAs in tandem, the self-filtering method is less effective than the ASE-filter method. This latter method allows greater flexibility in the choice of operational wavelength(s) and is easier to handle for channel level uniformity on WDM systems. Reference 10 advises that care should be taken in the selection of filter characteristics. It should be noted that the cascaded-filter passband is narrower than a single filter, unless the passband has a rectangular shape. Conventional filters with a full width at half-maximum (FWHM) on the order of 3 nm could be used on long-distance single-channel systems.

6.6.2 Self-Filtering Effect

There is a characteristic profile of the overall gain spectrum (or ASE noise spectrum) as a result of ASE noise accumulation due to OFAs in tandem. This spectrum tends to have a peak, and this peak's spectral linewidth narrows as the number of OFAs increases after it finally saturates with some number of OFAs. This may result in a 2 to 3 nm wide spectral linewidth after several tens of OFAs. This effect is called *self-filtering*.

Reference 10 reports that the self-filtering effect is determined by the spectral shape of the emission and absorption cross sections and by the degree of inversion of OFAs. The self-filtering wavelength may change with changes in the host-glass composition, input optical power, or interamplifier loss and their dependence on wavelength, pump wavelength, and the length of the doped fiber. Designers consider that the self-filtering effect can generally be desirable in single-channel systems but undesirable in WDM systems.

6.6.2.1 Impairment Effects on Transmission. When there are few OFAs in tandem, the spectral width of the self-filtering gain peak remains broad and does not reduce ASE noise accumulation. It does not even do so when the operational wavelength is adjusted to the self-filter peak wavelength. It may be tricky to try to take advantage of the self-filtering effect due to numerous OFAs in tandem. The SNR can be high but can degrade if the operational wavelength shifts from the self-filtering wavelength. Typically, this shift can occur after system reconfiguration or repair because of changes in interamplifier loss.

In the case of WDM/DWDM systems with cascaded OFAs, there can be a power level variation among channels that exponentially increases with the number of OFAs in tandem (the number of OFAs is the exponent). Reference 10 gives the example of the power spread for a five-channel system of ~ 3 dB after the first EDFA, and it increased to ~ 15 dB after the sixth EDFA. For WDM systems, changes in the total number of channels result in gain-spectral changes that perturb other channels. Besides, saturation-induced spectrally dependent gain also generates gain variation among the channels.

It is recommended that the link budget be carefully designed to accommodate such EDFA-gain spectral changes in both filtered and nonfiltered systems. It should be noted that ASE filters tend to eliminate the problem. This technique is described in Section 6.6.2.2.

6.6.2.2 Dealing with the Self-Filtering Effect. Using the self-filtering method to improve SNR is most effective when the optimum dispersion wavelength, the self-filtering wavelength, and the operational wavelength are the same. This method does not require the use of ASE noise filters, which can bring degradation associated with PDL in the filter. This is particularly true in transoceanic submarine systems. As we mentioned above, reliance on the self-filtering effect complicates system design, reconfiguration, and repair because the operational wavelength must always be adjusted to be the same as the self-filtering wavelength, which is changeable. The spectral characteristics of both OFA gains and interamplifier losses should be as uniform as possible. Otherwise, the self-filtering gain peak may not become sufficiently narrow, thus degrading the desired SNR improvement. It also makes it more difficult to prevent ASE noise-induced saturation in long-haul systems.

The ASE-filter method is the alternative that can avoid these disadvantages in reducing ASE noise accumulation with the filter passband adjusted to the operational wavelength. Narrow-band filters with FWHM < 1 nm are now commercially available. Using the ASE-filter method allows ASE noise accumulation to be minimized. The system is thus freed from restrictions such as complexities of system design, reconfiguration, and repair, and it is also freed from the requirements of short OFA span and uniform performance of each OFA.

In WDM systems, the interchannel power spread due to self-filtering effects can be avoided by amplifying each channel in a physically separate OFA. It is pointed out that this method demands a costly demultiplexer, a separate OFA, and a multiplexer. An alternative is to provide optical channel power equalization at each network node, even if this method requires additional control devices and a more critical power budget through the network. Another is to use a less saturated or less strongly inverted OFA, because this makes the attenuation less wavelength-dependent and, accordingly, reduces interchannel power spread. In a strongly inverted EDFA cascade, however, the ASE grows with the number of EDFAs at the first gain peak of $\sim$ 1530 nm and needs to be eliminated with a short wavelength eliminating filter. In addition, the preemphasis method minimizes SNR differences for all channels by adjusting the transmitter optical powers for each channel based on the received signal information from each end terminal.

Sections 6.4 through 6.6 are based on Appendix II, ITU-T Rec. 663, Ref. 10.

6.7 SELECTING OPTICAL FIBER FOR ITS DISPERSION CHARACTERISTICS

Throughout the text, we have mentioned four types of "specialized single-mode fiber":

1. Conventional single-mode fiber
2. Dispersion-shifted fiber
3. Nonzero-dispersion-shifted fiber (nz-dsf)
4. Dispersion-flattened fiber

Dispersion in single-mode fiber consists of two subsets:

1. Material dispersion (D_M) caused by the wavelength dependence of the refractive index
2. Waveguide dispersion (D_W) that results from the wavelength dependence of the light distribution of the fundamental mode over the core and cladding glass and therefore the refractive index difference.

Chromatic dispersion is the sum of these two subsets.

We remember that for wavelengths greater than 1300 nm the two dispersions in fused silica glass have opposite signs. Through the use of dopants, the material dispersion can only be changed slightly. On the other hand, waveguide dispersion is greatly influenced by the use of a different profile structure of the refractive index.

As we will remember from Chapters 1 and 2, the refractive index profile of an ordinary single-mode fiber is a step index profile with the refractive index difference Δ. For this profile structure of conventional single-mode fiber, the chromatic dispersion (sum of material and waveguide dispersions) is zero at about 1310 nm.

Since the waveguide contribution, D_W, depends on such fiber parameters as core radius a and the refractive index difference Δ, it is possible to move the zero dispersion wavelength to other wavelengths. One of the most desirable wavelengths is 1.550 μm because of its low loss characteristics. This type of fiber is covered by ITU-T Rec. G.653, whereas ITU-T Rec. G.652 specifies conventional single-mode fiber.

The dispersion shifting technique also can produce the flattened or dispersion compensated fiber where the dispersion values are very low for the entire wavelength range from 1300 to 1600 nm. Figure 6.7 shows chromatic dispersion as a function of wavelength for (1) conventional single-mode fiber, (2) dispersion-shifted fiber, and (3) dispersion-flattened fiber. Dispersion-flattened fiber is covered by ITU-T Rec. G.655.

To shift the zero-dispersion wavelength toward another wavelength, the waveguide dispersion and the profile structure of the optical fiber must be changed. This leads to *multistep* or *segmented index profiles*. Figure 6.8a shows the conventional index profile (i.e., simple step index or matched cladding) and Figure 6.8b illustrates the depressed index profile with reduced refractive index in the cladding (depressed cladding). Figure 6.9 illustrates the profile designs of optical fiber with dispersion shifting. Figure 6.9a shows a segmented profile with a triangular core (segmented core), Figure 6.9b illustrates a triangular profile, and Figure 6.9c shows

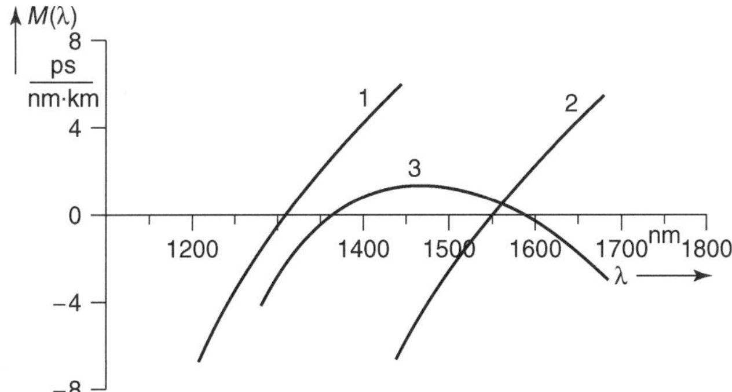

Figure 6.7. Chromatic dispersion as a function of wavelength. (Courtesy of Siemens, Ref. 9, Figure 4.11, page 48.)

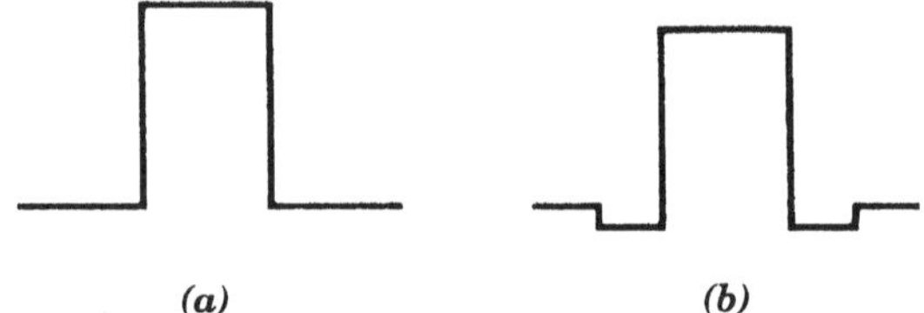

Figure 6.8. Profile design of optical fibers without dispersion shifting (conventional single-mode fiber). (a) Standard step index; (b) step index profile with reduced refractive index. (Courtesy of Siemens, Ref. 9, Figure 14.12, page 49.)

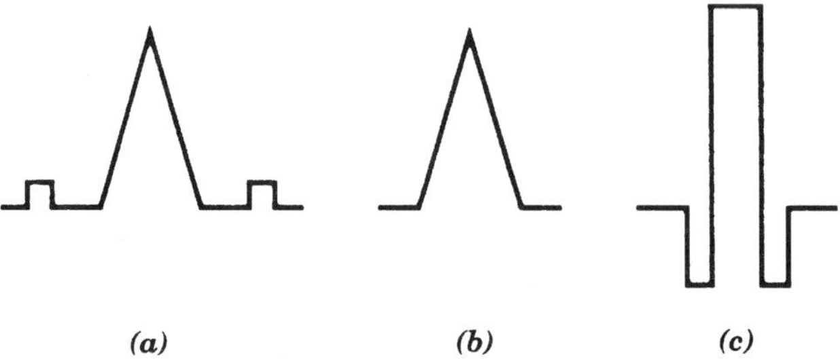

Figure 6.9. Profile design of optical fibers with dispersion shifting. (a) Segmented profile with triangular core (segmented core); (b) triangular profile; (c) segment profile with double-step refractive index in the cladding (double cladding). (Courtesy of Siemens, Ref. 9, Figure 4.13, page 49.)

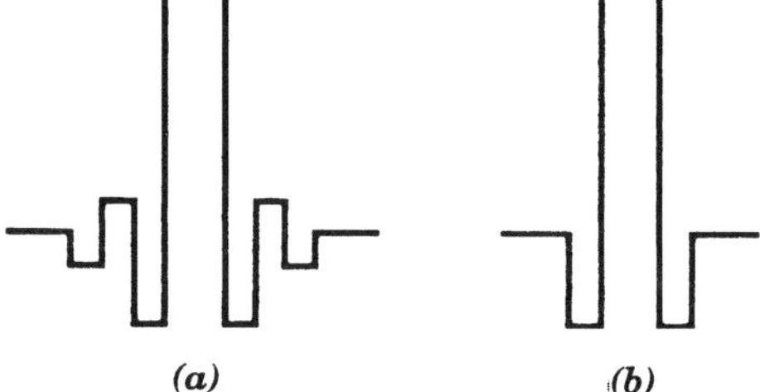

Figure 6.10. Profile design of optical fibers with dispersion flattening. (a) Segmented profile with a fourfold step in refractive index of the cladding (quadruple clad); (b) *W* profile (double clad). (Courtesy of Siemens, Ref. 9, Figure 4.14, page 50.)

a segment profile with a double-step refractive index in the cladding, also called double clad. Figure 6.10 shows the profile design for dispersion flattening. A segmented profile with a fourfold step in refractive index of the cladding (quadruple clad) is shown in Figure 6.10a. Figure 6.10b illustrates a *W* profile (double clad).

6.7.1 Salient Characteristics of ITU-T Rec. G.652 Single-Mode Fiber (Conventional, Single-Mode Fiber)

ITU-T Rec. G.652 (Ref. 5) describes a single-mode optical fiber that has the zero-dispersion wavelength around 1310 nm and that is optimized for use in the 1310-nm region. It can also be used in the 1550-nm wavelength region, where the fiber is not optimized. The salient characteristics of ITU-T Rec. G.752 are given in Table 6.2.

TABLE 6.2 Salient Characteristics of ITU-T G.652 Single-Mode Fiber

Characteristic	Value	Comments
Mode field diameter at 1310 nm	8.6–9.5 μm	10 μm used for matched cladding designs and 9 μm for depressed cladding designs.
Cladding diameter	125 μm $\pm$ 2 μm	
Mode field concentricity error at 1310 μm	Not to exceed 1 μm	
Cladding noncircularity	< 2%	
Maximum cable cutoff wavelength	1260 or 1270 nm	
1550-nm loss performance deployed; 1310 nm optimized for 1550 nm	< 1.0 dB	Loss increase of 100 turns with 37.5-mm radius and measured at 1550 nm, loosely wound.
Attenuation coefficient	< 0.5 dB/km in 1310-nm region and < 0.4 dB/km in the 1500-nm region	
Chromatic dispersion coefficient	Zero dispersion wavelength between 1300 nm and 1324 nm	Max. value $S_{0\,\max} = -0.093$ ps(nm$^2 \cdot$ km), zero dispersion slope.
Maximum chromatic dispersion coefficient		
1288–1339 nm	3.5 ps/(nm · km)	
1271–1360 nm	5.3 ps/(nm · km)	
Polarization mode dispersion coefficient	Generally < 0.5 ps/km$^{1/2}$	Lower bit–distance products (under 10 Gbps/400 km) can tolerate higher OMD coefficients value without impairment.

Source: Based on information derived from ITU-T Rec. G.652, Ref. 5.

To estimate maximum dispersion for the 1550-nm region, calculate $D_1(\lambda)$:

$$D_1(\lambda) = S_{0\,\max}/4[\lambda - (\lambda^4 0\min/\lambda^3]$$

where S_0 is the zero-dispersion slope.

6.7.2 Salient Characteristics of ITU-T Rec. G.653 Dispersion-Shifted Single-Mode Optical Cable

ITU-T Rec. G.653 (Ref. 3) describes a dispersion-shifted single-mode fiber that has a nominal zero-dispersion wavelength close to 1550 nm, along with a dispersion coefficient that is monotonically increasing with wavelength. The fiber is optimized for use at wavelengths in the region between 1550 nm and 1600 nm, but also may be used at around 1310 nm subject to the constraints outlined in the referenced recommendation. The salient characteristics of G.653 optical cable are given in Table 6.3.

TABLE 6.3 Salient Characteristics of G.653 Optical Cable

Characteristic	Value	Comments
Mode field diameter at 1550 nm	7.8–8.5 μm	Deviation not to exceed ±10%
Cladding diameter	125 μm	Deviation not to exceed ±2 μm
Mode field concentricity error at 1550 nm	Not to exceed 1 μm	Some jointing techniques and loss requirements, tolerances up to 3 μm
Cladding noncircularity	< 2%	
Maximum cutoff wavelength	1270 nm	Recommended
1550-nm bend performance	< 0.5 dB	Loss increase for 100 turns fiber loosely wound with 37.5-mm radius, measured at 1550 nm
Attenuation coefficient at 1550 nm	< 0.35 dB/km	In the 1300-nm region < 0.55 dB/km
Chromatic dispersion coefficient	$D_{max} = 3.5$ ps(nm · km)	Between 1525 and 1575 nm
Chromatic dispersion slope	$S_{0\,max} \leq 0.085$ ps(nm^2 · km)	$D(l) = (l - l_0)S_0$, where λ is the wavelength of interest
Nominal zero-dispersion wavelength l_0	1550 nm	

Source: Based on information derived from ITU-T Rec. G.653, Ref. 3.

TABLE 6.4 Salient Characteristics of G.654 Single-Mode Fiber-Optic Cable

Characteristic	Value	Comments
Mode field diameter (MFD) at 1550 nm	10.5 μm	MFD deviation not to exceed ±10%
Cladding diameter	125 μm ± 2 μm	
Mode field concentricity error	Not to exceed 1 μm	At 1550 nm
Cladding noncircularity	< 2%	
Cutoff wavelength (λ_c = fiber cutoff)	< 1600 nm	When lower limit > 1350 nm
Cable cutoff wavelength (λ_{cc})	Max value 1530 nm	Recommended
1550-nm bend loss performance	0.5 dB	Loss increase for 100 turns of fiber, loosely wound with 37.5-mm radius, measured at 1550 nm
Attenuation coefficient	0.22 dB/km	In the 1550-nm region
Chromatic dispersion coefficient (D)	20 ps(nm · km)	At 1550 nm
Maximum dispersion slope S_{1550}	Around 0.07 ps/(nm^2 · km)	
Polarization mode dispersion (PMD) coefficient	< 0.5 ps/km$^{1/2}$	Corresponds to transmission distance about 400 km for 10-Gbps system; lower bit-rate-distances product system can tolerate higher PMD coefficient values

Source: Based on information derived from ITU-T Rec.G.654, Ref. 8.

TABLE 6.5 Salient Characteristics and Requirements of G.655 Single-Mode Fiber-Optic Cable

Characteristic	Value	Comments
Mode field diameter at 1550 nm	8–11 μm	Mode field deviation not to exceed ±10%
Cladding diameter	125 μm ± 2 μm	
Mode field concentricity error at 1550 nm	Not to exceed 1 μm	
Cladding noncircularity	Not to exceed 2%	
Cutoff wavelength for fiber cable	1480 nm	Jumpers 1480 nm; worst-case length and bends, 1470 nm
1550-nm bend performance	Not to exceed 0.5 dB	Loss increase for 100 turns of fiber, loosely wound with 37.5-mm radius
Attenuation coefficient	< 0.35 dB/km	
Chromatic dispersion coefficient	Between 0.1 ps/nm-km and 6.0 ps/nm-km between 1530 nm and 1565 nm	
Polarization mode dispersion coefficient	$< 0.5\ \text{ps/km}^{1/2}$	Corresponds to 400 km of STM-64 with 1-dB penalty

Source: Based on information derived from ITU-T Rec. G.655, Ref. 6.

6.7.3 Characteristics of Cutoff-Shifted Single-Mode Optical Fiber Based on ITU-T Rec. G.654

ITU-T Rec. G.654 (Ref. 8) describes a single-mode fiber that has the zero-dispersion wavelength around 1300 nm which is cutoff-shifted and loss-minimized at a wavelength around 1550 nm and is optimized for use in the 1500- to 1600-nm region. Table 6.4 gives the basic characteristics and requirements for G.654 fiber-optic cable.

6.7.4 Salient Characteristics of Nonzero-Dispersion-Shifted Single-Mode G.655 Fiber-Optic Cable

ITU-T Rec. G655 (Ref. 6) describes a single-mode fiber whose chromatic dispersion (absolute value) is required to be greater than some nonzero value throughout the wavelength range of anticipated use. This dispersion suppresses the growth of four-wave mixing, a nonlinear effect that can be particularly deleterious to DWDM. The fiber is optimized for use at wavelengths in between 1500 and 1600 nm. The characteristics and requirements of ITU-T Rec. G.655 are given in Table 6.5.

REFERENCES

1. Govind P. Agrawal, *Fiber-Optic Communication Systems*, 2nd ed., John Wiley & Sons, New York, 1997.
2. *The IEEE Standard Dictionary of Electrical and Electronic Terms*, 6th ed., IEEE, New York, 1996.

3. *Characteristics of a Dispersion-Shifted Single-Mode Optical Fiber Cable*, ITU-T Rec. G.653, ITU Geneva, April 1997.
4. *Guide to WDM Technology Testing*, 2nd ed., EXFO Electro-Optical Engineering, Inc., Quebec City, Canada, 2000.
5. *Characteristics of Single Mode Optical Fiber Cable*, ITU-T Rec. G.652, ITU Geneva, April 1997.
6. *Characteristics of Non-Zero Dispersion Shifted Single-Mode Optical Fiber Cable*, ITU-T Rec. G.655, ITU Geneva, October 1996.
7. Andre Girard, Handling Special Effects: "Non-Linearity, Chromatic Dispersion, Soliton Waves," *Lightwave*, July 2000.
8. *Characteristics of a Cut-off Shifted Single-Mode Optical Fiber Cable*, ITU-T Rec. G.654, ITU Geneva, April 1997.
9. G. Mahlke and P. Goessing, *Fiber Optic Cables*, 3rd ed., Siemens Berlin–Munich, John Wiley & Sons, New York, 1997.
10. *Applications Related Aspects of Optical Fiber Amplifier Devices and Subsystems*, ITU-T Rec. G.663, ITU Geneva, October 1996.

7

REGENERATORS AND FIBER-OPTIC AMPLIFIERS

7.1 CHAPTER OBJECTIVE AND SCOPE

In the 1980s, fiber-optic systems began to be widely deployed. Each fiber carried a single stream of pulses representing binary 1s and 0s. A 1980s model of such a system would include a light source, connecting fiber-optic cable, and a light detector some distance away. The maximum distance between source and detector depended on the output of the laser source, connector losses, splice losses, fiber loss, bit rate, and the sensitivity of the light detector. If we wanted to extend the link still further, a regenerator was installed. Following this methodology, a connectivity could be established across the continent. Furthermore, the connectivity could have transmission capacity of hundreds of megabits per second. The entire large capacity would be carried all on one fiber in one direction; a second fiber would carry companion digital traffic in the other direction for full-duplex connections. As we added capacity, regenerator sections became shorter and shorter. The number of active components in a setup like this notably degraded system availability. There also was jitter buildup. A regenerator at that time was no more than a light receiver back-to-back with a light transmitter.

When I started in the fiber-optics business, we firmly believed that a light signal could not be amplified. Then, around 1989, after years of research, the rare-earth fiber-optic amplifier came on the scene. This brought several new dimensions to fiber-optic transmission. With the ~ 20-dB gain of these amplifiers, a fiber-optic link could be made much longer before introducing a regenerator, particularly if several of these amplifiers were used on the link. It also allowed the practical application of wavelength division multiplexing, and finally it made optical switching practical.

In this chapter the goal is to discuss light amplification and, in particular, the applications of erbium-doped fiber amplifiers (EDFAs) and Raman amplifiers. The chapter starts with a review of regenerators, their application to lightwave transmission systems, and their interface with an optical network management system.

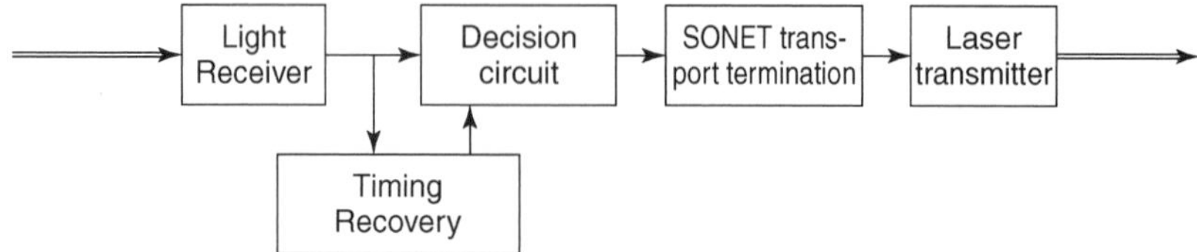

Figure 7.1. A simplified block diagram of a digital light regenerator.

7.2 THE APPLICATION OF REGENERATORS TO A LIGHTWAVE SYSTEM

A regenerator takes a corrupted light signal at its input and delivers a near-perfect replica of the signal as it was transmitted by the previous transmitter. The regenerated signal is practically free from distortion. The regeneration function is carried out by an all-digital light transmitter and receiver. The device we will discuss here is a stand-alone regenerator. Light amplifiers do not regenerate a digital light signal. Figure 7.1 illustrates a digital light regenerator.

The fiber-optic system engineer uses the link budget technique to position the regenerator. That point will be where the accumulated loss on the link would start to degrade its error performance plus a certain margin. (See Chapter 10 for application and methodologies of a link budget.)

Telcordia states in Ref. 8 that a regenerator reshapes, retimes, and retransmits a light signal. A number of texts say that regenerators amplify a light signal. This is not true unless a light amplifier is incorporated at the output of the regenerator's laser transmitter.

Turn to Figure 7.1, moving from left to right. The input to the regenerator consists of a distorted and corrupted light signal due to the loss of the fiber section just traversed and the accumulated dispersion of that section of interconnecting fiber. The light pulses that represent binary 1s are converted to electrical 1s, and the bit positions with little or no light signals are converted to binary 0s. This electrical baseband of signals is passed through the receiver's electrical circuits where retiming is carried out. The actual decision as to whether a binary 1 or binary 0 exists in a bit position is determined in the demodulation process of the light signal. In modern systems all these functions are carried out in an integrated PIN receiver.

The binary baseband signal is passed to the SONET transport termination. Here access is gained to the SONET transport overhead. This allows status of the regenerator and quality of the bit stream to be passed to a network operations control center.

The electrical baseband signal of the SONET transport termination is now passed to a laser transmitter that generates equivalent light pulses from the incoming bits. The laser transmitter light level coupled into the outgoing fiber often is in the range of 0 to +3 dBm. However, if the fiber-optic link design engineer desires longer spans,[1] he/she may opt to place a fiber amplifier at the regenerator laser output where the signal may be boosted as much as 20 or 25 dB.

[1]*Span:* On an optical network, the horizontal distance between adjacent regenerators, between a regenerator and an ADM, or between a regenerator and an optical terminal.

A regenerator has two advantages that an amplifier does not. An amplifier does not *regenerate* a digital signal, whereas a regenerator does. The advantage here is that a digital signal with its accumulated forms of distortion is applied to the input of an amplifier. That digital signal is output from the amplifier containing the same distortion plus the added distortion and noise imparted to the signal by the amplifier. Quite to the contrary, a regenerator cleans up much of the distortion or impairments on a digital signal and outputs a new "squared-up" serial bit stream. The second advantage of a regenerator is that it can easily access SONET or SDH overhead OA&M (operations, administration, and maintenance) fields to provide status of the regenerator and the bit stream passing through it. The status is passed to the responsible network operations center. This provides the network operator with an excellent monitoring capability and trouble-shooting tool. An amplifier does not have easy access to the bit stream baseband because it does not demodulate and remodulate the bit stream as a regenerator does.

We described a stand-alone regenerator in this section. Keep in mind that every light receiver and transmitter carries out a regeneration function. However, one might call the role of a light transmitter one of *generation* rather than one of *regeneration*.

Remote regenerators are powered by one of two general methods:

1. They can be powered by a wire pair deriving from the prime power system of the nearest ADM or terminal site
2. They can be powered locally by the local power company. They should have a no-break battery backup, or they should be powered locally by solar cells, small gas turbine, or wind generator, all with battery backup (see Chapter 13, Section 13.8).

Remote optical fiber amplifiers (OFAs) receive their prime power in a similar manner.

7.2.1 SONET Regenerators

The SONET regenerator described in this section follows the specification found in Bellcore (now Telecordia) TR-NWT-000917, Issue 1, Dec. 1990 (Ref. 9). This regenerator carries out the basic regenerator functions as described in Section 7.2 above. In addition, the regenerator has a specific interface with SONET OA&M overhead. Bellcore calls it section terminating equipment. This means that it terminates, and in most cases reinitiates, the section overhead. Section overhead is discussed in Chapter 9, Section 9.2.3. In Some instances, the SONET RGTR (Bellcore terminology) may relay some or all of the section overhead through the RGTR. The overhead bytes that are relayed are transmitted as received. The section overhead byte designations for STS-1[2] are shown in Figure 7.2.

With reference to Figure 7.2, each byte-interleaved STS-1, as part of an STS-*N* frame, has two framing bytes (A1 and A2) and an STA-1 identification byte (C1). The framing bytes contain a unique pattern used to identify the beginning of a frame.

[2]STS-1 stands for synchronous transport signal 1. SONET terminology discussion may be found in Chapter 9, Section 9.4.

3 BYTES

3 rows section overhead			
	Framing A1	Framing A2	STS-1 ID C1
	BIP-8 B1	Orderwire E1	User F1
	Data Com D1	Data Com D2	Data Com D3

Figure 7.2. Byte designations of SONET section overhead for STS-1. (From Ref. 12, page 2-33.)

The section bit interleaved parity-8 (BIP-8) code (B1), section orderwire (E1), section user channel (F1), and the section data communications channel (DCC), which consists of bytes D1, D2, and D3, are also part of the section overhead. The SONET RGTR uses the BIP-8 to detect errors, whereas the user channel (F1) is reserved for the network provider's applications. For sending and receiving operation information, typically OA&M information, the SONET RGTR relies on the

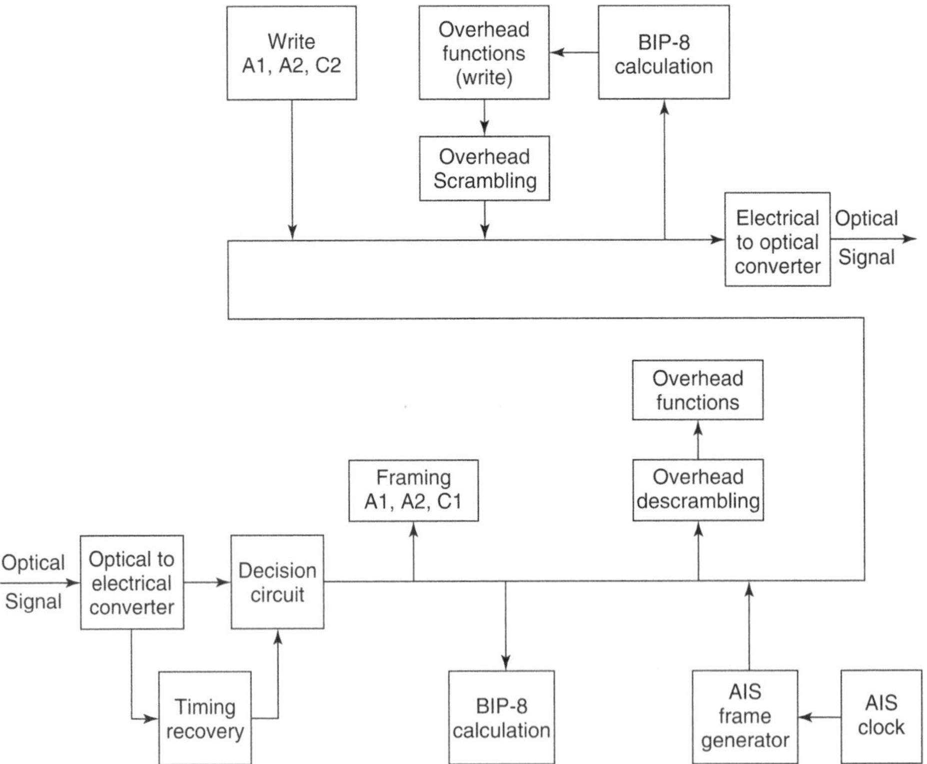

AIS = Alarm indication signal, A1, A2, C1, framing bytes and STS-1 identification byte respectively.

Figure 7.3. Functional block diagram of a SONET STE regenerator. (Based on Refs. 9 and 13.)

section DCC to provide 192-kbps embedded operations channel (EOC). Finally, the section orderwire channel provides voice communication for craftsperson activities via the E1 byte.

Figure 7.3 is a functional block diagram of a section terminating equipment (STE) regenerator (RGTR). From the figure, moving from left to right, the O/E block represents the optical-to-electrical conversion producing an electrical signal of 1s and 0s from which timing is recovered. The "decision" circuit is the actual regenerator determining whether the incoming pulse (or pulse position) is a binary 1 or binary 0. This is followed by the framer, which establishes the frame boundaries; overhead descrambling can be performed, and bytes B1, E1, F1, D1, D2, and D3 can be read. The section B1 bytes are read and compared with the value calculated from the previous frame. This supports the performance monitoring. Next, the section overhead bytes B1, E1, F1, D1, D2, and D3 can be written and rescrambled. *Note:* A new BIP-8 calculation is performed for the frame and is written in the overhead B1 byte of the next frame in the SONET STE RGTR. Finally, an electrical-to-optical conversion (E/O) is performed, generating the outgoing optical bit stream.

For completeness, Figure 7.3 shows the line AIS (alarm indication signal) function. The AIS frame generator generates framing bytes (A1 and A2), an STS-1 identification byte (C1), and a section BIP-8 (BI byte). The AIS frame generator generates valid E1, F1, and D1 through D3 bytes, if appropriate. The AIS frame generator also generates a scrambled all-1s pattern for the remainder of the STS-*N* signal. (For definition of STS-*N*, see Chapter 9, Section 9.2.1.2.)

7.3 OPTICAL FIBER AMPLIFIERS

There are three generic types of optical fiber amplifiers (OFAs) that have been developed for use in fiber-optic systems. These are the *laser-diode amplifiers*, *doped-fiber amplifiers*, and *Raman amplifiers*. At present the doped amplifiers dominate the market. The element, erbium, is used for doping, and the amplifier is called the *erbium-doped fiber amplifier* (EDFA). The competing laser-diode amplifiers have until recently been eclipsed by the EDFA because of their expense of manufacture and the coupling of light into the fiber pigtail, their polarization sensitivity, and their high level of crosstalk.

Figure 7.4 is a simplified functional block diagram of an EDFA. It contains just one active component, the pump. The pump is usually a laser-based light source much like we would find in a light transmitter. For production types of EDFAs, the pump provides an output at either 980 nm or 1480 nm.

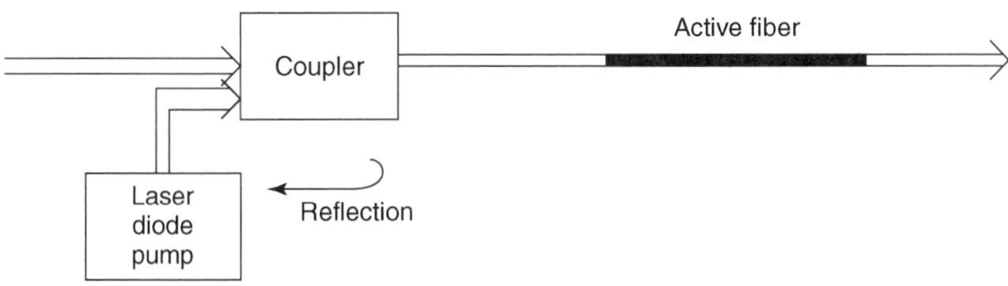

Figure 7.4. A simplified schematic drawing of an erbium-doped fiber amplifier.

The configuration we have in Figure 7.4 is an elementary wave-division multiplexer, where the light coupler does the multiplexing—that is, simply combining the pump light signal with the operational light signal. The two signals pass through the active area where the actual amplification takes. The active area consists of specially prepared optical fiber that has been properly doped with erbium, a rare-earth element. In the most elementary EDFA designs the necessary amplification occurs over a relatively narrow wavelength band, from about 1525 to 1565 nm. However, what appears to be a narrow range of wavelength provides sufficient spectral space for many DWDM channels.

One big advantage an OFA has over the regenerator is that in a multichannel WDM system a regenerator is required for each channel. On the other hand, only one OFA is required to amplify the entire WDM configuration. For example, say that a certain system has a 16-channel WDM configuration. Then 16 regenerators would be required if this option were selected. In the case of an OFA, only one OFA would be required. Still further, an OFA is transparent to bit rate; a regenerator is designed for a single, particular bit rate. On very long systems (e.g., $>$ 700 km) at least one regenerator is required to mitigate dispersion and to square-up the signal.

7.3.1 Types of Optical Fiber Amplifiers

There are two types of fiber-optic amplifiers: the laser-diode amplifier and the rare-earth-doped amplifier. There are also Raman and Brillouin amplifiers. Our discussion will concentrate on the first two types, especially on the rare-earth-doped amplifier. The practical realization of this amplifier group is the EDFA. However, other types of these amplifiers are in development, some of which are described below.

7.3.1.1 Laser-Diode Amplifiers. There are three types of laser-diode amplifiers: the injection-lock, Fabry–Perot, and traveling wave (TW). The first two are distinguished by lasing threshold. We have an *injection-locked* type when a conventional laser diode is biased above the lasing threshold value and used as an amplifier; it is a *Fabry–Perot* type when used as an amplifier that is biased below the lasing threshold. It is called a TW (traveling wave) amplifier when both facets of the semiconductor laser are given an antireflection coating. Recently, the TW amplifier has become predominant in this group principally because of its superior performance as an amplifier and the progress made in development of antireflective coating techniques. Unlike the rare-earth-doped amplifier, laser-diode amplifier can be made to work well at any wavelength band where lasers (i.e., transmitters) can operate.

The maximum signal gain of a laser-diode amplifier at an injection current of 80 mA is 19 dB; its 3-dB bandwidth is approximately 50 nm. This broad-gain bandwidth is one of the major advantages of semiconductor laser amplifiers. Even a broader-gain spectrum can be expected if a multiple quantum well (MQW) laser is used due to their peculiar band structure. The noise values of these amplifiers range from 5 dB to 7 dB.

Source: Ref. 1.

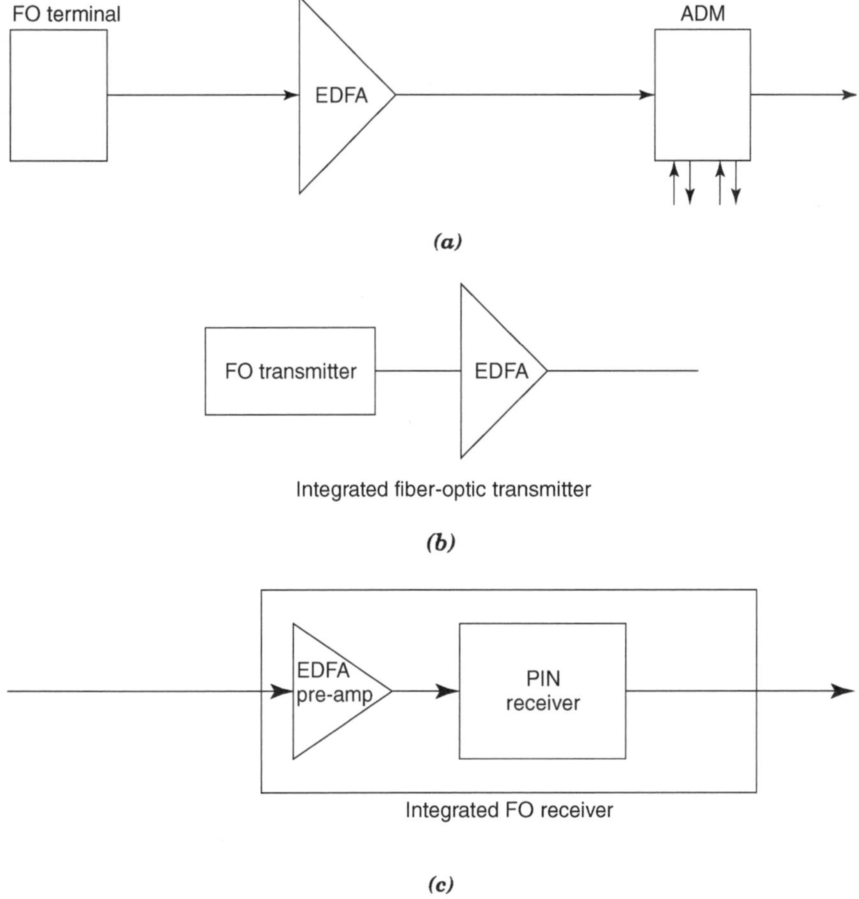

Figure 7.5. Three different application configurations of fiber-optic amplifiers. (a) In-line amplifier. (b) Power amplifier. (c) Preamplifier.

There are three different ways that fiber amplifiers can be deployed, as illustrated in Figure 7.5. Figure 7.5a shows these amplifiers used as *in-line amplifiers*. In this case the amplifiers are installed at strategic points along a long fiber span to increase signal level sufficiently so it will fall into the desired range of sensitivity of the far-end ADM or terminal receiver.

Figure 7.5b shows a power amplifier. In this configuration, the fiber amplifier is placed just after the light transmitter to boost signal level to +15 to +20 dBm. Such signal levels are necessary either for long fiber spans or when a number of lossy passive components are employed—for example, as an input to a WDM configuration that is lossy.

Figure 7.5c shows a fiber amplifier used as a *preamplifier*. In this case a fiber amplifier is placed so its output directly feeds the far-end receiver. In most cases, in this type of configuration, a fiber amplifier is integrated with the fiber receiver. Here the fiber amplifier lowers the sensitivity level of a receiver, in a manner of speaking. It takes low-level signals from a long fiber run or where lossy passive components are employed, and it brings the signal level up to the useful range level of the receiver.

Laser-diode amplifiers can fulfill all three amplifier applications illustrated in Figure 7.5. They can operate in the 1310-nm wavelength band where EDFAs have less than satisfactory performance. The EDFA is the fiber amplifier of choice for the 1550-nm wavelength band.

When a diode laser amplifier is used as a preamplifier (Figure 7.5c), the resulting signal level applied to the optical receiver is so high that the receiver performance is shot noise-limited rather than thermal noise-limited. These preamplifiers also degrade the signal-to-noise ratio through spontaneous-emission noise. A relatively high noise figure (i.e., 5–7 dB) of the typical laser diode amplifier makes them less than ideal as a preamplifier. Even so, they can improve the receiver sensitivity considerably.

When a laser diode amplifier is used as a power amplifier (Figure 7.5b), its power output is limited, usually < 10 mW. This is because of its relatively small value (approximately 5 mW) of output saturation power.

Semiconductor fiber amplifiers have several drawbacks that make their use as in-line amplifiers impractical. Among these drawbacks are polarization sensitivity, interchannel crosstalk (for WDM systems), and large coupling losses. EDFAs do not suffer from these problems, but may only be employed in a portion of the 1550-nm band as we have said.

7.3.1.2 Erbium-Doped Fiber Amplifier (EDFA). The EDFA is by far the most practical fiber-optic amplifier. As we have said above, its application is limited to the 1550-nm band. Its use makes wavelength division multiplexing a reality.

Amplification takes place in a length of low-loss fibers doped with a rare-earth metal. Many different rare-earth ions, such as erbium, holmium, neodymium, samarium, thallium, and ytterbium, can be used to produce fiber amplifiers operating in different wavelength bands from 0.5 μm to 3.5 μm. Figure 7.6 is a detailed block diagram of an EDFA.

The EDFA shown in Figure 7.6 consists only of two active components: the active fiber doped with Er^{3+} and a suitable pump. A pump is no more than a semiconductor-laser transmitter. At least one coupler is required to couple the pumped signal into the fiber. The typical EDFA operational spectrum is shown in Figure 7.7.

Pumping wavelengths may be either 0.98 μm or 1.48 μm. EDFAs can also be pumped using the wavelength range of 0.6–0.7 μm. GaAs laser diodes are preferred where pumping efficiencies as high as 11 dB/mW have been achieved.

Source: Ref. 1.

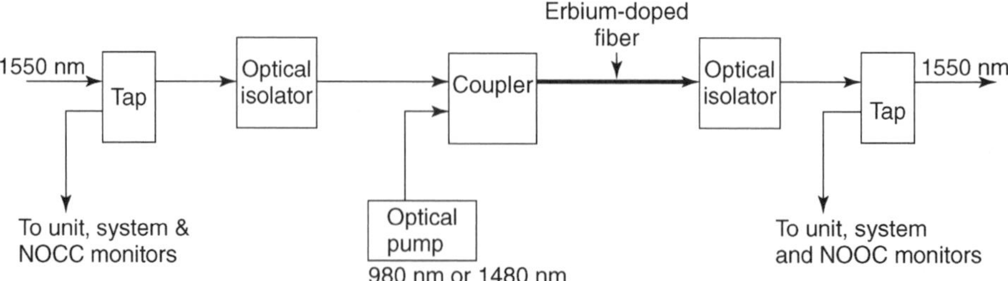

Figure 7.6. Detailed block diagram of an erbium-doped fiber amplifier.

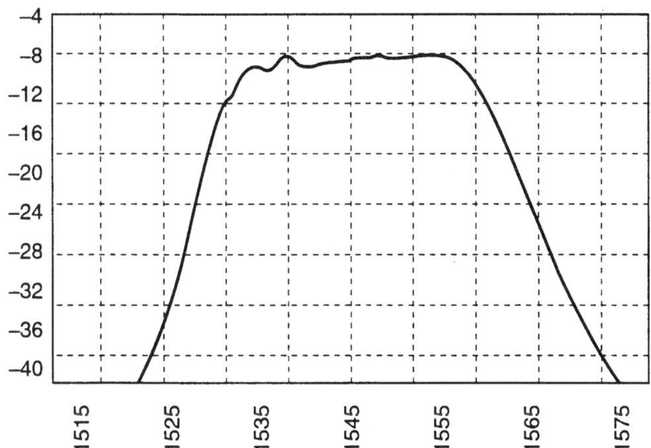

Figure 7.7. Typical EDFA optical spectrum. About 40 nm of bandwidth are available, of which only 30 nm are useful. (Courtesy of EXFO, from Figure 2.21, page 26, Ref. 2.)

There are several different configurations of EDFAs. One configuration is illustrated in Figure 7.4, where the pump and signal beams propagate in the same direction. Figure 7.8 illustrates four pumping configurations. Figure 7.8a (similar to Figure 7.4) shows a single pump source pumping in the forward direction. Figure 7.8b shows backward pumping with a single pump source. The performance is nearly the same in the two single-laser pumping configurations when the signal power is small enough for the amplifier to remain unsaturated. In the saturation regime, the power conversion efficiency is generally better in the backward pumping configuration, mainly because of the important role played by the amplified spontaneous emission (ASE). If noise is a major concern, forward pumping is recommended.

There is also the bidirectional, two-pump configuration shown in Figure 7.8c, where the amplifier is pumped in both directions simultaneously. Of course, one pump operates at 1.480 μm, usually in the backward direction, and the other pump operates at 0.980 μm and pumps in the forward direction. This makes the best of use of each of their strengths. The 1.480-μm pump has a higher quantum efficiency but also a somewhat higher noise figure, whereas the 0.980-μm pump can provide a near-quantum-limited noise figure.

Typically, a single-stage pumped EDFA provides about +16-dBm output power in the saturation region and provides a noise figure of 5–6 dB in the small-signal region. When both pumps are used simultaneously, one can expect higher output power, up to +26 dBm. Lower, near-quantum-limited noise figures needed for many preamplification applications can be attained by a multistage design. With such a design, an isolator is placed immediately after the first amplifying stage (which basically determines the noise figure) to prevent degradation of the first stage performance due to ASE that may propagate backwards from the second stage.

Source: Ref. 2.

Figure 7.8d illustrates reflection pumping where an optical circulator couples the light to the EDFA amplifier.

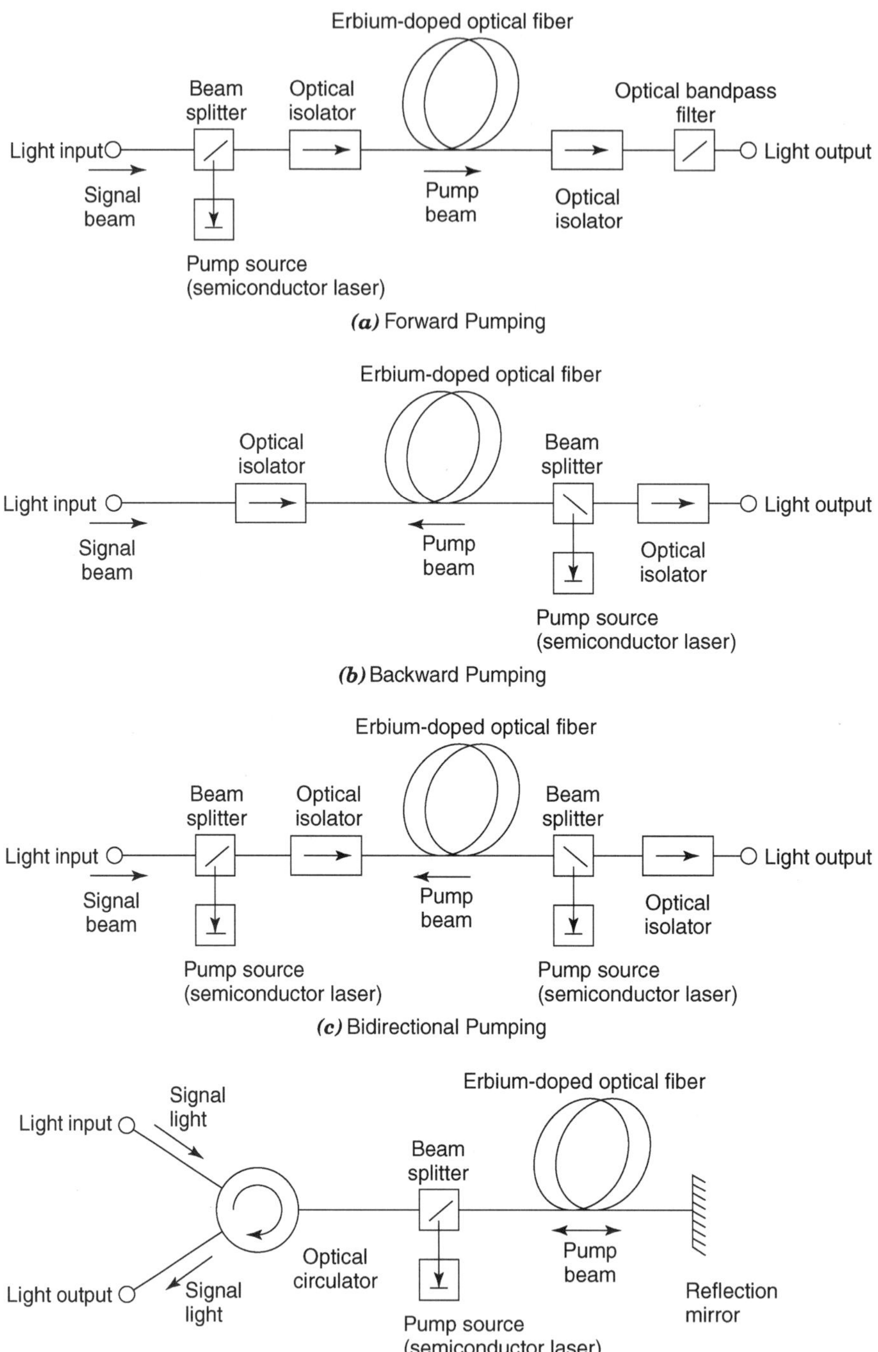

Figure 7.8. Construction of erbium-doped fiber amplifiers. (Reprinted with permission from Ref. 3, Figure 5.7, page 97.)

TABLE 7.1 Comparison of Pump Wavelengths for EDFAs

Wavelength	1.48 μm	0.98 μm
Light source	InGaAsP/InP FP-LD	InGaAs strained-layer superlattice LD
Gain efficiency	5 dB/mW	10 dB/mW
Noise index	~ 5.5 dB	3–4.5 dB
Saturation output[a]	+20 dBm	+5 dBm
Pump wavelength range	Broad (1.47–1.49 μm) 20 nm	Narrow (0.979–0.981 μm) (2.5 nm)
Beam splitter	Difficult	Simple
Pump output	50–200 mW	10–20 mW
Reliability	○	Δ

[a]Depends on pump output. Currently, higher output is easier to achieve with a 1.48-μm LD.

Source: Reprinted with permission from Ref. 3, Table 5.5, page 106.

A major consideration in the design of EDFAs is the selection of the pump wavelength to be employed, either 0.980 μm or 1.480 μm. A comparison of these two pump wavelengths is provided in Table 7.1.

7.3.1.3 Derivative EDFA-Type Amplifiers. There are two EDFA-type amplifiers that are commercially available to the fiber-optic system designer: the silica-based amplifier described above and the fluoride-based amplifier. They are very similar and differ only in the type of host fiber employed. They cover the same wavelength bandwidth from about 1525 nm to 1560 nm, with essentially the same gain characteristics. They differ only in their output signal response curve, with the fluoride EDFA having a somewhat flatter response as shown in Figure 7.9.

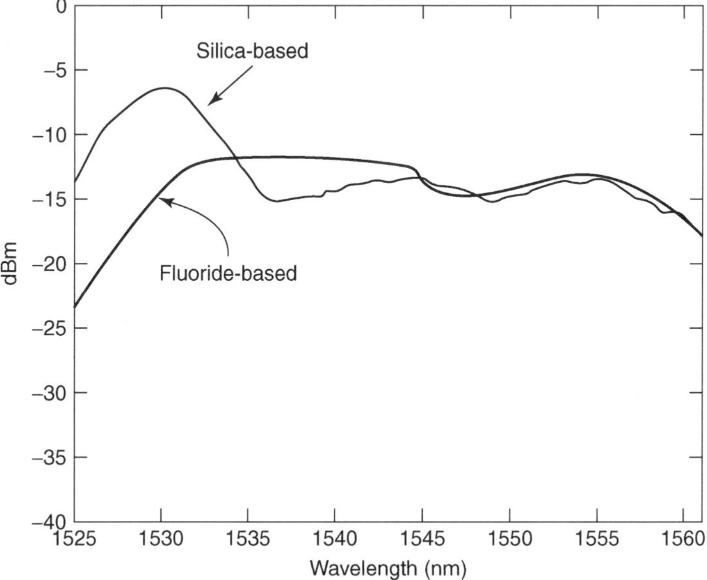

Figure 7.9. Output response curves for the silica-based and fluoride-based EDFAs. (Based on Refs. 4 and 14.)

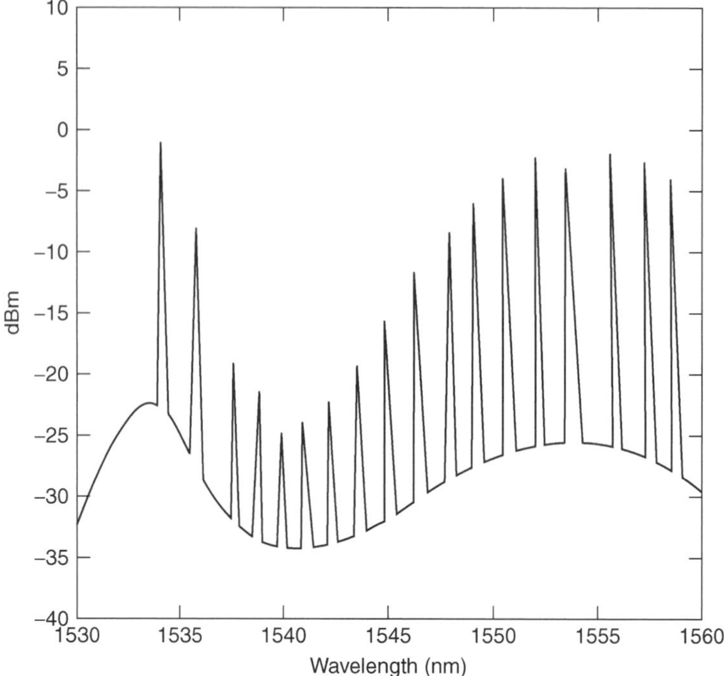

Figure 7.10. An unbalanced output response curve of a silica-based EDFA carrying a DWDM channel grouping. Note the dip at about 1540 nm. (Based on private communication with Corning, Inc., Ref. 12.)

The output response is very important when multichannel WDM is deployed. Conventional silica-based EDFAs were force-fitted in one way or another such that each WDM carrier had more or less the same output amplitude. One way of doing this was to reduce the bandwidth of the amplifier, using only the longer wavelength portion of the band (see Figure 7.9). This was done by filtering out the 1530- to 1542-nm region of the band. One result of this was tighter WDM channel spacing. This caused greater vulnerability to several nonlinear effects such as four-wave mixing, discussed in Chapter 6. Another flattening technique is to selectively attenuate each optical channel at the amplifier input to produce a flatter response curve. This is a labor-intensive operation.

To reduce or eliminate this complex operation of adjusting channel level, vendors are integrating self-optimization algorithms into the network element. An unbalanced response curve of a silica-based EDFA carrying a DWDM signal is illustrated in Figure 7.10.

The major advantage of the fluoride-based EDFA is that its response curve around the 1540-nm region is considerably flatter than its silica-based counterpart. There is one performance drawback using the fluoride-based amplifier. Its noise figure is higher as a result of being pumped at 1480 nm, whereas silica-based EDFAs use 980 nm for pumping. The 980-nm wavelength is inefficient for fluoride-based amplifiers due to excited-state absorption. But this is the price that must be paid to derive a flatter response curve so that the entire 1550-nm passband of the amplifier may be used.

Source: Based on Ref. 12.

7.3.1.4 Other Rare-Earth-Doped Fiber Amplifiers. An erbium-doped silica-based amplifier provides about 35 nm of bandwidth in the 1550-nm band. There is about 200 nm of bandwidth available with less than 0.25 dB per kilometer loss. If we were to allow an attenuation of up to 0.35-dB/km, some 400 nm would be available for DWDM operation. Other types of rare-earth-doped amplifiers are required to exploit more of this wide bandwidth. One possibility is the erbium-doped tellurite-based fiber amplifier (EDTFA) that can increase the bandwidth to 90 nm.

The 1550-nm usable passband for amplifiers can be extended still further by employing thulium- or praseodymium-doped floride-based fiber amplifiers (TDFFAs and PDFFAs). Using the thulium-doped fluoride-based amplifier, the 1550-nm amplifier passband can be extended from 1470 to 1650 nm. The praseodymium-doped amplifier operates in the 1310-nm band.

Japan's NTT Laboratories have experimented with an EDTFA, using an equalizer. The amplifier had a comparatively flat response from 1561 to 1611 nm realizing a usable bandwidth of 50 nm. The amplifier had a gain of $\sim$ 25-dB and a noise value of < 6 dB. Since this experiment, NTT has developed a gain-flattened EDTFA that is operational from 1570 to 1617 nm.

Source: Lightwave, November 1999, pages 56–64, Ref. 5.

7.4 CRITICAL PERFORMANCE PARAMETERS OF EDFAs

7.4.1 Gain and Output Response

As we have seen in Section 7.3.1, the output response of a silica-based EDFA varies with wavelength. Intuitively, we then can say that EDFA gain varies with wavelength. Gain also varies with input polarization state and power. The gain will change, depending on the relative input power of each WDM/DWDM channel. Thus, the effect of a temporary redistribution of the input power, typically when a channel is being added or dropped, must be characterized and controlled in multi-signal applications. Gain of an EDFA can be calculated given the average input power (P_{in}) and average output power (P_{out}). Note that in the equation the power measurement is related to the wavelength of the signal in question:

$$G_{dB} = 10\log[\{P_{out}(\lambda_c) - P_{ase}(\lambda_c)\}/P_{in}(\lambda_c)] \qquad (7.1)$$

where P_{ase} is the power level of the amplified spontaneous emission.

Note that the ASE power component has been removed from the gain (G_{dB}) calculation in equation (7.1).

The gain of an optical amplifier is very dependent on the input signal level. Curiously, the amplifier exhibits a large signal gain for weak input signals. For

TABLE 7.2 EDFA Amplification Characteristics Comparison at 1536 nm and at 1550 nm

Item-wavelength	At 1535–1536 nm	At 1550–1554 nm
Gain	38–43 dB	38–41 dB
Bandwidth of operation	2 nm	4 nm
Noise figure	5.0 dB	4.25 dB
Saturation output	+15 dBm	+15.8 dBm

This table reflects operating characteristics of typical INO EDFAs
Source: From INO, Quebec, Canada, (Ref 13).

example, a gain greater than 30 dB can be expected for an input signal less than −20 dBm (Ref. 2). Thus gain compression is an important parameter for large signal input levels.

To characterize EDFA gain, testing should include small-signal gain, the 3-dB compression, and saturated output power. These three parameters will vary with the wavelength of the input signal. The following critical EDFA parameters are defined.

Profile is a term used to describe wavelength dependence of a particular characteristic. Noise gain is expressed in decibels for one wavelength, where the noise gain profile characterizes how gain, for a particular amplifier, varies with wavelength.

Gain flatness is the maximum difference among individual channel gains at the output of a fiber amplifier when their input powers are equal.

Signal gain is the principal factor that determines the operating point of an amplifier. Noise gain, on the other hand, is the gain that applies to a small signal with a gain that has no impact on the amplifier's operating point, while another large signal is driving the amplifier into saturation.

Gain cross-saturation is the change in gain of a specific channel when the input level of another channel (or several channels) is changed by a specific amount.

Source: Ref. 2.

Table 7.2 shows a comparison of operational characteristics at two wavelengths: 1536 nm and 1550 nm. This latter wavelength corresponds to the zero-dispersion wavelength (shifted).

7.4.2 Amplified Spontaneous Emission

The principal source of noise in an optical amplifier is *amplified spontaneous emission* (ASE). Its spectral density is nearly constant similar to white noise or thermal noise. The effect of spontaneous emission is to add fluctuation to the amplified power which are converted to current fluctuations during the photodetection process. It turns out that the dominant contribution to the receiver noise derives from beating of spontaneous emission with the signal. This beating phenomenon is similar to heterodyne detection in that the spontaneously emitted radiation mixes with the amplified signal at the photodetector and produces a heterodyne component of the photocurrent. The beating of the spontaneous emission with the signal produces a noise current.

The amplifier noise figure (F_n) relates the amplifier gain (G) and *spontaneous emission factor* (n_{sp}) by the following:

$$F_n = 2n_{sp}(G - 1)/G \approx 2n_{sp} \tag{7.2}$$

where

$$n_{sp} = N_2/(N_2 - N_1) \tag{7.3}$$

N_1 and N_2 are the atomic populations for the ground and excited states.

Consider equation (7.2) for a moment. The equation tells us that the SNR (signal-to-noise ratio) of an amplified signal is degraded by 3 dB, even for a "perfect" amplifier for which $n_{sp} = 1$. For most practical amplifiers, the noise figure F_n must exceed 3 dB and can be as high as 6–8 dB.

Source: Ref. 2.

ASE power in a signal passing through the amplifier can be calculated using the following relationship:

$$N_{out}(\lambda) = (N_{in}(\lambda) \times G) + \text{ASE} \tag{7.4}$$

Two measurements are required: (1) the level of the input signal ($N_{in}(\lambda)$) and (2) the total noise level of the output signal ($N_{out}(\lambda)$).

In saturation or gain compression, the ASE contribution is small. Thus, we can say that the gain (G) is nothing more than the ratio of the output power to the input power where ASE is not taken into account.

Source: Ref. 2.

There are five contributors to the noise portion of the SNR relation in a light transmission system employing EDFAs. These are:

- Shot noise
- Spontaneous–spontaneous beat noise
- Signal-spontaneous beat noise
- Interference noise
- Excess noise

Most of us are familiar with shot noise, which is peculiar to light emission. Shot noise comes about from the random fluctuations in arrival times of the photons which make up a light signal. The IEEE (Ref. 6) defines *shot noise* as follows: Noise caused by current fluctuations due to the discrete nature of charge carriers and random or unpredictable (or both) behavior of charged particles from an emitter.

Spontaneous–spontaneous beat noise, is also termed ASE–ASE beat noise. It is noise developed between ASE signals within the bandwidth of the amplitude-modulated signal. As gain is increased in a fiber amplifier, ASE noise decreases with output power, or when saturation conditions are being approached. Thus, for fiber amplifiers being employed as power amplifiers, this type of noise is ignored. It can be very important where low signal levels are involved such as in preamplifiers unless narrow filters are used.

Signal-spontaneous beat noise is generated when the desired signal mixes (heterodynes) with white ASE noise. This noise cannot be removed either optically or electrically by filtering because the noise resides in the bandwidth of the modulated information carrier. However, it should be measured. The noise figure of an EDFA is usually defined in terms of signal-spontaneous beat noise effect.

7.5 RAMAN AMPLIFIERS

From Ref. 10, "Raman amplification occurs when higher-energy (shorter-wavelength) pump photons scatter off the vibrational modes of a material's lattice matrix (optical photons) and coherently add to lower-energy (longer wavelength) signal photons." In its practical implementation, it is called *Raman-assisted transmission* (D-RAT); the pump light is launched into the fiber in-line amplifier sites opposite to the signal direction. In this configuration a Raman amplifier acts as a low noise premaplifier. A major advantage of this low-level Raman amplification is that no additional fiber nonlinearities are incurred.

Figure 7.11 is a schematic of a distributed Raman amplifier. As Ashiqur Rahman of Corning states in Ref. 11: "Two orthogonally polarized pump diode lasers are polarization multiplexed and combined with a WDM to provide backward-pumping pump power in the transmission fiber. As a result, the forward-propagating signals achieve Raman gain in the fiber. The use of backward pumping reduces the impact of the pump noise in the signals."

The gain performance of a distributed Raman amplifier depends on the transmission fiber properties such as pump absorption, effective area, and the Raman gain coefficient. In one practical example given in Ref. 11, the gain was only 3.75 dB.

Gain flatness is an important parameter for OFAs, particularly when WDM/DWDM is involved. In the case of a Raman amplifier, gain for a particular

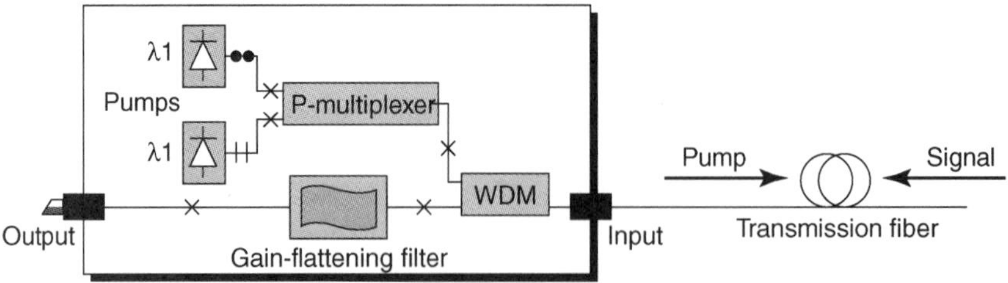

Figure 7.11. Schematic diagram of a distributed Raman amplifier. (Courtesy of Corning, Inc., Ref. 15.)

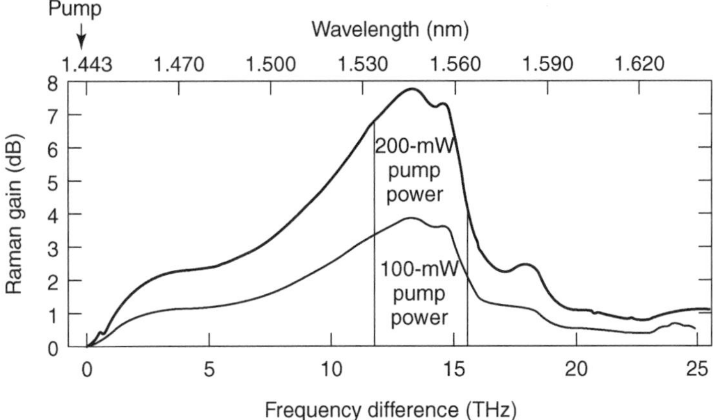

Figure 7.12. Raman gain spectra in a long fiber pumped by 1443-nm pump at 100 mW and with 200-mW pump power. For a 30-nm (~ 4-THz) C-band-signal window, 1530–1560 nm is also indicated in the figure. (Based on Refs. 10, 11, and 15.)

signal depends on the frequency difference of the pump and signal. Figure 7.12 shows the small-signal Raman gain in a long fiber. The gain/bandwidth is more than 20 THz, with its peak around 13.2 THz. Different signals achieve different amounts of gain, depending on their frequency difference with the pump. Thus, any signal-band window will have some gain ripple. For the 200-mW pumped case in Figure 7.12, the maximum gain is 7.78 dB and the gain ripple (meaning maximum gain–minimum gain) in 3.5 dB. The actual ripple, defined as (gain ripple in dB)/(maximum gain in dB) = 3.5 dB/7.78 dB = 0.45 in the C-band window, is shown in Figure 7.12.

In system engineering a fiber-optic span carries WDM formations; a combination of a distributed Raman amplifier and an EDFA in tandem makes an excellent combination to reduce ASE buildup.

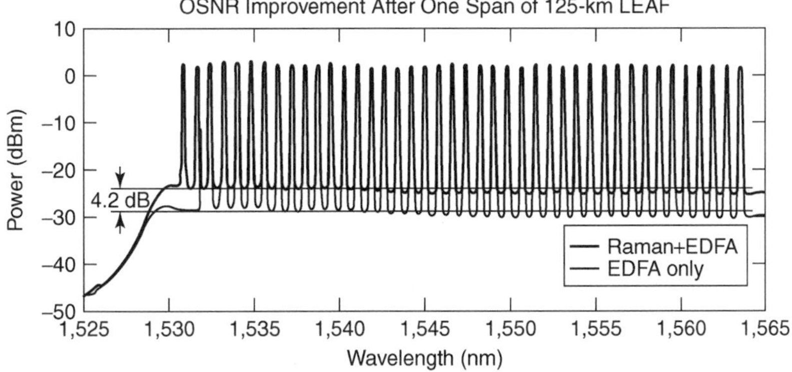

Figure 7.13. Optical spectrum of 32 wavelengths through a span of fiber amplified with a conventional EDFA and then with a Raman–EDFA hybrid. The peak power is the same, while the noise floor is 4.2 dB lower in the hybrid case. (Courtesy of Corning, Inc., Refs. 10 and 11.)

Figure 7.13 illustrates the optical spectrum for 32 DWDM wavelengths showing OSNR (optical signal-to-noise ratio) for a fiber length of 125 km with the hybrid Raman–EDFA preamplifier and with EDFA only. There is a 4.2-dB improvement with the hybrid scheme over an EDFA alone.

REFERENCES

1. Govind P. Agraval, *Fiber-Optic Communication Systems*, 2nd ed., John Wiley & Sons, New York, 1997.
2. *Guide to WDM Technology Testing*, EXPO Electro-Optical Engineering Inc., Quebec City, Canada, 2000.
3. S. Shimada and H. Ishio, eds., *Optical Amplifiers and Their Applications*, John Wiley & Sons, Chichester, England, 1992.
4. Thomas Fuerst, Today's Optical Amplifiers Enable Tomorrow's Optical Layer, *Lightwave*, July 1997.
5. Makoto Shimizu, Non-Silica-Based Fiber Amplifiers Open New Wavelength Regions for WDM, *Lightwave*, November 1999.
6. *The IEEE Standard Dictionary of Electrical and Electronic Terms*, 6th ed., IEEE Std-100-96, IEEE Press, New York, 1996.
7. *Application-Related Aspects of Optical Fiber Amplifier Devices and Subsystems*, ITU-T Rec. G.663, ITU Geneva, October 1996.
8. *Telcordia Notes on Dense Wavelength Division Multiplexing (DWDM) and Optical Networks*, Special Report, SR Notes, Series 02, Issue 1, Telecordia, Piscataway, NJ, May 2000.
9. *SONET Regenerator (SONET RGTR) Equipment Generic Criteria*, Technical Reference TR-NWT-000918, Issue 1, Piscataway, NJ, December 1990.
10. Alan Evans, Raman Amplification Key to Solving Capacity, System-Reach Demands, Corning, Inc., from *Lightwave*, August 2000, page 69.
11. Ashiqur Rahman, Design Issues of Distributed Raman Amplifiers for Reduced Noise Accumulation in Long-Haul, Repeatered Transmission, *Lightwave*, August 2000, page 70.
12. *Introduction to SONET*. Hewlett-Packard seminar, Burlington, MA 1993.
13. *Synchronous Optical Network (SONET)—Basic Description including Multiplex Structure, Rates and Formats.* ANSI T1.105-1995, ANSI, New York, 1995.
14. Private communication, Pierre Talbot, Inc., Quebec City, Canada, April 4, 2002.
15. Private communication, Dr. Alan Evans, Corning, Inc., Corning, NY, April 4, 2002.

8

WAVE-DIVISION MULTIPLEXING

8.1 THE DEMAND FOR BIT-RATE CAPACITY

The demand for bit-rate transmission capacity is estimated to quadruple every 18 months (Ref. 1). The fiber-optic medium is the only one in our toolbox that can fulfill this need. There are two ways to meet the requirement:

1. Add new fiber cable or install fiber cable with more strands of fiber; use available dark fiber.
2. Employ wave-division multiplexing (WDM) on existing fiber or in new fiber installations.

In nearly every instance, WDM will be the technology of choice due to cost considerations of adding fiber, particularly on longer systems. The objective of this chapter is familiarize the reader with WDM and its many ramifications.

In the early days of fiber-optic transmission, each operational fiber strand[1] carried a single bit stream. The technology was such that the 1310-nm band matured before the 1550-nm band. Thus, the bit stream operated at some wavelength in the 1310-nm band. Demand began to grow for bit-rate capacity, and the technology developed as well. When the 1550-nm band opened up, a second bit stream was placed on that band. Thus we had a crude but effective form of early WDM, a two-channel system. This system is shown conceptually in Figure 8.1.

In Figure 8.1 the coupler is an optical combiner. For simplicity's sake, the coupler could be a passive splitter in reverse using the biconic fused tapered coupler technology discussed in Section 3.5. The splitter on the right-hand side of Figure 8.1 could be the same device used in reverse. Both ports of the splitter contain signal energy of λ_1 and λ_2.

The filters, based on thin-film technology, block out the unwanted signal energy and allow the desired signal energy to pass through. Thin film technology is discussed in Section 8.6. The importance of filters in DWDM operation cannot be overemphasized. It is for this reason we have devoted over three-quarters of the chapter to filter technology.

[1]We hasten to add that in nearly all fiber systems, spare fibers (called *dark fibers*) are installed. Sometimes an astute system owner would install many dark fibers in view of leasing them out to other users on some future date or dates.

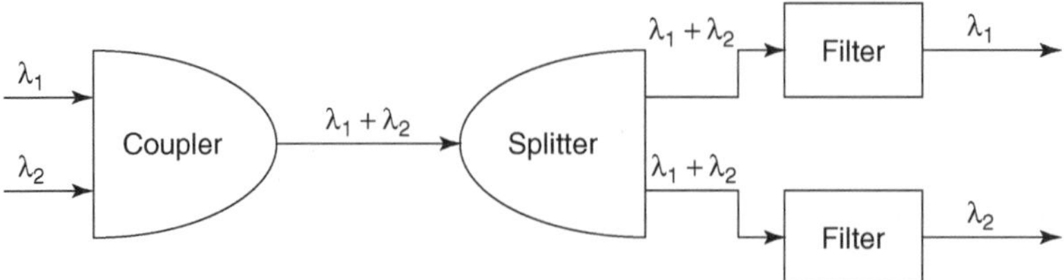

Figure 8.1. Conceptual block diagram of a two-wavelength WDM system. In an early layout, λ_1 was 1310 nm and λ_2 was 1550 nm.

8.2 BASES OF WDM SYSTEMS

The multiplexer–demultiplexer pair form the basis of a WDM system. As illustrated in Figure 8.1, a multiplexer can be simply a combining device of multiple wavelengths. A demultiplexer is a different matter. It must separate individual wavelengths from an aggregate of wavelengths. Thus demultiplexers require a wavelength selective mechanism. These mechanisms can be classified into two broad categories: *diffraction-based demultiplexers* and *interference-based demultiplexers*.

A diffraction-based demultiplexer uses an angularly dispersive element, such as a diffraction grating, which disperses incident light spatially into various wavelength components. The grating concept is illustrated in Figure 8.2. An interference-based demultiplexer makes use of devices such as directional couplers and optical filters. Due to the inherent reciprocity of optical waves in dielectric media, the same device can be used as a multiplexer or demultiplexer. This will depend on the direction of propagation.

The EDFA (erbium-doped fiber amplifier) is a necessary element in a WDM system. The useful gain of an EDFA (i.e., 20–25 dB) compensates for the losses incurred by the passive elements of the multiplexer and demultiplexer, including many of the devices described below. If we allow a $\geq$ 6-dB loss for a multiplexer and a 6-dB loss for the demultiplexer, we have a 12-dB loss. As the number of WDM channels increases, losses start to increase radically. The loss of a splitter to derive just two channels is 3 dB + ; four channels, 6 dB + ; and so on. The secondary effect is that we are stuck with the band 1530–1565 nm, the band of operation for the EDFA. The ITU has developed for this band a standardized

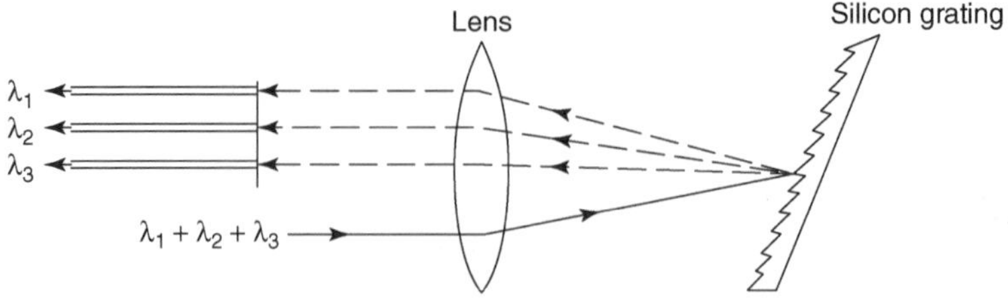

Figure 8.2. Concept of an angled grating and focusing lens used as a basis for demultiplexing.

wavelength grid for 200/100/50 and extended to 25-GHz channel spacing. The ITU grid is based on the relationship

$$F = 193.1 \pm m \times 0.1 \text{ THz} \tag{8.1}$$

where 193.1 THz is the reference frequency and m is an integer (see Section 8.9).

Even being limited to 1530–1565 nm, we expect to see 160 DWDM channels on one fiber, each transmitting 40 Gbps in 2004.

8.3 THE FABRY–PEROT INTERFEROMETER

The Fabry–Perot (FP) interferometer is an interference-type device and is based on the interference of multiple reflections of a light beam by two surfaces of a thin plate. This concept is illustrated in Figure 8.3. There is an interference maximum for each wavelength which is expressed mathematically by

$$m\lambda = 2d \cos \alpha \tag{8.2}$$

where m is an integer and d is the distance between plates.

This interferometer makes use of the multiple reflections between two closely spaced partially silvered surfaces. Part of the light is transmitted each time the light reaches the second surface, resulting in multiple offset beams that can interfere with each other. The large number of interfering rays produces an interferometer with extremely high resolution. This is somewhat like the multiple slits of a diffraction grating that increase its resolution.

The *FP resonator* is a derivative device of the FP interferometer. It is an arrangement of two parallel plates that reflect light back and forth. *Finesse* is an indication of how many wavelength channels can simultaneously pass without severe interference among them. The finesse is a measure of the energy of wavelengths within the cavity relative to the energy lost per cycle. The higher the finesse, the narrower the resonant line width. Think of finesse as the equivalent of the Q of electrical filters.

The FP interferometer makes an excellent optical fiber. Filter tuning is accomplished by changing the length of the gap between the two mirrors. With a more

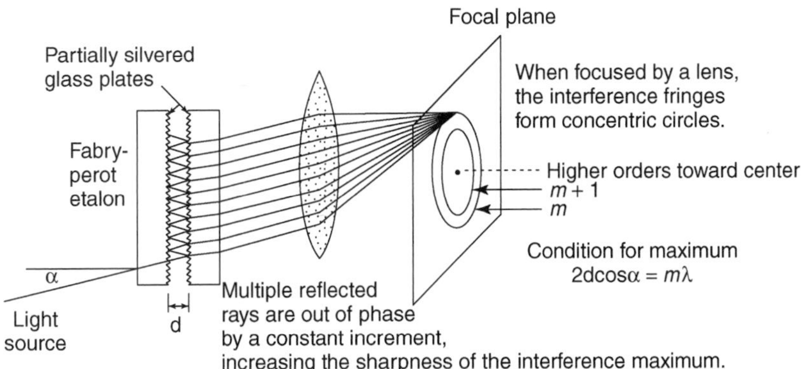

Figure 8.3. Fabry–Perot interferometer, conceptual drawing. (Sketch based on Ref. 2.)

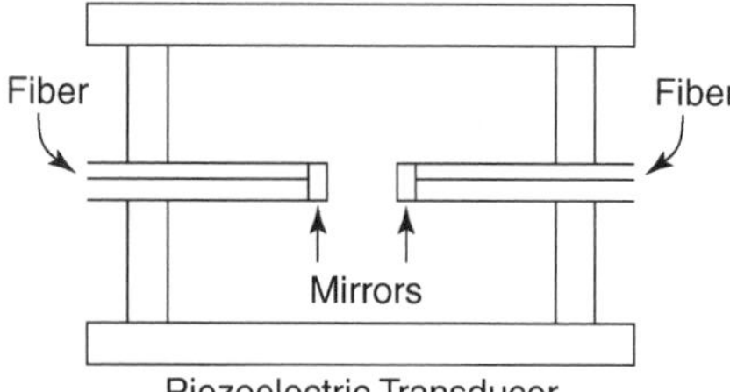

Figure 8.4. A practical implementation of an FP filter.

sophisticated FP interferometer, the entire structure is enclosed in a piezoelectric chamber so that the gap length can be changed electronically for tuning and selecting a specific channel. The advantage of fiber FP filters is that they can be integrated within the system without incurring coupling losses. The number of channels is limited in the range of 50–100 because of a limited finesse of a practical FP filter ($F = 100$ for 97% mirror in tandem, which increases the effective finesse to $F \sim 1000$. Figure 8.4 is an outline drawing of a practical FP filter.

Source: Ref. 3.

8.4 MACH–ZEHNDER FILTERS

A Mach–Zehnder (MZ) interferometer can be made by connecting the two output ports of a 3-dB coupler to the two input ports of another 3-dB coupler as shown in Figure 8.5. The first coupler splits the optical signal equally in two parts where each part acquires a different phase (when the arm lengths are different before the interference of one split signal to the other in the second coupler).

Of course the relative phase is wavelength-dependent, and transmittivity $T(\upsilon)$ also depends on wavelength. It may be calculated by the simple formula:

$$T(\upsilon) = \cos^2(\pi \upsilon \tau_m) \tag{8.3}$$

where τ is the relative delay in the two arms of the MZ interferometer and υ is the frequency. A cascaded chain of such MZ interferometers with relative delays suitably adjusted acts as an optical filter that can be tuned by adjusting the arm lengths slightly.

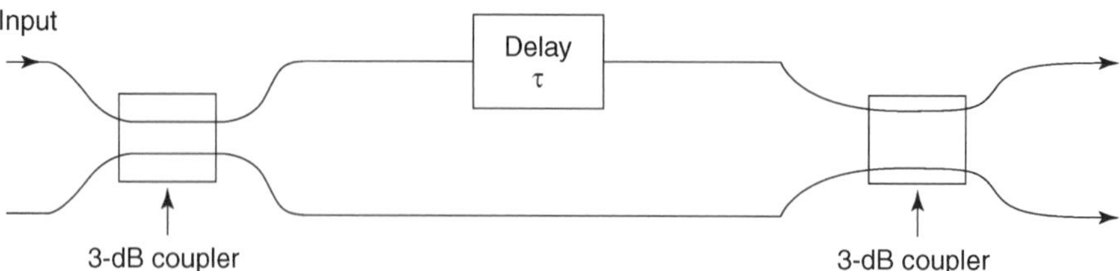

Figure 8.5. A Mach–Zehnder optical interferometer.

Reference 3 reports that a commonly used method implements the relative delays τ_m such that each MZ stage blocks the alternate channels successively. This scheme requires that $\tau_m = (2^m \, \Delta\upsilon_{ch})^{-1}$ for a channel spacing of $\Delta\upsilon_{ch}$. The resulting transmittivity of a 10-stage MZ chain has a channel selectivity as good as that offered by an FP filter having a finesse of 1600. In addition, such a chain is capable of selecting closely spaced channels. A chained group of MZ interferometers provides the WDM system designer still another filter technology to choose from.

Source: Ref. 2.

8.5 BRAGG GRATINGS AND FIBER BRAGG GRATINGS

Figure 8.6 is a model which we will use to describe the *Bragg grating concept*. A Bragg grating is an arrangement of parallel semireflecting plates. The plates, one from the next, are separated by distance *d*. Light, consisting of several or multiple wavelengths, enters from the left. Depending on distance *d*, there will be a strong reflection for one or several particular wavelengths. This reflected light will exit from the left, and the remainder of the group of wavelengths of light will exit from the right. The condition of strong reflection of *Bragg condition* is

$$d = -n\lambda_B/2 \tag{8.4}$$

where n is an arbitrary number and λ_B is the wavelength of the reflected channel; d represents the Bragg spacing or periodicity or grating period and should be an integer multiple of a half-wavelength. The negative sign denotes reflection, and n denotes the order of the Bragg grating. When $n = 1$ (first order) we have $d = \lambda/2$, and when $n = 2$ (second order) we have $d = \lambda$. The Bragg grating makes an excellent bandpass filter.

A *fiber Bragg grating* (FBG) consists of a length of fiber whose index of refraction varies periodically with its length. Variations of the refractive index constitute

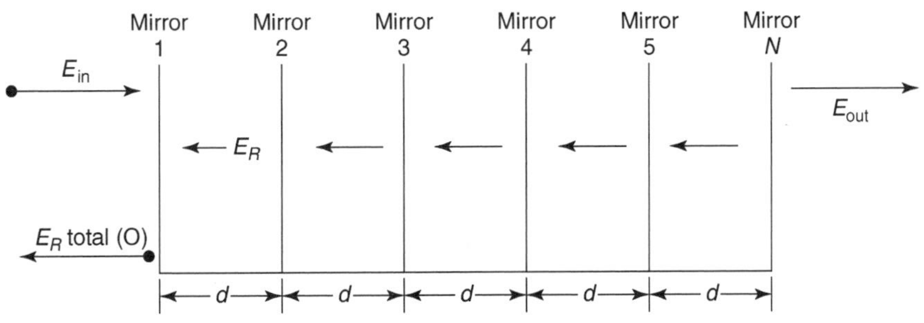

Figure 8.6. A Bragg grating model. (Based on concepts from Refs. 2 and 3.)

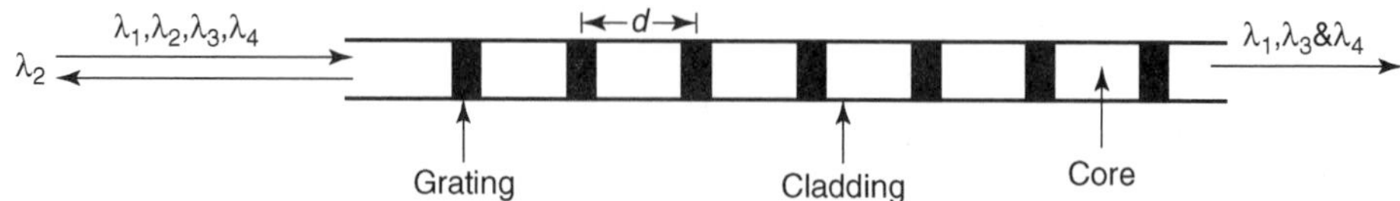

Figure 8.7. Illustrative model of a fiber Bragg grating. For the 1550-nm band, *d* will be in the range of 1–10 μm.

discontinuities that emulate a Bragg structure. A common method of fabricating an FBG is to subject the fiber to an intense ultraviolet optical interference pattern that has a periodicity equal to the periodicity of the grating to be fabricated. When the germanium-silicate core of the fiber is exposed to the intense ultraviolet pattern, structural defects are formed and thus a permanent variation of the refractive index is brought about. It will have the same periodicity as the ultraviolet pattern.

Figure 8.7 illustrates an FBG from another perspective. Several of the dimensions are exaggerated to better illustrate composition and structure.

Another method of constructing a Bragg grating reflector is based on a stacked-dielectric structure made up of $\lambda/4$-thick layers. This is known as a *photonic lattice*, each with a different index of refraction. These reflectors reflect wavelengths over all possible angles of incidence, and they do not absorb incident energy as mirror-based reflectors do.

Source: Ref. 3.

An FBG is commonly used with an optical circulator, typically in an optical add–drop multiplexer (OADM) where an FBG reflects back only the wavelength it is designed for. The remaining aggregate of wavelengths can now be passed to another circulator–FBG combination to extract another wavelength, and so on. Figure 8.8 illustrates this concept. FBGs may be used as passband filters, interference filters, and chromatic dispersion compensators, as well as to flatten EDFA output levels.

FBGs are temperature-sensitive regarding periodicity and bandwidth. They are usually mounted in a temperature-controlled enclosure.

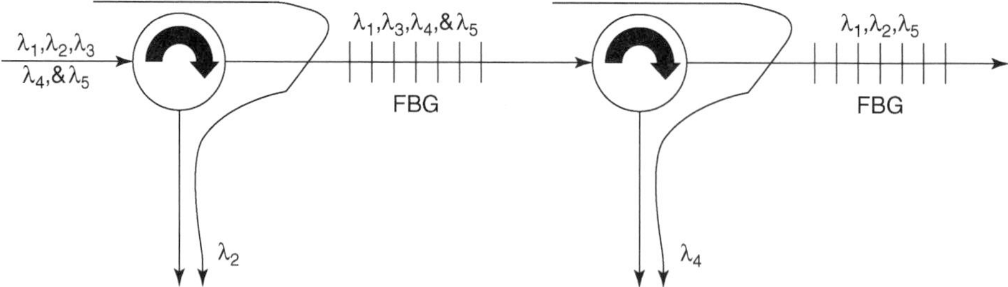

Figure 8.8. A circulator–FBG combination extracts a single wavelength channel from an aggregate of channels, followed by a second circulator FBG combination separating another wavelength. FBG = fiber Bragg grating.

TABLE 8.1 Specifications for Fiber Bragg Grating Bandpass Filters

	Type A (100 GHz)	Type A (50 GHz)	Type B	Type C
Application	WDM filter	WDM filter	External cavity	ASE rejection
Wavelength range	1530–1560 nm	1530–1560 nm	980-nm band 1310-nm band 1480-nm band 1550-nm band	1525–1545 nm
Wavelength accuracy	±0.05 nm	±0.05 nm	±0.05 nm	±0.05 nm
Reflectivity	≥ 99%	≥ 99%	≤ 1% to ≥ 99%	
Optical bandwidth	0.6 nm (FWHM)	0.3 nm (FWHM)	0.6 nm (FWHM)	≥ 10 dB
Crosstalk transmission	≥ 30 dB	≥ 30 dB ≥ 20 dB[a]	≥ 30 dB	≥ 30 dB
Crosstalk reflection	≥ 30 dB	≥ 30 dB ≥ 20 dB[a]	≥ 30 dB	≥ 30 dB
Package	(1) Coated UV acrylate, (2) glass case, (3) temperature-compensated			

[a]Adjacent channel.

Source: Courtesy of Sumitomo Electric Lightwave Corp., grating for de-mux application, Ref. 4.

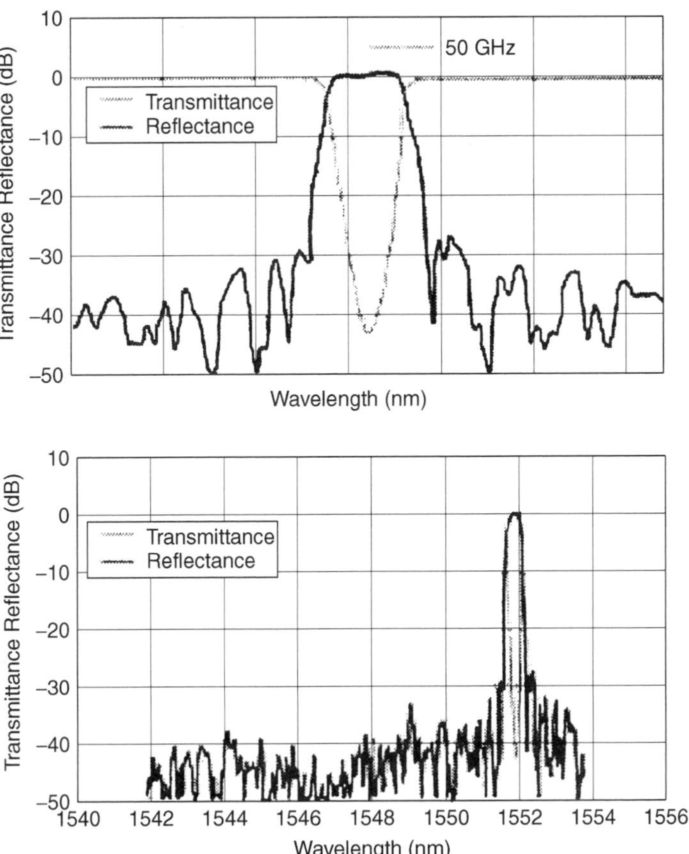

Figure 8.9. Fiber Bragg grating filter bandpass characteristics. (Courtesy of Sumitomo Electric Lightwave Corp., Ref. 4.)

Table 8.1 gives typical parameters and characteristics of FBG commercially available, off-the-shelf. In a DWDM configuration, our greatest concern would be crosstalk. Figure 8.9 shows typical bandpass characteristics in a demultiplex application. At 50-GHz spacing, adjacent channel interference is down almost 30 dB.

8.5.1 Several Specific Applications of Fiber Bragg Gratings

8.5.1.1 Unique Filtering Properties of FGBs. There are many optical filtering applications for FGBs because of their wavelength selectivity. In its simplest form, a fiber grating acts as a refraction filter whose central wavelength can be controlled by changing the grating period, and whose bandwidth can be tailored by changing the grating strength and by chirping the grating period slightly.

The stabilizer filter is another application. Erbium-doped fiber amplifier (EDFAs) make the practical implementation of WDM a reality. The various devices used for lightwave multiplexing and demultiplexing are passive and lossy, even very lossy (e.g., > 20 dB per device). EDFAs compensate for this loss. Even better, they can compensate for the aggregate of WDM channels, where a regenerator must work on a per-channel basis.

Suppose we had a formation of 32 WDM channels. If we used a regenerator, which we must do on some very long circuits, we would require a regenerator for each channel or a total of 32 regenerators. The good news is that if we use an EDFA, only one would be required. The bad news is that at the output of the EDFA we would like each of the 32 channels to have the same level (within reason).

The performance of an EDFA is affected by the characteristics of the pumped-laser diodes. Pump-laser performance is often adversely affected by unwanted external reflections back into the laser cavity and by the temperature and injection current fluctuations. The introduction of FBG-based pump stabilizer filters resolves this problem by locking the emission wavelengths of 980-nm and 1480-nm pump lasers and providing immunity to spectral-mode hopping due to changes in temperature, drive current and optical feedback. The key characteristic of an FGB-based pump stabilizer filter is its bandwidth and reflectivity with respect to the pump laser's characteristics and the particular application.

8.5.1.2 Flattening Filters. The flattening filter is another FBG application. Because EDFA gain tends to vary across the 1530- to 1560-nm spectrum of its output, it is necessary to equalize the gain across the spectrum with the use of FGB-based gain-flattening filters. In the three basic wavelength bands of interest, these filters can reduce the nonuniform gain variation across the full EDFA window, thereby improving performance and simplifying the design of WDM systems.

8.5.1.3 FBGs for Chromatic Dispersion Compensation. Chromatic dispersion is one of the two major overriding impairments to gigabit light transmission. It has been discussed at length in Section 6.3. The common approach to compensate for this distortion is achieved by passing the light signal through a device with equal distortion but with opposite sign, thus canceling out the accumulated chromatic

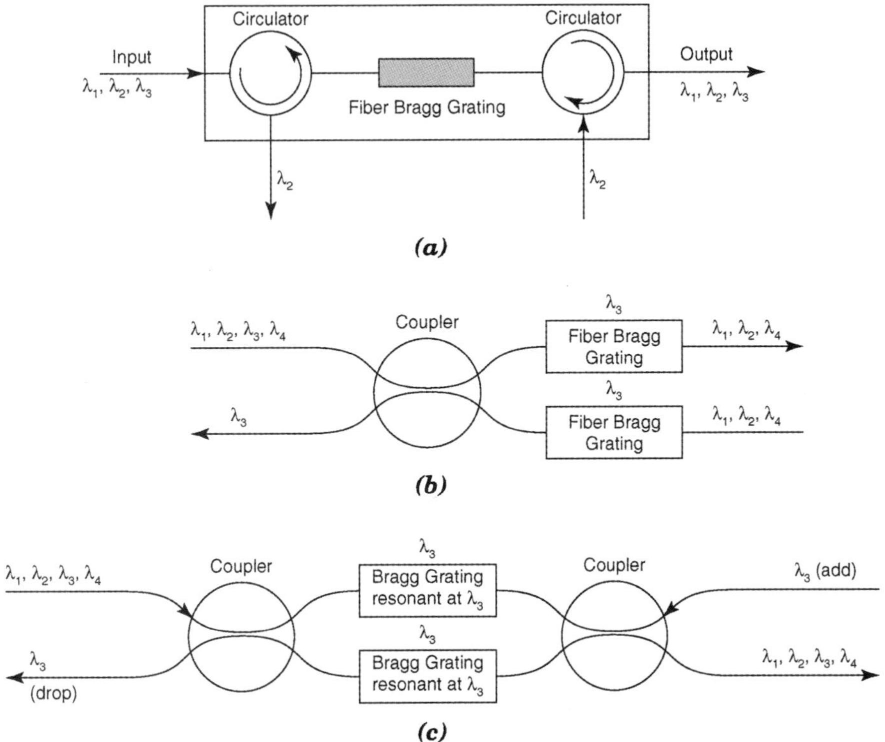

Figure 8.10. Application of DWDM filters based on fiber Bragg gratings. (a) Michelson interferometer, (b) Mach–Zehnder, and (c) add–drop configuration. (From *Lightwave*, Figure 2, page 191, Ref. 10.)

dispersion. The compensation can be done on an individual optical channel basis or over the entire DWDM band using a very long chirped grating.

Figure 8.10 shows DWDM filters based on FB gratings in add–drop and multiplexing configurations.

8.6 THIN-FILM FILTERS

A thin-film filter uses multiple coatings of dielectric filters. When the multiwavelength signal enters the filter, each layer of the film causes the various wavelengths of the signal to be reflected or passed through, depending on the wavelength and filter design. Each wavelength then contributes constructively or destructively to the entering signal. Each wavelength of the signal is canceled out or otherwise survives, then is passed to the output port (Ref. 1).

These filters usually have a large number of thin-film layers such that their bandpass characteristics can be fairly tightly controlled. This allows the transmission (passing through) of narrow ranges of wavelengths, or even a single wavelength.

The loss through a 1×2 splitter is about 4 dB, 3 dB of which is the loss due to splitting the power in half. The remaining decibel is the insertion loss of the device. As the number of output ports grow, the splitting loss grows accordingly.

A 1 × 16 splitter has an input port to output port loss of 14.5 dB. These losses can be approximated by the following formula:

$$\text{Loss}_{\text{dB}} = 0.5 + 3.5 \log_2 N \tag{8.5}$$

where N is the number of output ports.

Source: Ref. 1.

Such an approach would be used on an incoming channel before detection in a receiver. Thus the link budget must accommodate this high loss. One way to compensate for this loss would be to use an EDFA on the aggregate signal just before demultiplexing.

Thin-film technology is not the technology of choice for dense packing of channels (e.g., 50-GHz separation). However, thin-film filters provide bandpasses narrow enough for use with multiplexers/demultiplexers in WDM systems of 16 or 32 channels. More densely spaced networks are now turning to other technologies.

8.6.1 Optical Filter Summary

Optical filters are generally small. They are available embedded in connectors, in adapters, and in the fiber itself. Some typical performance values for an embedded filter specified for the band 1550–1625 nm are

Insertion loss: $<$ 1.0 to 1.5 dB

Wavelength isolation: $\geq$ 35 dB

Return loss: 40 dB

Operating temperature: −20°C to +70°C

8.7 ARRAYED WAVEGUIDE GRATING

An arrayed waveguide grating (AWG) is an integrated approach to demultiplexing. It consists of a phased array of optical waveguides that acts as a grating. This type of grating can be fabricated using InGaAsP/InP technology, permitting integration of these wave guides within a WDM transmitter or receiver. An illustration of an AWG appears in Figure 8.11. As shown in the figure, an incoming WDM signal is coupled into an array of planar waveguides after passing through a coupling section. During signal passage in each waveguide, each wavelength experiences a different phase shift because of the different lengths of the waveguides. Because of the frequency dependence of the mode-propagation constant, the phase shifts are wavelength-dependent. As a result, different channels focus on different spatial spots when the output of waveguides diffracts through another coupling section. As can be seen, an AWG acts as a conventional diffraction grating. Its efficiency can be close to 100% with proper design. A number of WDM components can be made using this technology such as waveguide-grating routers.

Figure 8.12 shows a typical application of an AWG used as an optical add–drop multiplexer.

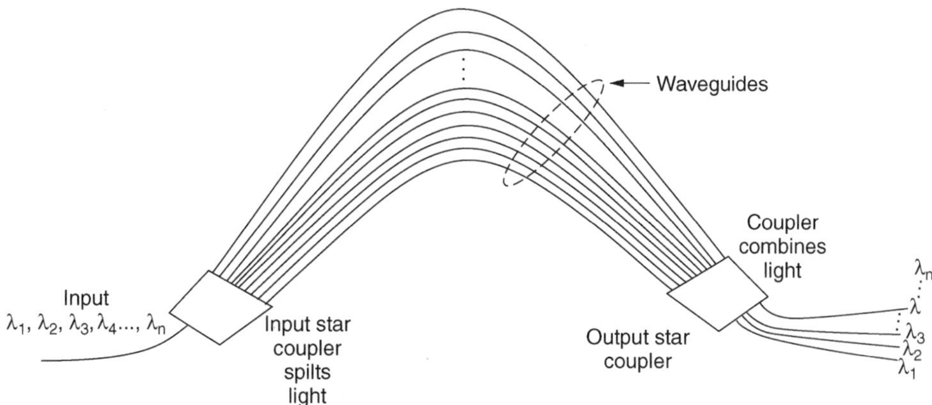

Figure 8.11. A conceptual drawing of an arrayed waveguide router.

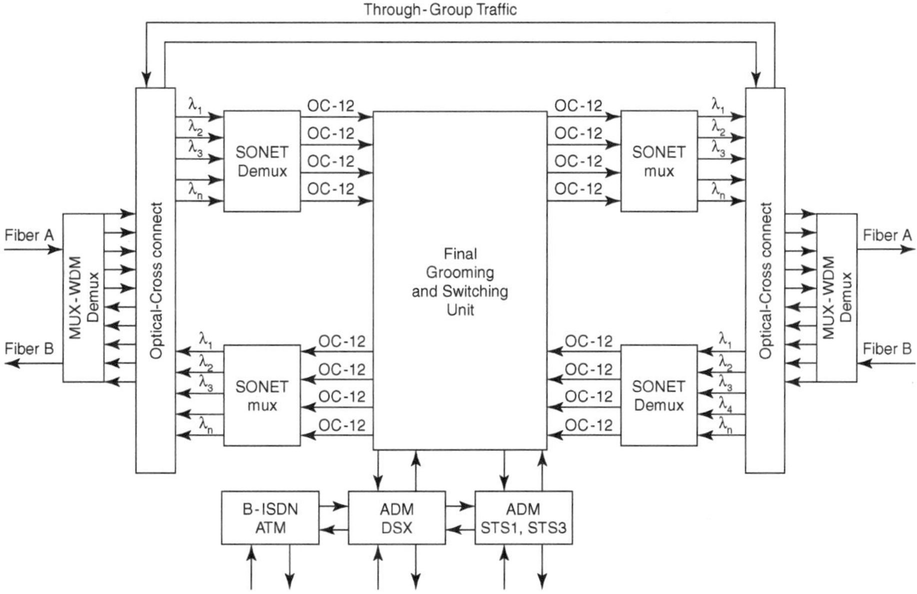

Figure 8.12. Typical transmission architecture using a multiplexer/demultiplexer and an add–drop configuration.

8.8 INTERLEAVING AND DEINTERLEAVING

An interleaver segregates a group of channels to be multiplexed into odd and even sets for a 1×2 configuration. In the simplest case, the interleaver combines two sets of channels into one densely packed set with half the channel spacing. In reverse, the deinterleaver routes the single input set of channels into two output streams with twice the channel spacing. Interleavers can be cascaded in a binary fashion. For example, a 1×4 deinterleaver takes a set of DWDM channels with 50-GHz spacing and routes them to four output fibers with 200-GHz channel

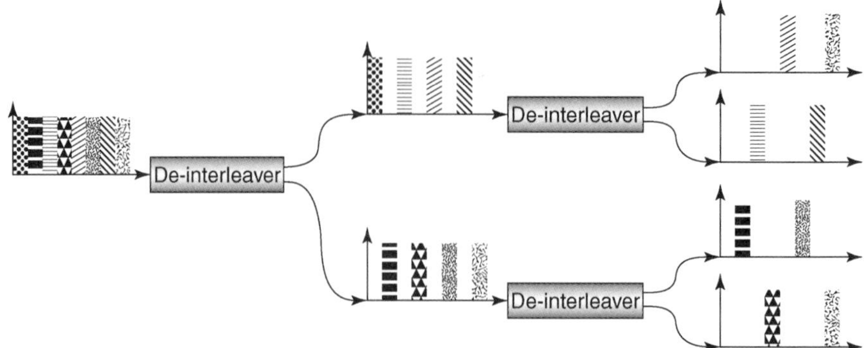

Figure 8.13. Illustrating the basic concept of a deinterleaver. Here a deinterleaver separates an input spectrum of periodically spaced wavelengths into two complementary sets at twice the original channel spacing. The component can be cascaded to create even wider channel separation and increases modularity of design. (Courtesy of Bob Shrine, Wavesplitters Technology, Fremont, CA.)

spacing. Interleavers allow us to achieve channel spacing we otherwise would not be able to accomplish. This basic concept is illustrated in Figure 8.13.

The general principle behind the operation of an interleaver is an interferometric overlap of two optical beams. The interference creates a periodic repeating output, as different integral multiples of wavelength pass through the interleaver. The desired channel spacing is set by controlling the fringe pattern. There are several approaches to achieve this fringe pattern such as employing fused fiber interferometers, liquid crystals, birefringent crystals, and others. Reference 9 describes one of the simplest designs in terms of raw material and technologies which is a fused-fiber Mach–Zehnder interferometer. In this design, the interference is created by using an unequal fiber path length between two 3-dB couplers. By carefully controlling the path-length difference, the channel spacing can be set to the desired value and be well-matched to the ITU grid (see Section 8.9). Because of the all-fiber design, this approach to interleaving has very low loss, uniform response over a wide range of wavelengths, low dispersion, and minimal polarization-dependent effects.

Liquid crystals and birefringent crystals take advantage of the different polarization states in the crystal (i.e., the ordinary and extraordinary ray), creating an effective path length difference for those different polarization states. The ordinary and extraordinary rays traveling through the crystal experience different indices of refraction and thus different retardations. When the input wavelength experiences an integral number of full-wave retardations between the two polarizations, the signal is then transmitted through the exit polarizer, while other wavelengths will be routed to a secondary output. One must realize that the input signal does not have a defined polarization, so extra components must be included to achieve the desired performance of any random polarization state. In either case, the output pattern of these devices exhibits a sine-squared waveform.

Often one can distinguish between interleavers/deinterleavers of different manufacturers by the general sine-squared interferometric pattern. The flat-top performance of one such device is shown in Figure 8.14.

The flat-top device is an all-fiber product and exhibits a low insertion loss, typically < 0.6 dB for a multiplexer and < 1.6 dB for a demultiplexer. The

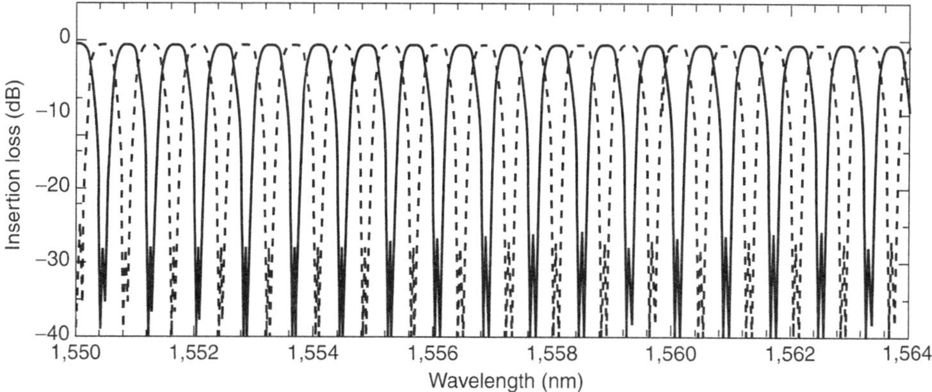

Figure 8.14. A spectrum showing the output for two arms of a deinterleaver. The example demonstrates some of the key performance parameters of an optical waveform at the output of a deinterleaver. Among these key performance parameters are: low insertion loss, flat-top passband, good crosstalk properties, and channel amplitude uniformity over a wide spectrum. (Courtesy of Bob Shrine, Wavesplitters Technology Inc., Fremont, CA, Ref. 9.)

TABLE 8.2 Performance Values of a State-of-the-Art, All-Fiber Interleaver/Deinterleaver

Parameter	Multiplexer	Demultiplexer
Channel spacing	50 GHz	50 GHz
Insertion loss	$<$ 0.6 dB ($<$ 0.4 typical)	$<$ 1.6 dB ($<$ 0.8 typical)
Ripple	$<$ 0.2 dB	$<$ 0.4 dB
0.5-dB bandwidth	$> \pm 11$ GHz	$> \pm 10$ GHz
3-dB bandwidth	$> \pm 20$ GHz	$> \pm 17.5$ GHz
Crosstalk	$>$ 12 dB (at ± 10 GHz)	$>$ 22 dB (at ± 10 GHz)
Dispersion	$<$ 10 ps/nm (at ± 10 GHz)	$<$ 10 ps/nm (at ± 10 GHz)
PDL	$<$ 0.1 dB	$<$ 0.1 dB

Source: Courtesy of Bob Shine, Wavesplitter Technology, Inc., Fremont, CA, Ref. 9.

waveform of the device is illustrated in Figure 8.14. Here the 0.6-dB insertion loss is achieved even with a 0.5-dB passband covering $> 50\%$ of the channel spacing. Additionally, the crosstalk performance maintains a level of < 25 dB, even allowing some drift of the signal wavelengths.

Performance parameter values of a typical interleaver/deinterleaver pair are shown in Table 8.2.

8.9 ITU RECOMMENDED WAVELENGTHS FOR WDM OPERATION

Table 8.3 gives the nominal central frequencies and equivalent wavelengths for 50- and 100-GHz spacings. This is known as the *ITU grid*. The ITU reference publication (G.692, Ref. 6), Annex A, adds the following note: To convert frequency to wavelength and vice versa, the speed of light, c, is taken as 2.99792458×10^8 m/s.

TABLE 8.3 Nominal Central Frequencies

Nominal Central Frequencies (THz) for Spacings of 50 GHz	Nominal Central Frequencies (THz for Spacings of 100 GHz and Above	Nominal Central Wavelengths (nm)
196.10	196.10	1528.77
196.05	—	1529.16
196.00	196.00	1529.55
195.95	—	1529.94
195.90	195.90	1530.33
195.85	—	1530.72
195.80	195.80	1531.12
195.75	—	1531.51
195.70	195.70	1531.90
195.65	—	1532.29
195.60	195.60	1532.68
195.55	—	1533.07
195.50	195.50	1533.47
195.45	—	1533.86
195.40	195.40	1534.25
195.35	—	1534.64
195.30	195.30	1535.04
195.25	—	1535.43
195.20	195.20	1535.82
195.15	—	1536.22
195.10	195.10	1536.61
195.05	—	1537.00
195.00	195.00	1537.40
194.95	—	1537.79
194.90	194.90	1538.19
194.85	—	1538.58
194.80	194.80	1538.98
194.75	—	1539.37
194.70	194.70	1539.77
194.65	—	1540.16
194.60	194.60	1540.56
194.55	—	1540.95
194.50	194.50	1541.35
194.45	—	1541.75
194.40	194.40	1542.14
194.35	—	1542.54
194.30	194.30	1542.94
194.25	—	1543.33
194.20	194.20	1543.73
!94.15	—	1544.13
194.10	194.10	1544.53
194.05	—	1544.92
194.00	194.00	1545.32
193.95	—	1545.72
193.90	193.90	1546.12
193.85	—	1546.52
193.80	193.80	1546.92
193.75	—	1547.32
193.70	193.70	1547.72
193.65	—	1548.11
193.60	193.60	1548.51
193.55	—	1548.91

TABLE 8.3 *(Continued)*

Nominal Central Frequencies (THz) for Spacings of 50 GHz	Nominal Central Frequencies (THz) for Spacings of 100 GHz and Above	Nominal Central Wavelengths (nm
193.50	193.50	1549.32
193.45	—	1549.72
193.40	193.40	1550.12
193.35	—	1550.52
193.30	193.30	1550.92
193.25	—	1551.32
193.20	193.20	1551.72
193.15	—	1552.12
193.10	193.10	1552.52
193.05	—	1552.93
193.00	193.00	1553.33
192.95	—	1553.73
192.90	192.90	1554.13
192.85	—	1554.54
192.80	192.80	1554.94
192.75	—	1555.34
192.70	192.70	1555.75
192.65	—	1556.15
192.60	192.60	1556.55
192.55	—	1556.96
192.50	192.50	1557.36
192.45	—	1557.77
192.40	192.40	1558.17
192.35	—	1558.58
192.30	192.30	1558.98
192.25	—	1559.39
192.20	192.20	1559.79
192.15	—	1560.20
192.10	192.10	1560.61

Note: The endpoints of this table are illustrative only. Future evolutions of multichannel systems are anticipated to include frequencies beyond those limits.

Source: Table A.1/G.692, ITU-T Rec. G.692, Ref. 6.

8.9.1 Selection of the Minimum Channel Spacing and Grid Reference Frequency for the WDM Plan

The ITU grid reference frequency is 193.10 THz. This frequency was not selected based on any known atomic standard AFR (absolute frequency reference) lines.

The minimum channel spacing was 100 GHz and has been reduced to 50 GHz. This minimum provides the flexibility of meeting the various ITU-T Rec. G.692 requirements. Multiples of the minimum channel spacing are compatible with respect to EDFA gain spectrum and capacity.

Technology limits, basically filter and light sources tolerances, for minimum channel spacing were taken into account. This approach makes the best use of technology and does not impose limitations associated with specific applications. Technology projections suggest that only a subset of ITU-T Rec. G.692 applications may be achievable.

The 193.10-THz AFR provides an optical signal with the necessary frequency accuracy and stability, where both can be verified by an ideal national/international frequency standard. Standards based on an iodine-stabilized He–Ne and a methane-stabilized He–Ne have been suggested.

AFRs may be used for the following applications:

- To calibrate WDM test equipment
- To provide a frequency reference for fabrication and calibration of WDM devices
- To directly provide reference frequency for multichannel systems
- To control and/or maintain optical source frequencies

Note: National/international AFR reference sources do not, as yet, have specified accuracy and stability values. These values are under study by the relevant ITU technical committee.

Source: Section 8.9.1 was based on Appendix II, ITU-T Rec. G.692, Ref. 6.

8.10 TYPICAL WDM PERFORMANCE

Table 8.4 shows the performance of ADC multiplexers; Table 8.5 gives typical ADC demultiplexer performance. Table 8.6 gives common performance values for ADC multiplexers/demultiplexers. Figure 8.15 shows an ADC 16-channel configuration.

TABLE 8.4 ADC DWDM Multiplexer Specifications

	Number of Channels			
	4	8	16	32
Channel spacing	200 GHz	100 GHz	100 GHz	100 GHz
1-dB Bandwidth (minimum)	0.7 nm	0.3 nm	0.3 nm	0.3 nm
3-dB Bandwidth (minimum)	0.8 nm	0.4 nm	0.4 nm	0.4 nm
Maximum insertion loss[a]	1.5 dB	2.3 dB	3.3 dB	5.3 dB

[a]Insertion loss does not include connectors.

Source: Information courtesy of ADC, Ref. 8.

TABLE 8.5 Specifications for ADC DWDM Demultiplexers

	Number of Channels			
	4	8	16	32
Channel spacing	200 GHz	100 GHz	100 GHz	100 GHz
1-dB Bandwidth (minimum)	0.7 nm	0.3 nm	0.3 nm	0.3 nm
3-dB Bandwidth (minimum)	0.8 nm	0.4 nm	0.4 nm	0.4 nm
20-dB Bandwidth (maximum)	1.5 nm	0.8 nm	0.8 nm	0.8 nm
Maximum insertion loss[a]	2.0 dB	2.8 dB	3.8 dB	5.8 dB
Minimum wavelength isolation[b]	30 dB	30 dB	30 dB	30 dB

[a]Insertion loss does not include connectors.

[b]The ratio of the optical power from the desired channel to the sum of the optical power from the blocked channels expressed in decibels.

Source: Information courtesy of ADC, Ref. 8.

TABLE 8.6 Common Specifications for ADC Multiplexer/Demultiplexer

Maximum in-band ripple	±0.1 dB
Central wavelength accuracy	±0.5 nm
Maximum wavelength variation with temperature	±0.0012 nm/°C
Maximum uniformity	1.0 dB
Maximum return loss	−45 dB
Maximum polarization-dependent loss	0.2 dB
Operating temperature	−5°C to 55°C
Fiber type	Corning SMF-28™

Source: Information courtesy of ADC, Ref. 8.

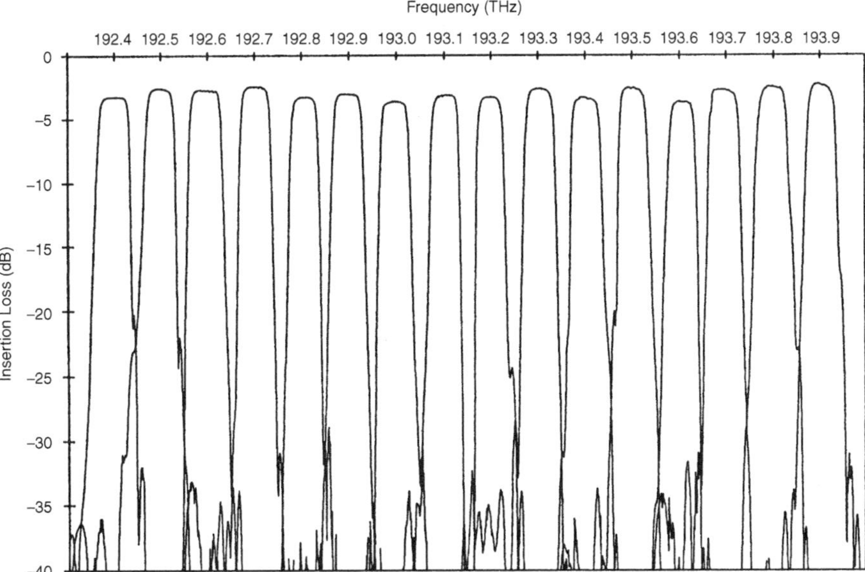

Figure 8.15. Sixteen-channel ADC demultiplexer performance. (Courtesy of ADC, Ref. 8.)

REFERENCES

1. *Telecordia Notes on Dense Wavelength-Division Multiplexing (DWDM) and Optical Networks*, Special Report, SR-NOTES-SERIES-02, Issue 1, Piscataway, NJ, May 2000.
2. Stamatios V. Kartalopoulos, *Introduction to DWDM Technology*, IEEE Press, New York, 2000.
3. Govind P. Agrawal, *Fiber-Optic Communication Systems*, 2nd ed., John Wiley & Sons, New York, 1997.
4. Technical promotional material from the Web at *www.sel-rtp.com/products/passives/Components/fbg.html*, Sumitomo Electric Lightwave Corp., Research Triangle, NC.
5. Lucent Technologies, from the Web at *www.lucent.com/micro/opto*, Lucent Technologies, Allentown, PA.

6. *Optical Interfaces for Multi-Channel Systems with Optical Amplifiers*, ITU-T Rec. G.692, ITU Geneva, October 1998.
7. *The IEEE Standard Dictionary of Electrical and Electronic Terms*, 6th ed., IEEE, New York, December 1996.
8. Dense Wavelength Division Multiplexing (DWDM), from the Web at www.adc.com, ADC Telecommunications, Minneapolis, MN, August 1998.
9. Bob Shine and Jerry Bautista, Interleavers Make High-channel-count Systems Economical (Wavesplitter Technologies), *Lightwave*, August 2000, page 140.
10. Franck Chatain, Fiber Bragg Grating Technology Passes Light to New Passive Components, *Lightwave*, March 2001, page 186.
11. Karen Liu and John Ryan, "The Animals in the Zoo: The Expanding Menagerie of Optical Components," *IEEE Communications Magazine*, July 2001, page 110.

9

SYNCHRONOUS OPTICAL NETWORK (SONET) AND SYNCHRONOUS DIGITAL HIERARCHY (SDH)

9.1 BACKGROUND AND INTRODUCTION

SONET and SDH are similar digital transport formats that were developed for the specific purpose of providing a reliable and versatile digital structure to take advantage of the higher bit-rate capacity of optical fiber. SONET is an acronym standing for *synchronous optical network*. In a similar vein, SDH stands for *synchronous digital hierarchy*. We could say that SONET has a North American flavor and that SDH has a European flavor. This may be stretching the point, because the two systems are very similar.

The original concept in developing a digital format for high-bit-rate-capacity optical systems was to have just one singular standard for worldwide application. This did not work out. The United States wanted the basic bit rate to accommodate DS3 (around 50 Mbps). The Europeans had no bit rates near this value and were opting for a starting rate around 150 Mbps. Another difference surfaced for framing alternatives. The United States perspective was based on a frame of 13 rows by 180-byte columns for the 150-Mbps rate reflecting what is now called STS-3 structure. Europe advocated a 9-row by 270-byte column STS-3 frame to efficiently transport the E1 signal (2.048 Mbps) using 4 columns of 9 bytes, based on 32 bytes/125 μs.

The ANSI T1X1 committee approved a final standard in August 1988, with CCITT following suit, and a global SONET/SDH standard was established. This global standard was based on a 9-row frame, wherein SONET became a subset of SDH (Ref. 1).

Both SONET and SDH use basic building-block techniques. As we mentioned above, SONET starts at a lower bit rate, at 51.84 Mbps. This basic rate is called STS-1 (synchronous transport signal layer 1). Lower rate payloads are mapped into the STS-1 format, while higher rate signals are obtained by byte interleaving N frame-aligned STS-1s to create an STS-N signal. Such a simple multiplexing approach results in no additional overhead; as a consequence the transmission rate of an STS-N signal is exactly $N \times 51.84$ Mbps, where N is currently defined for the values 1, 3, 12, 24, 48, and 192.

Source: Ref. 2.

The basic building block of SDH is the synchronous transport module level 1 (STM-1) with a bit rate of 155.52 Mbps. Lower rate payloads are mapped into STM-1, and higher rate signals are generated by synchronously multiplexing N STM-1 signals to form the STM-N signal. Transport overhead of an STM-N signal is N times the transport overhead of an STM-1, and the transmission rate is $N \times 155.52$ Mbps. Presently, only STM-1, STM-4, STM-16, and STM-64 have been defined by the ITU-T Organization.

Source: Ref. 5.

Both with SONET and SDH, the frame rate is 8000 per second resulting in a 125-μs frame period. There is high compatibility between SONET and SDH. Because of the different basic building-block size, they differ in structure: 51.84 Mbps for SONET and 155.52 Mbps for SDH. However, if we multiply the SONET rate by 3, developing the STS-3 signal, we do indeed have the SDH initial bit rate of 155.52 Mbps. Figures 9.1 and 9.2 compare multiplexing structures of each. Table 9.1 lists and compares the bit rates of each standard.

Besides differing in basic building-block bit rates, SONET and SDH differ with respect to overhead usage. These overhead differences have been grouped into two broad categories: format definitions and usage interpretation. As a result, we have opted to segregate our descriptions of each.

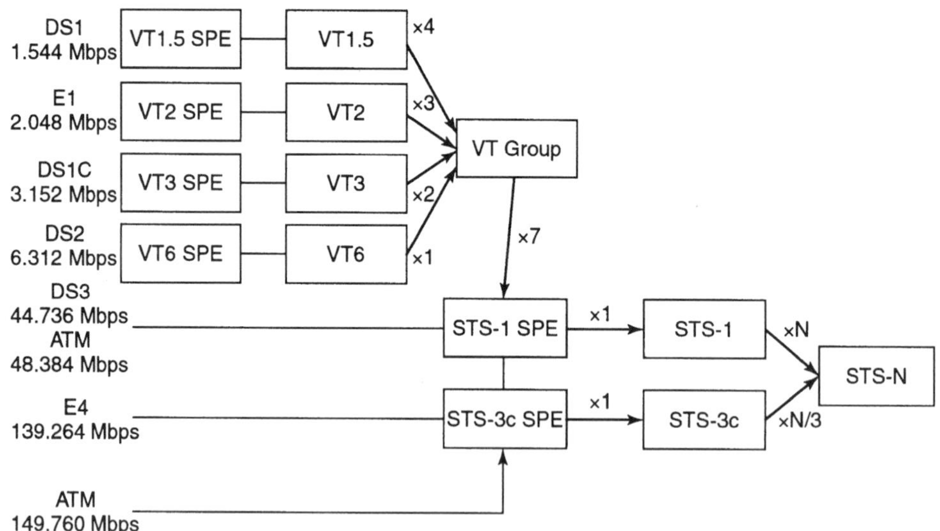

Figure 9.1. SONET multiplexing structure. (From *Synchronous Optical Network, Synchronous Digital Hierarchy: An Overview of Synchronous Network*, Ref. 1, Figure 3, page 4.)

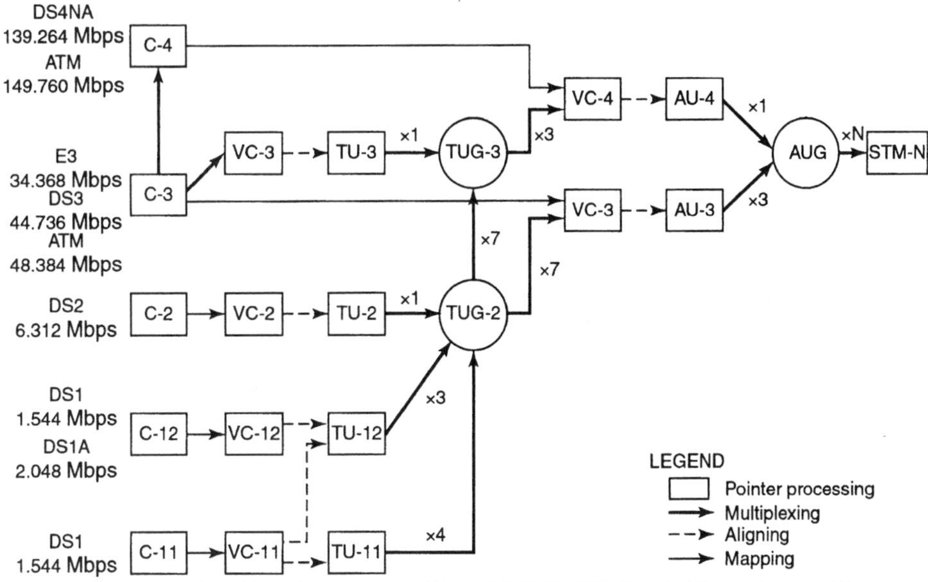

Figure 9.2. SDH multiplexing structure. (From *SONET/SDH*, Ref. 3, Figure 4, page 4.)

TABLE 9.1 SONET and SDH Transmission Rates

SONET Optical Carrier Level OC-*N*	SONET Electrical Level STS-*N*	Equivalent SDH STM-*N*	Line Rate (Mbps)
OC-1	STS-1	—	51.84
OC-3	STS-3	STM-1	155.52
OC-12	STS-12	STM-4	622.08
OC-24	STS-24	—	1244.16
OC-48	STS-48	STM-16	2488.32
OC-192	STS-192	STM-64	9953.28
OC-768	STS-768	STM-256	39,813,120[a]

[a]Initial testing as of this writing (Refs. 2–5).

We must dispel a concept that has developed by the unfortunate use of words and terminology. Some believe that because SONET stands for synchronous optical network, it will only operate on optical fiber light-guide. This is patently incorrect. Any transport medium that can provide the necessary bandwidth (measured in hertz, as one would expect) will transport the requisite SONET and SDH line rates. For example, digital LOS microwave, using heavy-bit-packing modulation schemes, readily transports 622 Mbps (STS-12 and STM-4) per carrier at the higher frequencies using a 40-MHz assigned bandwidth.

The objective of this section is to provide an overview of these two standards and some of their challenging innovations that make them interesting, such as the payload pointer. Section 9.2 deals with SONET, the North American standard, and Section 9.3 covers SDH. Section 9.4 presents a summary of the two standards in tabular form.

It should be noted that we keep the customary nomenclature in this work such that ITU recommendations will be identified by the CCITT or CCIR logo if that document was issued prior to January 1, 1993. If the document was issued after that date, where it originated from the Telecommunications Standardization Sector of the ITU, it will be referred to as an ITU-T Recommendation; or if it originated from the Radiocommunications Sector, it will be called an ITU-R Recommendation.

9.2 SYNCHRONOUS OPTICAL NETWORK (SONET)

9.2.1 Synchronous Signal Structure

SONET is based on a synchronous digital signal comprised of 8-bit octets, which are organized into a frame structure. The frame can be represented by a two-dimensional map comprising N rows and M columns, where each box so derived contains one octet (or byte). The upper left-hand corner of the rectangular map representing a frame contains an identifiable marker to tell the receiver if is the start of frame.

SONET consists of a basic, first-level structure called STS-1, which is discussed in the following. The definition of the first level also defines the entire hierarchy of SONET signals because higher-level SONET signals are obtained by synchronously multiplexing the lower-level modules. When lower-level modules are multiplexed together, the result is denoted STS-N (STS stands for synchronous transport signal), where N is an integer. The resulting format can be converted to an OC-N (OC stands for optical carrier) or STS-N electrical signal. There is an integer multiple relationship between the rate of the basic module STS-1 and the OC-N electrical equivalent signals (i.e., the rate of an OC-N is equal to N times the rate of an STS-1). Only OC-1, OC-3, OC-12, OC-24, OC-48, and OC-192 are supported by today's SONET.

9.2.1.1 *The Basic Building Block.* The STS-1 frame is shown in Figure 9.3. STS-1 is the basic module and building block of SONET. It is a specific sequence of 810 octets (6480 bits) which includes various overhead octets and an envelope capacity for transporting payloads.[1] STS-1 is depicted as a 90-column, 9-row structure. With a frame period of 125 μs (i.e., 8000 frames per second), STS-1 has a bit rate of 51.840 Mbps. Consider Figure 9.3. The order of transmission is row-by-row, from left to right. In each octet of STS-1 the most significant bit (MSB) is transmitted first.

As illustrated in Figure 9.3, the first three columns of the STS-1 frame contain the transport overhead. These three columns have 27 octets (i.e., 9×3), 9 of which are used for the *section overhead* and 18 for the *line overhead*. The remaining 87 columns make up the STS-1 envelope capacity, as shown in Figure 9.4.

[1]The several reference publications use the term *byte*, meaning, in this context, an 8-bit sequence. We prefer the term *octet*. The reason is that some argue that byte is ambiguous, having conflicting definitions.

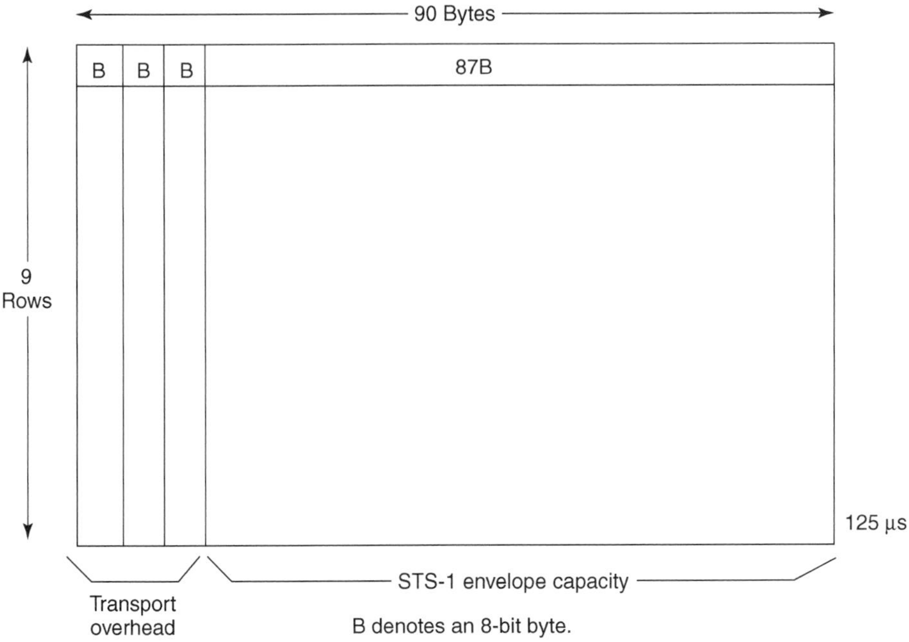

Figure 9.3. The STS-1 frame.

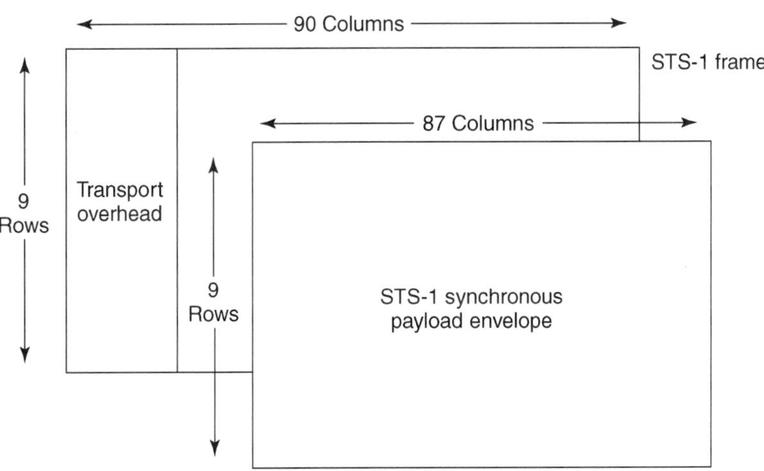

Figure 9.4. STS-1 synchronous payload envelope.

The STS-1 synchronous payload envelope (SPE) occupies the STS-1 envelope capacity. The STS-1 SPE consists of 783 octets and is depicted as an 87-column by 9-row structure. In that structure, column 1 contains 9 octets and is designated as the *STS path overhead* (POH). In the SPE, columns 30 and 59 are not used for payload but are designated *fixed stuff* columns and are undefined. However, the values used as stuff in columns 30 and 59 of each STS-1 SPE will produce even parity when calculating BIP-8 of the STS-1 path BIP value. The POH column and fixed stuff columns are shown in Figure 9.5. The 756 octets in the remaining 84 columns are used for the actual STS-1 payload capacity.

The STS-1 SPE may begin anywhere in the STS-1 envelope capacity. Typically, the SPE begins in one STS-1 frame and ends in the next. This is illustrated in Figure 9.6. However, on occasion, the SPE may be wholly contained in one frame. The *STS payload pointer* resides in the transport overhead. It designates the location of the next octet where the SPE begins. Payload pointers are described in the following paragraphs.

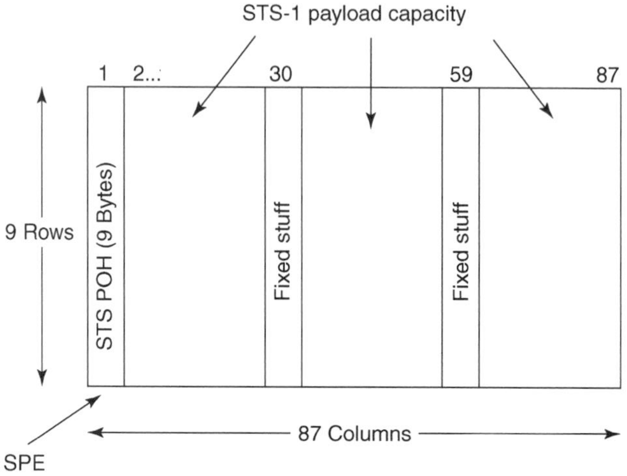

Figure 9.5. Path overhead (POH) and the STS-1 payload capacity within the STS-1 SPE. Note that the net payload capacity of the STS-1 is only 84 columns. (Courtesy of Agilent Technologies, Ref. 7.)

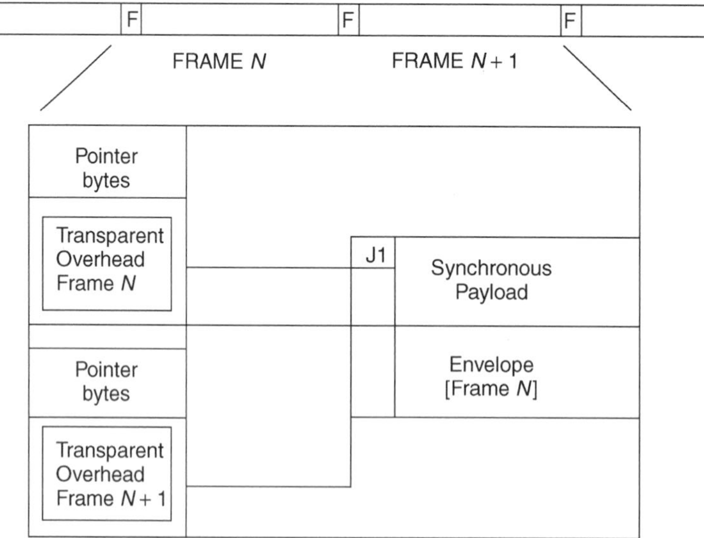

Figure 9.6. STS-1 SPE typically located in STS-1 frames. (Courtesy of Agilent Technologies, Ref. 7.)

The STS POH is associated with each payload and is used to communicate various pieces of information from the point where the payload is mapped into the STS-1 SPE to the point where it is delivered. Among the pieces of information carried in the POH are alarm and performance data.

9.2.1.2 STS-N Frames. Figure 9.7 illustrates the structure of an STS-N frame. The frame consists of a specific sequence of $N \times 810$ octets. The STS-N frame is formed by octet-interleaved STS-1 and STS-M ($3 \leq M < N$) modules. The transport overhead of the associated STS SPEs are not required to be aligned because each STS-1 has a payload pointer to indicate the location of the SPE or to indicate concatenation.

9.2.1.3 STS Concatenation. Superrate payloads require multiple STS-1 SPEs. FDDI and some B-ISDN payloads fall into this category. Concatenation means the linking together. An STS-Nc module is formed by linking N constituent STS-1s

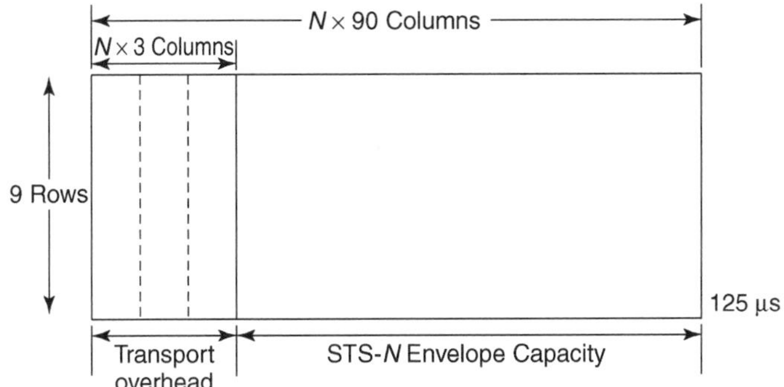

Figure 9.7. STS-N frame.

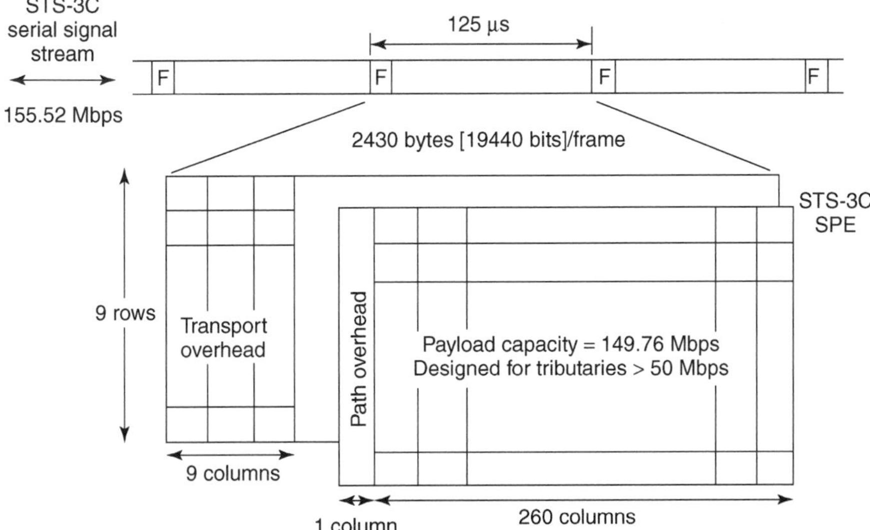

Figure 9.8. STS-3c concatenated SPE. (Courtesy of Agilent Technologies, Ref. 7.)

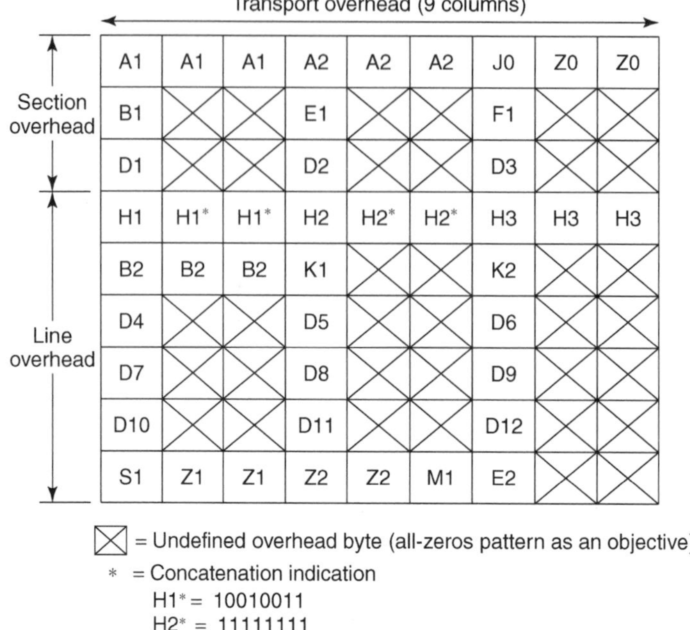

Figure 9.9. Transport overhead assignment showing OC-3 carrying an STS-3c SPE. (Courtesy of Telcordia Technology, Figure 3-8, page 3, GR-253-CORE, Issue 3, [Ref. 6].)

together in a fixed phase alignment. The superrate payload is then mapped into the resulting STS-*N*c SPE for transport. Such STS-*N*c SPE requires an OC-*N* or STS-*N* electrical signal. Concatenation indicators contained in the second through the *N*th STS payload pointer are used to show that the STS-1s of an STS-*N*c are linked together.

There are $N \times 783$ octets in an STS-*N*c. Such an STS-*N*c arrangement is illustrated in Figure 9.8 and is depicted as an $N \times 87$ column by 9-row structure. Because of the linkage, only one set of STS POH is required in the STS-*N*c SPE. Here the STS POH always appears in the first of the *N* STS-1s that make up the STS-*N*c (Ref. 10).

Figure 9.9 shows the assignment of transport overhead of an OC-3 carrying an STS-3c SPE.

9.2.1.4 *Structure of Virtual Tributaries.* The SONET STS-1 SPE with channel capacity of 50.11 Mbps has been designed specifically to transport a DS3 tributary signal. To accommodate sub-STS-1 rate payloads such as DS1, the VT structure is used. It consists of four sizes: VT1.5 (1.728 Mbps) for DS1 transport, VT2 (2.304 Mbps) for E1 transport, VT3 (3.456 Mbps) for DS1C transport, and VT6 (6.912 Mbps) for DS2 transport. The virtual tributary concept is illustrated in Figure 9.10. The four VT configurations are illustrated in Figure 9.11. In the 87-column by 9-row structure of the STS-1 SPE, the VTs occupy 3, 4, 6, and 12 columns, respectively.

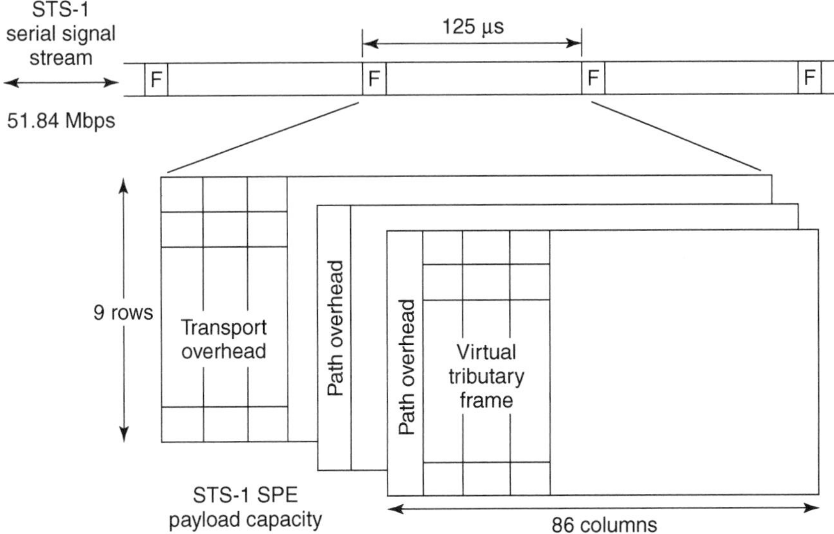

Figure 9.10. The virtual tributary (VT) concept. (Courtesy of Agilent Technologies, Ref. 7.)

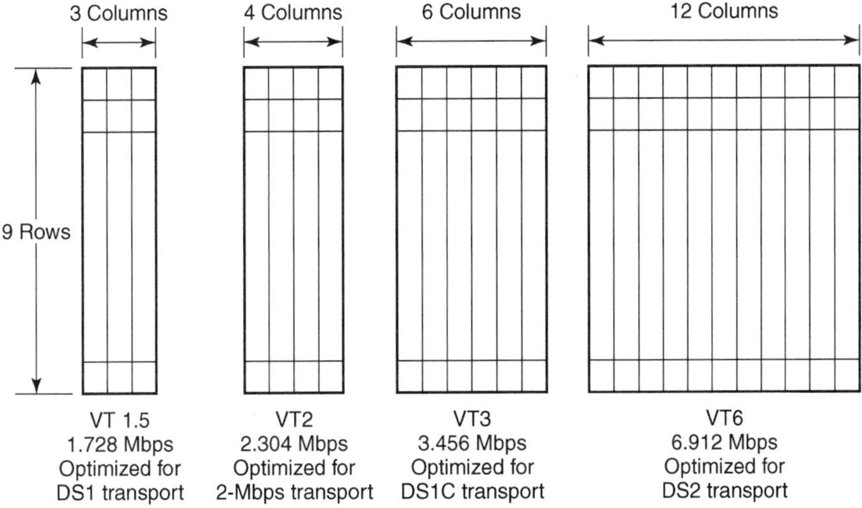

Figure 9.11. The four sizes of virtual tributary frames. (Courtesy of Agilent Technologies, Ref. 7.)

9.2.2 Payload Pointer

The STS payload pointer provides a method for allowing flexible and dynamic alignment of the STS SPE within the STS envelope capacity, independent of the actual contents of the SPE. SONET, by definition, is intended to be synchronous. It derives its timing from the master network clock.

Modern digital networks must make provision for more than one master clock. Examples in the United States are the several interexchange carriers that interface with local exchange carriers (LECs), each with its own master clock. Each master

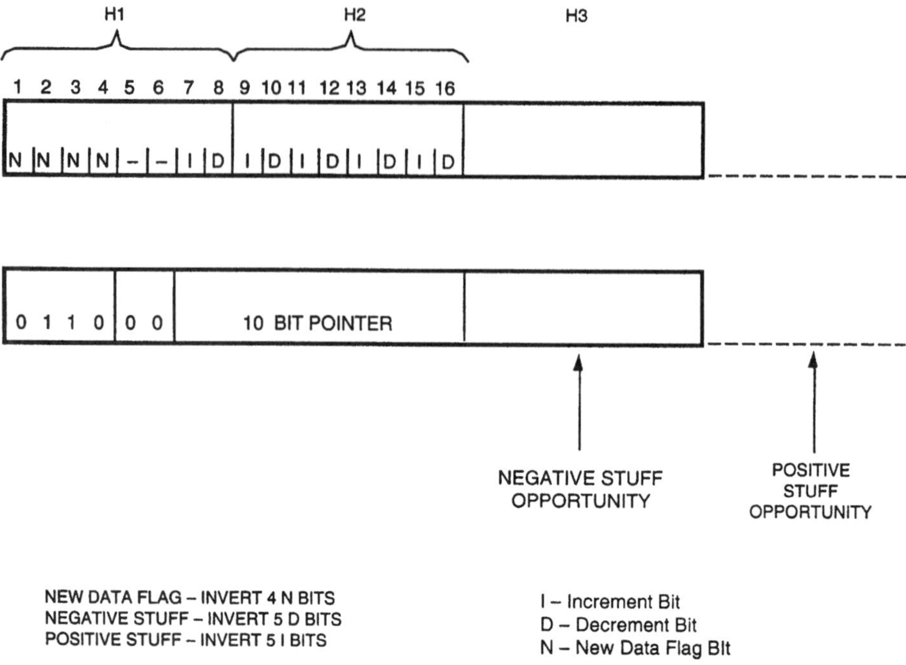

Figure 9.12. STS-1 payload pointer (H1, H2) coding. (Based on Refs. 1, 2, and 6.)

clock (stratum 1) operates independently. Also, each of these master clocks has excellent stability (i.e., better than 1×10^{-11} per month), yet there may be some small variance in time among the clocks. Assuredly, they will not be phase-aligned. Likewise, SONET must take into account loss of master clock or a segment of its timing delivery system. In this case, network switches fall back on lower-stability internal clocks.[2] The situation must be handled by SONET. Therefore, synchronous transport is required to operate effectively under these conditions, where network nodes are operating at slightly different rates.

To accommodate these clock offsets, the SPE can be moved (justified) in the positive or negative direction one octet at a time with respect to the transport frame. This is accomplished by recalculating or updating the payload pointer at each SONET network node. In addition to clock offsets, updating the payload pointer also accommodates any other timing phase adjustments required between the input SONET signals and the timing reference at the SONET node. This is what is meant by *dynamic alignment*, where the STS SPE is allowed to float within the STS envelope capacity.

The payload pointer is contained in the H1 and H2 octets in the line overhead (LOH) and designates the location of the octet where the STS SPE begins. These two octets are illustrated in Figure 9.12. Bits 1 through 4 of the pointer word carry the *new data flag* (NDF), and bits 7 through 16 carry the pointer value. Bits 5 and 6 are undefined.

[2]It is general practice in digital networks that switches provide timing supply for transmission facilities.

Let us discuss bits 7 through 16, the pointer value. It is a binary number with a range of 0 to 782. It indicates the offset of the point word and the first octet of the STS SPE (i.e., the J1 octet). The transport overhead octets are not counted in the offset. For example, a pointer value of 0 indicates that the STS SPE starts in the octet location that immediately follows the H3 octet, whereas an offset of 87 indicates that it starts immediately after the K2 octet location. Note that these overhead octets are shown in Figure 9.9.

Payload pointer processing introduces a signal impairment known as *payload adjustment jitter*. This impairment appears on a received tributary signal after recovery from an SPE that has been subjected to payload pointer changes. The operation of the network equipment processing the tributary signal immediately downstream is influenced by this excessive jitter. By careful design of the timing distribution for the synchronous network, payload jitter adjustments can be minimized, thus reducing the level of tributary jitter that can be accumulated through synchronous transport.

9.2.3 The Three Overhead Levels of SONET

The three embedded overhead levels of SONET are as follows:

1. Path (POH)
2. Line (LOH)
3. Section (SOH)

These overhead levels, represented as spans, are illustrated in Figure 9.13. One important function carried out by this overhead is the support of network operation, administration, and maintenance (OA&M).

The path overhead (POH) consists of nine octets and occupies the first column of the SPE, as pointed out previously. It is created by and included in the SPE as part of the SPE assembly process. The POH provides the facilities to support and

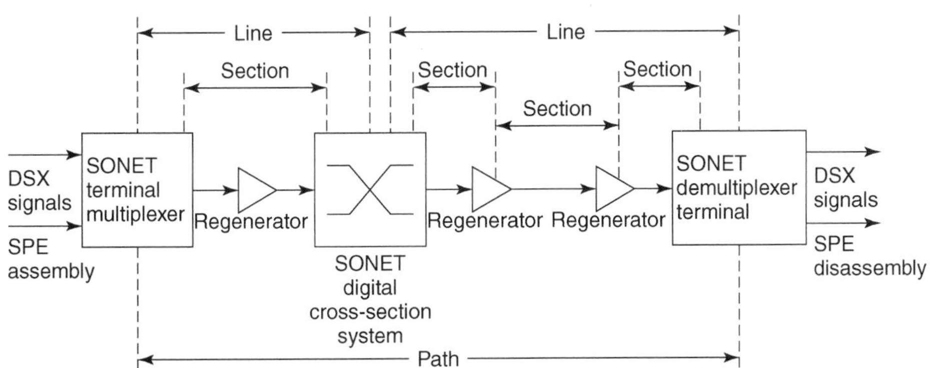

Figure 9.13. SONET section, line, and path definitions. (Based on Ref. 7, page 2-28.)

maintain the transport of the SPE between path terminations, where the SPE is assembled and disassembled. Among the POH specific functions are:

- An 8-bit wide (octet B3) BIP (bit-interleaved parity) check calculated over all bits of the previous SPE. The computed value is placed in the POH of the following frame.
- Alarm and performance information (octet G1).
- A path signal label (octet C2); gives details of SPE structure. It is 8 bits wide, which can identify up to 256 structures (2^8).
- One octet (J1) repeated through 64 frames can develop an alphanumeric message associated with the path. This allows verification of continuity of connection to the source of the path signal at any receiving terminal along the path by monitoring the message string.
- An orderwire for network operator communications between path equipment (octet F2).

Facilities to support and maintain the transport of the SPE between adjacent nodes are provided by the line and section overhead. These two overhead groups share the first three columns of the STS-1 frame. The SOH occupies the top three rows (total of 9 octets), and the LOH occupies the bottom six rows (18 octets).

The overhead line functions include the following:

- Payload pointer (octets H1, H2, and H3) (each STS-1 in an STS-*N* frame has its own payload pointer)
- Automatic protection switching control (octets K1 and K2)
- BIP parity check (octet (B2)
- 576-kbps data channel (octets D4 through D12)
- Express orderwire (octet E2)

A *section* is defined in Figure 9.13. Among the section overhead functions are:

- Frame alignment pattern (octets A1 and A2)
- STS-1 identification (octet C1): a binary number corresponding to the order of appearance in the STS-*N* frame, which can be used in the framing and deinterleaving process to determine the position of other signals
- BIP-8 parity check (octet B1): section error monitoring
- Data communications channel (octets D1, D2, and D3)
- Local orderwire channel (octet E1)
- User channel (octet F1)

Sources: Refs. 8 and 9.

9.2.4 SPE Assembly/Disassembly Process

Payload mapping is the process of assembling a tributary signal into an SPE. It is fundamental to SONET operation. The payload capacity provided for each individual tributary signal is always slightly greater than that required by that tributary

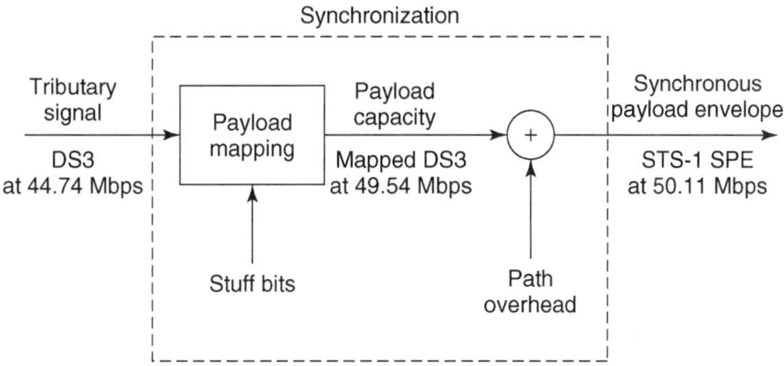

Figure 9.14. The SPE assembly process. (Courtesy of Agilent Technologies, Ref. 7.)

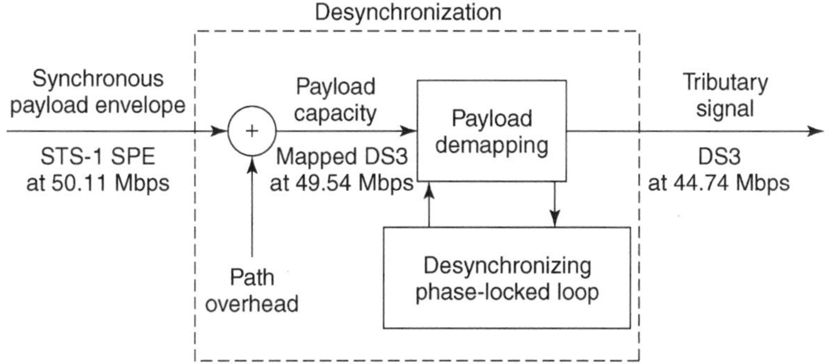

Figure 9.15. The SPE disassembly process. (Courtesy of Agilent Technologies, Ref. 7.)

signal. The mapping process, in essence, is to synchronize the tributary signal with the payload capacity. This is achieved by adding stuffing bits to the bit stream as part of the mapping process.

An example might be a DS3 tributary signal at a nominal bit rate of 44.736 Mpbs of 49.54 Mbps provided by an STS-1 SPE. The addition of path overhead completes the assembly process of the STS-1 SPE and increases the bit rate of the composite signal to 50.11 Mbps. The SPE assembly process is shown graphically in Figure 9.14. At the terminus or drop point of the network, the original DS3 payload must be recovered, as in our example. The process of SPE disassembly is shown in Figure 9.15. The term used in this case is *payload demapping.*

The demapping process desynchronizes the tributary signal from the composite SPE signal by stripping off the path overhead and the added stuff bits. In the example, an STS-1 SPE with a mapped DS3 payload arrives at the tributary disassembly location with a signal rate of 50.11 Mbps. The stripping process results in a discontinuous signal representing the transported DS3 signal with an average signal rate of 44.74 Mbps. The timing discontinuities are reduced by means of a desynchronizing phase-locked loop, which then produces a continuous DS3 signal at the required average transmission rate of 44.736 Mbps.

Sources: Refs. 8–10.

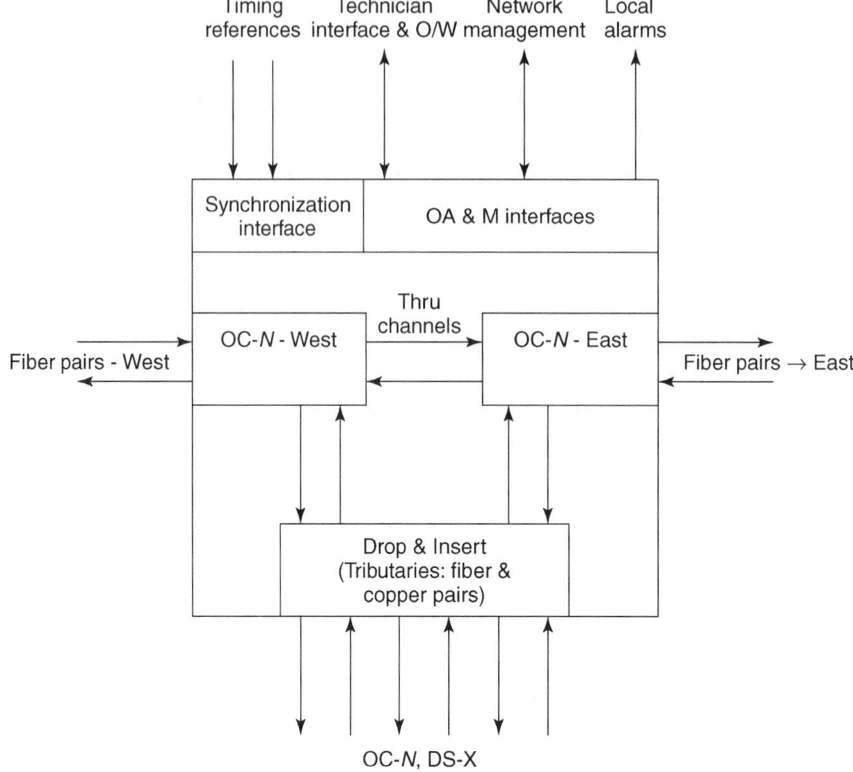

Figure 9.16. SONET ADM add–drop configuration example.

9.2.5 Add–Drop Multiplex (ADM)

A SONET ADM multiplexes one or more DS-n signals into a SONET OC-N channel. In its converse function, a SONET ADM demultiplexes a SONET STS-N configuration into its component, DS-N, to be passed to a user or to be forwarded on a tributary bit stream. An ADM can be configured for either the add–drop or terminal mode. In the ADM mode, it can operate when the low-speed DS1 signals terminating at the SONET derive timing from the same or equivalent source (i.e., synchronous) as the SONET system it interfaces with, but does not derive timing from asynchronous sources.

Figure 9.16 is an example of an ADM configured in the add–drop mode with DS1 and OC-N interfaces. A SONET ADM interfaces with two full-duplex OC-N signals and one or more full-duplex DS1 signals. It may optionally provide low-speed DS1C, DS2, DS3, or OC-M (where $M \leq N$). There are non-path-terminating information payloads from each incoming OC-N signal, which are passed to the SONET ADM and transmitted by the OC-N interface on the other side.

Timing for transmitted OC-N is derived from either (a) an external synchronization source, (b) an incoming OC-N signal, (c) each incoming OC-N signal in each direction (called through-timing), or (d) its local clock, depending on the network

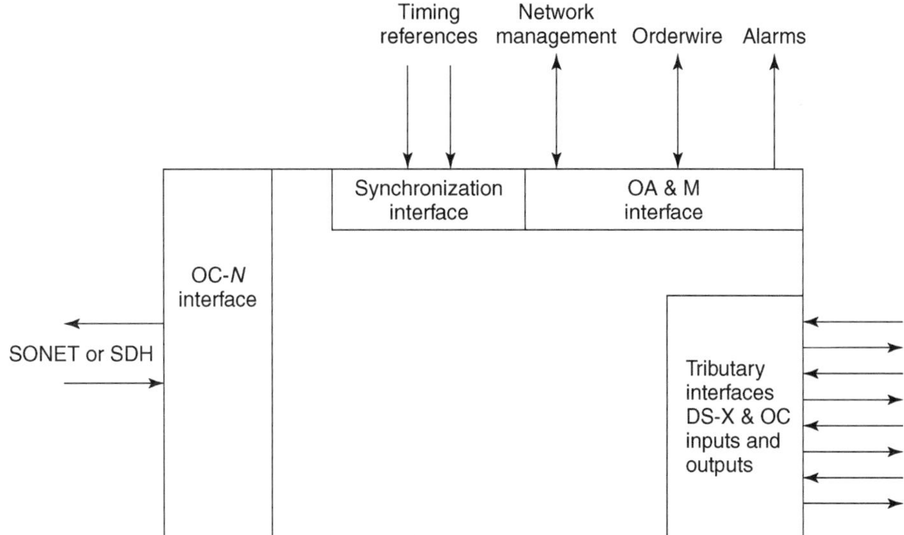

Figure 9.17. An ADM in a terminal configuration.

application. Each DS1 interface reads data from an incoming OC-*N* and inserts data into an outgoing OC-*N* bit stream as required. Figure 9.16 also shows a synchronization interface for local switch application with external timing and an operations interface module (OIM) that provides local technician orderwire,[3] local alarm, and an interface to remote operations systems. A controller is part of each SONET ADM, which maintains and controls ADM functions, to connect to local or remote technician interfaces, and to connect to required and optional operations links that permit maintenance, provisioning, and testing.

Figure 9.17 shows an example of an ADM in the terminal mode of operation with DS1 interfaces. In this case, the ADM multiplexes up to $N \times 28$DS1 or equivalent signals into an OC-*N* bit stream.[4] Timing for this terminal configuration is taken from either an external synchronization source, the received OC-*N* signal (called loop timing), or its own local clock, depending on the network application (Ref. 11).

9.2.6 Automatic Protection Switching (APS)

First, we distinguish $1 + 1$ protection from $N + 1$ protection. These two SONET linear APS options are shown in Figure 9.18. APS can be provided in a linear or ring architecture. SONET NEs (network elements) that have *line termination equipment* (LTE) and terminate optical lines may provide *linear* APS. Support of linear APS at STS-*N* electrical interfaces is not provided in the relevant Telcordia and ANSI standards.

Linear APS—and, in particular, the protocol for the APS channel—is standardized to allow interworking between SONET LTEs from different vendors. There-

[3]An *orderwire* (also known as service channel) is a voice or keyboard/printer/display circuit for coordinating setup and maintenance activities among technicians and supervisors.

[4]This implies a DS3 configuration because it contains 28 DS1s.

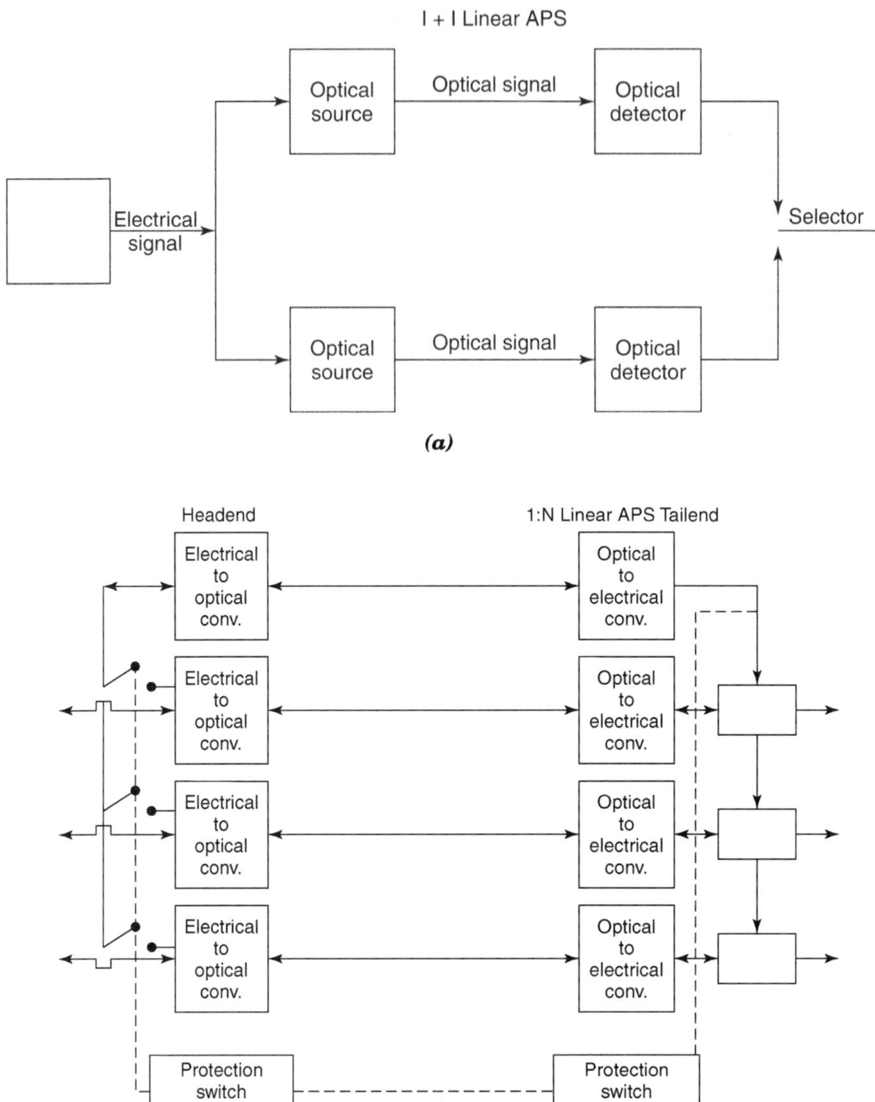

Figure 9.18. Linear SONET APS options: 1 + 1 protection, N + 1 protection.

fore, all the STS SPEs carried in an OC-N signal are protected together. The ANSI and Telcordia standards define two linear APS architectures:

1. 1 + 1
2. N + 1 (also called 1:n/1:1)

The 1 + 1 is an architecture in which the head-end signal is continuously bridged to working and protection equipment so the same payloads are transmitted identically to the tail-end working and protection equipment (see top of Figure 9.18).

At the tail end, the working and protection OC-N signals are monitored independently and identically for failures. The receiving equipment chooses either the working or protection signals as the one from which to select traffic, based on the switch initiation criteria [e.g., loss of signal (LOS), signal degrade, etc.]. Because of the continuous head-end bridging, the 1 + 1 architecture does not allow an unprotected extra traffic channel to be provided.

1 + 1 protection is very effective to achieve full redundancy. This type of configuration is widely employed usually with a ring architecture. In the basic configuration of a ring, the traffic from the source is transmitted simultaneously over both bearers, and the decision to switch between main and standby is made at the receiving location. In this situation, only LOS or similar indications are required to initiate changeover, and no command and control information needs to be passed between the two sites. It is assumed that after the failure in the main line, a repair crew will restore it to service. Rather than have the repaired line placed back in service as the "main line," it is designated the new "standby." Thus only one line interruption takes place, and the process of repair does not require a second break in service.

The best method of configuring 1 for 1 service is to have the standby line geographically distant from the main line. This minimizes common mode failures. Because of the simplicity of this approach, it ensures fastest restoral with the least requirement for sophisticated monitoring and control equipment. However, it is more costly and less efficient with regard to equipment usage than an $N + 1$ approach. It is inefficient in that the standby equipment sits idle nearly all the time, not bringing in any revenue.

The $N + 1$ (also denoted as 1:n or one for n) protection is an architecture in which any of n working lines can be bridged to a single protection line (see bottom of Figure 9.18). Permissible values for n are 1 to 14. The APS channel refers to the K1 and K2 bytes in the line overhead (LOH) used to accomplish head-end to tail-end signaling. Because the head-end is switchable, the protection line can be used to carry an extra traffic channel. Some texts call a subset of $N + 1$ architecture the 1:1 architecture.

The $N + 1$ link protection method makes more efficient use of standby equipment. It is just an extension of the 1 + 1 technique described above. With the excellent reliability of present-day equipment, we can be fairly well assured that there will not be two simultaneous failures on a route. This makes it possible to share the standby line among N working lines.

$N + 1$ link protection makes more cost-effective use of equipment but requires more sophisticated control and cannot offer the same level of availability as 1-for-1 protection. Diverse routing of main and standby lines is also much harder to achieve.

9.2.7 SONET Ring Configurations

A ring network consists of network elements connected in a point-to-point arrangement that forms an unbroken circular configuration as shown in Figure 9.19. As we must realize, the main reason for implementing path-switched rings is to improve network survivability. The ring provides protection against fiber cuts and equipment failures.

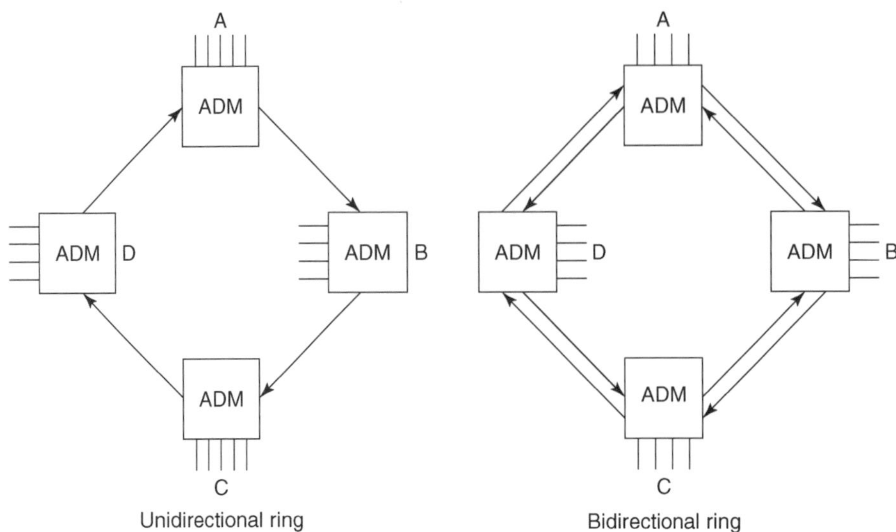

Figure 9.19. Ring definitions—working path direction. [From Figure 5-6, Ref. 1.]

A variety of names are used to describe path switched-ring functionality: for example, *unidirectional path-protection-switched* (*UPPS*) *rings*, *unidirectional path-switched rings* (*UPSR*), *uni-ring*, and *counterrotating rings*.

Ring architectures may be considered a class of their own, but we analyze a ring conceptually in terms of 1 + 1 protection. Usually, when we think of a ring architecture, we think of route diversity; and there are two separate directions of communications. The ring topology is most popular in the long-haul fiber-optics community. It offers what is called *geographical diversity*. Here we mean that there is sufficient ring diameter (e.g., > 10 miles) that there is an excellent statistical probability that at least one side of the ring will survive forest fires, large floods, hurricanes, earthquakes, and other force majeure events. It also means that only one side of the ring will suffer an ordinary nominal equipment failure or "backhoe fade."

There are some forms of ring topology used in CATV HFC systems, but more for achieving efficient and cost-effective connectivity than as a means of survivability. Rings are not used in building and campus fiber routings.

There are two basic SONET self-healing ring (SHR) architectures: the unidirectional and the bidirectional. Depending on the traffic demand pattern and some other factors, some ring types may be better suited to an application than others.

In a unidirectional SHR, shown on the left side of Figure 9.19, working traffic is carried around the ring in one direction only. For example, traffic going from node A to node D would traverse the ring in a clockwise direction, and traffic going from node D to node A would also traverse the ring in a clockwise direction. The capacity of an unidirectional ring is determined by the total traffic demands between any two node pairs on the ring.

In a bidirectional SHR (shown on the right in Figure 9.19), working traffic is carried around the ring in both directions, using two parallel paths between nodes (e.g., sharing the same fiber cable sheath). Using a similar example as above, traffic going from node A to node D would traverse the ring in a clockwise direction through intermediate nodes B and C, and traffic going from node D to node A would return along the same path also going through intermediate nodes B and C.

In a bidirectional ring, traffic in both directions of transmission between two nodes traverse the same set of nodes. Thus, unlike a unidirectional ring time slot, a bidirectional ring time slot can be reused several times on the same ring, allowing better utilization of capacity. All the nodes on the ring share the protection bit-rate capacity, regardless the number of times the time slot has been reused. Bidirectional routing is also useful on large rings, where propagation delay can be a consideration, because it provides a mechanism for ensuring that the shortest path is used (under normal conditions) against failures affecting both working and protection paths as well as node failures.

Telcordia points out in Ref. 6 that the term *unidirectional* has a dual meaning as follows: ion. Therefore, switching only affects one (uni) direction of a bidirectional circuit. Because of this, the various nodes through which any paths affected by a fiber cut or other failure pass do not need to communicate with each other. This makes UPSR architecture much simpler than a bidirectional ring.

Note: The term "unidirectional" in the UPSR refers to the traffic direction around the ring and should not be confused with the fact that UPSR is designed for *bi*directional symmetric transmission.

9.3 SYNCHRONOUS DIGITAL HIERARCHY (SDH)

9.3.1 Introduction

SDH resembles SONET in most respects. It uses different terminology, often for the same function as SONET. It is behind SONET in maturity by several years. History tells us that SDH will be more pervasive worldwide than SONET and is/will be employed in all countries using an E1-based PDH (plesiochronous digital hierarchy).

9.3.2 SDH Standard Bit Rates

SDH bit rates are built upon the basic rate of STM-1 (synchronous transport module 1) of 155.520 Mbps. Higher-capacity STMs are formed at rates equivalent to N times this basic rate. STM capacities for $N = 4$, $N = 16$, and $N = 64$ are defined. Table 9.2 shows the bit rates presently available for SDH (Rec. G.707 Ref. 5) with its SONET equivalents. The basic SDH multiplexing structure is shown in Figure 9.2.

TABLE 9.2 SDH Bit Rates

SDH Level	Equivalent SONET Level	Bit Rate (kbps)
1	STS-3 / OC-3	155,520
4	STS-12 / OC-12	622,080
16	STS-48 / OC-48	2,488,320
64	STS-192 / OC-192	9,953,280

Source: Based on data from Ref. 5.

9.3.3 Definitions

Synchronous Transport Module (STM). An STM is the information structure used to support section layer connections in the SDH. It consists of information payload and section overhead (SOH) information fields organized in a block frame structure that repeats every 125 μs. The information is suitably conditioned for serial transmission on the selected media at a rate that is synchronized to the network. As mentioned above, a basic STM is defined at 155.520 Mbps. This is termed STM-1. Higher-capacity STMs are formed at rates equivalent to N times the basic rate. STM capacities are currently defined for $N = 4$, $N = 16$, and $N = 64$. Figure 9.20 shows the frame structure for STM-1.

The STM-1 comprises a single administrative unit group (AUG) together with the section overhead (SOH). The STM-N contains N AUGs together with SOH. Figure 9.21 illustrates an STM-N.

Virtual Container-n (VC-n). A virtual container is the information structure used to support path layer connections in the SDH. It consists of information payload and path overhead (POH) information fields organized in a block frame structure

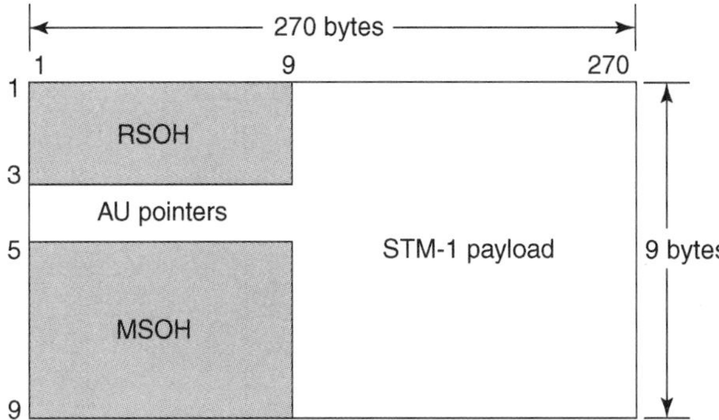

Figure 9.20. STM-1 frame structure. RSOH, regenerator section overhead; MSOH, multiplex section overhead.

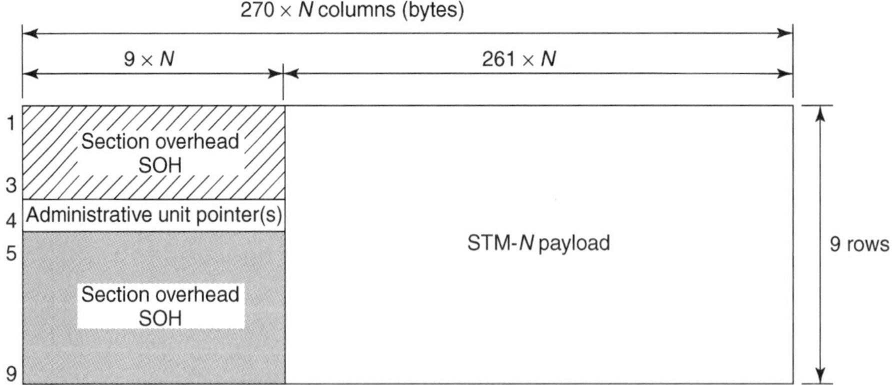

Figure 9.21. STM-N frame structure.

which repeats every 125 or 500 μs. Alignment information to identify VC-n frame start is provided by the server network.

Two types of virtual containers have been identified:

1. Lower-order virtual container-n: VC-n ($n = 1, 2, 3$). This element comprises a single container-n ($n = 1, 2, 3$) plus the lower-order virtual container POH appropriate to that level.
2. Higher-order virtual container-n: VC-n (n-3, 4). This element comprises either a single container-n ($n = 3$) or an assembly of tributary groups (TUG-2s or TUG-3s), together with virtual container POH appropriate to that level.

Administrative Unit-n (AU-n). An administrative unit is the information structure which provides adaptation between the higher-order path layer and the multiplex section layer. It consists of an information payload (the higher-order virtual container) and an administrative unit pointer which indicates the offset of the payload frame start relative to the multiplex section frame start.

Two administrative units are defined. The AU-4 consists of a VC-4 plus an administrative unit pointer that indicates the phase alignment of the VC-4 with respect to the STM-N frame. The AU-3 consists of a VC-3 plus an administrative unit pointer that indicates the phase alignment of the VC-3 with respect to the STM-N frame. In each case the administrative unit pointer location is fixed with respect to the STM-N frame.

One or more administrative units occupying fixed, defined positions in an STM payload are termed an administrative unit group (AUG). An AUG consists of a homogeneous assembly of AU-3s or an AU-4.

Tributary Unit-n (TU-n). A tributary unit is an information structure that provides adaptation between the lower-order path layer and the higher-order path layer. It consists of an information payload (the lower-order virtual container) and a tributary unit pointer that indicates the offset of the payload frame start relative to the higher-order virtual container frame start.

The TU-n ($n = 2, 2, 3$) consists of a VC-n together with a tributary unit pointer.

One or more tributary units, occupying fixed defined positions in a higher-order VC-n payload is termed a tributary unit group (TUG). TUGs are defined in such a way that mixed-capacity payloads made up of different-size tributary units can be constructed to increase flexibility of the transport network.

A TUG-2 consists of a homogeneous assembly of identical TU-1s or a TU-2.

A TUG-3 consists of a homogeneous assembly of TUG-2s or a TU-3.

Container-n ($n = 1$–4). A container is the information structure that forms the network synchronous information payload for a virtual container. For each of the defined virtual containers there is a corresponding container. Adaptation functions have been defined for many common network rates into a limited number of standard containers. These include all of the rates defined in CCITT Rec. G.702.

Pointer. A pointer is an indicator whose value defines the frame offset of a virtual container with respect to the frame reference of the transport entity which is supported.

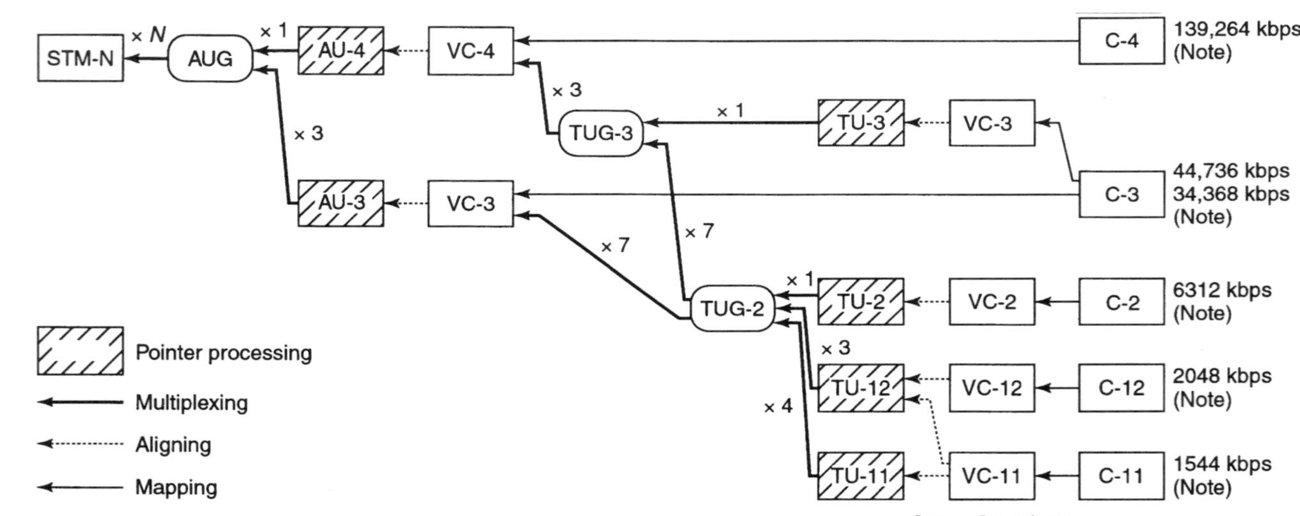

Note: G.702 tributaries associated with containers C-*x* are shown. Other signals, e.g. ATM, can also be accommodated.

Figure 9.22. Multiplexing structure overview. (From Figure 6-1/G.707, page 6, ITU-T Rec. G.707, Ref. 13.)

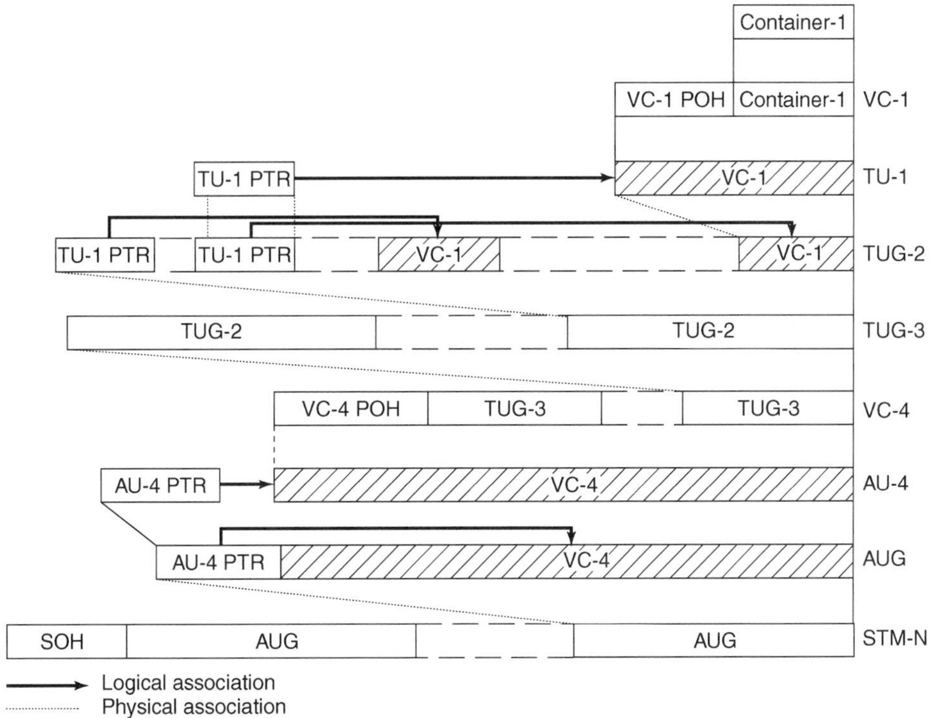

Figure 9.23. Multiplexing method directly from Container-1 using AU-4. (From Figure 6-2/G.707, page 7, ITU-T Rec. G.707, Ref. 13.)

9.3.3.1 Conventions. The order of transmission of information in all diagrams and figures in this section (9.3) is first from left to right and then top to bottom. Within each byte (octet) the most significant bit is transmitted first. The most significant bit (bit 1) is illustrated at the left in all the diagrams, figures, and tables in this section.

9.3.4 Basic SDH Multiplexing

Figure 9.22 illustrates the relationship between various multiplexing elements that are defined below. It also shows common multiplexing structures.

Figures 9.23–9.27 illustrate examples of various signal formations that are multiplexed using the multiplexing elements shown in Figure 9.22.

9.3.4.1 Administrative Units in the STM-N. The STM-*N* payload can support *N* AUGs where each AUG may consist of

- One AU-4 or
- Three AU-3s

The VC-*n* associated with each AU-*n* does not have a fixed phase with respect to the STM-*N* frame. The location of the first byte of the VC-*n* is indicated by the

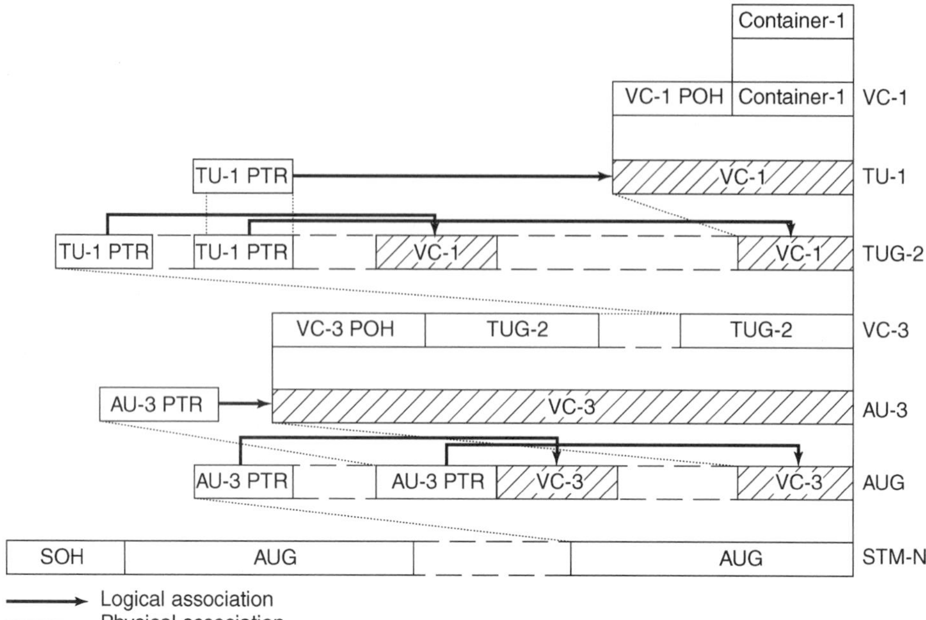

Figure 9.24. Multiplexing method directly from container-1 using AU-3. (From Figure 6-3/G.707, page 7, ITU-T Rec. G.707, Ref. 13.)

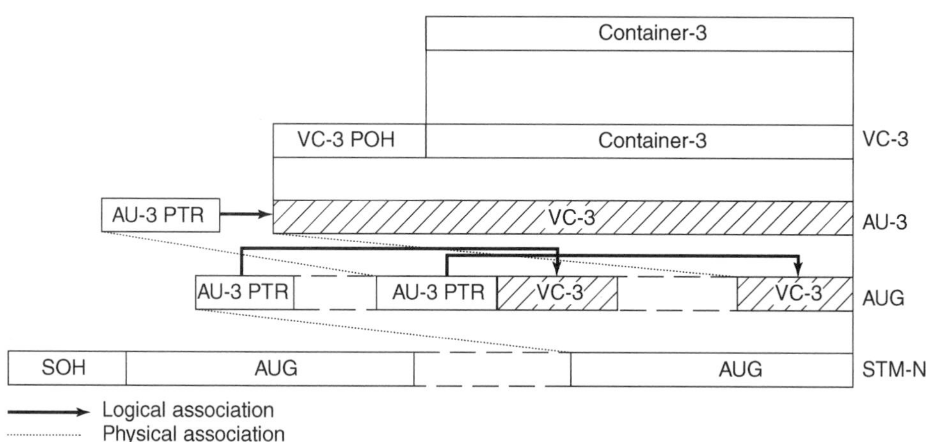

Figure 9.25. Multiplexing method directly from container-3 using AU-3. (From Figure 6-4/G.707, page 9, ITU-T Rec. G.707, Ref. 13.)

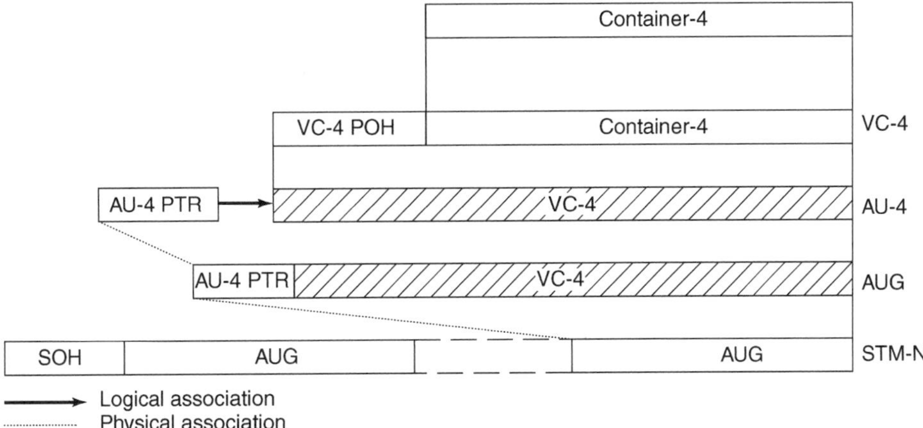

Figure 9.26. Multiplexing method directly from container-4 using AU-4. (From Figure 6-5/G.707, page 10, ITU-T Rec. G.707, Ref. 13.)

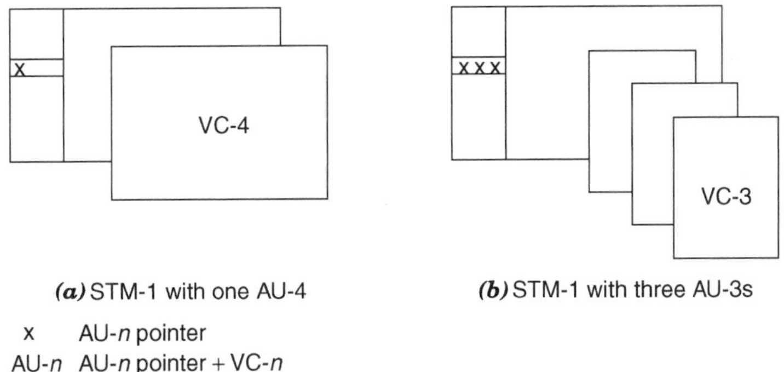

Figure 9.27. Administrative units in an STM-1 frame. (a) STM-1 with one AU-4. (b) STM-1 with three AU3s. (From Figure 6-7/G.707, page 12, ITU-T Rec. G707, Ref. 13.)

AU-*n* pointer. The AU-*n* pointer is in a fixed location in the STM-*N* frame. Examples are illustrated in Figures 9.21 and 9.23–9.28.

The AU-4 may be used to carry, via the VC-4, a number of TU-*n*s ($n = 1, 2, 3$) forming a two-stage multiplex. An example of this arrangement is illustrated in Figures 9.27a and 9.28a. The VC-*n* associated with each TU-*n* does not have a fixed phase relationship with respect to the start of the VC-4. The TU-*n* pointer is in a fixed location in the VC-4, and the location of the first byte of the VC-*n* is indicated by the TU-*n* pointer.

The AU-3 may be used to carry, via the VC-3, a number of TU-*n*s ($n = 1, 2$) forming a two-stage multiplex. An example of this arrangement is illustrated in Figures 9.27b and 9.28b. The VC-*n* associated with each TU-*n* does not have a fixed phase relationship with respect to the start of the VC-3. The TU-*n* pointer is in a fixed location in the VC-3 and the location of the first byte of the VC-*n* is indicated by the TU-*n* pointer.

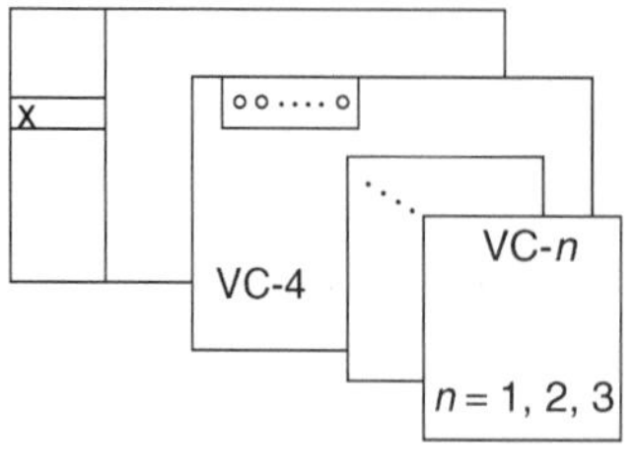

(a) STM-1 with one AU-4 containing Tus

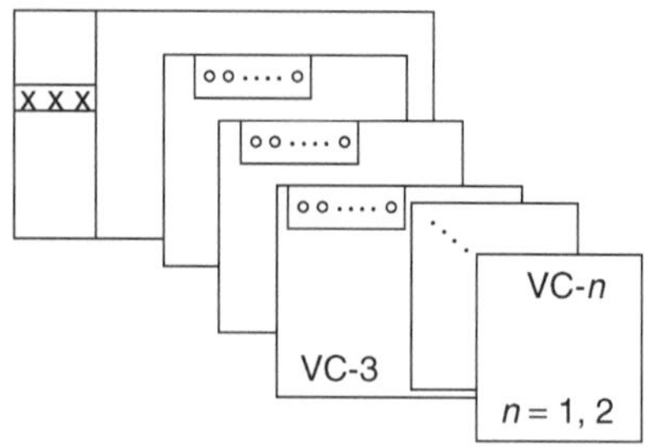

(b) STM-1 with three AU-3s containing TUs

x AU-*n* pointer
o TU-*n* pointer
AU-*n* AU-*n* pointer + VC-*n*
TU-*n* TU-*n* pointer + VC-*n*

Figure 9.28. Two-stage multiplex. (a) STM-1 with one AU-4 containing TUs. (b) STM-1 with three AU-3s containing TUs. (From Figure 6-8/G.707, page 12, ITU-T Rec. G.707, Ref. 13.)

9.3.4.2 *Interconnection of STM-Ns.* SDH is designed to be universal, allowing the transport of a large variety of signals including those specified in CCITT Rec. G.702. However, different structures can be used for the transport of virtual containers. The following two interconnection rules will be used (Ref. G707):

1. The rule for interconnecting two AUGs based upon two different types of administrative unit, namely AU-4 and AU-3, will be to use the AU-4 structure. Therefore, the AUG based upon AU-3 will be demultiplexed to the VC-3 or TUG-2 level according to the type of payload, and remultiplexed within an AUG via the TUG-3/VC-4/AU-4 route. This is illustrated in Figure 9.29a and 9.29b.
2. The rule for interconnecting VC-11s transported via different types of tributary unit, namely TU-11 and TU-12, will be to use the TU-11 structure. This is illustrated in Figure 9.29c. VC-11, TU-11, and TU-12 are described below. This SDH interconnection rule does not modify the interworking rules defined in ITU-T Rec. G.802 for networks based on different PDHs and speech encoding laws.

9.3.4.3 *Scrambling.* Scrambling ensures sufficient bit timing content (transitions) at the NNI to maintain synchronization and alignment. Figure 9.30 is a functional block diagram of the frame synchronous scrambler. The generating polynomial for the scrambler is $1 + X^6 + X^7$.

9.3.5 Frame Structure for 51.840-Mbps Interface

Low/medium-capacity SDH transmission systems based on radio and satellite technologies that are not designed for the transmission of STM-1 signals may operate at a bit rate of 51.840 Mbps across digital sections. However, this bit rate does not represent a level of the SDH or an NNI bit rate (Ref. 13).

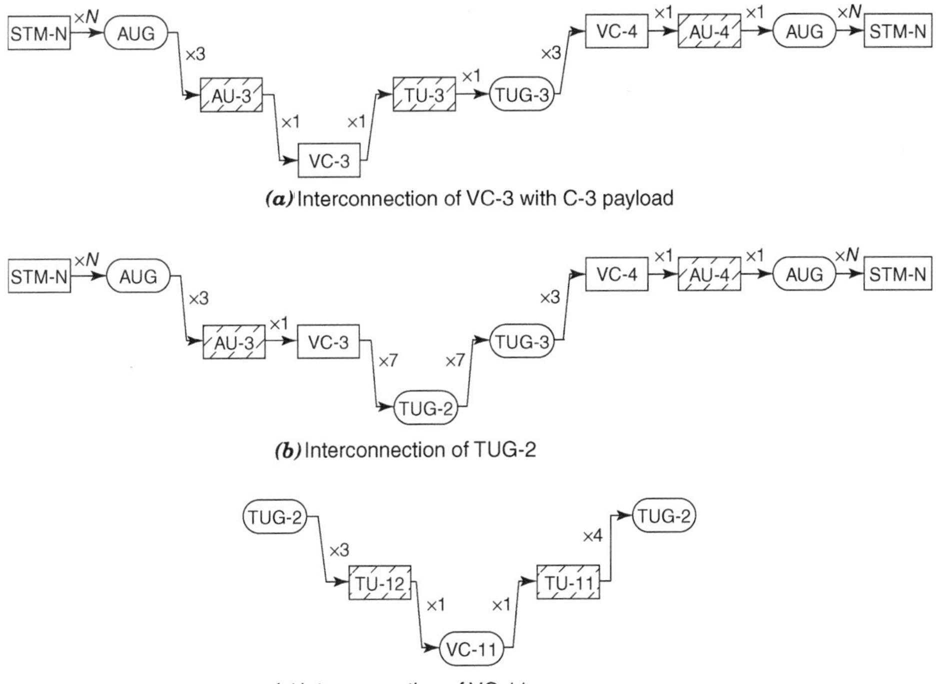

Figure 9.29. Interconnection of STM-*N*s. (a) Interconnection of VC-3 with C-3 payload. (b) Interconnection of TUG-2. (c) Interconnection of VC-11. (From Figure 6-9/G.707, page 16, ITU-T Rec. G.707, Ref. 13.)

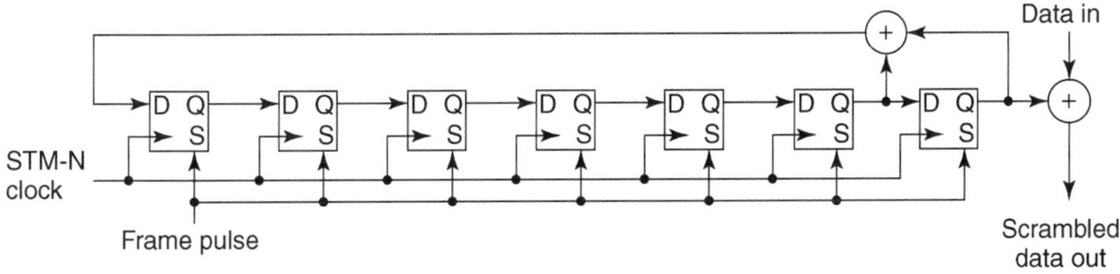

Figure 9.30. Functional block diagram of a frame-synchronous scrambler. (From Figure 6-10/G.707, page 17, ITU-T Rec. G.707, Ref. 13.)

The recommended frame structure for 51.840 Mbps signal for satellite (ITU-R Rec. S.1149) and LOS microwave application (ITU-R Rec. F.750) is shown in Figure 9.31.

9.3.6 SDH Multiplexing Methods

9.3.6.1 Multiplexing of Administrative Units into STM-N

Multiplexing of Administrative Unit Groups (AUGs) into an STM-N. The arrangement of *N* AUGs multiplexed into an STM-*N* is illustrated in Figure 9.32. The AUG is a structure of 9 rows by 261 columns plus 9 bytes in row 4 (for the AU-*n*

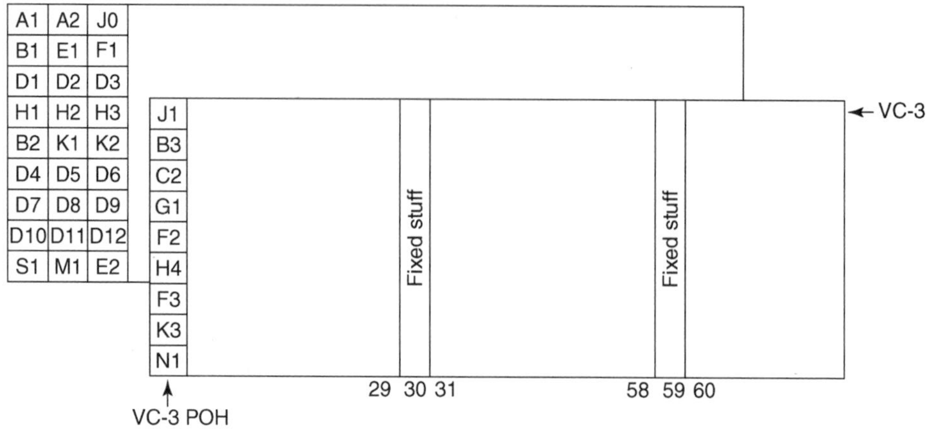

Notes

1 M1 position is not the same position (9, 3N+3) as in a STM-N frame.
2 Fixed stuff columns are not part of the VC-3.

Figure 9.31. Frame structure for 51.840 Mbps (SDH) operation. (From Figure A.1/G.707, page 89, ITU-T Rec. G.707, Ref. 13.)

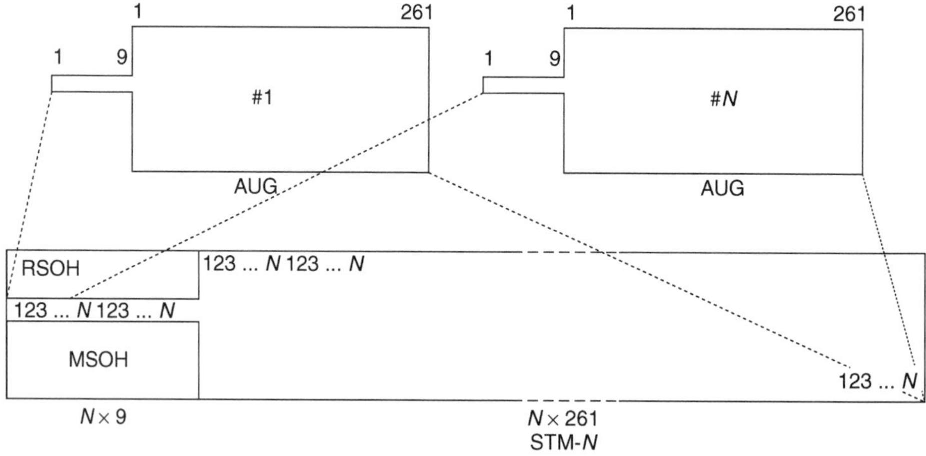

Figure 9.32. Multiplexing of *N* AUGs into an STM-*N*. (From Figure 7-1/G.707, page 18, ITU-T Rec. G.707, Ref. 13.)

pointers). The STM-*N* consists of an SOH described below and a structure of 9 rows by $N \times 261$ columns with $N \times 9$ bytes in row 4 (for the AU-*n* pointers). The *N* AUGs are one-byte-interleaved into this structure and have a fixed phase relationship with respect to the STM-*N*.

Multiplexing of an AU-4 via AUG. The multiplexing arrangement of a single AU-4 via the AUG is illustrated in Figure 9.33. The 9 bytes at the beginning of row 4 are assigned to the AU-4 pointer. The remaining 9 rows by 261 columns is allocated to the virtual container-4 (VC-4). The phase of the VC-4 is not fixed with respect

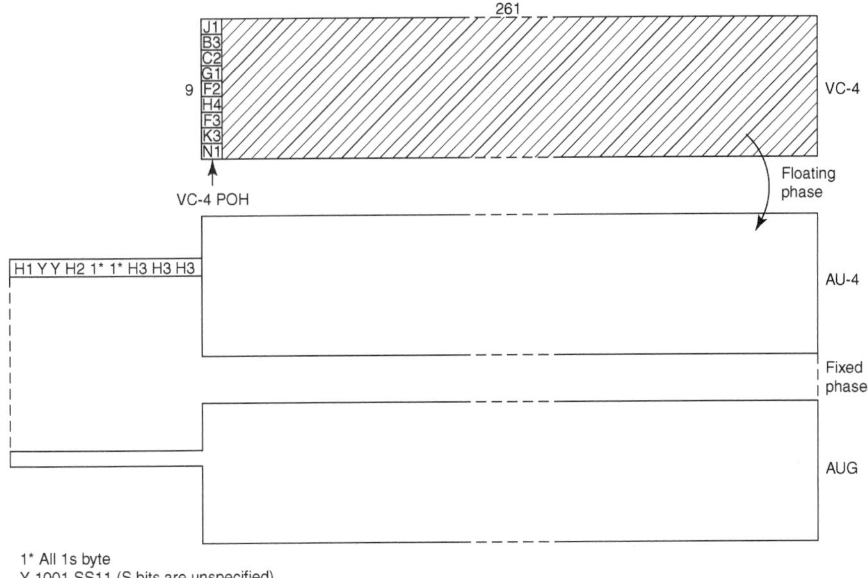

Figure 9.33. Multiplexing of AU-4 via AUG. (From Figure 7-2/G.707, page 19, ITU-T Rec. G.707, Ref. 13.)

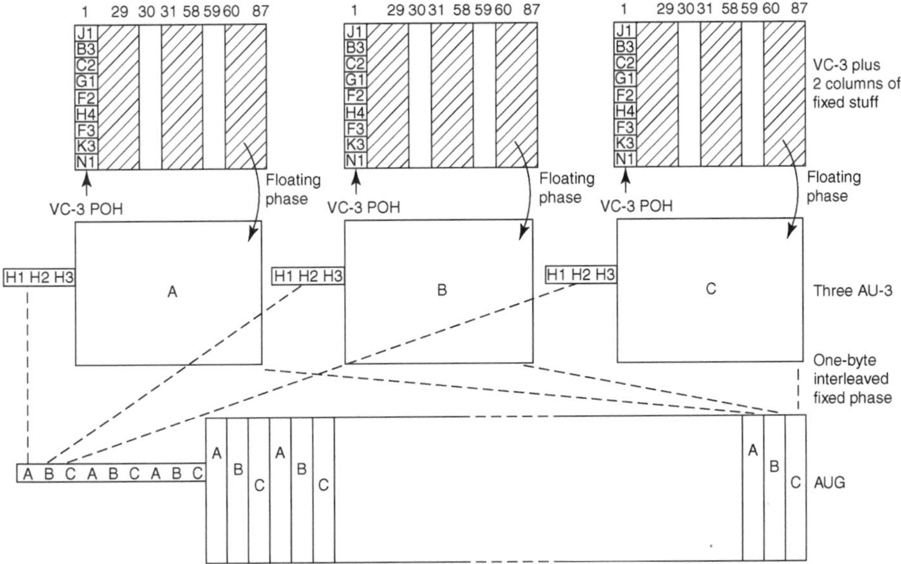

Note: The byte in each row of the two columns of fixed stuff of each AU-3 shall be the same.

Figure 9.34. Multiplexing of AU-3s via AUG. (From Figure 7-3/G.707, page 20, ITU-T Rec. G.707, Ref. 13.)

to the AU-4. The location of the first byte of the VC-4 with respect to the AU-4 pointer is given by the pointer value. The AU-4 is placed directly in the AUG.

Multiplexing of AU-3s via AUG. The multiplexing arrangement of three AU-3s via the AUG is shown in Figure 9.34. The 3 bytes at the beginning of row 4 are assigned to the AU-3 pointer. The remaining 9 rows by 87 columns is allocated to the VC-3 and two columns of fixed stuff. The byte in each row of the two columns of fixed stuff of each AU-3 shall be the same. The phase of the VC-3 and the two columns of fixed stuff is not fixed with respect to the AU-3. The location of the first byte of the VC-3 with respect to the AU-3 pointer is given by the pointer value. The three AU-3s are one-byte-interleaved in the AUG.

9.3.6.2 Multiplexing of Tributary Units into VC-4 and VC-3

Multiplexing of Tributary Unit Group-3s (TUG-3s) into a VC-4. The arrangement of three TUGs multiplexed in the VC-4 is illustrated in Figure 9.35. The TUG-3 is a 9-row by 86-column structure. The VC-4 consists of one column of VC-4 POH, two columns of fixed stuff, and a 258-column payload structure. The three TUG-3s are single-byte-interleaved into the 9-row by 258-column VC-4 payload structure and have a fixed phase with respect to the VC-4. As described in the previous subsection on multiplexing of AUGs into an STM-*N*, the phase of the VC-4 with respect to the AU-4 is given by the AU-4 pointer.

Multiplexing of a TU-3 via a TUG-3. The multiplexing of a single TU-3 via the TUG-3 is shown in Figure 9.36. The TU-3 consists of the VC-3 with a 9-byte VC-3 POH and the TU-3 pointer. The first column of the 9-row by 86-column TUG-3 is assigned to the TU-3 pointer (bytes H1, H2, H3) and fixed stuff. The phase of the VC-3 with respect to the TUG-3 is indicated by the TU-3 pointer.

Multiplexing of TUG-2s via a TUG-3. The multiplexing format for the TUG-2 via the TUG-3 is illustrated in Figure 9.37. The TUG-3 is a 9-row by 86-column structure with the first two columns of fixed stuff.

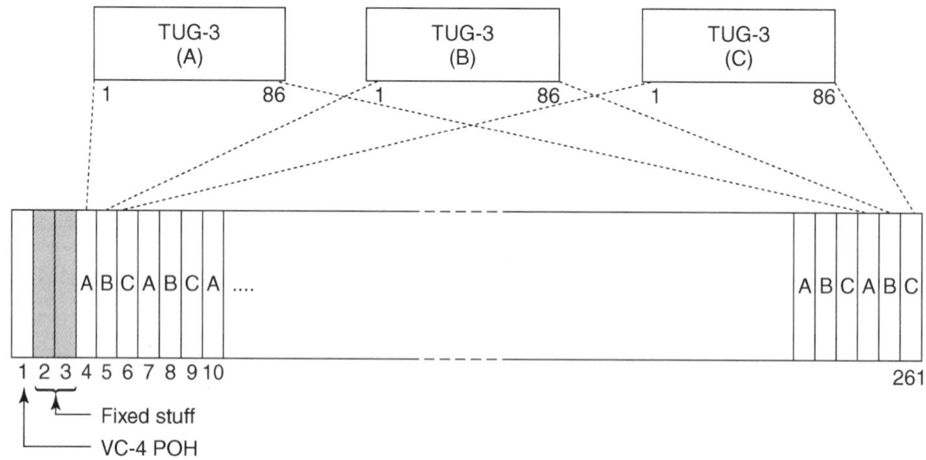

Figure 9.35. Multiplexing of three TUG-3s into a VC-4. (From Figure 7-4/G.707, page 21, Ref. 13.)

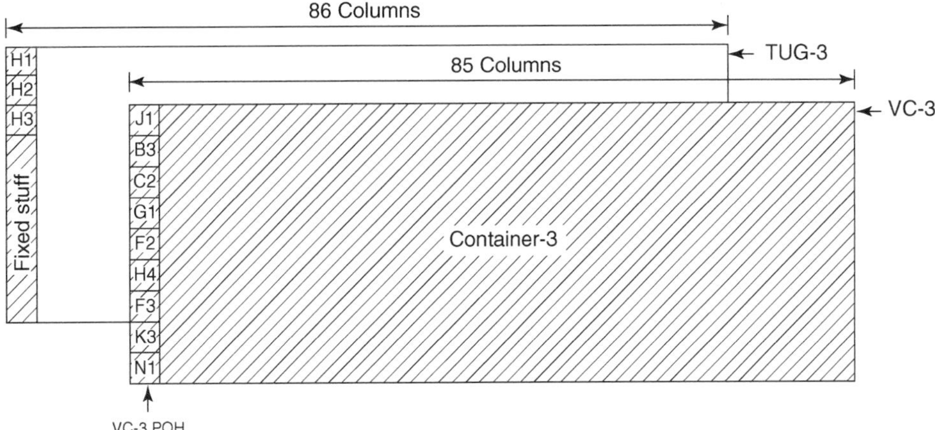

Figure 9.36. Multiplexing a TU-3 via a TUG-3. (From Figure 7-5/G.707, page 21, ITU-T Rec. G.707, Ref. 13.)

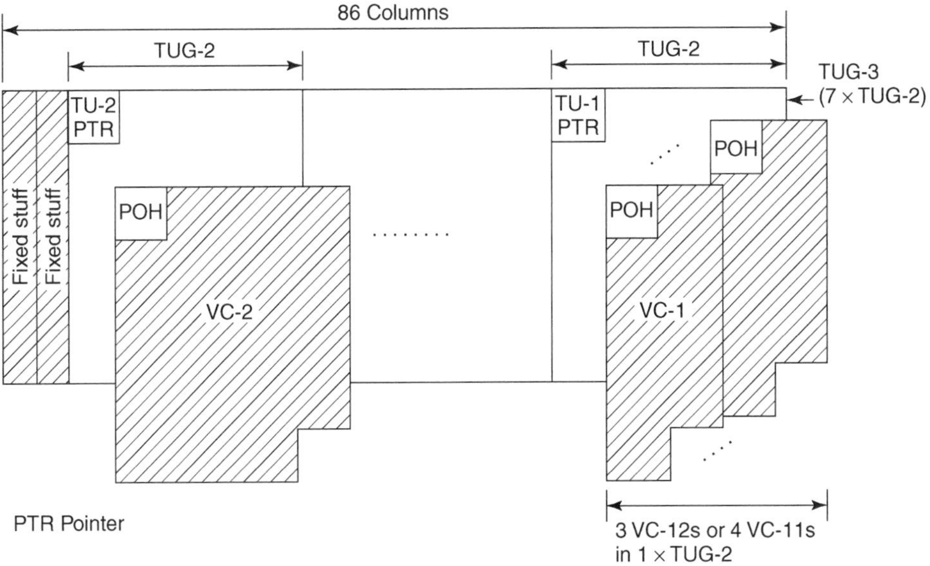

Figure 9.37. Multiplexing of TUG-2s via a TUG-3. (From Figure 7-6/G.707, page 22, ITU-T Rec. G.707, Ref. 13.)

Multiplexing of TUG-2s into a VC-3. The multiplexing structure for TUG-2s into a VC-3 is shown in Figure 9.38. The VC-3 consists of VC-3 POH and a 9-row by 84-column payload structure. A group of seven TUG-2s can be multiplexed into the VC-3.

9.3.7 Pointers

9.3.7.1 AU-n Pointer. The AU-*n* pointer provides a method of allowing flexible and dynamic alignment of the VC-*n* within the AU-*n* frame. Dynamic alignment means that the VC-*n* is allowed to "float" within the AU-*n* frame. Thus, the

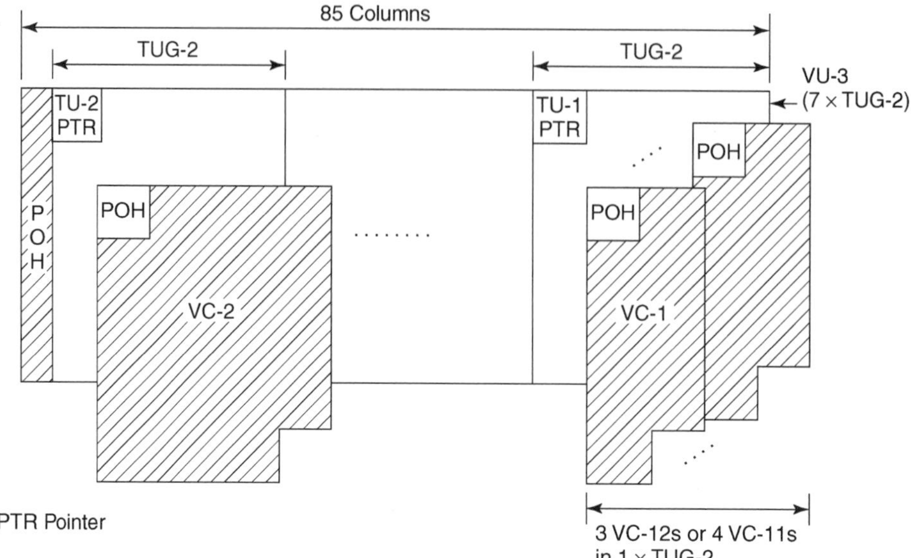

Figure 9.38. Multiplexing seven TUG-2s into a VC-3. (From Figure 7-8/G.707, page 24, ITU-T Rec. G.707, Ref. 13.)

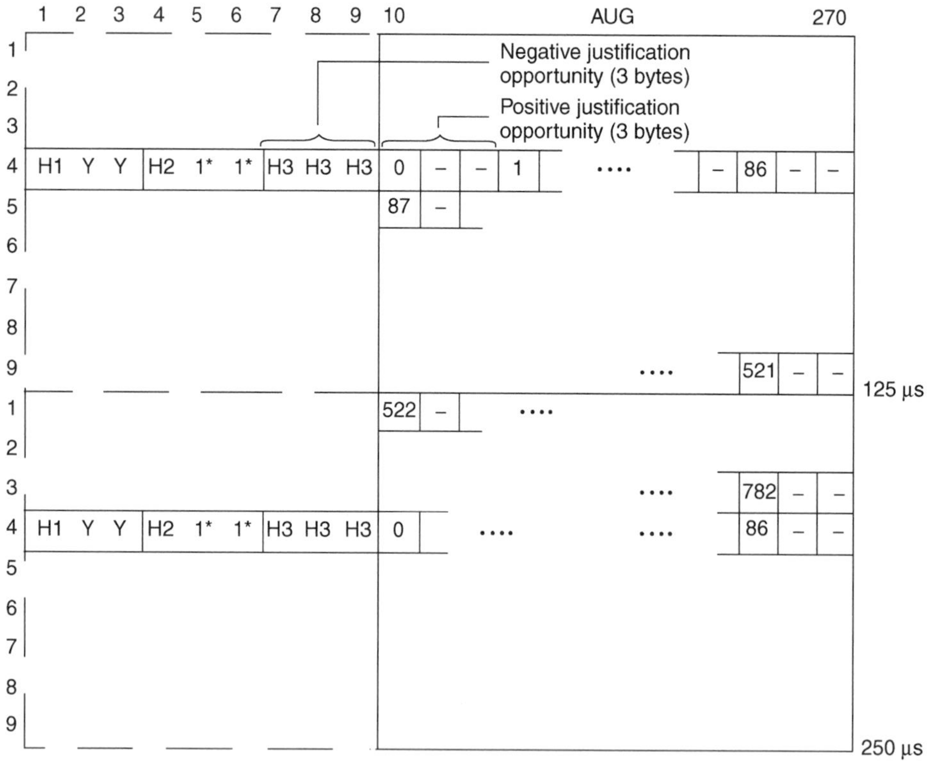

Figure 9.39. AU-4 pointer offset numbering. (From Figure 8-1/G.707, page 34, ITU-T Rec. G.707, Ref. 13.).

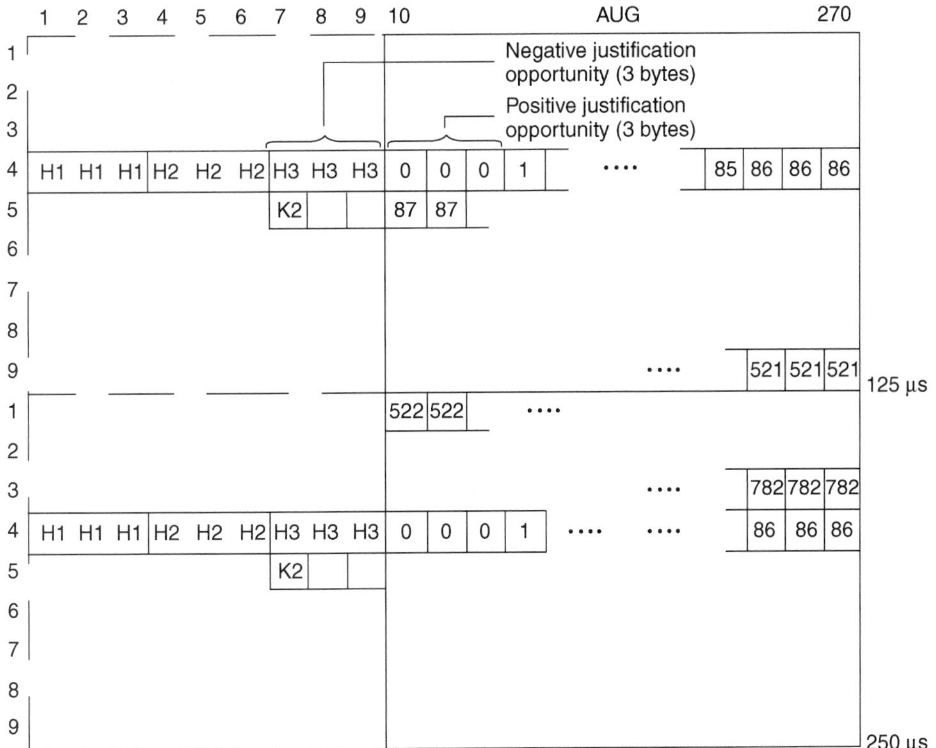

Figure 9.40. AU-3 pointer offset numbering. (From Figure 8-2/G.707, page 325, ITU-T Rec. G.707, Ref. 13.)

pointer is able to accommodate differences, not only in the phases of the VC-*n* and the SOH but also in the frame rates.

AU-n Pointer Location. The AU-4 pointer is contained in bytes H1, H2, and H3 as shown in Figure 9.39. The three individual AU-3 pointers are contained in three separate H1, H2, and H3 bytes as shown in Figure 9.40.

AU-n Pointer Value. The pointer contained in H1 and H2 designates the location of the byte where the VC-*n* begins. The two bytes allocated to the pointer function should be viewed as one word, as illustrated in Figure 9.41. The last 10 bits (bits 7–16) of the pointer word carry the pointer value.

As illustrated in Figure 9.41, the AU-4 pointer value is a binary number with a range of 0 to 782 which indicates the offset in three-byte increments, between the pointer and the first byte of the VC-4 (see Figure 9.42). Figure 9.41 also indicates one additional valid pointer, the concatenation indication. The concatenation indication is designated by binary 1001 in bits 1–4, bits 5–6 are unspecified, and by 10 binary 1s in bit positions 7–16. The AU-4 pointer is set to concatenation indication for AU-4 concatenation.

As shown in Figure 9.41, the AU-3 pointer value is also a binary number with a range of 0 to 782. Because there are three AU-3s in the AUG, each AU-3 has its own associated H1, H2, and H3 bytes.

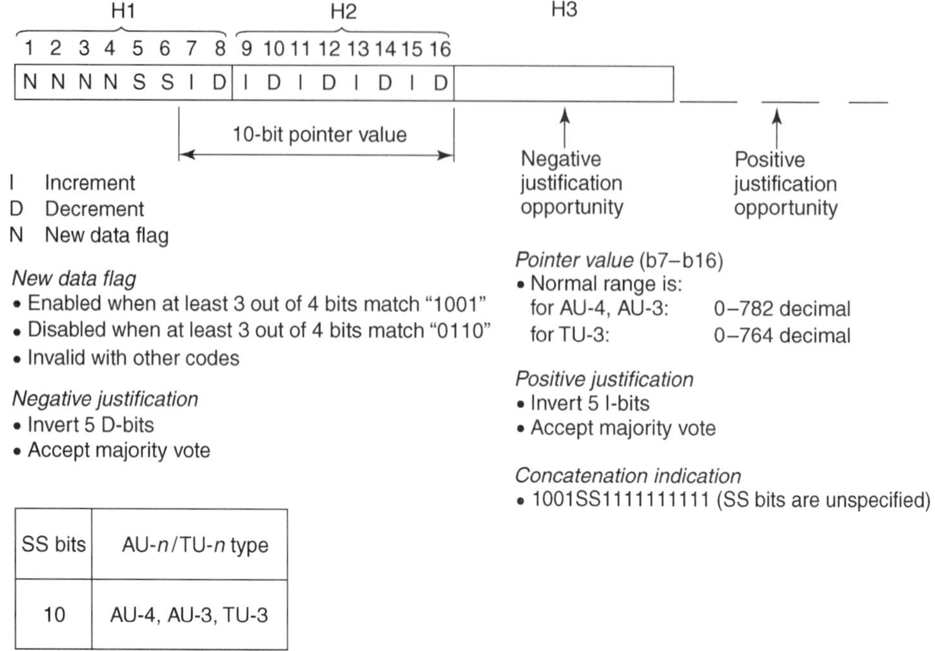

SS bits	AU-n/TU-n type
10	AU-4, AU-3, TU-3

Note: The pointer is set to all "1"s when AIS occurs.

Figure 9.41. AU-*n*/TU-3 pointer (H1, H2, H3) coding. (From Figure 8-3/G.707, page 36, Ref. 13.)

Note that the H bytes are shown in sequence in Figure 9.39. The first H1, H2, H3 set refers to the first AU-3, the second set refers to the second AU-3, and so on. For the AU-3s, each pointer operates independently.

In all cases, the AU-*n* pointer bytes are not counted in the offset. For example, in an AU-4, the pointer value of 0 indicates that the VC-4 starts in the byte location that immediately follows the last H3 byte, whereas an offset of 87 indicates that the VC-4 starts three bytes after the K2 byte.

Frequency Justification. If there is a frequency offset between the frame rate of the AUG and that of the VC-*n*, then the pointer value will be incremented or decremented as needed, accompanied by a corresponding positive or negative justification byte or bytes. Consecutive pointer operations must be separated by at least three frames (i.e., every fourth frame) in which the pointer value remains constant. If the frame rate of the VC-*n* is too slow with respect to that of the AUG, then the alignment of the VC-*n* must periodically slip back in time and the pointer value must be incremented by one. This operation is indicated by inverting bits 7, 9, 11, 13, and 15 (I-bits) of the pointer word to allow 5-bit majority voting at the receiver. Three positive justification bytes appear immediately after the last H3 byte in the AU-4 frame containing inverted I-bits. Subsequent pointers will contain the new offset. This is depicted in Figure 9.42.

For AU-3 frames, a positive justification byte appears immediately after the individual H3 byte of the AU-3 frame containing inverted I-bits. Subsequent pointers will contain the new offset.

If the frame rate of the VC-*n* is too fast with respect to that of the AUG, then the alignment of the VC-*n* must periodically be advanced in time and the pointer

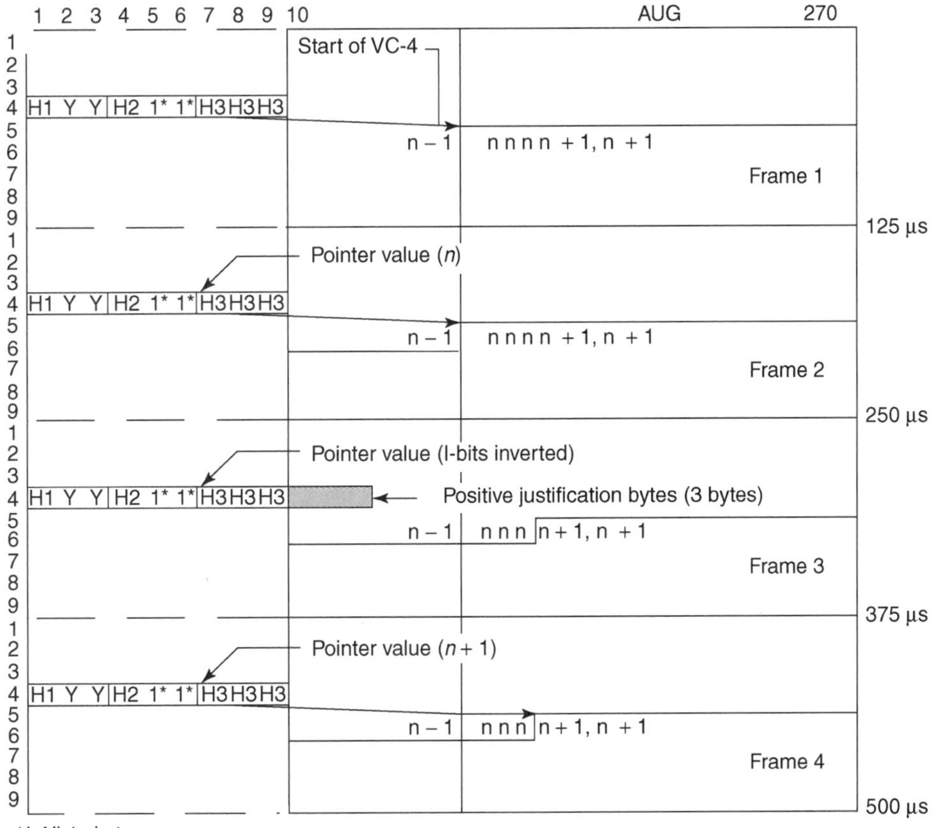

Figure 9.42. AU-4 pointer adjustment operation, positive justification. (From Figure 8-4/G.707, page 37, ITU-T Rec. G.707, Ref. 13.)

value must be decremented by one. This operation is indicated by inverting bits 8, 10, 12, 14, and 16 (D-bits) of the pointer word to allow 5-bit majority voting at the receiver. Three negative justification bytes appear in the H3 bytes in the AU-4 frame containing inverted D-bits. Subsequent pointers will contain the new offset.

For AU-3 frames, a negative justification byte appears in the individual H3 byte of the AU-3 frame containing inverted D-bits. Subsequent pointers will contain the new offset.

Pointer Generation. The following summarizes the rules for generating the AU-*n* pointers.

1. During normal operation, the pointer locates the start of the VC-*n* within the AU-*n* frame. The NDF is set to binary 0110. The NDF consists of the N-bits, bits 1–4 of the pointer word.
2. The pointer value can only be changed by operation 3, 4, or 5 (below).
3. If a positive justification is required, the current pointer value is sent with the I-bits inverted, and the subsequent positive justification opportunity is filled with dummy information. Subsequent pointers contain the previous pointer

value incremented by one. If the previous pointer is at its maximum value, the subsequent pointer is set to zero. No subsequent increment or decrement operation is allowed for at least three frames following this operation.

4. If a negative justification is required, the current pointer value is sent with the D-bits inverted and the subsequent negative justification opportunity is overwritten with actual data. Subsequent pointers contain the previous pointer value decremented by one. If the previous value is zero, the subsequent pointer is set to its maximum value. No subsequent increment or decrement operation is allowed for at least three frames following this operation.
5. If the alignment of the VC-*n* changes for any reason other than rules 3 or 4, the new pointer value should be sent accompanied by the NDF set to 1001. The NDF only appears in the first frame that contains the new values. The new location of the VC-*n* begins at the first occurrence of the offset indicated by new pointer. No subsequent increment or decrement operation is allowed for at least three frames following this operation.

Pointer Interpretation. The following summarizes the rules interpreting the AU-*n* pointers.

1. During normal operation, the pointer locates the start of the VC-*n* within the AU-*n* frame.
2. Any variation from the current pointer value is ignored unless a consistent new value is received three times consecutively or it is preceded by one of the rules 3, 4, or 5. Any consistent new value received three times consecutively overrides (i.e., takes priority over) rules 3 and 4.

TABLE 9.3 Summary of SDH/SONET Payloads and Mappings[a]

	SDH				SONET			
Payload	Container	Actual Payload Capacity	Payload and POH	Mapping AU-3/AU-4 Based	Container SPE	Actual Payload Capacity	Payload and POH	Mapping
DS1 (1.544)	VC-11	1.648	1.728	(AU-3), AU-4	VT1.5	1.648	1.664	STS-1
	VC-12	2.224	2.304	AU-3, AU-4				
E1 (2.048)	VC-12	2.224	2.304	(AU-3), AU-4	VT2	2.224	2.240	STS-1
DS1C (3.152)					VT3	3.376	3.392	STS-1
DS2 (6.312)	VC-2	6.832	6.912	(AU-3), AU-4	VT6	6.832	6.848	STS-1
E3 (34.368)	VC-3	48.384	48.960	AU-3, AU4				
DS3 (44.736)	VC-3	48.384	48.960	(AU-3), AU-4	STS-1	49.536	50.112	STS-1
E4 (139.264)	VC-4	149.760	150.336	(AU-4)	STS-3c	149.760	150.336	STS-3C
ATM (149.760)	VC-4	149.760	150.336	(AU-4)	STS-3c	149.760	150.336	STS-3C
ATM (599.040)	VC-4-4c	599.040	601.344	(AU-4)	STS-12c	599.040	601.344	STS-12c
FDDI (125.000)	VC-4	149.760	150.336	(AU-4)	STS-3c	149.760	150.336	STS-3c
DQDB (149.760)	VC-4	149.760	150.336	(AU-4)	STS-3c	149.760	150.336	STS-3C

[a]Numbers are in Mbp/s unit.

Source: Reprinted with permission from *SONET/SDH*, Table 2, page 64, Ref. 3.

TABLE 9.4 Overhead Summary, STM-1 and STS-3c

Framing A1	Framing A1	Framing A1	Framing A2	Framing A2	Framing A2	STS-1 1D C1	STS-1 1D C1	STS-1 1D C!		Trace J1	
BIP-8 B1			Order-wire E1			User F1			for STS-3c only, not included in STM-1	BIP-8 B3	
Data Comm D1			Data Comm D2			Data Comm D3				Sig Label C2	
Pointer H1	1001 ss11	1001 ss11	Pointer H2	1111 1111	1111 1111	Ptr Action H3	Ptr Action H3	Ptr Action H3		Path Stat G1	
BIP-8 B2	BIP-8 B2	BIP-8 B2	APS K1			APS K2				User F2	
Data Comm D4			Data Comm D5			Data Comm D6				Multi-frame H4	Path Overhead
Data Comm D7			Data Comm D8			Data Comm D9				Growth Z3	
Data Comm D10			Data Comm D11			Data Comm D12				Growth Z4	
Growth Z1	Growth Z1	Growth Z1	Growth Z2	Growth Z2	Growth Z2	Order-wire E2				Growth Z5	

Transport Overhead | Payload Capacity

Source: Reprinted with permission from *SONET/SDH*, Table 3, page 64, Ref. 3.

3. If the majority of the I-bits of the pointer word are inverted, a positive justi-fication operation is indicated. Subsequent pointer values shall be incremented by one.
4. If the majority of the D-bits of the pointer word are inverted, a negative justification operation is indicated. Subsequent pointer values shall be decremented by one.
5. If the NDF is interpreted as enabled, the coincident pointer value shall replace the current one at the offset indicated by the new pointer value unless the receiver is in a state that corresponds to a loss of pointer.

9.4 SONET/SDH SUMMARY

Table 9.3 summarizes SDH/SONET payloads and mappings. Table 9.4 reviews the various overhead for SDH STM-1 and SONET STS-3c.

REFERENCES

1. *Synchronous Optical Network (SONET)—Basic Description Including Multiplex Structure, Rates and Formats*, ANSI T1.105-1995, ANSI, New York, 1995.
2. Roger L. Freeman, *Reference Manual for Telecommunications Engineering*, 3rd ed., John Wiley & Sons, New York, 2001.
3. C. A. Siller and M. Shafi, eds., *SONET/SDH*, IEEE Press, New York, 1996.
4. Roger L. Freeman, *Telecommunication Transmission Handbook*, 4th ed., John Wiley & Sons, New York, 1998.
5. *Network–Node Interface for the Synchronous Digital Hierarchy (SDH)*, ITU-T Rec. G.707, ITU Geneva, March 1996.
6. *Synchronous Optical Network (SONET), Transport Systems, Common Generic Criteria*, Telecordia GR-253-CORE, Issue 3, Rev. 2, Piscataway, NJ, September 2000.
7. Introduction to SONET, Seminar, Hewlett-Packard Co., Burlington, MA, November 1993.
8. *SONET Add–Drop Multiplex Equipment (SONET ADM) Generic Criteria*, Bellcore, TR-TSY-000496, Issue 2, Bellcore, Piscataway, NJ, 1989.
9. *Automatic Protection Switching for SONET*, Telecordia Special Report SR-NWT-001756, Issue 1, Piscataway, NJ, October 1990.
10. *SONET Dual-Fed Unidirectional Path Switched Ring (UPSR) Equipment Generic Criteria*, Telcordia GR-1400-CORE, Issue 2, Piscataway, NJ, January 1999.
11. *SONET Bidirectional Line-Switched Ring Equipment Generic Criteria*, Telcordia GR-1230-CORE, Issue 4, Piscataway, NJ, December 1998.
12. *Telcordia Notes on the Synchronous Optical Network (SONET)*, Special Report, SR-NOTES, Series 01, Issue 1, Piscataway, NJ, December 1999.
13. *Network Node Interfaces for Synchronous Digital Hierarchy*, ITU-T Rec. G.707, ITU Geneva, March 1996.

10

LINK ENGINEERING OF LIGHTWAVE SYSTEMS

10.1 INTRODUCTION TO THE LINK BUDGET

The link budget on a fiber-optic link is very similar to link budget exercises the design engineer carries out for a loss-of-signal (LOS) microwave link, a troposcatter link, or a satellite link. It is a working aide to determine link parameters such as

- Light source output
- Fiber loss
- Connector, splice, and patch cord losses
- Penalties (see Section 4.5)
- Margin

A detector threshold value in dBm is based on the desired bit error rate (BER). In addition, we might have

- Amplifier net gain(s)
- Filter loss
- WDM passive losses (Ref. 11)
- Splitter loss
- Isolator loss

The handy decibel is employed almost exclusively with the link budget.

A link design is based on the link budget. Either we assign parameter values for the items listed above or we calculate a specific value. The objective is to have the most cost-effective design possible to meet our design objectives.

One of the first steps in carrying out a link budget exercise is to determine if the link in question is dispersion-limited or loss-limited. In general, when using G.653 (Ref. 1) or Corning LEAF fiber, links will be loss-limited up to about 1-Gbps transmission rate. One difference between a fiber-optic link budget and a radio/wireless system budget is the *rise-time budget*. The end point of this budget determines if a link is dispersion-limited. The budget also gives us clues on what we can do about it. For the first part of this chapter, we deal with budgets in the power domain, where a link is power-limited. In Section 10.4, we deal with the dispersion-limited problem calculating system rise time, where we are given or we estimate component rise times.

10.2 THE LINK MARGIN

Link margin may be defined as those extra decibels that are added to the link budget as a safety margin. It may well be a safety margin to compensate for the possibility that we did not estimate the link penalties correctly; we guessed on the low side, again to save money. Certain active components tend to have deteriorating performance as time passes (typically LEDs). Not all fusion splices will have extremely low losses. A link margin compensates for these shortfalls. We recommend a 6-dB margin on very long systems. Each decibel of margin costs the vendor of the system money. Many will lower this value to 4.8 or 3 dB. We would advise against this action.

ITU-T in Rec. G.957 (Ref. 2) allocates 2–4 dB to account for the end-of-life of equipment. It should be noted that reliability specialists [see Telecordia GR-468-CORE (Ref. 3) and TR-NWT-000357 (Ref. 4)] give equipment three periods in a life span:

1. Infant, where we can expect infant mortality
2. Operational, where equipment meets performance and reliability requirements.
3. End-of-life, where equipment begins to wear out and we cannot expect it to meet performance requirements

One can assume to read a parenthetical note on end-life degradation—namely, that in all probability the equipment in question should be replaced well before end-of-life because it has become obsolete.

ITU-T Rec. G.957 (Ref. 2) states:

> "Attenuation specifications are assumed to be worst-case values including losses due to splices, connectors, optical attenuators (if used) or other passive optical devices, and any additional cable margin to cover allowances for:
>
> (1) future modifications to the cable configuration (additional splices, increased cable lengths, etc.);
> (2) fiber cable performance variations due to environmental factors; and
> (3) degradation of any connector, optical attenuators (if used) or other passive optical devices between points S and R, when provided."

ITU-T Rec. G.957 defines point S as a reference point on the optical fiber just after the transmitter connector (C_{TX}), and point R is a reference point on the optical fiber just before the receiver optical connector (C_{RX}). These ITU fiber-optic system reference points are illustrated in Figure 10.1.

This ITU recommendation defines three types of fiber-optic links based on their length as shown in Table 10.1.

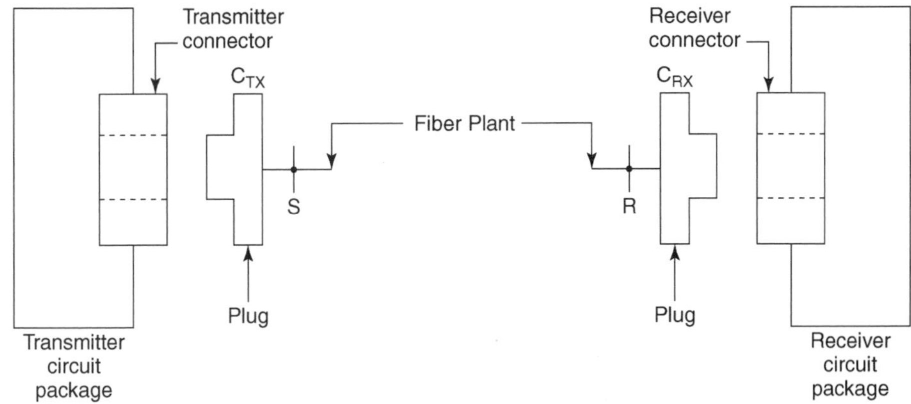

Figure 10.1. Reference fiber-optic link drawing showing reference points S (send) and R (receive) and the connector identifiers C_{TX} and C_{RX}. (From ITU-T Rec. 957, Figure 1/G.957, page 8, Ref. 2.)

TABLE 10.1 Link Types, Distances, Wavelengths, and Margins

Type	Intra-Office	Short-Haul (1)	Short-Haul (2)	Long-Haul (1)	Long-Haul (2)	Long-Haul (3)
Wavelength	1310 nm	1310 nm	1550 nm	1310 nm	1550 nm	1550 nm
Fiber type	Single mode	Single mode	Single mode	Single mode	Single mode and cutoff-shifted single mode	Dispersion-shifted single mode
Distance (km)	≤ 2	~ 15	~ 15	~ 40	~ 80	~ 80
Application code [a]	I-1	S-1.1	S-1.2	L-1.1	L-1.2	L-1.3
155-Mbps threshold [b]	−23 dBm [c]	−28 dBm	−28 dBm	−28 dBm	−28 dBm	−28 dBm
622-Mbps threshold [b]	−23 dBm	−28 dBm	−28 dBm	−28 dBm	−28 dBm	−28 dBm
2.5-Gbps threshold [b]	−18 dBm	−18 dBm	−18 dBm	−27 dBm	−28 dBm	−28 dBm
Attenuation range (dB)	0–7	0–12 dB	0–12 dB	10–28 dB	10–28 dB	10–28 dB
Receiver margin (dB)	3	3	3	4.0	4.8	4.8

[a] ITU-T application code typically ITU-T Recs. G.957 and G.662.

[b] *Threshold* refers to the detector threshold at reference point R. This threshold is set for a BER of 1×10^{-10}. For a BER of 1×10^{-12} the threshold will be 1 dB less sensitive, or algebraically add 1 dB. Thus a −23-dBm threshold becomes a −22-dBm threshold.

[c] Requires attenuator or reduced light source output to avoid receiver overload.

Sources: Based on Tables 2–4, ITU-T Rec. G.957 (Ref. 2) and from the *Telecommunication Transmission Handbook* (Ref. 5).

10.2.1 Check Tables

Table 10.2 through 10.4 are included so that the reader may cross-check his/her values in the link budget being prepared for a particular fiber-optic project. The values in the tables provided, taken from ITU-T Rec. G.957 (Ref. 2) Tables 2 through 4, are very conservative. For example, mean launched power for column 6 may be +3 or +6 dBm; and with an OFA (optical fiber amplifier), the power may reach +20 dBm or more, depending on circumstances.

TABLE 10.2 Link Budget Examples/Optical Interfaces for SDH STM-1 (155 Mbps)

	Unit	Values									
Digital signal		STM-1 According to Recommendation G.707									
Nominal bit rate	kbps	155,520									
Application code (Table 10.1)		1-1		S-1.1	S-1.2		L-1.1		L-1.2	L-1.3	
Operating wavelength range	nm	1260[a]–1360		1261[a]–1360	1430–1576	1430–1580	1263[a]–1360		1480–1580	1534–1566/ 1523–1577	1480–1580
Transmitter at reference point S											
Source type		MLM	LED	MLM	MLM	SLM	MLM	SLM	SLM	MLM	SLM
Spectral characteristics											
Maximum RMS width (s)	nm	40	80	7.7	2.5	—	3	—	—	3/2.5	—
Maximum −20-dB width	nm	—	—	—	—	1	—	1	1	—	1
Minimum side mode suppression ratio	dB	—	—	—	—	30	—	30	30	—	30
Mean launched power											
Maximum	dBm	−8		−8	−8		0		0	0	
Minimum	dBm	−15		−15	−15		−5		−5	−5	
Minimum extinction ratio	dB	8.2		8.2	8.2		10		10	10	
Optical path between S and R											
Attenuation range[b]	dB	0–7		0–12	0–12		10–28		10–28	10–28	
Maximum dispersion	ps/nm	18	25	96	296	NA	246	NA	NA	246/296	NA
Minimum optical return loss of cable plant at S, including any connectors	dB	NA		NA	NA		NA		20	NA	
Maximum discrete reflectance between S and R	dB	NA		NA	NA		NA		−25	NA	
Receiver at reference point R											
Minimum sensitivity[b]	dBm	−23		−28	−28		−34		−34	−34	
Minimum overload	dBm	−8		−8	−8		−10		−10	−10	
Maximum optical penalty	dB	1		1	1		1		1	1	
Maximum reflectance of receiver, measured at R	dB	NA		NA	NA		NA		−25	NA	

[a]Some administrations may require a limit of 1270 nm.
[b]See clause 6 of reference document.

Source: Table 2/G.957, ITU-T Rec. G.957, page 5, Ref. 2.

TABLE 10.3 Link Budget Examples/Optical Interfaces for STM-4 (622 Mbps)

	Unit	Values							
Digital signal		STM-4 According to Recommendation G.707							
Nominal bit rate	kbps	622,080							
Application code (Table 10.1)		I-4		S-4.1	S-4.2	L-4.1		L-4.2	L-4.3
Operating wavelength range	nm	1261[a]–1360		1293–1334/1274–1356	1430–1580	1300–1325/1296–1330	1280–1335	1480–1580	1480–1580
Transmitter at reference point S									
Source type		MLM	LED	MLM	SLM	MLM	SLM	SLM	SLM
Spectral characteristics									
Maximum RMS width (σ)	nm	14.5	35	4/2.5	—	2.0/1.7	—	—	—
Maximum −20-dB width	nm	—	—	—	1	—	1	< 1[b]	1
Minimum side mode suppression ratio	dB	—	—	—	30	—	30	30	30
Mean launched power									
Maximum	dBm	−8		−8	−8	+2		+2	+2
Minimum	dBm	−15		−15	−15	−3		−3	−3
Minimum extinction ratio	dB	8.2		8.2	8.2	10		10	10
Optical path between S and R									
Attenuation range[c]	dB	0–7		0–12	0–12	10–24		10–24	10–24
Maximum dispersion	ps/nm	13	14	46/74	NA	92/109	NA	*b*	NA
Minimum optical return loss of cable plant at S, including any connectors	dB	NA		NA	24	20		24	20
Maximum discrete reflectance between S and R	dB	NA		NA	−27	−25		−27	−25
Receiver at reference point R									
Minimum sensitivity[c]	dBm	−23		−28	−28	−28		−28	−28
Minimum overload	dBm	−8		−8	−8	−8		−8	−8
Maximum optical penalty	dB	1		1	1	1		1	1
Maximum reflectance of receiver, measured at R	dB	NA		NA	−27	−14		−27	−14

[a] Some administrations may require a limit of 1270 nm.
[b] See clause 6.2.2 of reference document.
[c] See clause 6 of reference document.

Source: Table 3/G.957, ITU-T G.957, page 6, Ref. 2.

TABLE 10.4 Link Budget Examples/Optical Interfaces for STM-16 (2.488 Gbps)

	Unit	Values					
Digital signal		STM-16 According to Recommendation G.707					
Nominal bit rate	kbps	2,488,320					
Application code (Table 10.1)		I-16	S-16.1	S-16.2	L-16.1	L-16.2	L-16.3
Operating wavelength range	nm	1266[a]–1360	1260[a]–1360	1430–1580	1280–1335	1500–1580	1500–1580
Transmitter at reference point S							
Source type		MLM	SLM	SLM	SLM	SLM	SLM
Spectral characteristics							
Maximum RMS width (σ)	nm	4	—	—	—	—	—
Maximum −20-dB width	nm	—	1	< 1[b]	1	< 1[b]	< 1[b]
Minimum side mode suppression ratio	dB	—	30	30	30	30	30
Mean launched power							
Maximum	dBm	−3	0	0	+3	+3	+3
Minimum	dBm	−10	−5	−5	−2	−2	−2
Minimum extinction ratio	dB	8.2	8.2	8.2	8.2	8.2	8.2
Optical path between S and R							
Attenuation range[c]	dB	0–7	0–12	0–12	10–24[e]	10–24[e]	10–24[e]
Maximum dispersion	ps/nm	12	NA	*b*	NA	1200–1600[b,d]	*b*
Minimum optical return loss of cable plant at S, including any connectors	dB	24	24	24	24	24	24
Maximum discrete reflectance between S and R	dB	−27	−27	−27	−27	−27	−27
Receiver at reference point R							
Minimum sensitivity[c]	dBm	−18	−18	−18	−27	−28	−27
Minimum overload	dBm	−3	0	0	−9	−9	−9
Maximum optical penalty	dB	1	1	1	1	2	1
Maximum reflectance of receiver, measured at R	dB	−27	−27	−27	−27	−27	−27

[a] Some administrations may require a limit of 1270 nm.
[b] See clause 6.2.2 of reference document.
[c] See clause 6 of reference document.
[d] The indicated dispersion range corresponds to the approximate worst-case dispersion for 80-km G.652/G.654 fiber over the wavelength range 1500–1580 nm: manufacturers shall ensure sufficient margins to guarantee proper operation over a target distance of 80 km.
[e] To meet 10-dB minimum attenuation instead of 12 dB, it will be required to decrease the maximum output power, to increase the minimum overload, to use optical attenuators, or a combination thereof.
Source: Table 4/G.957, ITU-T G.957, page 7, Ref. 2.

The ITU test links from which they built up Tables 2 through 4 of Rec. G.957 used binary NRZ optical line coding, scrambled and thus meeting the requirements of ITU-T Rec. G.707. (See Section 4.7 of this text.) We stress that the selection of line coding is a very important part of fiber-optic link design. If our concern was to recover timing, we may want to opt for the RZ coding regime. If we wish minimum transitions on the link, NRZ coding would be our choice. Most systems use NRZ coding or possibly Manchester coding. RZ systems require greater bandwidth capacity than NRZ-coded links, which may be a distinct disadvantage.

10.2.2 ITU-Prepared Practice Tables

Table 10.2 gives the parameters as specified by the ITU for STM-1 (155 Mbps) optical interfaces. Table 10.3 is similar to Table 10.2 but applies to STM-4 (622 Mbps) optical interfaces. Table 10.4 gives data on STM-16 (2.488 Gbps) optical interfaces. It should be noted that *STM-xxx* is SDH nomenclature. SDH is discussed in Chapter 12.

10.3 LINK BUDGETS: EXAMPLES

Input information for these examples comes from either Table 10.1 or information provided in Chapters 4, 5, and 6.

10.3.1 General Guidelines

The starting point, in nearly every case, is the distant-end light detector or receiver. The vendor of the receiving equipment will usually give one or several thresholds in its technical documentation. Even more appreciated by the link budget engineer is a family of sensitivity curves, usually a curve for each standard bit rate. Each curve will have a plot of BER versus input level to the receiver, usually in −dBm. We would expect the bit rates to be related to either SONET or SDH. Refer to Table 9.3.

Table 10.1 provides threshold levels for a BER of 1×10^{-10} for STS-3/STM-1, STS-12/STM-4, and STS-48/STM-16. These have been taken from ITU-T Rec. 957 (Ref. 2), Tables 2–4. We find the threshold levels on the conservative side.

10.3.2 Example 1

We desire to set up a 100-km fiber-optic link operating at 155 Mbps (STS-3/STM-1) meeting the G.826 (Ref. 6) performance requirement. There will be no regenerators or amplifiers in the link. Select the fiber you would use and the transmission wavelength. Use threshold values and requisite margins from Table 10.1.

10.3.2.1 Analysis. We would use single-mode fiber and would select a wavelength in the 1550-nm band. Actually, 1550 nm is the point of minimum loss for single-mode fiber (ITU-T G.654) fiber (Ref. 7). By inspection, the link would be loss-limited. Use a value of 0.25 dB/km as suggested in Chapter 6, Table 6.3. Assume there is a splice every 2 km with an insertion loss of 0.03 dB; there is a connector on each bitter end of the cable (reference points S and R), with an insertion loss of 0.5 dB per connector for a total of 1.0 dB. There is a 1.0-dB dispersion penalty on the link. Assume a mean launched power from an MLM laser of 10 dBm.

PRACTICE TABLE 10.5

Item or Parameter	Value	Comments or Advice
Light source output	0 dBm	Use most cost-effective LD transmitter. Link is loss limited by inspection.
Light detector/receiver threshold	−35 dBm	Alcatel PIN diode receiver
Decibels assignable to link budget	35 dB	Algebraic subtraction
Link losses		
Connector losses	1.0 dB	0.5-dB connector insertion loss each end, at reference points S and R
Fiber loss at 0.25 dB/km, 100 km; includes splice losses	25 dB	ITU-T G.654 fiber, λ-1550 nm
Dispersion penalty	1.0 dB	
Margin	4.8 dB	
Total link losses	31.8 dB	Includes margin
Excess margin	3.2 dB	Apply the excess margin to cable slack which was not accounted for. Add 5% additional loss for cable slack.[a]

[a]It is advisable to have a certain amount of slack in the cable which can be accommodated in hand-holds and man-holes along the cable route. Add 5% to the cable loss value to accommodate losses due to cable slack. The reasons we would want this cable slack are covered in Chapter 11.

Given a −28-dBm[1] threshold at the light detector and 0-dBm output at the near-end laser source, the link can support 28 dB of loss. Assigning loss:

- Connector losses: 1.0 dB
- Fiber and splice losses: 25.0 dB (100 km × 0.25 dB/km = 25.0 dB)
- Dispersion penalty: 1.0 dB

Total: 27.0 dB

This leaves only a 1.0-dB margin. We believe this value is insufficient.

However, Section 6.4 of ITU-T Rec. G.957 states the sensitivity values given in Tables 2–4 are worst case, end-of-life value. This allows about 2-dB additional margin until the end-of-life period is reached. Then at that time we would expect the link to begin to show deteriorating performance.

It is recommended at this point that the link designer investigate whether the light detector in question is a PIN receiver or an APD receiver. Turn to Table 5.2 herein. An Alcatel PIN receiver has a −35-dBm threshold for a BER of 1×10^{-10}. If we were to employ this value and assume it to be a nominal value, add +3 dB for end-of-life deterioration; we have picked up 4 dB of additional margin (i.e., the value of −28 dBm in G.957 Tables 2–4 and −35 dBm published performance of Alcatel PIN diode for the bit rate given and BER).

We suggest setting the values we have discussed into a tabular form, as one would expect in a link budget. The table should have three columns. Column 1 would be "Item or Parameter," column 2 would be "Value" of the item or parameter, and column 3 would be "Comments or Advice," where the designer

[1]A −28-dBm threshold in Tables 2–4, G.957, is for a BER of 1×10^{-10}; we would add 1 dB to comply with the BER requirement of 1×10^{-12}. However, we believe that the −28-dBm value will be valid for 1×10^{-12}. See Section 5.5.3 and Table 5.2 of this text. Most PIN diode detectors run in the range of −33 to −38 dBm for the desired BER and bit rate.

PRACTICE TABLE 10.6

Item or Parameter	Value	Comments or Advice
SLM laser diode with EDFA	+17 dBm	Call the EDFA a power amplifier.
PIN receiver threshold with EDFA front end	−45 dBm	
Decibels assignable to the link	62 dB	
Link losses		
Connector loss	2.0 dB	Four connectors
Fiber loss 160 km + 0.05 × 160 = 168 km	42 dB	5% length added for slack; loss at 0.25 dB/km
Splice loss at 0.1 dB/splice	2.1 dB	Fiber is on 2-km reels; 21 splices
Dispersion penalty	2.0 dB	
Margin	4.8 dB	
Total losses	52.9 dB	
Excess margin	9.1 dB	

would comment on any reservations he/she may have or other such advice. We will call this table Practice Table 10.5.

10.3.3 Example 2

The proposed route is 160 km long. The bit rate is 622 Mbps. No regenerator or amplifiers are desired along the route. However a power amplifier may be incorporated with the light source; a preamplifier may be incorporated in the far-end light receiver. The amplifiers are based on EDFA technology implying 1550-nm operation. The amplifiers have a gain of 17 dB. In other words, we expect the laser transmitter rated at 0-dBm output would then have a +17-dBm output due to the insertion of an EDFA. There will be little impact of four-wave mixing because there is only one light carrier on the fiber. The receiver has a threshold of −28 dBm (Table 10.1) for 622-Mbps operation. The gain from the EDFA extends the threshold 17 dB for a new threshold at −45 dBm. Subtracting −45 dBm from the +17-dBm value, we derive 62 dB assignable to the link. We go directly to a tabular calculation. See Practice Table 10.6.

Several decibels of surplus margin are advantageous, but 9.1 dB of surplus margin are excessive. We would recommend cutting back the output power of the transmitter–amplifier combination. The higher power can bring about several undesirable nonlinearities such as self-phase modulation, cross-phase modulation, and others. If we were building a long system, its length could be extended 40 km using the excessive margin.

10.3.4 Example 3

A SONET ring is to be installed around Municipalberg. The ring circumference is 36 km and will consist of 12 fibers: 4 spares, 4 traffic fibers upstream, and 4 traffic fibers downstream. The initial installation will transport STS-3 bit streams. However, with a minimum of field replacement, the traffic must be able to be upgraded to STS-24 (1244 Mbps). Six ADMs (add–drop multiplex) feed corresponding nodes. Consider an ADM simply as a regenerator. Assume equal length segments of 6 km. Likewise, consider the SONET switch as a regenerator in that we have to drop to electrical baseband in order to access SONET frame and OA&M over-

PRACTICE TABLE 10.7

Item or Parameter	Value	Comments or Advice
Transmit output power coupled to fiber	−3 dBm	MLM laser. Modulation bandwidth of ELED cannot support bit rate.
Threshold level BER = 1×10^{-12}	−25 dBm	This is the threshold for a PIN diode, bit rate 1244 Mbps.
Decibels allocated to the link budget	22 dB	
Link losses		
Connector insertion loss at points S and R	1.0 dB	
Splice loss	1.0 dB	
Fiber loss 1310 nm at 0.5 dB/km, 6 km	3.0 dB	*Note:* We only consider the distance between adjacent ADMs.
Dispersion penalty	1.0 dB	
Margin	3.0 dB	
Optical attenuator[a]	13 dB	
Alternate routing switch insertion loss	0.0 dB	

[a]An option would be to reduce the power output of the laser transmitter by 10–12 dB and eliminate the cost of the attenuator.

head and to have switching in the electrical domain. Thus, a switch, in this case, has no loss or gain. If it were a light switch, there would be considerable insertion loss.

Several observations can be made by simple inspection:

- The design bit rate will be 1244 Mbps.
- Each link will be loss-limited because the links are so short.
- The first attempt link budget should use LED as standardized source. Note that we may be reaching the limit of transmission rate for an LED. This should be investigated. If a laser diode source is contemplated, the link budget should include the PIN-diode receiver maximum input power. If the power input to the PIN receiver is higher than this maximum, either reduce the output power of the source, if this is possible, or install light attenuators as required.
- Fiber should be the most cost-effective type. Multimode fiber may even be a candidate.

Consult Practice Table 10.7 for link budget exercises.

10.3.5 Example 4

We desire to install a fiber-optic system across a single link 100 km long. The link shall carry 8 WDM channels supporting STM-16 (2.5 Gbps) each. The system will use the ITU grid with 200-GHz spacing. We will estimate fiber loss using ITU-T Rec. G.654 fiber, with 0.25 dB/km including splice losses and 20 ps/(nm-km) for chromatic dispersion accumulation.

WDM equipment insertion losses will be considerable. To estimate these losses, turn to ITU-T Rec. G.671 (Ref. 8), which deals with passive fiber components. The recommendation offers the following guidelines:

PRACTICE TABLE 10.8

Item or Parameter	Value	Comments or Advice
SLM laser diode power output	+3 dBm	Maximum advisable output. See Table 4/G.957.
PIN diode receiver threshold + EDFA preamp gain −28 dBm −17 dB = −45 dBm	−45 dBm	Bit rate 2.5 Gbps, BER = 1×10^{-10}. See Table 4/G.957
Link losses		
Budgetary loss	48 dB	+3 dBm − (−28 dBm) − 17 dB = 48 dB
G.654 fiber and splice losses	26.25 dB	100 km + 0.05 × 100 for slack or 105 × 0.25 dB/km = 26.25 dB
Connector losses	3.0 dB	Six connectors in tandem including WDM connector[a]
WDM +filter losses	10.5 dB	
Margin	4.8 dB	
Total link losses	44.55 dB	
Excess margin	3.45 dB	

[a]Connectors as follows: output of LD transmitter, input WDM multiplexer, composite output of WDM multiplexer to fiber line. Fiber line input to demultiplexer, demultiplexer to filter, filter output to pre-amp EDFA. The EDFA is integrated with the receiver.

WDM Multiplexers and Demultiplexers. Insertion loss = $1.5 \log_2 n$, where n is the number of ports [i.e., channels, either input to a multiplexer or output from a demultiplexer (our interpretation of Ref. 8)].

In the case for this example, there are eight input ports for the multiplexer and eight output ports for the demultiplexer. Thus, the insertion loss for each is 4.5 dB. Filter insertion loss is 1.5 dB. The total insertion loss for the eight-channel WDM equipment including filters is then 4.5 dB + 4.5 dB + 1.5 dB or 10.5 dB.

By inspection, each channel is carrying 2.5 Gbps on a span of 100 km, and we conclude that the link has to be analyzed whether it is loss-limited or dispersion-limited. This analysis will be demonstrated in Section 10.4. We will assume at this juncture that the link is loss-limited. The link budget may be found in Practice Table 10.8.

Note that the 200-GHz channel separation just about precludes problems with four-wave mixing. Assume the EDFA preamplifier has 17-dB gain.

10.4 LINK BANDWIDTH, RISE TIMES, AND CUMULATIVE DISPERSION

10.4.1 The Rise-Time Budget

The system rise time is calculated to assure ourselves that the fiber-optic link in question can handle the bit rate. Rise time and fall time are defined by the *IEEE Standard Dictionary of Electrical and Electronic Terms* (Ref. 9) as follows: "The rise time is the time for the light intensity to increase from the 10 to 90% intensity points. The fall time is the time for the light intensity to fall from the 90 to 10% intensity points."

The following equation relates light system rise time T_r to the rise times of the components that make up the system:

$$T_r^2 = T_{\text{tr}}^2 + T_{\text{fiber}}^2 + T_{\text{rec}}^2 \tag{10.1}$$

where T_{tr}, T_{fiber}, and T_{rec} are the rise times of the light transmitter, the fiber cable and the receiver, respectively.

Source: Ref. 10.

We can relate bandwidth, Δf, to rise time, T_r, by the following equation:

$$T_r = 2.2/2\pi\Delta f = 0.35/\Delta f \tag{10.2}$$

This just tells us that the inverse relationship between bandwidth and rise time holds for a linear system. The value 0.35 of the product of $T_r\Delta f$ should be handled with care. In the design of a fiber-optic link it is common practice to use this value (0.35) to keep the design on the conservative side.

As one can imagine, the relationship between bit rate, B, and bandwidth, Δf, depends on the format of the digital bit stream, whether it is an RZ format or NRZ format. (See Section 4.7 for a discussion of digital formats.) The following guidelines are useful when designing a fiber-optic link. If we follow these guidelines, we can be assured that the system bandwidth is sufficient to handle the bit rate, B.

T_r must be below the maximum value as follows:

For an RZ format: $T_r \leq 0.35/B$
For an NRZ format: $T_r \leq 0.70/B$

Source: Ref. 10.

The following are rise-time estimates that a design engineer might find useful when calculating system rise time, T_r, and for applying equation (10.2).

Transmitter LED: ~ 2 ns
Transmitter LD: ~ 0.1 ns

When given the 3-dB bandwidth of the fiber-optic receiver front end, the receiver rise time can be calculated by equation (10.2), where Δf is the receiver front end 3-dB bandwidth as follows:

$$T_{rec} = 0.35/\Delta f \tag{10.3}$$

and, using equation 10.1:

$$T_r^2 = T_{tr}^2 + T_{fiber}^2 + T_{rec}^2 \tag{10.4}$$

Now the practical design engineer can calculate system rise time allowing a 10% degradation factor; thus

$$T_r = 1.1\left(T_{tr}^2 + T_{fiber}^2 + T_{rec}^2\right)^{1/2} \tag{10.5}$$

T_{fiber} may be calculated from

$$T_{fiber}^2 = T_{modal}^2 + T_{GVD}^2 \tag{10.6}$$

where GVD is the group velocity dispersion.

When using single-mode fiber, T_{modal} is zero. Then $T_{fiber} = T_{GVD}$, and T_{GVD} may be calculated from the following approximate relationship:

$$T_{GVD} \approx DL\Delta\lambda \tag{10.7}$$

where D is the "real value of D" (which is the dispersion parameter), $\Delta\lambda$ is the FWHM (full width at half maximum) width of the optical source, and L is the length of the link in km. $\Delta\lambda$ values can be taken from Tables 2–4 of ITU-T Rec. G.957 (Ref. 2). It should be noted that the dispersion parameter D may change along a fiber line if different fiber sections have different dispersion characteristics. Thus, we should expect that the value of D is an average value.

Sources: Refs. 10 and 11.

Example 1. Calculate the system rise time given the following parameter values (see Section 10.3.5):

Link length: 100 km, G.654 fiber
LD transmitter rise time: 0.1 ns
Source spectral width (Table 10.4): 1 nm
Receiver front-end bandwidth: 10 GHz

The digital format is NRZ. The bit rate is 2.5 Gbps.

- G.654 fiber: For the exercises here, use 20 ps/nm-km. For more accurate results, turn to Section 6.2 of ITU-T Rec. G.654.
- G.653 fiber: D_{max} between 1525 and 1575 nm is 3.5 ps/nm-km. For more accurate results, turn to Section 6.2, of ITU-T Rec. G.653.

$$\begin{aligned} T_{fiber} &= T_{GVD} \approx DL\Delta\lambda \\ T_{fiber} &= 20 \text{ ps/nm-km} \times 100 \text{ km} \times 1 \text{ nm} \\ &= 2000 \text{ ps or } 2 \text{ ns} \end{aligned}$$

The receiver 3-dB bandwidths of its front end is 10 GHz.

$$\begin{aligned} T_{rec} &= 0.35/10 \text{ GHz} \\ T_{rec} &= 0.35/10 \times 10^9 = 3.5 \times 10^{-11} = 35.0 \times 10^{-12} \text{ or } 35 \text{ ps or } 0.035 \text{ ns} \\ 0.035^2 &= 0.001225 \\ T_r &= 1.1(0.01 \text{ ns} + 4 \text{ ns} + 0.000005 \text{ ns})^{1/2} \\ &= 1.1 \times 4.01^{1/2} = 1.1 \times 2.024 \text{ ns} = 2.203 \text{ ns} \end{aligned}$$

We now give the system rise-time value the test to see if it will support 2.5 Gbps by using the approximate relationship:

$$\begin{aligned} \text{System rise time } T_r &\leq 0.70/2.5 \times 10^9 \text{ bps} \\ 2.2363 \text{ ns} &\leq 0.70/2.5 \times 10^9 \text{ bps} \\ 2.203 \text{ ns} &\leq 2.8 \times 10^{-10} = 0.28 \text{ ns} \end{aligned}$$

It does not meet the test. What can be done about it? The biggest contributor to the rise time is the fiber rise time. We can shorten the link; we can use G.653 fiber with its 3.5-ps/nm-km value for chromatic dispersion.

$$T_{\text{fiber}} = 3.5 \text{ ps/nm-km} \times 100 \times 1 = 350 \text{ ps}$$

$$T_r = 1.1(0.01 \text{ ns} + 0.1225 \text{ ns} + 0.001225 \text{ ns})^{1/2}$$

$$= 1.1(0.361225)^{1/2} = 0.4004 \text{ ns}$$

$$0.4004 \text{ ns} \leq 0.28 \text{ ns}$$

It still does not meet the test. Now we shorten the link to 50 km.

Turning again to T_{fiber}:

$$T_{\text{fiber}} = 3.5 \text{ ps/nm-km} \times 50 \times 1$$

$$= 175 \text{ ps or } 0.175 \text{ ns}$$

$$T_r = 1.1(0.01 \text{ ns} + 0.175 \text{ ns} + 0.001225 \text{ ns})^{1/2} = 1.1(0.2015)$$

$$= 0.2217 \text{ ns}$$

$$0.2217 \text{ ns} \leq 0.28 \text{ ns}$$

It meets the test.

We would like to raise another red flag, or maybe just a yellow flag for caution. A fiber link of any length is made up of fiber sections, commonly of 2-km reels, but reels from 1 to 10 km of fiber may be used. The fiber may differ reel by reel such that the dispersion characteristics may vary sufficiently to throw our calculations off. It may turn out that a workable link on paper may not be a workable link in practice.

Example 2. A fiber link uses G.653 fiber (chromatic dispersion accumulating at 3.5 ps/nm-km) and transmits STS-24 (1244 Mbps), and we will assume that the link *may* be dispersion-limited. Can we stretch it 200 km without a regenerator? An EDFA, which does not contribute to accumulated dispersion, will provide sufficient gain to reach a comfortable receiver threshold level. Our concern here is that the link may be dispersion-limited. The link uses an NRZ waveform. The source spectral width is 1 nm. For convenience, the unit of time will be the nanosecond.

Now then, for the link to be power-limited, the following relationship must hold, namely,

$$T_r \leq 0.70/B, \qquad \text{where } B \text{ is 1244 Mbps thus}$$

$$T_r \leq 0.562 \text{ ns}$$

LD transmitter rise time is 0.1 ns. Receiver front-end rise time = 0.35/10 GHz = $0.35/10 \times 10^9$ Hz = 3.5×10^{-11} s or 35 ps or 0.035 ns.

$$T_r = 1.1\left(0.01 + T_{\text{fiber}}^2 + 0.001225\right)^{1/2}$$

$$T_{\text{fiber}} = 3.5 \text{ ps} \times 200 \times 1 = 0.0035 \text{ ns} \times 200 \times 1 = 0.7 \text{ ns}$$

$$T_r = 1.1(0.01 + 0.7 + 0.001225)^{1/2} = 0.843 \text{ ns}$$

The test: 0.562 has to be larger than 0.843; it is not. It failed the test.

In the T_r equation above, the largest value (factor) is the fiber contribution. Thus, it is that value that deserves our attention. Remember the factors that are used in the equation where we created the value $T_{\text{fiber}} = DL\Delta\lambda$ [from equation (10.7)]. With this relationship in mind, we can reduce or cut back on

- D chromatic dispersion, measured in ps/km
- L length of the link
- $\Delta\lambda$, the bandwidth, which is the spectral width of the light emission of the LD

We reduce the source spectral width to half of the original value by using a better light transmitter. Assume that the new spectral width is 0.5 nm, then

$$T_{\text{fiber}} = 3.5 \text{ ps} \times 200 \times 0.5 \text{ (ps)} = 350 \text{ ps or } 0.350 \text{ ns}$$

Calculate T_r:

$$\begin{aligned} T_r &= 1.1(0.01 \text{ ns} + 0.1225 \text{ ns} + 0.001225)^{1/2} \\ &= 0.402 \text{ ns} \end{aligned}$$

The value 0.562 ns is now indeed larger than 0.402 ns. We met the test simply by improving the spectral characteristics of the emitted light wave from the transmitter.

10.5 DERIVATION OF OPTICAL POWER LEVELS

10.5.1 Channel Power

10.5.1.1 Minimum Channel Power. In this section we describe how to derive the end-of-life minimum channel power that is required to maintain the desired optical signal-to-noise ratio (OSNR). First we must relate OSNR to BER. It is important that we realize BER will be different for amplified systems and for nonamplified systems. The relationship to the BER is a receiver characteristic, which has yet to be included in the design methodology. The resulting minimum channel optical power is independent of the number of channels (i.e., wavelengths) and can be used for both single-channel and multiple-channel systems. The section also described how ASE effects limit the minimum channel power for both single- and multiple-channel optically amplified systems.

The ASE power per unit frequency interval for an optical amplifier is given by

$$P_{\text{ASE}} = 2N_{\text{SP}}(G - 1)hv \qquad (10.8)$$

where $N_{\text{SP}} \geq 1$ is the spontaneous noise factor, G is the internal gain, h is Planck's constant, and ν is the optical frequency. The external amplifier noise figure is given (in dB) by

$$\text{NF} = 10\log\left[2N_{\text{SP}} - \frac{2N_{\text{SP}} - 1}{G}\right] + \eta_{\text{in}} \qquad (10.9)$$

where η_{in} is the input coupling loss for the amplifier in dB. If we make the simplifying assumptions that the total output power (including accumulated ASE power) is equal after each amplifier and that the gain G is much greater than 1, then the OSNR is given approximately by

$$\text{OSNR} = P_{\text{out}} - L - \text{NF} - 10 \log N - 10 \log[h\upsilon\Delta\upsilon_0] \tag{10.10}$$

where P_{unit} is the output power (per channel) in dBm, L is the span loss between amplifiers in dB, NF is the external noise figure in dB, $\Delta\upsilon_0$ is the optical bandwidth, N is the number of spans in the chain, and we have assumed that all the span losses are equal. In the 1.55-μm band, $10 \log(h\upsilon\Delta\upsilon_0) = -58$ dBm at 0.1-nm optical bandwidth. This approach could still be applied in a system where the span losses differ, by assuming that all losses are equal to or less than L, and would yield a worst-case estimate of the OSNR.

The above relationship provides a practical and useful prediction, because the OSNR at the input to the receiver (point $R_{(N)}$ in Figure 10.1) is the result of an rms average of N effective noise sources, such that small differences in span losses on output powers tend to average out. $G \gg 1$ assumption is true for most amplifier systems (Ref. 12).

Equation (10.10) can be used to estimate the minimum optical power (out) that is required to maintain a desired OSNR. The minimum output power would be measured at the output of the amplifiers. Because it is a limit on the minimum power (per channel) and is independent of the number of channels, it can be used for both single- and multiple-channel systems.

In the case where individual channel powers vary, if all channel powers are greater than or equal to the minimum power, then all OSNRs will also be greater than or equal to the minimum required value.

In a real WDM system, the output channel powers will probably not be equal because of unequal gain, and the noise figures may also differ among the amplifiers and among the channels. In addition, the span losses will likely not be equal. Nevertheless, equation (10.10) is useful in the establishment of minimum channel optical power levels because only the worst case needs to be considered (i.e., all span losses are taken to be at their highest value, and we consider the channel with the lowest output power).

10.5.1.2 Maximum Channel Power. Limitations on maximum optical power levels can be based on either fiber nonlinear effects or laser safety consideration. If the maximum total output power (including ASE) is fixed at the Class 3A laser limit, $P_{3\text{A}}$, then the maximum nominal channel power, $P_{\text{ch max}}$ is related to the channel number as given in

$$P_{\text{ch max}} = P_{3\text{A}} - 10 \log(M) \tag{10.11}$$

where M is the number of channels in operation. This equation is given for illustrative purposes because output power may vary among individual channels as long as the total output power is less than $P_{3\text{A}}$. This limitation would be valid for systems with and without line amplifiers as described in ITU-T Rec. G.692 (Ref. 12).

In some cases, fiber nonlinearities impose more restrictive limits on output power level than laser safety considerations. In particular, self-phase modulation (SPM), cross-phase modulation (XPM), and stimulated Brillouin scattering (SBS) place limits on the maximum channel power. The limits on optical power level imposed by SPM and SBS are independent of the number of channels present, and, in the case of SPM, only systems on G.652 and G.655 fiber are affected. However, XPM affects only multichannel systems and is more significant for systems that adopt narrow channel spacings. The impairments from XPM are more significant in G.652 fiber systems relative to G.653 and G.655 fiber systems. Maximum permitted channel output powers due to SPM-imposed or XPM-imposed limitations will vary between application codes, and they depend on the number of spans and the span target distance.

The limits on maximum channel optical power introduced by SBS are for further study by the ITU-T Organization and are not considered herein. Four-wave mixing (FWM) affects only multichannel systems and does not present a practical limitation to G.652- or G.655-type fiber systems. Stimulated Raman scattering also does not present a practical limitation to multichannel systems on G.652 fibers as described in this section and in ITU-T Rec. G.692. The impact of stimulated Raman scattering on some multichannel systems with unequal channel spacing using G.653 fiber is also open for further study.

10.5.1.3 Maximum Range for Channel Power. The three power level limits defined above in this section define the maximum range for channel power levels. The minimum channel power is independent of the number of channels present, whereas the maximum channel power depends on the number of channels present. As an example, for eight channels present, the maximum level is determined by the safety limit, whereas for one channel present, the maximum power level is dictated by the SPM limit applicable for the application code. A relatively high channel power is obtained when only a few channels are present, whereas the channel power would drop to a lower level when channels are being added. This, however, depends on the OFA implementation.

Source: Ref. 12.

10.5.2 Maximum Total Power

The total required output power from the optical amplifiers can be estimated by

$$P_{\text{tot}} = \sum P_{\text{out}} + N \cdot BW_{\text{eff}} \cdot h\upsilon \cdot 10^{(\text{NF}+L)/10} \tag{10.12}$$

Here, NF and L are given in dBs, and all other terms are expressed in linear units. The last term is the total accumulated ASE power, and BW_{eff} is the effective ASE bandwidth defined as total ASE power divided by the ASE power density. This bandwidth is about 20–30 nm for one amplifier and about 15 nm for a chain of up to 10 amplifiers, as long as the signal gain is close to the maximum spectral gain of the amplifier. This approximation is sufficient as long as the total power is dominated by the signal power.

Sources: Refs. 12 and 13.

REFERENCES

1. *Characteristics of a Dispersion-Shifted Single-Mode Optical Fiber Cable*, ITU-T Rec. G.653, ITU Geneva, April 1997.
2. *Optical Interfaces for Equipments and Systems Relating to the Synchronous Digital Hierarchy*, ITU-T Rec. G.957, ITU Geneva, June 1999.
3. *Generic Reliability Assurance Requirements for Optoelectronic Devices Used in Telecommunication Equipment*, Telcordia GR-468-CORE, Issue 1, Piscataway, NJ, December 1998.
4. *Generic Requirements for Assuring the Reliability of Components Used in Telecommunication Systems*, Telcordia TR-NWT-000357, Issue 2, Piscataway, NJ, October 1993.
5. Roger L. Freeman, *Telecommunication Transmission Handbook*, 4th ed., John Wiley & Sons, New York, 1998.
6. *Error Performance Parameters and Objectives for International, Constant Bit Rate Digital Paths at or Above the Primary Rate*, ITU-T Rec. G.826, ITU Geneva, February 1999.
7. *Characteristics of Cut-Off Shifted Single-Mode Optic Fiber Cable*, ITU-T Rec. G.654, ITU Geneva, April 1997.
8. *Transmission Characteristics of Passive Optical Components*, ITU-T Rec. G.671, ITU Geneva, November 1996.
9. *The IEEE Standard Dictionary of Electrical and Electronic Terms*, 6th ed., IEEE, New York, 1996.
10. Govind P. Agrawal, *Fiber-Optic Communication Systems*, 2nd ed., John Wiley & Sons, New York, 1997.
11. Donald J. Sterling, Jr., *Technician's Guide to Fiber Optics*, 3rd ed., Delmar, Albany, NY, 2000.
12. *Optical Interfaces for Multichannel Systems with Amplifiers*, ITU-T Rec. G.692, ITU Geneva, October 1998.
13. Karen Liu and John Ryan, "All the Animals in the Zoo: The Expanding Menagerie of Optical Components," *IEEE Communications Magazine*, page 110, July 2001.

11

OUTSIDE PLANT CONSIDERATIONS

11.1 CHAPTER OBJECTIVE

The outside plant portion of the fiber network may be buried or pole-supported in an aerial configuration. In this chapter we place emphasis on buried plant. The outside plant sector of the fiber cable system may also include (a) remote devices such as regenerators and optical amplifiers or (b) passive components such as signal splitters. Of great concern to the system design engineer is reliability and built-in aids to maintainability. Active remote devices require prime power. This power may be provided locally or carried in some metallic portion of the fiber cable.

11.2 BURIED PLANT

11.2.1 Standard for Physical Location

The proper design of a fiber-optic cable below-ground route is important, this being the first step in avoiding damage to the cable by future work operations performed in the area.

The following guidelines will be helpful for the outside plant engineer responsible for cable planning and placement:

1. Plans for the location and installation of below-ground fiber-optic cable should be made using information obtained from a field survey.
2. Installation plans will identify the cable route, placing and depth information, and information that is sufficient to locate subsurface structures. In line with this, Table 11.1 gives the standard uniform color code of underground facilities that might be encountered in this cable lay project.
3. In recognition of possible right-of-way congestion, the route should take into account future interfering structures as well as the present competing underground facilities.
4. Land acquisition rights and permissions should be obtained before actual installation is started. These will include right-of-way permits that recognize right of access; they will also include equipment enclosures and the work area involved for future maintenance activities.

TABLE 11.1 Standard Uniform Color Code—Below-Ground Markings

Color	Facility
Red	Electric power lines and conduit
Yellow	Gas, oil, steam, and petroleum lines
Blue	Water, irrigation, and slurry lines
Green	Sewer and drain lines
Orange	Communication lines, including fiber-optic cable
White	Proposed excavation
Fluorescent pink	Temporary survey markings

Source: From EIA/TIA-590-A, Table 1, page 4, Ref. 1.

5. A preconstruction meeting should be held with involved local government agencies, contractors, and other utilities to cover construction plans, schedules, sequence of operations, and other concerns.
6. The facility owner should conduct inspections as necessary to ensure that the installation is in accordance with approved plans.
7. Those all-important as-built location records and files are to be maintained by the facility owner. The location records should be made available for reference when government agencies or other entities are planning work in the area to allow them to plan to avoid damage or conflicts with the fiber-optic cable facilities being installed. The as-builts cannot be expected to reflect subsequent changes in public works, landscape/landmarks, or foreign underground structures.

Source: Guidelines 1–5 are from Ref. 1.

11.2.2 Depth of Plant

Buried or conduit plant should meet the depth requirements indicated in Table 11.2. These are requirements for fiber-optic cable. TIA/EIA-590-A (Ref. 1) also gives guidance on "Joint Construction" and depth of cover for power cables. The standard goes on to say that power cable depth is governed by the *National Electrical Safety Code* (*NESC*), Rule 353D (Ref. 2). When fiber cable is buried jointly with electrical cable, the minimum depth of cover is governed by Table 11.2 or Table 11.3, whichever depth is greater.

TABLE 11.2 Depth of Fiber-Optic Plant

Facility	Minimum Cover, mm (in.)
Toll, trunk cable	750 (30)
Feeder, distribution cable	600 (24)
Service / drop lines	450 (18)
Underground conduit	750 (30)

Source: From EIA/TIA-590-A, Table 2, page 6, Ref. 1.

TABLE 11.3 Depth of Electrical Power Cable

Maximum Voltage Phase-to-Phase, Volts	Depth of Cover, mm (in.)
0–600	600 (24)
601–50,000	750 (30)
50,001 and above	1070 (42)

Source: From EIA/TIA-590-A, Table 3, page 6, Ref. 1.

11.2.3 Marking the Facility Route

Both permanent above-ground markers and underground warning tape are recommended to identify the general location of the facility route. These devices, however, cannot be relied upon as markers for the precise location of the underground fiber cable lay.

The reference standard recommends that permanent markers be placed within line of sight of each other so the direction and location of the route is clearly indicated. Each marker so placed should be visible from the adjacent marker in each direction. The maximum recommended separation is 300 m (1000 ft) or less, if land use permits. Markers are identified by facility name, owner, and appropriate telephone number.

Warning tape should be buried at least 300 mm (12 in.) above the cable and should not deviate more than 450 mm (19.8 in.) from the outside edge of the facility. These warning tapes should have sufficient tensile strength and elongation properties so that when encountered in excavating, they are not easily broken and will stretch significantly before breaking.

Source: Ref. 1.

11.2.4 Riser Poles

When fiber-optic cable is placed on riser poles, the cable should have mechanical protection such as a duct or U-guard on the pole extending from the ground approximately 2.5 m (8 ft). The duct or U-guard protection should extend below ground via a conduit bend to the specified burial depth of the cable as given in Table 11.2. Risers on the pole should be placed in the safest position with respect to possible traffic damage and climbing space. When added protection is required, the fiber-optic cable may be placed in innerduct which extends above the U-guard up and onto the supporting aerial strand. From an underground conduit, the innerduct may be run from the manhole, through subsidiary duct and U-guard onto the supporting aerial strand.

11.2.5 Building Entrances

Cable entry to a building may be made above ground or at burial depth. The fiber cable in question should be secured to the building and mechanically protected by innerduct or U-guard.

11.2.6 Cables Crossing Water

Usually special permits are required for water crossings. The US Army Corps of Engineers through the Corps' Regional District Engineer will advise applicants for water crossing permits on what exactly is required regarding these permits. The Corps has published a pamplet entitled *Regulatory Program—Applicant Information*, which is available and gives permit information.

11.2.7 Railroad Crossing

It is advisable to notify the appropriate railroad headquarters' chief engineer of the planned crossing of railroad tracks. The railroad chief engineer will advise on the approved methods of crossing a railroad track. The Association of American Railroads in Washington, DC, is a good source of information and additional details for engineering and design for fiber-optic cable crossing railroad tracks.

11.2.8 Bridge Crossings

Each bridge crossing must be individually designed to conform to local conditions and physical constraints imposed a the bridge location. Likewise, local regulations will vary from place to place. Conduit would normally be used to provide the structure and mechanical protection of these cable crossings.

11.2.9 Highway Designs

In the design of long-haul fiber-optic systems, city and town streets and main and secondary highways are very handy to use for fiber cable right-of-way. All states and many counties have statutes or regulations that permit and define the use and occupancy of public highways and streets. There may be franchise agreements in place that specify the legal rights covering placement of utility in the highway path.

The fiber cable system owner should consult *A Guide for Accommodating Utilities Within Highway Right-of-Way*. This document is issued by the American Association of State Highway and Transportation Officials.

Highway design and type, soil conditions, traffic levels and patterns, and zoned land use restrictions will affect the ultimate fiber-optic cable installation accommodations along specific highway rights-of-way.

11.2.10 Excavation and Damage

Most states have laws that are intended to ensure safe working operations and to minimize any possibility of damaging existing subsurface facilities. These laws vary from state to state. Expect these laws to cover such things as required participation in a *one call bureau*, time of advanced notification to facility owners before excavation starts, size of tolerance zone, specifying use of the Utility Location and Coordination Council (ULCC) uniform color code for temporary facility location marking, and facility owner registration at a local government office.

Guidance should be taken from OSHA (Occupational Safety and Health Administration), *Code of Federal Regulations*, title 29, Chapter XVII, subpart P,

Excavations, Section 1926.651. Here it states that "the estimated location of utility installations, such as sewer, telephone, fuel electric, water lines, or any other underground installations that reasonably may be expected to be encountered during excavation work, shall be determined prior to opening and excavation." This regulation also advises that utilities shall be advised of the proposed work before the start of actual excavation.

Obviously, excavators (contractors) and facility owners should be knowledgeable about the specific laws and regulations governing damage prevention methods and procedures for their operational areas. It is urged that both parties follow not only the letter but the intent of such laws to minimize or avoid accidental damage to subsurface fiber-optic cable facilities and thereby reduce liability exposure of the excavators and/or contractors and service interruptions.

11.2.11 Damage Restoration

Facility owners (i.e., owners of the fiber-optic system to be installed) restore facilities damaged during excavation procedures. There should be advanced preparations for such possible events. Unfortunately, each damage case presents different situations, circumstances, and conditions that should be handled and coordinated for rapid service restoration.

The excavator or facility contractor should be prepared to carry out restoration work, which will include the following generic restoration work items and/or procedures:

- Network records, maps, installed-facility measurement data, requirements, and availability needed for rapid and effective restoration of service.
- Spare cable requirements, including metallic and fiber cable, for restoration and repair works. This item should include lengths of cable, types, quality, inventory, and availability, based on network layouts and design.
- Trained facility personnel including trained splicers and splicing jigs, test sets, and splicing restoration kits.
- Supply site protection as needed for temporary restorations.
- Be prepared to make facility test measurements.

11.3 PLANNING AND INSTALLATION OF OPTICAL FIBER SYSTEMS—WITH EMPHASIS ON OUTSIDE PLANT

11.3.1 Optical Fiber Cable—Special Considerations

There are literally perhaps millions of man-years of experience in the installation of wire pair and coaxial cable systems in the telecommunication plant. Of course, it would be very desirable to use these same installation methods on fiber-optic cable systems. However, particular attention should be paid to fiber cables' very low stress limits, its critical bending characteristics, the long installation lengths possible, and the effects of ambient conditions.

11.3.2 Installation Planning

We can rely on many of the procedures used in the installation of metallic cable plant, but certain aspects of the fiber-optic plant must be given special consideration. This includes, but is not limited to, the following:

- The effect of splices and connectors on span length.[1]
- The longer lengths of cables that can be installed.
- The low tolerance of the cable, in transmission terms, to additional splices and/or connectors.
- The different construction of optical fiber (when compared to its metallic cable counterpart). In planning and installation, care must be taken of the cable's low stress limits, bending characteristics, and ambient temperature ranges.
- Route construction, conditions, and access in term of both (a) installation and in-service factors and (b) the value of local information.
- The importance of information and training as part of the planning of optical cable installations.

11.3.2.1 Routing. The importance of the route survey cannot be overemphasized. The geometry and condition of an existing plant must be taken into account, and the longer-length spans in particular access arrangements must be carefully considered. Ducts should be in sound condition, and the generally smaller diameter of optical fiber cables provides an opportunity for considering subduct systems to provide better duct utilization, a clear and clean installation track, and better maintenance procedures. In large duct systems containing several cables, a mid-position is preferable. In aerial systems it is important to minimize in-service cable movement and cable strain, and to maximize the stability of the pole route. The element that should be borne in mind is the use of proper optical-fiber-type pole fittings. The greater traffic-carrying capacity of optical fibers indicates that they should be placed at the top position on poles. Other elements where care must be exercised are the special requirements for underground or overhead structures in terms of optical fibers' smaller diameter, longer lengths and critical bending limits, large splice configurations, and movement and strain limitations.

11.3.2.2 Total Link or Span Length. This length is based on the link budget discussed in Chapter 10 of what some texts call the maximum outside plant (OSP) loss. It should be taken into account that this loss is affected by many factors, such as cabled fiber loss and the number of connectors and splices. A route that contains a large amount of aerial plant may be shorter due to allowances for extreme temperature losses.

A conservative approach to speedier restoration after a link failure is to have on hand a spare length of operational cabled fiber that is equal in length to the longest length placed in the duct sections of the route. If the core of the cable is

[1] We define *span length* as the distance from a fiber-optic terminal or add–drop multiplexer (ADM) to an adjacent regenerator or between adjacent regenerators.

unfilled, the fibers in the cable may break over a length of hundreds of meters at a dig-up; but for filled core cables and cables with discrete blocking, the breaks are usually confined to near the sheath break. Therefore, for unfilled core cable the ITU (Ref. 2) recommends that it is good practice to have a spare length of cable for a route that equals or exceeds the longest cable lengths independent of where it is placed: duct plant, direct-buried, or aerial.

11.3.2.3 Considerations for Total Cable Length. The total cable length is taken from the final approved plans. To this value we add an additional length for each splice or connector. This should include one complete turn of spare cable around the splice/connector manhole plus spare cable within the splice/connector housing. The total cable length also must include the building lead-in cable from the first or last external splice/connector housing to the optical fiber distribution frame located near the fiber terminal equipment, or the in-house cable from the cable distribution room to the fiber distribution frame.

The spare length can be in the range of 4 m (only spare fibers in the splice/connector housing) to 12 m.

11.3.2.4 Reel or Drum Length. Compared to its metallic cable counterpart, longer fiber lengths can be placed on a reel. Equivalent fiber is considerably smaller and lighter than twisted pair or coaxial cable. However, we would want a fiber reel size and weight to be optimized for easy handling in the field. It should also be pointed out that fiber cable manufacturers supply reel lengths generally not much longer than several kilometers where, after this waypoint length, the cost meter begins to increase because of lower manufacturing yields.

For long routes, the maximum separation of splices depends on the physical characteristics of the route (e.g., mountainous, rocky terrain versus comparative rock-free rolling plains). This maximum separation, of course, also depends on maximum reel lengths. Cable lengths that are to be pulled into ducts will be shorter because of limited pulling-in tension and the actual position of manholes. Drum lengths, depending on cable diameter, vary from 1 km up to possibly 10 km.

When calculating drum length, the planning engineer should take into account the following. For duct routes, drum lengths should be determined as follows, once splice positions are identified:

- Distance between splices $= x$ (m).
- Length tolerance allowance $= 2\%$ of x (m).
- Splicing + measurement allowance at 10 m each end = 20 m.
- Thus, a drum length for a duct length of x meters is $(1 + 0.02)x + 20$ m.

Especially for long regenerator spans in ducts, it is important to take into consideration that the minimization of the number of joints should be consistent with the ability to install the resultant cable lengths. The cable length allowance above should be sufficient for an extra splice if the cable installation proves more difficult than anticipated (Ref. 2).

For ploughed or direct-burial routes of cables pulled into separate pipes, drum lengths are determined as follows, once splice locations have been firmly calculated:

- Distance between joints, as measured or taken from firm planning documents: x m.
- Connector tolerance allowance + measurement allowance: 10 m each end.
- Thus, the drum length of a direct-buried route of x meters is $(x + 20)$ m.

11.3.2.5 Number and Location of Splices/Connectors. The number of splices is controlled by the length of the optical fiber cable that is available on a reel and the location and physical constraints of the selected route. Generally speaking, on easy terrain, the longer the length of cable on a reel, the fewer the splices will be on a link. Another consideration is the condition and location of the duct system. It may be such that long lengths of cable cannot be placed. Similarly, for direct-buried routes, there may be obstacles in the route that require cutting the cable and placing the cable from two directions to the obstacle. Typically we may expect splices every one to three kilometers.

Splices are located in manholes in duct systems, as well as in handholes[2] in directly buried plant, and are attached to the support strand in aerial plant. In a subscriber distribution system or HFC system,[3] splices may be located in junction boxes above ground at sites where network rearrangements may be planned.

11.3.2.6 Right-of-Way. Right-of-way is one of the most important cost factors in outside plant design for fiber-optic cable systems. In general, guidance can be taken based on installation factors and right-of-way of metallic counterparts. Cables are generally placed on public rights-of-way such as along highways, and the duct systems are under or adjacent to suburban streets or roads. Other convenient rights-of-way are along railroad tracks, power transmission line rights-of-way, and natural gas and petroleum pipeline rights-of-way. Another possibility is over agricultural land. Here obtaining a right-of-way may be more complex and more difficult. However, if the fiber cable route is where deep plowing is practical, the cable depths may be deeper over such agricultural land.

This planning phase must be followed by a field survey to find out whether the planned route can in fact be installed—for example, if public roads are involved, if private property is involved, if facilities of other institutions (i.e., an electric power company) pose a problem, if cable ducts already installed by other entities can be used, or if houses, towers, and so on are endangered by the installation of the fiber cable system.

A final routing plan is formulated after all of the above factors have been considered and consultation has been held with the various federal, state, town, and private entities involved.

[2] *Handholes* are small, buried boxes.

[3] *HFC system* is a hybrid fiber coax system, found in a CATV plant. See Chapter 13.

11.3.2.7 Pipes, Manholes, Ducts, and Subducts. When there are existing duct lines and where there are insufficient spare tubes to allow the fiber-optic cable to solely occupy one of them, subducting should be considered. It is recommended that all subducts should be installed at the same time and in such a way that the configuration does not tend to spiral as it is pulled in.

The following are some reasons why cables cannot be direct-buried:

- Mechanical protection is required.
- Presence of roads or other obstacles.
- Possible expansion at a later date.
- Protection required against rodents.

As a preparatory measure, one or more spare pipes may be installed, either plowed or direct-buried. The spare pipes consist of individual sections 2500 m in length (Ref. 2) which are joined together by fittings to yield a single pipe spanning the entire length of the route. Later, based on the actual drum length or pulling-in length of the fiber cable, the fiber-optic cable is pulled through the pipe. At points of splicing (jointing), the pipe is open at the appropriate position and subsequently sealed. Where jointing (splicing) is to be carried out in a manhole, the initial route survey will verify if there is sufficient space in each manhole to accommodate the facilities needed such as a mounting table and light, and so on.

The duct or pipe material is normally PVC or HDPE (high-density polyethylene) or an epoxy fiberglass compound. Inside and outside walls for ducts and innerducts are available with longitudinal or corrugated ribs. These ribs help to reduce pulling tensions during installation.

Source: Ref. 3.

Ducts and innerducts have a minimum bending radius. The cable pulled through these ducts/innerducts should not be tighter than the bending radius value. This radius can be specified as supported or unsupported. The supported radius should only be used when the duct is bent around a supporting structure, such as another duct or a reel.

Duct and innerduct can be ordered from the manufacturer with pulling tape, which aids in the pulling procedure. Likewise, they can be prelubricated, which can greatly reduce pulling tensions required. Afterwards, installation ducts and innerducts should have end plugs installed to prevent water leakage. It is important to keep the duct installation dry and free of debris. When sizing ductwork, a fill ratio of 40% is a good rule of thumb. For instance, a 1-in. duct will take a 0.6-in. fiber cable (outer diameter). Calculate fill ratio as follows:

$$\text{Fill ratio}(\%) = \left(d^2/D^2\right) \times 100$$

where d is the outside diameter of the fiber-optic cable and D is the inside diameter of the duct in question. A larger duct size should be used on long runs to ease pulling procedures. Standard duct sizes vary from 3 in. to 8 in., and innerduct size varies from 0.75 in. to 2 in. (Ref. 3).

For cable pulls of any comparatively long length, the use of lubricant is mandatory, in our opinion. Based on Ref. 3, lubricant should be applied at all cable feed locations and intermediate pull locations, and whenever possible just before bends. The lubricant should be used with a lubricant collar and pump. There should lubricant covering the inner wall of the duct for its entire length.

The lubricant's coefficient of friction should be less the 0.25. The lubricant should have no effect on a cable's jacket, conduit duct, or innerduct throughout the life of the installation.

After a cable span has been installed, an OTDR (optical time-domain reflectometer) test should be carried out from each bitter end of the cable and on each fiber strand (see Chapter 15).

11.3.2.8 Plowing or Direct-Buried Cable (Spare Pipe) Sections. Optical fiber cables may be buried by trenching or plowing. Trenching is preferred because it is less demanding and provides a more gentle cable lay. However, it is more costly than plowing. Trenching is used in rocky soil where access is difficult. After layback of the soil in a trenching procedure, the large rocks and sharp stones should be removed from the refill soil to avoid damaging the cable.

Plowing is less costly. It is a satisfactory method where the terrain is gentle and along prepared rights-of-way. Where conditions of hard soil are encountered, it is advisable to rip the route with an empty plow ahead of the plow containing the cable. It is also possible to plow more than one cable in a single pass or to plow in a cable and a single duct simultaneously. The spare duct can be used at a later time for placing a second cable.

11.3.2.9 Use of an Aerial Cable. Should aerial cable be considered for a specific route? An aerial installation has the following advantages:

- It can use existing pole lines.
- It is independent of soil conditions.
- It probably is faster to install.
- There are possibilities of long-length cabling.
- It is easy to maintain, when cables run along public roads.

The following are disadvantages of an aerial installation:

- Shortened lifetime because of environmental stresses.
- Possible excessive cable strain in special conditions such as wind or ice loading or overlength spans.
- Aesthetic considerations.

11.3.3 Cable Installation

The fiber cable installer can use the same general methods with metallic cables as with fiber cables. However, certain aspects of fiber-optic cable must be taken into consideration. These include (a) overstrain of the fiber and (b) provision of proper bending and guiding methods, and one should expect longer lengths to be installed (compared to its metallic cable counterpart). The principal objective is to place the fiber-optic cable in as near a stress-free condition as possible. The manufacturer's recommended physical limitations of the cable must be adhered to to the letter.

11.3.3.1 Duct Installations. When installing a fiber-optic cable, care must be taken because of its small size and relatively low strength (when compared to metallic pair or coaxial cable). Factors that limit the length of a fiber cable that can be pulled into a duct include:

- Number and degree of bends
- Configuration changes at manholes
- Level changes between manholes
- Duct material
- Duct misalignment, damaged or repair section, and the general condition of the duct

There are several techniques that can mitigate these factors and maximize the distance between splicing positions:

- Consider pulling-in from a midpoint in both directions after fleeting or figure-eighting the cable before the second pull.
- Pull-in from one direction, looping out at intermediate manholes where the ductline makes a sharp turn or on each side of a known "difficult" section.
- Pull downhill rather than uphill.
- Use adequate lubrication.
- Thoroughly clean and check clearance in each duct.
- Use intermediate pulling points and use adequate cable guide equipment.

Where a fiber-optic cable is to be installed in the same duct as other big cables such as power cables, the optical fiber cable should have a minimum diameter to avoid wedging.

The use of pulling winches should be considered. These are capable of continuous monitoring of length and pulling tension. The winches are mounted on a trailer and fully self-contained. They are available for transfer where and when required.

11.3.3.2 Direct Burial. With the trenching technique, fiber cables may be directly buried. Trenching is used in difficult areas where the terrain is rocky. In congested areas it is mandatory. The cable is normally handled manually from a motorized reel carrier. It is important to maintain trench alignment as straight as possible.

The backfill, which consists of fines,[4] should be placed around the cable and carefully compacted. If the backfill does not contain fines, suitable fines should be

[4]*Fines:* Very finely ground material; in this case, very fine sand or equivalent.

trucked in and packed around the cable. Cable burial depth should be as recommended in Table 11.2 (750 mm or 30 in.) or dependent on relevant local regulations.

The spare pipes and directly buried cables should be marked by marker tape (e.g., PE foil) positioned approximately 30 cm (8.5 in.) above the cable or pipe bearing a warning of the presence of a telecommunication cable.

11.3.3.3 *Plowing.* Plowing is more cost-effective than trenching. It should be considered whenever there are long, obstruction-free cable runs. As we are reminded, installation cable tension in the laying of fiber-optic cable as well as avoidance, if possible, of short-term tension transients should be major considerations.

Installed cable tension can be reduced by a low-friction insert in the plow and by the use of large-bend radii curves. One way is to have a large-diameter power capstan mounted on the rear of the bulldozer which is used to pull the cable from the drum and present it to the plow chute under minimal tension. High transient tensions caused by overspeeding, as well as whiplash of the cable drum caused by rapid speed or attitude changes of the bulldozer, must be avoided. Marker tapes are plowed in at the same time, some 30 cm (8.5 in.) above the cable or spare pipe. The depth of burial, based on Table 11.2, is 750 mm (30 in.).

11.3.3.4 *Aerial Installation.* There are two methods of aerial installation depending on cable construction: self-supporting or lash construction.

Self-Supporting. The cable is first laid out along the pole line on cable rollers located at the pole positions. This can be done by installation vehicles, depending on terrain conditions, or by pulling the cable out by hand.

Spans of different lengths may be used if sag is selected properly keeping in mind excess load. It is recommended that the aerial fiber line take top position on the poles to avoid the problem of high vehicles passing underneath.

Lash Structure. With such a structure a supporting wire is used. The fiber cable is attached (lashed) to the supporting wire either on the ground or once the supporting wire is in the air. The supporting wire is pretightened before lashing to avoid-excess elongation of the cable. During the procedure, care must be taken to avoid fiber cable damage, especially because of too high tension of the lashing wire on the cable. Sags in the fiber cable must be equalized before lashing if the supporting wire is not first attached to the poles.

11.3.3.5 *Specific Problem Areas Along a Fiber Cable Route*

River Crossing. When the cable route is taken across a river or lake or other body of water, the following should be adhered to:

- For the water-crossing section of the route, special cable should be used that is constructed for underwater application. It should have wire-armor protection as well.

- Avoid underwater splices by using a long continuous section for the water-crossing.
- Avoid any form of cable movement in any direction by embedding the cable in the bottom soil of the crossing.
- Often cable laying in this circumstances is performed using a barge or some other type of boat. In this case, a small percentage of the length must be given over for cable slack.
- If a mid-span splice is required, provisions should be made for maintaining the armor strength across the splice, and the closure of the armor over the splice must withstand the water pressure without leaking.

Vertical Runs for Buildings, Towers, Bridges, and So On. The approach for bringing fiber-optic cables into buildings is very similar to the techniques used for metallic pair cables. The same types of fasteners should also be used.

Epoxy anchor blocks are inserted to hold the fibers and cable structure as one, where the cable runs vertically in a building, bridge abutments, or similar structures. The following guidelines are suggested for spacings:

- Internal terminating cable—every 3 m or less
- Filled cable—every 30 m or less

Locate the first anchor block at the top of the vertical section. Provisions should be made to hold each anchor block mechanically in the vertical riser. It is recommended that fiber splices be as far as practicable from the top and bottom of the vertical section.

Splice Location and Protection. Joints/splices in manholes are placed at the top of the manhole and as close to the wall as possible in a position offering the best possible protection against mechanical loads such as a craftsperson stepping on the splice/joint. Splice/joint housings of direct-buried cables or of cables pulled into spare pipes should also be directly buried.

Splice/joint housings should be located as close as possible to the cable route. Splice/joint housings are designed to withstand direct burial as well as manhole installation. A plastic mesh is used to cover the splice/joint housing. There should also be a coil of excess cable provided with physical protection from hand tools when reexcavating.

11.3.3.6 Pulling Tension of Fiber Cable. There is a limit to the pulling tension placed upon fiber-optic cable when it is being installed, pulled through conduit, along a trench or along any pathway. Table 11.4 gives guidance on maximum pulling tension for typical fiber-optic cable. The maximum pulling tension should never be exceeded. Excessive force will cause the cable to permanently elongate. Elongation may cause the optical fiber to fail by fracturing.

Notes on Cable Pulling. Tail Loading is the tension in the cable caused by the mass of the cable on the reel and reel brakes. Tail loading can be minimized by using little or no braking during the payoff of the cable from the reel. At times, no

TABLE 11.4 Maximum Pulling Tension

OSP Optical Fiber Cable Type	Max. Pulling Tension, lb / newtons
Fiber feeder, dielectric	400 / 1800
Fiber feeder, armored	400 / 1800
Fiber feeder, self-support	1000 / 4500
Central tube, dielectric	600 / 2700
Central tube, armored	600 / 2700
Loose tube, dielectric	600 / 2700
Loose tube, armored	600 / 2700

Source: From table appearing on page 7.2, CommScope *HFC Upgrade Manual*, Ref. 7. Chart is based on CommScope Optical Reach fiber-optic cables.

braking is preferred. Tail loading can also be minimized by rotating the reel in the direction of the payoff, but being careful not to let the reel overspin.

Dynamometers are used to measure the dynamic tension on the cable. They allow continuous review of pulling tension. Sudden increases in pulling tension, caused by factors such as cable binding at an entry/exit point or a tight bend, can be detected immediately.

Breakaway swivels are used alone or in conjunction with dynamometers to ensure that the maximum pulling tension is not exceeded. A swivel with a break tension equal to that of the pulling tension of the cable is placed between the cable puller and the pulling grip. One breakaway swivel should be used for each cable being pulled.

11.3.3.7 Minimum Bending Radius. Minimum bending radius was quantified in Chapter 2, Table 2.2b. Fiber-optic cables are often routed around corners during placement. A more flexible cable (i.e., one with a smaller bending radius) will require less pulling tension to get it through a bend in the route.

Care must be taken during installation that the minimum bending radius is never exceeded. Overbent cable may deform and damage the fiber inside and cause a jump in attenuation.

Table 11.5 gives values for minimum bending radii for various optical fiber cable types. For optical fiber cable, bending radius is given as *loaded* and *unloaded.* Loaded means that the cable is under pulling tension and is being bent simultaneously. Unloaded means that the cable is under no tension or up to a residual tension of around 25% of its maximum pulling tension (Table 11.4). The unloaded bending radius is also the radius allowed for storage purposes.

The loaded bending radii of cables during the construction process are controlled by technique and equipment. Installation lubricants help decreased the pulling tension required to pull the cable through a duct or a conduit.

Excess or Slack Loops. Excess or slack cable is pulled out and looped inside the manhole or vault to facilitate splicing or the future relation of the cable section. Normally, an additional 5% of the total cable span is stored at regular intervals during installation. Loops should be placed at every manhole or vault. The radius of the loop should be no smaller than the minimum bending radius of the cable.

TABLE 11.5 Minimum Bending Radii

OSP Optical Fiber Cable Type/Max. Fiber Count	Min. Bending Radii in./cm	
	Loaded	Unloaded
Fiber feeder, dielectric/18	7.3/18.5	3.6/9.2
Fiber feeder, armored/18	7.3/18.5	3.6/9.2
Central tube, dielectric/48	9.8/24.9	4.9/12.4
Central tube, armored/48	10.0/25.4	5.0/12.6
Central tube, armored/96	11.4/29.0	5.7/14.5
Loose tube, dielectric/72	9.8/24.9	4.9/12.4
Loose tube, armored/72	10.9/27.7	5.5/13.9
Loose tube, armored/216	14.4/36.6	7.2/18.4

Source: From table appearing on page 7.3, CommScope *HFC Upgrade Manual*, Ref. 7.

11.4 OUTSIDE PLANT TEST AND ACCEPTANCE

After the outside plant portion of the fiber cable system has been installed, the following tests and procedures are recommended for quality assurance:

- Total attenuation of the cable including splices and connectors
- Chromatic dispersion
- Metallic parts inspection
- Splice quality
- Verification of maintenance margin by testing and calculation
- Inspection of connectors

11.4.1 Total Attenuation

After the fiber cable has been installed, the total end-to-end insertion loss of each optical fiber is measured. The wavelength of the light source used in the attenuation measurement should be representative of the operational wavelengths of the system. The decibel loss value should be compared to the calculated value of the link budget. The two values should not vary by more than 1 dB. The attenuation measurement can be carried out by a calibrated light source and a light detector at the far end, which is also calibrated. Alternatively, an OTDR may be used. We discuss the applications and uses of an OTDR in Chapter 15. The OTDR is particularly useful in detecting irregularities in the attenuation coefficient, and it can also detect defects and breaks of fiber joints/splices/connectors. In certain cases where the installation acceptance technician decides, it may be useful to measure attenuation at each end of a link.

11.4.2 Dispersion

When a fiber is to transport bit rates in excess of 1 Gbps, dispersion may heavily influence system performance, especially chromatic dispersion. As we said in Chapter 1, a fiber-optic link may be loss-limited or dispersion-limited. Given the manufacturer's specifications for the fiber, we can calculate the accumulated dispersion because we know the total length of the link in question and the fiber's dispersion coefficient. On the other hand, a manufacturer may say that the fiber meets all the requirements of ITU-T Rec. G.652, G.653, G.654, or G.655 (see Chapter 6, Section 6.6). The testing of fiber-optic components and systems is covered in Chapter 15.

11.4.3 Error Performance Testing

The final end-point objective of a fiber-optic system is bit error performance. ITU-T Rec. G.955 and other ITU-T recommendations in the 600 and 900 series recommend a BER of 1×10^{-10}. However, in several documents a BER of 1×10^{-12} is suggested. We hold with the improved BER, as does US Sprint.

How long must we wait for the first random error to appear when we carry out tests for such stringent BERs? At the 10-Gbps SONET or SDH rate, 10 Gbps gives us for the BER value 1×10^{-10} 1 bit error in 10^{10} bits per second; thus, in theory, the first error would probably occur every second. For the 10^{-12} BER value, we would have to wait some 100 s for the first error to occur.

We now slow the bit rate down to, say, 2.5 Gbps and assume 1×10^{-10} BER; we now will have to wait 4 s for the first random error. At a 1-Gbps rate, 10 s; 500 Mbps, 20 s; 1 Mbps; 10,000 s. The message here is be careful when using such excellent BERs; they may be hard to measure. Then, of course, use some value more reasonable.

The ITU-T recommendations use a BER of 1×10^{-3} as a threshold. This BER value is based on telephone signaling. With a BER poorer than 1×10^{-3}, the telephone channel associated with that signaling will drop out, and the user will now get a dial tone. Because, statistically, less than 50% of PSTN network traffic is voice, we question that BER value. Wouldn't a value such as a BER of 1×10^{-6} be more reasonable? It certainly would tend to better satisfy data users.

11.5 SUBMARINE SYSTEMS

Long-reach undersea fiber-optic systems present ticklish problems for the system designer. Leaving aside the fact that special cable is required under these circumstances to prevent harm to the cable due to the corrosive effects of sea water and the tearing up of the cable by the odd trawler, the principal thrust of design is system availability. In Chapter 12 we discuss *availability* and the means and measures we have to take to maximize system *uptime* to values on the order of 99.99% and better. However, our discussion there centers basically on terrestrial systems (meaning overland). Undersea systems present very special problems.

First off, access to a submarine fiber-optic system is difficult and expensive. A specialized ship is required. Access could take many, many days. Furthermore, there are not much more than 40 fiber cable ships serving the entire world.

11.5.1 Measures to Improve Availability

The following are some of the measures that must be taken to bring system availability up to "many nines" (e.g., 99.9999+):

- For the entire undersea portion of the system, improve component, subsystem, and system reliability. Also, use hi-rel[5] parts. Rel testing must take into account the undersea environment (e.g., water pressure, salinity).
- Use redundancy at all strategic locations.
- Employ dual routing wherever possible. Consider arrangements with competing systems to lease space or dark filter in case of failure of the principal system. Another possibility is to alternate route critical via satellite.
- Use ring architectures wherever possible with switchover regimes in the millisecond range or less.

The ITU fiber-optics handbook (Ref. 2) recommends the following. Irrespective of specific cable characteristics from one design to another, there are several overriding criteria with which all submarine optical fiber cables must comply:

(a) To have a system life of over 25 years.
(b) To offer a power feed conductor to the repeaters which has a low DC resistance and a high-voltage insulation.
(c) To have sufficient strength and degree of protection appropriate to the service environment to enable safe laying and recovery of the fiber cable in adverse weather conditions using conventional shipboard handing methods.
(d) To withstand the anticipated hazards of the submarine environment.
(e) To provide a cable construction that gives protection for the fibers against the effects of excessive strain, pressure, water penetration, and hydrogen.
(f) To ensure that in the event of complete cable break, the cable is designed to restrict water ingress and minimize hydrogen generation.
(g) To enable the quick deployment of efficient and reliable cable repair techniques at sea.

Source: Ref. 1.

REFERENCES

1. *Standard for Physical Location and Protection of Below-Ground Fiber Optic Cable Plant*, TIA/EIA-590-A, Telecommunications Industry Association, Washington, DC, January 1997.
2. *Optical Fibres Systems Planning Guide*, CCITT–ITU, Geneva, 1989.
3. Bob Chomyez, *Fiber Optic Installer's Field Manual*, McGraw-Hill, New York, 2000.
4. *Digital Line Systems Based on 1544 kbps and 2048 kbps Hierarchy on Optical Fibre Cables*, ITU-T Rec. G.955, ITU Geneva, November 1995.
5. *General Features of Optical Fibre Submarine Cable Systems*, ITU-T Rec. G.971, ITU Geneva, November 1996.
6. *Characteristics of a Single-Mode Optical Fibre Cable*, ITU-T Rec. G.652, ITU Geneva, April 1997.
7. CommScope *HFC Upgrade Manual*, Vol. 2, *Fiber*, CommScope of North Carolina, 1999.

[5] *Rel* stands for reliability.

12

SYSTEM AVAILABILITY AND SURVIVABILITY

12.1 THE IMPORTANCE OF AVAILABILITY AND SURVIVABILITY

In the early days of optical fiber cable use, we pictured a buried cable passing through the farmer's field. The farmer is plowing the field, and his plow hooks the cable and snaps it. Service is lost due to the severed cable. Restoration will take at least a day. What to do?

Today, the cable will probably not pass through the farmer's field due to right-of-way considerations. It is still vulnerable to being severed, perhaps by a backhoe of a construction crew. For those of us in the industry the common term used for severing a fiber-optic cable is "*suffering from a backhoe fade*." Now, instead of cutting hundreds of trunks, we may well be cutting tens of thousands of trunks—a dreadful loss to the community, to public safety, to the revenue stream, and of customer goodwill.

Another way of cutting or degrading connectivity of fiber-optic cable is part failure or degradation of a component or subassembly in the cable system such as loss of a fiber optic amplifier. If active components in a fiber-optic system lose their prime power source, there is a failure. Another important item we tend to lose sight of is system synchronization and timing. For example, if there is a timing offset of more than half a bit duration on the principal bit stream, for all intents and purposes the system is down.

Our objective here is to be able to assure the customer of the installed fiber system that her/his system will remain operational, meeting performance objectives with some specified probability. This probability will be based on our capability and the capability of the system to be operational through some very high percentage of time. This time takes into account the probability of part failure, how fast the system can be repaired after a part failure, and how fast it can be returned to normal service after a catastrophic failure (severing of the cable).

12.1.1 Definitions of Availability and Survivability

Let us define *survivability* as how well the system can withstand a catastrophic event. Here we mean severing the cable in one place. *Availability* means the percentage of total time (e.g., in a year) that the system is up and operational. More formally we define *availability* as "the ability of an item—under combined aspects of its reliability, maintainability, and maintenance support—to perform its required function at a stated instant of time or over a stated period of time and as the ratio of uptime to uptime plus downtime" (Ref. 1).

Availability is usually expressed as a percentage (e.g., the system availability is 99.993%).

Another term we will run into is *unavailability*. If the availability of a certain fiber-optic system is 99.997%, what is its unavailability? This is simply 1 − availability—or, in this case, 1 − 0.99997, which is 0.00003 or 0.003%. If this is the unavailability, what is the *downtime* over 1 year? One year represents 8760 hours. $8760 \times 0.00003 = 0.2628$ hours or 15.768 minutes.

We often express availability with the following formula:

$$A\% = \text{MTBF}/(\text{MTBF} + \text{MTTR}) \times 100 \tag{12.1}$$

MTBF is mean time between failures, measured in hours; and MTTR is the mean time to repair, measured in hours.

The following example will clarify the use of equation 12.1. Assume that the MTBF of a certain fiber-optic amplifier is 10,000 hours and that the MTTR is 60 minutes or 1 hour. Calculate the availability of the amplifier:

$$A\% = 10{,}000/(10{,}000 + 1) \times 100 = 0.99990001 \times 100 = 99.990001\%$$

MTBF is a popular measure of reliability. Another measure of reliability is the *FIT*, which stands for "failure in time." This is defined as failures in 10^9 hours.

In formula (12.1), MTTR is the most difficult parameter to quantify with some high probability of being correct. Included in MTTR is (a) travel time to the site in question and (b) troubleshooting time to determine the defective part, subassembly, or circuit card. Then the service technician has to determine the location of a replacement circuit card or subassembly (in local stores, in a central storeroom). For a central storeroom, we will have to include in MTTR time to deliver the part or card to site, time to replace and test, then time to cutover on line. Suppose the replacement card or subassembly was not available in central stores and had to be ordered from the manufacturer. The replacement could take days or even weeks to arrive on site.

We will cover MTBF, availability, and MTTR more extensively further along in our discussion.

There are many ways to make the whole process more efficient. One could (1) reduce the different types of spares required making greater use of common parts, circuit boards, and subassemblies, or (2) provide redundant circuitry with fail-safe cutover wherever there are circuit boards or subassemblies with a poor repair record. One could also use LED displays or other visual indication to show the circuit board on line and a red neon light or color LED display for a replaceable redundant circuit board that is defective and that should be pulled and replaced. Most of these BITE (built-in test equipment) circuits are "go," "no-go."

12.1.2 Survivability and Force Majeure

A fairly large number of system acquisition and installation contracts are written with a paragraph allowing for "force majeure exception." *Force majeure* is a legal term meaning "act of God." Among such "acts of God" we include forest fires, hurricanes, earthquakes, volcano eruptions, floods, and avalanches. These severe disturbances can take a fiber-optic cable system out in a big way. The resulting damage may require weeks or months of major repair work to place the system back in operation. Many entities require a system availability study, but allow for a force majeure exception. In other words for the availability calculation, outage due to force majeure is not counted against availability. It is a lovely "out" for the contractor, because such calamitous events cannot be forecast.

There are many actions a contractor can take to either lessen the effect of force majeure events or even eliminate the damaging results. Buried cable withstands these events much better than does aerial-mounted cable. Even so, handhole slack must be provided for buried cable to allow strain relief on the cable. This is particularly necessary during an earthquake event. Other measures can be taken to lessen the effects of earthquakes. In equipment room locations or remote shelters, racks are mounted with both floor and ceiling reinforcement often using rubber mountings to absorb the shock. The buildings and shelters themselves must be made earthquake-resistant.

Another important measure is to add special mechanical reinforcement at water crossings for spring freshets, flood conditions, and ice movement. The installer should trench/plow on high ground where flooding can be expected.

12.1.3 Channel Outage Definition

The Telcordia TSGR (Ref. 6) defines *channel outage* as follows: "A digital transmission channel is considered unavailable, or in a complete downtime condition, when its error performance falls below a given threshold and remains below that threshold for some period of time." The definition is based on ITU-T Rec. G.821 (also see ITU-T Rec. G.826) and is quoted in part below:

> A period of *unavailable time* begins when the bit error ratio (BER) in each second is worse than 1×10^{-3} for a period of 10 consecutive seconds. These 10 seconds are considered unavailable time.

The unavailable time begins from the first severely errored second in the sequence. In addition:

> The period of unavailable time terminates when the BER in each second is better than 1×10^{-3} for a period of 10 consecutive seconds. These 10 seconds are considered to be available time

Available time starts at the first fault-free second in the sequence. Thus, unavailability or downtime of a channel begins when the first of 10 consecutive severely errored seconds to consecutive non-SESs occurs.[1] An SES is a second in which the BER is greater than 1×10^{-3}. This definition applies to all causes of degradation affecting the channel error performance, including (unprotected) equipment failures and other factors that contribute to poor performance.

Source: Ref. 6.

12.1.4 Short-Haul Availability Objective (Telcordia)

The short-haul Telcordia availability objective for a 250-mile two-way broadband channel is 99.98% availability, or 0.02% unavailability. When expressed as downtime, it is 105 minutes of downtime per year. For systems of less than 250 route miles, the downtime is reduced linearly by 0.42 minutes/year/mile.

For a DS3 configuration, the availability objective for a DSX-3 to a DSX-3 interface is 75% of the above objective for a DS1 channel: 79 minutes/year/DS3 channel for a 250-mile system.

For a DS3 configuration that is shorter than 250 miles, the downtime objective is decreased linearly by 0.32 minutes/year/mile. Allocate 75% of the downtime to transmission media and 25% to terminals and regenerators.

12.1.5 Electronic Equipment and System Reliability References

In the "reliability community" there are three accepted references with accompanying models. These include MIL-STD-217E (Ref. 2) and Telcordia TR-332 (Ref. 3) for equipment reliability, and GR-929-CORE (Ref. 4) deals with measuring reliability/availability of telecommunication systems. For fiber-optic systems, include ITU-T Rec. G.911 (Ref. 5).

12.1.6 Transmission Systems Generic Requirements (TSGR) Equipment Reliability Prediction

There are three accepted methods to calculate reliability of equipment:

Method I is the "parts count" method, where the unit failure rate is the sum of the device failure rates. Various stress levels (see MIL-STD-217E, Ref. 2) maybe applied. These are called *device quality factors* by Telcordia. Device failure rates are further modified by operating temperatures and electrical stress factos. If actual device temperatures and stress levels are not known, the devices are assumed to operate at 40°C and 50% of the rated electrical stress. An environmental factor is applied to the unit failure rate for applications other than a standard switching environment.

Method II combines data deriving from laboratory testing with predictions from Method I. The failure rate is calculated as a weighted average of measured failure rate and parts count generic failure rate, with the weighting determined by the confidence level of the laboratory data. The reliability prediction program (RPP) gives such information as how many devices or units should be tested, how long they should be tested, and how they should be tested.

[1]In the North American Network, carrier group alarms occur after a 2-s loss of synchronization at the DSI level. Affected channels then remain out of service for approximately 10 s (Ref. 6).

Method III is based on statistical estimates of actual in-service field data of a reliability study program. Such elements that are essential to a failure rate estimate are carefully controlled in this reliability program, which depends on the reliability experience of the actual equipment operating in the field. These elements include the number of units in the study, the number of units declared to have failed, the length of time each unit was in the study, and an estimate of the statistical precision of the failure rate estimate. As one would expect, the study population and environment must be representative of the product in-service. If not, an analytical basis must be available for making an appropriate adjustment. Throughout the study, each study unit must be accounted for. When reporting a failure, care must be taken to verify each failure. We must distinguish among initial defective units, infant mortality units, and long-term, in-service failed units, and we must ensure that all failures have been reported.

We recommend ITU-T Rec. G.911 (Ref. 5) to calculate availability of fiber-optic systems. However, our own approach is quite simplified. It is described in the next section.

12.2 RELIABILITY RELATIONSHIPS

FITs is an acronym meaning "failures in time." Its definition is "failures in 10^9 hours." To convert FITs to failures in 10^6 hours, multiply the FITs value by 10^{-3}. In most texts the notation λ is the failure rate of a unit, device, or entire system. If we measure λ as failure per million hours, then we can just invert this value to arrive at the equivalent MTBF. If devices are connected in a series configuration, the failure rate of the configuration is the sum of the failure rates of those devices.

If the failure rate is measured in FITs (F), then the MTBF (M) is

$$M = (1.14 \times 10^5)/F \tag{12.2}$$

where M is measured in years per failure.

For example, calculate the MTBF of a circuit pack. We sum the failure rate value for each of the elements comprising the circuit pack. The circuit-pack consists of:

1 integrated package	Unit FIT rate = 1500	Total FITs = 1500
6 ICs	Unit FIT rate = 300	Total FITs = 1800
5 resistors	Unit FIT rate = 123	Total FITs = 615
8 capacitors	Unit FIT rate = 57	Total FITs = 456
1 connector	Unit FIT rate = 27	Total FITs = 27
1 PCB/print ckt board	Unit FIT rate = 27	Total FITs = 27
		TOTAL 4425

IC = integrated circuit
PCB = printed circuit board
ckt = abbreviation for *circuit*

The circuit pack MTBF = $1.14 \times 10^5/4425 = 25.76$ years.

This second example is a simple series calculation for a fiber-optic system. Again we sum the failure rate for each of the plug-in circuit packs (and other equipment that is not on plug-in circuit packs) that comprise the system. In this example we have:

6 channels packs	Unit FIT rate = 8000	Total FITs = 48,000
5 power supplies	Unit FIT rate = 6500	Total FITs = 32,500
5 regenerators	Unit FIT rate = 12,050	Total FITs = 60,250
1 microprocessor board	Unit FIT rate = 12,300	Total FITs = 12,300
1 monitor board	Unit FIT rate = 3400	Total FITs = 3400
		TOTAL 156,450 FITs

MTBF $= 1.14 \times 10^5/1.56450 \times 10^5 = 0.7286673$ years or $0.72866 \times 8760 = 6383$ hours.

Suppose we thought that 6383 hours was not an acceptable value. What could we do? One approach is to add a full redundant line. Once we do that with fail-safe switching, we can now square the MTBF value and it becomes 40,744,291.76 hours, quite a respectable value.

Another, more economical approach is to make the device with the worst failure rate redundant on a 1 + 1 basis. In this case it is the system regenerators; there are five of them. In the example, a regenerator has a FIT rate of 12,050. Without going into the complexity of a Markov chain and stochastic processes, convert the FIT rate to MTBF, square the calculated value, and calculate the equivalent FIT rate.

Calculate the equivalent MTBF for a FIT rate of 23,050:

$$\text{MTBF} = 10^9 \text{ hours/FITs} = 10^9 \text{ hours}/12{,}050 = 82.987 \text{ hours}$$

$$(82{,}987 \text{ hours})^2 = 6.886933765 \times 10^9 \text{ hours}$$

$$F = 10^9/6.8869 \times 10^9 = 1/6.8869 = 0.1452$$

Recalculate the MTBF value with redundant regenerators, substituting 5×0.1452 for the regenerator value used. The total FITs for the system reduces to 96,201. The new value for MTBF is then

$$\begin{aligned}\text{MTBF} &= (1.14 \times 10^5)/0.96201 \times 10^5 \\ &= 1.1875 \text{ years, which is marginally better than the previous value.}\end{aligned}$$

If MTBF is measured in hours and we wish to convert this value to FITs, then

$$\text{FITs} = 10^9/\text{MTBF}_{\text{hours}} \tag{12.3}$$

Conversely, if we are given a failure rate in FITs and want the equivalent value of MTBF, measured in hours, then:

$$\text{MTBF}_{\text{hours}} = 10^9(\text{hours})/\text{FITs} \tag{12.4}$$

Remember, there are 8760 hours in a year (24×365).

12.3 CALCULATION OF SYSTEM AVAILABILITY

In our calculation of system availability, force majeure events are not included in the outage time.

12.3.1 Availability and Unavailability

Let us assume that a fiber-optic system is specified to have an availability of 99.0%. That means over the course of a year I can expect the system to be operational, meeting its BER requirements, for $0.99 \times 8760 = 8672.4$ hours. What about the remainder of the time or $8760 - 8672.4$ hours = 87.6 hours—that is, the period of time we would expect the system not to be meeting its BER requirements or to be just nonoperational (down) or *unavailable*. The 87.6 hours represents the system's *unavailability*. If we let A be availability and let U represent unavailability, given one value, we can always calculate the other as follows:

$$U = 1 - A \tag{12.5}$$

If the unavailability is 0.01%, what is the equivalent availability? Convert the percentage to a decimal, 0.0001, and then

$$0.0001 = 1 - A$$
$$A = 1 - 0.0001, \text{ or } 99.99\%$$

TABLE 12.1 Varying MTTR (MTBF Held Constant at 20,000 Hours)

MTTR (hours)	Availability	Comments
0.5 (i.e., 30 minutes)	0.999975	Technician on duty 24 hours a day, spares immediately available, advanced BITE system installed.[a]
1.0	0.99995	Technician on duty 24 hours a day, spares readily available, good maintainability.[b]
2.0	0.999900	TSGR recommended value at manned locations such as switching center.[c]
3.0	0.99985	
4.0	0.99980	TSGR recommended value for unmanned location—allow 3 hours for travel.[c]
6.0	0.99970	

[a]BITE stands for *built-in-test equipment*. What this means is that there is a system of "go, no-go" circuits allowing a technician to be able to spot the cause of outage down to the PCB (card) level. These circuits can be remoted to a centralized readout platform or to a network control center. Another maintainability enhancement is to have all testing by front access.

[b]MTTR = 1 hour. From 30 minutes for the previous entry, which allows extra time for troubleshooting. We can assume that the level of maintainability is less in this situation.

[c]TSGR is an abbreviation for Telcordia "transport system generic requirements" (Ref. 6). This document sets MTTR values as 2 hours for switching centers including the transmission equipment housed in the same building as the switch(es). An MTTR of 4 hours is assigned for remote equipment. On the fiber-optic side, this may be an ADM feeding a community dial office, a regenerator, or a fiber amplifier that are out along the fiber-optic cable line.

The standard equation for availability is [equation (12.1)]

$$A\% = \text{MTBF}/(\text{MTBF} - \text{MTTR})100$$

Suppose we assume that the MTBF of a fiber-optic terminal multiplexer is 50,000 FITs or $10^9/5 \times 10^4 = 20{,}000$ hours and we vary MTTR 30 minutes and 1, 2, 4, 6, and 8 hours. See Table 12.1 for availability figures.

12.4 NETWORK ARCHITECTURE AND SURVIVABILITY

Network architecture can have a major impact on system availability. Ring networks are nearly universally employed on long-haul fiber-optic cable systems. The ring will consist of fibers in groups of two. In one case, one of the fibers funnels traffic upstream while the other fiber carries traffic downstream. On these long-haul systems, often supporting the PSTN community, traffic is symmetrical. Some fair percentage of the traffic is voice serving telephony subscribers. These subscribers have full-duplex service. Other services are Internet data, enterprise network data, and leased data connectivities carrying frame relay, IP, and ATM traffic. Expect these to be full duplex and symmetrical. Certain video circuits may be one-way or with an asymmetrical imbalance. Conference television is usually two-way. Section 12.4.7 describes the basic ring configurations deployed by the fiber-optics community.

12.4.1 Automatic Protection Switching (APS)

Fiber-optic cable-based transport systems nearly always are connected in some sort of a ring configuration to protect against facility failures. In the ring architecture, which is the most commonly encountered, the traffic can be routed in either (or both directions) around a closed loop containing multiple nodes. This ring topology provides protection against intermediate node failures, although traffic originating or terminating at a failed node cannot be restored unless node redundancy or diversity is provided. If such facilities are provided, it may be possible to restore 100% of the traffic following any single failure in the ring. *Automatic protection switching* (APS) enables the restoral of traffic flow.

APS has also been defined both for systems that monitor for degradations and failures on individual channels or paths, as well as for systems that monitor the entire signal. When degradations or failures are detected, all or a portion of the connectivity, such as one channel, of the full capacity may be switched to a standby facility.

12.4.2 Switch Activation

Protection switching on a fiber-optic cable system activates when the BER of the signal on an active fiber (either a working line or a protection line) is greater than the switch initiation threshold and a standby fiber, fiber group, or cable (either a protection configuration or the original working configuration) with a better BER is available. Generally, the activation threshold occurs when the BER is in the range of 10^{-6} to 10^{-9}. To accommodate most cases, the threshold value can be selectable. Usually, in addition, the system can support a second switch initiation

TABLE 12.2 Default Frame Structure Based on the K1 and K2 Bytes

K1 Byte								K2 Byte							
X	X	X	X	0	0	0	0	0	0	0	0	X	X	X	X

X, any value.

Sources: Based on ANSI T1-105.01 (Ref. 8) and Telcordia GR-1230 (Ref. 7).

threshold so that severely degraded lines can be given priority. The second switch initiation threshold is usually set in the BER range of 10^{-3} to 10^{-4}.

Detection time is the time that the equipment takes to determine that a *hard* failure[2] had occurred or that a BER threshold is being exceeded, and to initiate a switching action if appropriate. The detection time for hard failures should be less than 10 ms.

Switch activation or deactivation is triggered by certain specific bit sequences in the K1 and K2 bytes of SONET overhead. See Table 12.2.

The time to complete a switching action once it is initiated should be under 50 ms. For the case of bidirectional switching, the switch action in both directions should be completed under 50 ms. Craftpersons can initiate a manual switchover. This means that traffic could be interrupted for as long as 50 ms during a manual switchover. Certain users may find the 50-ms value for switchover time excessive. Telcordia therefore has the following conditional requirements for manual switchover action. In certain cases, the switch action should be error-free. Manual switchover actions preclude a detection time, thus switchover actions may be much less than 50 ms for completion time.

12.4.3 Restoration

When the BER of the working line improves, the traffic may continue to be carried on the protection line (called *nonrevertive switching*), or it may be switched back to the working line (called *revertive switching*). Telcordia recommends that a restoral threshold be provided, with a value of $T/10$ (where T is the switch initiation threshold, discussed above).

After the BER on the working line meets the restoral threshold, a wait-to-restore (WTR) period of 5–12 minutes should elapse before revertive switching is performed, unless the protection line is needed for other purposes such as protecting another degraded line. If the protection line is needed for other purposes, the revertive switch is initiated as soon as the BER on the working line meets the restoral threshold.

The rationale for the WTR period is to eliminate frequent oscillations between working and protection lines resulting from intermittent failures. It only applies to automatically initiated switching caused by failures or degradations on working lines. It does not apply after manually initiated switches are cleared, or after failures or degradations on protection lines.

12.4.3.1 Enabling the APS Function. Upon a failure, either a fiber cut or nodal, the APS function is activated by the K1 and K2 bytes in the SONET overhead. Table 12.2 shows the default structure of the K1 and K2 byte frame in

[2] Hard failure is a failure of hardware (vs. a software failure or "bug").

the default mode. The APS codes are developed through the bits in these two bytes (octets).

The K1 and K2 bytes for rings may be in either the *idle state* or the *switching state*. When K1/K2 are in the idle state for rings, the bits take on the following formation:

K1 (bits 1–4)	0000 (No request code)
K1 (bits 5–8)	Destination NODE ID
K2 (bits 1–4)	Source NODE ID
K2 (bit 5)	0 (Short path code)
K2 (bits 6–8)	000 (Idle code)

When K1/K2 bytes are in a switching state formation as sourced by a ring node, expect the following formation:

K1 (bits 1–4)	Bridge request (status) code
K1 (bits 5–8)	Destination NODE ID
K2 (bits 1–4)	Source NODE ID
K2 (bit 5)	1/0 (Short/long path code)
K2 (bits 6–8)	STATUS code

We selected the following grouping of bit assignments in byte K2. This is included under *general criteria for the K1 and K2 bytes*. Only K2 data is shown as an example.

K2 Bit Assignments

Bits 1–4	Indicate the node ID of the node souring the request
Bit 5	Indicates if the bridge request in byte K1 bits 1–4 is a short path request (0) or a long path request (1)
Bits 6–8	111 = Line AIS
	110 = Line RDI
	101 = Reserved for future use
	100 = Reserved for future use
	011 = Extra traffic (ET) on protection channel
	010 = Bridged and switched (Br & Sw)
	001 = Bridged (Br)
	000 = Idle

Note: AIS = Alarm indication signal
RDI = Remote defect indicator

Sources: Based on Refs. 7 and 8.

12.4.4 Protection Switch Reliability and Availability

12.4.4.1 Silent Failures. Not all failures are alarmed. Those nonalarmed failures of the protection switching equipment that prevent switching to a protection line are called *silent failures*. These failures cause service outages, and their detection is important. There can be other alarmed failures of protection lines or associated control functions. These failures are assumed to be promptly corrected.

There are three types of silent failures:

1. Common equipment failures that prevent switching for all *N* service channels or working lines in an *MXN* protection switching system.
2. Equipment failures that prevent switching for a given service channel or working line, but can be detected by completing the final switching stage.
3. Equipment failures that prevent switching for a given service channel or working line and that can be detected without completing the final switching stage.

12.4.4.2 Exercising. Exercising of protection switching equipment is a form of preventive maintenance. The goal is to detect switching troubles before they can cause an outage. Protection switches are exercised all the way up to the final switching stage. Telcordia (in Ref. 6) states that in digital fiber-optic cable systems, outage caused by silent failures should be no more than 10% of the total outage.

The following exercising requirements apply:

1. The exercise frequency should be user-changeable.
2. The exercise routine should include a check that a transfer has been made at the head-end.
3. The exercise routine should include a check that an acceptable signal is being provided to the final transfer switch.
4. The switching system should drop an exercise routine to service failed lines.
5. If a switch exercise fails, a distinctive alarm should be produced locally, and it should be possible to disable this feature either locally or remotely.
6. If an exerciser has an option to complete the final switching stage, it should be possible to disable this feature either locally or remotely.
7. If a service failure results from the completion of the final switching stage, the system should return to normal operating conditions within 50 ms.

Source: Ref. 6.

12.4.5 Options with Line Protection

12.4.5.1 One-for-One Link Protection (1 + 1). 1 + 1 protection is very effective to achieve full redundancy as we will show in this section. This type of layout is widely employed usually with a ring architecture. In the basic configuration of a ring, the traffic from the source is transmitted simultaneously over both bearers, and the decision to switch between main and standby is made at the receiving location. In this situation, only *loss of signal* or similar indications are required to initiate changeover, and no command and control information needs to be passed between the two sites. It is assumed that after the failure in the main line, a repair crew will restore it to service. Rather than have the repaired line placed back in service as the "main line," it is designated the new "standby." Thus only one line interruption takes place, and the process of repair does not require a second break in service.

The best method of configuring 1-for-1 service is to have the standby line geographically distant from the main line. This minimizes common mode failures.

Because of the simplicity of this approach, it ensures fastest restoral with the least requirement for sophisticated monitoring and control equipment. However, it is more costly and less efficient of equipment usage than an $N + 1$ approach. It is inefficient in that the standby equipment sits idle nearly all the time, not bringing in any revenue.

12.4.5.2 The N + 1 Approach. The $N + 1$ link protection method makes more efficient use of standby equipment. It is just an extension of the 1 + 1 technique described above. With the excellent reliability of present-day equipment, we can be fairly well assured that there will not be two simultaneous failures on a route. This makes it possible to share the standby line among N working lines.

$N + 1$ link protection makes more cost-effective use of equipment but requires more sophisticated control and cannot offer the same level of availability as 1 for 1 protection. Diverse routing of main and standby lines is also much harder to achieve.

Source: Ref. 8.

12.4.6 SONET Self-Healing Rings (SHRs)

A SONET self-healing ring consists of four or more network elements tied together in series by either two or four fibers. The formation of NEs (network elements) is tied back on itself to form a ring. SONET NEs are actually ADMs (add–drop multiplexers) permitting access to the ring by PDH/SDH sources. There is a limit on the number of ADMs that can be interconnected by a ring. That value is usually 16. SONET SHRs have been previously introduced in Chapter 9, Sections 9.2.6 and 9.2.7.

There are two principal types of SONET SHRs: line-switched rings and path-switched rings. In a line-switched ring, incoming traffic is only assigned in one direction around the ring unless a switch action occurs (whether 1 + 1 or $1 + N$ style). Switching is coordinated by the nodes on either side of the failure condition in the ring via a special signaling protocol. The default configuration results in a bidirectional traffic flow. In a path-switched ring, add traffic (as in add–drop multiplex) is always routed in both directions around the ring (1 + 1 configuration). Protection switching is the responsibility of the *selector*, as the name of the device indicates. The selector makes its decisions on a per-path basis. The resulting default configuration gives rise to a unidirectional traffic flow; but after protection switches have been activated, the traffic for some connections will be bidirectional. If nonrevertive switching is used, then it will continue to be bidirectional.

12.4.6.1 Line-Protection Switched SHR. The line protection-switched SHR uses SONET line layer OAM indications to trigger protection switching action. This switching action is carried out at the line layer to recover from failure conditions on the ring. It does not involve path layer OAM. Line layer OAM includes line layer defects such as loss-of-frame and includes maintenance signals such as AIS-L.[3] Signaling messages are exchanged between nodes to affect a coordinated line protection switching action.

[2]AIS-L stands for alarm indication signal—line.

12.4.6.2 Path Protection Switched SHR. This type of SHR architecture uses SONET path layer OAM to trigger the path protection switching action. The switching action is carried out at the STS or VT path layer to recover from failure conditions. It does not involve line layer OAM indications. Path layer occurrences include such path layer defects as LOP-P and maintenance signals such as AIS-P. Path switching of a specific path is independent of any other path's status.

Here we deal with a dual-fed unidirectional path-switched ring (UPSR) implementation (Ref. 10). In this type of SHR, both directions of working traffic between any ring node A and any other ring node B travel on a single fiber ring in the same direction around the ring, arriving at their destinations by different paths. This means that different ring nodes are traversed for each path. A second ring connectivity carries a duplicate copy of path signals in the opposite direction around the ring. Thus we are dealing with a "dual-fed" or "1 + 1 protection" approach because duplicate information is transmitted in both directions on the ring. This means that the full capacity of one ring protects the full capacity of the other ring. At an exit node we can expect to find two signals (one from each direction) being available for signal selection. If a break occurs in the ring, such as in the span between nodes X and Y in Figure 12.1, exit node W switches to the standby path for path restoration. Exit node Y is not affected by the break, so no path switching is necessary.

One should take note that path switching in a UPSR is single-ended. What is meant here is that protection switching is only performed at the exit node—without any type of coordination with, or notification at, the entry node. Furthermore, it should be noted that this architecture changes from a ring to a linear network if a break in the ring (in both directions) occurs. However, service is maintained following such a break. That is, the ring nodes do not remain in a ring topology followlowing a break in the ring or loss of a node. When such a defect occurs, the ring nodes are now actually operating in a bidirectional manner. Besides, it should

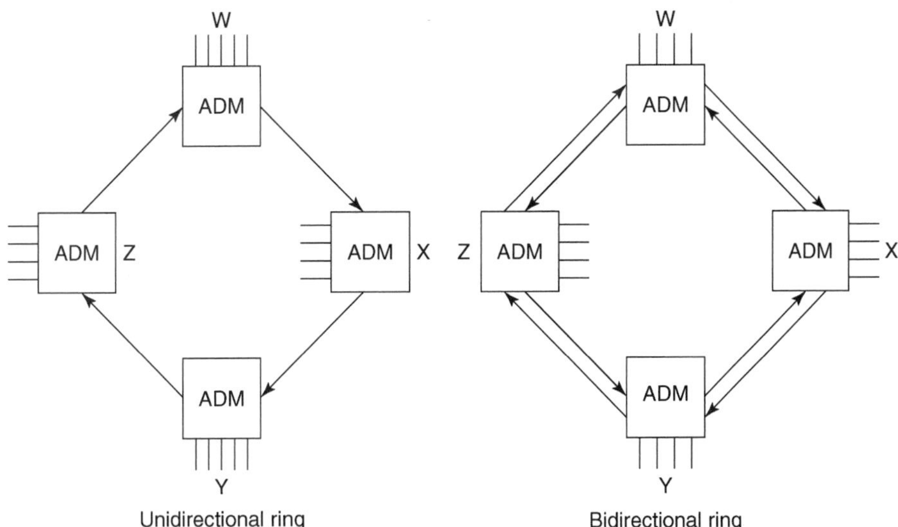

Figure 12.1. (a) Unidirectional and (b) bidirectional SONET rings distinguished. ADM stands for add–drop multiplex.

be noted that the ring could be provisioned for bidirectional working signal under normal conditions and have a unidirectional routing during a failure.

12.4.7 Ring Configurations

A ring network consists of network elements connected in a point-to-point arrangement that forms an unbroken circular configuration as shown in Figure 12.1. As we must realize, the main reason for implementing path switched rings is to improve network survivability. The ring provides protection against fiber cuts and equipment failures.

A variety of names are used to describe path-switched ring functionality—for example, *unidirectional path protection switched (UPPS) rings, unidirectional path switched rings (UPSR), uni-rings, and counterrotating rings.*

Ring architectures may be considered a class of their own, but we analyze this conceptually in terms of 1 + 1 protection. Usually, when we think of a ring architecture, we think of route diversity; and there are two separate directions of communications going around the ring. The ring topology is most popular in the long-haul fiber-optics community. It offers what is called *geographical diversity.* Here we mean that if there is sufficient ring diameter (e.g., > 10 miles), there is an excellent statistical probability that at least one side of the ring will survive forest fires, large floods, hurricanes, earthquakes, and other force majeure events. It also means that we assume that only one side of the ring will suffer an ordinary nominal equipment failure or "backhoe fade."

There are some forms of ring topology used in CATV HFC fade systems, but more for achieving efficient and cost-effective connectivity than as a means of survivability. Rings are not used in building fiber cabling systems and are not mandatory in campus fiber routings.

There are two basic SONET self-healing ring (SHR) architectures: the unidirectional and the bidirectional. Depending on the traffic demand pattern and some other factors, some ring types may be better suited to an application than others. Figure 12.1 distinguishes the unidirectional and bidirectional rings. With the unidirectional ring, operational traffic is carried around the ring in one directional only. For instance, traffic bound from node W to node Z would traverse the ring in a clockwise direction, and traffic bound from node Z to node W would also traverse the ring in a clockwise direction. The working capacity of a unidirectional ring is determined by the total traffic demand between any two node pairs.

In the case of a bidirectional ring, traffic goes around the ring in both directions using two parallel paths between nodes. (These paths travel on different fibers in the same fiber sheath). Based on the example above, traffic bound from node W to node Z traverses the ring in a clockwise direction through intermediate nodes X and Y, and traffic bound from node Z to node W returns along the same path which also passes through the intermediate nodes X and Y.

Another attribute of the bidirectional ring is that traffic in both directions of transmission between two nodes traverses the same set of nodes. Unlike a unidirectional ring time-slot, a bidirectional ring time-slot can be reused several times on the same ring, which better utilizes the ring's bit rate capacity. All nodes on the ring share the same protection bandwidth (bit rate capacity), regardless of the number of times a time-slot has been reused. With large rings where propagation delay may be a problem, there is a mechanism to ensure that the shortest path is used (under normal routing circumstances) in each direction of traffic flow.

12.4.7.1 SONET Unidirectional Path-Switched Rings (UPSR). A SONET UPSR is illustrated on the left side of Figure 12.1. The UPSR provides redundant bit rate capacity to protect message service nodes against failure or signal degradation. The basis of UPSR protection is *bridging*, at the entry node, identical STS/VT signals in both directions around the ring. It then selects at the exit node the best of the two paths. The decision on which signal to select is based on STS and VT path layer indications. In other words, the UPSR uses the 1 + 1 APS mechanism, but at the path layer. Thus the name *path-switched.* Linear 1 + 1 APS described in Section 12.4.5.1 operates in the line layer. Path layer indications to force switching include equipment degradation and defects.

In our case here, *unidirectional* has two meanings:

1. The initial setup of UPSR is such that the forward and return paths traverse the ring in the same direction. Suppose a certain path originates at node Y in Figure 12.2a. It will travel clockwise and be dropped at node Z in Figure 12.2a. Because we are dealing with full-duplex circuits here, the return path from Z to Y will normally carry reverse direction traffic.
2. Path switching in a UPSR is single ended. When a protection switch is operated at an exit node, there normally is no forcing of the opposite end to perform a protection switching operation. As a result, switching only affects only one end of a circuit. Thus the other meaning of *unidirectional* derives. There is one advantage here. The various nodes through which any paths affected by a fiber or equipment failure do not need to communicate with each other. This makes UPSR architecture much simpler than BLSR.

Such a ring network is illustrated in Figure 12.1. This is the two-fiber *bidirectional line-switched ring* (BLSR). Such a ring architecture also goes by the names *unidirectional path switched rings* (*UPSR*), *uni-rings, and counterrotating rings.*

12.4.7.2 Bidirectional Line-Switched Rings (BLSRs). There are two-fiber BLSRs and four-fiber BLSRs as illustrated in Figure 12.2. The four-fiber BLSR configuration is illustrated in Figure 12.2b. This may be viewed as two separate, but parallel, transmission rings. The working traffic is carried on one ring, and the protection traffic is carried on the other ring. The separation of working and protection traffic onto different fibers permits the use of *span switching.* A span consists of a set of SONET lines between two adjacent nodes on a ring.

Span Switching. In a four-fiber BLSR, the working and protection channels are carried over different lines. Thus, a four-fiber ring may allow protection similar to 1 + 1 point-to-point protection switching on individual spans. For failures that affect only the working channels, such as a single-fiber cut, the restoration can be performed by switching the working channels to a different line carrying the protection channels on the same span. The actual protocol for span switching is part

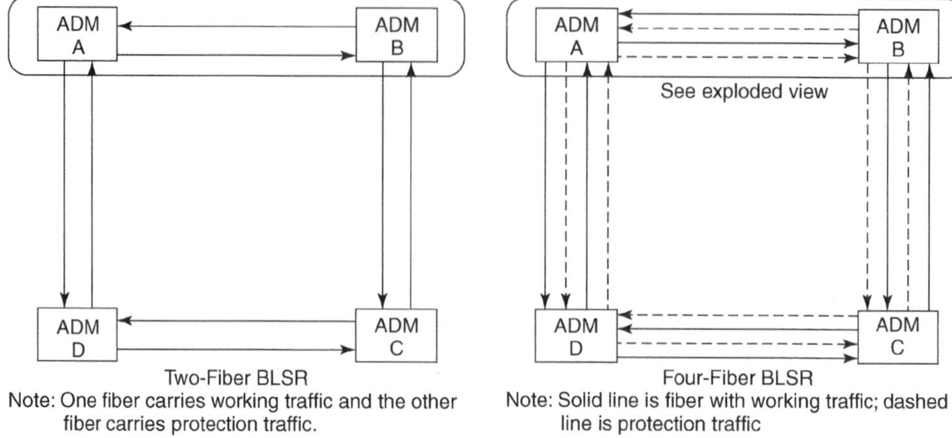

Figure 12.2. (a) Two-fiber and (b) four-fiber BLSRs.

of the BLSR protocol and differs from the protocol used in point-to-point APS systems. Span switching is not applicable for two-fiber BLSRs.

A four-fiber BLSR operating at an OC-N rate has a span capacity of OC-N, whereas for a two-fiber BLSR the span capacity is OC-($N/2$). However, the four-fiber BLSR requires more electronic equipment to operate.

For bidirectional rings both directions of transmission use the same set of nodes under normal conditions. A protection switching action is initiated based on performance conditions at a particular NE. Abnormal performance conditions sensed are loss of signal (LOS) or degraded BER. For both BLSRs and UPSRs there is a disabling feature for providing unprotected traffic to traverse the ring. In this case, end-to-end protection (e.g., between CPEs) can be used to provide protection.

A system designer may select a BLSR configuration as the most economical for applications that have cyclic or mesh traffic patterns. In a cyclic pattern, demand will exist only between adjacent nodes. In a mesh pattern, all nodes are similarly treated from the demand perspective, and demand exists between all nodes. Keep in mind that the capacity of a two-fiber BLSR may well be twice that for a UPSR (at the same data rate). The capacity of a four-fiber BLSR may well be four times that of a UPSR, depending on specific traffic patterns of the situation. A two-fiber BLSR is illustrated in Figure 12.3.

In the case of the two-fiber BLSR (2F-BLSR), neither fiber is exclusively dedicated for protection. Switching is performed by using a form of time-slot selection where each working time-slot is preassigned to a protection time-slot in the opposite direction and is not user-settable.

A 2F-BLSR can provide maximum restoration where there is 100% restorable traffic for single failures, when 50% of a ring's capacity is reserved for protection. This results in a span capacity of OC-($N/2$). Consider the example of a two-fiber OC-48 BLSR. It will have 24 time-slots between nodes W and X for working traffic and 24 time-slots between nodes W and X for protection traffic.

The intelligence to activate protection switches is carried in SONET overhead via bytes K1 and K2 on the protection lines. When a failure occurs, the nodes adjacent to the failed segment carry out the ring-switching function. A failed seg-

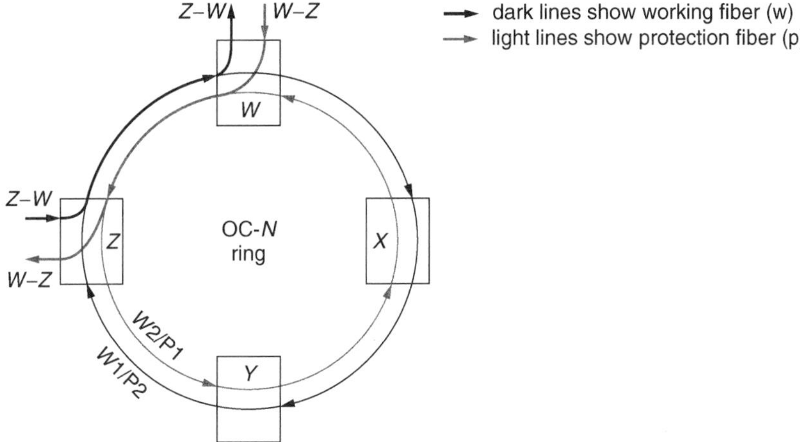

Figure 12.3. A typical two-fiber bidirectional line-switched ring.

ment may consist of a span or one or multiple nodes. In the case of a 2F-BLSR, which is operating at an OC-N rate, time-slots 1 through $N/2$ of the SONET multiplex input carry working channels. Then time-slot numbers $(N/2) + 1$ through N of the multiplex input are reserved for the protection channels. Time-slot number X of the first fiber is protected using time-slot number $X + (N/2)$ of the second fiber in the opposite direction, where X is an integer between 1 and $(N/2)$.

Figure 12.4 shows a 4F-BLSR. This type of BLSR consists of a set of nodes interconnected with two pairs of fiber. The configuration may include fiber-optic amplifiers and/or regenerators to form a closed loop. For the case of 4F-BLSR, the working and protection channels are on separate fibers. This allows the choice of two forms of switching to achieve improved service availability.

STS-level traffic protection is initiated based on line-level conditions detected by the NE. In a two-fiber bidirectional ring, protection is provided by reserving some of the bit-rate capacity on each fiber. Switching is performed by using a form of time-slot selection where each working time-slot is preassigned to a protection time-slot in the opposite direction. In SONET BLSRs, when the NEs perform a switching operation, all protected traffic at a site starts traversing the ring in the opposite direction. Bidirectional rings typically route traffic to the NE adjacent to the fiber cut.

To further clarify two-fiber BLSR, as the name implies, only two fibers are required for each span of the ring. Each fiber carries both working channels and protection channels. Thus, on each fiber, half of the channels are defined as working channels and half are defined as protection channels. The working channels on one fiber are protected by the protection channels traveling the opposite direction around the ring as illustrated in Figure 12.3. This permits the bidirectional transport of working traffic. Only one set of overhead channels is used on each fiber.

We introduce the term *nonpreemptible unprotected traffic*, often abbreviated "NUT." At system designers' discretion, certain selected channels of the working bit-rate capacity and their corresponding protection channels may be designated nonpreemptible unprotected channels. The remaining working channels are still

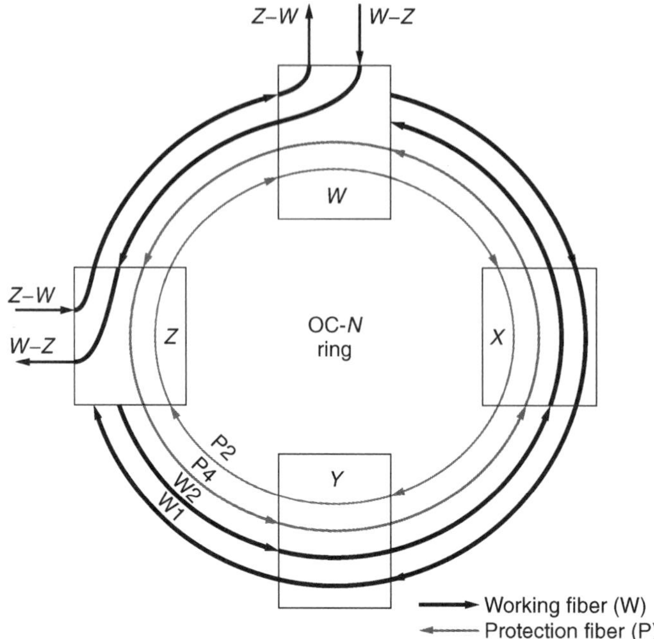

Figure 12.4. A typical four-fiber BLSR (4F-BLSR), line-switched.

protected by the corresponding protection channels. The NUT channels will have no BLSR APS protection.

Two-fiber BLSR supports ring switching only. When a ring switch is activated, the time-slots that carry the working channels are switched to the time-slots that carry the protection channels in the opposite direction.

Four-fiber BLSR requires four fibers for each span of the ring. This is shown in Figure 12.4, which illustrates that working and protection channels are carried over different fibers. Two of the fibers transmit in opposite directions carrying the working channels while the other two fibers, also transmitting in opposite directions, are carrying the protection channels. Such a concept permits the bidirectional transport of working traffic. The overhead of a fiber is dedicated to either working or protection channels because working and protection channels are not transported over the same fibers.

If NUT is supported, then on each span, selected channels on the working bit rate capacity and their corresponding protection channels are provisioned as non-preemptible unprotected channels. The remaining working channels are protected by their corresponding protection channels. NUT channels have no BLSR APS protection on the provisioned span. On other spans, the same channel (if not provisioned as NUT) has only span switching available to it.

In review, the line-switched architecture uses SONET line layer indications to trigger the protection switching action. Switching action is performed only at the line layer to recover from failures, and it does not involve path layer indications. Line-layer indications include line-layer failure conditions and signaling messages that are sent between nodes to effect a coordinated line protection switching action.

For the case of the path-switched ring, the trigger mechanism for the protection switch is derived from information in the SONET path layer overhead.

12.5 NETWORK OPERATIONS CONTROL CENTER (NOCC)

The NOCC is the nerve center of the fiber-optic network. It provides an operator with a visual presentation by computer display of network connectivities, related network nodes, operational circuits, standby circuits, and circuits that are down due to one or more faults. The network faults are broken out by location, and the affected equipment is broken out by nomenclature (possibly serial number) and by the portion of the equipment (down to the card level) where the fault occurred. Much or all of this status information is brought back to the NOCC via SONET overhead. Thus, in this argument it is assumed that the system transport is either SONET- or SDH-based.

12.5.1 Bringing Performance and Status Information to the NOCC

SONET and SDH as well as other systems of interest consist of network elements (NEs). An NE may consist of a transmitter assembly, its laser diode component, a PINFET assembly, and so on. Each element has a BITE structure from which we can derive performance information in a binary configuration. Thus, all we have to do is collect this information and ship it to the NOCC for display. The information is generally transported on DCCs using the TL1 network management language.

12.5.2 The TL1 Language

The TL1 protocol has been designed to carry network performance data, fault information, and status of network elements. TL1 interfaces NEs with network management command elements, usually found at the NOCC. TL1 is formally known as a network management protocol. It is used, for example, with the Megasys *Telenium System*, a typical network management and display system found on large fiber-optic networks. The protocol is governed by Telecordia GR-831 (Ref. 14).

12.6 SONET PERFORMANCE AND ALARM REPORTING

Through SONET (or SDH) overhead bits, fiber-optic/multiplex system faults are detected in NEs and forwarded through a maintenance data channel to the NOCC.

SONET/SDH overhead is quite extensive. It was discussed in Section 9.3.2. Figure 9.19 illustrates each byte of the transport overhead. Our concern here is the function of the various bytes that deal with fault reporting and maintenance performance monitoring and APS activation. Figure 12.5 illustrates SONET OA&M overhead and is a review of the information contained in Figure 9.19. The

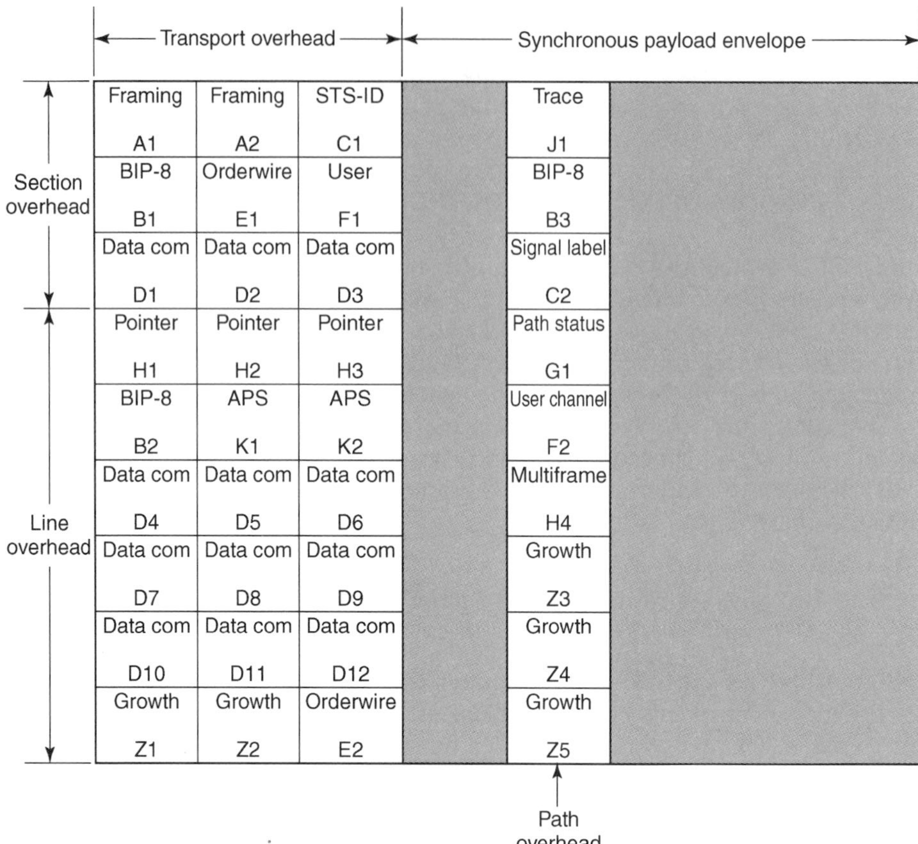

Figure 12.5. The SONET frame showing overhead byte (octet) assignments.

discussion that follows covers the minimum OA&M bytes that should have readout in the NOCC. Therefore, in this section, however, we wish to highlight the following:

- Maintenance capabilities and fault alarms
- Performance monitoring
- APS activation (see Section 12.5.3.1)
- Data links derived from this overhead

12.6.1 Overhead Structure by Layers

As we discussed in Chapter 9, the SONET (and SDH) electrical signal consists of three overhead layers: section, line, and path. These are illustrated in Figure 12.5, and the termination of each of these overheads is shown in Figure 12.6. A brief overview of each layer is given below. A more detailed description is provided in Section 12.6.1.1.

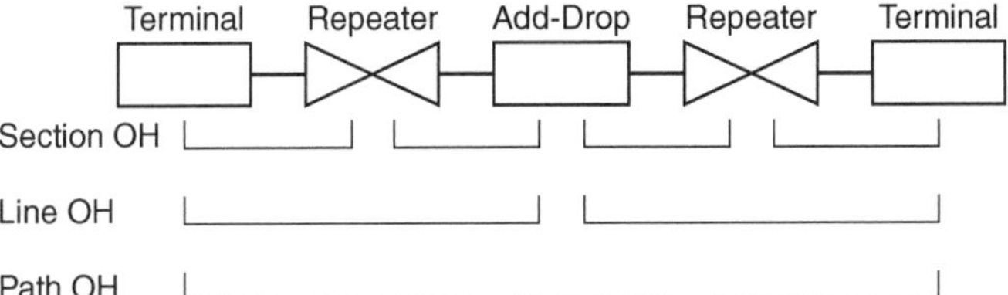

Figure 12.6. Termination points for the three overhead types (line, section, and path). OH = overhead.

- *Section Layer*. This provides signal framing and basic level of performance monitoring of the payload. In SDH this is referred to as a regenerator section.
- *Line Layer*. This layer provides protection switching and multiplexing functions for the information payload. In the SDH hierarchy this is called a multiplex section with an administrative unit.
- *Path Layer*. This provides signal labeling and tracing for end-to-end payload management.

12.6.1.1 Overhead Layer Descriptions. A SONET network is managed in a hierarchical fashion. Figure 12.6 shows the generic convention used in the SONET specification (e.g., GR-253-CORE). Each network element that performs the path, line, and section termination functions on the payload must be capable of handling the overhead information for signal processing and network management support.

Section Overhead. This overhead grouping must be processed by each NE to accomplish the basic transport functions of framing on the payload signals. There is also a basic level of performance monitoring data, a local orderwire (O/W) channel, a user channel, and a 192-kbps data communications channel. Every network element must process these overhead layers and the derived management features which are basic to all SONET offerings.

Line Overhead. The functions that are processed by line termination and multiplex elements in this overhead grouping support management of the payload. These include payload pointer storage and automatic protection switching (APS) commands. Additional functions covered in the line overhead include additional performance monitoring data, an express orderwire channel, and a 576-kbps data communications channel.

Path Overhead Groups. Path overhead (POH) provides end-to-end management of the payloads at the service terminating location (see Figure 12.6) of a SONET network. POH carries performance monitoring data, signal labeling, status feedback, a user channel, and a tracing function. There is also virtual tributary (VT) or virtual container (VC) path overhead. This overhead also carries performance monitoring data, status feedback, and signal labeling.

12.6.2 Performance Monitoring

Here we deal with data that give a measure of the performance of a SONET network. Prior to the implementation of SONET, this information was limited basically to alarms and status. ANSI and the ITU, the two primary standards agencies for SONET and SDH, have defined performance monitoring primitives, parameters, and failure criteria for various signal rates provided by NEs. The reader is encouraged to review ITU-T Rec. G.784 (Ref. 11) and ANSI T1.231-1997 (Ref. 12).

12.6.2.1 Primitives for Performance Monitoring and Network Defects. To provide meaningful management information, there is a basic group of primitives and defects that shapes the parameter and that is used to measure the performance of the network. This group includes the following:

- *Bit Interleaved Parity (BIP).* A parity error code generated for comparison at the receiver to determine transmission integrity.
- *Path Far-End Performance Report (FEPR).* This is the path status message sent from receiver to transmitter.
- *Loss of Signal (LOS).* Occurrence of no transitions on the incoming signal for a defined period of time.
- *Section Severely Errored Frame (SEF).* Four consecutive error frame alignment signals followed by two successive error-free frames.
- *Section Loss of Frame (LOF).* When the occurrence of a severely errored frame defect persists for a period of 3 ms.
- *Loss of Pointer (LOP).* Occurrence of a valid pointer not being detected in eight contiguous frames or when eight contiguous frames are detected with a new data flag set.
- *Alarm Indication Signal (AIS).* Defect occurs with the reception of an AIS signal for a set number of frames, five at the line layer and three at the path layer.
- *Remote Defect Indication (RDI).* Defect occurs with the receiving of the RDI signal for five frames defined at each layer. At the line layer, a remote failure indication (RFI) is derived with the persistence of the RDI signal at the path layer, previously referred to as a path yellow signal.

There are a number of optional primitives and defects that may be specific for a particular manufacturer. Among these optional signals indicating performance or performance defects are laser bias current, optical power transmitted and received, and protection switching events.

12.6.2.2 Performance Parameters. By processing of performance primitives and defects, we can derive performance parameters. These parameters are counts of various impairment events accumulated by the network in 15-minute intervals. It is such parameters which are usually employed in measuring QoS (Quality of

Service) on transport systems. The parameters involved here are described briefly in the following:

- *Coding Violations for Section (CV-S), Line (CV-L), Payload Path (CV-P), and VT/VC Path (CV-V).* These violations are BIP errors detected on incoming signals at each layer.
- *Errored Seconds for Section (ES-S), Line (ES-L), Payload Path (ES-P), and VT/VC Path (ES-V).* This consists of a 1-second interval during which at least one coding violation has occurred at that layer.
- *Severely Errored Seconds for Section (SES-S), Line (SES-L), Payload Path (SES-P), and VT/VC Path (SES-V).* This consists of a 1-second interval with a variable number of coding violations with the variable value set in relation to the layer having the errors.
- *Severely Error Framing Seconds for Section (SEFS-S).* A count of 1-second intervals containing one or more SEF events.
- *Unavailable Seconds for Line (UAS-L), Far-End Line (UAS-LFE), Payload Path (UAS-P), Far-End Payload Path (UAS-PFE), VT/VC Path (VAS-V), and Far-End VT/VC Path (UAS-VFE).* A count of 1-second intervals where the associated layer is not available.
- *Alarm Indication Signal Second for Line (AISS-L).* A count of 1-second intervals containing one or more AIS defects for the line layer.
- *Alarm Indication Signal/Loss of Pointer Second for Payload Path (ALS-P) and VT/VC Path (ALS-V).* A count of 1-second intervals containing one or more AIS or LOP defects at a particular layer.

12.6.3 Maintenance Signals

Standardized maintenance signals require that signal-terminating elements make decisions on the conditions of received payloads, as well as on the specific actions taken by the NEs to be reported as status information to the management system/NOCC. The information made available consists of the following:

- *Remote Failure Indication (RFI).* This shows receipt of AIS to upstream network elements of the same layer peer level.
- *Remote Defect Indication (RDI).* This defect indication message is returned to the transmitting network element of the AIS being received or loss of pointer defect.
- *Alarm Indication Signal (AIS).* This all-important indicator shows loss of signal condition on upstream network elements.
- *Unequipped Indications.* Provides status reporting on partially equipped network elements.

The use of these maintenance signals is illustrated in Figure 12.7.

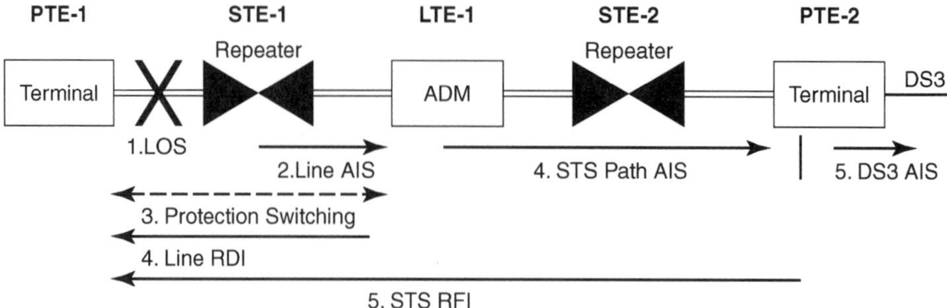

1. STE-1 detects LOS caused by laser failure.
2. STE-1 generates line AIS to LTE-1 and reports LOS alarm to OSS via section DCC.
3. LTE-1 attempts but fails protection switching.
4. LTE-1 generates STS path AIS, sends line RDI to PTE-1, and reports switch failure and AIS received to the operations system.
5. PTE-2 converts STS path AIS to DS3 AIS for termination reporting and generates STS RFI to PTE-1.

Figure 12.7. Illustrative examples of SDH/SONET maintenance signals. RFI stands for remote failure indication; PTE, STE, and LTE stand for path, section, and line termination equipment, respectively. (From *SONET/SDH*, Figure 3, page 262, Ref. 13.)

12.6.4 Orderwire Channel

Orderwire channels in the context of SONET/SDH are used for voice communications. See Figure 12.5, byte (octet) E1 in the section overhead and byte E2 in the line overhead. The application for the orderwire in the section overhead is voice communication between/among network elements which identifies this orderwire as local. The orderwire contained in the line overhead is not terminated at section terminating equipment (e.g., STEs or regenerators). In this case the orderwire is denoted as an express orderwire.

12.6.5 User Channels

These are 64-kbps channels that are contained in the section and path overheads. They provide termination at each NE and at the path terminating equipments. In the section overhead, byte F1 is dedicated as a user channel; in the path overhead, it is byte F2. These channels are employed at the manufacturer's option. They may be used to backhaul additional alarm information or as additional voice channels.

12.6.6 Data Communication Channels (DCCs)

Two data channels are defined by ANSI for SONET operations, bytes D1, D2, and D3 in the section overhead providing a single 192-kbps channel. In the line overhead, there are six 64-kbps channel slots reserved for data communications. They provide a combined bit rate of 576 kbps. The typical 576-kbps data channel uses a protocol stack based on the seven-layer ISO reference model. The channel serves as transport for transmit alarm, status, control, and performance information to other network elements. This high-speed channel only terminates at line terminating elements (LTEs).

12.7 STOCKING OF SPARES

From equation (12.1), availability can be improved by increasing the MTBF or by reducing the value of MTTR. We can improve equipment reliability either through the use of hi-rel (high reliability) parts or by redundancy, or both. Either way, it costs money.

The second factor (redundancy) can be equally important, or in some cases much more important. If a part fails and it is not available in the storeroom, MTTR can grow to weeks, a very undesirable situation. It would follow, then, that we better stock all parts, subassemblies, some complete assemblies, and replaceable printed circuit boards. This can become a major investment that could well be inefficiently used. For example, a printed circuit board may sit in stores for years where usage is zero. It may never be used through the life of the system.

Guidance may be found in Section 7 of Telcordia SR-TSY-000385, the *Reliability Manual* (Ref. 15). Its principle is based on a *service continuity objective* (SCO), where we probably would require an availability of at least 99.99% for a main line, long-haul fiber-optic system and the Telcordia SCO value is 99.9%. It is geared more for very large numbers of equipment (e.g., 160,000 repeaters); our numbers may well be smaller.

Mathematically, the *Reliability Manual* uses the product of NRT, where

N stands for in-service population of units
R stands for replacement rate per hour per unit
T stands for average lead time (hours)

Our concept for spare parts should mandate minimizing the number of *different* part types. For example, on long-haul systems, all transmitter assemblies should be the same and interchangeable, using the same type of laser diode; all receivers should be the same; and so on. We can use the stocking figures in Section 7 of the Telcordia *Reliability Manual* for the number of, say, transmitters to be stocked given the SCO value.

For fiber-optic systems covering large geographical areas, maintenance centers should be established where our maintenance personnel will be housed. The idea here is to reduce travel time to a site or location of failure. Each maintenance center will have a small storeroom. We should know the failure rate for each part or board over time. From these data, we base the level of parts stocking in each maintenance center. Expect each maintenance center to be connected to every other one by telephone and by an Ethernet-type data line. A maintenance center should be co-located with a major fiber link facility such as an ADM connecting a ranking switching center. For remote facilities, technician travel time must be included in the MTTR value. Again, our objective MTTR values are as follows:

- Two hours for a facility where a technician resides or is stationed. There is little time for travel in this MTTR value.
- Four hours for nearby facilities
- Six hours for remote facilities

When the technician departs a maintenance facility, he/she should know what the failed part or board is and have a spare with him/her. The site where a failure took place is known, and the unit with the failure and which board or subassembly failed have been identified from information provided by the NOCC. The NOCC plays a vital role in reducing MTTR values.

REFERENCES

1. *The IEEE Standard Dictionary of Electrical and Electronic Terms*, 6th ed., IEEE, New York, 1996.
2. *Reliability Prediction of Electronic Equipment*, MIL-STD-217E, US Department of Defense, Washington, DC, 1986.
3. *Reliability Prediction Procedure for Electronic Equipment*, Telcordia Technical Reference TR-332, Issue 5, Piscataway, NJ, December 1997.
4. *Reliability and Quality Measurements for Telecommunication System (RQMS-Wireline)*, Telcordia GR-929-CORE, Issue 5, Piscataway, NJ, December 1999.
5. *Parameters and Calculation Methodologies for Reliability and Availability of Fibre Optic Systems*, ITU-T Rec. G.911, ITU Geneva, April 1997.
6. *Transport Systems Generic Requirements (TSGR): Common Requirements*, Telcordia GR-499-CORE, Issue 2, Piscataway, NJ, December 1998.
7. *SONET Bidirectional Line-Switched Ring Equipment Generic Criteria*, Telcordia GR-1230-CORE, Issue 4, Piscataway, NJ, December 1998.
8. *Synchronous Optical Network (SONET)—Automatic Protection Switching*, ANSI T1.105.01-1998 (prepared by the Alliance for Telecommunications Industry Solutions), ANSI, New York, 1998.
9. *Telcordia Notes on the Synchronous Optical Network (SONET)*, Telcordia Special Report SR-NOTES-Series-01, Issue 1, Piscataway, NJ, December 1999.
10. *SONET Dual-Fed Unidirectional Path-Switched Ring (UPSR) Equipment Generic Criteria*, Telcordia GR-1400-CORE, Issue 2, Piscataway, NJ, January 1999.
11. *Synchronous Digital Hierarchy Management*, ITU-T Rec. G.784, ITU Geneva, 2000.
12. *Digital Hierarchy—Layer 1 In-Service Digital Transmission Performance Monitoring*, ANSI T1.231-1997, ANSI, New York, 1997.
13. *SONET / SDH: A Sourcebook of Synchronous Networking*, Curtis A. Siller, Jr., and Mansoor Shafi, eds., IEEE Press, New York, 1996; see the following articles: Rodney J. Boehm, Progress in Standardization of SONET, IEEE LCS, May 1999. Rony Holter, SDH/SONET—A Network Management Viewpoint, IEEE Network, November 1990.
14. *OTGR Section 12.1: Operations Application Messages—Language for Operations Application Messages*, Telcordia GR-831, Piscataway, NJ, November 1996.
15. *Reliability Manual*, (Bellcore) Telcordia SR-TSY-000385, Piscataway, NJ, June 1986.

13

POWERING OPTIONS TO IMPROVE AVAILABILITY

13.1 NO-BREAK POWER

A telecommunication system that loses its prime power source ceases to work. Many such communication systems are vital for the health and welfare of the community, state, and federal government, for the people as a whole. The public switched telecommunication network (PSTN) stands right at the top of this list along with public safety and the armed forces networks. In fact it was the PSTN people who really pioneered no-break power. Have you looked in on a PSTN switching center, called a local office in the United States? There are very large battery banks where the batteries are conventional lead–acid and look like big tubs. It has not changed much over the last many years. No-break power designs have been modernized, and the resulting product is certainly more efficient. The concept has been extended to other applications where it is vital that a certain device or group of devices require what some call "keep-alive" power. In other applications, such as telecommunication networks, such systems must be kept up and running at full power 100% of the time.

In this section we will describe no-break power systems that might be applied to fiber-optic networks at active nodes and terminals. Many of these systems use direct current (DC), and voltages that are very common in the field are −48 VDC or +24 VDC. If an AC voltage is required, an inverter can be used for DC-to-AC conversion.

The objective is to assemble a backup power system with reasonable parameters. Reserve time is probably the one single parameter that will govern cost. We define *reserve time* as that time the battery supply can carry the full load presented to it without a charging voltage present. Often the quantifying of reserve time is a judgment call by the system designer.

A basic static no-break power system consists of a battery charger, a battery or bank of batteries, and the load. A counter-emf cell may be placed in series between the battery and load. A simple functional diagram of this system is shown in Figure 13.1.

In my consulting business, I recommend a very common practice. When the basic electrical power system is installed, there should be two buses. One carries

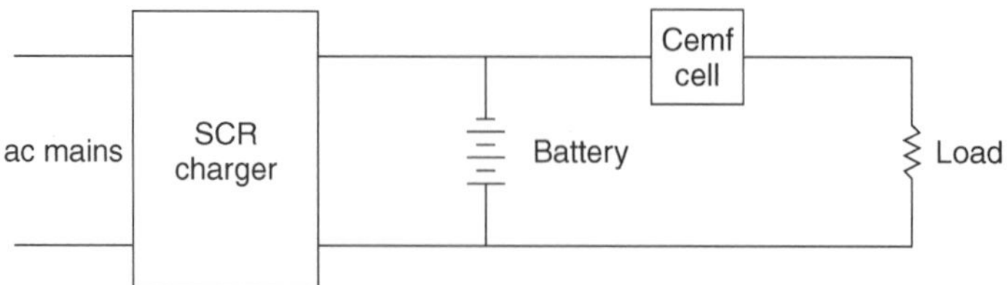

Figure 13.1. Simplified diagram of a static no-break power system. Cemf stands for counter emf.

critical load, and the other bus carries the regular load. Critical load should carry telecommunications line equipment, emergency lights, and other critical feature equipment. The telecommunication equipment commonly draws its load from a −48 VDC battery source. Only the critical load is supported by the no-break system. This allows considerable economy across the board in both first cost and recurring cost. It will also permit more leeway for the system designer regarding reserve time.

We would expect that the no-break power system would be backed up by a motor generator or even two motor generator sets. The design should be such that the motor generator will be cranked up, settle down, and be on line within five minutes. We recommend, in this case, that there be at least an hour of reserve time or even more.

Sources: Refs. 1 and 2.

13.2 FLYWHEEL SYSTEMS

Flywheel systems are coming back in vogue. Their principal drawback is that if there is a flaw in the wheel structure, the system will fly apart and shrapnel pieces of the wheel will be ejected at a tremendous velocity tangent to the turning angle of the wheel at that moment in time. On the bright side of the picture is that the hassle of any battery care is removed. Batteries also have finite lives, whereas flywheels do not. The estimated life of one flywheel system is 50 years. Batteries can also be dangerous. I've squirted sulfuric acid in my eyes; there can be explosion from accumulated gases.

A flywheel system can be described as a generator on a common shaft with a gas- or diesel-driven motor. However, between the generator and the motor there is a rather large flywheel. Between the flywheel and the gasoline or diesel motor there is an electrically controlled clutch. During normal operation when commercial mains are supplying power normally, current flows through a coil, keeping the clutch unengaged. When commercial prime power is lost, current stops flowing, thereby releasing the clutch and connecting the main shaft with the motor shaft. The inertia of the flywheel continues the shaft in motion, thereby cranking the gas or diesel engine and bringing the motor up to running condition. That inertia also keeps the generator turning, supplying voltage to the line. The concept of the flywheel is the kinetic energy stored in the rapidly turning flywheel. As we remember

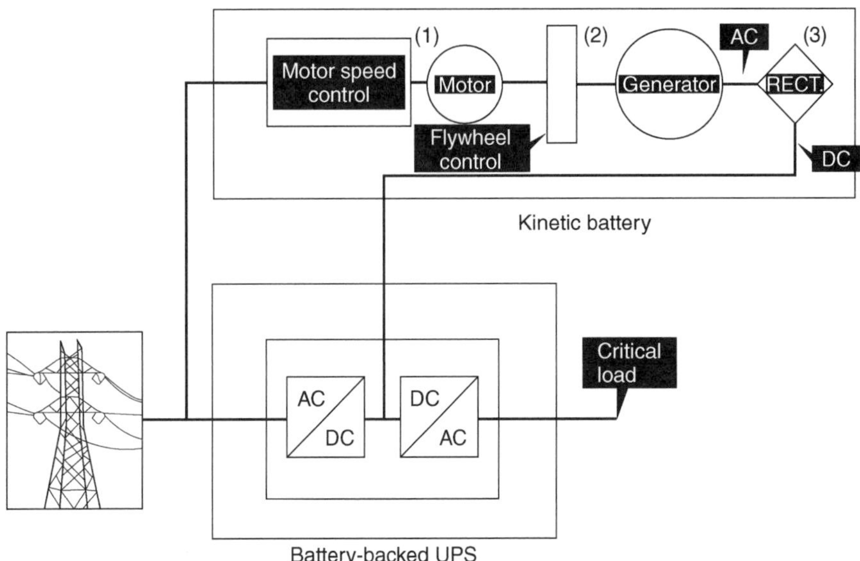

Figure 13.2. An illustrative block diagram of a dynamic energy storage system. (Reprinted with permission of International Computer Power, Ref. 3.)

from freshman physics, the amount of kinetic energy of a turning device such as a flywheel is a function of the weight of the wheel and the velocity of turning.

The reserve time equivalent of a flywheel is called *ride-through*. One commercial[1] kinetic energy no-break power system has a specified ride-through of 15 seconds. The device used to store the kinetic energy and to crank the motor/generator is called "Flywheel-Coupled Induction Motor-Generator." Of course, in this time period we expect the motor generator to settle down and start delivery of a uniform voltage to the load. Figure 13.2 is an illustrative block diagram of a *dynamic energy storage system* (DESS).

In the figure, a solid-state motor control circuitry (1) operates the motor at maximum continuous rated speed, increasing the kinetic energy in the flywheel (2) and allowing full use of the generator's capabilities. The output of the flywheel-driven generator is then rectified (3) into a regulated DC voltage permitting the flywheel to spin down to less than 50% of its normal running speed while still delivering "battery" power to the UPS inverter.

13.3 CONVENTIONAL STATIC NO-BREAK POWER SYSTEMS

Figure 13.3 is a simplified block diagram of a static no-break power system. It consists of a rectifier and a battery bank that feeds the load. The rectifier is fed by the AC mains. There will often be a switch that is connected to a backup AC gas- or diesel motor driven generator. When the AC power mains lose the prime power

[1]International Computer Power (Ref. 3).

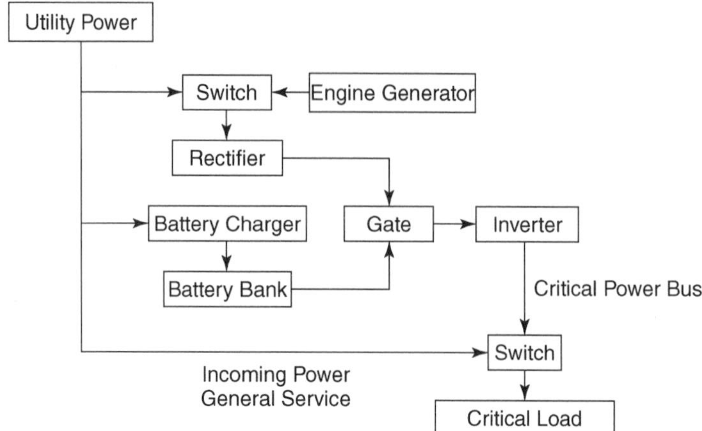

Figure 13.3. Functional block diagram of a complete static no-break UPS system. Note that if the telecommunication system supply voltage is a DC voltage (e.g., −48 VDC), then there is no need of the inverter shown in the figure.

source, the motor generator is automatically cranked to start up. Once the generator settles down, it takes over the battery charger AC inputs until the AC commercial mains are returned to their normal AC supply state. There are no switching transients because the load is always supplied by the battery bank, whether the load is DC or AC. If the load requires 110 or 220 VAC, the load is then supplied by a DC–AC inverter. The DC input to the inverter is provided by the battery bank. The switch connecting the motor-generator output to the AC input of the battery chargers may be autoactivated or manually activated.

13.3.1 Specifying a Conventional Static No-Break Power System

Still today, chargers are based on silicon-controlled rectifier (SCR) technology. Often these rectifiers are called *thyristors*. High-frequency chopping techniques were introduced years ago on AC–DC and DC–DC converters. Now these same techniques are being used on chargers. The results are much smaller filter rectifier components and less ripple. Ripple should be specified at 0.10% or below. To ensure a low hum level, 50 mV or less is a good figure to specify for maximum ripple voltage.

Noise from the battery bank source should not exceed −70 dBmp (100 pWp). Both ripple and noise measurements should be made during full-load conditions, with AC mains connected.

Given modern telecommunication plant life on the order of 15 years, lead–acid batteries remain very attractive. Older lead–antimony cells are still used. However, calcium–lead cells, costing somewhat near 10–15% more than conventional lead–acid batteries, compete with the neater, yet still more expensive, Nicad (nickel–calcium) cells.

The charging circuit should be able to withstand nominal variations of incoming AC line voltage of at least ±10%. The charger should also provide the required DC output voltage under widely varying load conditions.

For some applications, the DC load can vary tremendously. For example, a local serving switch (local office) in a business area, with a 10,000-line capacity, may draw 800 A at 48 V during the busy hour peaks. At about 2 A.M, it may only draw 15% of that amount, or about 120 A. Fiber-optic regenerator sites may require 5 A, and will display that load 24 hours per day.

We would expect the charger to use thyristors as rectifiers or another form of SCR because of their light weight, simplicity, and excellent operating characteristics. Load sharing should also be included in the charger specification. Load sharing permits two or more chargers, not necessarily of equal capacity, to provide output to a common bus and supply load to that bus according to their rated capacity. This is done by the proper interconnection of the DC control signals and adjustment of the load-sharing potentiometers.

To properly size a battery and charger, the following expression must be understood. Battery capacity (C_{ah}) in ampere-hours (A-hr) is expressed as

$$C_{\mathrm{ah}} = I_L \times T_R \tag{13.1}$$

where I_L is load current (A) and T_R is reserve time (hr).

Reserve time is the time that the system can operate off the battery supply exclusively, when the AC mains are shut down for whatever reason. Reserve time is usually given as 8, 12, 24, 48, and 72 hours. The time selected is a design decision. As reserve time increases, the cost of the UPS installation increases. Reserve time depends on the length of time an expected outage may last, or how long it would take to bring a spare motor generator setup on line and cutover. This time should include a technician's travel time to unmanned sites. However, in some circumstances, the standby generator is brought up and placed on line automatically, and the technician on site is unnecessary unless something goes wrong.

Under certain circumstances, when the installing contractor is trying to cut corners, he/she may assume that the battery installation is designed for, say, 8 hours reserve time. The contractor believes that 8 hours is sufficient time for the prime power utility to bring its basic utility power back up on line. We are of the opinion that such a practice is dangerous. There may be an excellent probability that utility service will be restored well inside the reserve time allotted. But suppose it does not. Thus, we are in the camp that says a standby generator should be mandatory. In theory, the standby generator will carry the plant as long as its fuel lasts. Some facilities not only carry an extra fuel tank, but a second generator as well in case the first generator fails. Our experience tells us that if a generator is going to fail, it will fail to start at first instance. That is why it is an excellent idea to cycle an emergency generator several times a year dumping its power onto a dummy load.

Source: Ref. 4.

The relationship to determine the load on the charger (I_{ch}) is

$$I_{\mathrm{ch}} = I_{\mathrm{ba}} + I_L \tag{13.2}$$

where I_{ba} is the battery current. These values of current are shown in the simplified drawing, Figure 13.4. Another parameter that must be considered is

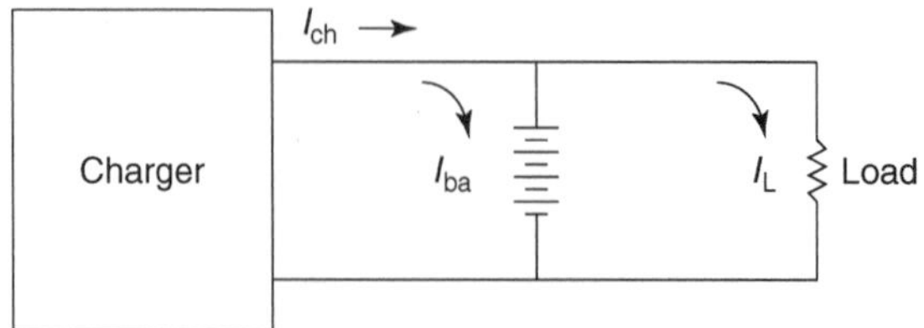

Figure 13.4. Simplified drawing showing charger, current, battery, and load.

recharge time T_{ch}, expressed in hours. The average current required to recharge the battery can now be expressed as follows:

$$I_{ba} = \frac{C_{ah}}{T_{ch}}$$

Combining these last two equations, we obtain

$$I_{ch} = \frac{C_{ah}}{T_{ch}} + I_L$$

Now combining this equation with equation (13.1), we obtain

$$I_{ch} = \frac{I_L \times T_r}{T_{ch}} + I_L$$

The energy required to recharge the battery is the term $(I_L \times T_r)/T_{ch}$ and assumes a battery that is 100% efficient. However, some energy is required to compensate for chemical and heat losses; usually 10% is a good figure. Thus

$$I_{ch} = I_L + 1.10\left(\frac{I_L \times T_r}{T_{ch}}\right)$$

Factoring gives

$$I_{ch} = \left(1 + \frac{1.10T_r}{T_{ch}}\right)$$

With this equation, charger current, I_{ch}, can be determined. I_L will be given, sometimes called the *house load* or *power budget*. Reserve hours and recharge time are design parameters for us to determine.

The standard nominal 2-V lead–acid storage cells are considered fully discharged when cell voltage reaches 1.75 V at 77°F (23°C). This is called the *final voltage* and is used as a standard figure in the industry. However, some power system designers use a more conservative figure, 1.84 V. The final voltage is valid for an 8-hr discharge rate (reserve hours = 8 hr).

To adjust this 8-hr reserve time capacity to other capacities, Table 13.1 will prove useful and introduces a new aspect in battery sizing.

TABLE 13.1 Conversion Table for Adjusting Reserve Time Capacity to Other Capacities

Rate (hr)	Capacity per Positive Plate (A-hr), Final Voltage 1.75	Capacity (%)
24	75	125
12	66	110
8	60	100
6	55	91
4	50	83
2	40	66

Source: Ref. 2, Table 34-1, page 3091.

TABLE 13.2 Terms Used When Specifying Battery/Charger Installations

Term	Voltage per Cell	Voltage for 24 Cells
Equalize[a]	2.30	55.20
Float[b]	2.17	52.08
AC-off (load not connected)[c]	2.05	49.20
AC-off, initial full load[d]	1.97	47.28
AC-off, average full load[e]	1.92	46.08
AC-off, final full load[f]	1.75	42.00

[a]The equalizing voltage is considerably higher than 2.05 VDC per cell and is applied for relatively short periods of time.
[b]The float condition permits a slow charge using a voltage just high enough to overcome internal resistance of the cell. Thus the charge voltage must be some tenths of a volt more than 2.05 VDC per cell.
[c]A fully charged lead\acid battery under no load has 2.05-V potential difference between negative and positive plates of the cell. Its specific gravity will be 1.215 (i.e., specific gravity of the electrolyte). Some automative batteries using high-density electrolytes have fully charged specific gravities up to 1.300.
[d]The initial full-load voltage is 2.05 VDC $- I_L \times R_{int}$, where R_{int} is the internal resistance of the cell in ohms.
[e]Average full-load voltage is the voltage that may be expected when the cell has reached a point in time halfway between full charge and the final voltage. A uniform discharge rate is assumed.
[f]Final voltage is a standard parameter.
Source: Ref. 2, Table 34-2, page 3091.

Batteries are sized (i.e., required ampere-hour capacity is established) in terms of single positive plates. Total cell rating is given by multiplying the single-plate rating by the number of plates in a cell. It should be kept in mind that there is always one more negative plate than positive plates in a cell.

When specifying battery/charger installations, the terms listed in Table 13.2 are used. The table shows the terms with *example* values given for a 48-VDC installation. The terms are given with equivalent declining voltage order.

On sizing batteries and chargers, the system designer must ensure that the equipment which will represent the load can withstand the usual $\pm 10\%$ DC input voltage variation. This figure should be verified on the specifications on the equipment that makes up the load.

Assume a 48-VDC supply. The high voltage would then be 52.8 VDC, and the low voltage would be 43.2 VDC. If the equalize charge were the 55.2 VDC from Table 13.2, then on equalize the charger would also employ a regulating element, the cemf (counter emf) cell shown in Figure 13.1. The voltage drop across that cell must be 55.2 − 52.8 VDC = 2.4 VDC. Provision is made in the cemf cell that it is dropped off line when AC power fails so that the load will get the full battery voltage. The cemf cell must also be specified to withstand the full-load current.

The following example will illustrate these concepts:

Example. Radiolink Repeater, 48 VDC

8 transmitter–receiver groups	20 A
2 orderwire	1
1 fault alarm system	1
Tower lights, strobe/control	10
Interior lighting	3.5
Heat and ventilation	14.5
Total	50 A

The site is unmanned and comparatively isolated. A reserve time of 12 hr is selected.

$$I_L = 50 \text{ A} \qquad T_R = 12 \text{ hr} \qquad T_{ch} = 48 \text{ hr}$$

$$\begin{aligned} I_C &= 50(1 + 1.10 \times 12/48) \\ &= 50(1 + 1.10 \times 0.25) \\ &= 63.75 \text{ A, the charger rating} \end{aligned}$$

Battery: 48 VDC or 24 cells.

Final cell voltage: 1.75 VDC.

From Figure 13.3 where 12 hr crosses the 1.75-V curve, 5.5 A per positive plate.

Number of positive plates: 50/5.5 = 9.9 rounded off to 10.

Number of negative plates: 10 + 1 = 11.

Total plates per cell: 21.

Figure 13.5 is a family of curves used to determine the amperes per positive plate of a typical battery. This information is vital when sizing a new battery installation.

13.3.2 Application Guidance for Secondary Cells

Secondary cells connected together make up a battery. These are called *stationary batteries*. There are two types of stationary batteries commonly used in the telecommunications industry: One is the lead–acid electrochemical couple, and the other is the nickel–cadmium electrochemical couple.

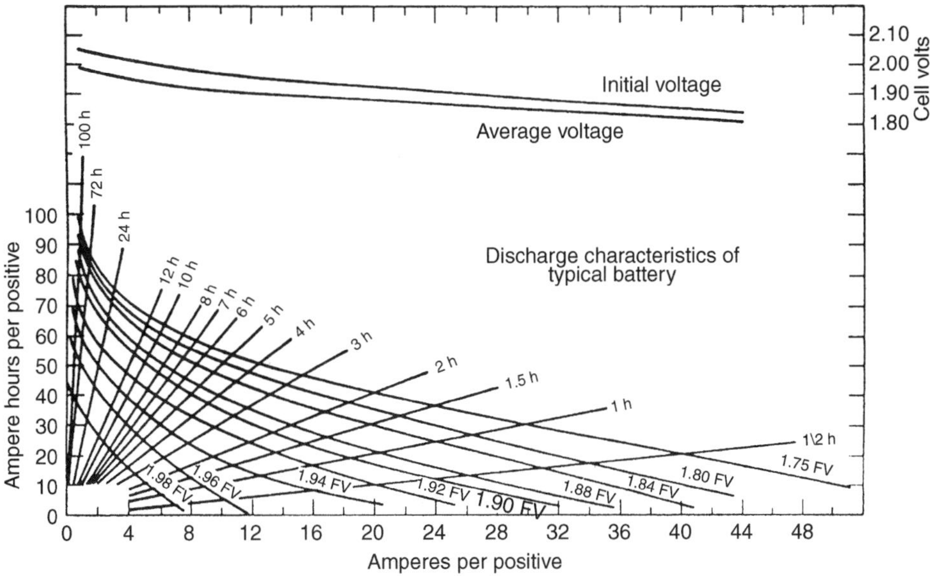

Figure 13.5. Curves to determine amperes per positive plate. (Courtesy of Warren G-V Communications, Ref. 2, page 3093, Figure 34-3.)

It can be briefly stated that the lead–acid battery is less expensive (first cost) than its nickel–cadmium counterpart. However, the initial capital cost may be offset in many applications because nickel–cadmium batteries generally exhibit a longer life, more rugged construction, and lower maintenance. The lower cost may be offset by the requirement of needing additional batteries to obtain the necessary voltage.

There are three basic designs employed in stationary lead–acid battery positive-plate construction: Fauré, Planté, and multitubular. The negative plate of the battery undergoes relatively little change, and virtually all manufacturers have standardized the Fauré construction for negative plates.

Fauré plates are made in two versions: lead–antimony and lead–calcium. In either version, an alloy of lead and antimony or calcium is pasted onto a flat lead grid. The advantage of the lead–antimony version is its ability to support prolonged and frequent discharges with a minimum of structural change; it has the disadvantage of requiring the frequent addition of water as the cell ages. The lead–calcium version requires very little watering throughout its lifetime, but frequent prolonged discharges can create structural growth that can shorten the battery's life.

Stationary nickel–cadmium batteries are normally constructed in a pocket plate design. Other designs, such as the sintered plate design, may have an undesirable feature referred to as a *memory*. The effect is described later.

The lead–acid electrochemical couple is nominally a 2-V couple. The nickel–cadmium couple is nominally 1.2 V. Therefore, more nickel–cadmium cells would be used to configure a given battery than would be required for a lead–acid battery.

TABLE 13.3 Number of Cells for Desired Voltage

Nominal battery voltage	120	48	32	24	12
Number of lead–acid cells	60	24	16	12	6
Number of nickel–cadmium cells	92	37	24	19	10
Equalize/recharge voltage	143	58	38	30	15.5
Float voltage	129	51	34	26	13
End voltage[a]	105	42	27	21	10.5
Voltage window	143–105	58–42	38–27	30–21	15.5–10.5

[a]The end voltage is a limit imposed by the manufacturer of the electrical equipment being powered. However, as a general rule of thumb, lead–acid cells should not be discharged below 75% of their nominal voltage (1.5 V per cell), and the pocket plate nickel–cadmium cells should not be discharged below 50% of their nominal voltage (0.6 V per cell). To avoid deep discharge, most battery systems include an under-voltage relay that automatically interrupts discharge at a preset end voltage.

Note: It is not uncommon to vary the number of cells for a specific application.

The number of cells in a battery for any specific system is a matter of adapting to suit the voltage available for charging and voltage required at the end of the discharge period (voltage window). The most frequently used systems encountered and the number of cells normally applied are given in Table 13.3.

13.3.3 Recharge/Equalizer Charge

In lead–acid batteries, even if the battery is not discharged, the individual cell voltage will begin to drift apart, and after approximately 60–90 days the lower-voltage cells will need to be brought back to full charge by increasing the charging voltage approximately 10% for 25–30 hours. This is referred to as *equalizing* the battery. Nickel–cadmium batteries have much less self-discharge; as a result, if the nickel–cadmium battery is not discharged with an external load, it will remain fully charged for many years at 1.4 V per cell. Therefore, nickel–cadmium cells do not need to be equalized.

However, it must be understood that nickel–cadmium batteries do need the dual-rate charging mode of the *float/equalize* battery charger.

Whether lead–acid or nickel–cadmium, both batteries need approximately 10% higher voltage to restore the discharged battery to a fully charged state.

Standby batteries are normally applied in *float*, where the battery, battery charger, and load are connected in parallel (see Figure 13.6). Charging equipment is sized to provide all of the power normally required by relatively steady loads

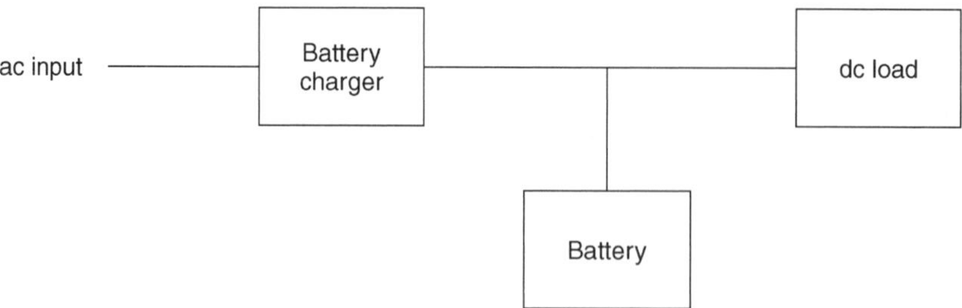

Figure 13.6. Battery "float" diagram.

(such as indicating lamps and relay holding coils) and minor intermittent loads, plus enough additional power to keep the battery at full charge. Relatively intermittent loads will draw power from the battery; the power will be restored by the charger when the intermittent load ceases.

When AC input power to the system is lost, the battery instantly assumes all of the connected load. If the battery and charger are properly matched to the load and to each other, there will be no discernible voltage dip when the system reverts to full-battery operation.

When an emergency load on the system ends and charging power is restored, the charger delivers more current than it would if the battery were fully charged. The charger must be properly sized to ensure that it can serve the load and restore the battery to full charge within an acceptable time. The increased current delivered by the charger during the restoration period will taper off as the battery approaches full charge. The charger controls will maintain the battery float voltage at the prescribed value when the battery is fully charged.

A single-rate float charger will adequately maintain a fully charged nickel–cadium battery until it is discharged by an external load. However, since the battery is discharged, it will not recharge to more than 85% at float voltage regardless of current capacity of the charger. With each successive discharge, the nickel–cadmium battery in such a charging circuit may continue to lose capacity. This phenomenon has been referred to as *memory effect*. It is simply the result of inadequately recharging any battery. It is even experienced in lead–acid batteries. However, usually before the loss of capacity is noted, the lead–acid battery is destroyed by sulfation of the positive plates, which is a rapid result in an undercharged lead–acid battery.

The charging rectifier, or battery charger, is a very important part of the emergency power system, and consideration should be given to redundant chargers on critical systems. Figure 13.7 shows this concept. A general formula for sizing the

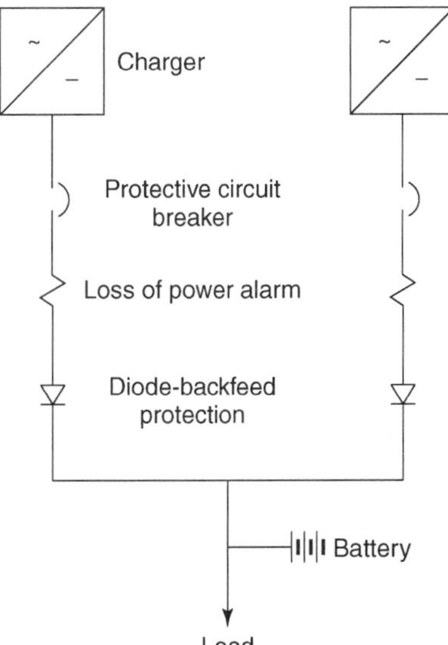

Figure 13.7. Typical redundant charger circuit. The redundancy, properly designed, greatly improves unit reliability, thus enhancing total system availability.

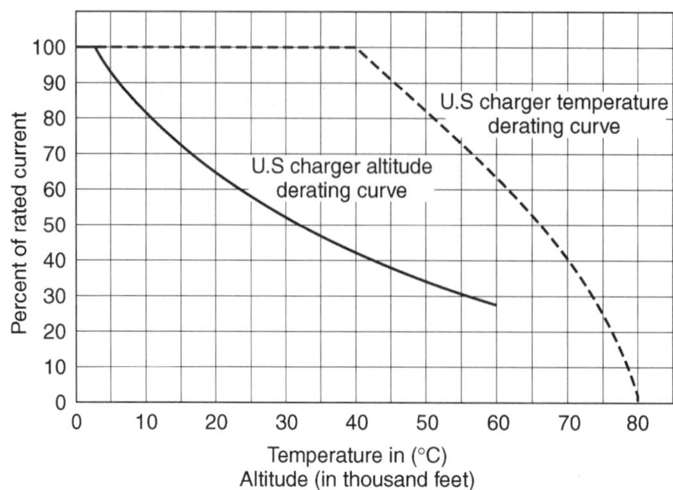

Figure 13.8. Derating curves for battery chargers due to altitude and temperature.

battery charger for an inverter system is as follows:

$$\begin{aligned}&\text{Battery charger (amperes)}\\&\quad= \text{inverter output (VA)} \times 100/(\text{voltage input} \times \text{conversion efficiency})\\&\qquad+ 1.15 \times \text{battery capacity } (\text{A} \times \text{H})/\text{desired recharge time (hours)}\end{aligned}$$

The battery charger output should be derated for both altitude and temperature. These requirements have to be recognized because the user frequently establishes these conditions. A larger-than-normal rating may be required to compensate for reduced capacity. A typical derating curve is shown in Figure 13.8.

13.3.4 Battery Sizing

To properly size any battery, the duty cycle should be defined with respect to the following:

- How many amperes?
- For how long?
- To what end voltage?
- At what temperature?

The size of the battery required depends not only on the size and duration of each load, but also on the sequence in which the loads occur.

The battery is sized to support the critical load until either the critical load can be shut down in an orderly manner, or the utility (line) power returns, or an alternate standby source (a motor–generator set) can be started and connected. Rather than purchasing larger battery capacity, an engine generator standby power might be considered.

The battery system should be sized from manufacturer's data for a particular application in a known operating temperature range. Most ampere-hour ratings are for a temperature of 77°F, and a reduction of ampere-hour capacity is usually necessary for operation at lower temperatures. Some manufacturers derate their lead–acid batteries by as much as 60% from the 77°F ratings for operation at 0°F.

Ampere-hour (AH) capacities decrease as the rate of discharge increases. Therefore, a simple summation of the various loads (area under the current–time curve) may yield an undersized battery. For varying loads, a summation of the various loads should be made as follows:

$$AH = A_1T_1 + A_2T_2 + \cdots + A_nT_n$$

where AH is ampere-hours, A is load (in amperes), and T is time (in hours).

A larger-capacity battery will be required if there is a large discharge rate at or near the end of the cycle. Therefore, to verify that the battery selected is adequate, it should be checked by starting at the beginning of the discharge cycle and subtracting the energy (AT) removed by each load in order to determine if adequate capacity still remains for the final load interval.

Experience has shown that the lead–calcium battery requires several days to weeks to return to full equal charge on all cells following a discharge to the rated end voltage.

Other batteries may be required where frequent prolonged power outages occur, since there may not be time to fully recharge between outages without raising the charging voltage beyond the rating of the connected load. Lead Planté and lead tubular batteries are also available, but their use is limited. For this reason they are not shown in Table 13.4.

Source: Ref. 4.

TABLE 13.4 General Differences for Various Battery Types

Battery Type	Physical Construction	Typical Characteristics
Lead–calcium	Pasted lead–calcium Positive plate Sulfuric acid Electrolyte	Life 12–15 years, poor in high temperatures or many or deep discharges. Lowest water loss of any lead battery. Lowest cost.
Lead–antimony	Pasted lead–antimony Positive plate Sulfuric acid electrolyte	Life 10–12 years, good for cycle applications. Medium cost.
Nickel–cadmium	Pocket plate construction Nickel positive plate Cadmium negative plate Potassium hydroxide Electrolyte	Life 20–23 years, good for high or low temperatures. Superior for short, fast discharges. Superior for deep discharges or for many cycles. Can be rapidly recharged. Highest cost.

Note: The end of battery life is defined as follows: When a rechargeable cell has been fully recharged and discharged in a load test and it fails to provide 80% of its original rated capacity, it has failed.

13.4 POWERING REMOTE LOCATIONS

There are many circumstances where we would wish to provide electric power at distant and remote locations where no electric utility is available to supply power. There is also the situation where we would rather provide our own power than use power from a nearby utility. There are several circumstances where this idea is not only feasible but also necessary if we are to provide dependable high-availability telecommunications. If a regenerator, optical amplifier, or ADM site loses power, the entire system goes down when in a linear configuration unless automatic protection switching has been designed into the configuration and has been invoked.

A remote site must provide its own power when it is not feasible to run a power line from the nearest utility termination to the site. It is also desirable for a site to generate its own power when the nearest utility power is of poor quality. By *poor quality* we mean an electrical service suffers many outages, has poor voltage regulation, and suffers many transients or other defects.

We may wish to power a fiber-optic cable system by bringing power along the cable itself by providing a fairly thick-gauge wire pair incorporated in the cable. This, of course, will defeat one of the advantages of fiber-optic cable, namely to be designed with no path for ground loops (i.e., no metal in the cable). The wire pair will add a ground loop path.

The method of powering a remote site must be simple, cost-effective, and with an MTBF measured in multiple years.

13.4.1 Gas-Turbine Power Generation

One such system is completely self-contained, consisting of a combustion system, which commonly is propane gas, a vapor generator, a turbo-alternator, an air-cooled condenser, a rectifier, alarms, and controls all housed in a shelter. A single unit will supply 400–3000 watts of filtered DC power on a continuous 24-hour-per-day basis for periods over 25 years with very low maintenance and no overhauls.

The unit utilizes a hermetically sealed Rankine cycle generating set that contains only one smoothly rotating part: the shaft on which the turbine wheel and brushless alternator rotor are mounted. The turbo-alternator shaft is supported by working fluid film bearings, which eliminate any metal-to-metal contact, resulting in years of trouble-free operation. The reliability of the unit at the 95% probability point is an MTBF of 200,000 hours of field experience on six continents.

To improve the reliability to a point better than 99.9999% availability, two gas turbine units are connected to the power bus in parallel, where each turbine unit carries half the load. If some sort of failure does occur, the other generator automatically assumes the full load.

Figure 13.9 is a cutaway drawing of the gas turbine unit. The figure shows the burner heating the organic working fluid in the vapor generator, where some of it vaporizes and expands through a turbine wheel and thereby produces shaft power to drive an alternator. The vapor then passes into a condenser where it is cooled, condensed back into the liquid state, and driven back to the vapor generator, thereby cooling the alternator on its way and lubricating the bearings. The cycle

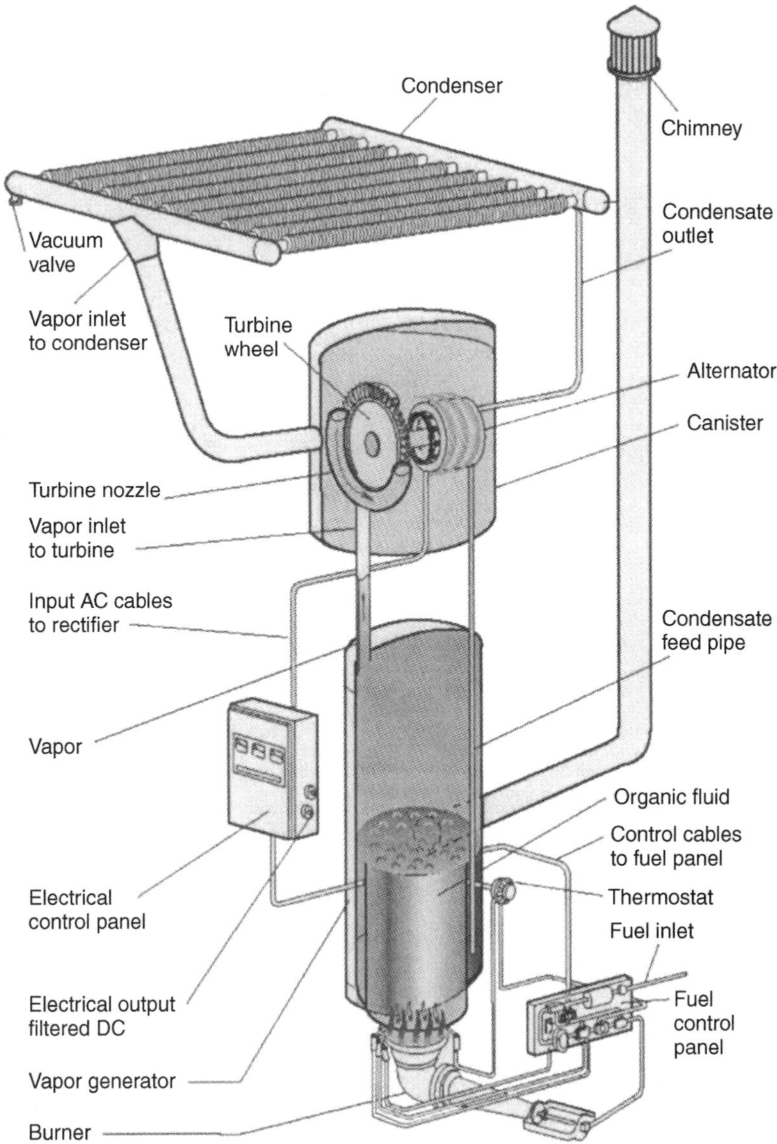

Figure 13.9. Cutaway drawing of an Ormat gas-turbine unit. (Reprinted with permission from Ormat Corp., Ref. 5.)

continues as long as heat is applied to the vapor generator. Because the liquid/vapor stainless steel envelope is sealed, none of the organic fluid is lost in the process.

The working fluid is immune to changing climatic conditions outside the sealed envelope. The turbo-alternator produces three-phase AC power, which is then rectified and filtered. The DC power is regulated for varying load conditions by automatically controlling the amount of fuel supplied to the burner.

The system is equipped with (a) a digital turbine control unit that provides for remote monitoring and control and (b) safety controls to protect it against any abnormalities.

Source: Ref. 5.

13.4.2 Fuel-Efficient Alternative

One alternative design for remote power is to combine solar cells, a no-break power system, and a gas turbine unit. The prime source would be the solar cells feeding the no-break battery bank. In parallel with the solar cells would be a gas turbine unit. Sizing of the battery bank in this configuration is vitally important. There will be a cost trade-off among the portion of the total electric power taken from the system, the portion of input power from solar cells, and the portion required from the gas turbine unit. The battery bank must carry the system during nighttime and on cloudy days. One configuration with a sufficiently large-capacity battery bank and replacement power sufficient from solar cells could possibly leave gas turbine service only for emergencies.

REFERENCES

1. Roger L. Freeman, *Telecommunication Transmission Handbook*, 1st ed., John Wiley & Sons, New York, 1975.
2. Roger L. Freeman, *Reference Manual for Telecommunication Engineering*, 3rd ed., John Wiley & Sons, New York, 2002.
3. International Computer Power, Dynamic Energy Storage System Extended Ride-through Kinetic Battery (Flywheel Systems), from the Web at www.rotoups.com (01/29/01).
4. *IEEE Recommended Practice for Emergency and Standby Power Systems for Industrial and Commercial Applications*, IEEE Std. 446-1987, IEEE, New York, 1987.
5. Ormat Corp. promotional material for OEC power systems, Ormat Corp., 980 Greg Street, Sparks, Nevada 89431-6039, February 1, 2001 (www.ormat.com).

14

HYBRID FIBER-COAX (HFC) SYSTEMS

14.1 THE OBJECTIVE OF THIS CHAPTER

Fiber-optic cable as a transmission medium has a very important niche in the transmission of CATV signals. Video transmission is the key material to be transmitted on a CATV service. The objective is to deliver a clean video signal to the CATV customer with a signal-to-noise ratio (S/N) greater than 46 dB. This S/N value is far in excess of what must be delivered for data or voice. We might say that if we make the video downstream work properly, then the other services will fall in line.

Optical fiber was introduced around 1990 as a transmission medium for CATV trunks and supertrunks. The rationale for its introduction is discussed in this section, along with various fiber network architectures. This is followed by an overview of the genesis of a CATV HFC network and its organization to provide the user with two-way service. The next topic that is covered is HFC system architectures and *last-mile connectivity.*

14.2 BACKGROUND

Figure 14.1 illustrates a CATV system design prior to the introduction of fiber. The figure is very simplified. The transmission medium was coaxial cable because of its broadband properties. Video/TV is essentially a broadband signal. The standard NTSC (National Television Systems Committee) television channel is employed. It is an RF channel 6 MHz wide as shown in Figure 14.2.

The CATV signal originates at the headend[1] and was delivered to the customer by coaxial cable. The cable TV system required broadband amplifiers to maintain the signal at a useful level all the way to the CATV user. Each amplifier develops its own noise and causes other degradations that corrupt the signal as it traverses the system to the end-user. The noise and degradations accumulate. Thus, there is

[1]In a CATV system, the headend is the point at which all programming (and other user signals) is collected and formatted for placement on the CATV cable. A headend can be identified by the several satellite antennas mounted around the CATV site.

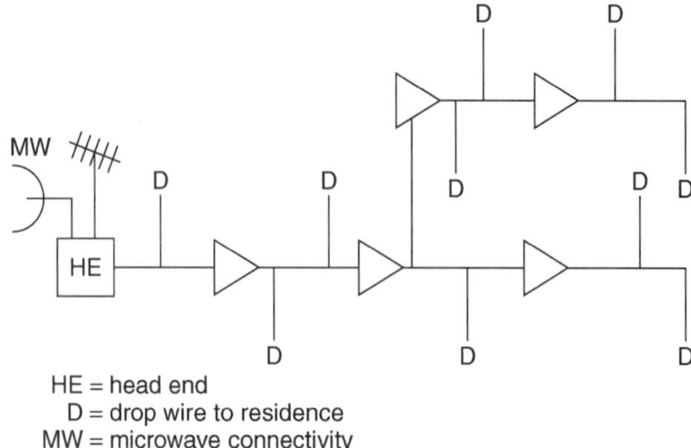

Figure 14.1. A conventional CATV system prior to 1970. The transmission subsystem is entirely based on coaxial cable/amplifier technology. The programming source at the headend relies on microwave extension, off-the-air, and magnetic tape. Satellite reception came later.

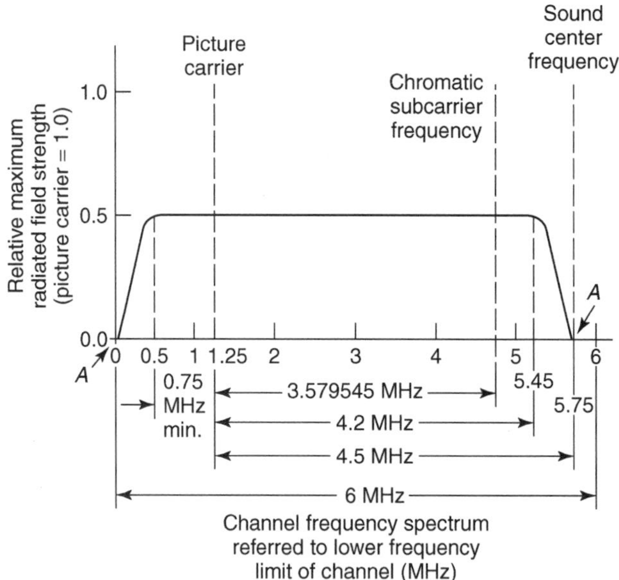

Figure 14.2. Amplitude characteristics of television video transmission. The bandwidth is 6 MHz. (From FCC Part 73.699, Figure 5, page 240, Ref. 4.)

a limit as to the number of amplifiers in series that the system may support before the "signal" is no longer useful to the user (consumer). Ciciora (Ref. 1) reports this number to be around 20. However, he states that the maximum number may range up to 50 amplifiers in tandem for narrower-band systems where the amplifiers have lower noise figures. In a standard TV system, expect some 2000 feet between amplifiers. That is roughly about half a mile. Thus the longest extension of an all-coaxial cable system connectivity may be up to 25 miles, probably less. This puts an upper limit for the service area of a CATV system based on just one headend.

We have found that optimum gain of a single broadband CATV amplifier is about 22 dB. Increasing the gain above this value allows intermods (intermodular products) to increase nonlinearly and soon to become excessive. A working number we can use in our calculations for conventional broadband amplifier noise figure is 8 dB (Refs. 1, 3, and 5), although 10 dB might be more conservative.

14.2.1 Rationale

Realize that coaxial cable presents an isolated RF environment. Because of its construction and electrical characteristics, coaxial cable has the ability to carry a separate frequency spectrum. It is separate from off-the-air, that is. One of its properties is that it can contain a frequency band that allows it to behave like an over-the-air spectrum. This means that a television receiver connected to a cable signal will behave as it does when connected to an antenna. Thus a television set owner can become a cable television subscriber without any additional expenditure on consumer electronics equipment. The subscriber can cancel his/her CATV subscription and is not left with any useless hardware.

With the cable spectrum tightly sealed inside the coaxial cable, a CATV system can use the frequencies assigned for other purposes than that of the over-the-air environment. This "other" usage takes place without interfering with conventional radiated TV. Of course we must take care that over-the-air radiated TV does not interfere with the cable signals, or does over-the-air radiated TV cause interference with the cable signals? Thus inside the coaxial cable of a CATV system we have a brand new spectrum with its own signals.

Whereas the conventional radiated TV spectrum is discontinuous, the CATV spectrum is continuous upwards in frequency from channel 2 (54 MHz). It is the cable television headend that fills in the normally vacant spectrum of radiated television using standard frequency translation equipment. These other 6-MHz channels derive from satellite feed or from LOS microwave which brings in distant TV radiated signals off-the-air.

One major disadvantage of coaxial cable is its loss, which does not increase linearly with increase in frequency. It is more nearly exponential. The attenuation of coaxial cable varies as the square root of the frequency. A $\frac{1}{2}$-in.-diameter aluminum cable has around 1 dB attenuation per 100 ft at 181 MHz; at 1-in. diameter the attenuation drops to 0.59 dB per 100 ft. The loss at 216 MHz (within channel 13) is twice that at 54 MHz (within TV channel 2) because the frequency is four times as great. The attenuation of cable at 216 MHz is twice that at 54 MHz. Thus cable equalizers are used to flatten out these characteristics. This is not so with optical fiber, where there is no increase in loss as we change wavelength (or frequency). As we have mentioned before, the loss of fiber in the 1310-nm window is about 0.35 dB/km and at the 1550 nm window it is about 0.25 dB/km. This is one good reason fiber optics is so attractive for long transmission runs for CATV. When using AM modulation where we convert to a coaxial transmission line, all that is required is a light receiver and an RF signal amplifier at the point of conversion. Along a fiber-optic line, little noise and distortion accumulate; along a

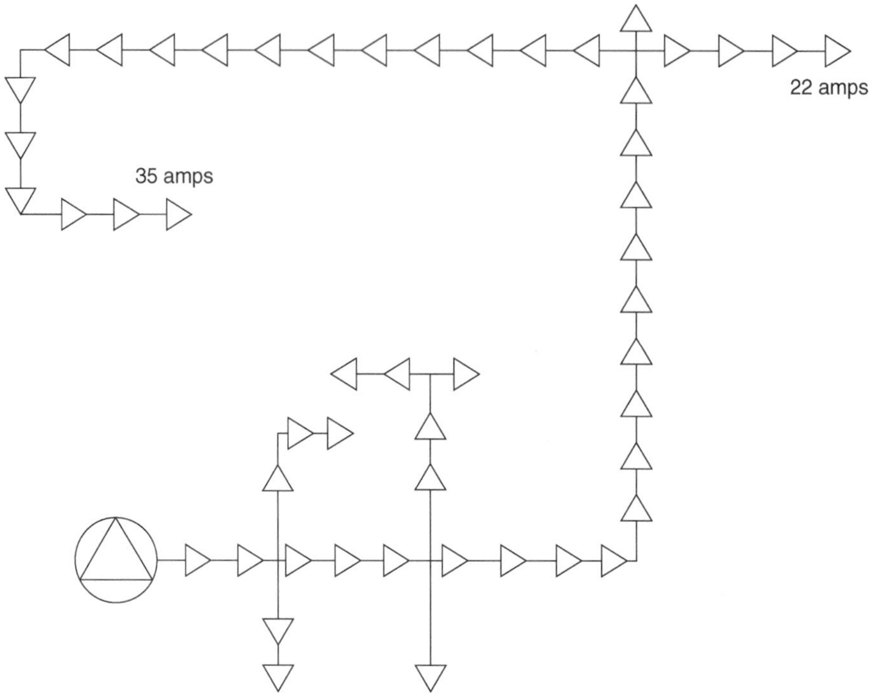

Figure 14.3. A conventional CATV system prior to about 1990. Note the plethora of RF amplifiers, as many as 35 in tandem. (Based on Refs. 8 and 9.)

coaxial line with amplifier, considerable noise and distortion accumulate. As a transmission medium, fiber wins hands down. A coaxial cable line is cheaper for the CATV case because, for fiber, a receiver demodulator would be required at each user's TV set. So the closer we can bring the fiber to the user's premises, the better the customer's picture.

Some Additional Notes. Active devices tend to have degraded MTBF values when compared to inactive devices. The more amplifiers in tandem, the more degraded the system MTBF becomes. This directly affects system availability.

If, somehow, we can reduce the number of broadband amplifiers in tandem, end-to-end, we help the noise performance of the system, thereby improving the signal-to-noise ratio of the picture at the customer receiver. We also improve the system reliability performance, increasing the system MTBF at the customer receiver. The system downtime can be notably less.

Fiber-optic cable was first introduced into the CATV plant on backbone trunk routes. It was found that a 20-mile fiber trunk route could be operational without repeaters or amplifiers. Thus, fiber-optic cable could replace up to some 30 broadband coaxial cable amplifiers on such a route. Figure 14.3 illustrates a CATV system based on coaxial cable signal delivery. The system is without fiber upgrades. Figure 14.4 illustrates this same system but with a fiber trunk network. Figure 14.5 shows one approach to optimum implementation of optical fiber in the CATV trunk plant. This approach to CATV system design was to say that no user was more than X amplifiers from the headend. In this case, $X = 3$.

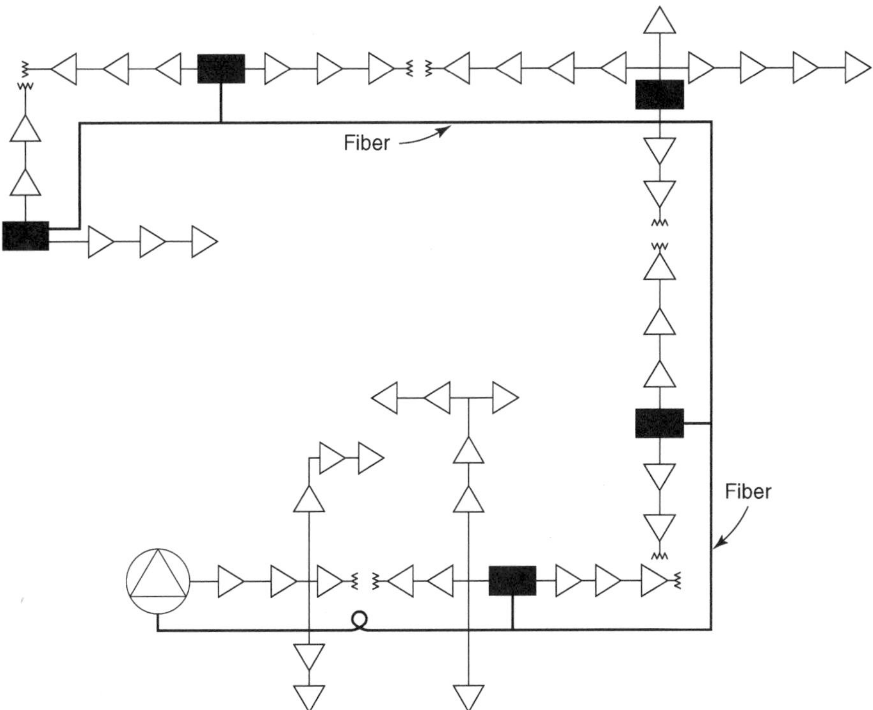

Figure 14.4. The same system as in Figure 14.3, but the primary distribution is carried out by a fiber-optic cable backbone. The span, on modern systems, will have two-fibers: one fiber for downstream and one for upstream. (Courtesy of ADC, Minneapolis, MN, Ref. 8.)

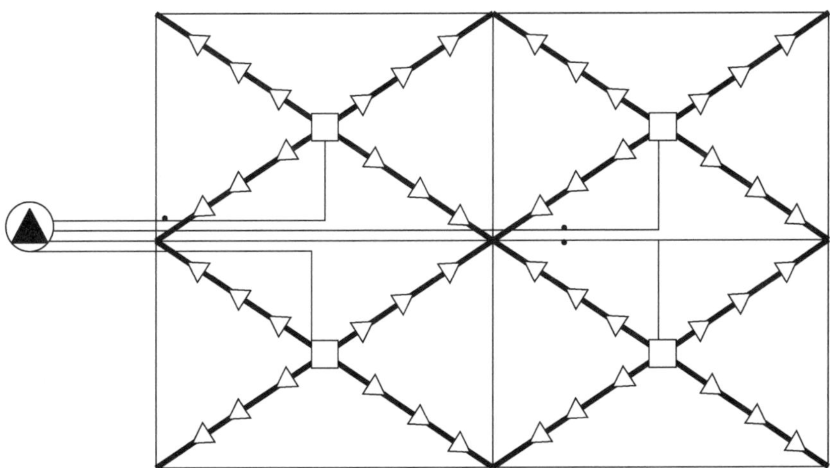

Figure 14.5. A CATV one-way HFC distribution architecture where there are no more than X RF amplifiers to any user. In this case, $X = 3$. The drawing is idealized. (Courtesy of ADC, Minneapolis, MN, Ref. 8.)

14.3 APPROACHES TO CATV TRANSMISSION OVER FIBER SPANS ON CATV SUPERTRUNKS

There are three ways to transmit a conventional analog TV signal on the fiber portion of a CATV transmission system. The most popular method is simple amplitude modulation. Here the 500-, 600-, or 800-MHz-wide CATV signal intensity modulates a laser transmitter. It is so popular because it requires practically no signal conversion.

The second method is called "FM" by the CATV technician. It is not FM (frequency modulation) in the purest sense. With this method, each CATV channel frequency modulates an independent RF carrier. Depending on the proprietary approach, these carriers are taken 8, 16, or 24 at a time and the aggregate grouping of subcarriers intensity (AM) modulates a laser diode transmitter. Each signal grouping will be coupled to a different fiber. If, for example, we wish to transmit 64 CATV channels to the node in the distribution plant and we are dealing with a system based on the 16-channel principle, then we will require four fibers for all 64 channels (i.e., $16 \times 4 = 64$). As one can see, the FM system is considerably more complicated than the AM system. This is because we will eventually have to convert back to the standard AM format compatible with the end-user's TV set.

Then why use FM in the first place? It is the only type of system that can meet the EIA/TIA-250C (Ref. 2) short-haul S/N requirement of 67 dB. EIA/TIA-250C is the governing standard in the United States for television QoS. There are two other advantages of FM systems. FM systems are much less prone to dispersion degradation on the fiber link, and longer links are possible when compared to an equivalent AM fiber span.

For our discussion here, we will focus on AM transmission over fiber.

The third method of transmission on optical fiber is the digital method. This is described below in Section 14.3.2.

14.3.1 AM Transmission of CATV Signals over a Fiber-Optic Span

Suppose our objective signal-to-noise (S/N) ratio value is 53 dB, but all we can readily measure on the signal output of our fiber PINFET receiver is carrier-to-noise ratio (C/N). In this case we use the relationship

$$S/N = C/N - 4.1 \text{ dB (unweighted)} \tag{14.1}$$

Adding a noise weighting improvement of 6.8 dB, we then calculate

$$S/N = C/N + 2.7 \text{ dB} \tag{14.2}$$

This is only valid on VSB-AM systems, and we assume a modulation index of 87.5%. When going through this exercise, remember that for S/N the signal level is measured peak-to-peak (sync tips) and the noise is an rms measurement. Regarding C/N, both C and N are rms measurements.

Substitute on the left side of equation (14.2) 53 dB and calculate on the right side the equivalent C/N value. C/N in this case is 50.3 dB.

Consider now the example link budget in Table 14.1. We need a 50.3-dB signal range to match the required 50.3 dB. This value shows we have little margin.

TABLE 14.1 Example AM Link Budget[a]

Item or Parameter	Value	Comment
Laser output (dBm)	+7 dBm	Pushing the laser
Receiver noise floor (dBm)	−58.7 dBm	
Receiver threshold for C/N = 50.3 dB	−8.4 dBm[b]	Derives a 53-dB S/N
Link budget overage	15 dB	
Leave 2-dB link margin	13 dB	Overage
Allow 0.25-dB/km loss, total fiber loss	10 dB	$\lambda = 1550$ nm; 2 dB for connectors and splices
Assume a 40-km link using 2-km sections, thus 19 splices at 0.08 dB/splice; splice loss, connector loss	1.52 dB	0.08-dB/splice insertion loss
Connector loss	0.48 dB	Total
	1.0 dB	Excess dB, add to link margin

[a]Based on a 40-km link with no regenerators or amplifiers.
[b]At first appearance this value for receiver threshold seems to be a very high level, whereas before in this text we dealt with threshold of some −23 dBm down to −40 dBm and below. The principal reason for this is that we would accept C/N values of around 15 or so dB. Here we want a C/N of 53 dB or more, a very big difference to achieve the desired S/N at the TV receiver.

The very large S/N (53 dB) with a 50.3 C/N leaves little room to extend link length or to increase margin (e.g., now only 2.0 dB). Often lasers are set for only 0 or +1 or +2 dBm output to improve lifetime duration. Of course, at the 40-km marker we can install an amplifier or a regenerator to extend link length.

With a C/N of 50 dB, the distortion objectives are as follows:

- Composite second order (CSO): −62 dBc
- Composite third order (CTB)(trible beats): −65 dBc

If we were to use signal splitters on the optical side of the link, an optical amplifier might be necessary to compensate for the insertion loss of the splitter. For example, if a 2:1 splitter were used, expect an insertion loss at the output of the splitter of about 3.5 dB.

14.3.2 Comments and Discussion on Fiber-Optic Link Budgets

It will be observed in Table 14.1 that there is a small margin, a margin that some experienced fiber engineers would call insufficient. We can achieve such a margin on this 40-km link by pushing the laser diode perhaps harder than we should. If the link designer will use FM rather than AM, perhaps a 60-km link could be maintained and possibly with a 4-dB margin.

We introduced three methods of transmitting CATV signals over fiber. The third method is digital. Because our digital bit stream has no compressed video, we are not so concerned about bit error rate (BER). Raw video is usually highly redundant with information. The bit stream can be regenerated. In fact, it is regenerated at each transmitter–receiver combination that the bit stream passes through. To extend the link(s) still further, regenerators may be considered.

TABLE 14.2 Link Budget for Model Fiber-Optic AM Link—A Different Approach

Item	Value	Comment
Laser output	+5 dBm	Λ = 1550 nm
Connector loss	1.0 dB	For two connectors, both sides
Fiber loss @0.35 dB/km	3.5 dB	For 10 km, include splice losses, G.652 fiber
PINFET receiver threshold	−12 dBm	Provides output S/N of 67 dB
Decibel range	17 dB	+5 dBm − (−12 dBm) = 17 dB
Dispersion penalty	1.0 dB	
Margin	11.5 dB	= 17 dB − 1 dB − 3.5 dB − 1.0 dB
The margin is excessive. Options: reduce laser output, extend link range.		Set new margin at 6 dB.

However, if the terminal end BER is 1×10^{-5} or better, often a wideband amplifier will suffice to bring the level of the bit stream to the desired value.

The major drawback to the use of a digital signal on fiber is expense, particularly in the A/D and D/A converters necessary to digitize the TV signals. We would opt for 10-bit coding rather than 8-bit to give better-quality pictures, especially for improved resolution.

Table 14.2 is a conventional link budget for a downstream fiber link in an AM HFC network. The laser diode transmitter puts out +5 dBm into the fiber. We would like to extend the fiber 10 km. Two popular fiber wavelengths will be checked for desirability: 1310 nm and 1550 nm. We could, for example, use poor man's WDM, by using each wavelength to carry independent CATV traffic. We set the C/N threshold at the input to the PINFET receiver as −12 dBm for a derived S/N of 67 dB. Using simple algebraic addition, the dB value left for engineering the link is +5 dBm (the laser output) and −12 dBm (the receiver threshold) or 17 dB (i.e., +5 − (−12) = 18). Table 14.2 shows these operations in tabular form.

We recommend expressing the link budget in tabular form as illustrated in Table 14.2. With this approach it is easier for a second or third person to spot methodology, catch arithmetical errors, and do a more presentable job. We do recommend laying the groundwork and setting guidelines in advance as we have done in the paragraph above.

14.4 PUSHING FIBER DEEPER INTO THE SUBSCRIBER PLANT

A *node* in a CATV network is a point of signal transformation. An optical signal may be transformed into a compatible electrical signal, or an electrical signal may be transformed into an optical signal. In our model, both transformations take place.

The goal is to push the fiber as close as possible to the CATV subscriber's TV set and still be cost-effective. There is some geographical point in the network where the "reach" of the electrical signal covers a certain number of residences. The number of residences may be from, say, 50 to over 10,000. Prior to conversion to two-way operation, the value was typically 1000 to 2000 residences. In these earlier systems, a node transformed the optical signal to an electrical signal which, by signal splitting, directly served many residences. The optical signal from the supertrunk termination could be split in the optical domain and carried still closer to groups of subscribers before being transformed to the electrical domain, compatible with the conventional CATV coaxial cable terminations. These nodes are the black boxes shown in Figure 14.4. The goal in Figure 14.4 was to have no more than a certain small number of amplifiers in tandem to reach any subscriber. In this case it was 3. This is a far cry from what we described earlier where the maximum number of wideband amplifiers to a subscriber was 35.

So far in the development of our argument, the objective was to reduce as much as possible the number of broadband amplifiers needed to reach any subscriber. As we mentioned earlier, two goals accrued:

1. The fewer the number of amplifiers in tandem, the smaller the accumulation of noise, and a better S/N at the subscriber equipment is achieved.
2. System reliability improves because there are considerably less amplifiers in tandem.

We now add a third improvement: two-way operation. Prior to this revolutionary change, CATV was simply a one-way entertainment system. The conversion to two-way operation made CATV a true competitor in the wideband digital marketplace, along with microwave, LMDS/MMDS, and ADSL.

14.5 TWO-WAY CATV

For two-way operation, we would want to employ as much of the plant-in-place as possible. The existing entertainment system was left in place. On the cable or AM fiber system, this involved TV channels spaced every 6 MHz from 54 MHz upwards to the upper limit, some 870 MHz. The conceptual designer eyed the band from about 5 MHz up to 40 MHz for the upstream link. The services envisioned first and foremost were the Internet, followed by other data services based on IP or ATM, VPN, interactive video, and telephony, among others.

Making maximum use of two-way services, did the band from 5 to 40 MHz provide sufficient spectrum to serve hundreds of users simultaneously? Suppose 100 users required service simultaneously and each user required 1 MHz of upstream spectrum for full period. That is 100 MHz, which we cannot provide. This is because in the band from 5 to 40 MHz (i.e., that unused segment), only a 35-MHz bandwidth exists. Following this thinking, only the needs of 35 users can be satisfied during the busy hour. We make the following observations on this scenario, and we assume that 100 residences will require upstream service.

(a) Not all 100 residences will require upstream service at once.
(b) We can roughly equate 1-MHz bandwidth to 1-Mbps user data rate. Allow that 1 Mbps upstream is a very high bit rate for Internet service in that direction. The high bit-rate requirement for Internet is downstream, and the upstream data rate can notably be reduced to no more than ~ 100 kbps.

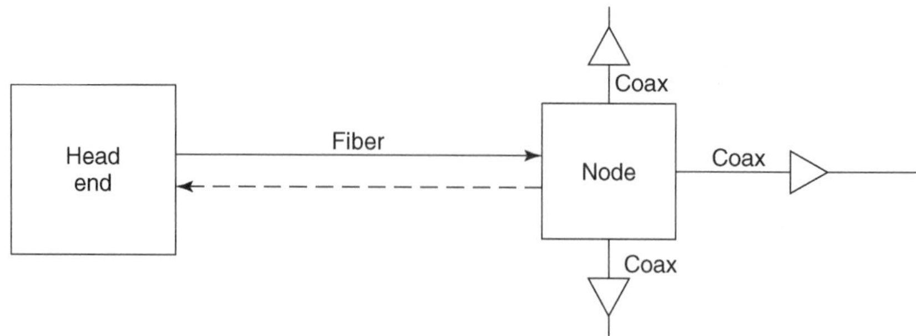

Figure 14.6. Functional diagram of a CATV node showing downstream service only.

(c) Employ TDMA (time division multiple access) in the upstream direction as the access method. This access technique can serve a larger user community considering the bursty nature of upstream Internet.

(d) The CATV two-way architecture should include a nodal capacity of no more than 500 residences. Of course, small numbers are more desirable but may have a negative impact on expense.

The downstream functions of a node are illustrated in Figure 14.6.

Transmission of traffic from user CPE (customer premises equipment) to the headend is introduced. For the sake of this argument, all traffic is digital. The anticipated traffic type and bit rates, along with expected direction of transmission (whether symmetrical or asymmetrical), are detailed in Table 14.3.

As a minimum, users will require voice telephony and Internet. For the sake of this discussion, the bit rate for voice telephony and Internet downstream is rounded to 2 Mbps, and upstream it is rounded off at 256 kbps. At this juncture, our interest is upstream. Allow 1 bit/Hz for spectral signal occupancy so each user will require 256 kHz. Based on Table 14.2, again assuming 1 bit/Hz, we can have

TABLE 14.3 Expected Traffic Type, Bit Rate, Direction of Transmission, and Whether Symmetrical or Asymmetrical, Continuous or Bursty

Traffic Type	Bit Rate	Direction of Trans	Symmetrical or Asymmetrical	Continuous or Bursty	Comment
Conventional TV	1.544 Mbps	Downstream	One way	Continuous	MPEG-2
HDTV	20 Mbps	Downstream	One way	Continuous	From Ciciora (Ref. 1)
Internet	2 Mbps 128 kbps	Downstream Upstream	Two-way asymmetrical	Continuous[a]	Bit rates, reasonable value
Virtual private networks (VPNs)	256 kbps	Downstream and upstream	Symmetrical	Bursty	—
B-ISDN/ATM	1024 kbps	Both ways	Symmetrical	Bursty	For various private networks
Voice telephony	64 kbps	Both ways	Symmetrical	Continuous	
Frame relay	1024 kbps	Both ways	Symmetrical	Bursty	LAN interconnect

[a]Continuous "on" for Internet. Most cable companies offer this continuous service at no charge.

130 users operating upstream at once. If a 50% duty cycle is allowed, 260 continuous users can be accommodated. Walt Ciciora (Ref. 1) states that conventional HFC systems should connect 1000–2000 users per node. This plan will not accommodate this large number of users during the Internet busy hour. The upstream is configured on a TDMA basis with four users per frequency segment. On this basis, we might expand to 800–1000 simultaneous users in the upstream band, 5–40 MHz, realizing that even during the busy hour all potential users will not demand service at once.

14.5.1 Segment Assignment on Upstream Fiber Trunk

Figure 14.7 shows the RF spectrum below some arbitrary frequency where frequency band assignments are for a model cable television service. Downstream service starts at 54 MHz, and the assignments proceed upward in 6 MHz segments. Upstream service begins at 5 MHz and proceeds upward to 40 MHz. The space between 40 MHz and 54 MHz is a guardband to perform isolation between upstream and downstream operations.

Let us consider residences in groups of 250, and each group is supported by a *node*. For upstream service, each node has its own upstream segment on a coaxial cable. That segment, as we discussed, resides between 5 and 40 MHz. In our model there are four nodes, each with its own upstream frequency segment between 5 and 40 MHz. We will call these nodes (or segments) A, B, C, and D. These nodal segments are brought to a central node that upconverts the aggregate band (5–200 MHz). The aggregate of four segments or blocks amplitude-modulates a laser diode, forming an optical signal. This block translation of four frequency segments, each 5–40 MHz, is illustrated in Figure 14.8. Each nodal upstream frequency aggregate grouping is carried to a *hub* on a comparatively short fiber link. A hub is a device that interfaces the local light signal from the node to the mainline fiber trunk connecting back to the headend. Another way of looking at a hub is as an *add–drop multiplexer* (ADM).

The mainline fiber trunk cable is often set up in a ring architecture to improve survivability and availability. This concept is illustrated in Figure 14.9.

Some of these two-way HFC installations use SONET or SDH as the underlying digital format to ride on the maintrunk fiber line in its ring architecture. Others may turn to DOCSIS or DAVIC, which provides the underlying fiber format (including MAC access techniques). SONET/SDH provides no access techniques.

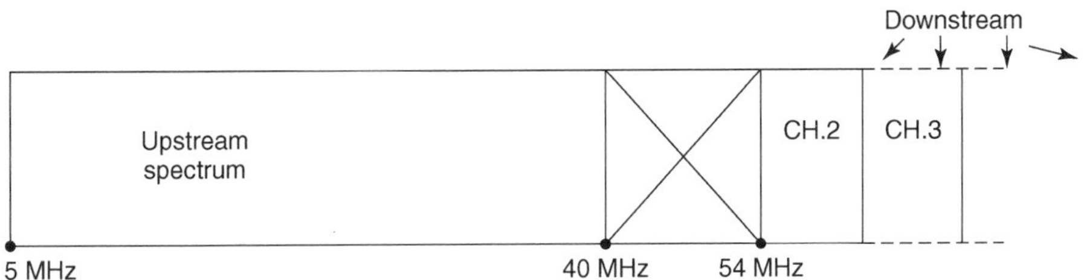

Figure 14.7. The lower RF spectrum in the CATV service.

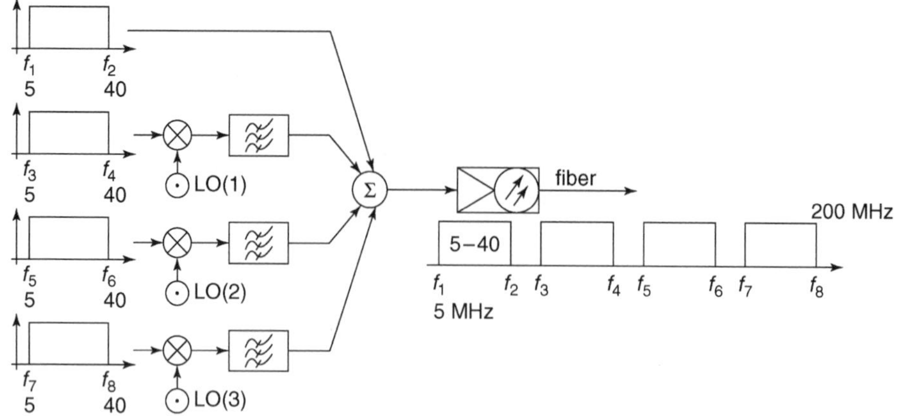

Figure 14.8. Block translation of four spectrum groups where each group segment is 35-MHz wide.

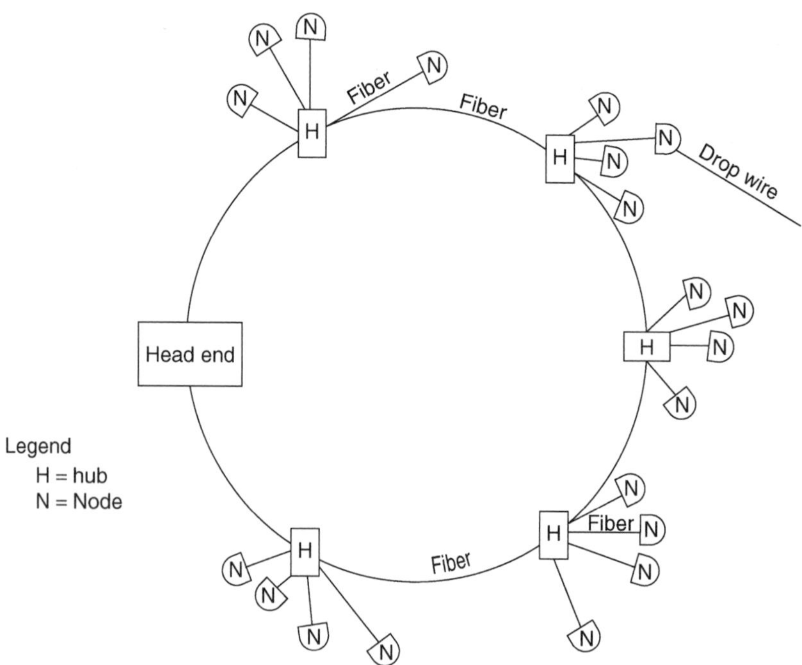

Figure 14.9. A fiber ring architecture connecting a CATV headend to various hubs supplying service via nodes to residence customers. H, hub; N, node.

Now let the subscriber density per node increase to 500 for a given geographical area such that the grouping of four nodes will serve about 2000 subscribers. The bit rate per carrier remains at 256 kbps, assuming, say, BPSK (binary phase-shift keying) modulation. In these rough calculations, we assume 1 bit per hertz and have a 256-kHz bandwidth. If we allow bandwidth for a raised cosine filter, then $\alpha = 0.25$. Our bandwidth per carrier now becomes 320 kHz. In this case, from the arithmetic above, the 35-MHz bandwidth can serve only 109 carriers at once. This boils down to about 100 simultaneous users with a 100% usage or duty cycle. These calculations allow no time for setup and takedown of a circuit. Note that we are assuming an FDMA (frequency-division multiple access) method, probably the most wasteful of bandwidth.

Keep in mind that most installed fiber-optic links use separate fibers for upstream and downstream communications; thus there are no restrictions on RF bandwidth occasioned by the need to share spectrum space. In the case of coaxial cable technology, however, there is a trade-off among bandwidth, signal loading, noise, and distortion. Thus the bottleneck is the coaxial cable portion of the link.

This section describes a means of increasing the number of users per node where we will deal principally with the upstream link. We will assign only 250 drops to each coaxial cable segment as a baseline. We recommend starting at this lower subscriber density to accommodate growth of new services offered to the CATV user.

The following measures may need to be considered to achieve more users per unit bandwidth:

- Access method: We would opt for TDMA; CDMA should also be considered. Timing and synchronization are important issues in both cases. Power control is a major consideration with CDMA.
- Use QPSK rather than BPSK, possibly 8-ary PSK, giving 3 bits/Hz theoretical. This assuredly is one way to increase the number of subscribers per unit bandwidth.
- Decrease the upstream bit rate per user with the growth of new services such as VPN, IP message service, and frame relay, and increase the upstream bit rate of Internet for faster uploading. Additional upstream bandwidth can be obtained by robbing some of the downstream bandwidth by placing these downstream channels in other segments of the frequency band. This reassignment will be in segments 6 MHz wide.

To achieve even greater bit rate capacity, we can resort to forms of bit packing. One cable television protocol, DOCSIS,[2] does just that. For downstream it uses 64- or 256-QAM; for upstream either QPSK or 16-QAM. For a discussion of bit packing refer to the *Telecommunication Transmission Handbook*, 4th edition (Ref. 6). Reference 7 contains a description of DOCSIS.

REFERENCES

1. Walter Ciciora, James Farmer, and David Large, *Modern Cable Television Technology*, Morgan Kaufmann Publishers, San Francisco, CA 1999.
2. *Electrical Performance for Television Transmission Systems*, EIA/TIA-250C, EIA/TIA Washington, DC, January 1990.
3. George Scherer, private communication, GI/Motorola, Hatsboro, PA, February 15, 2001.

[2]DOCSIS stands for data over cable service interface specification.

4. *Code of Federal Regulations*, 47, Parts 73.600 and 76, US Government Printing Office, Washington, DC, revised October 1996.
5. Eric Schweitzer, "Return Path Technologies," *Communications Technology*, February 2001.
6. Roger L. Freeman, *Telecommunication Transmission Handbook*, 4th ed., John Wiley & Sons, New York, 1998.
7. Roger L. Freeman, *Reference Manual for Telecommunication Engineers*, John Wiley & Sons, New York, 2002.
8. HFC Virgraph presentation, ADC Communications, Minneapolis, MN, 1993.
9. Private communication, Chuck Grothaus, ADC Communications, Jan 31, 2002 (permission to publish).

15

ON-PREMISES WIRING OF BUILDINGS—FIBER OPTICS

15.1 THE INTENT OF THIS CHAPTER

When a building is built for business or industry, its design should include basic telecommunication wiring and fiber-optic cabling. Sufficient space should be allowed for future expansion and modification. The definition of *sufficient* will be a difficult task for the responsible electrical engineer.

The intent of this chapter is to provide help and guidance for the fiber-optics cabling portion of the task. How will the fiber-optic cabling be employed? What will we ask it to do? Once we have that settled, we can define the cable to be used. Even more important, we must run a trade-off between fiber and copper. It may be much more economic to use a copper wire pair for a signal lead carrying 64 kbps than to assign a fiber strand for that purpose. On the other hand, if we need connectivity on a 1-Gbps CSMA/CD system for a 350-ft run, certainly a single-mode fiber strand is called for. Some of the questions to be covered in this chapter are as follows: What will the fiber be used for? What is the most cost-effective fiber solution?

15.2 RANGE OF APPLICATIONS

Prior to about 1975, the range of telecommunication applications for a business or industrial enterprise was limited to analog telephony with UTP (unshielded twisted pair) interconnectivity via a PABX and the outside world. Today the range is far greater, with most media systems being digital and involving data. Among these digital systems, we can expect to find Ethernet (CSMA/CD),[1] FDDI, token ring, TCP/IP, video, SONET/SDH transport, ATM transport, and, possibly, fiber channel. A 62.5-μm optical fiber (see Chapter 2, Section 2.1.2) with the bandwidth specified as 160 MHz/km at 850 nm and 500 MHz/km at 1300-nm multimode fiber is sufficient to carry these signals. Key in our thinking of on-premises

[1]CSMA/CD, more commonly called Ethernet, is by far the most popular of the LAN types. At last count, there are 22 versions of CSMA/CD. One version covers a 1-Gbps Ethernet and a 10-Gbps version is being readied for customer usage.

fiber-optic facilities is that distances are short, generally no more than 300 m for a single building. In the campus environment or even extending to a metropolitan area, distances are no greater that 60,000 m when we base design on ANSI/EIA/TIA-568-B.1. (Ref. 1).

According to Corning Cable Systems (Ref. 2), the following data rates may be expected: Data applications at 62.5-μm multimode fiber operating in the 850-nm band will satisfy the requirements of 155 Mbps or less for a fiber run up to 2 km. ATM (asynchronous transfer mode) running at 622 Mbps is based on the use of 62.5-μm fiber-optic cable for runs up to 300 m. It is felt that this type of fiber should meet the needs of most building requirements. Corning (Ref. 2) believes that 2.5 Gbps (OC-48) can be supported in a 62.5-μm fiber for a 100-m horizontal distance. The needed flexibility for all premises applications can be satisfied with a hybrid cable design made up of single fiber and 62.5-μm multimode fiber. It is advisable to use single-mode fiber in addition to 62.5 μm for campus backbone applications when distances exceed 300 m with data rates in excess of 622 Mbps.

Of course the 62.5-μm fiber can be operated at yet greater bit rates than those indicated above. For horizontal cable runs, a 90-m length limit standard has been set. The reason for this is that usually the fiber is cabled with UTP, and 90 m is its limit. Thus the limit carries over to the fiber that is cabled with it. Also, it should be kept in mind that ANSI/EIA/TIA-568B.1-2001 recommends the use of LEDs as the light source (transmitter) in the premises environment. This choice is based

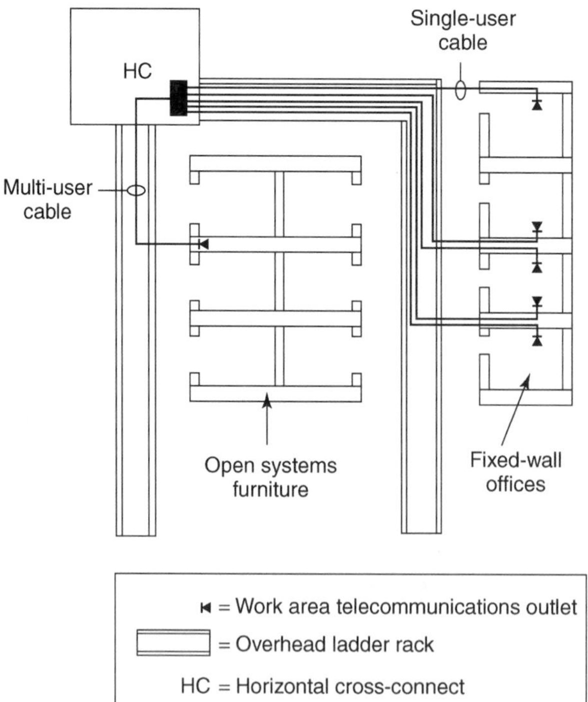

Figure 15.1. An example of horizontal cabling. (Reprinted with permission from Corning Cable Systems, Ref. 2, Figure 2.8, page 2.6.)

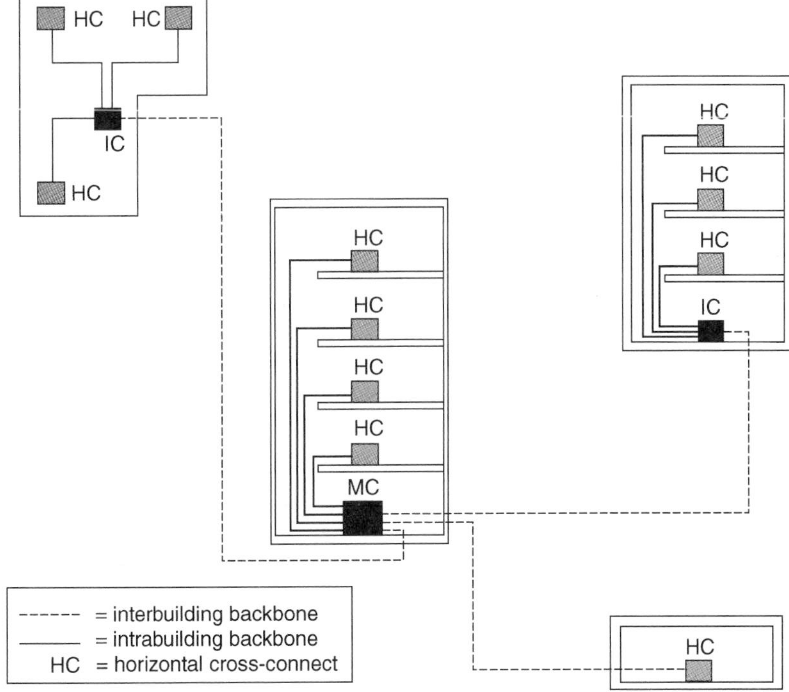

Figure 15.2. An example of backbone cabling. (Reprinted with permission from Corning Cable Systems, Ref. 2, Figure 2.7, page 2.6.)

on economy more than anything else. Thus, the use of 62.5/125-μm fiber is an ideal match for LED transmitters because of the fiber's larger NA. Simply said, this larger fiber collects more light from the transmitter than single-mode fiber would.

15.2.1 Building Backbone and Horizontal Cabling

There are two major components of premises wiring and cabling: horizontal cabling and backbone cabling. An overall premises wiring plan is illustrated in Figure 15.1. As shown in the figure, horizontal cabling connects work areas and equipment to a horizontal cross-connect (HC). We generally associate horizontal cabling with that covering work areas on one floor. Backbone cabling runs between cross-connects among floors and is illustrated in Figure 15.2. To further elaborate: The purpose of the building backbones is to connect the main cross-connects in the building with each of the telecommunication closets within the building. Figure 15.3 is a model of a typical campus environment with multiple buildings. This figure ties together the entire premises cabling scheme. Fiber has emerged as the medium of choice for intrabuilding data transmission because of its ability to support multiple, high-speed links in a much smaller cable without concern for crosstalk. More and more users are adopting fiber to support voice and PABX applications by installing small PABX shelves on each floor.

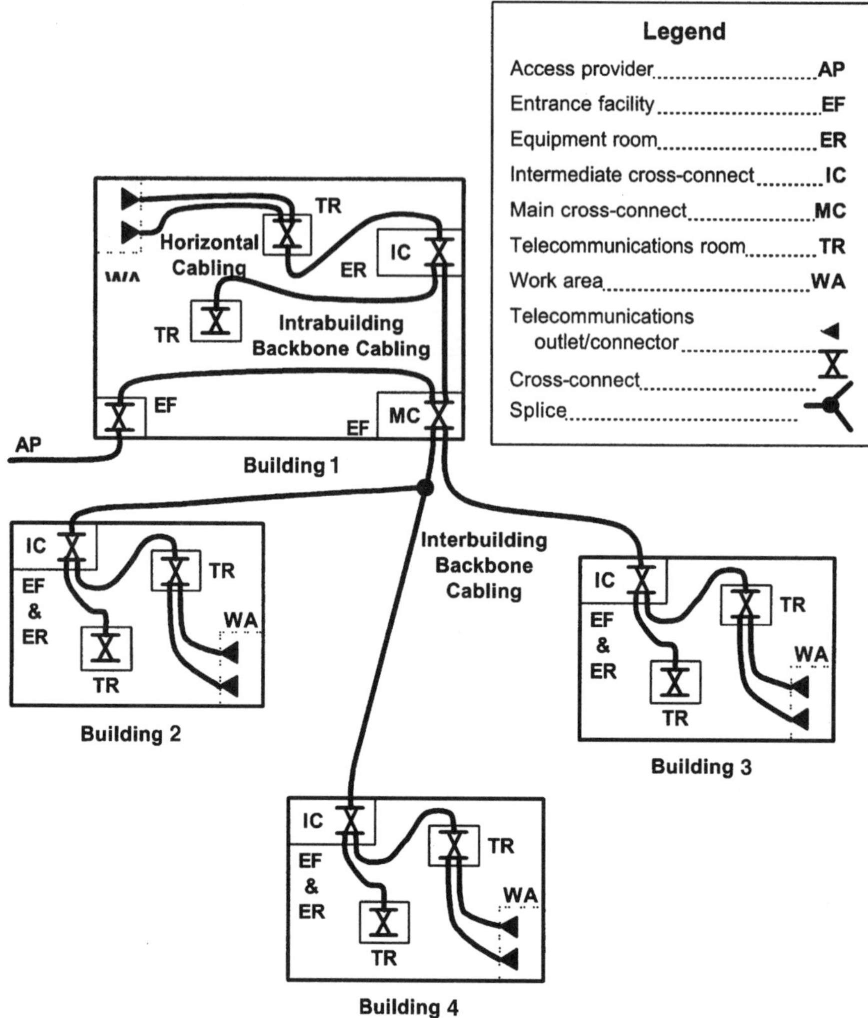

NOTES

1 – This figure is not meant to be an all-inclusive representation of the telecommunications cabling system and is provided only as a typical example.

2 – All cross-connects located in the telecommunications rooms (TRs) in this figure are horizontal cross-connects (HCs).

Figure 15.3. Typical telecommunications cabling system. (Reprinted with permission from ANSI/EIA/TIA-568-B.1, Ref. 1, Figure 1-1, page 3.)

15.3 NETWORK TOPOLOGIES

15.3.1 Campus Backbone Network

A campus network consists of telecommunication connectivity among a group of buildings with a single owner. Military bases, universities, hospitals, and large industrial facilities fall into this category. The modern communication elements nec-

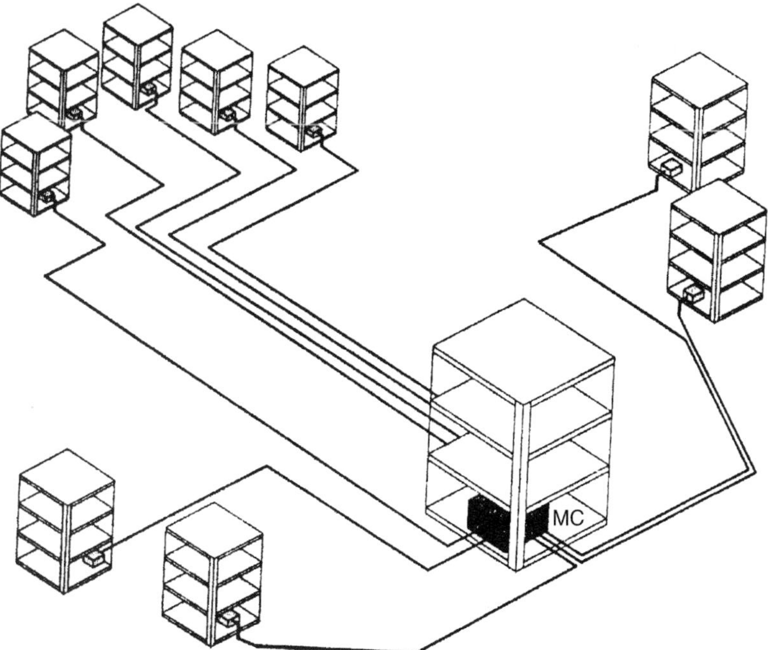

Figure 15.4. Illustrative drawing of a campus setting with a one-level hierarchy. MC = main cross-connect. (Reprinted with permission from Corning Cable Systems, Ref. 2, Figure 2.12, page 2.6)

essary in today's environment to satisfy the needs of the tenants are complex and costly. The total system breaks down in a rough hierarchical manner as follows:

1. Campus
2. Building
3. Floor of a building
   a. Telecommunication closet(s) (TCs)
   b Work areas (WAs)

Figure 15.3 illustrates the three levels. There is a fiber cross-connect or patch panel associated with each level. There is at least one telecommunication closet per floor. The fiber and wire-pair cross-connects should be physically separate yet have parallel purposes.

The MC is the master cross-connect and is usually found in the basement of one of the largest buildings. All buildings are linked to it via a single- or two-level hierarchy.

Figure 15.4 shows one method of interconnecting buildings in a campus network based on a single-level hierarchy. This is a campus backbone with one level of hierarchy. It has nothing to do with the three-level hierarchy—campus, building, floor—discussed above. There is also a two-level hierarchy for connecting buildings as well. It is used on large campus settings in a larger geographical area—for example, large military bases and larger universities that cover tens or even hundreds of square miles. In a two-level hierarchy design, all buildings are not linked directly to the MC, but rather via an IC (intermediate cross-connect). This two-level design used in large networks often translates to more effective use of electronics such as multiplexers, routers, or switches to better use the bit-rate capacity capabilities of the fiber or to segment the network.

There are several advantages accruing from a single hierarchical star-type network:

- It provides a single point of control for system administration.
- It allows the easy addition of future campus backbone links.
- It allows graceful change.
- It allows easy maintenance for security against unauthorized access.
- It allows testing and reconfiguration of the system's topology and applications from the MC.
- It provides for simple isolation for centralized testing.

ICs are illustrated in Figure 15.3. In the case of a two-level hierarchy, selected ICs will serve a number of buildings. These ICs are then linked to the MC.

15.4 NOTES ON FIBER INSTALLATIONS

Section 10.3 of ANSI/TIA/EIA-568-B.1 (Ref. 1) recommends the use of 62.5/125-μm optical fiber cable in the horizontal cable section of the premises installation, and both 62.5/125-μm and single-mode fiber for use on backbone installations.

In the case of 62.5/125-μm cable, as a minimum there will be two such fibers enclosed by a protective sheath. The bandwidth of this cable is 1 GHz for the 90 m (295 ft) maximum distance specified in the discussion on horizontal cabling. The fiber type is multimode graded index fiber. The fiber should comply with ANSI/EIA/TIA-492AAAA (Ref. 3).

The optical fiber employed in backbone cabling consists of 62.5/125-μm optical fibers or single-mode optical fibers, or both, typically formed in groups of 6 or 12 fibers each. The groups are assembled to form a single compact core and the core is covered by a protective sheath. The sheath consists of an overall jacket and may contain an underlying metallic shield and one or more layers of dielectric material applied over the core.

The transmission performance of 62.5/125-μm fiber is shown in Figure 15.5 and the transmission performance of single-mode fiber is shown in Figure 15.6. The operational wavelength in the case of the multimode 62.5-μm fiber, the transmitter is an LED operating at 1300-nm wavelength. The transmitter used for Figure 15.6 measurements was a laser operating at a wavelength of 1310 nm.

The ANSI/EIA/TIA standard 568-B.1 states that while it is recognized that the capabilities of single-mode optical fiber may allow backbone link distances of up to 60 km (37 miles), this distance is generally considered to extend outside the scope of the reference standard.

It should also be pointed out that system bit rate capacity is not only a function of the fiber but also of distance and transmitter characteristics, specifically center

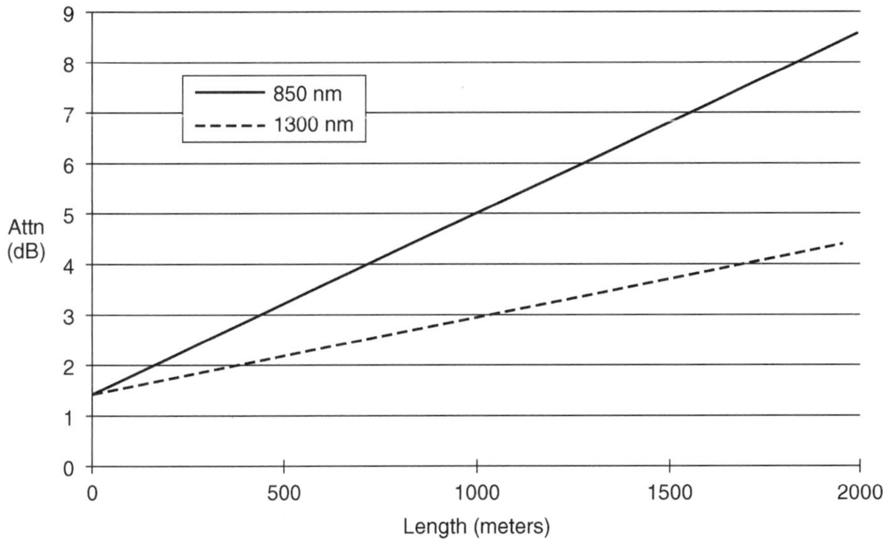

Figure 15.5. 62.5/125 μm or 50/125 μm backbone cabling link attenuation based on distance. (Reprinted with permission from ANSI/TIA/EIA-568-B.1, Ref. 1, Figure 11-6, page 56.)

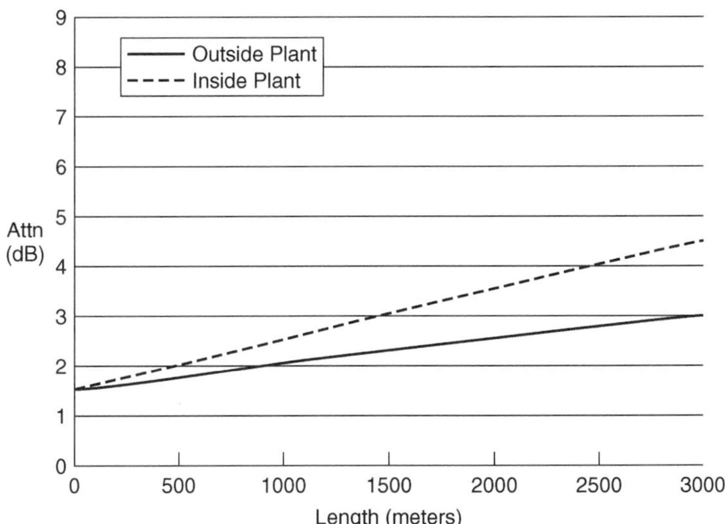

Figure 15.6. Single-mode backbone cabling link attenuation based on distance. (Reprinted with permission from ANSI/TIA/EIA-568-B.1, Ref 1, Figure 11.7, page 57.)

wavelength, spectral width, and optical rise time. Thus in the case of Figures 15.5 and 15.6, only typical system bit rate capacity values have been given.

Minimum Bend Radius. A most important factor in fiber-optic cable installation is to maintain the cable's minimum bend radius. If the installer bends the cable tighter than the minimum bend radius, doing so may result in markedly increased attenuation and broken fibers. If the elements of the cable (i.e., buffer tube or sheath) are not damaged when the bend is relaxed, the attenuation should return to normal. Cable manufacturers specify a minimum bend radii for cables under tension and for long-term installation. Table 15.1 shows a table of typical bend radius specifications.

TABLE 15.1 Typical Bend Radius Specifications[a]

		Minimum Bend Radius			
		Loaded		Unloaded	
Application	Fiber Count	cm	in.	cm	in.
Campus backbone	2–84	22.5	8.9	15.0	5.9
	86–216	25.0	9.9	20.0	7.9
Building backbone	2–12	10.5	4.1	7.0	2.8
	14–24	15.9	6.3	10.6	4.2
	26–48	26.7	10.5	17.8	7.0
	48–72	30.4	12.0	20.3	8.0
	74–216	29.4	11.6	19.6	7.7
Horizontal cabling	2	6.6	2.6	4.4	1.7
	4	7.2	2.8	4.8	1.9

[a]Specifications are based on representative cables for applications. Consult manufacturer for specifications of specific cables.

Note: For open-systems furniture applications where cables are placed, Siecor's 2- and 4-fiber horizontal cables have a one-inch bend radius.

Source: Reprinted with permission from Corning Cable Systems, Ref. 2, Figure 2.25, page 2.16.

TABLE 15.2 Typical Tensile Ratings[a]

		Maximum Tensile Load			
		Short Term		Long Term	
Application	Fiber Count	*N*	lbs	*N*	lbs
Campus backbone	2–84	2700	608	600	135
	86–216	2700	608	600	135
Building backbone	2–12	1800	404	600	135
	14–24	2700	608	1000	225
	26–48	5000	1124	2500	562
	48–72	5500	1236	3000	674
	74–216	2700	600	600	135
Horizontal cabling	2	750	169	200	45
	4	1100	247	440	99

[a]Specifications are based on representative cables for applications. Consult manufacturer for specific information.

Source: Reprinted with permission from Corning Cable Systems, Ref. 2, Figure 2.26, page 2.17.

Maximum Tensile Rating. Tensile rating means how tightly the installer can pull the cable before breaking it. The greatest possibility of this happening is during installation. Tension on the cable should be monitored when a mechanical pulling device is used.

Hand pulling does not require monitoring. Circuitous pulls can be accomplished through the use of backfeeding or centerpull techniques. For indoor installations, pull boxes can be used to allow cable access for backfeeding at every third 90° bend. Table 15.2 gives typical manufacturer's tensile ratings.

15.4.1 Recommended Connector for Building Fiber-Optic Cable Installations

The 568SC connector is recommended by ANSI/TIA/EIA-568-B.1 (Ref. 1) for fiber building installations because of its ability to establish and maintain correct polarization of transmit and receive optical fibers in two optical fiber transmission systems while still allowing transmission systems using other optical fiber counts. While the connector on the cabling side of the patch panel or telecommunications outlet/connector will use either two simplex SCs or the 568SC, the optical fiber patch cord will use the 568SC for two-fiber applications. The 568SC adapter will be either two simplex SC adapters or a duplex SC adapter of a one-piece construction.

Transmission Characteristics of Connector. The insertion loss of the 568SC connector will be no greater than 0.75 dB when installed. The total optical attenuation through a cross-connect from any terminated optical fiber to any other terminated optical fiber will not exceed 1.5 dB. The return loss of the connector will be greater than or equal to 26 dB on single-mode optical fiber. An optical fiber splice will have an insertion loss no greater than 0.3 dB.

The connecting hardware for patch panels for fiber-optic cable is installed at the following:

(a) The main cross-connect (MC)
(b) Intermediate cross-connect (IC)
(c) Horizontal cross-connect (HC)
(d) Horizontal cable transition points
(e) Telecommunications outlet/connectors

Typical cross-connect facilities consist of cross-connect jumpers or patch cords and patch panels that are connected directly to horizontal or backbone optical fiber cabling.

15.4.2 Cabling Practices—Polarization

To ensure a 568SC connection maintains the correct polarization throughout the cabling system, the correct adapter orientation and optical fiber cabling is followed. Once the installation is complete and the correct polarization is verified, the optical fiber cabling system will maintain that polarization of transmit and receive fibers and will not be of any concern to the end user.

The installation is such that the backbone and horizontal premises cabling is installed so as to pair an odd-numbered fiber with the next consecutive even-

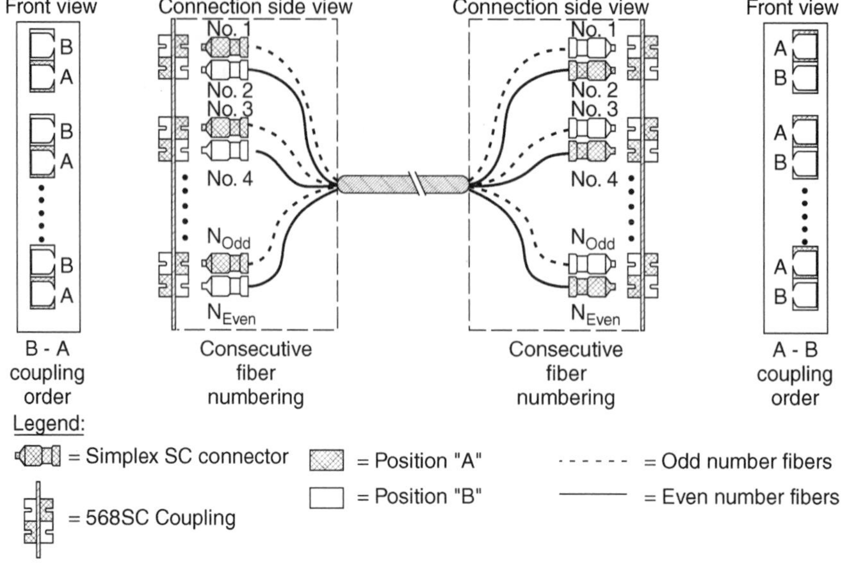

Figure 15.7. Specified optical fiber cabling for proper polarity (wall-mount hardware shown). (Reprinted with permission from ANSI/TIA/EIA-568-B.1, Ref. 1, Figure 10-1, page 36.)

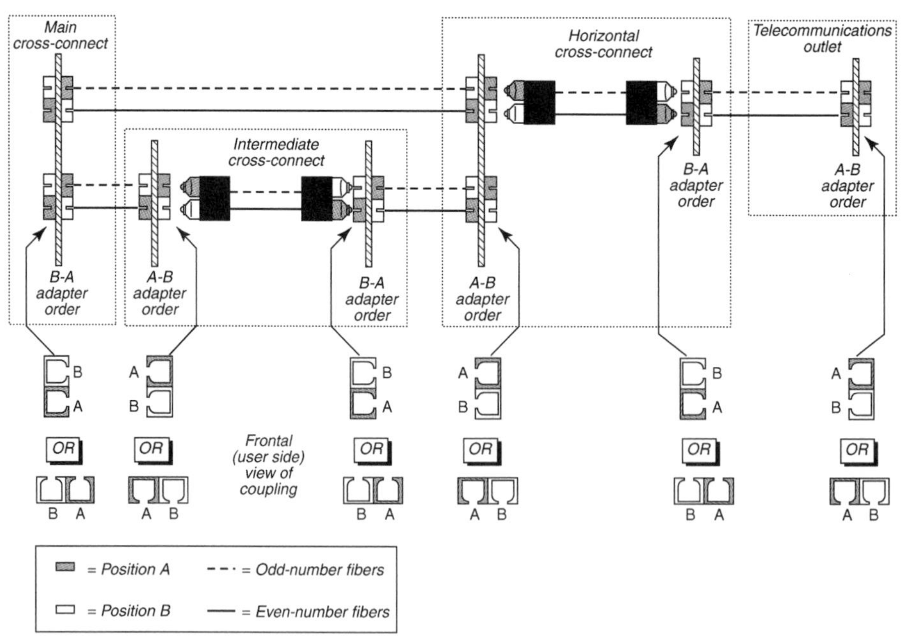

Figure 15.8. Optical fiber cabling plan for premises cabling. (Reprinted with permission from ANSI/TIA/EIA-568-B.1, Ref. 1, Figure 10-2, page 37.)

numbered fiber (i.e., fiber 1 with 2, 3 with 4, and so forth) to form two-fiber transmission paths. Each premises cabling segment will be installed with a pair-wise cross-over orientation such that odd-numbered fibers are Position A at one end and Position B at the other end while the even-numbered fibers are Position B at one end and Position B at the other end.

The cross-over is achieved by using consecutive fiber numbering (i.e., 1, 2, 3, 4 . . .) on both ends of an optical fiber line, but the 568SC adapters are installed in opposition ways on each end (i.e., A-B, A-B . . . on one end and B-A, B-A . . . on the other). Figure 15.7 shows optical fiber cabling for proper polarity.

Figure 15.8 shows the fiber cabling plan for premises cabling. It also shows the rules for the proper orientation of the 568SC adapter at the MC, IC and HC and telecommunications outlet/connector to insure proper polarization of an optical fiber system from the main cross-connect to the telecommunications outlet/connector. Optical fiber links not shown in the figure simply ensure that one end of the 568SC adapter be installed in the A-B orientation and the other in the B-A orientation.

15.5 FIBER SELECTION AND APPLICATION

As we stated above, there are two types of optical fiber that have application for premises wiring/cabling. These are 62.5/125-μm multimode fiber, which is best suited for short links typical of building applications, and single-mode fiber for longer links applicable for a campus setting or for very tall buildings.

The reason we are attracted to multimode fiber is that it allows us to use relatively inexpensive LED and VCSEL transmitters[2] and lower-cost connectors. LED, and to some lesser extent, VCSEL transmitters have broad natural spectral widths. The VCSEL also has a comparatively round beam of emitted light which couples better into round fibers. Multimode fibers have higher numerical apertures that will gather more light than their single-mode counterparts.

Table 15.3 is an applications summary table for 62.5/125-μm multimode fiber cable, and Table 15.4 is a similar table for single-mode fiber.

Corning Cable Systems and ANSI/EIA/TIA-568-B.1 recommend the use of hybrid cables consisting of both single-mode and 62.5/125-μm multimode fiber in the campus setting. This allows the use of lower-cost light sources and detectors where distances are 2000 m or less and for data rates up to 155 Mbps. For data rates of 622 Mbps and greater, use single-mode fiber for distances up to 40 km.

Premises fiber-optic systems operate at 850 or 1300 nm. A light signal behaves differently at these two wavelengths. The attenuation at 850 nm is higher than at 1300 nm. Also, the bandwidth tends to be smaller at 850 nm. Multimode fiber has better performance at 1300 nm, making longer system lengths possible. However, if bandwidth and length are not issues, 850-nm equipment is generally cheaper than its 1300-nm counterpart.

[2] LED and VCSEL transmitters are discussed in Chapter 4.

TABLE 15.3 62.5/125-μm Multimode Optical Fiber Applications Summary

		Cabling Scope and Technology Combinations					
		Horizontal Cabling < 100 meters		Building Cabling < 300 meters		Campus Backbone < 2000 meters	
Application	Bit Rate (Mbps)	Media	Tx Tech	Media	Tx Tech	Media	Tx Tech
10BASE-F	20	mm	S	mm	S	mm	S
Token ring	32	mm	S	mm	S	mm	S
100VG-AnyLAN	120	mm	S	mm	S	mm	LE
100BASE-F	125	mm	**S**	mm	**S**	mm	LE
FDDI	125	mm	**S**	mm	**S**	mm	LE
1000BASE-F	1250	mm	SL/LE	mm	SL/LE	sm	LL
Fiber channel	133	mm	**S**	mm	**S**	mm	**LE**
	266	mm	SL/LE	mm	SL/LE	sm	LL
	531	mm	SL/**LE**	mm	SL/**LE**	sm	LL
	1062	mm	SL	**mm**	SL	sm	LL
SONET-ATM	52	**mm**	**S**	**mm**	**S**	**mm**	**LE**
	155	mm	**S**/LE	mm	**S**/LE	mm	**SL**/LE
	622	mm	SL/LE	mm	SL/LE	sm	LL
	1244	**mm**	**SL**	**mm**	**SL**	sm	LL
	2488	**mm**	**SL**	**mm**	**SL**	sm	LL

Legend:
mm = 62.5/125-micron fiber with 160/500 MHz · km bandwidth
sm = standard single-mode fiber
S = short λ (850-nm window) LED or short λ laser equivalent
SL = short λ (850-nm window) laser
LE = long λ (1300-nm window) LED
LL = long λ (1300-nm window) laser
Note: Multiple entries are ordered for readability and uniformity only. Bold, underlined entries indicate where modifications to existing or draft standards are needed to support the most cost-effective approach.
Source: Reprinted with permission from Corning Cable Systems based on Ref. 2, Figure 3.4, page 3.6.

TABLE 15.4 Single-Mode Optical Fiber Applications Summary

Application/ Specification	Maximum Distance (m)	Data Rate (Mbps)	Attenuation (dB/km)	Wavelength (nm)
Cabling ANSI/TIA/EIA-568-B.1	60,000[a]	—	0.5/0.5[b]	1310/1550
Gigabit ethernet[c]	3000	1000	0.5	1310
FDDI-SMF	60,000	100	0.5[d]	1300
Fiber channel	10,000	531[e]	0.5	1300
Fiber channel	10,000	1063[e]	0.5	1300
ATM/SONET	55,000	155	Not specified	1300
ATM/SONET	50,000	622	Not specified	1300

[a]Beyond the scope of ANSI/TIA/EIA-568-B.1.
[b]Maximum attenuation for outside optical fiber cable. Maximum attenuation of inside optical fiber cable is 1.0/1.0 dB/km at 1310/1550 nm.
[c]1000BASE-LX specification values. Single-mode fiber is not supported by 1000BASE-SX specifications.
[d]Typical. (End-to-end attenuation is specified up to 32 dB.)
[e]MBaud.
Source: Reprinted with permission from Corning Cable Systems, Ref. 2, Figure 3.5, page 3.7.

TABLE 15.5 Network Applications and Number of Fibers Required

Network System	Fiber Required to Support Network	Fiber Type
ATM	2 Fibers	Single-mode or 62.5/125 μm
10BASE-F (ethernet)	2 Fibers	Multimode, 62.5/125μm
100BASE-F (ethernet)	2 Fibers	Multimode, 62.5/125 μm
1000BASE-F (ethernet)	2 Fibers	Single-mode or 62.5/125 μm
Token ring	4 Fibers	Multimode, 62.5/125 μm
FDDI	4 Fibers (DAS), 2 Fibers (SAS)	Multimode, 62.5/125 μm
Fiber channel	2 Fibers	Multimode, 62.5/125 μm
SONET[a]	4 Fibers	
Voice (two-way)	2 Fibers	Multimode, 62.5/125 μm
Video (broadcast)	1 or 2 Fibers	Single-mode or 62.5/125 μm
Video (security)	1 Fiber	Multimode, 62.5/125 μm
Video (interactive)	2 Fibers	Multimode, 62.5/125 μm
Telemetry	1 or 2 Fibers	Multimode, 62.5/125 μm

[a]SONET and video are listed here due to system design requirements to incorporate these applications into existing communication networks.

Note: DAS is an FDDI term for dual attachment station, SAS is an FDDI term for single attachment station.

Source: Reprinted with permission from Corning Cable Systems, Ref. 2, Figure 4.4, page 4.3.

Number of Fibers Required. Most applications for premises wiring requires full duplex operation; thus two fibers, as a minimum, are required. It should be noted, however, that video operation may be just one-way, and only one fiber would be required in that case. Broadcast or CATV video is rapidly changing from the traditional one-fiber source broadcast to receiver station interaction. In this case, two fibers are required. With the advances already in place today, all CATV applications will require two fibers.

Traditional telemetry requires only one fiber. But this is changing, too. Certain types of telemetry are interactive; thus two fibers are required, one for transmit and one for receive.

Data communications employ two fibers. Nevertheless, four-fiber systems are also in use. The determining factor is the type of network application being implemented. Table 15.5 lists some of the most current network applications and the number of fibers required.

Additional electronic equipment can be traded off for fiber count. Electronic equipment such as bridges and routers can be used to collect and distribute signals. Likewise, the fiber count can be increased, allowing for a decrease of electronic equipment. There is an interesting array of options available to the designer. There is a large variety of data services, which may or may not also include voice. For example, ATM accepts voice, data, and video over the same network. There is a form of FDDI that accepts voice transmission as well. Ethernet varieties carry voice.

The designer must decide whether to employ additional electronics equipment and reduce fiber count or minimize electronic equipment and use more fibers in a cable. Figure 15.9 shows two generalized setups for data communications operation between two buildings. With the *standard* installation, 12 fibers are required between the buildings. In an *enhanced* design, using the same equipment but adding a router, only four fibers are required to support the same operation.

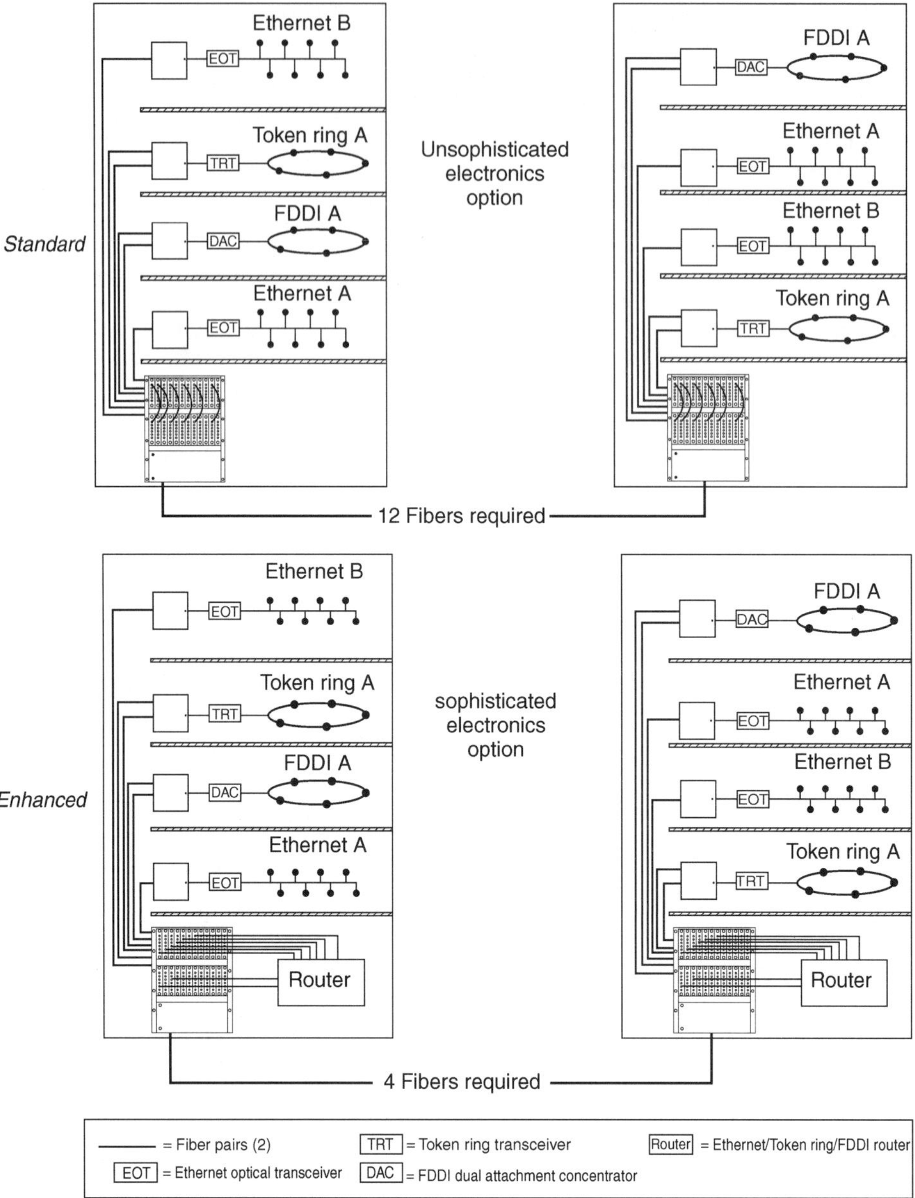

Figure 15.9. Number of fibers versus electronic complexity. (Reprinted with permission from Corning Cable Systems, Ref. 2, Figure 4.5, page 4.5.)

Network complexity and availability may also play a role here. Thus, users may often conclude that the most cost-effective networks are the simpler ones that use more fibers and less electronics.

Spare Fibers. Early on, the network designer should make a decision on spare fibers. The network designer can determine spare fiber count by examining future growth in existing area, through knowledge of new systems to be installed or planned new construction. If little or no change is expected over, say, 5 years in the future, there should be at least 25% greater number of fibers than those currently required. A 50% increase is recommended where future network applications are defined and potential demand for unanticipated service is high. In those cases where future additions can be anticipated where there is also a potential demand for unanticipated service, and especially where expansion in fiber count in the future is uncertain, where installation will be extremely difficult, a 100% addition is recommended to the number of fibers currently required. Also, the planner should consider standard fiber count in cables—usually in multiples of 6 for modest installations and in multiples of 12 for large installations.

A typical network consists of hubs or concentrators distributed to each telecommunications room, keeping the fiber requirements minimal in the building backbone. The designer should determine the level of the use of bridges and routers to help establish the fiber count on the building backbone. A table similar to Table 15.5 should be constructed to ensure that all data equipment requirements are fulfilled. One suggestion is to install a 30-fiber, 62.5-μm cable to each horizontal cross-connect if fiber-to-the-desk is contemplated. The 30-fiber count seems excessive unless one includes probable expansion. The inclusion of single-mode fiber is not recommended. Unless some very special circumstances dictate, the entire fiber population in a building should consist of 62.5-μm multimode cable mainly due to economy. However, this type of fiber can support transmission rates in excess of 1 Gbps for comparatively short distances, typically less than 300 meters. A minimum of a 24-fiber building backbone should be considered in support of distributed electronics.

Centralizing Backbone Cabling. Because fiber optics does not normally require the high-speed modem electronics of copper-pair systems, fiber-optic systems do not require the use of electronics in telecommunication rooms on each floor and, therefore, allow for a total centralized cabling network. Based on TIA document TSB72 (Ref. 5), a methodology is outlined where the end-user can place all of the data electronics in one room in the building. Compare this with the use of multiple rooms spread throughout the building when the backbone medium is exclusively copper. This greatly simplifies the management and maintenance of fiber-optic networks and provides the more efficient use of ports on electronic hubs, routers, and switches.

This design can be implemented in various ways. It can employ pull-through cables from outlets or multiuser outlets to the single central telecommunication room in the building. Another way of achieving the same thing is to place s splice between the horizontal cable and the higher fiber count backbone cable in the room. A third option is to use a passive interconnect in each telecommunications room. All three options have benefits, though the user will usually opt for the second alternative. The principal reasons are that it allows for simpler cable placement and offers an acceptable degree of flexibility for adding users at reasonable cost.

It is recommended that the building backbone cable be sized with sufficient spare capacity to permit additional building backbone cables. Of course the building backbone fiber count should be sized to deliver present and future applications to the maximum work area density within the area served by the TC.

Corning Cable Systems suggests, in its *Design Guide* (Ref. 2), that two fibers should be employed for each application delivered to a work area, so that a room serving 72 users with single applications would require a 144-fiber cable to be installed. Additionally, there may be application, such as voice and SCADA or telemetry that would be served by electronics in the room requiring an additional 12 or more fibers. Also recommended is a minimum two-fiber 62.5-μm cable for data communications and a separate category 3 UTP for voice telephony.

15.6 PERFORMANCE TESTING OF AN ENTERPRISE NETWORK

Network performance has to be measured. That is the hard way and the only way. The easy way is to take a vendor's word, a brochure specification, a magazine article review, or word-of-mouth. Many data networks are installed without acceptance testing. Breakdowns, unforeseen bottlenecks, and substandard performance after cutover to the customer can cause the telecommunication manager great stress and mental anguish, and cost the user great sums of money to fix the problems.

15.6.1 Enterprise Network Performance Issues

Consider the following six enterprise network performance issues:

1. *Throughput*. The network must faithfully deliver packets of data to all user nodes at the specified data rate and inside the system bit error rate.
2. *Latency*. Time to deliver, or how long it takes to deliver data packets (or frames) from originator to destination. Some traffic types are more urgent than others.
3. *Jitter*. Time-varying phase of serial bit stream relative to reference phase (derived from Ref. 8).
4. *Integrity*. Data should not be corrupted as it traverses the network.
5. *Order*. In some data systems, packet or frame order must be maintained from originator to destination.
6. *Priority*. If congestion occurs, the system may discard some packets or frames. The network should be capable of distinguishing between packets of differing priority and, under stress, discard packets (frames) of lower priority (typically, frame relay and ATM).

Network test plans should take these six issues into account when formulating a methodology of network performance verification and troubleshooting. We add the term *troubleshooting* as an important requirement because the test technician must be able to pinpoint the problem causing the network to underperform or just plain fail.

15.6.2 Preparation of Test Plans and Test Methodologies

We should be able to test for compliance each of the six listed performance issues or performance requirements. Reference 8 uses *throughput* as an example. There are numerous definitions of throughput. Reference 9 defines throughput as "the total capability of equipment to process or transmit data during a specified time period." Another definition is "the receipt of error-free frames per unit time." For a device the IETF defines it "as the maximum rate at which none of the offered frames are dropped by the device."

One approach to throughput compliance of a device is to stress it by increasing the input frame rate to the device until frames begin to be dropped. That is easier said than done on an enterprise network that runs with a minimum of test equipment and specialized device test positions on the device itself. Just to measure frame rate requires special test equipment.

Of course, to determine whether the network element actually dropped frames, the tester would have to bridge or terminate its output ports and account for every packet or frame put into the device. As we said, this is a big job. For example, each test packet (or frame) would have to be uniquely identified so that a receiving port can isolate and count it. Even when an underlying operating system is very sophisticated and fast, it cannot generate and count test packets on optical networks running at gigabit-per-second speeds.

Enterprise-network device testing thus should have specialized hardware with the following characteristics:

- Bit error rate test set (BERT) operational up to 1 Gbps. Design objective: 10 Gbps, BER $< 1 \times 10^{-12}$, Agilent 81250 (Ref. 10).
- Router test set, Agilent E5203A (router-tester) (Ref. 10).
- Protocol simulator/test set. Must include all protocols employed in the enterprise network.
- Ethernet frame generator; ATM cell generator; frame relay layer 2 frame generator and layer 3 health and welfare frames test. FDDI frame generator. Proprietary frame generators, if required. Agilent 86100A (Ref. 10).
- Capability to measure jitter and latency. Advisor LAN, Agilent p/o 5980-0990E (Ref. 10).
- Test equipment and techniques discussed in Chapter 16.

It should be noted that there may be some duplication of test/performance procedures with the equipment list provided above.

Reference 7 describes a throughput test described in IETF document RFC 2544 (Ref. 9) to determine maximum throughput of a network element.

Most test equipment vendors provide capability for the throughput test described in Ref. 7. Yet a network designer still needs to be aware of the various considerations that become important when dealing with some of the subtler issues. For example, a router on one end of a T1 link stores frames that are input at an equivalent 1-Gbps rate. It is only transmitting frames at 1.536 Mbps. Soon the router buffer fills at about the end of the test. However, the router is still bleeding off test frames to the other end, finally catching up. In such instances, the throughput algorithm would need to be modified such that the actual frames being counted are embedded in a series of frames not counted. Of course it would be more suitable to use test equipment built to operate at the 1-Gbps rate.

Reference 8 describes the latency performance measurement as one with issues that can become quite complex as the test evolves. Latency is defined for bit-forwarding devices as "the time interval starting when the end of the first bit of the input frame reaches the input port and ending when the start of the first bit of the output frame is seen on the output port." Here latency is defined with regard to the network element. Again we follow the methodology suggested in RFC 2544 (Ref. 9). First determine the maximum throughput of a network element, then transmit a continuous stream of frames at the throughput rate for 120 seconds with a "signature" frame transmitted at 60 seconds. The signature frame carries a time stamp, and the latency is determined by the transmitted time stamp. Several runs are recommended, and the final value is the average of the results. Reference 7 reports that this approach works well if the tester is trying to acquire a single number that is somewhat indicative to the network element's ability to forward frames.

REFERENCES

1. *Commercial Building Telecommunications Cabling Standard*, ANSI/TIA/EIA-568B.1, TIA, Arlington, VA, April 12, 2001.
2. *Design Guide*, Release 4, Corning Cable Systems, Hickory, NC 28603-0489, March 1999.
3. *Detail Specification for 62.5 μm Core Diameter/125 μm Cladding Diameter Class Ia Graded-Index Multimode Optical Fibers*, ANSI/TIA/EIA-492AAAA-A, TIA, Arlington, VA (no date).
4. *Optical Fiber Cabling Components Standard*, ANSI/TIA/EIA-568-B.3, TIA Arlington, VA, April 2000.
5. *Centralized Optical Fiber Cabling Guidelines*, ANSI/TIA TSB72, TIA, Arlington, VA, October 1995.
6. *Fiber Selection Guide for Premises Networks*, Corning WP1160, Corning Fiber Systems, Corning, NY, May 1998.
7. Dan Schaefer, Ixia, "Taking Stock of Premises-Network Performance," *Lightwave*, page 70, April 2001.
8. *The IEEE Standard Dictionary of Electrical and Electronic Terms*, 6th ed., IEEE, New York, 1996.
9. *Benchmarking Methodology for Network Interconnection Devices*, RFC 2544, March 1999, from the Internet.
10. Agilent Technologies Test and Measurement Catalog 2001, Agilent, Palo Alto, CA, 2001.

16

TOOLS FOR TROUBLESHOOTING FIBER-OPTIC SYSTEMS

16.1 SCENARIO

We are given a fully equipped fiber-optic network that has excessive outages or does not meet performance requirements based on BER measurements or is completely inoperative, or we want to verify performance parameters prior to final cutover.

There are four troubleshooting tools at our disposal. The first is the network operations control center (NOCC), which can provide certain noninvasive on-line performance tests (see Section 12.6). The second, third, and fourth test operations are *invasive*, meaning that the flow of revenue-bearing traffic is interrupted during the period of active testing.

In this chapter we will discuss the important test device, the *power meter*, which is in the second group of test procedures. The third group of tests are carried out with the versatile *OTDR* (optical time-domain reflectometer), which can serve to measure many of the important parameters of a fiber-optic line, typically used to locate a break or other discontinuity in the fiber line. The fourth group of tests deals with measuring BER of portions of the system or of the entire system. We also briefly cover the optical spectrum analyzer or OSA. At the end of the section, the possibility of having an optical service channel in a WDM aggregate of channels is discussed.

16.2 TEST EQUIPMENT

The test technician or craftsperson should have available the test equipment listed in Table 16.1.

16.3 TEST PROCEDURES USING AN OPTICAL POWER METER

16.3.1 Measurement of Fiber Strand Breaks in a Fiber-Optic Cable Using an Optical Power Meter

Fiber strand breaks are fairly common in a fiber-optic installation. Such breaks usually happen during installation. The symptoms are easy; there is no signal at the

TABLE 16.1 Test Equipment List

Equipment	Purpose
Digital voltmeter	Track operating levels; adjust, restore electrical levels.
Optical spectrum analyzer	Measure carriers and flatness of spectrum.
Optical power meter	Measure optical power at transmitter and at receiver input; measure segments of link from transmitter outwards.
True rms voltmeter	Measure signal-to-noise at electrical baseband.
NTSC or other relevant waveform generator	TV signals, measure differential phase/gain. Relevant waveform test sequences.
Waveform monitor	Measure video and other parameters.
OTDR (optical time-domain reflectometer)	Identify fiber discontinuities and their location; measure fiber attenuation, return loss at connector/splice.
BERT (bit error rate test set)	Bit error tests, localize faults.
Vectorscope	For TV, measure differential phase and gain.
Light signal source, four wavelength bands with resident power monitor	Often test setup includes power meter, which is connected at other side of component under test.

receiver end of the cable. The troubleshooter asks himself/herself: What are the possibilities? Here are several:

- There is no light signal out of the transmitter.
- There is a break in the cable strand feeding the far-end receiver.
- The receiver is nonoperational in that it does not carry out its function.

This test to determine a break has relevance with an electrical continuity test that many of us are familiar with. It is done with a volt-ohm meter. On a fiber line, some technicians just use a flashlight or other simple light source and check to see if light is coming out the other end of the fiber. Red light is preferred.

When the test is carried out with a power meter, we will know about how much light is available at the distant end of the fiber. Similar to the "flashlight" test, we need a light source. Ideally, it should have a calibrated source, and the source should be wavelength selectable. The desired wavelengths are 850 nm, 1300 or 1310 nm, and 1550 nm. In addition, it would be desirable have two other wavelengths: 780 nm and 1630 nm. This simple test setup is illustrated in Figure 16.1.

In the case at hand, we disconnected the transmitter from the cable and connected its output to the power meter. The reading is normal. In other words the light power coming out of the transmitter meets the specification (i.e., inside tolerance limits). We go to the point of access nearest the cable exit port and check the power level again. We get a reading on the power meter (usually in dBm). We note that the level is about 0.7 dB below the reading at the transmitter output. From the relevant link budget, this is the expected loss between the transmitter output and the access point (an optical patch panel or cross-connect). We connect a power meter at the distant end, at the output of the receiver connector. There is no reading of light power. The overwhelming probability is that we have a cable/fiber break somewhere along the fiber line between the transmitter

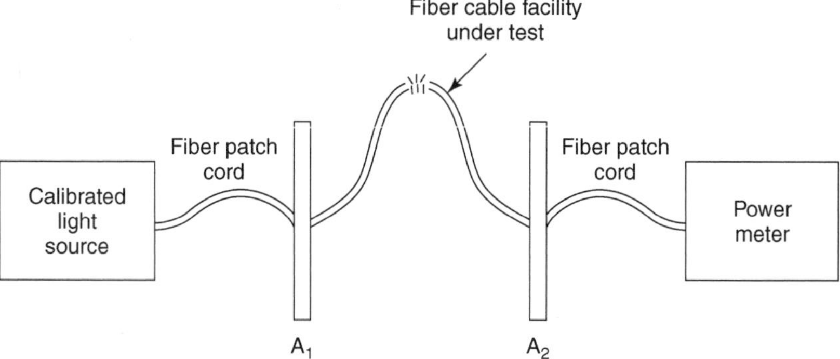

Figure 16.1. Basic test setup with a power meter. A_1 and A_2 are fiber patch panels.

and the far bitter end of the cable at the receiver connector. The big question now is, Where did the break occur?

If there are access points to the fiber strand, similar power tests may be carried out until we reach a point where we find an *expected* light level reading. We consult the link budget again and estimate the loss of fiber, connectors, and splices. Then subtract that loss (in dB) from the light level at the transmitter output. Compare this to the reading on the power meter. They should be around $\pm 5\%$ of each other. We continue down the fiber, taking level readings at sequential access points. We do this until we reach a point where the level reading is 0. We now can approximately identify the area along the line where the break probably occurred.

Following this method and doing no more tests, we would have to pull up that entire length of fiber and replace it. A lot of money can be saved on spare fiber if we use an OTDR so we can pinpoint within several feet where the break occurred. With further tests, we identify essentially the point of break, pull some slack, and splice.

16.4 INTRODUCTION TO AN OPTICAL TIME-DOMAIN REFLECTOMETER (OTDR)

In Section 16.3 we were left hanging, so-to-speak. We knew there was a break in the fiber, but we did not know where it was along the fiber line. Bring on an OTDR, and it will tell us where the break is.

An OTDR provides the user with a visual representation of an optical fiber's characteristics along its length. The OTDR gives a plot on a screen, in graph format, where distance is represented on the horizontal or x-axis and attenuation is on the vertical or y-axis. With the proper setup, an OTDR display can provide the user with a lot of information such as

- Anomaly location (along the fiber line)
- Fiber loss

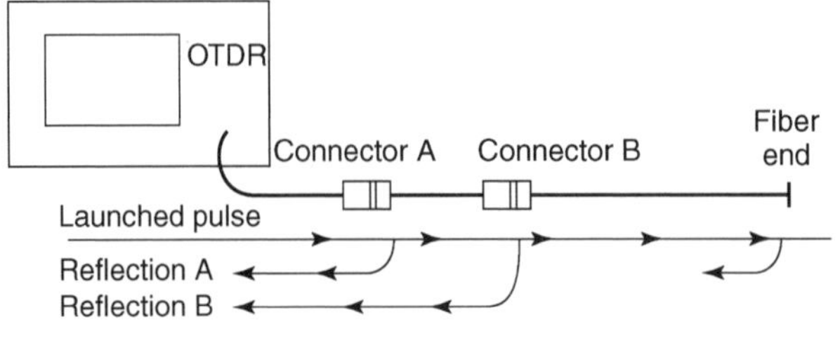

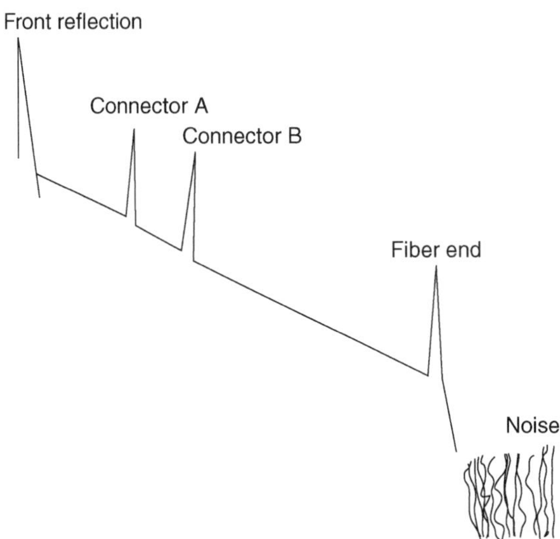

Figure 16.2. A typical OTDR setup illustrating a simple fiber line with two connectors installed. The upper half of the sketch shows an OTDR with a fiber line connected. The lower part of the sketch shows the resulting returns that may be seen on the OTDR display.

- Splice loss, connector loss
- Optical return loss (ORL) of components in the optical fiber line

It provides the best representation of overall fiber integrity.

The operation of an OTDR is similar to that of radar. It sends out a short pulse of light and then measures the time it takes for the reflection of the pulse to return. The pulse may be reflected off a ship or aircraft. In a similar fashion, an OTDR emits a light pulse that travels along the fiber until it meets an obstacle. The returned signal consists of (a) backscattered light along the fiber and (b) reflected light from *events* such as refractive index discontinuities at fiber joints, breaks, and ends. The optical loss between two points on the fiber can be indirectly determined by measuring the difference in the returned backscatter power between the two points in question. A typical fiber setup is illustrated in Figure 16.2.

A basic task of an OTDR is to measure the distance to a break or other anomaly in a fiber line. Knowing the fiber's index of refraction (n) of the core and the time required for the reflections to return (T in seconds), the OTDR computes the distance (meters) to an event based on the following equation:

$$\text{Distance}_{(\text{meters})} = 3 \times 10^{8}_{(\text{seconds})}(T)/2 \times n \tag{16.1}$$

Often the reflected light signal is very weak and is buried down in the noise of the receiver portion of the OTDR. This is one reason why an OTDR sends out multiple pulses, a chain of pulses. The receiver now has numerous pulses that are averaged such that it can compute and display a trace on the screen. A sketch of such a trace is shown in Figure 16.2.

OTDRs present the results of their measurement in the form of a fiber trace on a graphical display. The displayed information is a sloping trace of the logarithm of the backscattered power received by the OTDR. It should be noted that the slope of the trace approximately equals the attenuation coefficient of the fiber. Where there is a splice, connector, or other discontinuity in the fiber, their approximate loss can be measured by the vertical offset of the two fiber slopes on either side of the joint or other discontinuity. To accurately measure splice, connector or other discontinuity loss, an average of two OTDR fiber measurements, one from each end of the link, is required. The two measurements are then averaged. We can see that the OTDR is a unique tool that allows individual fiber events to be measured both in change of optical power and in distance from the OTDR. Figure 16.3 is a

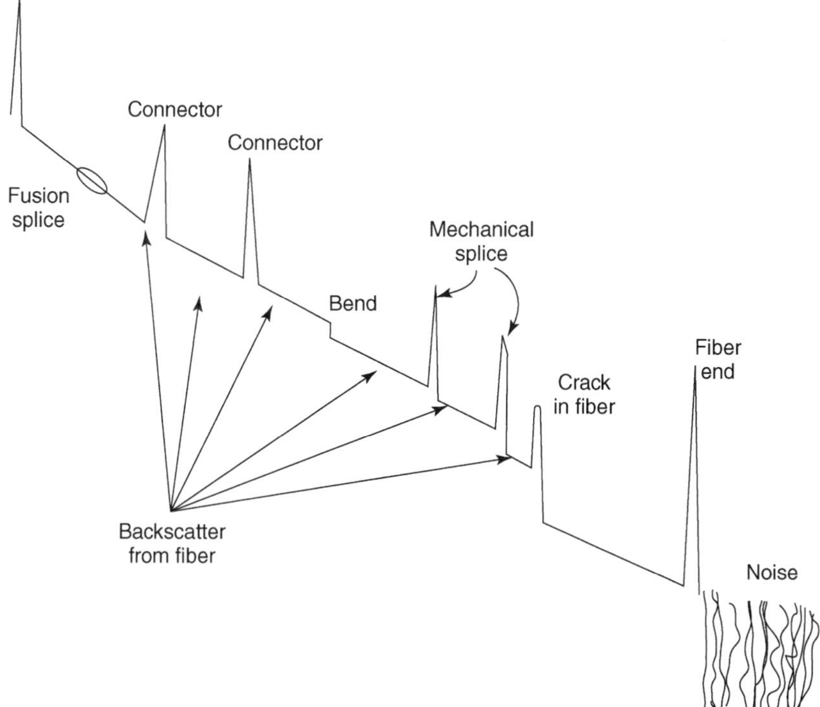

Figure 16.3. Sketch of a fiber trace as seen on the display of an OTDR. Discontinuities are identified.

sketch of a trace seen on the display of an OTDR. The various discontinuities along the fiber's trace are identified.

The entire length of the fiber reflects some light giving the impression of a straight line with a constant slope. The constant slope represents accumulating fiber attenuation as a pulse passes up the fiber. The straight line represents the *backscatter* resulting from scatter off minute imperfections of the fiber. Steps will be seen along the trace. A *step* is an instantaneous change of power. Most, or nearly all, of the steps seen along the fiber trace represent fusion splices. A step can also be a result of reflected light from microbends in the fiber. In this situation the fiber may be squashed or strongly bent (macrobends) such that some light may escape from the fiber resulting in a loss.

Peaks or spikes are seen along the trace. These represent reflections caused by a glass-to-air or air-to-glass transitions. One can expect to see these spikes where there are mechanical splices or at pairs of connectors. At the fiber end there is light return from noise. This noise derives most from the receiver, which has, of course, a finite sensitivity. An important parameter of an OTDR is its *dynamic range*. It may be defined as the difference between the power at the beginning of the backscatter shown as the trace and the peak of the noise at the far bitter end of the fiber.

There is another limit on the OTDR display. This is the *attenuation deadzone* or simply the *deadzone*. This limit describes the distance between the beginning of the peak and the point where it is almost back to backscatter level. The deadzone limits the resolution of measurement. Both of these parameters depend on the behavior of the receiver and on the pulse launched down the fiber. It turns out that the better the dynamic range, the worse the deadzone, and vice versa.

There are four types of test equipment now available in the OTDR family. These are:

1. Full-feature OTDR
2. Mini-OTDR
3. Fiber-break locator
4. Remote test unit (RTU)

Full-feature OTDRs are conventional optical time-domain reflectometers that are feature-rich. They are usually heavy and less portable than either mini-OTDRs or fiber-break locators. The full-feature OTDR is more commonly found in the laboratory or for very difficult troubleshooting procedures being carried out in the field.

Mini-OTDRs and fiber-break locators are highly portable devices used for fiber network troubleshooting. As the term implies, mini-OTDRs are intended OTDRs with reduced features. The mini-OTDR is inexpensive, can carry out many, if not most, of the functions of the full-feature OTDR, and is light and small such that a field technician can carry one around in his/her field kit. On the other hand, the fiber-break locator is nothing more than an optoelectronic tape measure that can only measure distance to catastrophic fiber events.

A full-feature OTDR should be able to operate on the following wavelength bands: 850 nm and 1300 nm (multimode fiber) and 1310 nm and 1550 nm (single-mode fiber).

We should expect an OTDR to provide us information on typical fiber-optic links such as:

- Fiber characteristics (typically attenuation coefficient, backscatter coefficient and group index)
- Insertion loss of individual events
- Reflectance of individual events
- Return loss of the links
- Distance between events
- End-to-end distance and loss between terminals

One or both of the following anomalies may appear on an OTDR screen:

Ghosts. Many of us have dealt with *ghosts* in TV reception. On an OTDR display, a ghost is an apparent event on its display which is not there in reality. Ghosts can be caused by both the fiber under test and the OTDR instrument itself. Ghosts are produced by multiple reflections in both directions along the fiber. Some of the light from ghosts is reflected back to the OTDR display, with extra delay appearing as an additional event. The intensity of ghosts can be reduced by

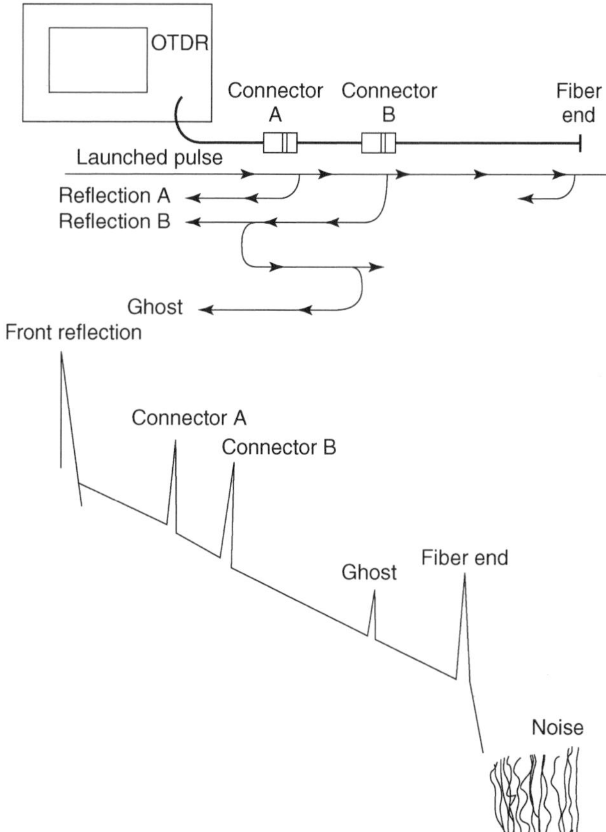

Figure 16.4. A typical spike that is a ghost.

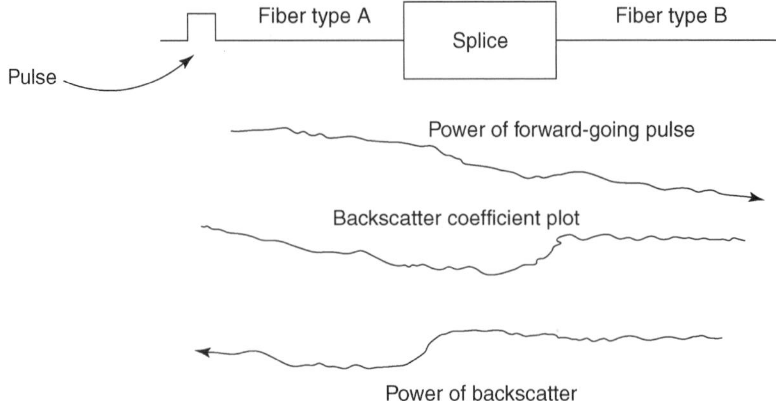

Figure 16.5. A sketch of an OTDR display showing a "gainer" splice. The joint seems to amplify light; thus it is called a gainer.

improving the return loss (also called reflectance) of connectors or by replacing connectors with splices. A typical ghost spike is shown in Figure 16.4.

We also can have ghosts when the OTDR has not been set up correctly. If pulse no. 1 is traveling down the fiber and pulse no. 2 is launched down the fiber before pulse no. 1 leaves the fiber at the distant end, a ghost will appear. This is because the OTDR receiver sees the echo from the first and second pulses at the same time and an event appears that does not exist. This can be cured by simply increasing the time between pulses. Several OTDRs on the market make the adjustment automatically.

Gainer. A *gainer* is another type of error in a fiber reading by an OTDR. When two different types of fiber are joined by a splice or a connector, a gainer can occur. Of course this appears at the point of splicing, and it is seen as a rise in amplitude instead of a drop in amplitude. This phenomenon is not at all helpful to the troubleshooting technician because he/she wishes to measure those losses quantitatively. Being there are two different types of fibers, each probably has different properties. Then one of the fibers at the joint sends back a stronger return than the other fiber, resulting in incorrect loss measurements. Figure 16.5 is a sketch of an OTDR display with a "gainer."

OTDR Range. An OTDR has a maximum range. That range is based on the "dynamic range" of the instrument. Suppose a manufacturer specifies that the dynamic range (D) of a certain OTDR type was 30 dB at 1550 nm and that the fiber attenuation [$L_{(\mathrm{dB/km})}$] at that wavelength was 0.25 dB/km, which includes all splices and connectors. What is the maximum range of the instrument?

$$\mathrm{Range}_{(\mathrm{km})} = D_{(\mathrm{OTDR})\,(\mathrm{dB})}/L_{(\mathrm{dB/km})} = 30\ \mathrm{dB}/0.25\ \mathrm{dB/km} = 120\ \mathrm{km} \quad (16.2)$$

To overcome the problem of the deadzone, bidirectional measurements are recommended. From one end of the fiber in question, the technician will measure loss to an identified event and record the reading. From the opposite end of the fiber, he/she will measure the loss to the same event and record reading. The next step is to sum the two readings and divide that sum by 2 for the average reading, which will be more accurate than just taking one reading alone.

Care must be exercised when loose tube fiber is employed in an installation. Generally, in this case, the fiber in the cable is longer than the cable itself. This variance will affect OTDR measurement accuracy. The excess fiber in the cable is due to the slight bunching of the loose tube fiber in the cable tubes and the tube's spiral wrap path around the cable's central strength member. The cable manufacturer should specify this excess amount of fiber in the cable as a percentage of the total cable jacket length. The OTDR should be adjusted for this excess fiber length.

The cable manufacturer should also supply the fiber's index of refraction accurate to four places. Before carrying out any measurements with an OTDR, the instrument should be adjusted, entering the correct index of refraction to the four-figure accuracy.

The following is a method for adjusting cable length for excess fiber.

Equation (16.3) shows the relationship between the cable distance D_{cable} to the event from the OTDR location and D_{fiber}, the actual fiber distance from the OTDR to the event.

$$D_{\text{cable}} = D_{\text{fiber}}/(1 + \alpha/100) \tag{16.3}$$

where α = excess amount (in percent) of fiber in the cable provided by the cable manufacturer.

Consider this example. A craftsperson read the distance to a fiber break as 48.36 km. The fiber manufacturer indicates that there is 5.4% excess fiber in a length of cable. For the craftsperson to go directly to the point on the cable where the break occurred, he/she will want to know the cable jacket distance to the break. Apply formula (16.3):

$$\begin{aligned} D_{(\text{cable})} &= 48.36/(1 + 0.054) \\ &= 45.88 \text{ km} \end{aligned}$$

Sources: Refs. 1 and 2.

16.5 BIT ERROR RATE TESTS AND OTHER ELECTRICAL TEST PROCEDURES

16.5.1 The Concept of a BERT

In a digital network the most important QoS parameter is bit error rate (BER). Outside of the world of CATV, in nearly every case a fiber network is transporting exclusively digital information. The only way we can measure BER is on the electrical equivalent of the light signal. So the first step for the BER test is to reduce the light signal to its electrical equivalent by passing it through a light detector/receiver.

First we review a BER test without the intercession of any light equipment. We can imagine our imaginary test as the test of a portion or the entire digital

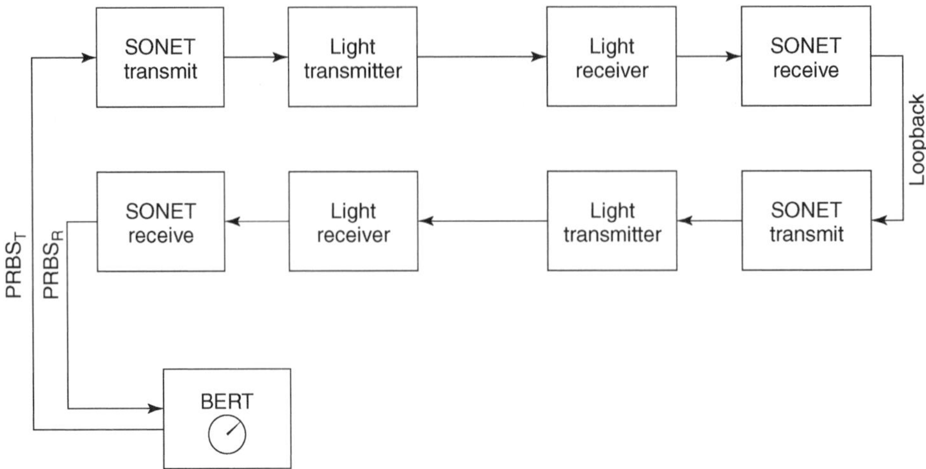

Figure 16.6. Conceptual layout and setup of a BER test using loopback procedure. Sketch represents a full-duplex channel (4-wire). PRBS stands for pseudorandom binary sequence.

network. The electrical signal will have the characteristics of a SONET, SDH, or PDH waveform (see Chapter 9). In this case it is a DS1 bit stream. The DS1 is taken out of service, and its bit stream is one generated by the BERT device at 1.544 Mbps. The most common test is a pseudorandom binary sequence (PRBS). If we were to break into the bit stream in the middle of the network and look at its bits, they would appear as some random arrangement of bits. It can be shown that a random bit stream is the most foolproof method of BERT testing.

At the distant-end DS1 demultiplexer output, a loopback test is invoked. This is simply taking the demultiplexed bit stream from the DS1 channel bank demultiplex port and feed stream right back into the multiplex section transmit port of the same channel. This procedure is illustrated in Figure 16.6.

The receive channel on the multiplex base equipment at the near end of the link should be carrying the bit stream for which we wish to measure BER. We use the word *should* because we are assuming that there are no faults on the link. The output of the demultiplex receive channel is connected to the BERT test set, and the BER reading is taken and recorded. Assuming that the link in question is performing correctly, then there are several performance parameters that can affect the BER reading. These are:

- Source transmitter power output
- Receiver threshold setting, far end
- Transmitter power output setting, far end
- Receiver threshold setting, near end
- Regenerator parameters, if there is a regenerator in the link
- Optical amplifier parameters, if there is an optical amplifier in the link

What BER is required by link specifications? Where will it be measured? The receiver thresholds are set for that value. It is advisable to compare these various readings with the values used on the link budget. Look for a BER in the vicinity of either 1×10^{-10} or 1×10^{-12}. High-bit-rate links may require even better BER values, in the range of 1×10^{-15}. By *high* we mean 10 Gbps, and 40 Gbps will be achieved during the early lifetime of this book. When setting up to run a BER test and such BERs of 1×10^{-15} are required, ask yourself, "How long will I have to wait statistically for the first error to appear?" Suppose a link was running at 1 Mbps and the BER required was 1×10^{-12}. How long must one wait statistically for the first error to appear (random errors assumed)? Answer (in seconds) $= 1 \times 10^{-6} - 1 \times 10^{-12}$ or 1 million seconds. Note that $1 \times 10^{6}/60 \times 60 = 277$ hours, or 11.57 days. That is a costly wait period. The BER value of 1×10^{-15} would seem kind of excessive, but the industry is beginning to talk in that value.

After consulting the link budget, the responsible engineer should ask, "How much of the link margin have I consumed to reach these values?" In the case where the link carries multiple light channels (WDM), each channel should be measured following identical criteria. The difference in level between channels should be greater than 2 dB.

Other loopbacks can be set up and BER measured. Remember that the BERT test is electrical, and we have to provide a light detector to bring the signal from the optical domain to the electrical domain. Other procedures allow us to use the link's own receiver. This, of course, is mandatory for the final link BER check.

One lesson here that with a little ingenuity, loopback testing is an excellent tool to isolate faults occurring on a fiber system.

Source: Ref. 3.

16.6 OPTICAL SPECTRUM ANALYZERS (OSAs)

An optical spectrum analyzer displays the spectrum of an optical signal or signals. The display of an OSA is calibrated in dBm along the y-axis (the vertical axis), and the x-axis (horizontal axis) displays wavelength in nanometers. The OSA is a most valuable instrument for debugging WDM systems. It is really the only device for measuring OSNR (optical signal-to-noise ratio), especially when working with DWDM. OSA measurements can include, besides OSNR, signal power, power levels of individual wavelengths in the case of WDM formations, spectral width of a light signal, wavelength, and channel spacing.

The OSNR is relatively simple to obtain using an OSA. It is the ratio (or difference, when the values are expressed in dBm) between the peak channel power and the noise power within the channel bandwidth. Many power measuring instruments will give the OSNR automatically. With DWDM systems an average value of OSNR of 18 dB is usually acceptable.

Source: Ref. 4.

The uniformity of transmitter power on WDM systems is another parameter of interest. It is generally quoted as the difference in level between the strongest and weakest channels. The value should not exceed 2 dB.

Figure 16.7 illustrates an OSA display of several WDM light signals.

An OSA in combination with a power meter and a tunable laser source (TLS) can be used to measure gain of an EDFA (erbium-doped fiber amplifier, see Chapter 7).

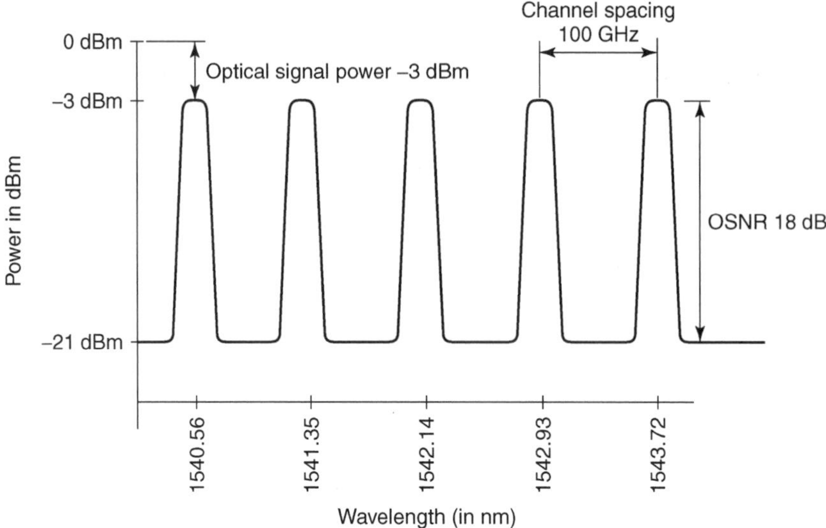

Figure 16.7. Several WDM channels are shown as they would appear on the display of an OSA.

Optical spectrum analyzers can be divided into three categories: diffraction-grating-based and two interferometer-based optical architectures, the Fabry–Perot and Michelson interferometer-based optical spectrum analyzers. Diffraction-grating-based optical spectrum analyzers are capable of measuring spectra of lasers and LEDs. The resolution of these instruments varies, typically ranging from 0.1 nm to 5 or 10 nm. Fabry–Perot-interferometer-based OSAs have a fixed narrow resolution, typically specified in frequency between 100 MHz and 10 GHz. This narrow resolution allows them to be used for measuring laser chirp, but can limit their measurement spans much more than the diffraction-grating OSAs. Michelson interferometer-based OSAs used for direct coherence-length measurements display the spectrum by calculating the Fourier transform of a measured interference pattern.

Sources: Refs. 2 and 6.

16.7 LIGHTWAVE SIGNAL ANALYZER

Lightwave signal analyzers help users measure important characteristics of optical fiber systems such as signal strength, modulation bandwidth, signal distortion, noise, and the effects of reflected light. When used in conjunction with a fiber-optic interferometer, light signal analyzers can also measure linewidth, chirp, and frequency modulation of single-frequency lasers.

Figure 16.8 shows the Agilent lightwave signal analyzer 70810A as a block diagram. It consists of a photoreceiver module along with other Agilent 70000 modular measurement system plug-in modules.

Modulated light enters the lightwave section of the receiver through a single-mode fiber front-panel connector. Here, the signal is expanded, collimated, and

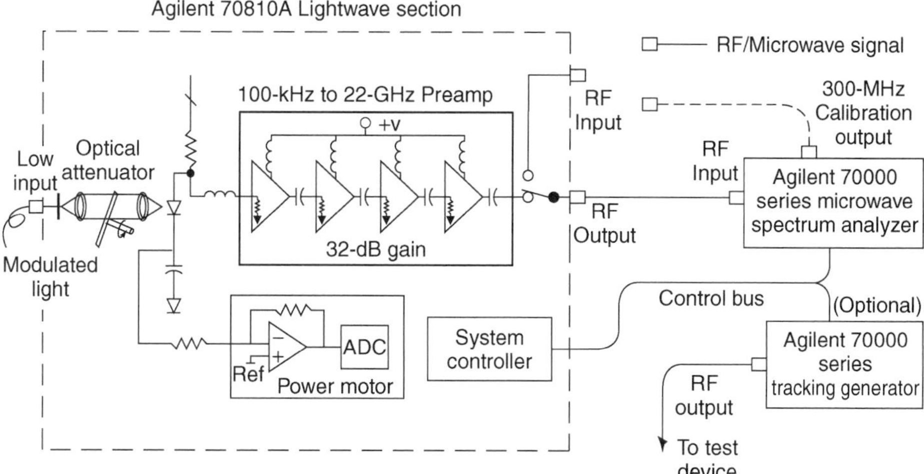

Figure 16.8. Lightwave signal analyzer system. (Based on Ref. 5, Figure 11, page 18.)

then focused onto the PIN photodetector. The collimated beam passes through an optical attenuator, which is controllable over a 30-dB range in 1-dB steps. The attenuator can be used to prevent front-end overload.

The receiver generates DC photocurrent that flows to the power-monitor circuit. This circuit measures the integrated, or average, optical power. The AC photocurrent flows to the 50-ohm, 100-kHz to 22-GHz preamplifier. The preamplifier amplifies the signal by 32 dB to improve the sensitivity of the analyzer. During production the combined frequency response of the photodetector, preamplifier, and spectrum analyzer is measured during production.

The calibration and correlation data at hundreds of frequency points from 100 kHz to 22 GHz are stored in the analyzer as correction factors. Using these correction factors, the system controller corrects each modulated power measurement at all frequencies from 100 kHz to 22 GHz before the information is displayed.

There is also a tracking generator that can be used to provide a swept modulation source whose frequency is synchronized with the sweep of the spectrum analyzer. With this tracking generator, the system can be used to conveniently generate frequency-response measurement for lightwave sources and detectors.

The lightwave analyzer in the lightwave mode allows the user to measure the following parameters:

- Optical or detected electrical average power
- Modulated optical or detected electrical power
- Relative power ($P_{\mathrm{MOD}}/P_{\mathrm{AVG}}$)
- Optical or electrical detected intensity noise
- Input noise-equivalent power (NEP) of the instrument
- Relative intensity noise, or RIN, independent of mode selection
- Bandwidth in optical or electrical dB

16.8 AN OPTICAL CHANNEL FOR SUPERVISION

An optical supervisory channel (OSC) is a dedicated channel for the continuous observation of operations and transmission efficiency. It is used to detect failures, power losses, or any significant change in system integrity. On conventional fiber-optic networks, many important tests disrupt revenue-bearing traffic. The OSC transmits appropriate tests and controls signals continuously. The OSC continuity on a link and network must be ensured because the channel carries control information. For this reason it is usually assigned a wavelength separate from the aggregate DWDM channel grouping.

The OSC is not used routinely to test system components or to issue operational reports. The concept behind the OSC is one of continuously monitoring of the system to keep an eye on network behavior. A reliable and functional OSC is mandatory and is vital for the system controller at the NOCC to guarantee the quality of network transmission for the most efficient use of network resources. When the OSC detects a fault or serious variation in performance level, it alerts the NOCC.

Because of the importance of maintaining continuity of the OSC, the wavelength selected is usually outside the passband of the EDFAs, whether at the top of the configuration or at the bottom such as 1525 nm or at 1610 nm. These two wavelengths lie outside the usual passband at 1550 nm, yet are close enough to carry the same fault phenomena. The monitoring test equipment to be supplied with the supervisory channel is vendor-specific. Various technician orderwire channels should be considered. Another idea is to include a SONET STS-1 configuration to provide dial telephone service up and down the entire fiber line. A 64-kbps channel should be available for SCADA activity, and SCADA access at each network element (NE) would be very helpful for a technician to monitor performance down to the card level. A 64-kbps channel could be made available for continuous BER checks of each link, another channel for system-wide BER. This would be in addition to the SONET/SDH self-monitoring capabilities (Ref. 2).

REFERENCES

1. *Beginner's Guide to Using the HP 8147 Optical Time Domain Reflectometer*, Product Note, Hewlett-Packard (Agilent Technologies), Santa Clara, CA, 1996.
2. Bob Chomycz, *Fiber Optic Installer's Field Manual*, McGraw-Hill, New York, 2000.
3. Roger L. Freeman, *Telecommunication System Engineering*, 3rd ed., John Wiley & Sons, New York, 1996.
4. *Guide to WDM Technology Testing*, EXFO, Quebec City, Canada, 2000.
5. *Agilent 71400 Lightwave Signal Analyzer*, Application Note 371, Agilent Technologies, Englewood, CO, 2000.
6. *Optical Spectrum Analysis*, Agilent Application Note 1550-4, Agilent Technologies, Englewood, CO, 2000.

17

OPTICAL NETWORKING

17.1 BACKGROUND AND CHAPTER OBJECTIVE

It is the old adage of supply and demand. The commodity is bit rate capacity, called bandwidth by IT people. Certainly the demand is there and probably doubling every year. In the supply chain the bits, now in great quantities, must be transported in some manner. The other requirement is that they be directed to or accepted from a user.

What we are talking about is information represented by bits. The only transport that can handle these great quantities is an optical link. At every node in the optical network the stream of bits must be returned to the electrical domain for switching/routing. The goal is the all-optical network except at the input/output points without the laborious return to the electrical domain.

Optical links are presently carrying 10 Gbps per bit stream per wavelength. With dense wavelength division multiplexing (DWDM) a single fiber can carry 8, 16, 32, 40, 80, 160, or 320 wavelengths. Within some years of the publication of this book we will have 40 Gbps per bit stream, and a single fiber will support in that same time frame 320 wavelengths or more. If each wavelength carries 40 Gbps, then a single fiber will have the capacity to carry 40×160 Gbps or 6400 Gbps.

As seen from the network provider, a major drawback in fiber networking technology is the costly requirement for repeated conversions from the optical domain to the electrical domain and back again, called OEO, at every regeneration point and for periodic signal monitoring along the line. Amplifiers replacing regenerative repeaters where the add–drop function is not necessary have improved the cost situation somewhat. When optical switching without OEO is employed in the network, the requirement for regenerative repeaters in the network will be vastly reduced.

For PSTN and fiber networks, which are primarily based on SONET and SDH infrastructures, the cost of regenerating optical signal can be very expensive, especially when it requires full SONET or SDH termination equipment at every ADM regeneration point. It has been found that even in these relatively homogeneous all-SONET environments, optical layer management can be a key factor in maintaining system integrity.

Even at locations along the fiber line where full conversion is not necessary, partial conversion at key points can be vitally important for monitoring to ensure circuit quality. At amplifier locations as in conversion signal points, active signal monitoring capabilities should be included. This will require optical signal splitting and optical-to-electrical conversion of some portion of the light signal.

The move toward direct deployment of gigabit Ethernet (GbE) over the MAN/WAN is a factor that may help to temporarily mitigate some of the push for all-optical switching because the cost of interface equipment for GbE fiber transport links is significantly less than for a SONET or SDH link. We see it as highly unlikely that GbE could completely replace SONET/SDH in the foreseeable future for anything other than very limited environments. However, the real-world impact of GbE deployment is likely to be increased traffic heterogeneity that will further drive the need for effective optical-layer management.

The ultimate goal behind DWDM deployment is the provision of more bit rate capacity. As a corollary then, MANs and WANs in the real world require optimization of the use of each wavelength's bit rate capacity. With the PSTN, where long-haul links traverse the network core, this goal of optimized utilization has typically been accomplished by pre-grooming all traffic such that uniform groups signals can be efficiently transported long distances with a minimum of intervening decision points along the way, However, for traffic traveling nearer the edge of the transport network, new-generation equipment needs to provide a higher level of traffic monitoring and grooming capabilities within the optical domain to achieve a balance of flexibility, performance, and capacity utilization.

We would argue that in most cases it would make little economic and practical sense to invest in DWDM and then map GbE connections across individual wavelength carriers on a one-to-one basis. Therefore, the push for aggregation of multiple connections can very quickly lead to a mix of heterogeneous nonconcatenated traffic traveling within a shared wavelength with a multitude of different end-point destinations.

The goal of the industry is to make the network all-optical except at the edge transition points. This point will be on a user's premises. By transition we mean the conversion of light to equivalent electrical information expressed as 1s and 0s.

This chapter's objective is to describe various steps to be taken toward what may now be called the "all-optical" network, its topology, and routing/switching in the optical domain.

17.2 NEW OPTICAL TECHNOLOGIES REQUIRED

The following is a list of new technology and radical approaches to make an all-optical network a reality:

- Optical switching
- Advanced wavelength demultiplexing, as well as multiplexing
- Tunable filters
- Stablized lasers
- New approaches to modulation
- Improved optical amplifiers with flat gain characteristics
- New and larger optical cross-connects (OXCs)

- Optical add–drop multiplexers (OADMs)
- Signaling techniques in the light domain

17.3 DISTRIBUTED SWITCHING

The new generation managed optical network is moving toward a *distributed switching* model in which lambda (λ) switches with intelligent Layer 1 cross-connect capability are distributed at various points along the network border. This concept is illustrated in Figure 17.1. Such an architecture provides seamless and efficient Layer 1 management of heterogeneous traffic types throughout the network, without sacrificing performance or flexibility in either the core or edge environments. This global distributed-switching architecture is equally adaptable to using dedicated wavelengths packed with homogeneous traffic for long-haul point-to-point transport or for flexibly managing heterogeneous traffic on dynamically allocated short-haul wavelengths.

With cross-connects along the edge of the network cloud, there is an emerging need for supporting a managed optical layer within a distributed optical switching environment. This outlines the crux of the matter and presents significant opportunities and challenges for both the semiconductor level and module-level developers and manufacturers. To achieve the required performance requirements, next-generation cross-connects need to be closer to the network by providing Layer 1 switching as opposed to the present traditional Layer 2 switching.

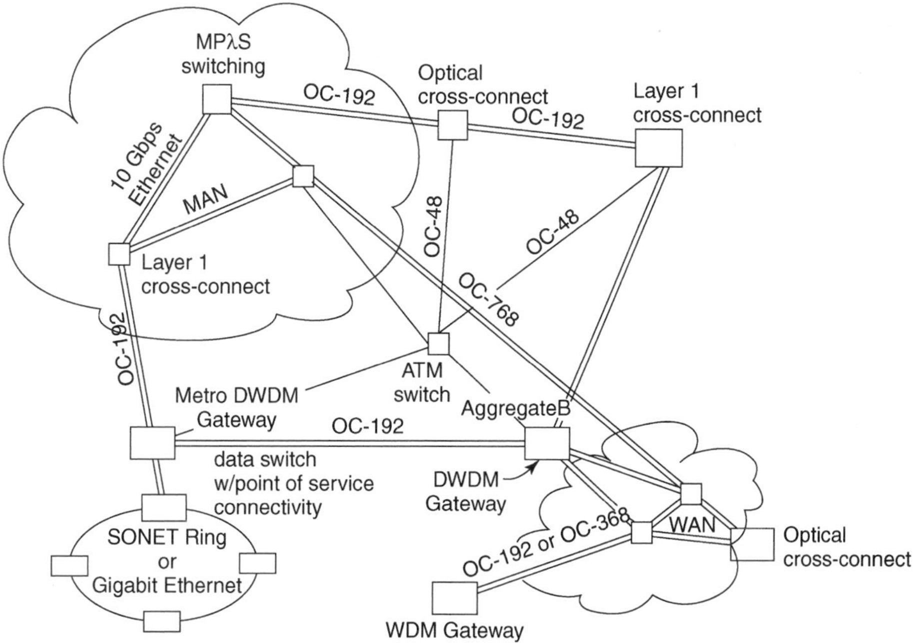

Figure 17.1. Distributed switching architectures. Note the combination of optical switches and Layer 1 cross-connects. (Based on references 1, 5, and 6.)

There will be two cross-point designs: asynchronous and synchronous. The higher-speed asynchronous cross-points enable heterogeneous MAN implementations to efficiently support different types of native-mode traffic within the same ring. In long-haul networks, there will be innovative uses of synchronous-switching cross-points that will provide the necessary performance requirements. These switches are seen as more of a *time–space–time* switch rather than the more rudimentary *space* switch. These new-generation synchronous cross-points will incorporate Layer 1 grooming capabilities that can selectively switch SONET (or SDH) or any other TDM (time-division multiplex) signals between any combination of inputs and outputs.

It is expected that these optical Layer 1 switching capabilities will use high-speed synchronous ICs. This next generation synchronous cross-point switching will offer the capability to selectively groom out and switch an STS-1 (STM-1) from within STS-48 (STM-16) or STS-192 (STM-64) bit streams. These devices will provide complete flexibility for provisioning IC-level managed optical cross-connects from any STS-1 input to any STS-1 output. Non-SONET traffic mapped to STS-N equivalent containers and protocol-independent wrapped traffic can be switched within the same cross-connects.

These high-density, high-speed grooming switches are deployed along the edge of the switching network cloud. They can optimize capacity utilization while efficiently making Layer 1 access decisions to partition out traffic to outlying Internet protocol, GbE, ATM, fiber channel, or other Layer 2 switches. Localized Layer 2 functions such as routing and policy management are appropriately handled by the outlying switches while Layer 1 access switches provide high-speed performance IC-level switching/grooming of DWDM wavelengths.

17.4 OVERLAY NETWORKS

Today's data networks typically have four layers:

1. IP for carrying applications
2. ATM for traffic engineering
3. SONET/SDH for transport
4. DWDM for capacity

This architecture has been slow to scale, making it ineffective for photonic networks. Multilayering architectures typically suffer from the lowest common denominator effect where any one layer can limit the scalability of the other layers and the entire network.

17.4.1 Two-Layer Networks Are Emerging

For the optical network developer, an absolute prerequisite for success is the ability to scale the network and deliver bit rate capacity where a customer needs it. Limitations of the existing network infrastructure are hindering movement to this service-delivery business model. It is the general belief in the industry that a new network foundation is required. This network foundation is seen as one that will easily adapt to support rapid change, growth, and highly responsive service delivery. What is needed is an intelligent, dynamic, photonic transport layer deployed in support of the service layer.

The photonic-network model divides the network into two domains: service and optical transport. The new architecture is seen as combining the benefits of photonic switching with advances in DWDM technology. It delivers a multigigabit bit rate capacity and provides wavelength-level traffic-engineered network interfaces to the service platforms. The service platform includes routers, ATM switches, and SONET/SDH add–drop multiplexers, which are redeployed from the transport layer to the service layer. The service layer is seen as relying completely on the photonic transport layer for the delivery of the necessary bit rate capacity where and when it is needed to peer nodes or to network elements (NEs). The bit rate capacity is provisional in wavelength granularity rather than in PDH TDM granularities. We expect exponential growth rate of the fiber network; and to meet these requirements, rapid provisioning is an integral part of the new architecture. While the first implementations of this model will support error detection, fault isolation, and restoration via SONET, these functions will gradually move to the optical layer.

Expect to see routers, ATM switches, and SONET/SDH ADMs to request bit rate capacity where and when needed via the provisioning capabilities of photonic switching with the traffic engineering capabilities of MPLS.[1] For the protocol designed for the optical domain, the name *MPλS* was selected. This protocol is designed to combine recent advances in MPLS traffic engineering control-plane techniques with emerging photonic switching technology to provide a framework for real-time provisioning of optical channels. This will allow the use of uniform semantics for network-management operations control in hybrid networks consisting of photonic switches, label-switched routers (LSRs), ATM switches, and SONET/SDH ADMs. While the proposed approach is particularly advantageous for data-centric optical internetworking systems, it easily supports basic transmission services. MPλS supports the basic network architectures, overlay and peer, proposed for designing a dynamically provisionable optical network.

Figure 17.2 illustrates the photonic network model. Here the network is divided into two domains: service and optical transport. The service platform includes routers, ATM switches, and SONET ADMs.

With the overlay model, there are two different control planes. One of these is in the core optical network and the other is the edge interface, variously called the UNI or user-network interface. The interaction between the two planes is virtually minimum. The derived network is very similar to our present IP/ATM networks. It can be dynamically set up through signaling or is statically configured. The internal operation of the network is transparent to the light carriers entering from the edge.

One drawback of an overlay network that has been envisioned is the amount of signaling and control traffic required due to the edge mesh of point-to-point connections. This excessive amount of routing protocol traffic results in limiting the number of edge devices in the network. For example, a single link-state advice

[1]MPLS stands for multiprotocol label switching. See Section 17.11 for a brief overview of MPLS.

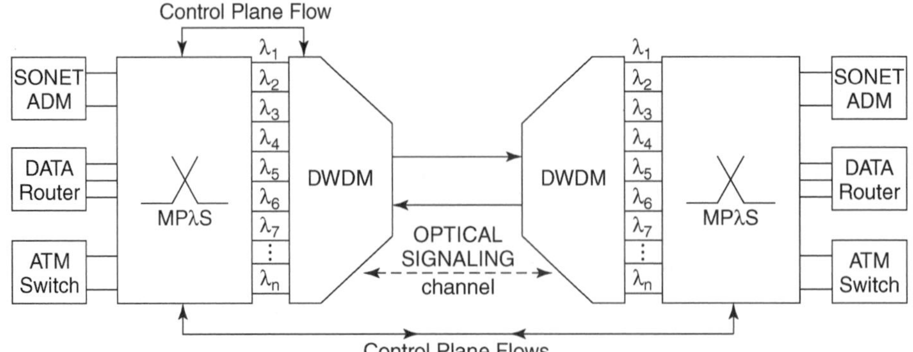

Figure 17.2. The photonic network model illustrating its two platforms: service and optical transport. The service platform at far left and far right shows individual routers, ATM switches, and SONET ADM capabilities. Inside the switch is the photonic transport layer, which consists of optical switches and DWDM equipment. There is a standardized control plane used to communicate between the various elements. (Based on references 2, 4, and 5.)

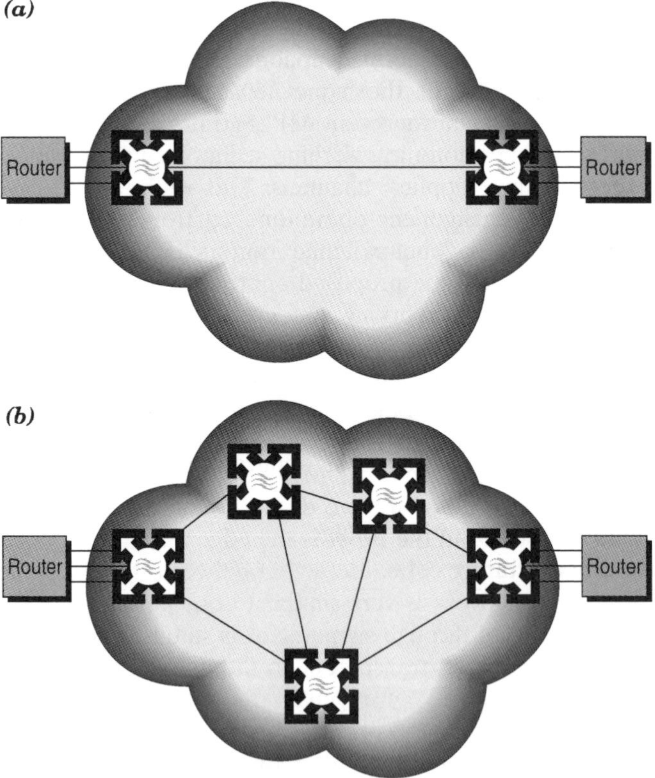

Figure 17.3. (*a*) Overlay and (*b*) peer network models. (Based on references 2, and 4. Courtesy of Calient Networks.)

flooding event is multiplied, creating a very large number of repetitive messages on the point-to-point mesh.

In the peer-to-peer model, a single action of the control plane spans both the core optical network and the surrounding edge devices as shown in Figure 17.3. Here we see the overlay and peer network models distinguished. In Figure 17.3a the overlay model hides the internal topology of the optical network, thereby producing the optical cloud. In Figure 17.3b the peer model allows edge devices to participate in the routing decisions and eliminates the artificial barriers between the network domains.

17.5 OPTICAL SWITCHING

The optical switch is one of the most important fiber-optic components that seamlessly maintains network survivability and is a flexible platform for signal routing. Today's switching in telecommunication systems is done electronically (in the electrical domain); however, as modern photonic networks evolve, routing of fiber-optic signals will be carried out completely in the optical realm.

The most common form of optical switches found off-the-shelf today are either electro-optical or optomechanical models. The electro-optical switch usually consists of optical waveguides with electro-optic crystal materials such as lithium niobate. 1×2 and 2×2 switch configurations are achieved using a Mach–Zehnder interferometer structure with a 3-dB waveguide coupler (see Sections 3.3 and 8.2). The differential phase between the two paths is controlled by a voltage applied to one or both paths. The resulting interference effect routes signals to the desired output as the drive voltage applied to one or both paths of the Mach–Zehnder interferometer structure changes the phase differential between them.

Electro-optical switches have many limitations such as the following:

- They have high insertion loss.
- They have high polarization-dependent loss.
- They have high crosstalk.
- They are sensitive to electrical drift.
- They are nonlatching, limiting their applications in network protection and reconfiguration.
- They require a high operating voltage.
- They have manufacturing costs.

The principal advantage of this type of switch is switching speed, that is in the nanosecond range.

Optomechanical switches rely on mechanically driven moving parts. They are the most widely used switches for optical application relying on mature optical technologies. Their operation is straightforward. Input optical signals are mechanically switched by moving fiber ends or prisms and mirrors to direct or reflect light to different output fibers in the switch. Switch movements must be precise for correct alignment and are usually solenoid-driven. Switching speeds, unfortunately, are in the millisecond range. However, these switches are widely used because of

TABLE 17.1 Typical Specifications of a 2 × 2 Optomechanical Switch

Parameter	Unit	Specification
Wavelength range	nm	1260–1600
Insertion loss	dB	0.6
Polarization-dependent loss	dB	0.05
Crosstalk	dB	−60
Switching speed	ms	5
Polarization mode dispersion	ps	0.1
Return loss	dB	55

Source: Courtesy of Yigun Hu, E-Tek Dynamics, Ref. 3.

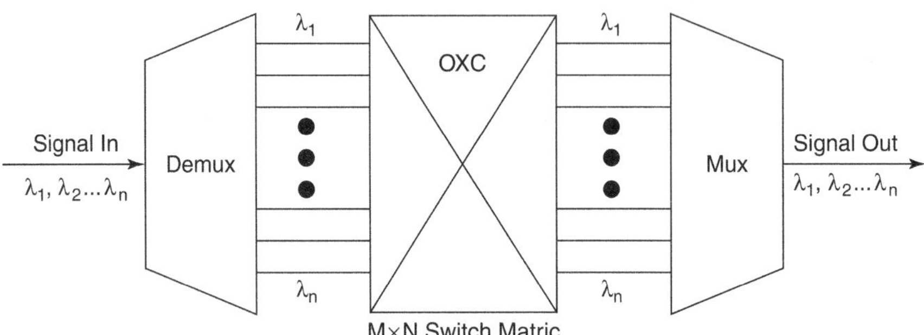

Figure 17.4. $M \times N$ switching by wavelength selection using optical cross-connects. (Courtesy of Yigun Hu, E-Tek Dynamics, Ref. 3.)

low cost, simplicity in design, and good optical performance. Simple configurations are readily available in 1 × 2 and 2 × 2 structures. Also small-scale matrix nonblocking $M \times N$ optical switches are easy to build. By using multistage configurations, switches such as 64 × 64 can be built which are partially nonblocking. However, larger matrix switches are complex and unwieldy. Table 17.1 gives typical specifications of a 2 × 2 optomechanical switch.

Switching in this regime will be wavelength switching. A DWDM configuration out of the multiplexer will be in formations of optical wavelengths consisting of two wavelengths up to 160 or more. Let us think that some wavelengths must be routed to point X, others to point Y, and still others to point Z. This concept is illustrated in Figure 17.4.

Wavelength division multiplexing and λ-switching (wavelength switching) are tightly tied together. In a DWDM aggregate, each wavelength must be clearly separable to minimize crosstalk.

17.5.1 MEMS Switching

MEMS is an abbreviation for *microelectromechanical systems*. There are two types of these switches being developed: mechanical and microfluidic. The mechanical variety utilizes an array of micromachined mirrors that can range into the hundreds of thousands on a single chip. The fluidic-type switch relies on the movement of liquid in small channels that have been etched into a chip. In the case of a mirror-based switch, an array of micromachined mirrors is fabricated in silicon. The incoming light signal is directed to the desired output port by means of control signals applied to the MEMS chip, which fixes the position of each individual mirror.

Mirror-based MEMS switches are classified by mirror movement. There are twodimensional (2-D) switches and three-dimensional (3-D) switches. In the case of the 2-D switches, a mirror takes on only one of two positions. This would be typically either up or down or side-by-side. With a 3-D switch, there are a wide variety of positions that a mirror can assume. A mirror can be swiveled in a wide variety of positions and in multiple angles.

Source: Ref. 7.

Reference 7 states that most of the major players in optical switching are following the mirror-based route. This does not include Agilent Technologies, which is taking advantage of the company's knowledge of microfluidics. It has developed a switch that is based on Hewlett-Packard's inkjet printing technology. This Agilent optical switch is quite unique in that it consists of intersecting silica waveguides, with a trench etched diagonally at each point of intersection. The trenches contain a fluid that, in the default mode, allows the light to travel through the switch. To activate the switch such that it switches light, bubbles are formed and removed hundreds of times per second in the fluid, which then reflect the light to the appropriate output port (Ref. 7).

17.5.1.1 Control of Mirrors and Bubbles. Reference 7 covers three types of actuation mechanisms being used in MEMS switching: electromagnetic, electrostatic, and thermal.

Electrostatic Actuators. This is the most common and well-developed method to actuate MEMS, because IC processes provide a wide selection of conductive and insulating materials. By using conductors as electrodes and insulators for electrical isolation of electrodes, electrostatic forces can be generated by applying a voltage across a pair of electrodes. This type of actuation requires lower power levels than other methods and is the fastest.

Magnetic Actuators. Magnetic actuators usually require relatively higher currents (resulting in higher power), which can limit their use. In addition, these mechanisms use magnetic materials that are not common in IC technology; and they often require some manual assembly, a distinct drawback. Thus the choice of magnetic material is limited to those that can be easily micromachined. However, electromagnetic microactuators can be actuated much faster and consume less power than their thermal actuator equivalents.

Thermal Actuators. These require heating that is accomplished by passing a current through the device. To its detriment, the heating elements can require high power consumption. Then, of course, the heated material has to cool down to return the actuator to its original position. Then the heat is dissipated in the surrounding structure. All of this takes time, limiting the switching speed of the device.

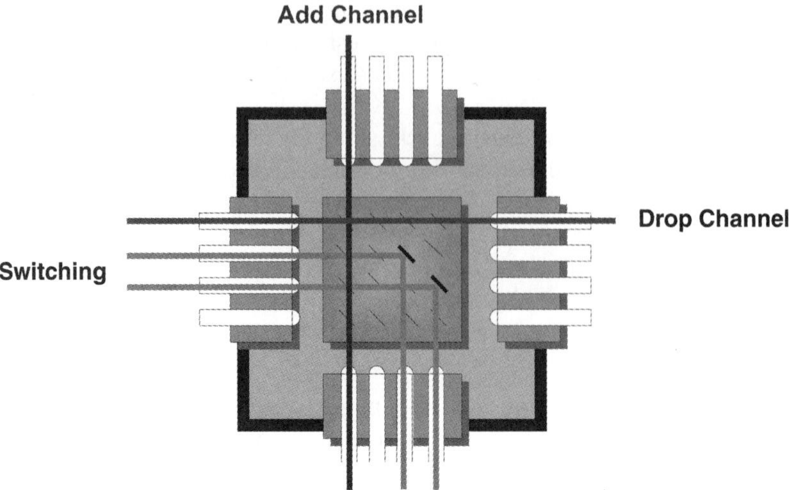

Figure 17.5. A 2-D MEMS optical cross-connect switch with additional third and fourth planes to add/drop functionality. (Courtesy of Zeke Kruglic, OMM, Inc., San Diego, CA, Ref. 13.)

LIGA (lithographic plating and molding) is a very promising fabrication method of MEMS devices. LIGA combines the basic process of IC lithography with electropating and molding to achieve depth. In LIGA, patterns are created in a substrate, which are then electroplated to create 3-D molds. The molds can be used to create the final product. However, a variety of materials can also be injected into them to produce a product. Here we have two distinct advantages to this technique: Materials other than silicon can be used (in particular, metal and plastic), and devices with very high aspect ratios can be built.

MEMS switches currently support 32 bidirectional ports (or fewer). The goal in MEMS switching is to develop devices with a 1000 × 1000 arrays. Some optical switching companies are tackling this task head-on. Other companies believe that scaling smaller switches together to make a larger array is the better approach (based on Ref. 7). Figure 17.5 is an example of a 2-D MEMS optical cross-connect switch with additional third and fourth planes to add/drop functionality.

Note: MEMS cross-connects are described in Section 17.8

17.6 A PRACTICAL OPTICAL ADD–DROP MULTIPLEXER

True optical ADMs will allow provisionable add–drop channels similar to time-slot assignment (TSA) and reassignment of optical channels similar to time-slot interchange (TSI) found in present-day electronic digital switches. Figure 17.6 is a block diagram showing the basic functions of a programmable optical ADM. Because these cross-connects will be provisioned on a wavelength basis, new sites requiring access to the network can be more easily added and the planning burden of

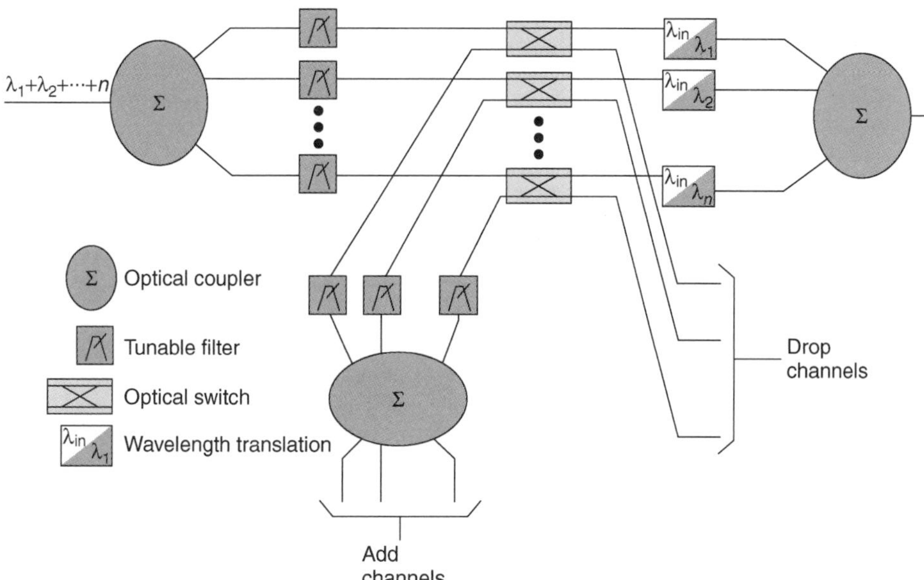

Figure 17.6. Provisionable optical ADMs. (Reprinted with permission from Alcatel, Ref. 4, Figure 3.)

network planners can be reduced. Migration to the all-optical layer provides new methods of protection for network restoration.

The evolution of optical networks will lead to more advanced systems that provide wavelength routing capability. As technology breakthroughs are made in the arena of optical gates and matrices, optical cross-connect systems will begin to appear on the scene. Figure 17.7 is a high level diagram of an optical cross-connect system (OCCS).

There are two basic types of cross-connect systems: line side and tributary side. The tributary side, or Type 1 OCCS, will provide functions similar to the broadband SONET cross-connects available today. The line side, or Type 2 OCCS, will support high-level network restoration and network reconfiguration on the high-speed transport system.

As new optical network services become available, there will be a dramatic increase in customer base and a similar increase in the demand to transport traffic. Up to now, electronic broadband cross-connects have met the requirements demanded by the network, but the complexity of these systems and their matrix sizes will eventually reach the limit of feasibility. Optical cross-connects can reduce the size and complexity of electronic digital cross-connect (DCSs) with a higher level of traffic loading and routing at the optical wavelength level. Signals can be routed at levels greater than STS-1 and can be efficiently handled at the optical layer. An optical matrix is inherently smaller than its electronic counterpart, requires less power, switches at higher speeds, and handles larger configurations of bit rate capacity with less complexity. Because a significant portion of the "bandwidth explosion" is due to customer-driven requirements for larger pipes, these connections can be more efficiently managed with an optical matrix rather than its electronic counterpart.

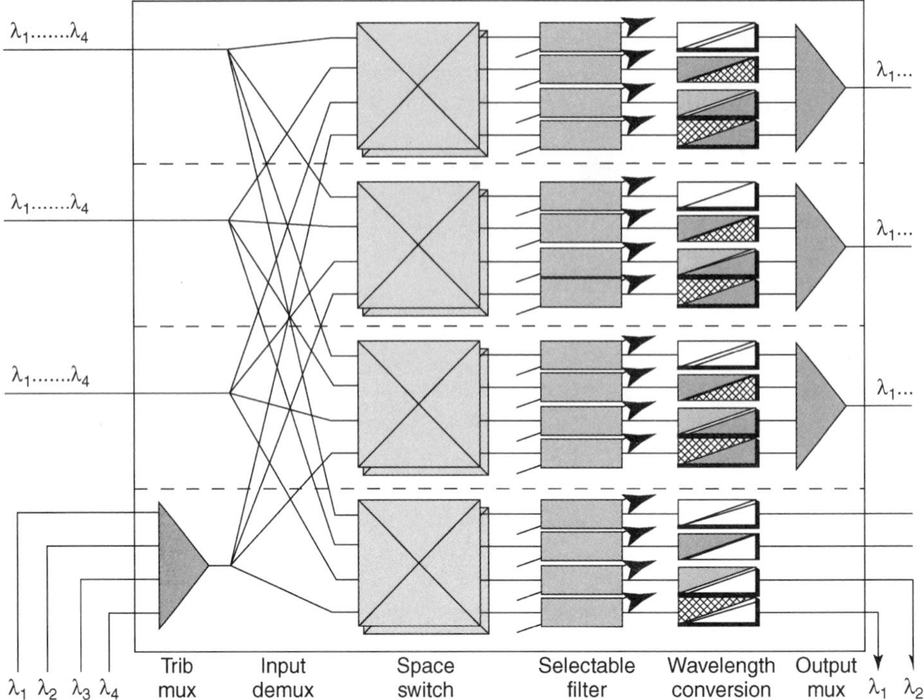

Figure 17.7. Block diagram of OCCS. (Reprinted with permission from Alcatel, Ref. 4, Figure 4, page 30.)

Currently, network restoration for fully restorable services is accomplished by two basic methods: mesh protection using digital cross-connects and ring protection by means of SONET/SDH multiplexes. Both methods have weak points such as in restoration time, cost, and management efficiency. Once the optical network is in place, one major efficiency will accrue over its electronic counterpart. Consider SONET (or SDH) rings. All ADMs in the ring must operate at the same data rate. Reference 4 suggests that this will lead to inefficiencies and additional cost in the transport network especially because some routes will be significantly more dense than others. With a ring in the all-optical network, some wavelengths can be operated at one rate, say OC-24, while others can be operated at OC-192 and still others at OC-48.

17.6.1 OXCs and OADMs Enhance Availability and Survivability

The major step toward the all-optical network is adding both OADM and OXC[2] elements. These network elements provide carriers with the ability to reconfigure the network traffic for optimal data transport. It will also have the capability of rapid restoration in the event of link failure, all within the optical layer.

[2]OADM stands for optical add–drop multiplex, and OXC stands for optical cross-connect.

OXCs are dynamic switching fabrics with connections between any of M fiber inputs and any of N fibers outputs in a DWDM network. Thus, optical cross-connect switches have a nonblocking one to many connections in a matrix configuration. OXCs provide good network survivability, lower network management cost, and reconfigurable paths for signal routing in the optical layer. These attributes help eliminate the need for complex and expensive digital switches operating in the electrical domain. Because they operate in the optical realm, optical cross-connects can potentially accommodate terabit data streams due to their wavelength, bit-rate, and protocol transparent characteristics.

Source: Ref. 3.

17.7 IMPROVEMENTS IN THE MANAGEMENT OF THE NEW NETWORK ARCHITECTURE

As discussed above, the network architecture will be two-layer. Both the IP router community and the optical community have agreed that the way to control both layers is by means of multiprotocol label switching (MPLS). MPLS for this application has been slightly modified; and, as covered above, it is appropriately called MPλS. Each control plane, the optical and IP router, has two phases in a switching routine. One phase sets up paths, and there is the steady-state traffic phase in which the state information has been set up at each node to define the paths and then forwards packets in a way that provides much of the missing QoS capability.

MPλS will replace the two current protocols operating in the lower layers with their many variants for several reasons. First, these traditional software families are very much vendor-dependent. Second, IP and SONET/SDH are very different one from the other, and third, they are very slow for the anticipated needs of restoration, provisioning, and protection.

There are two communities, MPLS and the MPλS, and there is only one disagreement—that is, whether the control entity within each set of IP routers forming the IP layer will be topologically aware of just what pattern of OXC traversals will constitute a lightpath across an optical network cloud, or whether the optical layer will set these up autonomously and then tell the IP layer where the endpoints are without saying which sequence of OXCs constitutes a lightpath. Paul Green (Ref. 5) believes that the latter will prevail, at least initially.

Protection switching, discussed in Section 12.5, was the first to receive attention in the optical layer integrity processes. To activate protection requires precanned algorithms similar to those used in SONET/SDH. Only a small portion of the network is affected when protection switching is invoked. This is an optical layer function, and the trigger to activate can be the loss of OSNR (optical signal-to-noise ratio).

Similar to present automatic protection switching, there is the restoration phase, which is the replacement of the failed optical path by another. Once repairs have been completed, this failed path, now capable of being operational, becomes the protection path.

Provisioning/reconfiguration becomes an interesting concept. Reference 5 describes the condition of *stranded bit rate capacity* where capacity on a fiber or on a cable lies fallow, unused. This bit rate capacity can now be brokered between service providers who might use this optical facility, setting up a rent-a-wavelength condition.

17.8 ALL-OPTICAL CROSS-CONNECTS

During the period of preparation of this book, large all-optical cross-connects arrived on the scene. These are microelectromechanical system (MEMS), which now have progressed from 2-D optical switches to 3-D devices. The concept of the 2-D MEMS switch derives from the old analog circuit crossbar switch. It involves N^2 popup mirrors to deflect collimated light from an input port to an output port. This is shown in Figure 17.8a. Figure 17.8b shows a MEMS switching machine consisting of only $2N$ mirrors, N of them directing the inputs toward distinct outputs and another N directing the outputs to connect to the inputs. Green (Ref. 5) states that the advantage of 3-D is therefore linear scalability with port count (compared to quadratic for 2-D), but at the expense of analog mirror tilt control versus binary for 2-D.

The 3-D OXC has other advantages. A large port-count device can be used for managing whole fibers as well as many wavelengths carried on a fiber. Its cost is

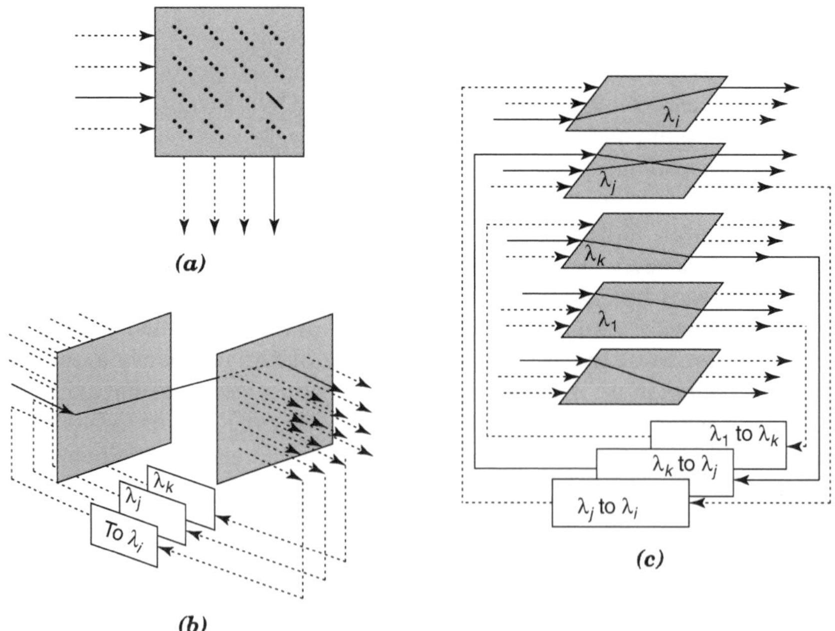

Figure 17.8. Optical cross-connects (OXCs). (a) 2-D using binary (popup) mirrors; (b) 3-D using mirrors with analog control, showing an attached wavelength converter pool; (c) multiplane architecture using multiple 2-D modules, showing an attached wavelength converter pool. (From *IEEE Communications Magazine*, Ref. 5, Figure 3.)

comparatively low and does not suffer the high attenuation due to the interconnections involved to accommodate large nonblocking $N \times N$ structures from small-N 2-D subcomponents. These subcomponents have achieved at most a 32×32 size.

Source: Ref. 5.

Another problem arises when an OXC is to be used just for WDM switching. By WDM switching we mean wavelength switching all optically that is protocol transparent. Within the OXC we would want a path between wavelength A and any output connected to wavelength B. As the number of wavelengths increases, the number of "from–to" paths goes up radically. The number of lasers in the wavelength converters can be decreased by using tunable lasers.

17.9 OPTIONS FOR OPTICAL LAYER SIGNALING

We assume that the optical network is *connection-oriented.* Thus circuits require setup and release algorithms. These signaling protocols are implemented in software leading to limitations on the call-handling capacity of a switch.

From the signaling and control perspective, two network models have evolved to provide interoperability between the IP and optical layers. There is the peer model, which is based on the premise that the optical-layer control intelligence can be transferred to the IP layer, which has assumed end-to-end control.

The second model is the client–server model. This model is based on the premise that the optical layer is independently intelligent and serves as an open platform for the dynamic interaction of multiple client layers. This includes IP.

In either case, we assume an optical mesh network. The control plane is IP-compatible based on the MPLS protocol discussed above. The routing protocols are IP and carry out topology discovery. MPLS signaling protocols are used for automatic provisioning. Expect to see the IP-based optical-layer control stack to be standardized as the model becomes adopted.

Applications are managed differently. The optical control plane will control *dynamic lambda provisioning* with routers on the edge of the network cloud and linked by optical subnetworks as illustrated in Figure 17.9.

When a router encounters congestion, either the network management system or the router itself will request the provisioning of a dynamic lambda, meaning an additional wavelength carrier. This requires that optical switches have the capability to create new or enhanced service channels such as OC-48 or OC-192 capacity to meet the needs of that router. This dynamic lambda provisioning can adapt to traffic flows.

The client–server model handles things differently. It will let each router communicate directly with the optical network using a well-defined UNI (user–network interface). The interconnection between subnetworks would be via an NNI (network-to-network interface). This permits each subnetwork to evolve independently.

In optical networks as with wire and radio networks, operators wish to take advantage of competition and thereby build multivendor networks. To do this, standardized interoperability is required.

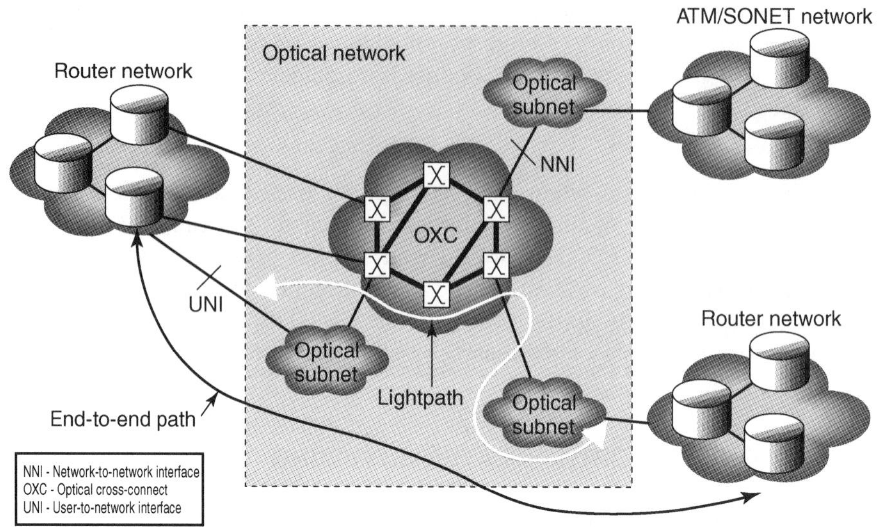

Figure 17.9. This sketch illustrates the client–server model. The optical layer has intelligence that controls lightwave links. The network is made up of subnetworks with well-defined interfaces. (Courtesy of Calient Networks, San Jose, CA, Ref. 12.)

When we compare the two models, the client–server model has a significant advantage over the peer model in that it has a faster path to interoperability. On the other hand, the client–server model is more direct and simplified. To manage end-to-end paths on optical links, additional communication between IP and optical layers is required. These additional communication connectivities will pervade across the entire network cloud.

17.10 FOUR CLASSES OF OPTICAL NETWORKS

17.10.1 Generic Networks

Whether optical or electrical, there are three general network types: Two are connection-oriented (CO), namely ATM and PSTN, and there is one connectionless (CL) network, which is IP. They can also be placed in categories of circuit-switching and packet-switching. The PSTN relies on circuit-switching, and ATM and IP can be classified as packet-switching networks.

Optical network designers have changed the definitions of circuit-switching and packet-switching to meet the special requisites of an optical network. *Circuit-switching* is position-based in that bits arriving in a certain position are switched to a different output position. The position is determined by a combination of one or more of three dimensions: space (port number), time, and wavelength. Packet switches are label-based in that they use intelligence in the header, which we call *labels*, to decide how and where to switch the packet. Note how these definitions differ from conventional ones. Still here in the case of data a circuit may be set up prior to the exchange of packets and thus meets the conventional definition, but

TABLE 17.2 Classes of Optical Networks Based on Components Employed

Types of Optical Communications Components	Classes of Optical Networks			
	Optical Link Networks	Broadcast-and-Select Networks	Wavelength-Routed Networks	Photonic Packet-Switched Networks
Nonswitching optical components	√	√	√	√
Tunable transmitters and/or tunable receivers	X	√	May or may not be present	May or may not be present
Optical circuit switches (OADMs and OXCs)	X	X	√	May or may not be present
Optical packet switches	X	X	X	√

Note: The check sign √ indicates nonswitching optical components and the X indicates switching components (author).

Source: Table 1, page 121, *IEEE Communications Magazine*, Ref. 6.

this setup is not a necessary attribute. Think of the permanent virtual circuit in X.25 and frame relay.

In the case of optical networks, it is also important to note whether a circuit is set up prior to data exchange or is a property of whether a circuit is CL or CO and not whether a circuit is circuit-switched or packet-switched. IP is a typical example of CL packet-switched networks, and ATM is a typical example of CO packet-switched networks to simplify discussion. We recognize that when resource reservation protocol (RSVP) and/or multiprotocol label switching (MPLS) add a CO mode of operation to IP networks, there may be confusion in semantics. Thus we hold with our ATM and IP examples.

We will briefly review four classes of optical networks based on the types of components used: optical link networks, broadcast-and-select (B&S) networks, wavelength-routed (WR) networks, and photonic packet-switched networks. Table 17.2 organizes these network types based on their optical components.

Optical link networks are defined as a network using electronic switches interconnected by optical links, either single-channel or multichannel. Multichannel links are derived from optical WDM multiplexers/demultiplexers at either end. WDM passive star couplers are used to create broadcast links for shared-medium operation. These two component types are not programmable; as a result, no reconfiguration is possible.

Figure 7.10 illustrates the classes of optical networks. There are (a), (b), and (c) columns, where column (a) lists optical link networks using all-electronic switching. Column (b) is a listing of single-hop B&S networks and photonic packet-switched networks. These network classes give examples using all-optical switching. Column (c) is a listing of multihop and WR networks. These network types use a hybrid switching optoelectronic types.

Single-hop B & S networks have optical transmitters and receivers that can be tuned on a packet-by-packet or call-by-call basis. All three networking techniques are theoretically possible with single-hop B&S networks as shown in Figure 17.10, column (b).

There are also *multihop B&S networks*. With this type of network, data are broadcast on all links. Effectively electronic switches provide wavelength conver-

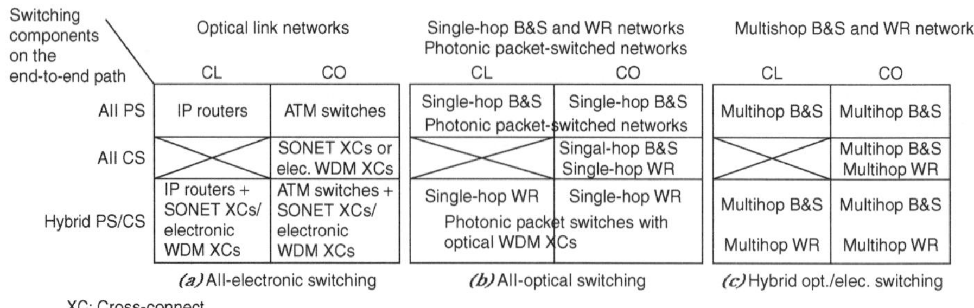

Figure 17.10. Classification of optical networks. B & S, broadcast and select; WR, wavelength routing; CL, connectionless, CO, connection-oriented; CS, circuit-switched; PS, packet-switched. (Based on Figure 3, page 120, *IEEE Communications Magazine*, Ref. 6.)

sion on the desired path between source and destination because not all nodes receive all wavelengths. In that the classification of such networks is B&S, the only optical switching takes place in the tunable transmitters and receivers. The electronic switches can be circuit switches or packet switches because the components can be tuned on a packet-by-packet basis or call-by-call basis. Multihop B&S networks can be operated in all categories of column (c) of Figure 17.10, except for the CS-CL category.

Wavelength-routed (*WR*) *networks* include optical circuit switches, which we will call OADMs and OXCs. These networks may also have, on an optical basis, tunable transmitters and receivers. WR networks can be single-hop or multihop. Single-hop networks use only switching components, and thus they are listed in column (b).

The final category of optical networks is *photonic packet-switched networks*. We can consider these networks as having packet switches and, optionally, circuit switches with tunable transmitters and receivers. Look for these networks in column (b).

Of all the networks listed in Figure 17.10, only the optical link network is feasible today and is presently in operation. Of the remaining three, the fiber-optic industry is directing its attention to WR networks. Multihop WR networks with electronic packet switches are the most common (Ref. 6). An example is a network of IP routers interconnected by optical circuit switches such as OADMs/OXCs.

17.11 BRIEF OVERVIEW OF MULTIPROTOCOL LABEL SWITCHING (MPLS)

Based on RFC 3031, from the Internet.

17.11.1 Introduction

MPLS is an outgrowth of IP and is very similar to that well-known protocol. It uses hop-by-hop source routing. It uses *labels* that are really addresses similar to those found in other protocols, like DLCI (data links connection identifier) block in frame relay, VPI/VCI (virtual path indicator/virtual channel indicator) in ATM, and so on. The labels used are native to the media employed.

We dedicate considerable space to MPLS because we believe it will become a major player in the optical network arena. It should be pointed out that the term *multiprotocol* implies that the techniques of MPLS are applicable to *any* network layer protocol.

17.11.2 Initial MPLS Terminology

Label—a short, fixed-length, physically contiguous identifier that is used to identify an FEC, usually of local significance.

Label merging—the replacement of multiple incoming labels for a particular FEC with a single outgoing label.

LDP—label distribution protocol.

LSP—label-switch path.

FEC—forwarding equivalence class.

LSR—label-switched router.

LER—label edge router (term is useful, but not found in standards).

17.11.3 MPLS Architecture

When dealing with a connectionless network layer protocol, a packet travels from one router to the next where each router involved makes an independent forwarding decision for that packet. All routing information is contained in a packet's header. And each router runs a network layer routing algorithm to determine a route for a particular packet. In this hop-by-hop routing, each router independently chooses a next hop for a packet, based on its analysis of the packet's header and the results of running the routing algorithm. The MPLS routing concept is built on standard IP as illustrated in Figure 17.11.

Packet headers contain considerably more information than is needed simply to choose the next hop. Choosing the next hop can therefore be thought of as the

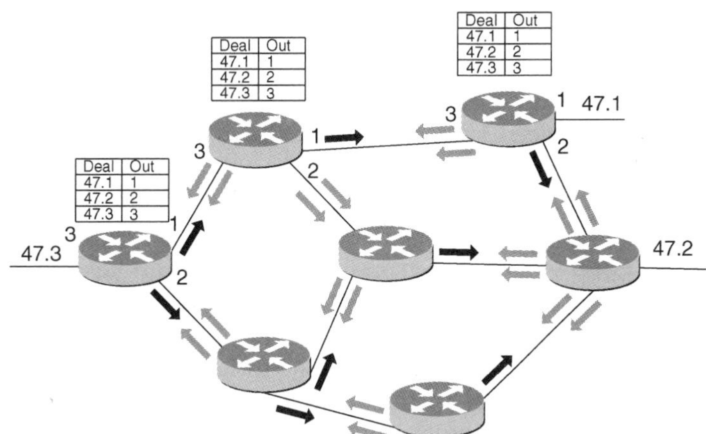

Figure 17.11. MPLS built on standard IP. Note the routing tables. (From Ref. 10.)

composition of two functions. The first function partitions the entire set of possible packets into a set of *forwarding equivalence classes* (*FECs*). The second maps each FEC to a next hop. Insofar as the forwarding decision is concerned, different packets that get mapped into the same FEC are indistinguishable. All packets that belong to a particular FEC and that travel from a particular node will follow the same path; or if certain kinds of multipath routing are in use, they will all follow one of a set of paths associated with the FEC.

In conventional IP forwarding, a particular router will typically consider two packets to be in the same FEC if there is some address prefix X in the router's routing tables such that X is the *longest match* for each packet's destination address. As the packet traverses the network, each hop in turn reexamines the packet and assigns it to an FEC.

In MPLS, the assignment of a particular packet to a particular FEC is done just once, as the packet enters the network. The FEC to which the packet is assigned is encoded as a short fixed-length value known as a *label*. When a packet is forwarded to its next hop, the label is sent along with it; that is, the packets are *labeled* before they are forwarded.

At subsequent hops, there is no further analysis of the packet's network layer header. Rather, the label is used as an index into a table that specifies the next hop, and a new label. The old label is replaced with a new label, and the packet is forwarded to its next hop.

In the MPLS forwarding paradigm, once a packet is assigned to an FEC, no further header analysis is done by subsequent routers; all forwarding is driven by the labels. This has a number of advantages over conventional network layer forwarding.

1. MPLS forwarding can be done by switches that are capable of doing label lookup and replacement, but are either not capable of analyzing the network layer headers or are not capable of analyzing network layer headers at adequate speed.
2. Because a packet is assigned to an FEC when it enters the network, the ingress router may use, in determining the assignment, any information it has about the packet, even if that information cannot be gleaned from the network layer header. For example, packets arriving on different ports may be assigned to different FECs. Conventional forwarding, on the other hand, can only consider information that travels with the packet in the packet header.
3. A packet that enters the network at a particular router can be labeled differently than the same packet entering that network at a different router; as a result, forwarding decisions that depend on the ingress router can be easily made. This cannot be done with conventional forwarding because the identify of a packet's ingress router does not travel with the packet.
4. The considerations that determine how a packet is assigned to an FEC can become ever more and more complicated, without any impact at all on the routers that merely forward the labeled packets.
5. Sometimes it is desirable to force a packet to follow a particular route that is explicitly chosen at or before the time the packet enters the network, rather than being chosen by the normal dynamic routing algorithm as the packet travels through the network. This may be done as a matter of policy, or to

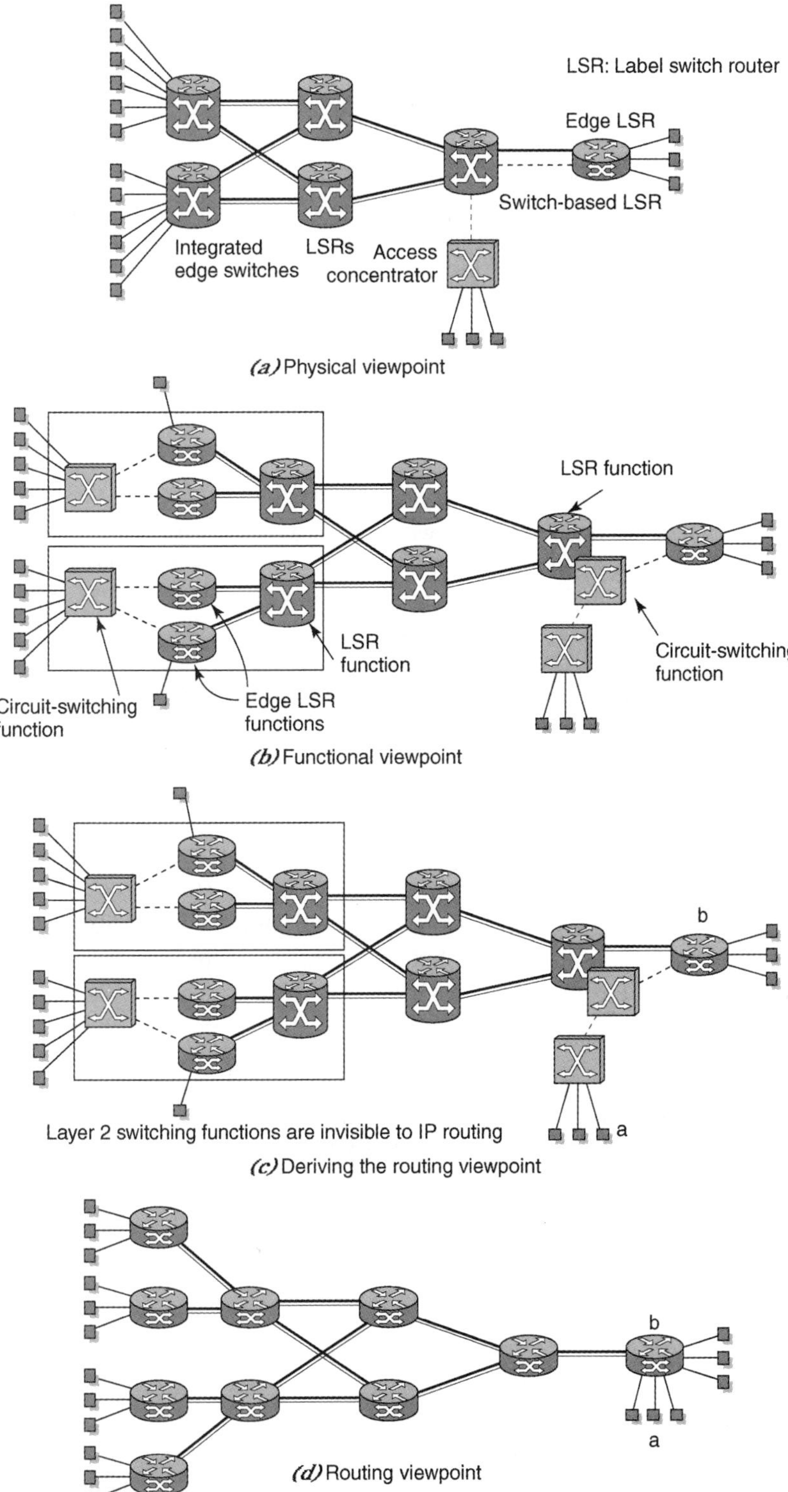

Figure 17.12. Viewpoints of an MPLS network. (From Figure 5, page 140, *IEEE Communications Magazine*, Ref. 8.)

support traffic engineering. In conventional forwarding, this requires the packet to carry an encoding of its route along with it (*source routing*). In MPLS, a label can be used to represent the route, so that the identity of the explicit route need not be carried with the packet.

Some routers analyze a packet's network layer header not merely to choose the packet's next hop, but also to determine a packet's *precedence* or *class of service*. They then may apply different discard thresholds or scheduling disciplines to different packets. MPLS allows (but does not require) the precedence or class of service to be fully or partially inferred from the label. In this case, one may say that the label represents the combination of an FEC and a precedence or class of service (Ref. 8).

Figure 17.12 shows that it is possible to have several viewpoints of an MPLS:

1. The physical viewpoint is shown in Figure 17.12a. This viewpoint represents the physical devices and links in a network.
2. The functional viewpoint is shown in Figure 17.12b. As we can see, where a device has several functions, these can be shown separately. For example, Figure 17.12 includes two MPLS edge devices of a type available today (Ref. 8). Each of the two edge devices includes two functionally separate edge LSRs and an LSR. In addition, each of the edge devices includes a permanent virtual circuit switching function that is functionally separate from the LSR function.
3. The routing viewpoint. This viewpoint is the network as it is seen by an IP routing protocol. It is derived from the functional viewpoints as follows:
   a. Layer 2 switches and permanent virtual circuits switching functions are invisible to IP routing. If a customer site is connected to a router by a permanent virtual circuit, the virtual circuit is seen by the IP routing as a single-hop direct connection. For example, note the sites labeled in Figure 17.12c, and assume that these are connected to edge LSR b. Then, in the routing viewpoint, the sites are directly adjacent to router b.
   b. Each edge LSR or LSR constitutes a router in the routing viewpoint.

Designing IP routing in an MPLS network is generally the same process as designing IP routing for an ordinary IP network. With reference to the routing viewpoint, a network can be divided into interior gateway protocol areas, route summarization can be designed, and so on (Ref. 8).

17.12 SUMMARY

A network consists of switches and transmission links connecting those switches. Switches and transmission links and devices involve hardware. In addition, a network requires a means of access and a method of routing messages. This portion of the network deals with protocols built with software. It would seem that we are describing conventional electronic networks. We are really dealing with optical networks. In this chapter we covered possible hardware and software combinations for feasible optical networks of the present and future. We addressed the question of how much of the network will remain electronic and how much can be implemented optically. We believe that the truly all-optical network remains illusive and a future goal. Optical signaling used for circuit setup in a connection-oriented network and in message headers for a connectionless-based network is going to be difficult to achieve with near-term technology.

REFERENCES

1. Michael Sluyski, AMCC, The Evolution of Crossconnects within the Emerging Managed Optical Layer, *Lightwave*, June 2000.
2. Luc Ceuppens, Multiprotocol Lambda Switching Comes Together, *Lightwave*, August 2000, page 80.
3. Ronnie Chua and Yiqun Hu, Optical Switches Are Key Components in High-Capacity, Data-Centric Networks, *Lightwave*, November 1999, page 43.
4. Tim Krause, *Migration to All-Optical Networks*, Alcatel Raleigh, July 24, 2000, from the Web at www.usa.alcatel.com/telecom.
5. Paul Green, Progress in Optical Networking, *IEEE Communications Magazine*, January 2001.
6. Malathi Veeraraghavan, Ramesh Karri, et al. Architectures and Protocols That Enable New Applications on Optical Networks, *IEEE Communications Magazine*, March 2001.
7. Marlene Bourne, MEMS Switching ... and Beyond, Cahners In-Stat Group, *Lightwave*, March 2001, page 204.
8. Jeremy Lawrence, Cisco Systems, Designing Multiprotocol Label Switching Networks, *IEEE Communications Magazine*, July 2001, page 134.
9. *MPLS Architecture*, E. Rosen et al., RFC 3031, January 2001. From the Internet.
10. Internet: www.nanog.org/mtg-9905/ppt/mpls, October 23, 2001.
11. Marc Fernandez and E. Kruglic, MEMS Technology Ushers in New Age in Optical Switching, *Lightwave*, August 2000, page 146.
12. Private communication, Luc Ceuppens, Calient Networks, San Jose, CA.
13. Private communication, Ezekial Kruglick, OMM Inc., San Diego, CA.

APPENDIX

ACRONYMS AND ABBREVIATIONS

2F-BLSR	two-fiber bidirectional switched ring
A	
AC	alternating current
A/D	analog-to-digital converter, conversion
ADC	analog-to-digital converter; a telecommunications company
ADM	add–drop multiplexer
ADSL	asymmetric digital subscriber line
AGC	automatic gain control
A-hr, AH	ampere-hour
AIS	alarm indication signal
AIS-L	alarm indication signal—line
AM	amplitude modulation
AMI	alternate mark inversion
AM/PM	amplitude modulation/phase modulation
Amp	amplifier
AM/VSB	amplitude modulation/vestigial sideband modulation
ANSI	American National Standards Institute
APC	angle PC or angle physical contact; automatic power control
APD	avalanche photodiode
APS	automatic protection switching
ASE	amplifier spontaneous emission
ATM	asynchronous transfer mode
AU	administrative unit
AUG	administrative unit group
A/W	ampere/watts
AWG	arrayed waveguide grating
B	
BER	bit error rate (ratio)
BERT	bit error rate test (set)

BIP	bit interleaved parity
B-ISDN	broadband integrated services digital network
BITE	built-in test equipment
BL product	bandwidth–distance product
BLSR	bidirectional line switched ring
BOL	beginning of life
BPSK	binary phase-shift keying
B&S	broadcast & select

C

CATV	community antenna television (cable television)
CCIR	International Consultive Committee on Radio
CCITT	International Consultive Committee on Telephone and Telegraph
CDMA	code division multiple access
Cemf	counter emf, counter electromotive force
CL	connectionless
C/N	carrier-to-noise (ratio)
CO	connection-oriented; central office
cpe, CPE	customer premise equipment
cpm	cross-phase modulation
CS	circuit switched
CSMA/CD	carrier sense multiple access/collision detection
cv	coding violation(s)
CW	carrier wave

D

D/A	digital-to-analog
DAVIC	Digital Audio Visual Council
dB	decibel
dBm	decibels referenced to a milliwatt
dBmp	decibels related to a milliwatt, psophometrically weighted
dBW	decibels related to a watt
DBR	distributed Bragg reflector
DC	direct current
DCF	dispersion-compensating fiber
DCS	digital cross-connect system
DFB	distributed feedback (laser)
DLCI	data link connection identifier
DOCSIS	Data Over Cable Service Interface Specification
DRIE	deep reactive ion etching
DS (DS1, DS1A, DS1c, DS2, DS3, etc.)	digital system, refers to North American PDH
dsf	dispersion-shifted fiber
DWDM	dense wavelength division multiplexing

E

ECL	external cavity laser
EDFA	erbium-doped fiber amplifier
EDTFA	erbium-doped tellurite-based fiber amplifier
EF	entrance facility
EIA	Electronic Industries Alliance
ELED	edge-emitting LED
EMC	electromagnetic compatibility
emf	electromotive force (voltage)
EOC, eoc	embedded operations channel
EOL	end of life
E/P	electrically pumped
ER	equipment room
eV	electron-volts

F

FBG	fiber Bragg grating
FCC	Federal Communications Commission
FDDI	fiber distributed data interface
FDM	frequency-division multiplex
FDMA	frequency-division multiple access
FEC	forwarding equivalence class
FEPR	far-end performance report
FET	field-effect transistor
FIT	failures in time (failures in 10^9 hours)
FM	frequency modulation
FP	Fabry–Perot
FWHM	full-width half-maximum
FWM	four-wave mixing

G

GaAs	gallium-arsenide
GbE	gigabit ethernet
Gbps	gigabits per second or 1×10^9 bps
GHz	gigahertz or 1×10^9 Hz
GVD	group velocity dispersion

H

HC	horizontal cross-connect
HDPE	high-density polyethylene
HEMT	high-electron mobility transistor
HFC	hybrid fiber coax(ial)

I

IC	integrated circuit; intermediate cross-connect
IEC	International Electrical Commission
IEEE	Institute of Electrical and Electronics Engineers

IETF	Internet Engineering Task Force
InGaAs	indium–gallium–arsenide
I/O	input–output (device)
IP	Internet protocol
ISI	intersymbol interference
IS-IS	intermediate system to intermediate system (intradomain routing)
ISP	Internet service provider
IT	information technology
ITU	International Telecommunication Union
IVD	inside vapor deposition (process)

K

kHz	kilohertz or Hz $\times 10^3$
km	kilometer

L

λ	Greek letter lambda, the symbol for wavelength
LAN	local area network
LD	laser diode
LDP	label distribution protocol
LEAF	large effective area fiber (trademark Corning)
LEC	local exchange carrier
LED	light emitting diode
LER	label edge router
LID	local injection and detection
LIGA	lithographic plating and molding
LMDS	local multipoint distribution system
LOF	loss of frame
LOH	line overhead
LOP	loss of pointer
LOS	loss of signal
LSP	label-switched path
LSR	label-switched router
LTE	line termination equipment

M

μm	micrometer, micron, meter $\times 10^{-6}$
MAC	medium access control
MAN	metropolitan area network
Mbps	megabits per second, bps $\times 10^6$
MC, mc	main cross-connect
MCVD	modified chemical vapor deposition (process)
MEMS	microelectromechanical system or switch
MFD	mode field diameter
MI	modulation instability
MLM	multilongitudinal mode (laser)
MMDS	microwave multipoint distribution system

MPEG	Motion Picture Experts' Group
MPLS; MPλS	multiprotocol label switching; multiprotocol λ (wavelength) switching
MPN	mode partition noise
MQW	multi-quantum well
MTBF	mean time between failures
MTTR	mean time to repair
MSR	mode suppression
mW	milliwatt
M–Z, MZ	Mach–Zehnder

N

NA	numerical aperture
NDF	new data flag
NE	network element
NEC	National Electric(al) Code
NEP	noise equivalent power
NESC	National Electric(al) Safety Code
NF	noise figure
nm	nanometer, meter $\times 10^{-9}$
NNI	network–node interface or network–network interface
NOCC	network operations control center
nom	nominal
NTSC	National Television Systems Committee
NTT	Nippon Telephone and Telegraph (Co.)
NUT	nonpreemptible unprotected traffic
nz-dsf	nonzero dispersion-shifted fiber

O

OADM	optical add–drop multiplexer
OAM, OA&M	operations and maintenance; operations, administration, and maintenance
OC	optical carrier
OCC	optical cross-connect, also see OXC
OEO	optical-electrical-optical
OFA	optical fiber amplifier
OH	hydroxyl ion
O/P	optically pumped
ORL	optical return loss
OSA	optical spectrum analyzer
OSC	optical supervisory channel
OSHA	Occupational Safety and Health Administration
OSP	outside plant
OTDR	optical time-domain reflectometer
O/W	orderwire
OXC	optical cross-connect, also see OCC

P

PABX	private automatic branch exchange
PC	physical contract
PCB	printed circuit board
pdh, PDH	plesiochronous digital hierarchy
PDFA, PDFBFA	praseodymium-doped fluoride-based fiber amplifier
PDL	polarization-dependent loss
PHB	polarization hole burning
p-HEMT	pseudomorphic high-electron mobility transistor
PIN	*p*-intrinsic-*n*
PMD	polarization mode dispersion
POH	path overhead
Pn junction	p-layer interface with n-silicon
POS	packet-over-SONET
prbs, PRBS	pseudorandom binary sequence
PS	packet-switched
PSK	phase-shift keying
PSTN	public-switched telecommunications network
PVC	polyvinyl chloride
pWp	picowatts psophometrically weighted

Q

QAM	quadrature amplitude modulation
QoS	quality of service
QPSK	quadrature phase-shift keying

R

RC	resistance–capacitance
RDI	remote defect indication
RF	radio frequency
RFC	request for comment
RFI	radio-frequency interference
rgtr, RGTR	regenerator
RIN	relative intensity noise
RIP	routing information protocol
RL	return loss
rms	root mean square
RPP	reliability prediction program
RSOH	regenerator section overhead
RSVP	resource reservation protocol
RTU	remote test unit
RZ	return to zero

S

SAM	separate absorption and multiplication
satcom	satellite communications
SCADA	supervisory control data acquisition (system)
SBS	simulated Brillioun scattering

SC	subscriber connector (name of connector)
SCO	service continuity objective
SCR	silicon-controlled rectifier
SDH	synchronous digital hierarchy
SEF	severely error frame
SES	severely errored second(s)
SG-DBR	sampled grating—DBR
SHR	self-healing ring
Si	silicon
SiO_2	silicon oxide
SLM	single longitudinal mode
S/N, SNR	signal-to-noise (ratio)
SOH	section overhead
SONET	synchronous optical network
SPE	synchronous payload envelope
SPM	self-phase modulation
SRS	simulated Raman scattering
STE	section terminating equipment
STM	synchronous transport module
STP	shielded twisted pair
STS	synchronous transport signal

T

Tbps	terabits per second, bps $\times 10^{12}$
TC	telecommunication closet
TCP/IP	transmission control protocol/Internet protocol
TDFFA	thalium-doped flouride-based amplifier
TDM	time division multiplex
TDMA	time division multiple access
TEC	thermoelectric cooler
THz	terahertz, Hz $\times 10^{12}$
TIA	Telecommunications Industry Association
TLS	tunable laser source
TOH	transport overhead
TR	telecommunication room
TSA	time-slot assignment
TSGR	transmission systems generic requirement(s)
TSI	time-slot interchanger
TU	tributary unit
TUG	tributary unit group
TW	traveling wave

U

UAS-L	unavailable seconds for line
UL	Underwriter's Laboratory
ULCC	Utility Location and Coordination Council
UNI	user–network interface
UPPS	unidirectional path-protection switched

UPS	uninterruptible power supply
UPSR	unidirectional path-switched ring(s)
UTP	unshielded twisted pair
UV	ultraviolet

V

va, VA	volt-amperes
VC	virtual container; virtual circuit
VCI	virtual circuit indicator
VCSEL	virtual cavity surface-emitting laser
VDC	volts, direct current
VOA	variable optical attenuator
VPI	virtual path indicator
VPN	virtual private network
VSB	vestigial sideband
VT	virtual tributary

W

WA	work area(s)
WAN	wide area network
WDM	wavelength division multiplex
WR	wavelength-routed
WTR	wait-to-restore

X, Y, Z

XC	cross-connect
XPM	cross-phase modulation
YIG	yttrium–iron–garnet

INDEX

Accounting for Non-Specialists

Michael Jones

Cardiff Business School

JOHN WILEY & SONS, LTD

Published by John Wiley & Sons Ltd
The Atrium, Southern Gate, Chichester,
West Sussex, PO19 8SQ, England
National 01243 779777
International (+44) 1243 779777

e-mail (for orders and customer service enquiries): cs-books@wiley.co.uk

Visit our Home Page on http://www.wiley.co.uk

Other Wiley Editorial Offices
John Wiley & Sons, Inc., 605 Third Avenue,
New York, NY 10158–0012, USA

Wiley-VCH Verlag GmbH
Pappelallee 3, D-69469 Weinheim, Germany

John Wiley & Sons (Australia) Ltd, 33 Park Road, Milton,
Queensland 4064, Australia

John Wiley & Sons (Asia) Pte Ltd, 2 Clementi Loop #02-01,
Jin Xing Distripark, Singapore 129809

John Wiley & Sons (Canada) Ltd, 22 Worcester Road,
Rexdale, Ontario, M9W 1L1, Canada

British Library Cataloguing in Publication Data
A catalogue record for this book is available from the British Library

ISBN 0-471-49572-7

Designed and typeset in Caslon and Helvetica by
Florence Production Ltd, Stoodleigh, Devon

Printed and bound in Great Britain by
Ashford Colour Press Ltd, Gosport, Hampshire.

This book is printed on acid-free paper responsibly manufactured from sustainable forestry in which at least two trees are planted for each one used for paper production.

Contents

About the Author

Michael Jones has taught accounting to non-specialists for 24 years, first at Hereford Technical College, then at Portsmouth Polytechnic (now University) and currently at Cardiff Business School. He has taught financial accounting at all levels from GCSE level to final-year degree course. He has published over 100 articles in both professional and academic journals. These articles cover a wide range of topics such as financial accounting, the history of accounting and international accounting. The author's main research interest is in financial communication. He is Professor of Financial Reporting at Cardiff Business School and Director of the Financial Reporting and Business Communication Unit. Michael is married and has one daughter, Katherine.

About the Book

Background

Accounting is a key aspect of business. All those who work for, or deal with, businesses, therefore, need to understand accounting. Essentially, understanding accounting is a prerequisite for understanding business. This book aims to introduce students to accounting and provide them with the necessary understanding of the theory and practice of financial and management accounting. The book, therefore, is aimed primarily at non-specialists and seeks to be as understandable and readable as possible.

The Market

This book is intended as a primary text for non-accounting students: either those following an undergraduate degree in a business school, or non-business studies students studying an accounting course. This will include, for example, engineers, physicists, hotel and catering, social studies and media-study students. The text aims to produce a self-contained, introductory, one-year course for non-specialists. However, it is also designed so that students can progress to more advanced follow-up courses in financial accounting or management accounting. The text should also be accessible to mainstream accounting graduates or MBA students as a supplementary text. In particular, MBA students should find the chapters on creative accounting, international accounting and strategic management accounting beneficial. The book should be particularly useful in reinforcing the fundamental theory and practice of introductory accounting.

Scope

The book sets down my acquired wisdom (such as it is) over 24 years and interweaves context and technique. It aims to introduce the topic of accounting to students in a student-friendly way. Not only are certain chapters devoted solely to context, but the key to each particular topic is seen as developing the student's understanding of the underlying concepts. This is a novel approach for this type of book.

The book is divided into 22 chapters within three sections. Section A deals with the context and techniques of basic financial accounting and reporting. Section B looks at the context of financial accounting and reporting. Finally, Section C provides an introduction to the context and techniques of management accounting.

Section A: Financial Accounting: The Techniques

In the first section, after introducing the context and background to accounting, the mechanics of financial accounting are explored, for example, bookkeeping and the preparation of financial statements (such as the profit and loss account and balance sheet). This section is presented using a tried and tested example, which starts from the accounting equation. A profit and loss account and balance sheet are then prepared. The section continues by explaining the adjustments to financial statements, different enterprises' financial statements, the cash flow statement and the interpretation of accounts.

Section B: Financial Accounting: The Context

The focus in this section is on exploring some wider aspects of external financial reporting. It begins by contextualising financial reporting by looking at the regulatory framework, measurement systems and the annual report. This is followed by an introductory overview of creative accounting and international accounting. This part of the book allows students to appreciate how accounting is rooted in a wider social and international context. Particularly novel is the inclusion of the regulatory framework, corporate governance, international accounting, and creative accounting. This material is included to give students a wider appreciation of accounting than is traditionally presented.

Section C: Management Accounting

This section begins with an exploration of the main concepts underpinning management accounting. Seven main areas are covered: costing, budgeting, standard costing, short-term decision making, strategic management accounting, capital investment and sources of finance. This section thus provides a good coverage of the basics of costing and management accounting as well as introducing strategic management accounting, a topic which has recently become more prominent. The aim of this section is to introduce students to a wide range of key concepts so that they gain a good knowledge base.

Coverage

The issues of double-entry bookkeeping, partnerships, manufacturing accounts, computers, internationalisation and the public sector are tricky ones for an introductory, non-specialist text. I have included double-entry bookkeeping, but tried to introduce a no-frills approach and focus on the essentials. This is because I strongly believe that double-entry bookkeeping is a fundamental stepping stone to understanding accounting. *The book is so designed that it is possible to miss out the detailed explanation of double-entry bookkeeping, but still follow the rest of the textbook.*

This book, after much consideration, focuses on three types of business enterprise: sole traders, partnerships and limited companies. These three enterprises comprise the vast bulk of UK businesses. Indeed, according to the Office of National Statistics (2000), in the UK, 39% of enterprises are companies, 36% sole traders and 23% partnerships.

The sole trader is the simplest business enterprise. The earlier chapters in Section A, therefore, primarily focus on the sole trader to explain the basics. However, later chapters in Section A are more concerned with companies.

Manufacturing accounts are excluded because I consider them unnecessarily complex for non-specialist students and also because of their diminishing importance within the UK economy. Indeed, in 2000, only 10% of UK businesses were in production (Office for National Statistics, 2000).

The impact of computers on accounting is covered in the text where appropriate. I consider the widespread use of computers makes it even more important than before to understand the basics of accounting.

Although the primary audience for this textbook is likely to be the UK, I have, where possible, attempted to 'internationalise' it. In fact, I have taught international accounting for 12 years. A synthesis of international accounting is thus provided in Chapter 14. However, I have also tried to integrate other international aspects into the book, for instance, drawing on the International Accounting Standards Board's Statement of Principles.

A decision was taken at an early stage to focus on the private sector rather than the public sector. So when the terms company, firm, enterprise and organisation are used, sometimes interchangeably, generally they refer to private sector organisations. There is some coverage of public sector issues, but in general, students interested in this area should refer to a more specialised public sector textbook.

Special Features

A particular effort has been made to make accounting as accessible as possible to students. There are thus several special features in this book which, taken together, distinguish it from other introductory textbooks.

Blend of Theory and Practice

I believe that the key to accounting is understanding. As a result, the text stresses the underlying concepts of accounting and the context within which accounting operates. The book, therefore, blends practice and theory. Worked examples are supplemented by explanation. In addition, the context of accounting is explored. The aim is to contextualise accounting within a wider framework.

Novel Material

There has been a conscious attempt to introduce into the book topics not usually taught at this level, for example, regulatory framework, corporate governance, creative accounting, international accounting and strategic management accounting. The regulatory framework is introduced to give the students the context within which accounting is based. Corporate governance aims to set accounting within the wider relationship between directors and shareholders. Creative accounting, I have found, is an extremely popular subject with students and demonstrates that accounting is an art rather than a science. It is particularly topical, given the collapse of the US company Enron and other recent US accounting scandals as WorldCom and Xerox. Creative accounting also enables the behavioural aspects of accounting to be explored. International accounting allows students to contextualise UK accounting within a wider international setting. This is increasingly important with the growth in international trade. Finally, strategic management accounting is a growing area of interest to management accountants. Its inclusion in this book reflects this.

Interpretation

I appreciate the need for non-specialist students to evaluate and interpret material. There is thus a comprehensive chapter on the interpretation of accounts. Equally important, the book strives to emphasise why particular techniques are important.

Readable and Understandable Presentation

Much of my research has been into readable and understandable presentation. At all times, I have strived to achieve this. In particular, I have tried to present complicated materials in a simple way.

Innovative Presentation

I am very keen to focus on effective presentation. This book, therefore, includes many presentational features which aim to enliven the text. Quotations, extracts from newspapers and journals (real-life nuggets), and extracts from annual reports (the company camera) convey the day-to-day relevance of accounting. I also attempt to use realistic examples. This has not, however, always been easy or practical given the introductory nature of the material. In addition, I have attempted to inject some wit and humour into the text through the use of, among other things, cartoons and soundbites. The cartoons, in particular, are designed to present a sideways, irreverent look at accounting which hopefully students will find not only entertaining, but also thought-provoking. Finally, I have frequently used boxes and diagrams to simplify and clarify material. Throughout the text there are reflective questions (pause for thought). These are designed as places where students may pause briefly in their reading of the text to reflect on a particular aspect of accounting or to test their knowledge.

End-of-Chapter Questions and Answers

There are numerous questions and answers at the end of each chapter which test the student's knowledge. These comprise both numerical and discussion questions. The discussion questions are designed for group discussion between lecturer and students. At the end of the book an outline is provided to, at least, the first discussion question of each chapter. This answer provides some outline points for discussion and allows the students to gauge the level and depth of the answers required. However, it should not be taken as exhaustive or prescriptive. The other discussion answers are to be found on the lecturers' area of the website.

The answers to the numerical questions are divided roughly in two. Half of the answers are provided at the back of the book for students to practise the techniques and to test themselves. These questions are indicated by the number being in blue. The other numerical answers are to be found on the lecturer area of the website. A further extensive supplementary set of over 100 questions and answers are also available to lecturers on their website.

Websites

In addition to the supplementary questions, there are Powerpoint slides available on the lecturers' website. On the student's website, there are 220 multiple-choice questions (ten for each chapter) as well as 22 additional questions with answers (one for each chapter).

Overall Effect

Taken together, I believe that the blend of theory and practice, focus on readable and understandable presentation, novel material, interpretative stance and innovative presentation make this a distinctive and useful introductory textbook. Hopefully, readers will find it useful and interesting! I have done my best. Enjoy!

Acknowledgements

In many ways writing a textbook of this nature is a team effort. Throughout the three years it took me to write this book I have consistently sought the help and advice of others in order to improve it. I am, therefore, extremely grateful to a great number of academic staff and students (no affiliation below) whose comments have helped me to improve this book. The errors, of course, remain mine.

Malcolm Anderson (Cardiff Business School)
Tony Brinn (Cardiff Business School)
Alex Brown
Mark Clatworthy (Cardiff Business School)
Alpa Dhanani (Cardiff Business School)
Mahmoud Ezzamel (Cardiff Business School)
Charlotte Gladstone-Miller (Portsmouth Business School)
Tony Hines (Portsmouth Business School)
Deborah Holywell
Carolyn Isaaks (Nottingham Trent University)
Tuomas Korppoo
Margaret Lamb (Warwick Business School)
Les Lumsdon (Manchester Metropolitan University)
Claire Lutwyche
Louise Macniven (Cardiff Business School)
Neil Marriott (University of Glamorgan)
Howard Mellett (Cardiff Business School)
Joanne Mitchell
Peter Morgan (Cardiff Business School)
Barry Morse (Cardiff Business School)
Simon Norton (Cardiff Business School)
Phillip O'Regan (Limerick University)
David Parker (Portsmouth Business School)
Roger Pegum (Liverpool John Moores University)
Maurice Pendlebury (Cardiff Business School)

Elaine Porter (Bournemouth University)
Neil Robson (University of the West of England)
Julia Smith (Cardiff Business School)
Aris Solomon (Cardiff Business School)
Jill Solomon (Cardiff Business School)
Tony Whitford (University of Westminster)
Jason Xiao (Cardiff Business School)

This book is hopefully enlivened by many extracts from books, newspapers and annual reports. This material should not be reproduced, copied or transmitted unless written permission is obtained from the original copyright owner. I am, therefore, grateful to all those who kindly granted the publisher permission to reproduce the copyright material. This includes the following companies and plcs: AstraZeneca, BG, British Petroleum, H.P. Bulmers (Holdings), Glaxo-Wellcome, Halifax, Heineken, Hyder, Laing, Manchester United, Marks and Spencer, Rentokil, Rolls-Royce, J. Sainsbury, Shell Transport and Trading, Tesco, Vodafone, J.D. Wetherspoon, and Woolwich. In addition, *Accountancy Age*, the *Economist*, *The Financial Times*, *The Guardian*, *Management Accounting*, the *New Scientist*, *Sunday Business* and *The Sunday Telegraph*. Each source is also specifically mentioned in the text.

Many of the Soundbites are drawn from the following texts. The full references are: H. Ehrlich (1998), *The Wiley Book of Quotations* (John Wiley & Sons, Inc., New York); R. Flesch (1959), *The Book of Unusual Quotations* (Cassell, London); J. Vitullo-Martin and J. Robert Moskin (1994), *The Executive's Book of Quotations* (Oxford University Press, New York); and E. Weber (1991), *The Book of Business Quotations* (Business Books, London).

Every effort has been made to trace the original copyright owner; if we have accidentally infringed any copyright the publishers offer their apologies.

I should also like to thank Steve Hardman and Sarah Booth of John Wiley for their help and support. Finally, last but certainly not least, I should like to thank Jan Richards for her patience and hard work in turning my generally illegible scribbling into the final manuscript.

Dedication

I would like to dedicate this book to the following people who have made my life richer.

- My parents, Donald and Lilian
- My wife, Christine
- My daughter, Katherine
- All my friends in Hereford, Cardiff and elsewhere
- All my colleagues
- And, finally, my past students!

Chapter 1

Introduction to Accounting

'One way to cheat death is to become an accountant, it seems. The Norfolk accountancy firm W.R. Kewley announces on its websites that it was "originally established in 1982 with 2 partners, one of whom died in 1993. After a short break he re-established in 1997, offering a personal service throughout."

He was, Feedback presumes, dead only for tax purposes.'

New Scientist, 1 April 2000, p. 96

Learning Outcomes

After completing this chapter you should be able to:

- ✔ **Explain the nature and importance of accounting.**
- ✔ **Outline the context which shapes accounting.**
- ✔ **Identify the main users of accounting and discuss their information needs.**
- ✔ **Distinguish between the different types of accountancy and accountant.**

In a Nutshell

- Accounting is the provision of financial information to managers or owners so that they can make business decisions.
- Accounting measures, monitors and controls business activities.
- Financial accounting supplies financial information to external users.
- Management accounting serves the needs of managers.
- Users of accounting information include shareholders and managers.
- Accounting theory and practice are affected by history, country, technology and organisation.
- Auditing, bookkeeping, financial accounting, financial management, insolvency, management accounting, taxation and management consultancy are all branches of accountancy.
- Accountants may be members of professional bodies, such as the Institute of Chartered Accountants in England and Wales.

Introduction

The key to understanding business is to understand accounting. Accounting is central to the operation of modern business. Accounting enables businesses to keep track of their money. If businesses cannot make enough profit or generate enough cash they will go bankrupt. Often accounting is called the 'language of business'. It provides a means of effective and understandable business communication. If you understand the language you will, therefore, understand business. However, like many languages, accounting needs to be learnt. The aim of this book is to teach the language of accounting.

Nature of Accounting

At its simplest, accounting is all about recording, preparing and interpreting business transactions. Accounting provides a key source of information about a business to those who need it, such as managers or owners. This information allows managers to monitor, plan and control the activities of a business. This enables managers to answer key questions such as:

- How much profit have we made?
- Have we enough cash to pay our employees' wages?
- What level of dividends can we pay to our shareholders?
- Should we expand our product range?

Pause for Thought 1.1

Some Accounting Questions

You are thinking of manufacturing a new product, the superwhizzo. What are the main accounting questions you would ask?

The principal questions would relate to sales, costs and profit. They might be:

- What price are rival products selling at?
- How much raw material will I need? How much will it cost?
- How many hours will it take to make each superwhizzo and how much is labour per hour?
- How much will it cost to make the product in terms of items such as electricity?
- How should I recover general business costs such as business rates or the cost of machinery wearing out?
- How much profit should I aim to make on each superwhizzo?

In small businesses, managers and owners will often be the same people. However, in larger businesses, such as large companies, managers and owners will not be the same. Managers will run the companies on behalf of the owners. In such cases, accounting information serves a particularly useful role. Managers supply the owners with financial information in the form of a profit and loss account, a balance sheet and a cash flow statement. This enables the owners to see how well the business is performing. In companies, the owners of a business are called the shareholders.

Essentially, therefore, accounting is all about providing financial information to managers and owners so that they can make business decisions (see Definition 1.1). The formal definition (given below), although dating from 1966, has stood the test well as a comprehensive definition of accounting.

Definition 1.1

Accounting

Working definition

The provision of information to managers and owners so that they can make business decisions.

Formal definition

'The process of identifying, measuring and communicating economic information to permit informed judgments and decisions by users of the information.'

American Accounting Association (1966), *Statement of Basic Accounting Theory*, p. 1

Importance of Accounting

Accounting is essential to the running of any business or organisation. Organisations as diverse as ICI, Barclays Bank and Manchester United football club all need to keep a close check on their finances.

At its simplest, money makes the world go round and accounting keeps track of the money. Businesses depend on cash and profit. If businesses do not make enough cash or earn enough profit, they will get into financial difficulties, perhaps even go bankrupt. Accounting provides the framework by which cash and profit can be monitored, planned and controlled.

Unless you can understand accounting, you will never understand business. This does not mean everybody has to be an expert accountant. However, it is necessary to know the language of accounting and to be able to interpret accounting numbers. In some respects, there is a similarity between learning to drive a car and learning about accounting. When you are learning to drive a car you do not need to be a car mechanic. However, you have to understand the car's instruments, such as a speedometer or fuel gauge. Similarly, with accounting, you do not have to be a professional accountant. However, you do need to understand the basic terminology such as income, expenses, profit, assets, liabilities, capital, and cash flow.

Pause for Thought 1.2

Manchester United

What information might the board of directors of Manchester United find useful?

Manchester United plc is both a football club and a thriving business. Indeed, the two go hand in hand. Playing success generates financial success, and financial success generates playing success. Key issues for Manchester United might be:

- How much in gate receipts will we get from our league matches, cup matches and European fixtures?
- How much can we afford to pay our players?
- How much cash have we available to buy rising new stars and how much will our fading old stars bring us?
- How much will we get from television rights and commercial sponsorship?
- How much do we need to finance new capital expenditure, such as building a new stadium?

Financial Accounting and Management Accounting

A basic distinction is between financial accounting and management accounting. Financial accounting is concerned with information on a business's performance and is targeted primarily at those outside the business (such as shareholders). However, it is also used internally by managers. By contrast, management accounting is internal to a business and used solely by managers. A brief overview is provided in Figure 1.1.

Figure 1.1 Overview of Financial and Management Accounting

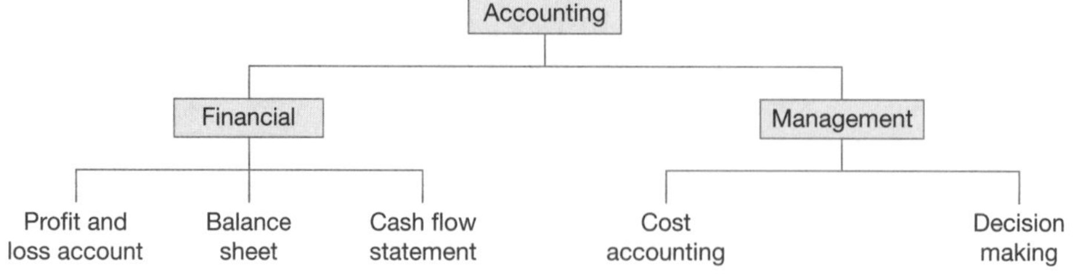

Financial Accounting

Financial accounting is the provision of financial information on a business's recent financial performance targeted at external users, such as shareholders. However, internal users, such as management, may also find it useful. It is required by law. Essentially, it is backward-looking, dealing with past events. Transactions are initially recorded using double-entry bookkeeping (see Chapter 3). Three major financial statements can then be prepared: the profit and loss account, the balance sheet and the cash flow statement (see Chapters 4–8). These are then interpreted using ratios (see Chapter 9) by users such as shareholders and analysts.

Soundbite 1.1

Investment Analysts

'I am not a professional security analyst. I would rather call myself an insecurity analyst.'

George Soros, *Soros on Soros*, Wiley (1995)

Source: *The Wiley Book of Business Quotations* (1998), p. 192

Management Accounting

By contrast, management accounting serves only the internal needs of the business. It is not required by law. However, all organisations generate some sort of management accounting: if they did not they would not survive long. Management accounting can be divided into cost accounting and decision making. In turn, cost accounting can be split into costing (Chapter 16) and planning and control: budgeting (Chapter 17) and standard costing (Chapter 18). Decision making is divided into short-term decisions (Chapter 19) and long-term decision making (Chapters 20–21). In Chapter 22 both short-term and long-term sources of finance are considered.

Users of Accounts

The users of accounting information may broadly be divided into insiders and outsiders (see Figure 1.2). The insiders are the management and the employees. However, employees are also outsiders in the sense that they often do not have direct access to the financial information.

Figure 1.2 Main Users of Accounting Information

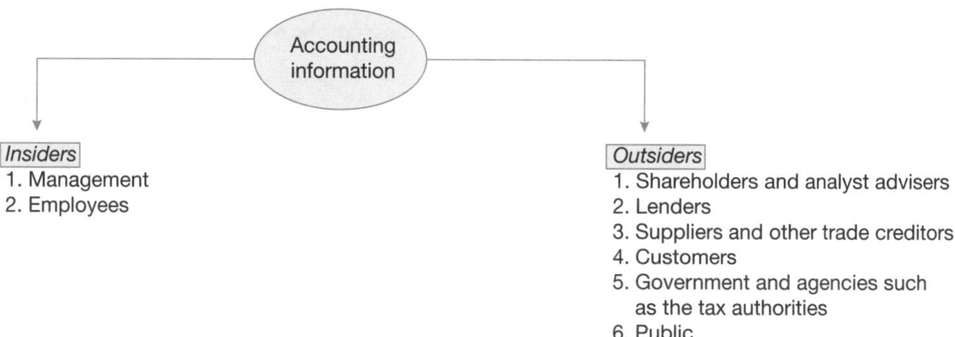

The primary user groups are the management and shareholders. Shareholders or investors are often advised by professional financial analysts who work for stockbrokers or big city investment houses. These financial analysts help to determine the share prices of companies quoted on the stock exchange. However, sometimes, they are viewed with mistrust (see Real-Life Nugget 1.1 on the next page).

Real-Life Nugget 1.1

Financial Analysts

Working with analysts is a little like being a member of the Magic Circle or the freemasons – those who know how to perform these masterful sleights of hand or arcane rituals are forbidden from ever revealing them to outsiders. Of course, the alternative is that I can't reveal them because I don't understand them myself – I'll let you decide.

Source: 'Showbiz values come to the City', James Montier, *The Guardian*, January 15, 2000, p. 5

The influence of other users (or stakeholders) is growing in importance. Suppliers, customers and lenders have a closer relationship to the company than the government and tax authorities, or the public.

These users all need accounting information to help them make business decisions. Usually, the main information requirements concern a company's profits, cash flow, assets and debt (see Figure 1.3).

Shareholders, for example, require information so that they can decide whether to buy, hold or sell their shares. The information needs of each group differ slightly and, indeed may conflict (see Pause for Thought 1.3).

Figure 1.3 User Information Requirements

User Group	*Information Requirements*
Internal Users	
1. Management	Information for costing, decision making, planning and control.
2. Employees	Information about job security and for collective bargaining.
External Users	
1. Shareholders and analyst advisers	Information for buying and selling shares.
2. Lenders	Information about assets and the company's cash position.
3. Suppliers and other trade creditors	Information about assets and the company's cash position.
4. Customers	Information about the long-term prospects and survival of the business.
5. Government and agencies such as tax authorities	Information to enable governmental planning. Information primarily on profits to use as a basis for calculating tax.
6. Public (e.g., individual citizens, or organisations such as Greenpeace)	Information about the social and environmental impact of corporate activities.

Pause for Thought 1.3

Conflicting Interests of User Groups

Can you think of an example where the interests of users might actually conflict?

Pause for Thought 1.3 (continued)

A good example would be in the payment of dividends to shareholders. The higher the dividend, the less money is kept in the company to pay employees or suppliers.

Another, more subtle, example is the interests of shareholders and analyst advisers. Shareholders own shares. However, they rely on the advice of analyst advisers such as stockbrokers. Their interests may superficially seem the same (e.g., selling underperforming shares and buying good performers). However, the analyst advisers live by the commission they make. It is in their interests to advise shareholders to buy and sell shares. Unfortunately, it costs money to buy and sell shares; therefore, this may not always be in the potential shareholders' best interests.

Accounting Context

It is important to realise that accounting is more than just a mere technical subject. Although it is true that at the heart of accounting there are many techniques. For example, as we will see in Chapter 3, double-entry bookkeeping is essential when preparing financial statements. However, accounting is also determined by the context in which it operates. Accounting changes as society changes. Accounting in medieval England and accounting today, for example, are very different. Similarly, there are major differences between accounting in Germany and in the United Kingdom. We can see the importance of context, if we look briefly at the effect of history, country, technology, and organisation (see Figure 1.4).

Figure 1.4 Importance of Accounting Context

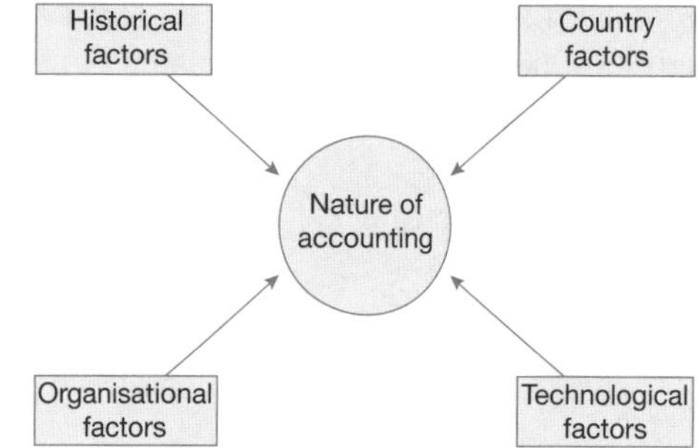

History

Accounting is an integral part of human society. Early societies had accounting systems which although appearing primitive to us today, served their needs adequately. The Incas in South America, for example, used knotted ropes, called quipus, for accounting. In medieval England notched sticks called tally sticks were used to record transactions.

Pause for Thought 1.4

The Term 'Accounting'

Why is accounting so called?

It is believed that accounting derives from the old thirteenth-century word *aconter*, to count. At its simplest, therefore, accounting means counting. This makes sense as the earliest accountants would have counted sheep or pigs!

Gradually, over time, human society became more sophisticated. A form of accounting called double-entry bookkeeping (every transaction is recorded twice) arose. Emerging from Italy in the fifteenth century, at the same time that Columbus discovered America, double-entry bookkeeping is now the standard way by which accounting transactions are recorded throughout the world.

International Accounting

Double-entry bookkeeping, the profit and loss account and the balance sheet are now routine in most major countries. The International Accounting Standards Board (IASB) is also making great efforts to harmonise the disclosure and measurement practices of listed companies worldwide. However, there is still great diversity in the broader context in which accounting is carried out. Accounting in the UK, for example, is very different from accounting in France despite the fact that both countries are members of the European Union. The UK prides itself on a flexible and self-regulated accounting system largely free of government control. By contrast, the French system is very standardised and largely government controlled. In general, as Soundbite 1.2 shows, accounting plays an important role globally.

Soundbite 1.2

'The significance of the accounting profession in the global marketplace is underestimated, particularly in the financial markets of the United States and Europe.'

Jim Schiro, quoted in the *Financial Times* (October 3, 1997)

Source: *The Wiley Book of Business Quotations* (1998), p. 192

There are also clear differences between the UK and the US. For example, in the UK, there are Companies Acts which applies to all UK companies whereas in the US only companies quoted on the stock exchange (i.e., listed companies) are subject to detailed and comprehensive Federal regulations. Unlisted companies are subject only to state regulations which vary from state to state.

Technology

A rapid change which has affected accounting is computerisation. Up until the advent of computers, accounting was done manually. This was labour-intensive work. Each transaction was entered into the books twice using double-entry bookkeeping. The accounts were then prepared by hand. Similarly, costing, budgeting and decision making were all carried out manually.

Today, most businesses use computers. However, they must be used with caution. For the computer, GIGO rules. If you put garbage in, you get garbage out. To avoid GIGO, one needs to understand accounting. In fact, computerisation probably makes it more, rather than less, important to understand the basics.

Organisations

The nature of accounting will vary from business to business. It will depend on the structure of the business and the nature of the business activity.

Structure

If we take the accounts of the three types of business enterprise with which this book deals, a sole trader's accounts will normally be a lot simpler than those of either a partnership or a company. Sole traders generally run smaller, less complicated businesses (for example, a small butcher's shop). Partnerships are multi-owned businesses typically larger in size than sole traders. The sole traders' and partnerships' accounts will normally be less complicated than company accounts as they are prepared for the benefit of active owner-managers rather than for owners who do not actually run the business.

Nature of the Business

Every organisation is different. Consequently, every organisation's accounts will differ in certain respects. For example, property companies will own predominantly more land and buildings than non-property companies. Manufacturing companies will have more stock than non-manufacturing companies.

It is clear from Figure 1.5 that the nature of sales varies from business to business. In some businesses, a service is provided (e.g., bank, football club, insurance company and plumber).

Figure 1.5 Nature of Sales

Business	***Nature of Main Sales***
Bank	Interest received from customers
Football club	Gate receipts
Insurance company	Premiums received
Manufacturing company	Sales of goods to retailers
Plumber	Sales of services and other goods
Shop	Sales of goods to customers

The Company Camera 1.1 shows the sales (often termed turnover) of Manchester United plc, mainly gate receipts, television and merchandising, which are generated by entertaining its customers.

The Company Camera 1.1

Sales or Turnover

Turnover

Turnover, all of which arises from the Group's principal activity, can be analysed into its main components as follows:

	2000	1999
	£'000	£'000
Gate receipts and programme sales	**36,626**	41,908
Television	**30,546**	22,503
Sponsorship	**18,513**	17,488
Conference and catering	**6,698**	7,189
Merchandising and other	**23,622**	21,586
	116,005	110,674

Source: Manchester United plc, 2000 Annual Report, p. 35

In other businesses, the sale is more tangible as goods change hands (for example, manufacturing companies and shops).

Overall, within the UK economy, services are becoming relatively more important and manufacturing industry is declining. In particular, there is an increase in information technology and knowledge-based industries. This trend is set to continue.

Types of Accountancy

We need to distinguish between the types of accountancy and the types of accountant. Accountancy refers to the process, while accountant refers to the person. In other words, accountancy is what accountants do! In this book, we primarily focus on bookkeeping, financial accounting and management accounting. However, accountants perform other roles such as auditing, financial management, insolvency, taxation and management consultancy. All of these are briefly covered below.

Auditing

Auditing is carried out by teams of staff headed by qualified accountants who are independent of the business. Essentially, auditors check that the financial statements, prepared by management, give a true and fair view of the accounts. Auditing is normally associated with company accounts. However, the tax authorities or the bank may request an audit of the accounts of sole traders or partnerships. For companies, auditors issue an auditor's report annually to shareholders. The Company Camera 1.2 (on the next page) provides an auditors' report for Shell Transport and Trading Company plc. This is issued after a thorough examination by the auditors of the accounting records and systems of the company. Auditors charge management an audit fee. For example, KPMG, a firm of auditors, charged HSBC Holdings £17.2 million in 2001.

Bookkeeping

Bookkeeping is the preparation of the basic accounts (bookkeeping is dealt with in more depth in Chapter 3). It involves entering monetary transactions into the books of account. A trial balance is then extracted, and a profit and loss account and balance sheet are prepared. Nowadays, most companies use computer packages for the basic bookkeeping function which is often performed by non-qualified accountants.

Financial Accounting

Financial accounting is a wider term than bookkeeping. It deals with not only the mechanistic bookkeeping process, but the preparation and interpretation of the financial accounts. For companies, financial accounting also includes the preparation of the annual report (a document sent annually to shareholders, comprising both financial and non-financial information). In orientation, financial accounting is primarily outward-looking and aimed at providing information for external users. However, monthly financial accounts are often prepared and used internally within a business. Within a company, financial accounting is usually carried out by a company's employees. Smaller businesses, such as sole traders, may use professionally-qualified independent accountants.

The Company Camera 1.2

Report of the Independent Auditors

To the Members of the "Shell" Transport and Trading Company, p.l.c.

We have audited the Financial Statements which comprise the Profit and Loss Account, the Statement of Retained Profit, the Balance Sheet, the Statement of Total Recognised Gains and Losses, the Statement of Cash Flows and the related notes. We have also examined the amounts disclosed relating to the emoluments, share options and pensions benefits of the Directors which form part of the Remuneration Report.

Respective responsibilities of Directors and Auditors

The Directors' responsibilities for preparing the Annual Report and the Financial Statements in accordance with applicable United Kingdom law and accounting standards are set out in the statement of Directors' responsibilities.

Our responsibility is to audit the Financial Statements in accordance with relevant legal and regulatory requirements, United Kingdom Auditing Standards issued by the Auditing Practices Board and the Listing Rules of the Financial Services Authority.

We report to you our opinion as to whether the Financial Statements give a true and fair view and are properly prepared in accordance with the Companies Act 1985. We also report to you if, in our opinion, the Directors' report is not consistent with the Financial Statements, if the Company has not kept proper accounting records, if we have not received all the information and explanations we require for our audit, or if information specified by law or the Listing Rules regarding Directors' remuneration and transactions is not disclosed.

We read the other information contained in the Annual Report and consider the implications for our report if we become aware of any apparent misstatements or material inconsistencies with the Financial Statements.

We review whether the Corporate Governance statement reflects the Company's compliance with the seven provisions of the Combined Code specified for our review by the Listing Rules, and we report if it does not. We are not required to consider whether the Board's statements on internal control cover all risks and controls, or to form an opinion on the effectiveness of the Company's corporate governance procedures or its risk and control procedures.

Basis of audit opinion

We conducted our audit in accordance with Auditing Standards issued by the Auditing Practices Board. An audit includes examination, on a test basis, of evidence relevant to the amounts and disclosures in the Financial Statements. It also includes an assessment of the significant estimates and judgements made by the Directors in the preparation of the Financial Statements, and of whether the accounting policies are appropriate to the Company's circumstances, consistently applied and adequately disclosed.

We planned and performed our audit so as to obtain all the information and explanations which we considered necessary in order to provide us with sufficient evidence to give reasonable assurance that the Financial Statements are free from material misstatement, whether caused by fraud or other irregularity or error. In forming our opinion we also evaluated the overall adequacy of the presentation of information in the Financial Statements.

Opinion

In our opinion, the Financial Statements give a true and fair view of the state of the Company's affairs at December 31, 2000 and of its profit and cash flows for the year then ended and have been properly prepared in accordance with the Companies Act 1985

PricewaterhouseCoopers
Chartered Accountants and Registered Auditors
London, March 15, 2001

Source: Shell Transport and Trading plc, Annual Report, 2000, p. 6

Financial Management

An area of growing importance for accountants is financial management. Some aspects of financial management fall under the general heading of management accounting. Financial management, as its name suggests, is about managing the sources of finance of an organisation. It may, therefore, involve managing the working capital (i.e., short-term assets and liabilities) of a company or finding the cheapest form of borrowing. These topics are briefly examined in Chapter 22. There is often a separate department of a company called the financial management or treasury department.

Insolvency

One of the main reasons for the rise to prominence of professional accountants in the UK was to wind up failed businesses. This is still part of a professional accountant's role. Professional accounting firms are often called in to manage the affairs of failed businesses, in particular to pay creditors who are owed money by the business.

Management Accounting

Management accounting covers the internal accounting of an organisation. There are several different areas of management accounting: costing (see Chapter 16), budgeting (see Chapter 17), standard costing (see Chapter 18), short-term decision making (see Chapter 19), strategic management accounting (see Chapter 20), capital investment appraisal (see Chapter 21), and management of working capital and sources of finance (Chapter 22). Essentially, these activities aim to monitor, control and plan the financial activities of organisations. Management uses such information for decisions such as determining a product's selling price or setting the sales budget.

Taxation

Taxation is a complicated area. Professional accountants advise businesses on a whole range of tax issues. Much of this involves tax planning (i.e., minimising the amount of tax that organisations have to pay by taking full advantage of the often complex tax regulations). This tax avoidance which operates within the law should be distinguished from tax evasion which is illegal. Professional accountants may also help individuals with a scourge of modern life, the preparation of their annual income tax assessment.

> *Soundbite 1.3*
>
> **Management Consultancy**
>
> **'[Definition of management consulting] Telling a company what it should already know.'**
>
> *The Economist* (September 12, 1987)
>
> *Source*: *The Wiley Book of Business Quotations* (1998), p. 359

Management Consultancy

Management consultancy is a lucrative source of income for accountants (see Real-Life Nugget 1.2 on the next page). However, as Soundbite 1.3 shows, management consultants are often viewed cynically. Management consultancy embraces a whole range of activities such as special efficiency audits, feasibility studies, and tax advice. Many professional accounting firms now make more money from management consultancy than from auditing. Examples of management consultancy are investigating the feasibility of a new football stadium or the costing of a local authority's school meals proposals.

Real-Life Nugget 1.2

Management Consultancy

CONSULTING: A RICH SOURCE OF BUSINESS (£M)

	Audit/business advisory/tax	Consulting and other services	UK total fee income
PwC*	977.3	611.8	1589.1
KPMG	569.0	298.0	867.0
Ernst & Young	383.8	249.1	633.0
Deloitte & Touche	291.4	271.8	563.2
Arthur Andersen*	356.4	108.6	465.0

Figures based on Accountancy Age Top 50

* estimates based on global figures

Source: *Accountancy Age*, 24 February, 2000, p. 7

Types of Accountant

There are several types of accountant. The most high-profile are those belonging to the six professionally qualified bodies. In addition to these six accountancy bodies, there are other accounting associations, the most important of which is probably the Association of Accounting Technicians.

Professionally Qualified Accountants

Chartered Accountants

There are six institutions of professionally qualified accountants currently operating in the UK (see Figure 1.6 on the next page). All jealously guard their independence and the many attempts to merge over the past few years have all failed (see Real-Life Nugget 1.3).

Real-Life Nugget 1.3

Professional Accountancy Bodies

There have been several attempts to persuade the UK's accountancy bodies to merge over the past few years, all without much success. Six accountancy bodies is rather a lot and the government understandably gets exasperated from time to time by six (and sometimes seven) different responses to a consultation paper. But the bodies' members have consistently refused merger initiatives, always citing differing training requirements as a major consideration – and not without justification.

Source: 'Big Five Pressure Gets Results', Elizabeth Mackay, *Accountancy Age*, 9 March, 2000, p. 18

Figure 1.6 **Main UK Professional Accountancy Bodies**

Body	*Main Activities*
Institute of Chartered Accountants in England and Wales (ICAEW)	Generally auditing, financial accounting, management consultancy, insolvency and tax advice. However, many work in industry.
Institute of Chartered Accountants in Ireland (ICAI)	Similar to ICAEW.
Institute of Chartered Accountants of Scotland (ICAS)	Similar to ICAEW.
Association of Chartered Certified Accountants (ACCA)	Auditing, financial accounting, insolvency, management consultancy, and tax advice. Many train or work in industry.
Chartered Institute of Management Accountants (CIMA)	Management accounting.
Chartered Institute of Public Finance and Accounting (CIPFA)	Accounting within the public sector and privatised industries.

There are three institutes of chartered accountants: the Institute of Chartered Accountants in England and Wales (ICAEW), the Institute of Chartered Accountants in Ireland (ICAI), and the Institute of Chartered Accountants of Scotland (ICAS). The largest of these three is the ICAEW. Its members were once mainly financial accountants and auditors, but now take part in a whole range of activities. Many leave the professional partnerships with which they train to join business organisations. In fact, qualifying as a chartered accountant is often seen as a route into a business career.

Association of Chartered Certified Accountants (ACCA)

The ACCA's members are not so easy to pigeonhole as the other professionally qualified accountants. They work both in public practice as auditors and as financial accountants. They also have an enormous number of overseas students. Many certified accountants train for their qualification in industry and never work in public practices.

Chartered Institute of Management Accountants (CIMA)

This is an important body whose members generally train and work in industry. They are found in almost every industry, ranging, for example, from coal mining to computing. They mainly perform the management accounting function.

Chartered Institute of Public Finance and Accountancy (CIPFA)

This institute is smaller than the ICAEW, ACCA or CIMA. It is also much more specialised with its members typically working in the public sector or the newly privatised industries, such as Railtrack. CIPFA members perform a wide range of financial activities within these organisations, such as budgeting in local government.

Second-Tier Bodies

The main second-tier body in the UK is the Association of Accounting Technicians. This body was set up by the major professional accountancy bodies. Accounting tech-nicians help professional accountants, often doing the more routine bookkeeping and costing activities. Many accounting technicians go on to qualify as professional accountants.

The different accountancy bodies, therefore, all perform different functions. Some work in companies, some in professional accountancy practices, some in the public sector. This diversity is highlighted in an original way in Real-Life Nugget 1.4.

Real-Life Nugget 1.4

A Sideways Look at the Accounting Profession

Thus, to take parallels from the Christian church, we have:

- **the lay priest**: the accountant working for a company;
- **the mendicant priest**: the professional accountant in a partnership;
- **the monastic priest**: the banker, who, while not strictly an accountant, serves much the same ends in a separate and semi-isolated unit;
- **the father confessor**: the auditing accountant to whom everything is (officially) revealed, and who then grants absolution.

Source: Graham Cleverly (1971) *Managers and Magic*, Longman Group Ltd, London, p. 47

Conclusion

Accounting is a key business activity. It provides information about a business so that managers or owners (for example, shareholders) can make business decisions. Accounting provides the framework by which cash and profit can be monitored and controlled. A basic distinction is between financial accounting (accounting targeted primarily at those outside the business, but also useful to managers) and management accounting (providing information solely to managers).

Accounting changes as society changes. In particular, it is contingent upon history, country, technology and the nature and type of the organisation. There are at least eight groups which use accounting information, the main ones being managers and shareholders. These user groups require information about, amongst other things, profits, cash flow, assets and debts. There are several types of accountancy and accountant. The types of accountancy include auditing, bookkeeping, financial accounting, financial management, insolvency, management accounting, taxation and management consultancy. The six UK professional accountancy bodies are the Chartered Association of Certified Accountants, the Chartered Institute of Management Accountants, the Chartered Institute of Public Finance and Accountancy, and the Institute of Chartered Accountants in England and Wales, the Institute of Chartered Accountants in Ireland and the Institute of Chartered Accountants of Scotland.

Discussion *Questions*

Questions with numbers in blue have answers at the back of the book.

Q 1 What is the importance, if any, of accounting?

***Q* 2** Can you think of three business decisions for which managers would need accounting information?

Q 3 What do you consider to be the main differences between financial and management accounting?

***Q* 4** Discuss the idea that as society changes so does accounting.

***Q* 5** 'Managers should only supply financial information to the 'current' shareholders of companies, no other user groups have any rights at all to information, particularly not the general public or government.' Discuss.

Section A

Financial Accounting: The Techniques

In this section, we look at the accounting techniques which underpin the preparation and interpretation of the financial statements. Chapter 2 sets the scene for this section explaining the essential background. It deals with the nature and importance of financial accounting and introduces some of the basic concepts and terminology.

In these initial chapters, we focus primarily on the financial statements of the sole trader as these are the most straightforward. In Chapter 3, the key accounting techniques of the accounting equation, double-entry bookkeeping and the trial balance, are introduced. The double-entry section of this chapter is self-contained and can be passed over by those students not wishing to study bookkeeping in depth. In Chapters 4 and 5, the essential nature, function and contents of the profit and loss account and balance sheet are discussed. This will enable students more fully to appreciate the importance of the profit and loss account and balance sheet before their preparation from the trial balance is explained in Chapter 6. In Chapter 7, we prepare the profit and loss account and balance sheet of partnerships and limited companies. The final two chapters look at the cash flow statement (Chapter 8) and the interpretation of accounts (Chapter 9) primarily from the perspective of limited companies. The cash flow statement is the third most important financial statement for a company. Finally, in Chapter 9 we look at 16 ratios commonly used to assess an organisation's performance.

Chapter 2

The Accounting Background

'Living up to his reputation, he brooks no nonsense, adds no frills. A murmured thank you to the chair, then: "Let us never forget that we are all of us in business for one thing only. To make a profit." The hush breaks, the apprehension goes. Audibly, feet slide forward and chairs ease back. Orthodoxy has been established. The incantation has been spoken. No one is going to be forced to query the framework of his world, to face the terrible question, why?'

Graham Cleverly (1971), *Managers and Magic*, Longman, pp. 25–6

Learning Outcomes

After completing this chapter you should be able to:

- ✔ Explain the nature of financial accounting.
- ✔ Appreciate the basic language of accounting.
- ✔ Identify the major accounting conventions and concepts.

In a Nutshell

- Financial accounting is about providing users with financial information so that they can make decisions.
- Key accounting terminology includes income, expenses, capital, assets and liabilities.
- The three major financial statements are the profit and loss account, the balance sheet and the cash flow statement.
- The most widely agreed objective is to provide information for decision making.
- The most important external users for companies are the shareholders.
- Four major accounting conventions are entity, money measurement, historic cost and periodicity.
- Four major accounting concepts are going concern, matching, consistency and prudence.

Introduction

Financial accounting is the process by which financial information is prepared and then communicated to the users. It is a key element in modern business. For limited companies, it enables the shareholders to receive from the managers an annual set of accounts which have been independently checked by auditors. For sole traders and partnerships, it allows the tax authorities to have a set of accounts which are often prepared by independent accountants. There are thus three parties to the production and dissemination of the financial accounts: the preparers, the users and the independent accountants. The broad objective of financial accounting is to provide information for decision making. Its preparation is governed by basic underpinning principles called accounting conventions and accounting concepts.

Financial Accounting

Once a year company shareholders receive through the post an annual report containing the company's annual accounts. These comprise the key financial statements as well as other financial and non-financial information. Sole traders and partnerships annually prepare a set of financial statements for the tax authorities. Managers will also use these financial statements to evaluate the performance of the business over the past year. The whole process is underpinned by a set of overarching accounting principles (i.e., accounting conventions and accounting concepts) and by detailed accounting measurement and disclosure rules.

Essentially, financial accounting is concerned with providing financial information to users so that they can make decisions. Definition 2.1 provides a more formal definition from the International Accounting Standards Board (a regulatory body which seeks to set accounting standards which will be used worldwide) applicable to all commercial, industrial and business reporting enterprises as well as a fuller definition from the Chartered Institute of Management Accountants.

Definition 2.1

Financial Accounting

Working definition

The provision of financial information to users for decision making.

Formal definitions

'The objective of financial statements is to provide information about the financial position, performance and change in financial position of an enterprise that is useful to a wide range of users in making economic decisions.'

International Accounting Standards Board (2000), *Framework for the Preparation and Presentation of Financial Statements*

'The classification and recording of the monetary transactions of an entity in accordance with established concepts, principles, accounting standards and legal requirements and their presentation, by means of profit and loss accounts, balance sheets and cash flow statements, during and at the end of an accounting period.'

Chartered Institute of Management Accountants (2000), *Official Terminology*

Financial accounting meets the common needs of a wide range of users. It does not, however, provide all the information users may need. An important role of financial accounting is that it shows the stewardship of management (i.e., how successfully they run the company). Shareholders may use this information to decide whether or not to sell their shares.

At a still broader level, accounting allows managers to assess their organisation's performance. It is a way of seeing how well they have done or, as Soundbite 2.1 shows, of keeping the score. The main financial statements for sole traders, partnerships and companies are the profit and loss account and the balance sheet. These contain details of income, expenses, assets, liabilities and capital. For companies (and often for other businesses), these two statements are accompanied by a cash flow statement which summarises a company's cash flows. Generally, the principal user is assumed to be the shareholder. These three statements are sent to shareholders once a year in a document called an annual report. However, managers need more detailed and more frequent information to run a company effectively. Monthly accounts and accounts for different parts of the business are, therefore, often drawn up.

Soundbite 2.1

Money

'Money never meant anything to us. It was just sort of how we kept the score.'

Nelson Bunker Hunt, *Great Business Quotations*, R. Barron and J. Fisk

Source: *The Book of Business Quotations* (1991), p. 155

Pause for Thought 2.1

Annual Financial Accounts

Why do sole traders, partnerships and limited companies produce financial accounts?

There are several reasons. First, those running the business wish to assess their own performance and regular, periodic accounts are a good way to do this. Second, businesses may need to provide third parties with financial information. In the case of sole traders and partnerships, the tax authorities need to assess the business's profits. Bankers may also want regular financial statements if they have loaned money. Similarly, companies are accountable not only to their shareholders, but to other users of accounts.

Language of Accounting

Accounting is a language. As with all languages, it is important to understand the basics. Five basic accounting terms (income, expenses, assets, liabilities and capital) are introduced here as well as the three main financial statements (profit and loss account, balance sheet and cash flow statement). These concepts are explained more fully later in the book.

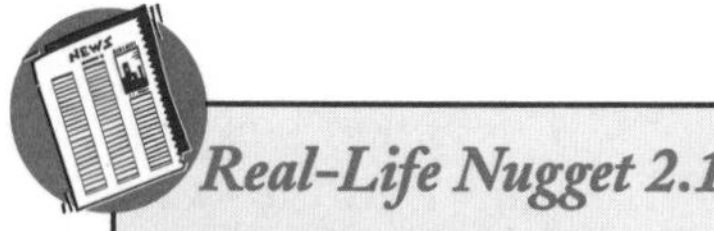

Real-Life Nugget 2.1

Football Club Expenses

Football clubs continue to spend most of their revenue on players' wages, but may end up paying them even more because of increased income from the new television deal.

Though revenue for all 20 Premiership clubs increased by 18% during the 1998-99 season, their total wage bill shot up by 31% to just over £397m. Salary costs now account for more than 59% of the clubs' income, with Manchester United the highest payers.

According to the survey of Premiership finances by the accountants Deloitte and Touche, gate receipts still generate the bulk of club revenue but television money is catching up fast.

Source: V. Chaudhary, TV deal to fuel Premiership wages hike, *The Guardian*, 9 May, 2000, p. 30

Income

Income is essentially the revenue earned by a business. Sales is a good example. Income is income, even if goods and services have been delivered but customers have not yet paid. Income thus differs from cash received.

Expenses

Expenses are the costs incurred in running a business. Examples are telephone, business rates and wages. The nature of expenses varies from business to business. Real-Life Nugget 2.1 shows that the biggest expense for football clubs is their wages. Expenses are expenses, even if goods and services have been consumed but the business has still not paid for them. Expenses are, therefore, different from cash paid.

Assets

Assets are essentially items owned (or leased) by the business which will bring economic benefits. An example might be a building or a stock of goods awaiting resale. Assets may be held for a long time for use in the business (such as motor vehicles) or alternatively be short-term assets (such as stock which is held for immediate resale). As Real-Life Nugget 2.2 shows, the asset base of modern companies is enormous.

Real-Life Nugget 2.2

Assets

New accounting rules behind explosion in UK asset base

By Gavin Hinks

Stricter accounting requirements could lead to the UK's commercial asset base growing by up to 33%, according to a new poll.

The survey, which questioned 300 finance managers, found over 90% of UK companies are predicting larger asset registers – some doubling in size by 2005. One reason for the growth was that auditors had advised finance managers to split major assets into separate accounting elements to cater for different depreciation methods.

It is now calculated that between now and 2005 there are likely to be an additional nine million assets on the registers of the UK's top 5,000 companies.

Mark Johnson, marketing manager at Britannia Software which carried out the survey, said: 'Nine million is a conservative estimate and only takes into account capitalised items.

'You must also take into consideration the potential explosion in non-capitalised or attractive items like low-cost IT equipment, mobile phones and software which many companies want to track and hold on a separate register.'

Source: *Accountancy Age*, 24 February 2000, p.5

Liabilities

Liabilities are amounts the business owes to a third party. An example might be money owed to the bank following a bank loan. Alternatively, the company may owe money to the suppliers of goods (known as creditors).

Capital

Capital represents the owner's interest in the business. In effect, capital is a liability as it is owed by the business to the owner. Owners may be sole traders, partners or shareholders. Capital is the assets of a business less its liabilities to third parties.

Profit and Loss Account

A profit and loss account, at its simplest, records the income and expenses of a business over time. Income less expenses equals profit. By contrast, where expenses are greater than income losses will occur. It is important (as Real-Life Nugget 2.3 shows) for even the world's largest companies to ensure that income (or revenue) exceeds expenses. The net profit (or net loss) in the profit and loss account is added (or subtracted) to capital in the balance sheet.

Real-Life Nugget 2.3

Revenues and Expenses

TROUBLED computer giant Compaq has warned of lower earnings from Europe amid growing concern about its key US direct-sales operation.

The world's second biggest computer company last week surprised markets when it forecast second-quarter losses, which are a result of higher expenses and lower-than-expected revenues.

Compaq is predicting losses of 15 cents a share, or $237m, the first quarterly loss in eight years.

Source: Compaq's troubles mount, *Sunday Business*, 20 June 1999, p. 15

Pause for Thought 2.2

Accounting Terms

Can you think of two examples of income, and three examples of expenses, assets and liabilities which a typical business might have?

These are many and varied. A few examples are given below:

Income	*Expenses*	*Assets*	*Liabilities*
Sales of goods	Telephone	Buildings	Bank loan
Sales of assets	Business rates	Motor cars	Creditors
Bank interest earned	Electricity	Furniture	Bills owing
	Repairs	Stock	
	Petrol consumed	Debtors	

The Company Camera 2.1 shows a reconstruction of the summary profit and loss account of Marks and Spencer plc. The actual profit and loss account is given in Appendix 2.1 on page 36. From The Company Camera 2.1 it can be seen that expenses are deducted from sales and other income to give profit before taxation. Once taxation and dividends have been deducted, we arrive at the retained profit (£407 million for 1998) or the retained loss (£41 million for 1999). The 1999 accounts were used as they are easier to understand than the 2000 accounts.

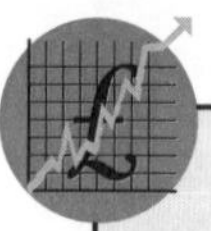

The Company Camera 2.1

Marks and Spencer Summarised Company Profit and Loss Account

Marks and Spencer
Summary profit and loss account for the year ended 31 March, 1999

	1999 £m	1998 £m
Sales	8,224	8,243
Add: Other income	34	104
	8,258	8,347
Less: Expenses	7,712	7,192
Profit before taxation	546	1,155
Taxation	(176)	(339)
Profit after taxation	370	816
Other	2	–
Dividends	(413)	(409)
Retained (loss)/Profit	(41)	407

Note: The profit and loss account has been simplified and reconstructed. The original summary profit and loss account can be found as Appendix 2.1 at the back of this chapter.

Source: Marks and Spencer, 1999, Annual Review and Summary Financial Statement, p. 20.

Balance Sheet

A balance sheet records the assets, liabilities and capital of a business at a certain point in time. Assets less liabilities will equal capital. Capital is thus the owners' interest in the business. The summary balance sheet for Marks and Spencer for 2000 is presented in The Company Camera 2.2. Here the assets are added together and then the liabilities are taken away. The net assets (i.e., assets less liabilities) equals the total capital employed by the business.

The Company Camera 2.2

Illustration of a Summarised Company Balance Sheet

SUMMARY BALANCE SHEET
AT 31 MARCH 2000

	2000	1999
	£m	£m
FIXED ASSETS		
Tangible and intangible assets	**4,243.4**	4,387.5
Investments	**55.0**	61.2
	4,298.4	4,448.7
CURRENT ASSETS		
Stocks	**474.4**	514.7
Debtors	**2,555.2**	2,355.7
Cash and investments	**687.5**	485.5
	3,717.1	3,355.9
CURRENT LIABILITIES		
Creditors: amounts falling due within one year	**2,162.8**	2,029.8
NET CURRENT ASSETS	**1,554.3**	1,326.1
TOTAL ASSETS LESS CURRENT LIABILITIES	**5,852.7**	5,774.8
Creditors: amounts falling due after more than one year	**804.3**	772.6
Provisions for liabilities and charges	**126.6**	105.0
NET ASSETS	**4,921.8**	4,897.2
SHAREHOLDERS' FUNDS (all equity)	**4,905.3**	4,883.9
Minority interests (all equity)	**16.5**	13.3
TOTAL CAPITAL EMPLOYED	**4,921.8**	4,897.2

Source: Marks and Spencer, 2000, Annual Review and Summary Financial Statement, p. 30

Cash Flow Statement

A cash flow statement shows the cash inflows and outflows of the business. The Company Camera 2.3 shows a summary cash flow statement for Marks and Spencer for 2000. All the cash flows from the day-to-day operations of the business (i.e., operating activities such as buying and selling goods or paying wages) are recorded. This gives a cash inflow from operating activities. Other cash flows (such as taxation and the payment of dividends) are then listed. Overall, Marks and Spencer has a cash outflow before funding (from external loans and so on) of £884 million in 1999 and £91 million in 2000.

The Company Camera 2.3

Illustration of a Summarised Company Cash Flow Statement

SUMMARY CASH FLOW INFORMATION
FOR THE YEAR ENDED 31 MARCH 2000

	2000 £m	1999 £m
OPERATING ACTIVITIES		
Received from customers	**7,989.9**	7,884.1
Payments to suppliers	**(5,357.1)**	(5,464.2)
Payments to and on behalf of employees	**(1,138.3)**	(1,153.9)
Other payments	**(803.8)**	(793.1)
Exceptional operating cash flows	**(49.2)**	(0.6)
CASH INFLOW FROM OPERATING ACTIVITIES	**641.5**	472.3
Returns on investments and servicing of finance	**15.2**	29.0
Taxation	**(145.7)**	(345.9)
Capital expenditure and financial investment	**(167.0)**	(628.1)
Acquisitions and disposals	**(21.1)**	1.0
Equity dividends paid	**(413.5)**	(412.6)
CASH OUTFLOW BEFORE FUNDING	**(90.6)**	(884.3)

Source: Marks and Spencer, 2000, Annual Review and Summary Financial Statement, p. 31

We will now look at three summary financial statements for a business called Gavin Stevens which we will meet in more depth in Chapter 3. Gavin Stevens runs a hotel and summary details of his income, expenses, assets, liabilities and capital are given below. At this stage, the financial statements for Gavin Stevens (Figure 2.1 below) and for Simon Tudent (Figure 2.2 on page 28) are drawn up using only broad general headings. More detailed presentation is covered in later chapters.

Figure 2.1 Preparation of Summary Profit and Loss Account, Balance Sheet and Cash Flow Statement for Gavin Stevens

Financial Information

Income	£8,930	Liabilities	£1,350
Expenses	£5,600	Opening capital	£200,000
Assets	£204,680	Closing capital	£203,330
Cash inflows	£204,465	Cash outflows	£117,550

(i) Profit and Loss Account

Here we are concerned with determining profit by subtracting expenses from income. We call the profit, net profit.

	£
Income	8,930
Less: *Expenses*	5,600
Net Profit	3,330

(ii) Balance Sheet

Here we deduct the assets from the liabilities to give net assets. This represents the capital employed in the business.

	£
Assets	204,680
Liabilities	(1,350)
Net Assets	203,330

	£
Opening capital employed	200,000
Add: Profit	3,330
Closing capital employed	203,330

(iii) Cash Flow Statement	£
Cash Inflows	204,465
Cash Outflows	(117,550)
Net cash inflow	86,915

We have a positive cash flow. In other words, our cash has increased by £86,915 over the year. Note that in accounting when figures such as liabilities and cash outflows are subtracted it is common to use brackets if the word 'less' is not used.

Student Example

In order to give a further flavour of the nature of the main accounting terms, this section presents the income and expenditure of Simon Tudent. Simon is a student who has just completed his first year at university.

Figure 2.2 A Student's Financial Statements

Simon has collected the following details of his finances for his last student year and possessions on 31 December. He has already divided them into income, expenses, assets and liabilities.

	£
Income	
Wages received from working in Student Union (38 weeks at £25 per week)	950
Wages owing from Student Union (2 weeks at £25 per week – also an asset because he is owed it)	50
Expenses	
Hall of residence fees	2,300
Money spent on books	200
Money spent on entertainment	500
Petrol used in car	300
Phone calls	50
Car repairs	200
General	150

Assets
Second-hand car worth probably £700*
Cash at bank at start of year £2,500*
Computer worth about £150*
CD player worth about £200*
* All possessions at start of year.

Liabilities
Owes parents £150 for loan this year
Student loan from Government £3,360

Note: Simon's opening capital is simply his opening assets less any opening liabilities. For simplicity, we assume that they are worth the same as at the start and as at the end of the year (except for cash at bank).

From the information shown above, we can present three financial statements, shown below and on the next page.

1. A profit and loss account (strictly, for a student we should call this an income and expenditure statement).
2. A balance sheet (strictly, for a student we should call this an assets and liabilities statement).
3. A cash flow statement.

S. Tudent

Profit and Loss Account (Income and Expenditure) Year Ended 31 December

Income	£	£
Student union wages (note: includes £50 owing)		1,000
Less *Expenses*		
Hall of residence fees	2,300	
Books	200	
Entertainment	500	
Petrol	300	
Phone calls	50	
Car repairs	200	
General	150	3,700
Net Deficit		(2,700)

Note that this statement deals with all income and expenses *earned* and *incurred*, not just with cash paid and received. We deduct all the expenses from the income and ascertain that S. Tudent has a net deficit. In business, this would be called a net loss.

Figure 2.2 A Student's Financial Statements (continued)

S. Tudent

Balance Sheet (Assets and Liabilities) as at 31 December

Assets	£	£
Cash at bank		
(balance from cash flow statement)		3,260
Second-hand car		700
Computer		150
CD player		200
Owed by student union		50
		4,360
Liabilities		
Parental loan	(150)	
Government loan	(3,360)	(3,510)
Net assets		850
Capital Employed		£
Opening capital employed		3,550*
Net deficit		(2,700)
Closing capital employed		850

Note: These two figures balance

*Opening possessions (£700 + £2,500 + £150 + £200)

We are simply listing the assets and liabilities. The assets less liabilities gives net assets. This also equals capital employed. The opening capital is simply opening assets less opening liabilities. Note that opening capital employed less the net deficit gives closing capital employed. A student loan is a liability because it is owed to the government. It is not income.

S. Tudent

Cash Flow Statement Year Ended 31 December

	£	£
Bank balance at start of year		2,500
Add *Receipts*:		
Student loan	3,360	
Student Union	950	
Loan from parents	150	4,460
		6,960
Less *Payments*:		
Books	200	
Entertainment	500	
Petrol used	300	
Phone calls	50	
Car repairs	200	
General	150	
Hall fees	2,300	3,700
Bank balance at end of year (balancing figure)		3,260

Note that we are simply recording all cash received and paid. We were not given the closing bank figure. However, it must be £3,260: opening cash of £2,500 plus receipts of £4,460 gives £6,960 less £3,700 payments.

Pause for Thought 2.3

Student Loan

Students are often granted loans by the government. Why would a student loan from the government or a loan from one's parents be a liability, but a gift from parents be income?

This is because the loans must be repaid. They are, therefore, liabilities. When they are repaid, in part or in full, the liability is reduced. By contrast, a gift will be income as it does not have to be repaid.

Why is Financial Accounting Important?

Financial accounting is a key control mechanism. All businesses prepare and use financial information in order to help them measure their performance. It is also useful in a business's relationship with third parties. It enables sole traders and partnerships to provide accounting information to the tax authorities or bank. For the limited company, it makes company directors accountable to company shareholders. For small businesses, the accounts are normally prepared by independent qualified accountants. In the case of large businesses, such as limited companies, the accounts are normally prepared by the managers and directors, but then audited by professional accountants. Auditing means checking the accounts are 'true and fair'. This term 'true and fair' is elusive and slippery. It is probably best considered to mean faithfully representing the underlying economic transactions of a business.

The independent preparation and/or auditing of the financial accounts by accountants and auditors is an essential task in the protection of the users. The tax authorities need to ensure that the sole traders and partnership accounts have been properly prepared by an expert. Similarly, the shareholders need to have confidence that the managers have prepared 'true and fair' accounts. For shareholders, this is particularly important as they are not directly involved in running the business. They provide the money, then stand back and allow the managers to run the company. So how can shareholders ensure that the managers are not abusing their trust? Bluntly, how can the shareholders make sure they are not being 'ripped off' by the managers. Auditing is one solution.

Pause for Thought 2.4

Directors' Self-Interest

How might the directors of a company serve their own interests rather than the interests of the shareholders?

Both directors and shareholders want to share in a business's success. Directors are rewarded by salaries and other rewards, such as company cars, profit-related bonuses, or lucrative pensions. Shareholders are rewarded by receiving cash payments in the form of dividends or an increase in share price. The problem is that the more the directors take for themselves, the less there will be left for the shareholders. So if directors pay themselves large bonuses, the shareholders will get smaller dividends.

Accounting Principles

There are several accounting principles which underpin the preparation of the accounts. For convenience, we classify them here into accounting conventions and accounting concepts (see Figure 2.3). Essentially, conventions concern the whole accounting process, while concepts are assumptions which underpin the actual accounts preparation. There are four generally recognised accounting conventions and four generally recognised accounting concepts.

Figure 2.3 **Accounting Principles**

Accounting Conventions	*Accounting Concepts*
• Entity • Money measurement • Historic cost • Periodicity	• Going concern • Matching (or accruals) • Consistency • Prudence

Accounting Conventions

Entity

The entity convention simply means that a business has a distinct and separate identity from its owners. This is fairly obvious in the case of a large limited company where shareholders own the company and managers manage the company. However, for a sole trader, such as a small baker's shop, it is important to realise that there is a theoretical distinction between personal and business assets. The business is treated as a separate entity from the owner. The business's assets less third party liabilities represent the owners' capital.

Monetary Measurement

Under this convention only items which can be measured in financial terms (for example, in pounds or dollars) are included in the accounts. If a company pollutes the atmosphere this is not included in the accounts, since this pollution has no measurable financial value. However, a fine imposed for pollution is measurable and should be included in the accounts.

Historical Cost

Businesses may trade for many years. The historical cost convention basically states that the amount recorded in the accounts will be the *original* amount paid for a good or service. In some countries, such as the US, the historical cost convention is still very closely followed. Nowadays, in the UK, there are some departures. For example, many companies revalue land and buildings. This reflects the fact that they have increased in value. The Company Camera 2.4 shows that Tesco plc prepared its accounts using the historical cost convention.

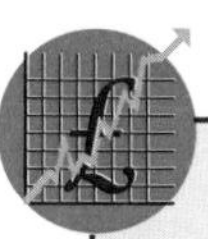

The Company Camera 2.4

Historical Cost Convention

Basis of financial statements

These financial statements have been prepared under the historical cost convention, in accordance with applicable accounting standards and the Companies Act 1985.

Source: Tesco plc, 2001 Annual Report, p. 22

Periodicity

This simply means that accounts are prepared for a set period of time. Audited financial statements are usually prepared for a year. Financial statements prepared for internal management are often drawn up more frequently. This means, in effect, that sometimes rather arbitrary distinctions are made about the period in which accounting items are recorded.

Accounting Concepts

There are four generally recognised accounting concepts.

Going Concern

This concept assumes the business will continue into the foreseeable future. Assets, liabilities, income and expenses are thus calculated on this basis. If you are valuing a specialised machine, for example, you will value the machine at a higher value if the business is ongoing than if it is about to go bankrupt. If it were bankrupt the machine would only have scrap value. In Company Camera 2.5 on the next page, we can see that J.D. Wetherspoon's directors have assured themselves that the company is a going concern.

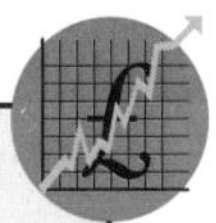

The Company Camera 2.5

Going Concern Concept

The Directors have made enquiries into the adequacy of the company's financial resources through a review of the company's budget and medium-term financial plan, which includes capital expenditure plans and cash flow forecasts, and have satisfied themselves that the company will continue in operational existence for the foreseeable future. For this reason, they continue to adopt the going concern basis in preparing the company's financial statements.

Source: J.D. Wetherspoon plc, 2000 Annual Report, p. 22

Matching

The matching concept (often known as the accruals concept) recognises income and expenses when they are accrued (i.e., earned or incurred rather than when the money is received or paid). Income is matched with any associated expenses to determine the appropriate profit or loss. A telephone bill owing at the accounting year end is thus treated as a cost for this year even if it is paid in the next year. If the telephone bill is not received by the year end, then the amount of telephone calls will be estimated.

Consistency

This concept states that similar items will be treated similarly from year to year. Thus consistency attempts to stop companies choosing different accounting policies in different years. If they do this, then it becomes more difficult to compare the results of one year to the next.

Prudence

This is the most contentious of the four accounting concepts. Prudence introduces an element of caution into accounting. Income and profits should only be recorded in the books when they are *certain* to result in an inflow of cash. By contrast, provisions or liabilities should be made *as soon as they are recognised*, even though their amount may not be known with certainty. Prudence is contentious because it introduces an asymmetry into the accounting process. Potential incomes are treated differently from potential liabilities.

Pause for Thought 2.5

Personal Finances

Draw up a set of financial statements for the last twelve months.
I hope they are not too gruesome!

Conclusion

Financial accounting, along with management accounting, is one of the two main branches of accounting. Its main objective is to provide financial information to users for decision making. Shareholders, for example, are provided with information to assess the stewardship of managers so that they can then make decisions such as whether to buy or sell their shares. Understanding the accounting language is a key requisite to understanding accounting itself. Four accounting conventions (entity, money measurement, historical cost and periodicity) and four accounting concepts (going concern, matching, consistency and prudence) underpin financial accounting.

Discussion *Questions*

Questions with numbers in blue have answers at the back of the book.

Q 1 What is financial accounting and why is its study important?

Q 2 'The objective of **financial statements** is to provide information about the **financial position, performance** and **change in financial position** of an enterprise that is useful to a wide range of **users** in making **economic decisions**.'

Discuss the key aspects *highlighted in bold* of this formal definition of the Objectives of Financial Statements as formulated by the International Accounting Standards Board in its *Framework for the Preparation and Presentation of Financial Statements*.

Q 3 Sole traders, partnerships and limited companies all have different users who need financial information for different purposes. Discuss.

Q 4 Classify the following as an income, an expense, an asset, or a liability:

(a) Friend owes business money
(b) Football club's gate receipts
(c) Petrol used by a car
(d) Photocopier
(e) Sales
(f) Telephone bill outstanding
(g) Long-term loan
(h) Cash
(i) Wages
(j) Capital

Q 5 State whether the following are true or false. If false, explain why.

(a) Assets and liabilities show how much the business owns and owes.
(b) The profit and loss account shows the income, expenses and thus the net assets of a business.

(c) Stewardship is now recognised as the primary objective of financial accounting.
(d) When running a small business the owner must be careful to separate business from private expenditure.
(e) The matching and prudence accounting concepts sometimes conflict.

Numerical *Questions*

Questions with numbers in blue have answers at the back of the book.

*Q*1 Sharon Taylor has the following financial details:

Sales	£8,000	Assets	£15,000
General expenses	£4,000	Liabilities	£3,000
Trading expenses	£3,000	Cash outflows	£12,000
Cash inflows	£10,000	Closing capital employed	£12,000
Opening capital employed	£11,000		

Required: Prepare Sharon Taylor's
(a) Profit and loss account
(b) Balance sheet
(c) Cash flow statement

*Q***2** Priya Patel is an overseas student studying at a British University. Priya has the following financial details:

	£		£
Tuition fees paid	6,840	Food paid	550
Hall of residence fees paid	2,000	General expenses paid	180
Money spent on books	160	Parental loan of £2,000 during year	
Money spent on entertainment	500		
Money earned at arts and crafts bazaar	1,800	*State of affairs at the start of the year:*	
		Cash at bank	8,600
Phone calls paid	100	Music system*	200

* Still worth £200 at end of year.

Required: Priya's:
(a) Profit and loss account (income and expenditure account)
(b) Balance sheet (assets and liabilities)
(c) Cash flow statement

Appendix 2.1: Illustration of a Summarised Company Balance Sheet

SUMMARY PROFIT AND LOSS ACCOUNT
FOR THE YEAR ENDED 31 MARCH 1999

	1999	1998
		As restated
	£m	£m
TURNOVER	**8,224.0**	8.243.3
OPERATING PROFIT		
Before exceptional operating (charges)/income	**600.5**	1,050.5
Exceptional operating (charges)/income	**(88.5)**	53.2
TOTAL OPERATING PROFIT	**512.0**	1,103.7
Profit/(loss) on sale of property and other fixed assets	**6.2**	(2.8)
Net interest income	**27.9**	54.1
PROFIT ON ORDINARY ACTIVITIES BEFORE TAXATION	**546.1**	1,155.0
Analysed between:		
Profit on ordinary activities before taxation and exceptional items	**634.6**	1,101.8
Exceptional operating (charges)/income	**(88.5)**	53.2
Taxation on ordinary activities	**(176.1)**	(338.7)
PROFIT ON ORDINARY ACTIVITIES AFTER TAXATION	**370.0**	816.3
Minority interest (all equity)	**2.1**	(0.4)
PROFIT ATTRIBUTABLE TO SHAREHOLDER	**372.1**	815.9
Dividends	**(413.3)**	(409.1)
RETAINED (LOSS)/PROFIT FOR THE YEAR	**(41.2)**	406.8
BASIC EARNINGS PER SHARE	**13.0p**	28.6p
ADJUSTED EARNINGS PER SHARE	**15.8p**	27.3p
DIVIDENDS PER SHARE	**14.4p**	14.3p

Source: Marks and Spencer, 1999 Annual Review and Summary Financial Statement, p. 20

'Old accountants never die, they just lose their balance'

Anon

Chapter 3

Recording: Double-Entry Bookkeeping

Learning Outcomes

After completing this chapter you should be able to:

- ✔ **Outline the accounting equation.**
- ✔ **Understand double-entry bookkeeping.**
- ✔ **Record transactions using double-entry bookkeeping.**
- ✔ **Balance off the accounts and draw up a trial balance.**

In a Nutshell

- Double-entry bookkeeping is an essential underpinning of financial accounting.
- The accounting equation provides the structure for double-entry.
- Assets and expenses are increases in debits recorded on the left-hand side of the 'T' (i.e., ledger) account.
- Income and liabilities are increases in credits recorded on the right-hand side of the 'T' (i.e., ledger) account.
- Debits and credits are equal and opposite entries.
- Initial recording in the books of account using double-entry, balancing off and preparing the trial balance are the three major steps in double-entry bookkeeping.
- The trial balance is a listing of all the balances from the accounts.
- The debits and credits in a trial balance should balance.

Real-Life Nugget 3.1

Kissing Your Accountant

It has been a week for displays of affection in the market. The darlings of the entertainment industry have long been accustomed to hugging and kissing in public – anyone who has watched the Oscars ceremony knows this only too well. Now, this tendency towards overtly physical contact has spread into the financial world. The latest way to convey to the market that a deal is truly wonderful is for the leading figures to share a moment of intimacy.

The chief executives of Time Warner and AOL were photographed cuddling up after announcing their tie up. Chris Evans was even seen planting a smacker on his accountant after pocketing a cool £75m, when Scottish Media Group bought out Ginger Productions. Of course, if I had made £75m I might even be tempted to kiss my accountant – despite the fact that he still hasn't completed my income tax return and the January 31 deadline is getting ever closer.

Source: James Montier, Showbiz values come to the City, *The Guardian*, January 15, 2000, p. 31

Introduction

Accounting is a blend of theory and practice. One of the key elements to understanding the practice of accounting is double-entry bookkeeping (see Definition 3.1). Although mysterious to the uninitiated, double-entry bookkeeping, in fact, is a mechanical exercise. It is a way of systematically recording accounting transactions into an organisation's accounting books.

Definition 3.1

Double-Entry Bookkeeping

Working definition
A way of systematically recording the financial transactions of a company so that each transaction is recorded twice.

Formal definition
'The most commonly used system of bookkeeping based on the principle that every financial transaction involves the simultaneous receiving and giving of value, and is therefore recorded twice.'

Chartered Institute of Management Accountants (2000), *Official Terminology*.

In a way it is like a business diary where all the financial transactions are recorded. Therefore, in essence, double-entry bookkeeping is the systematic recording of income, expenses, assets, liabilities and capital. Its importance lies in the fact that the profit and loss account and balance sheet are prepared only after the accounting transactions have been recorded.

The Accounting Equation

The reason why each transaction is recorded twice in the accounting books is not that accountants like extra work. In effect, it is a method of checking that the entries have been made correctly. At the heart of the double-entry system is the accounting equation. This starts from the basic premise that assets equal liabilities. It then logically builds up in complexity, as follows:

Step 1. **Assets = Liabilities**

If there is an asset of £1, then somebody must be owed £1. This can either be a third party (such as the bank) or the owner of the business. There is thus a basic equality. For every asset, there is a liability.

Step 2. It should be appreciated that capital is a distinct type of liability because it is owed to the owner of a business. If we expand our accounting equation to formally distinguish between third party liabilities and capital we now have:

Assets = Liabilities + Capital

Pause for Thought 3.1

The Accounting Equation

John decides to start a business and puts £10,000 into a business bank account. He also borrow £5,000 from the bank. How does this obey the accounting equation?

The asset here is easy. It is £15,000 cash. There is also clearly a £5,000 liability to the bank. However, the remaining £10,000, at first glance, is more elusive. A liability does, however, exist. This is because of the entity concept where the business and John are treated as different entities. Thus, the business owes John £10,000. We therefore have:

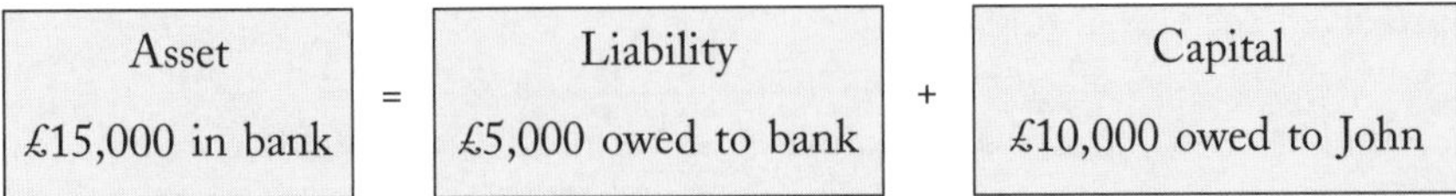

Where a business has a liability to its owner, this is known as capital.

In a sole trader, the capital is initially invested by the owner. In a company, like Manchester United (see The Company Camera 3.1), the capital will have been invested by shareholders.

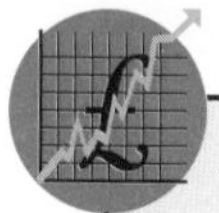

The Company Camera 3.1

Share Capital

Share capital

	Company	
	2000	1999
	£'000	£'000
Authorised:		
350,000,000 ordinary shares of 10p each	**35,000**	35,000
Allotted, called up and fully paid:		
259,768,040 ordinary shares of 10p each	**25,977**	25,977

Note: Authorised share capital is the amount Manchester United is allowed to issue; allotted, called up and fully paid share capital is that actually issued.

Source: Manchester United plc, 2000 Annual Report, p. 44, note to the accounts 24

Step 3. **Assets = Liabilities + Capital + Profit**

When an organisation earns a profit its assets increase.

Profit is on the same side as liabilities because profit is owed to the owner. Profit thus also increases the owner's share in the business.

If a loss is made assets will decrease, but the principle of equality still holds.

Step 4. **Assets = Liabilities + Capital + (Income – Expenses)**

All we have done is broken down profit into its constituent parts (i.e., income less expenses). We have still maintained the basic equality.

Step 5. **Assets + Expenses = Liabilities + Capital + Income**

We have now rearranged the accounting equation by adding expenses to assets. We have still preserved the accounting equation.

Step 6. In accounting terms, **assets** and **expenses** are recorded using **debit entries** and **income**, **liabilities** and **capital** are recorded using **credit entries**. Each page of each book of account has a debit side (left-hand side) and a credit side (right-hand side). This division of the page is called a 'T' account, with debits being on the left and credits on the right. Thus:

'T' Account (ledger account)	
Assets and expenses on the left-hand side	Incomes, liabilities and capital on the right-hand side
DEBIT	CREDIT

Pause for Thought 3.2

Illustration of Accounting Equation

A firm starts the year with £10,000 assets and £10,000 liabilities (£6,000 third party liabilities and £4,000 owner's capital). During the year, the firm makes £5,000 profit (£9,000 income, £4,000 expenses). Show how the accounting equation works.

	Accounting equation	*Transaction*
(1)	Assets = Liabilities	£10,000 = £10,000
(2)	Assets = Liabilities + Capital	£10,000 = £6,000 + £4,000
(3)	Assets = Liabilities + Capital + Profit	£15,000 = £6,000 + £4,000 + £5,000
(4)	Assets = Liabilities + Capital + (Income − Expenses)	£15,000 = £6,000 + £4,000 + (£9,000 − £4,000)
(5)	Assets + Expenses = Liabilities + Capital + Income	£15,000 + £4,000 = £6,000 + £4,000 + £9,000

'T' Account		'T' Account	
Assets + Expenses	Liabilities + Capital + Income	£15,000 + £4,000	£6,000 + £4,000 + £9,000

The 'T' account is central to the concept of double-entry bookkeeping. In turn, double-entry bookkeeping is the backbone of financial accounting. As Helpnote 3.1 below shows, it is underpinned by three major rules.

Helpnote 3.1

Basic Rules of Double-Entry Bookkeeping

1 For every transaction, there must be a **debit and a credit entry**.
2 These debit and credit entries are **equal** and **opposite**.
3 In the **cash book** all accounts **paid in** are recorded on the **debit** side, whereas all amounts **paid out** are recorded on the **credit** side.

In practice, there are many types of asset, liability, capital, income and expense. Figure 3.1 provides a brief summary of some of these.

Figure 3.1 Summary of Some of the Major Types of Assets, Liabilities and Capital, Income and Expenses

Four Major Types of Items

1. **Assets**
2. **Liabilities and capital**
3. **Income**
4. **Expenses**

---

1. **Assets**
Essentially items owned or leased by a business which will bring economic benefits. Two main sorts of tangible (i.e., assets with a physical existence):

I. ***Fixed assets***
These are infrastructure assets *not* used in day-to-day trading. They are assets in use usually over a long period of time.
- i. Motor vehicles
- ii. Land and buildings
- iii. Fixtures and fittings
- iv. Plant and machinery

II. ***Current assets***
These are assets used in day-to-day trading
- i. Stock
- ii. Debtors
- iii. Cash

2. **Liabilities**
Essentially these can be divided into:
I. Short-term and long-term third party liabilities; and
II. Capital which is a liability owed by the business to the owner.

I. ***Third party liabilities***

(a) ***Short-term***
- (i) Creditors
- (ii) Bank overdraft
- (iii) Proposed dividends (companies only)
- (iv) Proposed taxation (companies only)

(b) ***Long-term***
- (i) Bank loan repayable after several years
- (ii) Mortgage loan

II. ***Capital***
Capital is a liability because the business owes it to the owner. Owner's capital is increased by profit, but reduced by losses.

3. **Income**
This is the day-to-day revenue earned by the business, e.g. sales.

4. **Expenses**
These are the day-to-day costs of running a business, e.g. rent and rates, electricity, wages.

Figure 3.1 does not provide an exhaustive list of all assets and liabilities. For instance, it only deals with tangible assets (literally assets you can touch). It thus ignores intangible assets (literally assets you cannot touch) such as royalties or goodwill. However, for now, this provides a useful framework. Intangible assets are most often found in the accounts of companies and are discussed later. More detail on the individual items in Figure 3.1 is provided in later chapters.

Pause for Thought 3.3

Debits and Credits

What do the terms 'debit' and 'credit' actually mean?

Debit and credit have their origins in Latin terms (*debeo*, I owe) and (*credo*, I make a loan). Debtor (one who owes, i.e., a customer) and creditor (one who is owed, i.e., a supplier) have the same origins. Over time, these terms have changed so that nowadays perhaps we have the following:

Debit = An entry on the left-hand side of a 'T' account. Records principally increases in either assets or expenses. However, may also record decreases in liabilities, capital or income.

Credit = An entry on the right-hand side of a 'T' account. Records principally increases in liabilities, capital or income. However, may also record decreases in assets or expenses.

HEALTH WARNING

Those students not wishing to gain an in-depth knowledge of double-entry bookkeeping can miss out pages 43–54.

Worked Example

If we now look at an example. Gavin Stevens has decided to open a hotel to cater for conferences and large functions.

1 January	G. Stevens invests £200,000 capital into a business bank account.	
2 January	Buys and pays for a hotel	£110,000
	Buys and pays for a second-hand delivery van	£3,000
	Buys cash purchases	£2,000
	Buys credit purchases from Hogen	£1,000
	Buys credit purchases from Lewis	£2,000
3 January	Returns goods costing £500 to Hogen	

3 January	Credit sales to Ireton £4,000 for a large garden party Credit sales to Hepworth £5,000 for a business conference	
4 January	Pays electricity bill Pays wages	£300 £1,000
5 January	Ireton returns a crate of wine costing £70 to G. Stevens	
7 January	G. Stevens pays half of the bills outstanding and half of the debtors pay him	

It is now time to enter the transactions for Gavin Stevens into the books of account. This will be done in the next section. As Figure 3.2 shows, there are three main parts to recording the transactions (recording, balancing off and the trial balance).

The bookkeeping role is an essential function of an accountant. It is a precursor of the arguably more challenging job of analysing and interpreting the information (see Real-Life Nugget 3.2 below).

Figure 3.2 Recording the Transactions

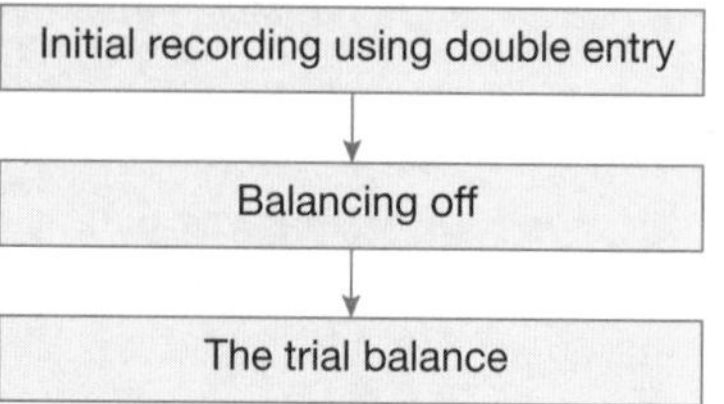

Real-Life Nugget 3.2

Keepers of the Books

As Keeper of the Books – a traditional priestly title – the accountant has a rational and useful role. But the accountant who sticks to it is a rare fish. Certainly the officers of the established institutions would claim that it was a minor part of their usefulness. After all, it doesn't offer much in the way of power or esteem to an ambitious or intelligent man – and Valerie Barden's studies showed accountants on the whole as more intelligent than the average manager.

Source: Graham Cleverly (1971), *Managers and Magic*, Longman, London, p. 39

Step 1 The Initial Recording Using Double-Entry Bookkeeping

The individual amounts are directly entered into two different ledger accounts, **one on each side**. Thus, on the 1st January we credit the capital account with £200,000, but debit the bank account with £200,000. This represents the initial £200,000 capital invested. We record three items (1) date of transaction, (2) **account** which is **equal and opposite** to complete the double-entry and (3) amount. Each 'T' account (i.e., ledger page in the books of account) is treated separately. So that the transactions are easier to follow, each separate transaction is given a letter. Thus, the first transaction (investing £200,000 capital) is recorded as A on the credit (i.e. right-hand side) in the capital account and also as A on the debit (i.e. left-hand side) in the bank account (*see Figure 3.3 on oppposite page*).

It must be remembered that all organisations will structure their accounts books in slightly different ways. In particular, very small businesses may not keep any proper books of accounts, just filing the original invoices and passing them on to their accountants. However, bigger businesses will keep day books (such as the sales day book and purchases day book) in which they will list their credit sales and credit purchases. Double-entry using day books is considered too complex for this book. Interested readers are referred to Alan Sangster and Frank Wood's *Business Accounting*.

Figure 3.3 Recording Gavin Steven's Entries Using Double-Entry Bookkeeping

Capital

	£			£	
			1 Jan. Bank	200,000	A

Bank

	£			£	
1 Jan. Capital	200,000	A	2 Jan. Hotel	110,000	B
7 Jan. Ireton	1,965	M	2 Jan. Van	3,000	C
7 Jan. Hepworth	2,500	N	2 Jan. Purchases	2,000	D
			4 Jan. Electricity	300	J
			4 Jan. Wages	1,000	K
			7 Jan. Hogen	250	O
			7 Jan. Lewis	1,000	P

Sales

	£			£	
			3 Jan. Ireton	4,000	H
			3 Jan. Hepworth	5,000	I

Purchases

	£			£	
2 Jan. Bank	2,000	D			
2 Jan. Hogen	1,000	E			
2 Jan. Lewis	2,000	F			

Sales Returns

	£			£	
5 Jan. Ireton	70	L			

Purchases Returns

	£			£	
			3 Jan. Hogen	500	G

Electricity

	£			£	
4 Jan. Bank	300	J			

Wages

	£			£	
4 Jan. Bank	1,000	K			

Hotel

	£			£	
2 Jan. Bank	110,000	B			

Van

	£			£	
2 Jan. Bank	3,000	C			

Ireton (debtor)

	£			£	
3 Jan. Sales	4,000	H	5 Jan. Sales returns	70	L
			7 Jan. Bank	1,965	M

Hepworth (debtor)

	£			£	
3 Jan. Sales	5,000	I	7 Jan. Bank	2,500	N

Hogen (creditor)

	£			£	
3 Jan. Purchases returns	500	G	2 Jan. Purchases	1,000	E
7 Jan. Bank	250	O			

Lewis (creditor)

	£			£	
7 Jan. Bank	1,000	P	2 Jan. Purchases	2,000	F

Helpnote 3.2

Note that each entry in Figure 3.3 appears on both sides (i.e., as a debit and a credit, equal and opposite) of different accounts. For example, capital of £200,000 is a credit in the capital account of £200,000 and a debit in the bank account of £200,000. Each account represents one page. So, for example, the sales account has nothing on the left-hand side of the page.

In most businesses, the initial transactions are now recorded using a computer system. The individual ledger accounts are then stored in the computer. However, it is necessary to appreciate the underlying processes involved. These are now explained, both for Gavin Stevens and more generally.

If we look at the double-entry in terms of debit and credit for Gavin Stevens, we have the following, see Figure 3.4:

Figure 3.4 Debit and Credit Table for Gavin Stevens

Account	Debit	Credit
Capital	–	Liability to owner increases by £200,000
Bank	Assets increase through capital introduced and money from debtors	Assets decrease through payments to suppliers and payments of expenses
Sales	–	Income increases through credit sales £9,000
Purchases	Expenses increase through £2,000 cash purchases and £3,000 credit purchases	–
Sales returns	Income reduced when customers return £70 goods	–
Purchases returns	–	Expenses reduced when £500 goods returned to supplier
Electricity	Expenses increase by £300	–
Wages	Expenses increase by £1,000	–
Hotel	Assets increase by £110,000	–
Van	Assets increase by £3,000	–
Ireton	Asset of debtor increases by £4,000	Asset reduced by returns of £70 and receipt of £1,965
Hepworth	Asset of debtor increases by £5,000	Asset reduced by receipt of £2,500
Hogen	Liability decreases by payment of £250 and returns of £500	Liability to third party increases by creditor of £1,000
Lewis	Liability decreases by payment of £1,000	Liability to third party increases by creditor of £2,000

Step 2 Balancing Off

Helpnotes 3.3 (on page opposite) and 3.4 (on page 48) provide a number of rules to guide us through double-entry book keeping. When all the entries for a period have been completed then it is time to balance off the accounts and carry forward the new total to the next period. We are, in effect, signalling the end of an accounting period. It is convenient to carry all the figures forward. These

Helpnote 3.3

Double-Entry Checks

Because of the way double-entry is structured, a number of rules can guide us when we make the **initial** entries.

1 **Sales and purchases**
There will **never** be a **debit** in a **sales account** or a **credit** in a **purchases account**. Assets and liabilities never pass through these accounts

2 **Returns**
Sales returns and purchases returns have their own accounts. You will **never** find a **credit** in a **sales returns account** or a **debit** in a **purchases returns account**

3 **Assets and expenses**
When making the **initial** entries you **never credit** a **fixed assets** or **expenses account**

4 The bank account represents cash at bank. We always talk of cash received, and cash paid. Given the number of cash transactions the normal business conducts, there would normally be a separate cash book. The totals from the cash book would be summarised and then transferred to the bank account.

Bank account

In	*Out*
Capital invested	Cash paid for purchases
Cash from sales	Cash paid for fixed assets such as cars
	Cash paid for expenses such as:
	Wages paid
	Rent paid
	Electricity paid
	Light and heat paid

Finally, if all else fails and you are still struggling with double-entry then Helpnote 3.4 on the next page may be useful.

Helpnote 3.4

The Bank Account

If you are having trouble remembering your double-entry, work back from the bank account; remember

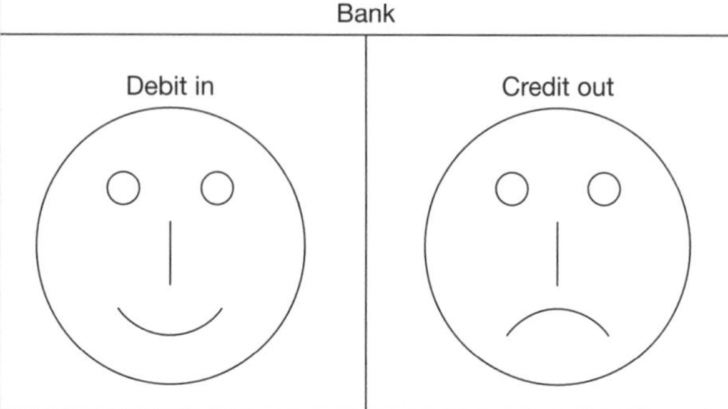

The smiling face represents money received so the business is better off. By contrast, the grumpy face represents money out, so the business is worse off.

carried forward figures will form the basis of the trial balance. However, the **sales, purchases, sales returns, purchases returns** and **expenses accounts** will then be **closed off** on the last day of the accounting period as they will **not** be carried forward to the next period. This is because in the next period all we are concerned with is that period's income and expenses. **The balance sheet items will be brought forward** on the first day of the new accounting period. This is because assets, liabilities and capital continue from accounting period to accounting period. Accounting periods may be weekly, monthly or annually. The aim of balancing off is threefold.

1 To prepare a trial balance from which we will prepare a balance sheet and trading, profit and loss account.
2 To close down income and expenses accounts which relate to the previous period.
3 To bring forward to the next period the assets, liabilities and capital balances.

First of all the accounts are balanced off. We then close off the revenue and expenses items to the trading account, and the profit and loss account. The trading account deals with purchases, purchases returns, sales and sales returns. The profit and loss account is concerned with expenses.

To illustrate balancing off we take three items from Gavin Stevens' accounts: sales (a trading account item), electricity (a profit and loss account item) and the bank account (a balance sheet item). We balance off after the first week. We will then draw up a trial balance.

Pause for Thought 3.4

The Trial Balance I

Why should the trial balance balance?

A trial balance is simply a list of all the balances on the individual accounts. If the double-entry process has been completed correctly, each debit will be matched by an equal and opposite credit entry. There will thus be equal amounts on the debit and credit sides. If the trial balance fails to balance – and it often will even for experts – then you know a mistake has been made in the double-entry process. You need, therefore, to find it. This process of trial and error is the reason why a trial balance is so called.

1. Trading account item

Sales

	£		£
		3 Jan. Ireton	4,000
7 Jan. Bal. c/f	9,000	3 Jan. Hepworth	5,000
	9,000		9,000
7 Jan. Transfer to trading account	9,000	7 Jan. Bal. b/f	9,000

Note that the balance (bal.) would be carried forward (c/f) above the totals and brought forward (b/f) below them.

2. Profit and loss account item

Electricity

	£		£
4 Jan. Bank	300	7 Jan. Bal. c/f	300
	300		300
7 Jan. Bal. b/f	300	7 Jan. Transfer to profit and loss account	300

3. Balance sheet item

Bank

	£		£
1 Jan. Capital	200,000	2 Jan. Hotel	110,000
7 Jan. Ireton	1,965	2 Jan. Van	3,000
7 Jan. Hepworth	2,500	2 Jan. Purchases	2,000
		4 Jan. Electricity	300
		4 Jan. Wages	1,000
		7 Jan. Hogen	250
		7 Jan. Lewis	1,000
		7 Jan. Bal. c/f	86,915
	£204,465		£204,465
8 Jan. Bal. b/f	£86,915		

Note that there is no need to close off the bank account because it will be used at the start of the next period. However, note **the date is the starting date for the next period**.

NOW GO BACK TO THE DOUBLE-ENTRY STAGE ON PAGE 45 AND COMPLETE THE 'T' ACCOUNTS SHOWN THERE. MAKE SURE YOU UNDERSTAND HOW TO BALANCE OFF PROPERLY. THEN CHECK YOUR ANSWER WITH THE ANSWER BELOW ON PAGES 50 AND 51.

Helpnote 3.5

Balancing Off

1. Always find which side has the greatest total (in the case below, £250) (a)
2. Record the greatest total twice, once on the debit side and once on the credit side (b)
3. Carry forward the balancing figure (bal. c/f) (c)
4. Bring forward the balancing figure (bal. b/f) (d)
5. For trading and profit and loss account items, only transfer the balance (e)

In this case, we use a telephone account for period to 31 January.

Balancing off a telephone account

	£		£
1 Jan. Telephone	250 (a)	31 Jan. Bal. c/f	250 (c)
	250 (b)		250 (b)
31 Jan. Bal. b/f	250 (d)	31 Jan. Transfer to profit and loss account	250 (e)

The completed 'T' accounts are shown in Figure 3.5 (below and on the next page). The *balancing off* is shown in *italics* for ease of understanding. We have abbreviated balance brought forward (to Bal. b/f) and balance carried forward (to Bal. c/f), transfer to profit and loss account (to To P&L) and transfer to trading account (to To trading a/c). Note that balance sheet items brought forward (Bal. b/f) are brought forward on the first day of the new accounting period, i.e. 8th January.

Figure 3.5 Completed Double-Entry Bookkeeping Entries for Gavin Stevens

Capital

	£		£
7 Jan. Bal. c/f	*200,000*	1 Jan. Bank	200,000
	200,000		*200,000*
		8 Jan. Bal. b/f	*200,000*

Bank

	£		£
1 Jan. Capital	200,000	2 Jan. Hotel	110,000
7 Jan. Ireton	1,965	2 Jan. Van	3,000
7 Jan. Hepworth	2,500	2 Jan. Purchases	2,000
		4 Jan. Electricity	300
		4 Jan. Wages	1,000
		7 Jan. Hogen	250
		7 Jan. Lewis	1,000
		7 Jan. Bal. c/f	*86,915*
	204,465		*204,465*
8 Jan. Bal. b/f	*86,915*		

Figure 3.5 Completed Double-Entry Bookkeeping Entries for Gavin Stevens (continued)

Sales

	£		£
		3 Jan. Ireton	4,000
7 Jan. Bal. c/f	*9,000*	3 Jan. Hepworth	5,000
	9,000		*9,000*
7 Jan. To trading a/c	*9,000*	*7 Jan. Bal. b/f*	*9,000*

Purchases

	£		£
2 Jan. Bank	2,000		
2 Jan. Hogen	1,000		
2 Jan. Lewis	2,000	*7 Jan. Bal. c/f*	*5,000*
	5,000		*5,000*
7 Jan. Bal. b/f	*5,000*	*7 Jan. To trading a/c*	*5,000*

Sales Returns

	£		£
5 Jan. Ireton	70	*7 Jan. Bal. c/f*	*70*
	70		*70*
7 Jan. Bal. b/f	*70*	*7 Jan. To trading a/c*	*70*

Purchases Returns

	£		£
7 Jan. Bal. c/f	*500*	3 Jan. Hogen	500
	500		*500*
7 Jan. To trading a/c	*500*	*7 Jan. Bal. b/f*	*500*

Electricity

	£		£
4 Jan. Bank	300	*7 Jan. Bal. c/f*	*300*
	300		*300*
7 Jan. Bal. b/f	*300*	*7 Jan. To P&L*	*300*

Wages

	£		£
4 Jan. Bank	1,000	*7 Jan. Bal. c/f*	*1,000*
	1,000		*1,000*
7 Jan. Bal. b/f	*1,000*	*7 Jan. To P&L*	*1,000*

Hotel

	£		£
2 Jan. Bank	110,000	*7 Jan. Bal. c/f*	*110,000*
	110,000		*110,000*
8 Jan. Bal. b/f	*110,000*		

Van

	£		£
2 Jan. Bank	3,000	*7 Jan. Bal. c/f*	*3,000*
	3,000		*3,000*
8 Jan. Bal. b/f	*3,000*		

Ireton (debtor)

	£		£
3 Jan. Sales	4,000	5 Jan. Sales Returns	70
		7 Jan. Bank	1,965
		7 Jan. Bal. c/f	*1,965*
	4,000		*4,000*
8 Jan. Bal. b/f	*1,965*		

Hepworth (debtor)

	£		£
3 Jan. Sales	5,000	7 Jan. Bank	2,500
		7 Jan. Bal. c/f	*2,500*
	5,000		*5,000*
8 Jan. Bal. b/f	*2,500*		

Hogen (creditor)

	£		£
3 Jan. Purchases Rets.	500	2 Jan. Purchases	1,000
7 Jan. Bank	250		
7 Jan. Bal. c/f	*250*		
	1,000		*1,000*
		8 Jan. Bal. b/f	*250*

Lewis (creditor)

	£		£
7 Jan. Bank	1,000	2 Jan. Purchases	2,000
7 Jan. Bal. c/f	*1,000*		
	2,000		*2,000*
		8 Jan. Bal. b/f	*1,000*

Note: We have closed off the trading and profit and loss accounts as at 7 January (sales, purchases, sales returns, purchases returns, electricity, wages) so that we can transfer these accounts to the trading and profit and loss account on that day. Balance sheet items are, however, brought forward on 8 January ready for the new period. In practice, we would probably not bring forward accounts with only one entry in, such as capital.

Step 3 Trial Balance

Definition 3.2

Trial Balance

Working definition

A listing of debit and credit balances to check the correctness of the double-entry system.

Formal definition

'A list of account balances in a double-entry accounting system. If the records have been correctly maintained, the sum of the debit balances will equal the sum of the credit balances, although certain errors such as the omission of a transaction or erroneous entries will not be disclosed by a trial balance.'

Chartered Institute of Management Accountants (2000), *Official Terminology*

The trial balance may now be prepared. *Prepare your own answer before you look at the 'answer' shown in Figure 3.6*

Figure 3.6 **Gavin Stevens: Trial Balance as at 7 January**

	Debit	Credit
	£	£
Hotel	110,000	
Van	3,000	
Sales		9,000
Purchases	5,000	
Capital		200,000
Sales returns	70	
Purchases returns		500
Bank	86,915	
Electricity	300	
Wages	1,000	
Ireton	1,965	
Hepworth	2,500	
Hogen		250
Lewis		1,000
	210,750	210,750

Soundbite 3.1

Trial Balance

'Nowadays, you hear a lot about fancy accounting methods, like LIFO and FIFO, but back then we were using the ESP method, which really sped things along when it came time to close those books. It's a pretty basic method: if you can't make your books balance you take however much they're off by and enter it under the heading ESP, which stands for Error Some Place.'

Sam Walton in *Sam Walton* p. 53

Source: The Executive's Book of Quotations (1994), pp. 3–4

Note that the **debit** items in Figure 3.6 (on the previous page) are either **assets** or **expenses** and the **credit** items are either **income** or **liabilities**. This conforms to the accounting equation:

Assets + Expenses = Liabilities + Capital + Income

Figure 3.7 **Analysis of Gavin Stevens' Trial Balance**

Assets	Hotel, van, bank, debtors (Ireton, Hepworth)
Expenses	Purchases, sales returns, electricity, wages
Income	Sales, purchases returns
Liabilities	Creditors (Hogen, Lewis)
Capital	Capital

Essentially, the trial balance is a check on the accuracy of the double-entry process. If the double-entry has been done properly the trial balance will balance. If it does not, as Soundbite 3.1 (on the page opposite) shows, there is an Error Some Place. Ideally, the books are then checked to find the error. Even when the trial balance balances, as Pause for Thought 3.5 shows, the accounts may not be totally correct.

Pause for Thought 3.5

The Trial Balance II

If the trial balance balances that is the end of all my problems, I now know the accounts are correct. Is this true?

Unfortunately, this is not true! Although you can be happy that the trial balance does indeed balance, you must still be wary. There are several types of error (listed below) which may have crept in, perhaps at the original double-entry bookkeeping stage.

1 **Error of omission**
You have omitted an entry completely. The trial balance will still balance, but your accounts will not be correct.

2 **Reverse entry**
If you have completely reversed your entry and entered a debit as a credit, and vice versa, then your trial balance will balance, but incorrectly.

3 **Wrong amount**
If the wrong figure (say £300 for the van instead of £3,000) was entered in the accounts, the books would still balance, but at the wrong amount.

Pause for Thought 3.5 (continued)

4 **Wrong account**
One of the entries might have been entered in the wrong account, for example, the van might be recorded in the electricity account.

5 **Compensating errors**
If you make errors which cancel each other out, your trial balance will once more wrongly balance.

HEALTH WARNING

Those students who did not wish to gain an in-depth knowledge of double-entry bookkeeping can restart here.

Computers

Double-entry bookkeeping is obviously very labour-intensive. Computerised packages, such as *Sage*, are very popular. Normally, an entry is keyed into the bank account (e.g., hotel £110,000 or into a debtor's or creditor's account (e.g., sales to Ireton £4,000). The computer automatically completes the entries. Sales, purchases and bank transactions are standard and, therefore, only require one entry. For non-standard items both a debit and a credit entry is recorded. The computer, after all the transactions have been input, can produce a trial balance, a balance sheet and a trading and profit and loss account.

At first sight, therefore, computerised accounting packages which perform the double-entry transactions and then prepare the trading and profit and loss account and balance sheet would seem to be a gift from heaven. However, there are problems. The principal one is that, as many organisations have found to their cost, if you put rubbish into the computer, you get rubbish out. In other words, computer operators who are unskilled in accounting can create havoc with the accounts. To enter items correctly, it is necessary to have an understanding of the accounting process. Otherwise, disasters may occur with meaningless accounts. Therefore, unfortunately, understanding double-entry bookkeeping is just as important in this computer age.

Conclusion

The double-entry process is a key part of financial accounting. Without understanding double-entry it is difficult to get to grips with accounting itself. However, there is no need to be scared of double-entry. Essentially, it means that for every transaction which is entered on one side of a ledger account, an equal and opposite entry is made in another ledger account. These entries are called debit and credit. All the debits will equal all the credits. This is proved when the accounts are balanced off.

The balance from each account is then listed in a trial balance. The trial balance shows that the double-entry process has been completed. However, a balanced trial balance does not necessarily guarantee that the double-entry has been correctly carried out (for example, there may be some errors of omission). Once the trial balance has been prepared it is possible to complete the trading and profit and loss account and balance sheet.

Discussion *Questions*

Questions with numbers in blue have answers at the back of the book.

Q 1 Why is double-entry bookkeeping so important?

Q 2 How do you think that the books of account kept by different businesses might vary?

Q 3 Computerisation means that there is no need to understand double-entry bookkeeping. Discuss.

Q 4 How much trust can be placed in a trial balance which balances?

Q 5 Why are there usually more debit balances in a trial balance than credit balances?

Q 6 State whether the following are true or false. If false, explain why.
- (a) We debit cash received, but credit cash paid.
- (b) Sales are debited to a sales account, but purchases are credited to a purchases account.
- (c) If a business purchases a car for cash we debit the car account and credit the bank account.
- (d) Purchases, hotel, electricity and wages are all debits in a trial balance.
- (e) Sales, rent paid and capital are all credits in a trial balance.

Numerical *Questions*

Questions with numbers in blue have answers at the back of the book.

Q 1 From the following accounting figures show the six steps in the accounting equation: opening assets £25,000, opening liabilities £25,000 (£15,000 third party and £10,000 capital), profit £15,000 (income £60,000, expenses £45,000).

***Q*2** Show the debit and credit accounts of the following transactions in the ledger and what effect (i.e., income/decrease) they have on assets, liabilities, capital, income and expenses. The first one is done as an illustration.

(a) Pay wages of £7,000

Debit effect	*Credit effect*
Wages: increases an expense	Bank: decreases an asset

(b) Introduces £10,000 capital by way of a cheque.
(c) Buys a hotel £9,000 by cheque.
(d) Pays electricity £300.
(e) Sales £9,000 cash.
(f) Purchases £3,000 on credit from A. Taylor.

***Q*3** You have the following details for A. Bird of transactions with customers and suppliers.

(a) 1 June credit purchases of £8,000, £6,000 and £5,000 from Robin, Falcon and Sparrow, respectively.
(b) 4 June A. Bird returns goods unpaid of £1,000 and £2,000 to Robin and Falcon, respectively.
(c) 6 June A. Bird makes credit sales of £4,000, £7,000 and £6,000 to Thrush, Raven and Starling, respectively.
(d) 7 June Starling returns £1,000 goods unpaid which are faulty.

Required:

(i) Write up the relevant ledger accounts.

(ii) Balance off the accounts on 7 June for the trading and profit and loss account. Bring them forward on 7 June, however, there is no need actually to transfer them to the trading and profit and loss account.

(iii) Bring forward balances for assets and liabilities on 8 June.

***Q*4** Balance off the following four accounts at the month end. Transfer the balances for sales and purchases to the trading account.

(i) Sales

	£		£
		8 June Bank	1,000
		9 June Brown	2,000

(ii) Purchases

	£		£
1 June Bank	500		
30 June Patel	9,500		

(iii) Bank

	£		£
3 June Cash Sales	500	1 June Wages	800
		7 June Rent	300
		8 June Purchases	200

(iv) R. Smith (debtor)

	£		£
1 June Sales	800	3 June Sales Rets.	1,000
		4 June Bank	3,000

Sales Rets. represents sales returns.

Q5 John Frier has the following transactions during a six month period to 30 June.

(a) Invests £10,000 on 1 January.
(b) Buys a motor van for £4,000 on 8 February by cheque.
(c) Purchases £8,000 goods on credit from A. Miner on 10 March.
(d) Pays A. Miner £3,000 on 12 April.
(e) Sells £9,000 credit sales to R. Army on 7 May.
(f) Receives £4,500 in cash on 10 June from R. Army.

Required:

On 30 June prepare John Frier's:

(i) Ledger accounts
(ii) Trial balance after balancing off the accounts. Bring forward sales and purchases on 30 June, but there is no need to transfer the trading and profit and loss account items.

Q6 Katherine Jones sets up a small agency that markets and distributes goods. She has the following regular credit customers (Edwards, Smith and Patel) and regular suppliers of credit goods (Johnston and Singh). She has the following transactions in the first week of July.

1 July	Invests £195,000 capital into the business.
2 July	Buys some premises for £75,000 by cheque.
	Buys office equipment for £9,000 by cheque.
	Buys goods from Johnston for £3,000 and from Singh for £1,000 both are on credit.
	Cash purchases of £7,000.
3 July	Sales of £10,000, £9,000 and £7,000 are made on credit to Edwards, Smith and Patel, respectively.
4 July	Returns £500 goods unpaid to Johnston as they were damaged.
5 July	Pays bills for wages £4,000, electricity £2,000 and telephone £1,000.
7 July	Settles half of the outstanding creditors and receives half of the money outstanding from debtors.

Required:

Prepare Katherine Jones':

(i) Ledger accounts. Balance off the trading and profit and loss account items on 7 July. However, there is no need to transfer the trading and profit and loss account items.
(ii) Trial balance after balancing off the accounts.

Q7 R. Poon was having a whale of a time trying to get a trial balance for his company Redwar. He had the following transactions during the month of May. He asks you to prepare the book entries and a trial balance.

1 May — Invests £8,000 in his business bank account.
2 May — Purchases £2,000 goods for cash from E. Skimo.
£4,000 of goods on credit from S. Eal.
£1,000 of goods on credit from P. Olar.
4 May — Sells £1,500 goods for cash to A.R.C. Tic.
£5,000 goods on credit to H. Unter.
£3,000 goods on credit to M. Dick.
7 May — M. Dick returns £2,000 goods unpaid saying that they were rotting.
8 May — R. Poon sends M. Dick's £2,000 goods (that he had originally purchased from S. Eal for £1,800) back unpaid because he can't stand the smell.
10 May — Purchases another £1,200 of goods on credit from S. Eal.
11 May — H. Unter buys on credit £1,600 of goods from R. Poon.
15 May — R. Poon receives bank interest of £250.
20 May — R. Poon pays S. Eal the amount owing.
20 May — R. Poon pays P. Olar £750.
21 May — R. Poon receives £2,000 from H. Unter
21 May — R. Poon receives £800 from M. Dick.
31 May — R. Poon pays by cheque wages £800, rent £500, electricity £75, and stationery £25.

Required:

Prepare R. Poon's:

(i) Ledger accounts.

(ii) Trial balance after balancing off the accounts. Bring forward the trading and profit and loss account items on 31 May, but there is no need actually to transfer the trading and profit and loss account items.

Q8 Jay Shah has the following balances as at 31 December in his accounts.

	£		£
Capital	45,300	Purchases returns	500
Motor car	3,000	Bank	3,600
Building	70,000	Electricity	1,400
Office furniture	400	Business rates	1,800
A. Smith (debtor)	250	Rent	1,600
J. Andrews (creditor)	350	Wages	3,500
T. Williams (creditor)	550	Long-term loan	9,000
G. Woolley (debtor)	150	Sales	100,000
		Purchases	70,000

Required:

(i) Jay Shah's trial balance as at 31 December.

(ii) An indication of which balances are assets, liabilities, capital, income or expenses.

Q9 Mary Symonds, a management consultant has the following balances from the accounts on 30 September.

	£		£
Office	80,000	Capital	28,150
Long-term loan	3,000	Electricity	1,600
Van	3,500	Telephone	3,400
H. Mellet (debtor)	650	Repairs	300
R. Edwards (debtor)	1,300	Business rates	900
P. Morgan (creditor)	1,400	Computer	3,000
Y. Karbhari (creditor)	600	Travel	4,000
Consultancy fees	70,000	Stationery	800
Cash at bank	3,700		

Required:
Mary Symonds' trial balance as at 30 September.

Q10 The following trial balance for Rajiv Sharma as at 31 December has been incorrectly prepared. Prepare a correct version.

	£	£
Shop		55,000
Machinery	45,000	
Car		10,000
Sales	135,000	
Purchases	80,000	
Opening stock		15,000
Debtors		12,000
Creditors	8,000	
Long-term loan	16,000	
General expenses		300
Telephone	400	
Light and heat	300	
Repairs		400
	284,700	92,700

Required:
A corrected trial balance as at 31 December.

Helpnote: If you rearrange the balances and the trial balance irritatingly still doesn't balance, remember capital.

Q 11 Rachel Thomas's trial balance balances as follows:

	£	£
Sales		100,000
Shop	60,000	
Van	50,000	
Purchases	60,000	
Capital		172,800
Bank	100,000	
General expenses	1,000	
Return inwards	300	
Repairs	800	
A. Bright (debtor)	2,000	
B. Dull (creditor)		1,600
Telephone	300	
	274,400	274,400

Unfortunately, Rachel Thomas's bookkeeper was not very experienced. The following transactions were incorrectly entered.

(a) Purchase of a computer for £3,000 cash completely omitted.
(b) Credit sales of £800 to A. Bright forgotten.
(c) A photocopier worth £800 wrongly debited to the shop account.
(d) £300 credited to B. Dull's account should have been charged to A. Bright as it was a sales return.
(e) The van really cost £5,000, but had wrongly been recorded as costing £50,000 in both the van and bank accounts.

Required:
A corrected trial balance as at 31 December.

Q 12 Which of the following balances extracted from the books on 31 December would be used as a basis for next year's accounts?

	£		£
Capital employed	9,200	Debtors	700
Rent and rates	1,000	Creditors	1,400
Buildings	7,000	Bank	600
Telephone	1,500	Stock	1,300
Sales	100,000	Purchases	80,000
Computer	1,000		

Chapter 4

Main Financial Statements: The Profit and Loss Account

'Around here you're either expense or you're revenue.'

Donna Vaillancourt, Inc. (March 1994), *The Wiley Book of Business Quotations* (1998), p. 90

Learning Outcomes

After completing this chapter you should be able to:

- ✔ Explain the nature of the profit and loss account.
- ✔ Understand the individual components of the profit and loss account.
- ✔ Outline the layout of the profit and loss account.
- ✔ Evaluate the nature and importance of profit.

In a Nutshell

- One of three main financial statements.
- Consists of income, cost of sales and expenses.
- Cost of sales is essentially opening stock plus purchases less closing stock.
- Gross profit is sales less cost of sales.
- Net profit is income less cost of sales less expenses.
- Profit is determined by income earned less expenses incurred **not** cash received less cash paid.
- Profit is an elusive concept.
- Capital expenditure (i.e., on fixed assets such as property) is treated differently to revenue expenditure (i.e., an expense such as telephone).
- Profit is useful when evaluating an organisation's performance.

Introduction

The profit and loss account is one of the three most important financial statements. Effectively, it records an organisation's income and expenses and is prepared from the trial balance. For companies, it is required by the Companies Act 1985. Profit and loss accounts seek to determine an organisation's profit (i.e., income less expenses) over a period of time. They are thus concerned with measuring an organisation's performance. Different organisations will have different profit and loss accounts. Indeed, they also have slightly different names. In this chapter, we focus on understanding the purpose, nature, contents and layout of the profit and loss account of the sole trader. The preparation of the profit and loss account of the sole trader from the trial balance is covered in Chapter 6. In Chapter 7 we investigate the preparation of the profit and loss accounts of partnerships and limited companies. The term 'profit and loss account' is used except when we are *specifically* discussing the sole trader. For sole traders, the more accurate term 'trading and profit and loss account' will be used.

Context

The profit and loss account, along with the balance sheet and cash flow statement, is one of the three major financial statements. It is prepared from the trial balance and presents an organisation's income and expenses over a period of time. This period may vary. Many businesses prepare a monthly profit and loss account for internal management purposes. Annual accounts are prepared for external users like shareholders or the tax authorities. From now on we generally discuss a yearly profit and loss account. By contrast, the balance sheet presents an organisation's assets, liabilities and capital and is presented at a particular point in time. The cash flow statement shows the cash inflows and outflows of a business.

The major parts of the profit and loss account are:

Income – **Expenses** = **Profit**

The profit figure is extremely important as it is used for a variety of purposes. It is used as an overall measure of performance and for more specific purposes, such as the basis by which companies distribute dividends to shareholders or as a starting point for working out taxation payable to the government. As Soundbite 4.1 shows, profits are central to evaluating an organisation's performance. An organisation's profit performance is closely followed by analysts and by the press. In Real-Life Nugget 4.1 on the next page, for example, there is discussion of Sage's 2000 financial results.

Soundbite 4.1

Profits

'You must deodorise profits and make people understand that profit is not something offensive, but as important to a company as breathing.'

Sir Peter Parker, quoted in the *Sunday Telegraph* (5 September 1976)

Source: *The Book of Business Quotations* (1991), p. 183

Real-Life Nugget 4.1

Profit Performance

Sage profits leap 52%

By Gavin Hicks

Accountancy software giant Sage has reported pre-tax profits up by 52% for the six months to 31 March this year.

The figures come to £54m compared to £35.5m for the same six months last year. Turnover also saw marked increases between the two periods with a 53% leap from 132m to 202.5m.

Company chairman Michael Jackson, said: "Our well-defined e-business products and services targeted at our 2.3 million customer base, coupled with recent acquisitions, will help us build on our existing business, the board remains confident about prospects for the year.'

The last six months have been busy for Sage with five completed acquisitions. Best Software, supplying asset management solutions, was bought in the US while Ubiquis SA of France, supplying web trading software, was acquired in France. Swiss accounting software supplier Sesam was bought while two British outfits, Hartley International and CSM were added to the stable.

Sage's strategy is firmly linked to the internet with the company believing its customers are looking for safe ways of completing e-transactions.

www.sage.com

Source: *Accountancy Age*, 18 May 2000, p. 19

Definitions

As Definition 4.1 indicates, a profit and loss account represents the income less the expenses of an organisation. Essentially, income represents money earned by the organisation (for example, sales), while expenses represent the costs of generating these sales (for example, purchases) and of running the business (for example, telephone expenses).

Definition 4.1

Definition of a Profit and Loss Account

Working definition

The income less the expenses of an organisation over a period of time, giving profit.

Formal definition

A key financial statement which represents an organisation's income less its expenses over a period of time and thus determines its profit so as to give a 'true and fair' view of an organisation's financial affairs.

Broadly, a working definition of income is the revenue earned by a business, while expenses are the costs incurred running a business. The International Accounting Standards Board formally defines income and expenses using the concept of an increase or a decrease in economic benefit. Incomes are thus increases in economic benefit (i.e., increases in assets or decreases in third party liabilities) which increase owner's capital. Expenses are decreases in economic benefit (i.e., decreases in assets or increases in third party liabilities) that decrease owner's capital. Both working and formal definitions are provided in Definition 4.2 on the next page.

Definition 4.2

Income

Working definition
Revenue earned by a business

Formal definition
'Increases in economic benefits during the accounting period in the form of inflows or enhancements of assets or decreases of liabilities that result in increases in equity.'

International Accounting Standards Board (2000), *Framework for the Preparation and Presentation of Financial Statements*

Expenses

Working definition
Costs incurred running a business

Formal definition
'Decreases in economic benefits during the accounting period in the form of outflows or depletions of assets or incurrences of liabilities that result in decreases in equity.'

International Accounting Standards Board (2000), *Framework for the Preparation and Presentation of Financial Statements*

Pause for Thought 4.1

Limited Companies' Profit and Loss Account

Why do you think limited companies produce only abbreviated figures for their shareholders?

The answer to this is twofold. First of all, many limited companies are large and complicated businesses. They need to simplify the financial information provided. Otherwise, the users of the accounts might suffer from information overload. Second, for public limited companies, there is the problem of confidentiality. Remember that a company is owned by shareholders. Anybody can buy shares. A competitor, for example, could buy shares in a company. Companies would not wish to give away all the details of their sales and expenses to a potential competitor. They, therefore, summarise and limit the amount of information they provide. Limited companies, therefore, publicly provide abridged (or summarised) accounts rather than full ones using all the figures from the trial balance.

Layout

The profit and loss account is nowadays conventionally presented in a vertical format (such as in Figure 4.1 below). Here we begin with sales and then deduct the expenses. The 1985 Companies Act sets out several possible formats. For sole traders or partners, a full profit and loss account is sometimes prepared using all the figures from the trial balance. Often, however, the profit and loss account is presented in summary form, with many individual revenues and expenses grouped together. For companies presenting their results to the shareholders in the annual report, the exact relationship to the original trial balance is often unclear. Examples of Marks and Spencer plc and J. Sainsbury plc profit and loss accounts are given in The Company Cameras 2.1 (on page 24 in Chapter 2) and 7.3 (on page 147 in Chapter 7) respectively.

By contrast the trading and profit and loss account of the sole trader is more clearly derived from the trial balance. In this section we explain the theory behind this in more detail. In the following section the main terminology is explained. In order to be more realistic, we use the adapted trading and profit and loss account of a real person, a sole trader who runs a public house. This is presented in Figure 4.1.

Figure 4.1 Sole Trader's Trading and Profit and Loss Account

R. Beer
Trading and Profit and Loss Account Year Ended 31 March 2002

	£	£
Sales		100,425
Less *Cost of Sales*		
Opening stock	3,590	
Add Purchases	58,210	
	61,800	
Less Closing stock	2,200	59,600
Gross Profit		40,825
Add *Other Income*		
Gaming machine		2,000
		42,825
Less *Expenses*		
Wages	8,433	
Rates and water	3,072	
Insurance	397	
Electricity	2,714	
Telephone	292	
Advertising	172	
Motor expenses	530	
Darts team expenses	1,865	
Repairs and renewals	808	
Laundry	1,174	
Music and entertainment	3,095	
Licences	604	
Guard dog expenses	385	
Garden expenses	1,716	
Sundry expenses	1,648	
Accounting	800	
Depreciation	6,770	34,475
Net Profit		8,350

The corresponding balance sheet for R. Beer is presented in the next chapter in Figure 5.2. In Chapter 6, we show how to prepare a trading and profit and loss account from the trial balance.

Figure 4.2 Overview of Trading and Profit and Loss Account

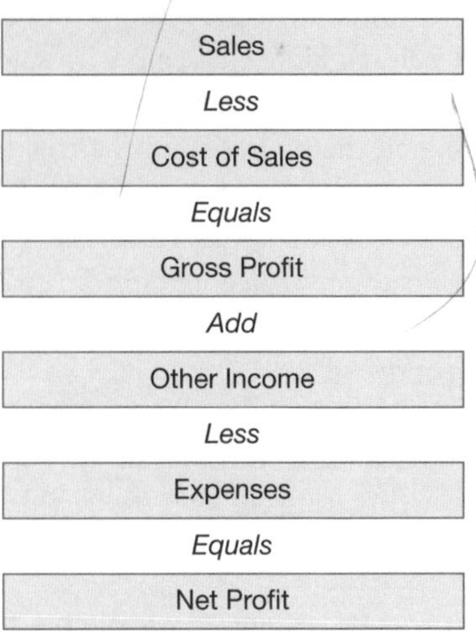

Main Components

Figure 4.2 shows the six main components of the trading and profit and loss account (sales, cost of sales, gross profit, other income, expenses and net profit) of a sole trader.

These six components are shown in Figure 4.3. In the first column, an overview definition is provided. This is followed by some general examples and then, whenever possible, a specific example. These components are then discussed in more detail in the text. The same order is used as for R. Beer's trading, and profit and loss account.

Figure 4.3 Main Components of the Trading and Profit and Loss Account

	Overview	*General Examples*	*Specific Examples*
Sales	Income earned from selling goods (may be reduced by sales returns, i.e. goods returned by customers)	Sales	e.g., Sales of beer
Cost of Sales	The items directly incurred in selling goods	1. Opening stock 2. Purchases (may be reduced by purchases returns, i.e., goods returned to supplier) 3. Closing stock	e.g., Barrels of beer a pub has at start of year e.g., Purchases of beer e.g., Barrels of beer a pub has left at end of year
Gross Profit	Sales less cost of sales	Measures gain of organisation from buying and selling	e.g., The direct profit a pub makes by reselling the beer
Other Income	Non-trading income which a firm has earned	1. Income from investments 2. Income from sale of fixed assets	e.g., Interest received from deposit account at bank e.g., Profit on selling a car for more than it was recorded in the accounts
Expenses	Items indirectly incurred in selling the goods	1. Light and heat 2. Employees' pay	e.g., Electricity e.g., Wages and salaries
Net Profit	Sales less cost of sales less expenses	Measures gain of organisation from all business activities	e.g., The profit a pub makes after taking into account all the pub's expenses

Sales

Generating sales is a key ingredient of business success. However, the nature of sales varies considerably from organisation to organisation. For companies, the word 'turnover' is often used for sales. Essentially, sales is the income that an organisation generates from its operations. For example, Brook Brothers in Real-Life Nugget 4.2 sold clothes.

These sales may be for credit or for cash. Credit sales create debtors, who owe the business money. Some businesses, such as supermarkets, have predominantly cash customers. By contrast, manufacturing businesses will have largely credit customers. A further distinction is between businesses that primarily sell goods (e.g., supermarkets which supply food) and those which supply services (for example, a bank). Sales are very diverse. In the modern world, both developed and developing countries have varied businesses with varied sales. The UK is an example of this (see Figure 4.4). It must be stressed that Figure 4.4 is for guidance only. The division between credit and cash, in particular, is very rough and ready.

Real-Life Nugget 4.2

Sales

Thoughtless Thieves

In 1964 the lower Manhattan branch of Brook Brothers was robbed, and the thieves got away with clothes worth $200,000. One clerk remarked: "If they had come during our sale two weeks ago, we could have saved 20 percent."

Source: Peter Hay (1988), *The Book of Business Anecdotes*, Harrap Ltd, London, p. 100

Figure 4.4 Ten Top Industrial Sectors for 2000

Sector	*Example of Sales*	*Credit/cash*	*Goods/services*
Agriculture	Farm produce	Credit	Goods
Manufacturing	Manufactured goods	Credit	Goods
Construction	Buildings	Credit	Goods
Motor trade	New cars or car repairs	Cash and credit	Goods
Wholesale*	Food wholesaler*	Credit	Goods
Retail	Supermarket	Cash and credit card	Goods
Hotels and catering	Hotels	Cash and credit card	Services
Transport	Taxi fares	Cash	Services
Finance	Interest earned	Not applicable	Services
Property and business services	Rents	Cash and credit	Services

*Wholesale businesses act as middlemen between manufacturers and customers. They buy from manufacturers and sell on to retailers

Source: Size Analysis of United Kingdom Businesses, 2000, Office for National Statistics, Table 1A

Sales are reduced by sales returns. These are simply goods which are returned by customers, usually because they are faulty or damaged.

The Company Camera 4.1 demonstrates the sales for Manchester United.

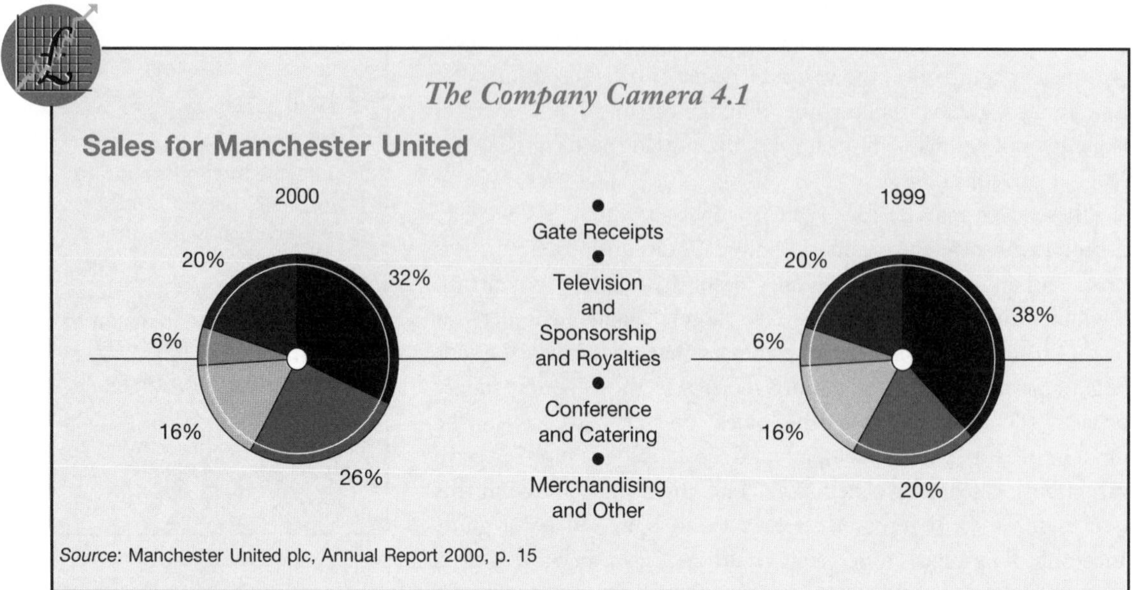

Source: Manchester United plc, Annual Report 2000, p. 15

Cost of Sales

Cost of sales is essentially the expense of directly providing the sales. Cost of sales are normally primarily found in businesses which buy and sell goods rather than those which provide services. The main component of cost of sales is purchases. However, purchases are adjusted for other items such as purchases returns (goods returned to suppliers), carriage inwards (i.e., the cost of delivering the goods from the supplier), and, of particular importance, stock. In businesses, which manufacture products, rather than buying and selling goods, the cost of sales is more complicated. It will contain, for example, those costs which can be directly related to manufacturing. The Company Camera 4.2 shows the cost of sales for J. Sainsbury plc, and Figure 4.5 on the next page presents a detailed example of cost of sales.

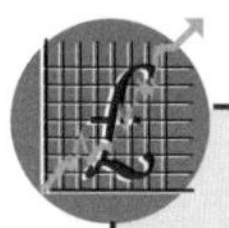

The Company Camera 4.2

Cost of Sales

Cost of sales consists of all costs to the point of sale including warehouse and transportation costs, all the costs of operating retail outlets and, in the case of Sainsbury's Bank plc, interest payable.

Source: J. Sainsbury plc Annual Report and Financial Statements 2001, p. 19

Figure 4.5 Cost of Sales

John Green, a retailer, runs a small business. He has the following details of a week's trading: Opening stock £2,000, carriage inwards £100, purchases £9,000, purchases returns £300, closing stock £3,800.
(i) What is his cost of sales? (ii) Why do we adjust for stock?

	£	£
Cost of Sales		
Opening stock		2,000
Add Purchases	9,000	
Less Purchases returns	300	
	8,700	
Add Carriage inwards	100	8,800
		10,800
Less Closing stock		3,800
		7,000

i. Cost of sales is thus £7,000. Note that we have adjusted (i) for goods returned to the supplier £300, (ii) for carriage inwards which is the cost of delivering the goods from the supplier (i.e., like 'postage' on goods), and (iii) for closing stock.
ii. We need to match the sales with the actual cost directly incurred in generating them. Essentially, opening stock represents purchases made last period, but not used. They were not, therefore, directly involved in generating last period's sales. Similarly, closing stock represents this year's purchases which have not been used. They will, therefore, be used to generate the next period's, rather than this period's, sales.

Helpnote: When presenting cost of sales we subtract purchases returns and closing stock. We indicate this using the word 'less'. As a result, we do not put these figures in brackets.

Gross Profit

Gross profit is simply sales less cost of sales. It is a good measure of how much an organisation makes for every £1 of goods sold. In other words, the mark-up an organisation is making. Mark-ups and margins are discussed in Pause for Thought 4.2.

Pause for Thought 4.2: Gross Profit

A business operates on a mark-up of 50% on cost of sales. If its cost of sales was £60,000, how much would you expect gross profit to be?

	Mark-up		Gross margin
	%	£	%
Sales	150	90,000	100
Cost of sales	100	60,000	67
Gross profit	50	30,000	33

Alternatively, a business might talk in terms of gross margin. This is a percentage of sales **not** of cost of sales. As we can see above, both mark-up and gross margin are different ways of expressing the same thing.

Businesses watch their cost of sales and their gross profit margins very closely. This is illustrated in The Company Camera 4.3 for Manchester United plc. Gross profit is determined in the trading account.

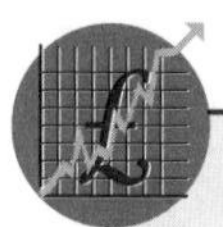

The Company Camera 4.3

Cost of Sales and Gross Margin

Cost of sales decreased by £3.8 million to £20.1 million as a result of the decrease in the number of home games and the cost of travel packages referred to above. The gross profit margin was 83 per cent compared to 78 per cent reflecting the improved sales mix with the higher margin television increasing and gate receipts decreasing.

Source: Manchester United plc, Annual Report 2000, p. 15

Pause for Thought 4.3

Trading Account

What actually is the trading account and why is it so called?

The trading account is the initial part of the trading and profit and loss account. In other words, sales less cost of sales. As we can see below, it deals with the sales and purchases of goods and gives gross profit.

	£	£	£
Sales			100,000
Less Sales returns			10,000
			90,000
Less *Cost of Sales*			
Opening stock		8,000	
Add Purchases	50,000		
Less Purchases returns	4,000	46,000	
		54,000	
Less Closing stock		2,000	52,000
Gross Profit			38,000

It is termed a trading account because it gives details of an organisation's direct trading income (i.e., buying and selling goods) rather than non-trading income (e.g. bank interest) or expenses. Nowadays, because of the growth of businesses which have little stock (e.g., service companies), the trading account is becoming less important.

Other Income

This is basically income from activities other than trading. So it might be interest from money in a bank or building society (interest received). Or it might be dividends received from an investment or profit on sale of a fixed asset. In the case of the R. Beer example it was income from the gaming machine. Sometimes organisations report operating profits. Operating profit is concerned with trading activities (e.g., sales, purchases and expenses). Other income would be excluded.

Expenses

Expenses are many and varied. They are simply the costs incurred in meeting sales. Some examples of expenses are given below. However, the list is far from exhaustive.

- Accountants' fees (A)
- Advertising (S)
- Insurance (A)
- Light and heat (A)
- Petrol consumed (S)
- Sales commission (S)
- Business rates (A)
- Rent paid (A)
- Repairs and renewals (A)
- Telephone bill (A)

Often, these expenses are grouped into broad headings. For instance, in UK companies' published accounts, expenses are divided into selling and distribution costs (i.e., costs of marketing, delivering or distributing goods or services) and administrative costs (general running of the business). These categories are often very subjective. In the list above S = selling and distribution, and A = administrative.

Most expenses *are* a result of a cash payment (e.g., rent paid) or *will* result in a cash payment (e.g., rent owing). However, depreciation is a non-cash payment. It represents the expense of using fixed assets, such as motor vehicles, which wear out over time. The topic of depreciation is dealt with in more detail in Chapter 5.

Net Profit

Net profit is simply the amount left over after cost of sales and expenses have been deducted. It is a key method of measuring a business's performance. It is often expressed as a percentage of sales, giving a net profit to sales ratio.

Profit

The concept of profit seems a simple one, at first. Take a barrow boy, Jim, selling fruit and vegetables from his barrow in Manchester. If he buys £50 of fruit and vegetables in the morning and by the evening has sold all his goods for £70, Jim has made a profit of £20.

However, in practice profit measurement is much more complicated and often elusive. Although there are a set of rules which guide the determination of income and expenses, there are also many assumptions which underpin the calculation of profit. The main factors which complicate matters are:

- the matching concept
- estimation
- changing prices
- the wearing out of assets

Matching Concept

It is essential to appreciate that the whole purpose of the profit and loss account is to match **income earned** and **expenses incurred**. This is the matching concept which we saw in Chapter 2. **Income earned** and **expenses incurred** are not the same as **cash paid** and **cash received**. We have already seen that depreciation is a non-cash item. However, it is also important to realise that for many other income and expense items the **cash received and paid** during the year are **not the same as income earned and expenses incurred.** When we are attempting to arrive at income earned and expenses incurred we have to estimate certain items such as amounts owing (known as accruals). We also need to adjust for items paid this year which will, in fact, be incurred next year (for example, rent paid in advance).

Estimating

Accounting is often about estimation. This is because we often have uncertain information. For example, we may estimate the outstanding telephone bill or the value of closing stock.

Changing Prices

If the price of fixed assets, such as property, rises then we have a gain from holding that asset. Is this gain profit? Well, yes in the sense that the organisation has gained. But no, in that it is not a profit from trading. This whole area is clouded with uncertainty. There are different views. Normally such fixed asset gains are only included in the accounts when the fixed assets are sold.

Wearing Out of Assets

Assets wear out and this is accounted for by the concept of depreciation. However, calculating depreciation involves a lot of assumptions, for example, length of asset life.

Profit is thus contingent upon many adjustments, assumptions and estimates. All in all, therefore, the determination of profit is more an art than a science. It is true to say that different accountants will calculate different profits. And they might all be correct! It is, however, also true that despite the assumptions needed to arrive at profits, profits are a key determinant by which businesses of all sorts are judged. Real-Life Nugget 4.3, for example, shows press comment on football clubs.

Real-Life Nugget 4.3

Football Clubs' Profits

Football clubs' profits drop

Profits made by football clubs in the English Premiership more than halved during 1998/99, according to Deloitte & Touche. While average club income stood at a record-breaking £33.5m, higher spending in the transfer market meant pre-tax profits shrunk by £22.7m to £18.7m compared to the previous season. Player transfer costs and in particular spiralling wages – which jumped by 31% in the 1997/1998 season – were highlighted as the key to the profits slump.

Source: *Accountancy Age*, 11 May 2000, p. 3

Capital and Revenue Expenditure

One example of the many decisions which an accountant must make is the distinction between capital and revenue expenditure. Capital expenditure is usually associated with balance sheet items, such as fixed assets, in other words, assets which may last for more than one year (e.g., land and buildings, plant and machinery, motor vehicles, and fixtures and fittings). By contrast, revenue expenditure is usually associated with profit and loss account items such as telephone, light and heat or purchases. This appears simple, but sometimes it is not. For example, to a student, is this book a capital or a revenue expenditure? Well, it has elements of both. A capital expenditure in that you may keep it for reference. A revenue expenditure in that its main use will probably be over a relatively short period of time. So we can choose! Often when there is uncertainty small-value items are charged to the profit and loss account. Interestingly, the WorldCom accounting scandal involved WorldCom incorrectly treating £2.5 billion of revenue expenditure as capital expenditure. This had the effect of increasing profit by £2.5 billion.

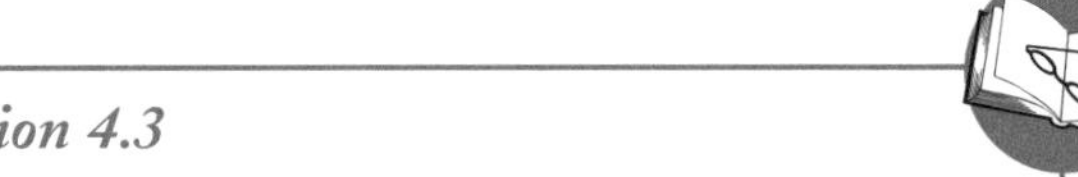

Definition 4.3

Capital and Revenue Expenditure

Capital expenditure
A payment to purchase an asset with a long life such as a fixed asset.

Revenue expenditure
A payment for a current year's good or service such as purchases for resale or telephone expenses.

Limitations

So does the profit and loss account provide a realistic view of the performance of the company over the year, especially of profit? The answer is, maybe! The profit and loss account does list income and expenses and thus arrives at profit. However, there are many estimates which mean that the profit figure is inherently subjective. At the end of a year, for example, there is a need to estimate the amount of phone calls made. This brings subjectivity into the estimation of profit. Another example is that the loss in value of fixed assets is not accurately measured. Similarly, the valuation of stock is very subjective.

Interpretation

The profit and loss account, despite its limitations, is often used for performance comparisons between companies and for the same company over time. These performance comparisons can then be used as the basis for investment decisions. Profitability ratios are often used to assess a company's performance over time or relative to other companies (for example, profit is often measured against sales or capital employed). Ratios are more fully explained in Chapter 9.

> *Soundbite 4.2*
>
> **Profits**
>
> **'Nobody ever got poor taking a profit.'**
>
> André Meyer, Reich, *Financier: André Meyer*, p. 119
>
> *Source*: *The Executive's Book of Quotations*, (1994), p. 233

Conclusion

The profit and loss account presents an organisation's income and expenses over a period. It allows the determination of both gross and net profit. In many ways, making a profit is the key to business success (as Soundbite 4.2 shows). Gross profit is essentially an organisation's profit from trading. Net profit represents profit after all expenses have been taken into account. It is important to realise that the profit and loss account is concerned with matching revenues earned with expenses incurred. It is not, therefore, a record of cash paid less cash received. Profit is not a precise absolute figure: it depends on many estimates and assumptions. However, despite its subjectivity, profit forms a key element in the performance evaluation of an organisation.

Discussion *Questions*

Questions with numbers in blue have answers at the back of the book.

Q 1 Why is the profit and loss account such an important and useful financial statement to such a variety of users?

Q 2 'There is not just one profit there are hundreds of profits.' Do you agree with this statement, taking into account the subjectivity inherent in calculating profit?

Q 3 The matching principle is essential to the calculation of accounting profit. Discuss.

Q 4 Over time, with the decline of the manufacturing company and rise of the service company, stock, cost of sales and gross profit are becoming less important. Discuss.

Q 5 State which of the following statements is true and which false? If false, explain why.
- (a) Profit is income earned less expenses paid.
- (b) Sales less cost of sales less expenses will give net profit.
- (c) Sales returns are returns by suppliers.
- (d) If closing stock increases so will gross profit.
- (e) The purchase of a fixed asset is known as a capital expenditure.

Numerical *Questions*

Questions with numbers in blue have answers at the back of the book.

Q 1 Joan Smith has the following details from her accounts year ended 31 December 2001.

Sales	£100,000	Purchases	£60,000	General expenses	£10,000
Opening stock	£10,000	Closing stock	£5,000	Other expenses	£8,000

Required:
Draw up Joan Smith's trading and profit and loss account for the year ended 31 December 2001.

Q 2 Dale Reynolds has the following details from his accounts for the year ended 31 December 2001.

	£		£
Opening stock	5,000	Closing stock	8,000
Purchases	25,000	Purchases returns	2,000
Sales	50,000	Sales returns	1,000
Carriage inwards	1,000		

Required:
Draw up Dale Reynolds trading account for the year ended 31 December 2001.

Q 3 Mary Scott has the following details for the year to 31 December 2001.

	£		£
Sales	200,000	Income from investments	3,000
Opening stock	5,000	Wages	4,000
Purchases	100,000	Insurance	2,500
Closing stock	3,000	Electricity	3,500
Advertising	3,500	Telephone	4,000
Motor expenses	1,500	Purchases returns	1,000
Repairs	800	Sales returns	2,000
Sundry expenses	2,400		

Required:
Draw up Mary Scott's trading and profit and loss account for the year ended 31 December 2001.

Q 4 Given the following different scenarios, calculate sales, gross profit and cost of sales from the information available.

(a) Sales £100,000, gross margin 25%.
(b) Sales £200,000, mark-up 30%.
(c) Cost of sales £50,000, gross margin 25%.
(d) Cost of sales £40,000, mark-up 20%.

Chapter 5

Main Financial Statements: The Balance Sheet

'[Creative accounting practices] gave rise to the quip, "A balance sheet is very much like a bikini bathing suit. What it reveals is interesting, what it conceals is vital."'

Abraham Briloff, *Unaccountable Accounting*, Harper & Row (1972). *The Wiley Book of Business Quotations* (1998), p. 351

Learning Outcomes

After completing this chapter you should be able to:

- ✔ Explain the nature of a balance sheet.
- ✔ Understand the individual components of a balance sheet.
- ✔ Outline the layout of a balance sheet.
- ✔ Evaluate the usefulness of a balance sheet.

In a Nutshell

- One of three main financial statements.
- Consists of assets, liabilities and capital.
- Assets are most often fixed (e.g., land and buildings, plant and machinery, motor vehicles, and fixtures and fittings) or current (e.g., stock, debtors, cash).
- Liabilities can be current (e.g., trade creditors, short-term loans) or long-term (e.g., long-term loans).
- A sole trader's capital is capital plus profit less drawings.
- In modern balance sheets, a vertical format is most popular.
- A balance statement's usefulness is limited by missing assets and inconsistent valuation.
- Different business organisations will have differently structured balance sheets e.g., sole traders and limited companies.
- Balance sheets are used as a basis for determining liquidity.

Introduction

The balance sheet, along with the profit and loss account and cash flow statement, is one of the most important financial statements. However, it is only in the last century that the profit and loss account has become pre-eminent. Before then, the balance sheet ruled supreme. The balance sheet is prepared from an organisation's trial balance. It consists of assets, liabilities and capital. For companies, it is required by the Companies Act 1985. Balance sheets seek to measure an organisation's net assets at a particular point in time. They developed out of concepts of stewardship and accountability. Although useful for assessing liquidity, they do not actually represent an organisation's market value. In this chapter, the primary focus will be on understanding the purpose, nature and contents of the balance sheet. The preparation of balance sheets of sole traders, partnerships and limited companies from the trial balance is covered, in depth, in Chapters 6 and 7, respectively.

Context

The balance sheet is one of the key financial statements. It is prepared from the trial balance at a particular point in time, which can be any time during the year. The balance sheet is usually prepared for shareholders at either 31 December or 31 March. The balance sheet has three main elements: assets, liabilities and capital. Essentially, the assets less the third party liabilities (i.e., net assets) equal the owner's capital. Thus,

Assets – **Liabilities** = **Capital**

The balance sheet and profit and loss account are complementary (see Figure 5.1). The profit and loss account shows an organisation's performance over the accounting period, normally a year. It is thus concerned with income, expenses and profit. A balance sheet, by contrast, is a snapshot of a business at a particular point in time. It thus focuses on assets, liabilities and capital. As Soundbite 5.1 shows, the balance sheet provides basic information which helps users to judge the value of a company.

Soundbite 5.1

Balance Sheets

'The market is mostly a matter of psychology and emotion, and all that you find in balance sheets is what you read into them; we've all guessed to one extent or another, and when we guess wrong they say we're crooks.'

Major L.L.B. Angas, *Stealing the Market*, p. 15

Source: *The Executive's Book of Quotations* (1994), p. 179

Figure 5.1 **Comparison of Profit and Loss Account and Balance Sheet**

	Profit and Loss Account	*Balance Sheet*
Preparation source	Trial balance	Trial balance
Main elements	Income, expenses, profit	Assets, liabilities, capital
Period covered	Usually a year	A point in time
Main focus	Profitability	Net assets

Definitions

As Definition 5.1 shows a balance sheet is essentially a collection of the assets, liabilities and capital of an organisation at a point in time. It is prepared so as to provide a true and fair view of the organisation.

Definition 5.1

Balance Sheet

Working definition

A collection of the assets, liabilities and capital of an organisation at a particular point in time.

Formal definition

'A statement of the financial position of an entity at a given date disclosing the assets, liabilities and accumulated funds such as shareholders' contributions and reserves prepared to give a true and fair view of the financial state of the entity at that date.'

Chartered Institute of Management Accountants (2000), *Official Terminology*

Broadly, assets are the things an organisation owns or leases; liabilities are things it owes. Assets can bring economic benefits by either being sold (for example, stock) or being used (for example, a car). Capital (sometimes known as ownership interest) is accumulated wealth. Capital is effectively a liability to the business because it is 'owed' to the owner: Standard setters more formally define assets in terms of rights to future economic benefits and liabilities as obligations. Capital (ownership interest) is what is left over. In other words, assets less third party liabilities equal owners' capital.

Pause for Thought 5.1

Net Assets

If the assets were £20,000 and the liabilities to third parties were £10,000, what would (a) net assets and (b) capital be?

The answer would be £10,000 for both. This is because net assets equals assets less liabilities and capital equals net assets.

By formally defining assets and liabilities, the balance sheet tends to drive the profit and loss account. In Definition 5.2 on the next page we present both formal and working definitions of assets, third party liabilities and capital (ownership interest).

Definition 5.2

Assets

Working definition
Items owned or leased by a business.

Formal definition
'An asset is a resource controlled by the enterprise as a result of past events and from which future economic benefits are expected to flow to the enterprise.'

International Accounting Standards Board (2000), *Framework for the Preparation and Presentation of Financial Statements*

Liabilities

Working definition
Items owed by a business.

Formal definition
'A liability is a present obligation of the enterprise arising from past events, the settlement of which is expected to result in an outflow from the enterprise of resources embodying economic benefits.'

International Accounting Standards Board (2000), *Framework for the Preparation and Presentation of Financial Statements*

Capital (Ownership Interest)

Working definition
The funds (assets less liabilities) belonging to the owner(s).

Formal definition
'Equity is the residual interest in the assets of the enterprise after deducting all its liabilities.'

International Accounting Standards Board (2000), *Framework for the Preparation and Presentation of Financial Statements*

Layout

Traditionally, the balance sheet was always arranged with the assets on the right-hand side of the page and the liabilities on the left-hand side of the page. However, more recently, the vertical format has become most popular. For companies, the use of the vertical format was set out by the 1985 UK Companies Act. Examples of Marks and Spencer plc and J. Sainsbury plc balance sheets are given in The Company Cameras 2.2 (on page 25 in Chapter 2) and 7.4 (on page 148 in Chapter 7), respectively. However, even organisations which are not companies now commonly use the vertical format. This sets out the assets and liabilities at the top. The capital employed is then put at the bottom. This modern format is the one which this book will use from now on. However, Appendix 5.1 (at the end of this chapter) gives an example of the traditional 'horizontal' format which readers may occasionally encounter.

In order to be more realistic, we use the adapted balance sheet of a real person, a sole trader who runs a public house. This is given in Figure 5.2. In Chapter 6, we show the mechanics of the preparation of the balance sheet from the trial balance. The purpose of this section is to explain the theory behind the presentation.

Figure 5.2 Sole Trader's Balance Sheet

R. Beer
Balance Sheet as at 31 March 2002

	£	£	£
Fixed Assets			
Land and buildings			71,572
Plant and machinery			3,500
Furniture and fittings			5,834
Motor car			3,398
Total fixed assets			84,304
Current Assets			
Stock	2,200		
Debtors	100		
Prepayments	50		
Bank	3,738		
Cash	340	6,428	
Current Liabilities			
Creditors	(3,900)		
Accruals	(91)		
Short-term loans	(1,000)	(4,991)	
Net current assets			1,437
Total assets less current liabilities			85,741
Long-term Creditors			(6,500)
Total net assets			79,241
Capital Employed			£
Opening capital			80,257
Add Net profit			8,350
			88,607
Less Drawings			9,366
Closing capital			79,241

Main Components

In Figure 5.3, on the next page, we now list the main components commonly found in the balance sheet of a sole trader. An overview of the structure of balance sheet is shown in Figure 5.4 on page 83.

In Figure 5.3, we provide an overview definition of the individual balance sheet components, some general examples and then, wherever possible, a specific example. The main components of the balance sheet are then discussed in the text. The same order is used as for R. Beer's balance sheet. Finally, in Figure 5.6, on page 86, we summarise the major valuation methods used for the main assets.

Figure 5.3 Main Components of the Balance Sheet

	Overview	General Examples	Specific Examples
Fixed Assets	Assets used to run the business long-term	1. Land and buildings 2. Plant and machinery 3. Fixtures and fittings 4. Motor vehicles	Public house, factory Lathe Computer, photocopier Car, van
Current Assets (i.e., short-term assets)			
(i) Stocks	Goods purchased and awaiting use or produced awaiting sale	1. Finished goods	Tables manufactured and awaiting sale
		2. Work in progress	Half-made tables
		3. Raw materials	Raw wood awaiting manufacture
(ii) Debtors	Amounts owed to company	Trade debtors	Customers who have received goods, but not yet paid
(iii) Prepayments	Amounts paid in advance	Prepayments for services	Insurance prepaid
(iv) Cash and bank	Physical cash Money deposited on short-term basis with a bank	Cash in till Cash and bank deposits	Petty cash Current account in credit
Current Liabilities (i.e., amounts falling due within one year)			
(i) Creditors	Money owed to suppliers	Trade creditors	Amounts owing for raw materials
(ii) Accruals	Amounts owed to the suppliers of services	Accruals for services	Amounts owing for electricity or telephone
(iii) Loans	Amounts borrowed from third parties and repayable within a year	Short-term loans from financial institutions	Bank loan
Long-term Creditors	Amounts borrowed from third parties and repayable after a year	Long-term loans from financial institutions	Loan secured, for example, on business property
Capital Employed	Originally, the money the sole trader introduced into the business. Normally represents the net assets (i.e., assets less liabilities)	The capital at the start of the year and the capital at the end of the year are generally known as opening and closing capital	The opening and closing capital represent the opening and closing net assets
(i) Profit	The profit earned during the year	Taken from the profit and loss account, represents income less expenses	Net profit for year
(ii) Drawings	Money taken out of the business by the owner. A reduction of owner's capital	Living expenses	Owner's salary or wages

Pause for Thought 5.2

Balance Sheets

Why does a balance sheet balance?

This is a tricky question! For the answer we need to think back to the trial balance. The trial balance balances with assets and expenses on one side and income, liabilities and capital on the other. Essentially, the balance sheet is a rewritten trial balance. It includes all the individual items from the trial balance in terms of the assets, liabilities and capital. It also includes all the income and expense items. However, all of these are included only as one figure 'profit' (i.e., all the income items less all the expense items). Incidentally, a balance sheet is probably called a balance sheet not because it balances, but because it is a list of all the individual balances from the trial balance.

Fixed Assets

These are the assets that a business uses for its continuing operations. There are generally recognised to be four main types of tangible assets (intangible assets, i.e. those which do not physically exist, are discussed in Chapter 7): land and buildings, plant and machinery, fixtures and fittings, and motor vehicles. Fixed assets are traditionally valued at historical cost. In other words, if a machine was purchased 10 years ago for £100,000, this £100,000 was originally recorded in the books. Every year of the fixed asset's useful life, an amount of the original purchase cost will be allocated as an expense in the profit and loss account. This allocated cost is termed depreciation. Thus, depreciation simply means that a proportion of the original cost is spread over the life of the fixed asset and treated as an annual expense. In essence, this allocation of costs relates back to the matching concept. There is an attempt to match a proportion of the original cost of the fixed asset to the accounting period in which the fixed assets were used up.

The most common methods of measuring depreciation are the straight-line method and the reducing balance method. The straight-line method is the one used in Figure 5.5 on the next page. Essentially, the same amount is written off the fixed asset every year. With reducing balance, a set percentage is written off every year. Thus, if the set percentage was 20%, then in Figure 5.5 £20,000 would be written off in year 1. This would leave a net book value of £80,000 (£100,000 − £20,000). Then 20% of the net book value of £80,000 (i.e., £16,000) would be written off in year 2, and so on).

Companies have great flexibility when choosing appropriate rates of depreciation. These rates should correspond to the lives of the assets. So,

Figure 5.4 Overview of a Balance Sheet

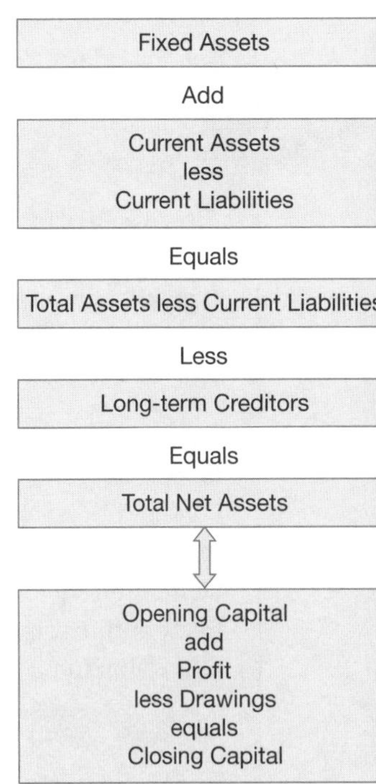

Figure 5.5 Illustrative Example on Depreciation

A machine was purchased 10 years ago for £100,000. Estimated useful life 20 years. We will assume the depreciation is equally allocated over 20 years. What would be the total depreciation (known as accumulated depreciation) after 12 years and at how much would the machine be recorded in the balance sheet?

Balance Sheet			
Fixed Assets	*Cost*	*Accumulated depreciation*	*Net book value*
	£	£	£
Plant and machinery	100,000	(60,000)	40,000

In the balance sheet, the original cost (£100,000) is recorded, followed by accumulated depreciation (i.e., depreciation over the 12 years: 12 × £5,000 = £60,000). The term 'net book value' simply means the amount left in the books after writing off depreciation. It is important to note that the net book value does not equal the market value. Indeed, it may be very different. In the profit and loss account only one year' depreciation (£100,00 ÷ 20 years) of £5,000 is recorded each year.

Note: This topic is discussed more fully in Chapter 6.

for example, if the directors believe the fixed assets have a life of five years they would choose a straight-line rate of depreciation of 20%. The Company Camera 5.1 gives the rates of depreciation used by Manchester United plc. Interestingly, the club uses the reducing balance method, which is relatively uncommon in the UK.

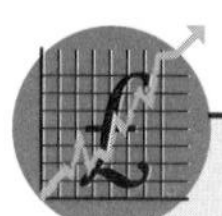

The Company Camera 5.1

Depreciation

Depreciation is provided on tangible fixed assets at annual rates appropriate to the estimated useful lives of the assets, as follows:

	Reducing Balance
Freehold land	Nil
Freehold buildings	1.33%
Assets in the course of construction	Nil
Computer equipment and software	33%
Plant and machinery	20% – 25%
General fixtures and fittings	15%

Source: Manchester United plc, 2000 Annual Report, p. 34

Finally, it is important to realise that nowadays, many businesses, regularly revalue some fixed assets such as land and buildings every five years. Businesses can also revalue their fixed assets whenever they feel it is necessary (or, alternatively, devalue them if they have lost value). Where revaluations occur, the depreciation is based on the revalued amount.

Current Assets

Current assets are those assets which a company owns which are essentially short-term. They are normally needed to perform the company's day-to-day operations. The five most common forms of current assets are stocks, debtors, prepayments, cash and bank.

1. Stocks

Stocks are an important business asset. This is especially so in manufacturing businesses. Stocks can be divided into three categories: raw material stocks, work-in-progress stocks and finished goods stocks (see, for example, The Company Camera 5.2).

The Company Camera 5.2

Stocks

	2000 $m	1999 $m
Raw materials and consumables	543	581
Stocks in process	768	699
Finished goods and goods for resale	794	876
	2,105	2,156

Source: AstraZeneca, 2000 Annual Report, p. 72

Each category represents a different stage in the production process.

- *Raw material stocks*. These are stocks a company has purchased and are ready for use. A carpenter, for example, might have wood awaiting manufacture into tables.
- *Work-in-progress*. These are partially completed stocks, sometimes called stocks in process. They are neither raw materials nor finished goods. They may represent partly manufactured goods such as tables with missing legs. Some of the costs of making the tables should be included.
- *Finished goods stock*. This represents stock at the other end of the manufacturing process, for example, finished tables. Cost includes materials and other manufacturing costs (e.g., labour and manufacturing overheads).

Stocks at the year end are often determined after a stock take. As Real-Life Nugget 5.1 shows, stocks can often represent a substantial percentage of a company's net assets.

Real-Life Nugget 5.1

The Importance of Stock Valuation

Company	Stock Value £m	Net Assets £m	Stock ÷ Net Assets %
Bulmers	31	85	36.4
Marconi	1,721	4,637	37.1
Laing	3	241	1.2
Marks and Spencer	472	4,661	10.1
Rolls-Royce	1,179	2,064	57.1
Vodafone	316	147,782	0.2
J.D. Wetherspoon	5	282	1.8

Source: 2000/2001 Annual Reports

Generally, stock is valued at the lower of cost or net realisable value (i.e., the value it could be sold for). In the case of work-in-progress and finished goods, cost could include some overheads (i.e., costs associated with making the tables). In Chapter 16, we will look at several different ways of calculating cost (e.g., FIFO and AVCO).

Figure 5.6 Summary of the Valuation Methods Used for Fixed Assets and Stock

	Valuation method
Fixed Assets	Normally valued at historical cost, or revaluation less depreciation. Historical cost is the original purchase price of the asset. Revaluation is the value of the fixed assets as determined, usually by a surveyor, at a particular point in time.
Stock	Stock is generally valued at the lower of cost (i.e., what a business paid for it) and the amount one would realise if one sold it (called net realisable value). For a business with work-in-progress or finished goods stock, an appropriate amount of overheads is included

2. Debtors

Debtors are sales which have been made, but for which the customers have not yet paid. If all transactions were in cash, there would be no debtors. Debtors at the year end are usually adjusted for those customers who it is believed will not pay. These are called bad and doubtful debts. Bad debts are those debts which will definitely not be paid. Doubtful debts have an element of uncertainty to them. Usually, businesses estimate a certain proportion of their debts as doubtful debts. These bad and doubtful debts are also included in the profit and loss account.

Pause for Thought 5.3

Bad and Doubtful Debts

Can you think of any reasons why bad and doubtful debts might occur?

There may be several reasons, for example, bankruptcies, disputes over the goods supplied or cash flow problems. Most businesses constantly monitor their debtors to ensure that bad debts are kept to a minimum.

3. Prepayments

Prepayments are those items where a good or service has been paid for in advance. A common example of this is insurance. A business might, for example, pay £1,000 for a year's property insurance on 1 October. If the accounts are drawn up to 31 December, then at 31 December there is an asset of nine months' insurance (January–September) paid in advance. This asset is £750.

4. Cash and bank

This is the actual money held by the business. Cash comprises petty cash and unbanked cash. Bank comprises money deposited at the bank or on short-term loan. As Soundbite 5.2 shows, money has long been the topic of humour.

Soundbite 5.2

Money

'They say money can't buy happiness, but it can facilitate it. I thoroughly recommend having lots of it to anybody.'

Malcolm Forbes (*Daily Mail*, 20.6.1988)

Source: *The Book of Business Quotations* (1991), p. 158

Current Liabilities

These are the amounts which the organisation owes to third parties. For a sole trader, there are two main types:

1. Creditors

These are the amounts which are owed to suppliers for goods received, but not yet paid (for example, raw materials). Like money, debts have often been a fertile subject for humourists (see Soundbite 5.3).

2. Accruals

Accruals is accounting terminology for expenses owed at the trial balance or balance sheet date. Accruals comply with the basic accounting concept of matching. In other words, because an expense has been incurred, but not yet paid, there is no reason to exclude it from the profit and loss account. Accruals are amounts owed, but not yet paid, to suppliers for services received. Accruals relate to expenses such as telephone or light and heat. For example, we might have paid the telephone bill up to 31 November. However, if our year end was 31 December then we might owe, say, another £250 for telephone. Importantly, accruals do **not** relate to purchases of trade goods owing: these are creditors.

Soundbite 5.3

Debts

'Debts shorten life.'

Joseph Joubert

Source: *The Book of Unusual Quotations* (1959), p. 56

3. Loans

Loans are the amounts which a third party, such as a bank, has loaned to the company on a short-term basis and which are due for repayment within one year.

The current assets less the current liabilities is, in effect, the operating capital of the business. It is commonly known as a business's working capital. Businesses try to manage their working capital as efficiently as possible (see Chapter 22). A working capital cycle exists (see Figure 5.7), where cash is used to purchase goods which are then turned into stock. This stock is then sold and cash is generated. Successful businesses will sell their goods for more than their total cost, thus generating a positive cash flow.

Figure 5.7 The Working Capital Cycle

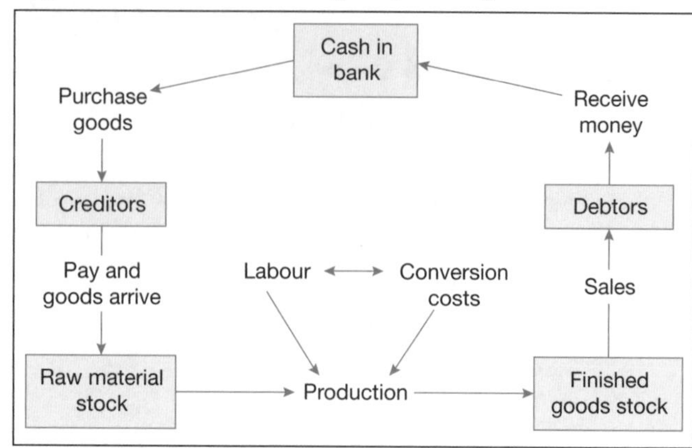

In some cases, a business may have more current liabilities than current assets. Instead of being net current assets, this section becomes net current liabilities. This is the situation with J.D. Wetherspoon plc in 2000, see The Company Camera 5.3.

The Company Camera 5.3

Current Assets and Current Liabilities

	2000	1999
Current assets	£000	£000
Investments	100	253
Stocks	4,686	3,845
Debtors due within one year	7,378	11,472
Debtors due after more than one year	5,588	5,588
Cash	41,685	62,578
	59,437	83,736
Creditors due within one year	(67,936)	(67,296)
Net current (liabilities)/assets	(8,499)	16,440

Source: J.D. Wetherspoon, 2000 Annual Report, p. 26

Long-term Creditors

These are liabilities that the organisation owes and must repay after more than one year. The most common are long-term loans. The total assets of a business less current liabilities and long-term creditors give the total net assets of the business. Total net assets represent the total capital employed by a business.

Capital Employed

For a sole trader the capital employed is opening capital plus profit less drawings. Thus for R. Beer it is:

	£
Opening capital	80,257
Add Profit	8,350
	88,607
Less Drawings	9,366
Closing capital	79,241

Opening capital is that capital at the start of the year (i.e., 1 April 2001). In essence, it represents the total net assets at the start of the year (i.e., all the assets less all the liabilities). If the business had made a loss, opening capital would have been reduced. It is important to realise that the profit (or loss) recorded under capital employed represents the net profit (or loss) as determined from the profit and loss account. It is, thus, a linking figure.

Pause for Thought 5.4

Capital

A chip shop owner, B. Atter has opening capital of £19,500. His income is £100,000 and expenses are £90,000. He has taken out £15,000 to live on. His financial position has improved over the year. True or false?

Unfortunately, for B. Atter, the answer is false. If we quickly draw up his capital employed.

	£
Opening capital	19,500
Add Profit (i.e., income less expenses)	10,000
	29,500
Less Drawings	15,000
Closing capital	14,500

His capital has declined by £5,000 over the year.

Profit is the profit as determined from the profit and loss account (i.e., sales less purchases and expenses). Drawings is the money that R. Beer has taken out of the business for his own personal spending. Finally, closing capital is capital at the balance sheet date. It must be remembered at all times that:

Assets – **Liabilities** = **Capital**

Soundbite 5.4

Capital

'Capital isn't scarce; vision is.'

Michael Milken (junk-bond creator)

Source: The Executive's Book of Quotations (1994), p. 47

Businesses need capital to operate. However, the capital needs to be used wisely (see Soundbite 5.4).

Limitations

To the casual observer, it looks as if the balance sheet places a market value on the net assets of the organisation. Unfortunately, this is wrong. Very wrong! To understand why, one must look at the major components of the balance sheet. The balance sheet is a collection of individual assets and liabilities. These individual assets and liabilities are not valued at a real-world, market value, so neither is the balance sheet as a whole. Taking the fixed assets, for example, only selected fixed assets are revalued. These valuations are neither consistent nor necessarily up-to-date. Also depreciation does not accurately measure the loss in value of fixed assets (nor is it supposed to). The balance sheet is thus a tangle of assets all measured in different ways.

Pause for Thought 5.5

Financial Position

Angela Roll, a baker, has the following assets and liabilities. Fixed assets £59,000, current assets £12,000, current liabilities £8,000 and long-term creditors £7,000. Is it true that her total net assets are £50,000?

Actually, she is better off. They are £56,000:

	£	£
Fixed assets		59,000
Current assets	12,000	
Current liabilities	(8,000)	
Net current assets		4,000
Total assets less current liabilities		63,000
Long-term creditors		(7,000)
Total net assets		56,000

Another problem, especially for companies, is that significant assets are not shown on the balance sheets. All the hard work of an owner to generate sales and goodwill will only be recognised when the business is sold. Many key items that drive corporate value, such as know-how and market share,

are also not recorded (see Real-Life Nugget 5.2). Neither is the value of human assets recorded! For example, what is the greatest asset of football clubs? You might think it was the footballers like David Beckham or Michael Owen. However, conventionally these players would not be valued as assets on the balance sheet unless they have been transferred from another club. Nor, in the case of zoos, would the animals bred in captivity be valued.

Real-Life Nugget 5.2

Missing Assets

In many respects, the current reporting model is more suited to a manufacturing economy than one based on services and knowledge. For example, it does a good job measuring historical costs of physical assets, like plant and equipment. But it ignores many of the key drivers of corporate value, such as know-how and market share. The result is incomplete, or distorted, information about a company's worth, and diminished relevance to investors.

Source: Dennis M. Nally, *The Future of Financial Reporting* (1999), International Financial Reporting Conference
From Web http://www.amazon.co.uk/exec/abidos/ASIN

Interpretation

Given the above limitations, any meaningful interpretation of amounts recorded in balance sheets is difficult. However, there are ratios which are derived from the balance sheet which are used to assess the financial position of a business. These ratios are dealt with in more depth in Chapter 9. Here we just briefly comment on liquidity (i.e., cash position) and long-term capital structure.

The balance sheet records both current assets and current liabilities. This can be used to assess the short-term liquidity (i.e., short-term cash position) of a business. There are a variety of ratios such as the current ratio (current assets divided by current liabilities) and quick ratio (current assets minus stock then divided by current liabilities) which do this. In addition, the balance sheet records the long-term capital structure of a business. It is particularly useful in determining how dependent a business is on external borrowings. In Soundbite 5.5, for example, excessive borrowing created a balance sheet where there was negative net worth.

Soundbite 5.5

Balance Sheet

'[MCI's balance sheet] looked like Rome after the Visigoths had finished with it. We had a $90-million negative net worth, and we owed the bank $100 million, which was so much that they couldn't call the loan without destroying the company.'

W.G. McGowgan, *Henderson Winners*, p. 187

Source: The Executive's Book of Quotations (1994), p. 81

Conclusion

The balance sheet is a key financial statement. It shows the net assets of a business at a particular point of time. The three main constituents of the balance sheet are assets (fixed and current), liabilities (current and long-term) and capital. Normally a vertical balance sheet is used to portray these elements. The balance sheet itself is difficult to interpret because of missing assets and inconsistently valued assets. However, it is commonly used to assess the liquidity position of a firm.

Discussion *Questions*

Questions with numbers in blue have answers at the back of the book.

Q 1 The balance sheet and profit and loss account provide complementary, but contrasting information. Discuss.

Q 2 What are the main limitations of the balance sheet and how can they be overcome?

Q 3 Is the balance sheet of any use?

Q 4 Are the different elements of the balance sheet changing over time, for example, as manufacturing industry gives way to service industry?

Q 5 State whether the following are true or false. If false, explain why.

(a) A balance sheet is a collection of assets, liabilities and capital.
(b) Stock, bank and creditors are all current assets.
(c) Total net assets is fixed assets plus current assets less current liabilities.
(d) Total net assets equals closing capital.
(e) An accrual is an amount prepaid, for example, rent paid in advance.

Numerical *Questions*

Questions with numbers in blue have answers at the back of the book.

Q 1 The following financial details are for Jane Bricker as at 31 December 2001.

	£		£
Capital 1 January 2001	5,000	Profit	12,000
Drawings	7,000		

Required:
Jane Bricker's capital employed as at 31 December 2001.

Q 2 Alpa Shah has the following financial details as at 30 June 2002.

	£		£
Fixed assets	100,000	Current liabilities	30,000
Current assets	50,000	Long-term creditors	20,000

Required:
Alpa Shah's total net assets as at 30 June 2002.

Q3 Jill Jenkins has the following financial details as at 31 December 2001.

	£		£
Stock	18,000	Cash	4,000
Debtors	8,000	Creditors	12,000

Required:
Jill Jenkins' net current assets as at 31 December 2001.

Q4 Janet Richards has the following financial details as at 31 December 2001.

	£		£
Land and buildings	100,000	Creditors	15,000
Plant and machinery	60,000	Long-term loan	15,000
Stock	40,000	Opening capital	200,000
Debtors	30,000	Net profit	28,000
Cash	20,000	Drawings	8,000
		Closing capital	220,000

Required:
Janet Richards' balance sheet as at 31 December 2001.

Appendix 5.1: Horizontal Format of Balance Sheet

R. Beer's Balance Sheet as at 31 March 2002 (presented in horizontal format)

	£	£		£	£
Capital Employed			**Fixed Assets**		
Opening capital		80,257	Land and buildings		71,572
Add Profit		8,350	Plant and machinery		3,500
		88,607	Furniture and fittings		5,834
Less Drawings		9,366	Motor vehicles		3,398
Closing capital		79,241	*Total fixed assets*		84,304
Long-term loan		6,500			
		85,741			
Current Liabilities			**Current Assets**		
Creditors	3,991		Stock	2,200	
Loan	1,000	4,991	Debtors	100	
			Prepayments	50	
			Bank	3738	
			Cash	340	6,428
		90,732			90,732

Chapter 6

Preparing the Financial Statements

'Mr Evans was the chief accountant of a large manufacturing concern. Every day, on arriving at work, he would unlock the bottom drawer of his desk, peer at something inside, then close and lock the drawer. He had done this for 25 years. The entire staff was intrigued but no one was game to ask him what was in the drawer. Finally, the time came for Mr Evans to retire. There was a farewell party with speeches and a presentation. As soon as Mr Evans had left the buildings, some of the staff rushed into his office, unlocked the bottom drawer and peered in. Taped to the bottom of the drawer was a sheet of paper. It read, "The debit side is the one nearest the window."'

R. Andrews (April 2000), C.A. Magazine, 'Funny Business', p. 26

Learning Outcomes

After completing this chapter you should be able to:

- ✔ **Show how the trial balance is used as a basis for preparing the financial statements.**
- ✔ **Prepare the profit and loss account and the balance sheet.**
- ✔ **Understand the post-trial balance adjustments commonly made to the accounts.**
- ✔ **Prepare the profit and loss account and balance sheet using post-trial balance adjustments.**

In a Nutshell

- The trial balance when rearranged creates a profit and loss account and a balance sheet.
- The profit and loss account presents sales, cost of sales, other income and expenses.
- The balance sheet consists of assets, liabilities and capital.
- Sales less cost of sales less expenses gives net profit.
- Net profit is the balancing figure in the balance sheet.
- There are five main post-trial balance adjustments to the accounts: closing stock, accruals, prepayments, depreciation and doubtful debts.
- All five adjustments are made twice to maintain the double-entry, first, in the profit and loss account and, second, in the balance sheet.

Introduction

A trial balance is prepared after the bookkeeping process of recording the financial transactions in a double-entry form. The bookkeeping stage is really one of aggregating and summarising the financial information. The next step is to prepare a profit and loss account and balance sheet from the trial balance. In essence, the profit and loss account sets out income less expenses and thus determines profit. Meanwhile, the balance sheet presents the assets and liabilities, including owner's capital. The two statements are seen as complementary. Chapters 4 and 5 discussed the nature and purpose of the profit and loss account and balance sheet. This present chapter looks at the mechanics of how we prepare the final accounts from the trial balance. It is thus a continuation of Chapter 3.

Main Financial Statements

Essentially the two main financial statements rearrange the items in the trial balance. The profit for the year effectively links the profit and loss account and the balance sheet. In the **profit and loss account profit is income less expenses paid,** while in the **balance sheet, closing capital less drawings and opening capital give profit.** Looked at another way profit represents the increase in capital over the year. Profit is a key figure in the accounts and plays a vital part in linking the profit and loss account to the balance sheet. Profit is often reported graphically by companies in their annual reports. Rentokil Initial plc's profit figure is given in The Company Camera 6.1.

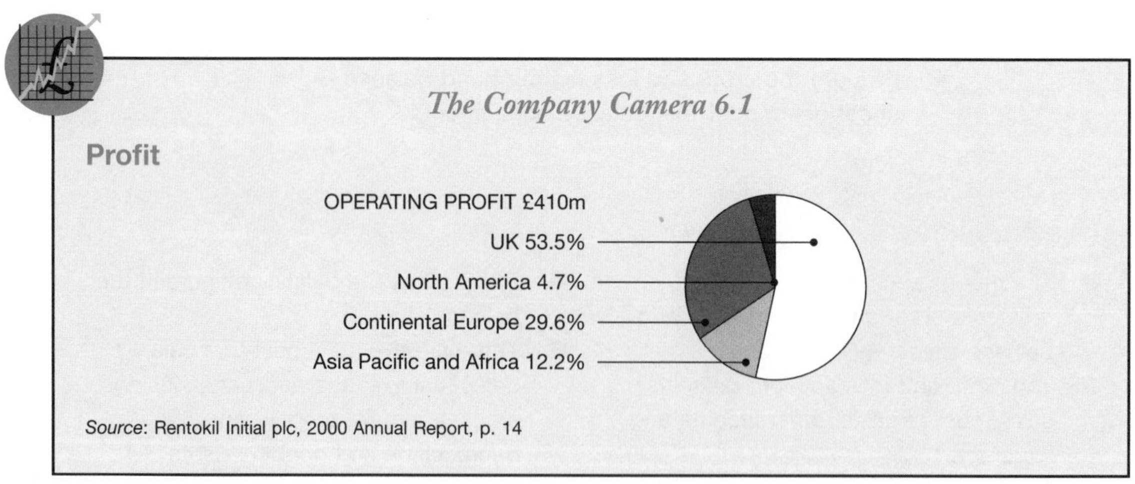

Pause for Thought 6.1

Accounting Equation and Financial Statements

What is the relationship between the accounting equation, profit and loss account and balance sheet?

In essence, the accounting equation develops into the financial statements. Remember from Chapter 3 that the expanded accounting equation was:

Assets + Expenses = Liabilities + Capital + Income

Rearrange thus:

Assets − (Liabilities + Capital) = Income − Expenses

Assets − (Liabilities + Capital) ↓ Balance sheet

Income − Expenses ↓ Profit and loss account

Trial Balance to Profit and Loss Account and Balance Sheet

The elements of the trial balance need to be arranged to form the profit and loss account and balance sheet. Different organisations (for example, sole traders, partnerships and limited companies) all have slightly different formats for their balance sheets. However, the basic structure remains the same. For the sole trader, we normally refer to the trading and profit and loss account, rather than the profit and loss account. For illustrative purposes, we continue the example of Gavin Stevens, a sole trader, who is setting up a hotel. The format used here is that followed by most UK companies and broadly adheres to the requirements of the UK Companies Acts. There is *no prescribed format* for sole traders, therefore, *for consistency this book broadly adopts the format used by companies*. The individual items were explained in more detail in Chapters 4 and 5. It is worth pointing out that the cash flow statement is usually derived from the profit and loss account and balance sheet once they have been prepared and *not* from the trial balance (see Chapter 8).

Gavin Stevens, Continued

In Chapter 3, we prepared the trial balance. We now can use this to prepare the profit and loss account, and balance sheet. However, when we prepare the financial accounts we also need to collect information on the amount of stock held and on any expenses which are yet to be paid or have been paid in advance. In this example, we set out more details about stock, amounts owing for telephone and amounts prepaid for electricity in the notes to the trial balance. It is important to realise that none of these items have so far been entered into the books. We deal with these adjustments in more depth later in this chapter.

Gavin Stevens
Trial Balance as at 7 January

		Debit	Credit
		£	£
Hotel		110,000	
Van		3,000	
Sales			9,000
Purchases		5,000	
Capital			200,000
Sales returns		70	
Purchases returns			500
Bank		86,915	
Electricity		300	
Wages		1,000	
Debtors	Ireton	1,965	
	Hepworth	2,500	
Creditors	Hogen		250
	Lewis		1,000
		210,750	210,750

Notes:

1. Gavin Stevens does not use all the catering supplies. He estimates that the amount of catering supplies left as closing stock is £50.
2. There is a special arrangement with the electricity company in which Gavin Stevens pays £300 in advance for a quarter. By 7 January he has used £50 which means he has prepaid £250. This is known as a prepayment.
3. There is a telephone bill yet to be received. However, Gavin Stevens estimates that he owes £100. This is known as an accrual.

We now prepare the trading and profit and loss account following the steps in Helpnote 6.1.

Helpnote 6.1

Presentational Guide to the Four Steps (Given as A–D in Gavin Stevens' Trading and Profit and Loss Account)

In terms of presentation, we should note:

- We must first determine cost of sales (Step A). This is, at its simplest, opening stock add purchases less closing stock. Purchases must be adjusted for purchases returns.
- Sales less cost of sales gives gross profit (Step B). Sales must be adjusted for sales returns.
- We list and total all expenses (Step C).
- We determine net profit (Step D) by taking expenses away from gross profit.

Gavin Stevens
Trading and Profit and Loss Account for Week Ended 7 January

	£	£	£		
Sales			9,000		Trading account
Less Sales returns			70		
			8,930		
Less *Cost of Sales*					
Opening stock		–			
Add Purchases	5,000				
Less Purchases returns	500	4,500			
Less Closing stock		50	4,450	A	
Gross Profit			4,480	B	
Less *Expenses*					Profit and loss account
Electricity		50			
Wages		1,000			
Telephone		100	1,150	C	
Net Profit			3,330	D	

Points to notice:

1. All the figures are from the trial balance except for closing stock £50, electricity £50 and telephone £100 (from notes) and profit £3,330 (calculated).
2. The first part of the statement down to gross profit (B) (sales less cost of sales (A)) is called the trading account. It deals with sales and purchases. Essentially sales returns and purchases returns are deducted from sales and purchases, respectively. Stock is simply unsold purchases.
3. Gross profit (B) is sales minus cost of sales.
4. Net profit (D) is sales minus cost of sales minus expenses (C).
5. Net profit of £3,330 is also found in the balance sheet. It is the balancing item, simply income minus expenses.

We now prepare the balance sheet following the steps in Helpnote 6.2.

Helpnote 6.2

Presentational Guide to the Five Steps (Given as A–E in Gavin Stevens' Balance Sheet)

In terms of presentation, we should note:

- All the fixed assets are added together (Step A).
- Current assets less current liabilities are determined next (i.e., net current assets). This is often termed the working capital and represents the day-to-day trading resources of the business (Step B).
- The total assets less current liabilities are determined (Step C).
- If there are any long-term creditors these are deducted to arrive at total net assets (Step D).
- Capital employed is determined. Effectively, net profit is added to opening capital to give closing capital (Step E).

Gavin Stevens
Balance Sheet as at 7 January

	£	£	£	
Fixed Assets				
Hotel			110,000	
Van			3,000	
			113,000	A
Current Assets				
Stock	50			
Debtors (£1,965 + £2,500)	4,465			
Electricity prepayment	250			
Bank	86,915	91,680		
Current Liabilities				
Creditors (£250 + £1,000)	(1,250)			
Telephone bill accrued	(100)	(1,350)		
Net current assets			90,330	B
Total net assets			203,330	C,D
Capital Employed				
Opening capital			200,000	
Add Net profit			3,330	
Closing capital			203,330	E

Points to notice:

1. All figures are from the trial balance except for closing stock £50, electricity prepayment £250, telephone bill (from notes) and profit £3,330 (balancing figure). Except for electricity all *these* figures are the same in the trading and profit and loss account. Electricity is different because in effect, we are splitting up £300. Thus:

   Total paid £300 = £50 used up as an expense in the profit and loss account, and £250 not used up recorded as an asset in the balance sheet.

2. Our statement is divided into fixed assets, current assets, current liabilities and capital employed.
3. The debtors are the trial balance figures for Ireton and Hepworth; the creditors those for Hogen and Lewis.
4. Our opening capital plus our net profit gives us our closing capital. In other words, the business owes Gavin Stevens £200,000 at 1 January, but £203,330 at 7 January.
5. Net profit of £3,330 is also found in the trading and profit and loss account. It is the balancing item. As we saw from Pause for Thought 6.1, the profit can be seen as the increase in net assets over the year. Or, alternatively, it can be viewed as assets less liabilities less opening capital.
6. The balance sheet balances. In other words total net assets equals closing capital.

Pause for Thought 6.2

Accounting Equation and Gavin Stevens

How does the Gavin Stevens example we have just completed fit the accounting equation?

If we take our expanded accounting equation:

Assets + Expenses = Liabilities + Capital + Income

and rearrange it,

Assets − (Liabilities + Capital) = Income − Expenses

Now, if we substitute the figures from Gavin Stevens, we have:

Fixed assets (£113,000) + Current assets (£91,680) − (Current liabilities (£1,350) + Opening capital (£200,000)) = Income (£8,930) − (Cost of sales (£4,450) + Expenses (£1,150))

∴ £113,000 + £91,680 − (£1,350 + £200,000) = £8,930 − (£4,450 + £1,150)

∴ £204,680 − £201,350 = £8,930 − £5,600

∴ £3,330 = £3,330

The £3,330 represents net profit. This net profit, therefore, provides a bridge between the balance sheet and the profit and loss account.

Adjustments to Trial Balance

The trial balance is prepared from the books of account and is then adjusted for certain items. These items represent estimates or adjustments which typically do not form part of the initial double-entry process. Five of the main adjustments are *closing stock*, *accruals*, *prepayments*, *depreciation*, and *bad and doubtful debts*. The mechanics of their accounting treatment is discussed here. However, the items themselves are discussed in more depth in Chapters 4 and 5 on the profit and loss account and balance sheet.

Stock

Stock is an important asset to any business, especially manufacturing businesses. Stock control systems in large businesses can be very complex and sophisticated. In most sole traders and in many other businesses it is normal not to record the detailed physical movements (i.e., purchases and sales of stock). Stock is not, therefore, formally recorded in these organisations in the double-entry process. However, at the balance sheet date stock is valued. To maintain the double-entry, the asset of stock is entered twice: *first* in the trading part of the trading and profit and loss account; and *second*, in the current assets section of the balance sheet. These two figures for accounting purposes cancel out and thus the double-entry is maintained. Last year's closing stock figure becomes The current year's opening stock figure. Hyder's closing stock is given in The Company Camera 6.2.

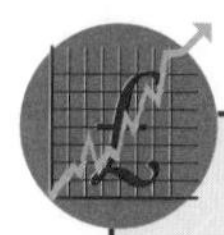

The Company Camera 6.2

Stocks

Stocks and work in progress

	Group 1999 £m	Grwp 1998 £m
Raw materials and consumables	8.8	9.4
Work in progress	6.5	4.4
Finished goods and goods for resale	0.7	0.3
	16.0	14.1

The replacement cost of stocks is not materially different from their carrying value.

Source: Hyder plc, 1999 Annual Report, p. 50

Accruals

Accruals are the amounts we owe to the suppliers of services such as telephone or light and heat. In small businesses, accruals are normally excluded from the initial double-entry process. Accruals appear in the final accounts in two places. *First*, the amount owing is included in the profit and loss account under expenses. *Second*, a matching amount is included under current liabilities in the balance sheet. The double-entry is thus maintained. Figure 6.1 gives an example of accruals and prepayments (payments in advance for services).

Figure 6.1 Accruals and Prepayments

From the following information calculate the trading and profit and loss, and balance sheet entries for:

1. Electricity

Mary Christmas has received three electricity bills for the year (£300, £400, £550). Another bill is due which is estimated at £600.

2. Rent

The quarterly rent for Mary Christmas's offices is £300 payable in advance. The first payment, when the business starts, is made on 1st January, the last on 31st December.

1. Electricity

Effectively, the total bill incurred for the year is the amount paid (£300 + £400 + £550) £1,250 and the amount due £600. Therefore, the total bill is £1,850; this is included in the profit and loss account. In the balance sheet, £600 is included.

2. Rent

The amount paid is 5 × £300 = £1,500.

However, only £1,200 relates to this year and should go in the trading and profit and loss account. The £300 balance is a prepayment in the balance sheet.

Trading and Profit and Loss Account		**Balance Sheet**	
Expenses	**£**	**Current Assets**	**£**
Electricity	1,850	Rent prepaid	300
Rent	1,200	**Current Liabilities**	
		Electricity owing	600

Prepayments

Prepayments represent the amount paid in advance to the suppliers of services, for example rent paid in advance. In many ways prepayments are the opposite of accruals. Whereas accruals must be added to the final accounts to achieve the matching concept, a prepayment must be deducted. The amount paid in advance is treated as an asset which will be used up at a future date. It is therefore excluded from the accounts.

Pause for Thought 6.3

Accruals and Prepayments

Can you think of four examples of accruals and prepayments?

We might have, for example:

Accruals	*Prepayments*
Electricity owing	Rent paid in advance
Business rates owing	Prepaid electricity on meter
Rent owing	Prepaid standing charge for telephone
Telephone owing	Insurance paid in advance

Depreciation

As we saw in Chapter 5, fixed assets wear out over time and depreciation seeks to recognise this. Depreciation in the accounts is simply recording this – twice. *First*, a proportion of the original cost is allocated as an expense in the profit and loss account. *Second*, an equivalent amount is deducted from the fixed assets in the balance sheet (see Figure 6.2. on the next page).

Rentokil Initial plc's fixed assets (as recorded in its 2000 balance sheet) are then given as an illustration in the Company Camera 6.3 on page 105. As you can see, they can be quite complex.

Bad and Doubtful Debts

This is the last of the five adjustments. Bad and doubtful debts are also considered in Chapter 5 on the balance sheet. Essentially, some debts *may* not be collected. These are termed doubtful debts.

The Company Camera 6.4

Doubtful Debts

Provisions for Bad and Doubtful Debts

Balance sheet	1999 £m	% of loan balances	1998 £m	% of loan balances
Residential property and other secured advances	405	0.43	398	0.48
Unsecured	150	7.04	156	8.42
Total	555	0.58	554	0.65

Figure 6.2 Example of Depreciation

A business has five assets shown at cost. Ten per cent of the original cost of each asset has been allocated as depreciation. Record the transactions in the accounts.

	£
Premises	80,000
Machine	75,000
Office furniture	12,000
Computer	1,500
Motor van	6,000

There are two parts to this. **First**, we record 10% of the original cost as an expense in the expenses section of the trading and profit and loss account.

Trading and Profit and Loss Account (Year 1)

Expenses	£
Depreciation on premises	8,000
Depreciation on machine	7,500
Depreciation on office furniture	1,200
Depreciation on computer	150
Depreciation on motor van	600

Second, we record the original cost, total depreciation (called accumulated depreciation) and net book value (original cost less depreciation) in the balance sheet. Note that accumulated depreciation is recorded in brackets – this shows it is being taken away.

Balance Sheet (Year 1)

Fixed Assets	£ *Cost*	£ *Accumulated depreciation*	£ *Net book value*
Premises	80,000	(8,000)	72,000
Machine	75,000	(7,500)	67,500
Office furniture	12,000	(1,200)	10,800
Computer	1,500	(150)	1,350
Motor van	6,000	(600)	5,400
	174,500	(17,450)	157,050

In next year's trial balance, the cost and accumulated depreciation figures would be recorded. The *accumulated depreciatio* is often called *provision for depreciation*. This example records only one year's depreciation. However, in future years, there will be more depreciation, which is why it is called accumulated depreciation.

In the following year, we also have 10% depreciation. The trading and profit and loss account is thus the same.

Trading and Profit and Loss Account (Year 2)

Expenses	£
Depreciation on premises	8,000
Depreciation on machine	7,500
Depreciation on office furniture	1,200
Depreciation on computer	150
Depreciation on motor van	600

In the balance sheet, however, we *add* this year's depreciation to the accumulated depreciation figure. We, therefore, have:

Balance Sheet (Year 2)

Fixed Assets	£ *Cost*	£ *Accumulated depreciation*	£ *Net book value*
Premises	80,000	(16,000)	64,000
Machine	75,000	(15,000)	60,000
Office furniture	12,000	(2,400)	9,600
Computer	1,500	(300)	1,200
Motor van	6,000	(1,200)	4,800
	174,500	(34,900)	139,600

In essence, we have merely added the two years' depreciation figures. Thus, for premises, our opening figure for accumulated depreciation was £8,000. We then added this year's depreciation to arrive at £16,000. The net book value is simply the residual figure.

For east of understanding these same figures are used in our comprehensive example, Live Wire, later in this chapter.

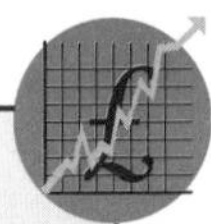

The Company Camera 6.3

Fixed Assets

Tangible fixed assets

	Land and buildings £m	Plant, equipment & tropical plants £m	Vehicles and office equipment £m	Total £m
Cost or valuation				
At 1st January 2000 as previously reported	232.8	1,065.3	414.5	1,712.6
Adjustment on adoption of FRS 15	(3.8)	–	–	(3.8)
At 1st January 2000 as restated	229.0	1,065.3	414.5	1,708.8
Exchange adjustments	1.4	31.3	1.5	34.2
Additions at cost	18.9	135.5	49.5	203.9
On acquisitions	12.0	18.5	8.6	39.1
Disposal of businesses	(14.5)	(651.1)	(69.1)	(734.7)
Disposals	(5.9)	(71.7)	(38.9)	(116.5)
Reclassification	1.3	(0.1)	(1.2)	–
At 31st December 2000	242.2	527.7	364.9	1,134.8
Aggregate depreciation				
At 1st January 2000 as previously reported	27.5	547.6	246.6	821.7
Adjustment on adoption of FRS 15	1.8	–	–	1.8
At 1st January 2000 as restated	29.3	547.6	246.6	823.5
Exchange adjustments	0.3	19.3	1.5	21.1
On acquisitions	7.6	8.9	4.2	20.7
Disposal of businesses	(3.1)	(310.9)	(48.4)	(362.4)
Depreciation	3.3	114.7	49.0	167.0
Disposals	(0.7)	(62.9)	(33.2)	(96.8)
Reclassification	1.0	(0.7)	(0.3)	–
At 31st December 2000	37.7	316.0	219.4	573.1
Net book value at 31st December 2000	204.5	211.7	145.5	561.7
Net book value at 31st December 1999 (restated)	199.7	517.7	167.9	885.3
Net book value of assets under finance leases at 31st December 2000	3.4	–	17.4	20.8

Source: Rentokil Initial plc, 2000 Annual Report, p. 45

Other debts *will definitely not* be collected; they are called bad debts. The accounting entries to record this are in the profit and loss account as an expense and in the balance sheet as a reduction in debtors. Some businesses, such as building societies (see Halifax's balance sheet in The Company Camera 6.4), typically carry a high level of bad and doubtful debts.

When considering the accounting treatment of bad and doubtful debts, it is crucial to distinguish between bad and doubtful debts.

1. Bad Debts

Bad debts are recorded as an expense in the profit and loss account and written off debtors in the balance sheet (see Figure 6.3).

Figure 6.3 Illustrative Example of Bad Debts

A business has debtors of £4,800, but estimates that bad debts will be £320.

Profit and Loss Account	£
Expenses	
Bad debts	320
Balance Sheet	
Current Assets	
Debtors less bad debts	
(£4,800 – £320)	4,480

For ease of understanding, these same figures are used in our comprehensive example, Live Wire, later in this chapter.

2. Provision for Doubtful Debts

A provision for doubtful debts is set up by a business for those debts it is dubious about collecting. This provision is *always* deducted from debtors in the balance sheet. However, only *increases or decreases* in the provision are entered in the profit and loss account. An *increase* is recorded as an *expense* and a *decrease* as an *income* (see Figure 6.4). *Where there are both bad and doubtful debts, the provision for doubtful debts is calculated first.*

Figure 6.4 Illustrative Example of Doubtful Debtors

A business has debtors of £4,800. There is a doubtful debt provision of 10% of debtors. Last year debtors were £2,400 and the doubtful debt provision was £240.

Profit and Loss Account	£
Expenses	
Increase in doubtful debt provision (£480 – £240)	*240*
Balance Sheet	
Current Assets	
Debtors less provision for doubtful debts	
(£4,800 – £480)	*4,320*

For ease of understanding, these same figures are used in our comprehensive example, Live Wire, later in this chapter.

Soundbite 6.1

Debts

'A small debt makes a man your debtor, a large one makes him your enemy.'

Seneca (Ad Lucilium xix)

Source: The Executive's Book of Quotations (1994), p. 80

It is now time to introduce some items commonly found in accounts. These are briefly explained here. Mainly we focus on the accounts of a sole trader. Goodwill and other intangible assets such as brands (i.e., assets that we cannot touch) are discussed in more detail in Chapter 12.

Figure 6.5 Introducing Common Items Found in the Final Accounts

Item	Explanation	Location
Bank overdraft	This is where the business owes the bank money. Too many students are in this position!	Balance sheet Current liabilities
Carriage inwards	This is usually found in manufacturing businesses. It is a cost of purchasing raw materials. It refers to the days when goods were brought in by horse-and-carriage.	Trading and profit and loss account Trading account Added to purchases
Carriage outwards	Similar to carriage inwards, except that it is an expense incurred by the business when selling goods.	Trading and profit and loss account Profit and loss account Expenses
Discount allowed	This is discount allowed by the business to customers for prompt payment. In other words, instead of paying, say £100, the customer pays £95. The sale of £100 is recorded as normal, but discount allowed is recorded separately.	Trading and profit and loss account Profit and loss account Expenses
Discount received	Similar to discount allowed. However, it is received by the business from the supplier for paying promptly. The business therefore, pays less. The purchase is recorded as normal, the discount received is recorded separately.	Trading and profit and loss account Proft and loss account Other income
Drawings	This is money which the sole trader or owner takes out for his or her living expenses. It is, in effect, the owner s salary. It is really a withdrawal of capital.	Balance sheet Capital employed *Note*: Do *not* put in expenses
Income receivable*	This is income which is received by the business from a third party. Examples, include dividends receivable from companies or interest receivable from the bank.	Trading and profit and loss account Profit and loss account Other income
Interest payable*	This is the reverse of income receivable. It is interest payable by the business to outsiders, especially on bank loans.	Trading and profit and loss account Profit and loss account Expenses
Long-term loan	This is a loan not repayable within a year. The loan may be with a bank or other organisation. Sometimes long-term loans are called debentures.	Balance sheet Long-term creditors

*Income receivable and interest payable are broader phrases than interest received and interest paid. They include interest earned yet to be received and interest incurred yet to be paid, respectively.

Comprehensive Example

We now close this chapter with a comprehensive example. This example includes most items normally found in the accounts of a sole trader. We use the example of a small engineering business run by Live Wire, which buys and sells electrical products. The trial balance is provided in Figure 6.6(a), and then in Figure 6.6(b) are the trading and profit and loss account, the balance sheet and the explanatory notes.

Figure 6.6 **Worked Example of a Sole Trader's Accounts**

(a)

Live Wire
Trial Balance as at 31 December 2001

	Debit £	Credit £
Sales		250,000
Sales returns	800	
Purchases	100,000	
Purchases returns		600
Carriage inwards	200	
Carriage outwards	300	
Discounts allowed	150	
Discounts receivable		175
Dividends receivable		225
Interest receivable		100
Drawings	26,690	
Advertising	1,250	
Telephone	2,150	
Wages	27,000	
Electricity	3,700	
Business rates	1,500	
Travelling expenses	1,200	
Van repairs	650	
Petrol	3,800	
General expenses	1,200	
Insurance	1,500	
Premises at cost	80,000	
Accumulated depreciation as at 1 January 2001		8,000
Machine at cost	75,000	
Accumulated depreciation as at 1 January 2001		7,500
Office furniture at cost	12,000	
Accumulated depreciation as at 1 January 2001		1,200
Computer at cost	1,500	
Accumulated depreciation as at 1 January 2001		150
Motor van at cost	6,000	
Accumulated depreciation as at 1 January 2001		600
Opening stock	9,000	
Cash at bank	7,050	
Bank overdraft		3,600
Debtors	4,800	
Provision for doubtful debts as at 1 January 2001		240
Creditors		9,800
Long-term loan		15,600
Interest payable	1,150	
Capital		70,800
	368,590	368,590

You have the following extra information:

(i) Closing stock at 31 December 2001 was £8,600.

(ii) Of the insurance £500 was paid in advance.

(iii) Live Wire still owes £450 for the telephone.

(iv) Depreciation is charged at 10% on the cost of the fixed assets (i.e., £8,000 for premises, £7,500 for machine, £1,200 for office furniture, £150 for computer and £600 for motor van). The accumulated depreciation is the depreciation charged to date.

(v) Bad debts are £320. They have not yet been written off. The provision for doubtful debts has increased from £240 to £480 and is 10% of debtors.

Required: Live Wire's Trading and Profit and Loss Account for year ended 31 December 2001 and Balance Sheet as at 31 December 2001

Figure 6.6 **Live Wire (*continued*)**

(b) **Live Wire**

Trading and Profit and Loss Account Year Ended 31 December 2001

	£	£	£
Sales			250,000
Less Sales returns			800
			249,200
Less *Cost of Sales*			
Opening stock		9,000	
Add Purchases	100,000		
Carriage inwards	200		
	100,200		
Less Purchases returns	600	99,600	
		108,600	
Less Closing stock (Note 2)		8,600	100,000
Gross Profit			149,200
Add *Other Income*			
Discounts receivable		175	
Dividends receivable		225	
Interest receivable		100	500
			149,700
Less *Expenses*			
Carriage outwards		300	
Discounts allowed		150	
Advertising		1,250	
Telephone (Note 3)		2,600	
Wages		27,000	
Electricity		3,700	
Business rates		1,500	
Travelling expenses		1,200	
Van repairs		650	
Petrol		3,800	
General expenses		1,200	
Insurance (Note 4)		1,000	
Interest payable		1,150	
Depreciation (Note 5)			
Premises		8,000	
Machine		7,500	
Office furniture		1,200	
Computer		150	
Motor van		600	
Bad debts (Note 6)		320	
Doubtful debts (Note 7)		240	63,510
Net Profit			86,190

Figure 6.6 Live Wire (*continued*)

(b)

Live Wire
Balance Sheet as at 31 December 2001

	£ *Cost*	£ *Accumulated depreciation (Note 5)*	£ *Net book value*
Fixed Assets			
Premises	80,000	(16,000)	64,000
Machine	75,000	(15,000)	60,000
Office furniture	12,000	(2,400)	9,600
Computer	1,500	(300)	1,200
Motor van	6,000	(1,200)	4,800
	174,500	(34,900)	139,600
Current Assets			
Stocks (Note 2)	8,600		
Debtors less bad and doubtful debts (Notes 6,7)	4,000		
Prepayments (Note 4)	500		
Cash at bank	7,050	20,150	
Current Liabilities			
Creditors	(9,800)		
Bank overdraft	(3,600)		
Accruals (Note 3)	(450)	(13,850)	
Net current assets			6,300
Total assets less current liabilities			145,900
Long-term Creditors			(15,600)
Total net assets			130,300
Capital Employed			£
Opening capital			70,800
Add Net profit			86,190
			156,990
Less Drawings			26,690
Closing capital			130,300

Notes:

1. All the figures in Live Wire's accounts, except for closing stock, telephone, insurance, depreciation, and bad and doubtful debts are as listed in the trial balance. The figures for profit and closing capital are calculated as balancing figures.
2. Closing stock of £8,600 is recorded in the trading account and in the balance sheet.
3. The telephone expense is adjusted for the £450 owing. In the profit and loss account, the expense increases to £2,600 (£2,150 + £450). In the balance sheet, the £450 owing becomes an accrual.
4. The insurance is adjusted for the £500 paid in advance. The expense in the profit and loss account thus becomes £1,000 (£1,500 – £500). The £500 is recorded as a prepayment in current assets in the balance sheet.
5. The depreciation for the year, which was the same as in Figure 6.2, has been recorded twice: *first* under expenses in the trading and profit and loss account, *second*, in the balance sheet under fixed assets. The double-entry is thus maintained.
6. The £320 for bad debts has been recorded twice: *first*, under expenses in the trading and profit and loss account, and, *second*, under current assets (debtors) in the balance sheet. The double-entry is thus completed (see also Figure 6.3).
7. Doubtful debts represents the increase in the provision for doubtful debts from £240 to £480. The increase of £240 is recorded twice. *First*, under expenses in the trading and profit and loss account, and, *second*, as part of the £480 deduction from debtors under current assets in the balance sheet. The double-entry is thus completed (see also Figure 6.4).

Conclusion

The two main financial statements – the trading and profit and loss account and the balance sheet – are both prepared from the trial balance. The trading and profit and loss account focuses on income, such as sales or dividends receivable, and expenses, such as telephone or electricity. The balance sheet, by contrast, focuses on assets, liabilities and owner's capital. In both financial statements, profit becomes the balancing figure.

After the trial balance has been prepared, the accounts are often adjusted for items such as closing stock, accruals (amounts owing), prepayments (amounts prepaid), depreciation (the wearing out of fixed assets), and bad and doubtful debts. In each case, we adjust the accounts twice: *first* in the trading and profit and loss account and *second* in the balance sheet. The double-entry and thus the symmetry of the accounts is thus maintained.

Discussion *Questions*

Questions with numbers in blue have answers at the back of the book.

Q 1 What is a sole trader and why is it important for the sole trader to prepare a set of financial statements?

Q 2 'Profit is the figure which links the profit and loss account and the balance sheet.' Discuss.

Q 3 Why do we need to carry out post-trial balance adjustments when we are preparing the final accounts?

Numerical *Questions*

The numerical questions which follow are graded in difficulty. Those at the start are about as complex as the illustrative example, Gavin Stevens. They gradually become more complex, until the final questions equate to the illustrative example, Live Wire.

Questions with numbers in blue have answers at the back of the book.

Q **1** Michael Anet has the following trial balance

M. Anet
Trial Balance as at 31 December 2001

	Debit	Credit
	£	£
Hotel	40,000	
Van	10,000	
Sales		25,000
Purchases	15,000	
Capital		51,900
Bank	8,000	
Electricity	1,500	
Wages	2,500	
A. Brush (Debtor)	400	
A. Painter (Creditor)		500
	77,400	77,400

Required:
Prepare Michael Anet's trading and profit and loss account for the year ended 31 December 2001 and balance sheet as at 31 December 2001.

Q **2** Paul Icasso has the following trial balance.

P. Icasso
Trial Balance as at 31 March 2002

	Debit	Credit
	£	£
Hotel	50,000	
Van	8,000	
Sales		35,000
Purchases	25,000	
Sales returns	3,000	
Purchases returns		4,000
Capital		60,050

Q 2 P. Icasso's trial balance (continued)

	Debit	Credit
	£	£
Bank	9,000	
Electricity	1,000	
Advertising	800	
Debtors Shah	1,250	
Debtors Chan	2,250	
Creditors Jones		1,250
	100,300	100,300

Required:
Prepare Paul Icasso's trading and profit and loss account for the year ended 31 March 2002 and a balance sheet as at 31 March 2002.

Q 3 Rose Ubens, buys and sells goods. Her trial balance is presented below.

R. Ubens
Trial Balance as at 31 December 2001

	Debit	Credit
	£	£
Opening stock	3,600	
Building	20,400	
Motor van	3,500	
Debtors	2,600	
Creditors		3,800
Cash at bank	4,400	
Electricity	1,500	
Advertising	300	
Printing and stationery	50	
Telephone	650	
Rent and rates	1,200	
Postage	150	
Drawings	7,800	
Capital		19,950
Sales		88,000
Purchases	66,000	
Sales returns	800	
Purchases returns		1,200
	112,950	112,950

Note:
1. Closing stock is £4,000

Required:
Prepare Rose Ubens trading and profit and loss account for the year ended 31 December 2001 and a balance sheet as at 31 December 2001.

Q 4 Clara Onstable has prepared her trial balance as at 31 December. She has the following additional information.

(a) She pays £240 rent per month. She has paid £3,600, the whole year's rent plus three months in advance.

(b) She has paid insurance costs of £480. However, she has paid £120 in advance.

Required:
Prepare the extracts for the final accounts.

Q 5 Vincent Gogh, a shopkeeper, has the following trial balance.

V. Gogh
Trial Balance as at 31 December 2001

	Debit	Credit
	£	£
Sales		40,000
Sales returns	500	
Purchases	25,000	
Purchases returns		450
Opening stock	5,500	
Debtors	3,500	
Creditors		1,500
Cash at bank	1,300	
Long-term loan		3,700
Motor car	8,500	
Shop	9,000	
Business rates	1,000	
Rent	600	
Electricity	350	
Telephone	450	
Insurance	750	
General expenses	150	
Wages	10,500	
Drawings	12,900	
Capital		34,350
	80,000	80,000

You also have the following additional information.

1. Closing stock as at 31 December 2001 is £9,000.
2. V. Gogh owes £350 for electricity.
3. £200 of the rent is paid in advance.

Required:
Prepare V. Gogh's trading and profit and loss account for the year ended 31 December 2001 and a balance sheet as at 31 December 2001.

Q 6 Leonardo Da Vinci, who sells computers, has extracted the following balances from the accounts.

L. Da Vinci
Trial Balance as at 30 September 2001

	Debit	Credit
	£	£
Sales		105,000
Sales returns	8,000	
Purchases	70,000	
Purchases returns		1,800
Opening stock of computers	6,500	
Drawings	8,500	
Debtors	12,000	
Creditors		13,000
Cash at bank	1,800	
Long-term loan		6,600
Discounts allowed	300	
Carriage inwards	250	
Business premises	18,000	
Motor van	7,500	
Computer	1,500	
Wages	32,500	
Electricity	825	
Telephone	325	
Insurance	225	
Rent	1,250	
Business rates	1,000	
Capital		44,075
	170,475	170,475

You have the following additional information.

1. Closing stock of computers as at 30 September 2001 is £7,000.
2. Da Vinci owes £175 for the telephone and £1,200 for electricity.
3. The prepayments are £25 for insurance and £250 for rent.

Required:
Prepare Da Vinci's trading and profit and loss account for the year ended 30 September 2001 and a balance sheet as at 30 September 2001.

*Q*7 Helen Ogarth is preparing her accounts for the year to 31 December 2002. On 1 January 2002 she purchased the following fixed assets.

	£
Buildings	100,000
Machine	50,000
Motor van	20,000

She wishes to write off the following amounts for depreciation.

	£
Buildings	10,000
Machine	3,000
Motor van	2,000

Required:
Prepare the appropriate extracts for the balance sheet and trading and profit and loss account.

*Q*8 Michael Atisse, a carpenter, has the following trial balance as at 31 December 2001.

	Debit	Credit
	£	£
Work done		50,000
Purchases of materials	25,000	
Opening stock of tools	650	
Motor expenses	3,550	
Debtors	1,000	
Creditors		4,000
Cash at bank	3,600	
Long-term loan		16,800
Building at cost	52,300	
Motor car at cost	8,000	
Computer at cost	6,300	
Office equipment at cost	10,200	
Business rates	1,300	
Electricity	900	
Interest on loan	1,600	
Drawings	5,200	
Telephone	1,200	
Capital		50,000
	120,800	120,800

Q8 Michael Atisse (continued)

You have the following additional information.

1. Closing stock of tools £4,500
2. Depreciation is to be written off the fixed assets as follows:

Buildings	£3,000
Motor car	£2,000
Computer	£1,400
Office equipment	£1,800

Required:

Prepare M. Atisse's trading and profit and loss account for the year ended 31 December 2001 and the balance sheet as at 31 December 2001.

Q9 Clare Analetto, an antique dealer, has the following trial balance as at 31 December 2001.

	Debit	Credit
	£	£
Sales		100,000
Sales returns	5,000	
Purchases	70,000	
Purchases returns		6,000
Opening stock of antiques	9,000	
Debtors	16,800	
Business rates	800	
Creditors		14,000
Cash at bank	17,100	
Long-term loan		12,000
Bank interest receivable		850
Rent	2,050	
Electricity	1,950	
Insurance	1,250	
Loan interest	1,200	
General expenses	1,025	
Motor van expenses	1,800	
Premises at cost	60,000	
Machinery at cost	16,500	
Office equipment at cost	1,750	
Motor car at cost	2,050	
Drawings	8,200	
Repairs to antiques	1,300	
Telephone	500	
Capital		85,425
	218,275	218,275

Q 9 Clare Analetto (continued)

You also have the following notes to the accounts.

1. Closing stock of antiques is £7,000.
2. Depreciation for the year is to be charged at 2% on premises, 10% on machinery, 15% on office equipment, and 25% on the motor car.

Required:

Prepare C. Analetto's trading and profit and loss account for the year ended 31 December 2001 and the balance sheet as at 31 December 2001.

Q 10 Michelle Angelo has debtors of £40,000 at the year end. However, she feels that £4,000 are bad.

Required:

Prepare the appropriate balance sheets and trading and profit and loss account extracts.

Q 11 Simon Eurat, who runs a taxi business, has the following trial balance.

S. Eurat

Trial Balance as at 30 June 2002

	Debit	Credit
	£	£
Cash overdrawn at bank		1,500
Long-term loan		3,550
Receipts		28,300
Diesel and oil	8,250	
Taxi repairs and service	3,950	
Radio hire	3,400	
Road fund licences	2,300	
Buildings at cost	68,000	
Taxis at cost	34,500	
Business rates	450	
Electricity	1,300	
Telephone	1,250	
Debtors	100	
Creditors for motor repairs		1,800
Insurance on buildings	1,300	
General expenses	850	
Drawings	9,600	
Bank interest	150	
Capital		103,250
Wages	3,000	
	138,400	138,400

Q **11** Simon Eurat (continued)

You have the following additional information.

1. 10% of the debtors are considered bad.
2. £800 of the insurance is prepaid.
3. There is £250 owing for electricity and £300 owing for telephone.
4. Depreciation on taxis is to be 25% on cost and on buildings 2% on cost.

Required:

Prepare S. Eurat's trading and profit loss account for the year ended 30 June 2002 and balance sheet as at 30 June 2002.

Q **12** Rebecca Odin has the following details of her fixed assets.

	Cost	*Accumulated depreciation as at 31 December 2000*
	£	£
Buildings	102,000	8,000
Machinery	65,000	6,500
Motor car	8,000	4,000
Computer	9,000	2,700

She charges depreciation at 2% per annum on cost for buildings, 10% per annum on cost for machinery, 25% per annum on cost for the motor car and 15% per annum on cost for the computer.

Required:

Prepare the appropriate extracts for

(a) the balance sheet as at 31 December 2000.

(b) the trading and loss account for year ended 31 December 2001 and for the balance sheet as at 31 December 2001.

Q **13** Deborah Urer owns a small bar. Her trial balance as at 30 June 2002 is set out below.

	Debit £	Credit £
Takings from sales		145,150
Purchases of beer, wine and spirits	83,250	
Discounts receivable		450
Dividends receivable		150
Drawings	26,400	
Advertising	3,600	
Motor expenses	1,750	
Telephone	2,800	
Electricity	1,250	
Insurance	1,900	
General expenses	2,250	
Repairs	350	
Premises at cost	20,340	
Accumulated depreciation as at 1 July 2001		6,300
Bar equipment at cost	8,200	
Accumulated depreciation as at 1 July 2001		2,500
Bar furniture at cost	5,600	
Accumulated depreciation as at 1 July 2001		1,800
Motor car at cost	3,600	
Accumulated depreciation as at 1 July 2001		2,000
Debtors	220	
Provision for doubtful debts		20
Loan interest	650	
Creditors		3,650
Cash at bank	4,350	
Long-term loan		6,500
Wages	2,560	
Business rates	2,450	
Opening stock	5,500	
Capital		8,500
	177,020	177,020

You also have the following additional information.

1. Closing stock is £6,250.
2. There is £200 owing for electricity and £300 of the insurance is prepaid.
3. Bad debts are £25.
4. The doubtful debt provision is to be increased to £25 on 30 June 2001.
5. There are the following depreciation charges:
   2% on premises
   10% on bar equipment and bar furniture
   25% on motor car

Required:

Prepare D. Urer's trading and profit and loss account for the year ended 30 June 2002 and the balance sheet as at 30 June 2002.

Q **14** Bernard Ruegel has drawn up a trial balance which is presented below.

B. Ruegel
Trial Balance as at 30 September 2002

	Debit	Credit
	£	£
Carriage inwards	350	
Carriage outwards	180	
Discounts allowed	80	
Discounts receivable		75
Sales		208,275
Sales returns	185	
Purchases	110,398	
Drawings	38,111	
Purchases returns		98
Interest receivable		790
Advertising	1,987	
Telephone	476	
Wages and salaries	10,298	
Electricity	1,466	
Rent	2,873	
Cash at bank	21,611	
Travelling expenses	1,288	
Van repairs	1,471	
Petrol	2,187	
Business rates	2,250	
General expenses	1,921	
Insurance	1,100	
Premises at cost	50,981	
Accumulated depreciation as at 1 October 2001		28,300
Machine at cost	21,634	
Accumulated depreciation as at 1 October 2001		8,200
Office furniture at cost	8,011	
Accumulated depreciation as at 1 October 2001		2,386
Computer at cost	2,980	
Accumulated depreciation as at 1 October 2001		1,200
Motor van at cost	2,725	
Accumulated depreciation as at 1 October 2001		1,725
Motor car at cost	5,386	
Accumulated depreciation as at 1 October 2001		1,980
Opening stock	11,211	
Loan interest	1,500	
Bank overdraft		8,933
Debtors	11,000	

Q **14** Bernard Ruegel (continued)

	Debit	Credit
	£	£
Provision for doubtful debts		850
Creditors		4,279
Long-term loan		14,811
Capital		31,758
	313,660	313,660

You have the following extra information:

1. Closing stock is £13,206
2. There are the following amounts owing: advertising £325, telephone £125, carriage outwards £20, general expenses £37, wages and salaries £560.
3. The following amounts are prepaid: travelling expenses £288, electricity £76, insurance £250.
4. It is business policy to treat 10% of total debtors as doubtful.
5. There is a bad debt of £600.
6. It has been decided to write down the fixed assets **to** the following net book value amounts as at 30 September 2002.

	£
Premises	21,081
Machine	12,434
Office furniture	4,151
Computer	125
Motor van	275
Motor car	1,599

Required:
Prepare, taking the necessary adjustments into account, the trading and profit and loss account for the year ended 30 September 2002 and the balance sheet as at 30 September 2002.

Chapter 7

Partnerships and Limited Companies

> *'Corporation, [i.e. Company] n. An ingenious device for obtaining individual profit without individual responsibility.'*
>
> Ambrose Bierce, *The Devil's Dictionary*, p. 29

Learning Outcomes

After completing this chapter you should be able to:

- ✔ **Explain the nature of partnerships and limited companies.**
- ✔ **Outline the distinctive accounting features of partnerships and limited companies.**
- ✔ **Demonstrate how to prepare the accounts of partnerships and limited companies.**

In a Nutshell

- Sole proprietors, partnerships and limited companies are the main forms of business enterprise.
- A partnership is more than one person working together.
- Partnership accounts must share out the profit and capital between the partners.
- Sharing out profit, capital and current accounts are special partnerships features.
- A limited company is based on the limited liability of the shareholders (i.e., they lose only their initial investment if things go wrong).
- A limited company's special features are taxation, dividends and capital employed split between share capital and reserves.
- In company accounts it is common to find intangible assets (i.e. assets you cannot touch) such as goodwill or patents.
- Limited companies may be private or public.
- A limited company's published accounts are standardised, abridged, supplemented by notes and incorporated in an annual report.
- Annual reports are sent to shareholders. They are also increasingly put on a company's website.

Introduction

The three most common types of business enterprise are sole traders, partnerships and public corporations. For example, in 2000, in the UK, there were 584,460 sole proprietors (or sole traders), 365,995 partnerships and 636,930 companies and public corporations (Office for National Statistics, 2000). So far, in this book, we have focused on sole traders. Sole traders are, typically, relatively small enterprises owned by one person. Their businesses and accounts tend to be less complicated than those of either partnerships or companies. They are thus ideal for introducing the basic principles behind bookkeeping and final accounts. In this chapter, we now look at the profit and loss accounts and balance sheets of partnerships and limited companies. The cash flow statements prepared by companies are covered in Chapter 8. Partnerships are normally larger than sole traders. However, basically their accounts are similar to those of the sole trader. This reflects the fact that partners, like sole traders, generally own and run their own businesses. For companies, however, the owners provide the capital, but the directors run the company. This divorce of ownership and management is reflected in the accounts of limited companies. In certain respects, particularly the capital employed, the accounts of limited companies thus appear quite different to those of partnerships and sole traders.

Context

In this section, we briefly set out the main features of sole traders, partnerships and limited companies. The main points are summarised in Figure 7.1 on the next page. In essence, the differences between these three types of business enterprise can be traced back to size and capital structure. In terms of size, sole traders are normally smaller than partnerships, which are usually smaller than companies. This greater size causes accounting to be more complicated for companies than for sole traders.

An important distinction between the three businesses is the capital structure. Sole traders and most partners own and run their businesses. They provide the capital, although they may borrow money. The main problem for partnerships is simply the fair allocation of both the capital and profit to the partners. For companies, the owners provide the capital whereas the directors run the company. The concept of limited liability for companies means that shareholders can only lose the money they initially invested.

> *Soundbite 7.1*
>
> **Partners**
>
> **'The history of human achievement is rich with stories of successful partnerships – while the history of human failure is rife with tales of fruitless competition and wilful antagonisms.'**
>
> Margaret E. Mahoney, *The Commonwealth Fund* (Annual Report 1987)
>
> *Source*: *The Executive's Book of Quotations* (1994), p. 212

Partnerships

Introduction

Partnerships may be seen as sole traders with multiple owners. Many sole traders take on partners to help them finance and run their businesses. As in all human relationships, when partners are well-matched partnerships can prove very successful businesses. However, when they are ill-suited problems can occur (see Soundbite 7.1). Except for certain occupations (such as firms of accountants or solicitors) the maximum number of partners in the UK is 20. An important aspect of both sole traders and partnerships is that liability is generally

Figure 7.1 **Sole Traders, Partnerships, and Limited Companies Compared**

Feature	*Sole Traders*	*Partnerships*	*Limited Companies*
Business			
(i) Owners	Sole traders	Partners	Shareholders
(ii) Run company	Sole traders	Partners	Directors
(iii) Statutory accounting legislation	No specific act	Partnership Act, 1890	Companies Acts
(iv) Number of owners	1	2–20 (but certain exceptions such as accountants, solicitors)	Private 1–50 Public 2 upwards
(v) Liability	Unlimited	Unlimited, except for limited partners	Limited
(vi) Number in UK in 2000*	584,460	365,995	636,930
*(vii) Size of turnover in UK**	(a) 66% under £100,000 (b) 1% over £1m	(a) 42% under £100,000 (b) 4% over £1 m	(a) 36% under £100,000 (b) 19% over £1m
Accounting			
(i) Main external users of accounts	Tax authorities, bank	Tax authorities, bank	Tax authorities for small, private companies, shareholders for public companies
(ii) Main financial statements	Trading and profit and loss account and balance sheet	Trading, profit and loss and appropriation account and balance sheet	Profit and loss account, balance sheet and cash flow statement
(iii) Main differences in profit and loss account from sole trader	–	Appropriation account shares out profit	Appropriation account has dividends and taxation
(iv) Main differences in net assets from sole trader	–	None	Companies, when in groups, may have goodwill. They are also likely to have other intangible assets such as patents or brands. In current liabilities, there are proposed dividends and taxation payable
(v) Main differences in owners' capital from sole trader	–	Capital and current accounts record partners' share of capital invested and profit	Capital essentially divided into share capital and reserves

****From Office for National Statistics, Size Analysis of United Kingdom Businesses, 2000***

unlimited. (There is an exception, a limited partnership. Limited partners can lose only the capital invested. However, they do not participate in running the company and there must be at least one unlimited liability partner.) In other words, if a business goes bankrupt the personal assets of the owners are *not* ring-fenced. Bankrupt partners may have to sell their houses to pay their creditors.

Sole traders and partners prepare accounts for their own personal internal management uses, but also for external users, such as the tax authorities and banks. The key issue which underpins partnerships is how the partners should split any profits. The ratio in which the profits are split is called the profit sharing ratio or PSR.

The allocation of profits between the partners is presented in the appropriation account. The appropriation, or 'sharing out', account appears after the calculation of net profit. In other words, we add a section at the bottom of the sole trader's trading and profit and loss account. So we now have the trading and profit and loss and appropriation account.

Real-Life Nugget 7.1

Partnerships

Occasionally partners have more in common than business interests: their names seem to complement each other. The following curious and apt names of partnerships were collected from old English signs and business directories: Carpenter & Wood; Spinage & Lamb; Sage & Gosling; Rumfit & Cutwell, and Greengoose & Measure, both tailors; Single & Double; Foot and Stocking, Hosiers. One is not quite sure whether to believe that Adam & Eve were two surgeons who practised in Paradise Row, London, though Byers & Sellers did have a shop in Holborn.

'Sometimes the occupation of persons harmonizes admirably with their surnames,' a nineteenth-century antiquarian continues:

> Gin & Ginman are innkeepers; so is Alehouse; Seaman is the landlord of the Ship Hotel, and A. King holds the 'Crown and Sceptre' resort in City Road. Portwine and Negus are licensed victuallers, one in Westminster and the other in Bishopsgate Street. Mixwell's country inn is a well-known resort. Pegwell is a shoemaker, so are Fitall and Treadaway, likewise Pinch; Tugwell is a noted dentist; Bird an egg merchant; Hemp a sherriff's officer; Captain Isaac Paddle commands a steamboat; Mr. Punt is a favorite member of the Surrey wherry [rowing] club; Laidman was formerly a pugilist; and Smooker or Smoker a lime burner; Skin & Bone were the names of two millers in Manchester; Fogg and Mist china dealers in Warwick street: the firm afterward became Fogg & Son, on which it was naturally enough remarked that the 'son had driven away the mist.' Mr. I. Came, a wealthy shoemaker in Liverpool, who left his immense property to public charities, opened his first shop on the opposite side of the street to where he had started as a servant, and inscribed a sign: 'I CAME from over the way.'
>
> Finally, Going & Gonne was the name of a well-known banking house in Ireland, and on their failure in business some one wrote:
>
> Going & Gonne are now both one
> For Gonne is going and Going's gone.

Source: Peter Hay (1988) *The Book of Business Anecdotes*, Harrap Ltd, London, pp. 119–20

The main elements of the basic appropriation account are salaries and the sharing of the profit. Salaries are allocated to the partners before the profit is shared out. Note that for partners, salaries are an appropriation *not* an expense. Figure 7.2 demonstrates the process. Two partners, A and B, share £18,000 net profit. The profit sharing ratio is 2:1. Their salaries are £5,500 and £3,500, respectively. The example is continued in Figure 7.3 on page 130.

Figure 7.2 Main Elements in the Appropriation Account

Main Elements	*Explanation*	*Layout*			
				£	£
Net Profit	Profit as calculated from trading and profit and loss account	Net profit before appropriation			18,000
Salaries	The amount which each partner earns must be deducted from net profit before profit sharing	Less: Salaries	A	5,500	
			B	3,500	9,000
					9,000
Residual Profit	The profit share for each individual	Profit	A	6,000	
			B	3,000	9,000

Partners' capital can be divided into two parts: capital accounts and current accounts. Each partner needs to keep track of his or her own capital.

Pause for Thought 7.1

Partners' Profit Sharing

Why are profits not just split equally between partners?

Superficially, it might seem that the partners might just split the profit equally between them. So if a partnership of two people earns £30,000; each partner's share is £15,000. Unfortunately, life is not so simple! In practice, profit sharing is determined by a number of factors, such as how hard each partner works, their experience and the capital each partner has contributed. It might, therefore, be decided that the profit sharing ratio or PSR was 2:1. In this case, one partner would take £20,000; the second would take £10,000.

Capital Accounts

These accounts represent the long-term capital invested into the partnership by the individual partners. When new partners join a partnership it is conventional for them to 'buy their way' into the partnership. This initial capital introduction can be seen as purchasing their share of the net assets of the business they have joined. This initial capital remains unchanged in the accounts unless the partners specifically introduce or withdraw long-term capital. It represents the amount which the business owes the partners.

Current Accounts

In contrast to the capital accounts, current accounts are not fixed. Essentially, they represent the partners' share of the profits of the business since they joined, less their withdrawals. In basic current accounts, the main elements are the opening balances, salaries, profit for year, drawings and closing balances. These elements are set out in Figure 7.3. This continues Figure 7.2, with drawings of £12,000 for A and £10,000 for B. It is important to realise that the salaries are credited (or added) to the partners' current account rather than physically paid. The partners physically withdraw cash which is known as drawings. Drawings are essentially sums taken out of the business by the partners as living expenses.

Figure 7.3 Main Elements in Partners' Current Accounts

Main Elements	*Explanation*	*Layout*		
			A £	B £
Opening Balances	Amount of profits brought forward from last year. Normally a credit balance, and is the amount the business owes the partner	Opening balances	7,000	6,000
		Add:		
Salaries	The amount which the partners earn by way of salary	Salaries	5,000	4,000
Profit Share	Amount of the profit attributable to partner. Determined by profit sharing ratio	Profit	5,500	3,500
			17,500	13,500
Drawings	Amount the partners take out of the business to live on	Less:		
		Drawings	12,000	10,000
Closing Balances	The amount of profits carried forward to next year. This is usually the balance owed to the partner	Closing balances	5,500	3,500

Pause for Thought 7.2

Debit Balances on Current Accounts

What do you think a negative or debit balance on a partner's current account means?

....

Pause for Thought 7.2 (continued)

This means that the partner owes the partnership money! A current account represents the partner's account with the business. It is increased by or credited with (i.e., the business owes the partner money) the partner's salary and share of profit. The account is then debited (or reduced) when the partner takes money out (i.e., drawings). If the partner takes out more funds than are covered by the salary and profit share a debit balance is created. It is, in effect, like going overdrawn at the bank. A partner with a debit balance owes rather than is owed capital.

Partnership Example: Stevens and Turner

Let us imagine that Gavin Stevens has traded for several years. He has now teamed up with Diana Turner. Both partners have invested £35,000 capital. Their salaries are £12,000 for Stevens and £6,000 for Turner. Their current accounts stand at £8,000 (credit) Stevens, £9,000 (credit) Turner. They share residual profits in Steven's favour 2:1. Their opening trial balance is below.

Stevens and Turner: Partnership Trial Balance as at 31 December 200X

		£	£
Capital accounts	Stevens		35,000
	Turner		35,000
Current accounts	Stevens		8,000
	Turner		9,000
Drawings	Stevens	18,000	
	Turner	13,500	
Hotel		110,000	
Vans		30,200	
Opening stock		5,000	
Debtors		15,000	
Creditors			20,000
Bank		20,300	
Electricity		1,850	
Wages		12,250	
Telephone		350	
Long-term loan			50,000
Sales			270,000
Purchases		195,000	
Other expenses		5,550	
		427,000	427,000

Notes:

1. Closing stock is £10,000.
2. For simplicity, we are ignoring all other post-trial balance adjustments such as depreciation, bad and doubtful debts, accruals and prepayments.
3. Salaries are £12,000 for Stevens and £6,000 for Turner. These are 'notional' salaries in that the money is not actually paid to the partners. Instead it is credited to their accounts.

Using this trial balance we now prepare, in Figure 7.4, the trading and profit and loss and appropriation account and the balance sheet

Figure 7.4 Stevens and Turner Partnership Accounts Year Ended 31 December 200X

Stevens and Turner
Trading and Profit and Loss and Appropriation Account for Year Ended 31 December 200X

		£	£
Sales			270,000
Less *Cost of Sales*			
Opening stock		5,000	
Add Purchases		195,000	
		200,000	
Less Closing stock		10,000	190,000
Gross Profit			80,000
Less *Expenses*			
Electricity		1,850	
Wages		12,250	
Telephone		350	
Other expenses		5,550	20,000
******************************			
Net profit before appropriation			60,000
Less Salaries:			
Stevens		12,000	
Turner		6,000	18,000
			42,000
Profits:			
Stevens	2	28,000	
Turner	1	14,000	
			42,000
******************************			

Note: The net profit is calculated as for a sole trader. The profit is then shared out between the partners. This appropriation is shown between the asterisks.

Stevens and Turner
Balance Sheet as at 31 December 200X

	£	£	£
Fixed Assets			
Hotel			110,000
Vans			30,200
			140,200
Current Assets			
Stock	10,000		
Debtors	15,000		
Bank	20,300	45,300	

Figure 7.4 **Stevens and Turner Partnership (*continued*)**

	£	£	£
Current Liabilities			
Creditors	(20,000)	(20,000)	
Net current assets			25,300
Total assets less current liabilities			165,500
Long-term Creditors			(50,000)
Total net assets			115,500

****

	Stevens £	Turner £	£
Capital Employed			
Capital Accounts	35,000	35,000	70,000
Current Accounts			
Opening balances	8,000	9,000	
Add:			
Salaries	12,000	6,000	
Profit share	28,000	14,000	
	48,000	29,000	
Less Drawings	18,000	13,500	
Closing balances	30,000	15,500	45,500
Total partners' funds			115,500

****

Note: The total net assets part of the balance sheet is drawn up as for a sole trader. The capital and current accounts show the amounts due to the partners. They are distinctive to partnership accounts and are shown between the asterisks.

Limited Companies

The Basics

Limited companies are a popular form of legal business entity. As Real-Life Nugget 7.2 on the next page indicates, in actual fact, a 'company' does not physically exist. The essence of limited companies lies in the fact that the shareholders' (i.e., owners') liability is limited. This means that owners are only liable to lose the amount of money they have initially invested. For example, if a shareholder invests £500 in a company and the company goes bankrupt, then £500 is all the shareholder will lose. All the shareholder's personal possessions (for example, house or car) are safe! This is a great advantage over partnerships and sole traders where the liability is unlimited.

Real-Life Nugget 7.2

The Company

A second major feature of the orthodox creed is the ascription of supernatural existence to the 'company'. Objectively and rationally speaking, the 'company' does not exist at all, except in so far as it is a heterogeneous collection of people. In order to facilitate the mutual ownership and use of assets, and for other technical reasons, the corporation is treated in law as a person. It is a legal fiction, but a fiction nonetheless.

Source: Graham Cleverly (1971), *Managers and Magic*, Longman Group Limited, London, p. 31

Pause for Thought 7.3

Limited Liability

For suppliers, and particularly lenders, limited liability can be bad news as their money may be less secure. Can you think of any ways they may seek to counter this?

A fact that is often overlooked is that in small companies limited liability is often not seen as a bonus to those who have close connections with the company. Suppliers may be less certain that they will be paid and bankers more worried about making loans. In many cases, unlimited liability is replaced by other control mechanisms. For example, suppliers may want to be paid in cash or have written guarantees of payment. Bankers will often secure their loans against the property of the business and, in many cases, against the personal assets of the owners. So the owners may not avoid losing their personal possessions in a bankruptcy after all.

The mechanism underpinning a limited liability company is the share. The total capital of the business is divided into these shares (literally a 'share' in the capital of the business). For instance, a business with capital of £500,000 might divide this capital into 500,000 shares of £1 each. These shares may then be bought and sold. Subsequently, they will probably be bought or sold for more or less than £1. For instance, Sheilah might sell 50,000 £1 shares to Mary for £75,000. There is thus a crucial difference between the face value of the shares (£1 each in this case) and their trading value (£1.50 each in this case). The face value of the shares is termed **nominal value**. The trading value is termed **market price**. If market price increases it is the individual shareholder *not* the company that benefits.

The risk for the shareholders is that they will lose the capital they have invested. The reward is twofold. First, shareholders will receive dividends (i.e., annual payments based on profits) for investing their capital. The dividends are the reward for investing their money in the company rather than, for example, investing in a bank or building society where it would earn interest. The second reward is any potential growth in share price. For example, if Sheilah originally purchased the shares for £50,000, she would gain £25,000 when she sold them to Mary.

It is important to realise that the shareholders own the company, they do not run the company. Running the company is the job of the directors. This division is known as the 'divorce of ownership and control'. In many small companies, however, the directors own most of the shares. In this case, although in theory there is a separation of ownership and control, in practice there is not.

There are two types of company in the UK. The private limited company and the public limited company. The main features are outlined in Figure 7.5.

Figure 7.5 Features of Private Limited Companies (Ltds) and Public Limited Companies (Plcs) in the UK

Feature	*Private Limited Company*	*Public Limited Company*
Names	Ltd after company name	Plc after company name
Number of shareholders	1 to 50	2 to unlimited
Share trading	Restricted	Unrestricted
Stock market listing	No	Usually
Authorised share capital	No minimum	At least £50,000
Size	Usually small to medium enterprises	Usually medium to large enterprises

The essential difference is that private limited companies are usually privately controlled and owned whereas public limited companies are large corporations usually trading on the stock market. The major companies world-wide such as British Petroleum, Toyota and Coca Cola are all, in essence, public limited corporations. (Even though laws vary from country to country, they are broadly equivalent.) In these large companies, the managers are, in theory, accountable to the shareholders. In practice, many commentators doubt this accountability.

There are several reasons why companies prepare accounts. First, the detailed accounts (normally comprising a profit and loss account, a balance sheet and cash flow statement) will be used by management for internal management purposes. They will often be prepared monthly. The published accounts which are sent to the shareholders will usually be prepared annually.

The second reason is to comply with the Companies Act 1985. This lays down certain minimum statutory requirements which are supplemented by accounting standards and stock exchange regulations. Companies following these accounting requirements will normally prepare and send their shareholders published accounts containing a profit and loss account, balance sheet and (except for small companies which are exempt) a cash flow statement. These account are prepared using a standardised format. For large companies, these accounts are sent out to shareholders as part of the annual reporting package. This is dealt with in detail in Chapter 12; however, it is briefly introduced later in this chapter. Companies, as part of their statutory reporting requirements, will also submit a set of accounts to the Registrar of Companies.

Third, as well as preparing accounts for shareholders, companies may also provide accounts to other users who have an interest in the company's affairs. Of particular importance is the role of corporation tax. The shareholder accounts are usually used as the starting point for assessing this tax, which was introduced in 1965, and is payable by companies on their profits. It is calculated according to a complicated, and often-changing, set of tax rules. 'Accounting' profit is usually adjusted to arrive at 'taxable' profit. Unlike partnerships and sole traders, who are assessed for income tax as individuals, companies are assessed for corporation tax as taxable entities themselves.

Finally, especially for large companies, there may be a wide range of potential users of the accounts such as employees, customers, banks and suppliers. Their information needs were discussed in Chapter 1. The accounts of medium and large companies are usually prepared by the directors and then audited by independent accountants. This is so that the shareholders and other users can be assured that the accounts are 'true and fair'. The auditors' report is a badge of quality.

A distinctive feature of many companies is that they are organised into groups. This book does not cover the preparation of group accounts (which are often complex and complicated and best left to more specialist textbooks.) Interested readers might try Alexander and Britton (2001), *Financial Reporting*. For now we merely note that many medium and large companies are not, in fact, single entities, but are really many individual companies working together collectively. There is, usually one overall company which is the controlling company. This is further discussed in Chapter 12.

Pause for Thought 7.4

Abridged Company Accounts

Companies often prepare a full, detailed set of accounts for their internal management purposes. Why would they not wish to supply these to their shareholders?

For internal management purposes detailed information is necessary to make decisions. However, in the published accounts, directors are careful what they disclose. In a public limited company, anybody can buy shares and thus receive the published accounts. Directors do not wish to give away any secrets to potential competitors just because they own a few shares. In actual fact, the Companies Acts requirements allow the main details to be disclosed in a way that is sometimes not terribly informative.

Distinctive Accounting Features

The essence of the profit and loss account of sole traders, partnerships and limited companies is the same. However, there are some important differences. In the case of a sole trader and partnerships the only important difference is that for partnerships the profit is divided (this is formally known as appropriated) between the partners. For a limited company, there is an abridged, standardised format set out by the 1985 Companies Act. An illustration of the formats for the three business entities is given in Figure 7.6 on the next page.

Be careful of the limited companies' format. There are several possible variations and the example in Figure 7.6 has been grossly simplified for comparison purposes.

Figure 7.6 Differences between Profit and Loss Accounts of Sole Traders, Partnerships and Limited Companies

Sole Traders: *Trading and Profit and Loss Account*		***Partnerships:*** *Trading and Profit and Loss and Appropriation Account*		***Limited Companies:*** *Profit and Loss Account*	
	£		£		£
Sales	200,000	Sales	200,000	Sales	200,000
Cost of Sales	(100,000)	Cost of Sales	(100,000)	Cost of Sales	(100,000)
Gross Profit	100,000	*Gross Profit*	100,000	*Gross Profit*	100,000
Other Income	10,000	Other Income	10,000	Other Income	10,000
	110,000		110,000		110,000
Expenses	(60,000)	Expenses	(60,000)	Expenses	(60,000)
Net Profit	50,000	*Net Profit*	50,000	*Profit before Tax*	50,000
				Taxation	(20,000)
		Partner A	25,000	*Profit after Tax*	30,000
		Partner B	25,000	Dividends	(15,000)
				Retained Profit	15,000

The special nature of limited companies leads to several distinctive differences between the accounts of limited companies and those of sole traders and partnerships, both in the profit and loss account and in the balance sheet. We now deal with three special features of a company: taxation, dividends and long-term capital. We also discuss intangible assets which can occur in the accounts of sole traders and partnerships, but are much more common in company accounts.

1. Taxation

As previously noted, companies pay corporation tax. As Soundbite 7.2 shows, taxation has never proved very popular. For companies, taxation is assessed on annual taxable profits. Essentially, these are the accounting profits adjusted to comply with taxation rules. The accounting consequences of taxation on the profit and loss account and balance sheet are twofold.

- In the profit and loss account, *the amount for taxation for the year is recorded.*
- In the balance sheet, under current liabilities, *the liability for the year is recorded as proposed taxation.*

2. Dividends

A reward for shareholders for the capital they have invested is the dividends they receive. The accounting treatment for dividends mirrors that of taxation.

- In the profit and loss account, record *all dividends for the year.*
- In the balance sheet, record *dividends proposed (i.e. final dividend) under current liabilities.*

Soundbite 7.2

Taxation

'The art of taxation consists in so plucking the goose as to obtain the largest possible amount of feathers with the smallest possible amount of hissing.'

Jean-Baptiste Colbert

Source: *The Book of Business Quotations* (1998), p. 257

Many companies, like H.P. Bulmers Holdings plc (the cider makers – see The Company Camera 7.1) pay interim dividends as payments on account during the accounting year and then final dividends once the accounts have been prepared and the actual profit is known.

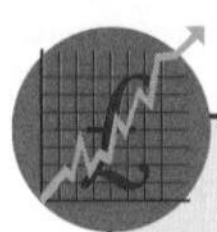

The Company Camera 7.1

Dividends

An interim dividend of 5.90 pence per share was paid on the ordinary shares on 19 February 2001. A final dividend of 12.60 pence per ordinary share has been recommended by the directors and, subject to approval at the Annual General Meeting, will be paid on 10 September 2001 to those shareholders whose names are on the register at 10 August 2001.

Source: H P Bulmer Holdings plc, 2000 Annual Report, p. 28

It is important to note that there is no direct record of taxation paid or dividends paid in profit and loss account or balance sheet. Taxation paid and dividends paid, however, appear in the cash flow statement.

Pause for Thought 7.5

Taxation and Dividends Paid

If you have details of the profit and loss charge for taxation and dividends and the opening and closing liabilities, how do you calculate the amount actually paid?

One becomes a detective. Take this example: opening taxation payable £800, closing taxation payable £1,000, profit and loss charge £3,000.

Our opening liability of £800 plus this year's charge of £3,000 equals £3,800. At the end of the year, however, we only owe £1,000. We must, therefore, have paid £2,800. This logic underpins the calculation of tax paid and dividends paid in the cash flow statement, covered in Chapter 8.

3. Long-Term Capital

The long-term capital of a company can be categorised into share capital (comprising ordinary and preference shares) and loan capital (often called debentures). We can see the differences between these in Figure 7.7 on the next page. This is portrayed diagrammatically in Figure 7.8 on the next page.

Essentially, ordinary shareholders own the company and, therefore, take the most risk and, potentially, gain the most reward. Preference shareholders normally receive a fixed dividend, whereas debenture holders receive interest. Debentures are long-term loans.

Figure 7.7 Different Types of Long-Term Capital of a Company

Features	*Ordinary Shareholders*	*Preference Shareholders*	*Debenture holders*
Type of capital	Share	Share	Loan
Ownership	Own company	Do not own company	Do not own company
Risk	Lose money invested first	Lose money invested after ordinary shareholder	Often loans secured on assets
Reward	Dividends	Usually fixed dividends	Interest
Accounting treatment	Under capital employed	Under capital employed	Deducted from net assets

Figure 7.8 Long-Term Capital Structure of a Company

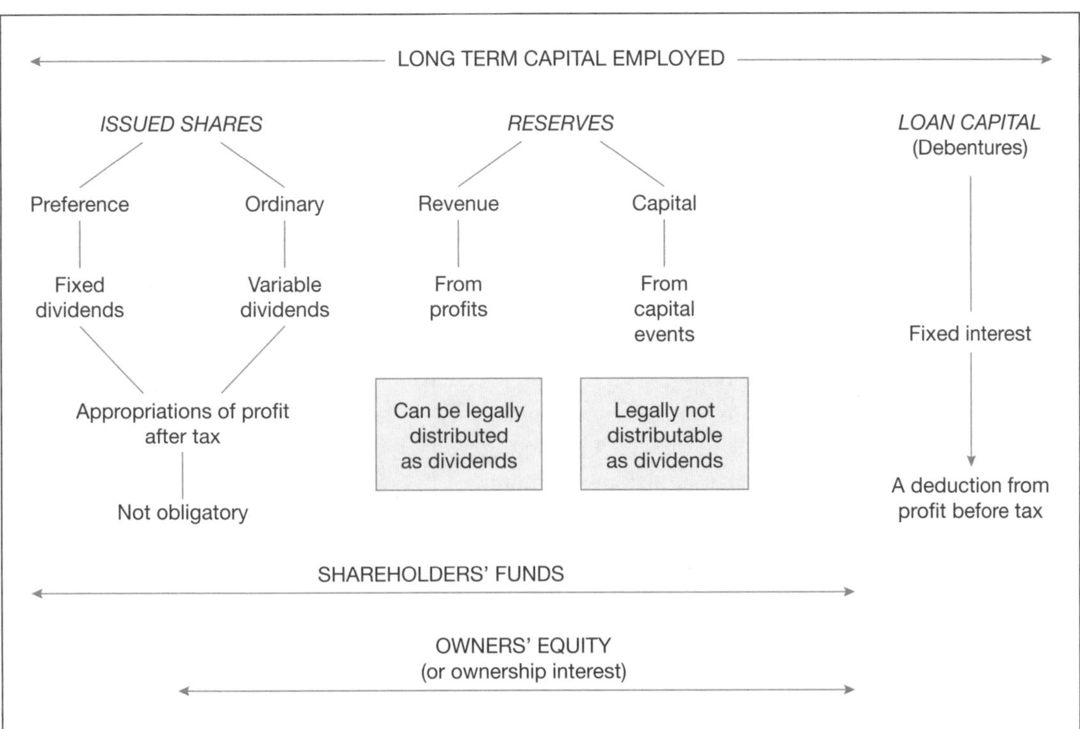

As Figure 7.9 below shows, the limited company's capital employed is presented differently from that of the sole trader or partnership. For a limited company, the capital employed is not adjusted for drawings. Excluding long-term capital, a company's capital employed is essentially represented by share capital and reserves. Unfortunately, there are many different types of share capital and reserves. Figure 7.10 on the page opposite provides a quick overview of these.

Figure 7.9 Owner's Capital Employed for Sole Trader, Partnerships and Limited Companies

Sole Trader		*Partnership*		*Limited Company*	
	£		£		£
Opening capital	100,000	Capital Accounts	100,000	Share Capital	100,000
Add Net Profit	50,000	Current Accounts	20,000	Reserves	20,000
	150,000	*Total partners' funds*	120,000	*Total shareholders' funds*	120,000
Less Drawings	30,000				
Closing capital	120,000				

Essentially, share capital represents the amount that the shareholders have directly invested. The amount a company is allowed (authorised share capital) to issue is determined in a company document called the Memorandum of Association. The amount actually issued is the issued share capital. It is important to emphasise that the face value for the shares is not its market value. If you like, it is like buying and selling stamps. The Penny Black, a rare stamp, was issued at one penny (nominal value), but you would have to pay a fortune for one today (market value). Shareholders' funds is an important figure in the balance sheet and is equivalent to total net assets (The Company Camera 7.2 shows the shareholders' funds for J.D. Wetherspoon).

Reserves are essentially gains to the shareholder. Capital reserves are gains from activities such as issuing shares at more than the nominal value (share premium account) or revaluing fixed assets (revaluation reserve). They cannot be paid out as dividends. By contrast, revenue reserves, essentially accumulated profits, are distributable.

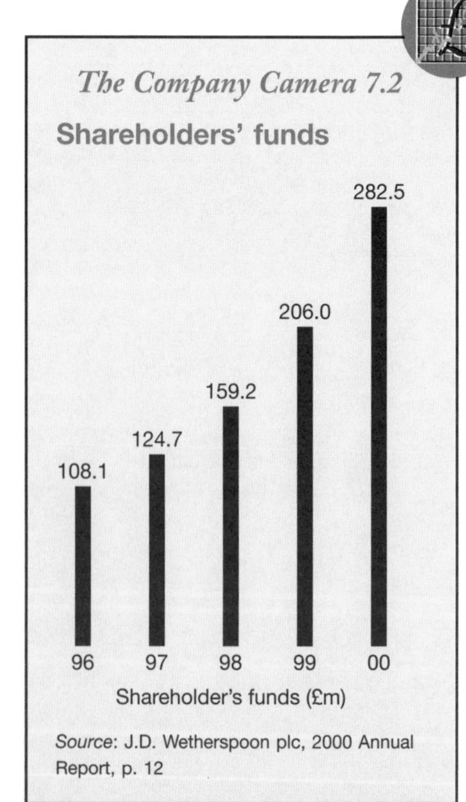

Source: J.D. Wetherspoon plc, 2000 Annual Report, p. 12

Figure 7.10 **Overview of the Main Terminology of a Limited Company's Share Capital and Reserves, and Loan Capital**

Term	*Explanation*
Share Capital	The capital of the company divided into shares.
Authorised share capital	The amount of share capital that a company is allowed to issue to its shareholders.
Called-up share capital	The amount of issued capital that has been fully paid to the company by shareholders. For example, a share may be issued for £1.50 and paid in three equal instalments. After two instalments are paid the called-up share capital will be £1.
Issued share capital	The amount of share capital *actually* issued.
Ordinary (equity) share capital	The amount of share capital relating to the shareholders who own the company and are entitled to ordinary dividends.
Preference share capital	The amount of share capital relating to shareholders who are not owners of the company and are entitled to fixed dividends.
Market value	The value the shares will fetch on the open market. This may differ significantly from their nominal value.
Nominal value	The face value of the shares, usually their original issue price.
Reserves	The accumulated profits (revenue reserves) or capital gains (capital reserves) to shareholders.
Capital reserves	Reserves which are not distributable to shareholders as dividends, for example, the share premium account or revaluation reserve.
General reserve	A reserve created to deal with general, unspecified contingencies such as inflation.
Profit and loss account	The accumulated profits of the company.
Revaluation reserve	A capital reserve created when fixed assets are revalued at more than the original purchase cost. The revaluation is a gain to the shareholders.
Revenue reserves	Reserves that are distributable to shareholders as dividends, for example, the profit and loss account, general reserve.
Share premium account	A capital reserve created when new shares are issued for more than their nominal value. For example, if shares were issued for £150,000 and the nominal value was £100,000, the share premium account would be £50,000.
Total shareholders' funds	The share capital and reserves which are owned by ordinary and preference shareholders.
Loan capital	Money loaned to the company by third parties. They are not owners of the company and are entitled to interest not dividends.
Debentures	Just another name for a long-term loan. Debentures may be secured or unsecured.
Secured and unsecured loans	Secured loans are loans which are secured on (or guaranteed by) the assets of the company. Unsecured loans are loans which are not secured on the assets.

Pause for Thought 7.6

The Nature of Reserves

Can you spend reserves?

No! Reserves are not cash. Reserves are in fact liabilities which the business owes to the shareholders. They represent either accumulated profits (profit and loss account) or gains to the shareholders such as an issue of shares above the nominal value (share premium account) or the revaluation of fixed assets (revaluation account). Reserves are represented by assets. However, these assets may be fixed assets or current assets. They do not have to be cash, and, if they are not they obviously cannot be spent.

4. Intangible Assets

Intangible assets are literally fixed assets one cannot touch. As opposed to tangible fixed assets such as land and buildings, plant and machinery, fixtures and fittings and motor vehicles. They can occur in all businesses, but are most common in companies.

Intangible assets are increasing in value and frequency as organisations become more knowledge-based. As the old manufacturing firms which are heavily dependent on tangible fixed assets decline and the new information technology businesses arise, the balance of assets in companies changes. There are now many types of intangible assets, such as goodwill. Goodwill is covered in more depth in Chapter 12.

Pause for Thought 7.7

Intangible Assets

Apart from goodwill, can you think of any other intangible assets?

There are many, but four of the most common are brands, copyright, patents and software development costs. Perhaps the most frequently occurring of these is patents. A patent is an intangible asset which represents the amount a firm has paid to register a patent or has paid to purchase a patent from another business or individual. A patent itself is the right of the patent's owner to exploit the invention for a period of time. Patents are recorded in the balance sheet under intangible assets.

Accounting Treatment

Profit and Loss Account

Essentially, the profit and loss account is calculated as normal. We then have to appropriate (or distribute) the profit to the government by way of tax, and to the shareholders by way of dividends. This is demonstrated in Figure 7.11 on the next page.

Figure 7.11 Limited Companies' Appropriation Account Format

The profit and loss account was prepared as normal. The profit was £100,000. The tax for the year was £40,000 and dividends are £25,000.

Limited Co. Ltd
Profit and Loss Account (extract)

	£
Profit before Taxation (calculated as normal)	100,000
Taxation	(40,000)
Profit after Taxation	60,000
Dividends for year	(25,000)
Retained Profit	35,000

Balance Sheet

In the balance sheet, the main differences from a sole trader or partnership are (i) that taxation payable and dividends payable are recorded under current liabilities; (ii) in the presentation of capital employed; and (iii) that limited companies are more likely than sole traders or partnerships to have intangible assets such as goodwill or patents. The current liabilities presentation is relatively straightforward, so only the capital employed format is presented here (see Figure 7.12).

Figure 7.12 Limited Companies' Capital Employed Format

Limited Co. Ltd has the following details of capital employed.

Ordinary share capital	*£100,000*	*Profit and loss*	*£15,000*
Preference share capital	*£50,000*	*Share premium account*	*£5,000*
		Revaluation reserve	*£6,000*

Share Capital and Reserves	£	£
Share Capital		
Ordinary share capital		100,000
Preference share capital		50,000
		150,000
Reserves		
Capital reserves		
Share premium account	5,000	
Other reserves		
Revaluation reserve	6,000	
Profit and loss account	15,000	26,000
Total shareholders' funds		176,000

Intangible assets are recorded in the balance sheet under fixed assets. The more conventional fixed assets, such as land and buildings, are then recorded as tangible assets. This is demonstrated in the comprehensive example which follows (see Figure 7.13 on page 145).

Limited Company Example: Stevens, Turner Ltd

In order to demonstrate a more comprehensive example of the preparation of a limited company's accounts, we now turn again to the accounts of Gavin Stevens. When we last met Gavin Stevens he had formed a partnership with Diana Turner and they were trading as a partnership, Stevens and Turner (see Figure 7.4 on page 137). We now assume that many years into the future the partnership has turned into a company. We will now prepare the company accounts (see Figure 7.13). **The distinctive elements of a limited company's account are bordered by asterisks.** It is important to realise that we are preparing here the **full accounts for internal management purposes.** In the following section, however, we show the published accounts of Stevens, Turner assuming they are a public limited company.

Stevens, Turner Ltd: Trial Balance as at 31 December 202X

	£000	£000
Ordinary share capital (£1 each)		300
Preference share capital (£1 each)		150
Share premium account		25
Revaluation reserve		30
General reserve		20
Profit and loss account as at 1 January 202X		50
Long-term loan		80
Land and buildings	550	
Patents	50	
Motor vehicles	50	
Opening stock	10	
Debtors	80	
Creditors		45
Bank	75	
Electricity	8	
Wages	50	
Telephone	7	
Sales		350
Purchases	150	
Loan interest	8	
Other expenses	12	
	1,050	1,050

Notes:

1. Closing stock is £25,000.
2. For simplicity, we are ignoring all other post-trial balance adjustments such as depreciation, bad and doubtful debts, accruals and prepayments.
3. The following are not recorded in the trial balance: (a) ordinary dividends payable £6,000; (b) preference dividends payable £3,000; (c) taxation payable £26,000.
4 Authorised ordinary share capital is £350,000; authorised preference share capital is £200,000.
5 Patents are included as an example of an intangible asset. They need to be recorded under fixed assets at the top of the balance sheet. Below these the other fixed assets are recorded as tangible assets.

Figure 7.13 Stevens, Turner Ltd: Accounts for Year Ended 31 December 202X

Stevens, Turner Ltd
Trading, Profit and Loss and Appropriation Account for Year Ended 31 December 202X

	£000	£000
Sales		350
Less *Cost of Sales*		
Opening stock	10	
Add Purchases	150	
	160	
Less Closing stock	25	135
Gross Profit		215
Less *Expenses*		
Electricity	8	
Wages	50	
Telephone	7	
Loan interest	8	
Other expenses	12	85
************		
Profit before Taxation		130
Taxation[1]		(26)
Profit after Taxation		104
Ordinary dividends[1]		(6)
Preference dividends[1]		(3)
Retained Profit		95
************		

Stevens, Turner Ltd
Balance Sheet as at 31 December 202X

	£000	£000	£000
Fixed Assets			
************			
Intangible Assets			
Patents			50
************			
Tangible Assets			
Land and buildings			550
Motor vehicles			50
Total fixed assets			650
Current Assets			
Stock	25		
Debtors	80		
Bank	75	180	
Current Liabilities			
Creditors	(45)		
************			
Taxation payable[1]	(26)		
Dividends payable[1]	(9)	(80)	
************			
Net current assets			100
Total assets less current liabilities			750
Long-term Creditors			(80)
Total net assets			670
************			

Figure 7.13 Stevens, Turner Ltd (*continued*)

	£000	£000	£000
Share Capital and Reserves			
Share Capital		*Authorised*	*Issued*
Ordinary share capital		350	300
Preference share capital		200	150
		550	450
Reserves			
Capital reserves			
Share premium account		25	
Other reserves			
Revaluation reserve		30	
General reserve		20	
Opening profit and loss account	50		
Retained profit for year	95		
Closing profit and loss account		145	220
Total shareholders' funds			670

****

Note 1:
In this case, the amount payable equals the charge for the year. This will not always be so.

Limited Companies: Published Accounts

The internal company accounts are not suitable for external publication. The published accounts which are sent to shareholders are incorporated in a special document called an annual report and have several special features: they are standardised, abridged and have supplementary notes.

- *Standardised.* Published accounts use special Companies Acts' formats. In actual fact, these formats are broadly used throughout this book for consistency and to aid understanding.
- *Abridged.* The Companies Acts' formats means that the details are summarised.
- *Supplementary notes.* Supplementary notes are used to flesh out the details of the main accounts.
- *Annual report.* In the case of public limited companies, an annual report is sent to shareholders. Increasingly, the annual report is also put onto a company's website. The website will include the financial statements, but also much more information. A separate chapter is devoted to the annual report (Chapter 12), given its importance, with Figure 12.4 on page 274 providing some corporate Web site addresses.

The Company Cameras 7.3 and 7.4 which follow show the profit and loss account and balance sheet for J. Sainsbury plc.

The Company Camera 7.3

Profit and Loss Account of a Public Limited Company

	Note	**1999** **56 weeks** **£m**	1998* 52 weeks £m
Group sales including VAT and sales taxes		**17,587**	15,496
VAT and sales taxes		**1,154**	996
Group sales excluding VAT and sales taxes	1	**16,433**	14,500
Cost of sales	1	**15,095**	13,289
Exceptional cost of sales – Texas Homecare integration costs	23	**21**	28
Gross profit		**1,317**	1,183
Administrative expenses	1	**406**	357
Year 2000 costs	1	**30**	20
Group operating profit before profit sharing	1	**881**	806
Profit sharing	2	**45**	44
Group operating profit		**836**	762
Associated Undertakings – share of profit		**12**	16
Profit on sale of properties		**11**	3
Profit/(loss) on disposal of an associate/subsidiary	3	**84**	(12)
Profit on ordinary activities before interest		**943**	769
Net interest payable	4	**55**	78
Profit on ordinary activities before tax	5	**888**	691
Tax on profit on ordinary activities	8	**292**	226
Profit on ordinary activities after tax		**596**	465
Minority equity interest		**2**	4
Profit for the financial year		**598**	469
Equity dividends	9	**294**	264
Retained profit	26	**304**	205
Earnings per share	10	**31.4p**	25.1p

Source: J. Sainsbury plc, 1999 Annual Report and Accounts, p. 36

Helpnote: Sainsbury's profit and loss account is presented in summary, abridged form. The detail is in the attached notes (not shown here). Value added tax (VAT) is a tax added to the cost of goods purchased and included in the final selling price. Exceptional cost of sales is a one-off cost incurred by Sainsbury to integrate the Texas Homecare chain into their group. Year 2000 costs were the costs of updating computer programmes for the new millennium. Associate and subsidiary are companies that are part of the Sainsbury group.

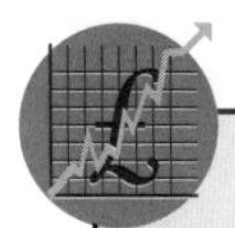

The Company Camera 7.4

Balance Sheet of a Public Limited Company

		Group		Company	
	Note	**1999 £m**	1998 £m	**1999 £m**	1998 £m
Fixed assets					
Tangible assets	11	**6,409**	6,133	**226**	228
Investments	12	**41**	151	**5,726**	5,023
		6,450	6,284	**5,952**	5,251
Current assets					
Stocks	15	**843**	743	**–**	–
Debtors	16	**249**	229	**72**	67
Investments	17	**17**	14	**–**	–
Sainsbury's Bank	18	**1,766**	1,584	**–**	–
Cash at bank and in hand		**725**	270	**1**	–
		3,600	2,840	**73**	67
Creditors: due within one year					
Sainsbury's Bank	18	**(1,669)**	(1,502)	**–**	–
Other	19	**(2,880)**	(2,499)	**(835)**	(518)
		(4,549)	(4,001)	**(835)**	(518)
Net current liabilities		**(949)**	(1,161)	**(762)**	(451)
Total assets less current liabilities		**5,501**	5,123	**5,190**	4,800
Creditors: due after one year	19	**(804)**	(949)	**(767)**	(926)
Provisions for liabilities and charges	23	**(8)**	(9)	**–**	–
Total net assets		**4,689**	4,165	**4,423**	3,874
Capital and reserves					
Called-up share capital	24	**480**	476	**480**	476
Share premium account	24	**1,359**	1,295	**1,359**	1,295
Revaluation reserve	25	**38**	38	**–**	–
Profit and loss account	26	**2,767**	2,318	**2,584**	2,103
Equity shareholders' funds		**4,644**	4,127	**4,423**	3,874
Minority equity interest		**45**	38	**–**	–
Total capital employed		**4,689**	4,165	**4,423**	3,874

Source: J. Sainsbury plc,1999 Annual Report and Accounts, p. 37

Helpnote: Sainsbury's balance sheet is presented in summarised, abridged form. The detail is in the attached notes (not shown here). Sainsbury has long-term investments which because of their long-term nature are treated as fixed assets. It is also noteworthy that Sainsbury operates a separate bank. This means that its current assets and current liabilities are very high compared to other companies.

The specific requirements for published company accounts are complex and beyond the scope of this book. However, Figure 7.14 on the next page is a summary of what Stevens, Turner Plc might look like.

Figure 7.14 **Stevens, Turner Ltd: Accounts Presented as Published Accounts**

Stevens, Turner Plc
Profit and Loss Account Year Ended 31 December 202X

	£000
Turnover[1]	350
Cost of Sales	(135)
Gross Profit	215
Administrative expenses	(85)
Profit before Taxation	130
Taxation	(26)
Profit after Taxation	104
Dividends	(9)
Retained Profit	95

Stevens, Turner Plc
Balance Sheet as at 31 December 202X

	Notes	£000	£000
Fixed Assets			
Intangible Assets			50
Tangible Assets	1		600
Total fixed assets			650
Current Assets			
Stock		25	
Debtors		80	
Bank		75	
		180	
Creditors: Amounts Falling Due within One Year[2]	2	(80)	
Net current assets			100
Total assets less current liabilities			750
Creditors: Amounts Falling Due after More than One Year[2]			(80)
Total net assets			670
			£000
Capital and Reserves			
Called-up share capital	3		450
Share premium account			25
Other reserves	4		50
Profit and loss account	5		145
Total shareholders' funds			670

Notes to the accounts

1. Tangible Assets	£000
Land and buildings	550
Motor vehicles	50
	600

2. Creditors: Amounts Falling Due within One Year	
Creditors	45
Taxation payable	26
Ordinary dividends payable	6
Preference dividends payable	3
	80

3. Called-up Share Capital[3]	£000
Ordinary share capital	300
Preference share capital	150
	450

4. Other Reserves	
Revaluation reserve	30
General reserve	20
	50

5. Profit and Loss Account	
Balance as at 1 January 202X	50
Retained profit for year	95
Balance as at 31 December 202X	145

We have thus summarised the accounts of Stevens, Turner Plc and supplemented them with notes to the accounts. The main figures can thus easily be identified. There are some points of interest, indicated by the superscript notes in Figure 7.14.

1 Turnover is the normal term for sales in published accounts.
2 These are the terms used in published accounts. They are equivalent to the terms current liabilities and long-term creditors we have used so far.
3 Called-up share capital has a technical meaning (see Figure 7.10). However, for convenience, it can be taken here as issued share capital.

Conclusion

Partnerships and limited companies are important types of business organisation. Partnerships are broadly similar to sole traders, except that there is the problem of how to divide the capital and profit between the partners. Limited companies, unlike partnerships or sole traders, are based on the concept of limited liability. The principal differentiating features in the accounts of companies are corporation tax, dividends and the division of capital employed into share capital and reserves. Limited companies may be either private limited companies or public limited companies. It is the latter which are quoted on the stock exchange.

Discussion *Questions*

Questions with numbers in blue have answers at the back of the book.

Q 1 Why do you think that three different types of business enterprise (sole traders, partnerships and limited companies) exist?

Q 2 Discuss the view that the accounts of partnerships are much like those of sole traders except for the need to share out the capital and profit between more than one partner.

Q 3 Distinguish between a private limited company and a public limited company. Is there any difference between the users of the accounts of each type of company?

Q 4 Why is the distinction between capital and revenue reserves so important for a company?

Q 5 State whether the following are true or false. If false, explain why.

(a) Drawings are an expense recorded in the partners' trading, profit and loss and appropriation account.
(b) Partner's current accounts report the yearly short-term movements in partners' capital.
(c) The nominal value of a company's shares is the amount the shares will fetch on the stock market.
(d) An unsecured loan is secured on specific assets such as the company's machinery.
(e) Reserves can be spent on the purchase of fixed assets.

Numerical *Questions*

These questions are separated into those on (A) partnerships and (B) limited companies. Within each section, they are graded in difficulty.

Questions with numbers in blue have answers at the back of the book.

A Partnerships

Q 1 Two partners, Peter Tom and Sheila Thumb, have the following details of their accounts for the year ended 31 December 2001.

			£	
Net profit before appropriation £100,000	Capital accounts:	Tom	8,000	
Profit sharing ratio: 3 Tom, 1 Thumb		Thumb	6,000	
Salaries: Tom £10,000, Thumb £30,000	Current accounts:	Tom	3,000	cr
Drawings: Tom £25,000, Thumb £30,000		Thumb	1,000	dr

Required:
Prepare the relevant trading, profit and loss and appropriation account, and balance sheet extracts.

Q 2 J. Waite and P. Watcher's trial balance as at 30 November 2001 is set out below.

		£	£
Capital accounts:	Waite		88,000
	Watcher		64,000
Current accounts:	Waite	2,500	
	Watcher		12,000
Drawings:	Waite	13,300	
	Watcher	6,300	
Land and buildings at cost		166,313	
Motor vehicles at cost		65,000	
Opening stock		9,000	
Debtors		12,000	
Creditors			18,500
Bank		6,501	
Electricity		3,406	
Wages		14,870	
Telephone		1,350	
Rent and business rates		6,660	
Long-term loan			28,000
Sales			350,000
Purchases		245,000	
Interest on loan		2,800	
Other expenses		5,500	
		560,500	560,500

Notes:

1. Closing stock is £15,000.
2. Salaries are £18,000 for Waite and £16,000 for Watcher.
3. There is £300 owing for rent.
4. Depreciation for the year is £2,000 on land and buildings and £3,000 on motor vehicles. The business was started on 1 December 2000.
5. The split of profits is 3 Watcher: 2 Waite.

Required:

Prepare the trading, profit and loss and appropriation account for year ended 30 November 2001 and the balance sheet as at 30 November 2001.

Q3 Cherie and Tony's trial balance as at 31 December 2001 is set out below.

		£	£
Capital accounts:	Cherie		30,000
	Tony		35,000
Current accounts:	Cherie		26,000
	Tony		18,500
Drawings	Cherie	21,294	
	Tony	18,321	
Land and buildings at cost		203,500	
Plant and machinery at cost		26,240	
Land and buildings accumulated depreciation as at 1 January 2001			12,315
Plant and machinery accumulated depreciation as at 1 January 2001			9,218
Debtors		18,613	
Creditors			2,451
Bank		25,016	
Electricity		1,324	
Wages		12,187	
Telephone		1,923	
Insurance		1,318	
Long-term loan			83,000
Sales			251,800
Sales returns		340	
Purchases		128,317	
Purchases returns			206
Other expenses		1,497	
Opening stock		8,600	
		468,490	468,490

Notes:

1. Closing stock is £12,000.
2. £197 of the other expenses was prepaid and £200 is owed for the telephone.
3. Salaries will be £12,000 for Cherie and £10,000 for Tony.
4. Depreciation is fixed at 2% on the cost of land and buildings and 10% on the cost of plant and machinery.
5. Profits are shared in the ratio 2 for Cherie and 1 for Tony.

Required:

Prepare the trading, profit and loss and appropriation account for the year ended 31 December 2001 and the balance sheet as at 31 December 2001.

*Q*4 Sister and Sledge are trading in partnership, sharing profits and losses in the ratio of 2:1, respectively. The partners are entitled to salaries of Sister £6,000 per annum and Sledge £5,000 per annum. There is the following additional information.

(1) Stock as at 31 December 2001 was valued at £8,800.
(2) Staff salaries owing £290.
(3) Advertising paid in advance £200.
(4) Provision for bad and doubtful debts to be increased to £720.
(5) Provision should be made for depreciation of 2% on land and buildings on cost, and for fixtures and fittings at 10% on cost.

Trial Balance as at 31 December 2001

	£	£
Capital accounts:		
Sister		12,500
Sledge		5,000
Current accounts:		
Sister		1,500
Sledge	600	
Drawings:		
Sister	9,800	
Sledge	6,700	
Long-term loan		40,250
Land and buildings at cost	164,850	
Stock as at 1 January 2001	9,500	
Fixtures and fittings at cost	12,500	
Purchases	126,000	
Cash at bank	3,480	
Sales		305,400
Trade debtors	9,600	
Carriage inwards	200	
Carriage outwards	300	
Staff salaries	24,300	
Trade creditors		26,300
General expenses	18,200	
Provision for bad and doubtful debts		480
Advertising	5,350	
Discounts receivable		120
Discounts allowed	350	
Rent and business rates	2,850	
Land and buildings accumulated depreciation as at 1 January 2001		9,750
Fixtures and fittings accumulated depreciation as at 1 January 2001		3,500
Electricity	4,500	
Telephone	5,720	
	404,800	404,800

Required:

Prepare the trading and profit and loss and appropriation account for the year ended 31 December 2001 and the balance sheet as at 31 December 2001.

B Limited Companies

Q 5 Red Devils Ltd has the following extracts from its accounts.

Red Devils Ltd
Trial Balance as at 30 November 2001

	£	£
Gross profit for year		150,000
7% Debentures		200,000
6% Preference share capital (£150,000 authorised)		150,000
£1 Ordinary share capital (£400,000 authorised)		250,000
Share premium account		55,000
Fixed assets	680,900	
General expenses	22,100	
Directors' fees	19,200	
Debtors	4,700	
Bank	5,300	
Creditors		12,200
Profit and loss account as at 1 December 2000		9,000
General reserve as at 1 December 2000		11,000
Stock as at 30 November 2001	105,000	
	837,200	837,200

Notes:

1. An audit fee is to be provided of £7,500.
2. The debenture interest for the year has not been paid.
3. The directors propose the following:
   (a) A dividend of 10p per share (10%) on the ordinary shares
   (b) To pay the preference dividend
   (c) To transfer £3,500 to the general reserve (Note: transfers are recorded in the appropriation account)
4. Corporation tax of £17,440 to be provided on the profit for the year.

Required:
Prepare for internal management purposes:
(a) The profit and loss account and appropriation account for the year ended 30 November 2001.
(b) The balance sheet as at 30 November 2001.

Q 6 Superprofit Ltd

Trial Balance as at 31 December 2001

	£000	£000
Ordinary share capital		210
Preference share capital		25
Share premium account		40
Revaluation reserve		35
General reserve		15
Profit and loss account as at 1 January 2001		28
Long-term loan		32
Land and buildings	378	
Patents	12	
Motor vehicles	47	
Opening stock	23	
Debtors	18	
Creditors		33
Bank	31	
Electricity	12	
Insurance	3	
Wages	24	
Telephone	5	
Light and heat	8	
Sales		351
Purchases	182	
Other expenses	26	
	769	769

Notes (all figures in £000s):

1. Closing stock is £26.
2. The following had not yet been recorded in the trial balance:
   (a) Dividends payable on ordinary shares £9 and on preference shares £3
   (b) Taxation payable £13
   (c) Interest on long-term loan £4
   (d) Auditors' fees £2
   (e) The authorised share capital is ordinary share capital £250, preference share capital £50.
3. The business started trading on 1 January 2001. Depreciation for the year is £18 for land and buildings and £7 for motor vehicles.

Required:

Prepare for *internal management purposes* the trading, profit and loss and appropriation account for year ended 31 December 2001 and the balance sheet as at 31 December 2001.

Q7 Lindesay Trading Ltd

Trial Balance as at 31 March 2001

	£000	£000
Ordinary share capital		425
Preference share capital		312
Share premium account		18
Revaluation reserve		27
General reserve as at 1 April 2000		13
Profit and loss account as at 1 April 2000		17
Long-term loan		87
Land and buildings at cost	834	
Patents	25	
Motor vehicles at cost	312	
Opening stock as at 1 April 2000	10	
Debtors	157	
Creditors		93
Bank	186	
Electricity	12	
Wages and salaries	183	
Telephone	5	
Sales		1,500
Insurance	6	
Purchases	750	
Other expenses	125	
Land and buildings accumulated depreciation as at 1 April 2000		25
Motor vehicles accumulated depreciation as at 1 April 2000		88
	2,605	2,605

Notes (all figures in £000s):

1. Closing stock is £13.
2. The following are not recorded in the trial balance:
   (a) Dividends payable on ordinary shares £19 and on preference shares £9.
   (b) Taxation payable £58.
3. Authorised share capital was £500 for ordinary share capital and £400 for preference share capital.
4. There was £45 owing for wages and salaries.
5. Debenture interest was £8.
6. Of the insurance £1 was prepaid.
7. The proposed auditors' fees are £3.
8. A transfer to the general reserve was made of £16.
9. Depreciation is to be £17 on land and buildings and £60 on motor vehicles.

Required:

Prepare for *internal management purposes* the trading, profit and loss and appropriation account for the year ended 31 March 2001 and the balance sheet as at 31 March 2001.

Q **8** The following trial balance was extracted from the books of Leisureplay Plc for the year ended 31 December 2001.

	£000	£000
Ordinary share capital (£1 each)		700,000
Preference share capital (£1 each)		80,000
Debentures		412,000
Profit and loss account as at 1 January 2001		68,000
Share premium account		62,000
Revaluation reserve		70,000
General reserve		18,000
Freehold premises at cost	1,550,000	
Motor vehicles at cost	18,000	
Furniture and fittings at cost	8,000	
Freehold premises accumulated depreciation as at 1 January 2001		102,000
Motor vehicles accumulated depreciation as at 1 January 2001		9,350
Furniture and fittings accumulated depreciation as at 1 January 2001		1,100
Stock as at 1 January 2001	5,000	
Cash at bank	183,550	
Provision for bad and doubtful debts		1,000
Purchases/sales	500,000	800,000
Debtors/creditors	28,900	7,000
Sales returns/purchases returns	3,500	3,800
Carriage inwards	60	
Carriage outwards	70	
Bank charges	20	
Rates	4,280	
Salaries	5,970	
Wages	3,130	
Travelling expenses	1,980	
Discount allowed	20	
Discount received		15
General expenses	8,100	
Gas, electricity	9,385	
Printing, stationery	1,850	
Advertising	2,450	
	2,334,265	2,334,265

Q **8** Leisureplay (continued).

Notes (all figures are in 000s):

(a) Stock as at 31 December 2001 is £12,000

(b) Depreciation is to be charged as follows:

(i)	Freehold premises	2% on cost
(ii)	Motor vehicles	10% on cost
(iii)	Furniture and fittings	5% on cost

(c) There is the following payment in advance:

General expenses	£500

(d) There are the following accrued expenses:

Business rates	£300
Advertising	£550
Auditors' fees	£250

(e) There are proposed dividends of £70,000 for ordinary shares and £4,800 for preference shares.

(f) Authorised ordinary share capital is £1,000,000 £1shares, and authorised preference share capital is 100,000 £1shares.

(g) Taxation has been calculated as £58,500.

(h) Debenture interest should be charged at 10%.

(i) Provision for bad and doubtful debts is increased to £1,600 and a bad debt of £400 is to be written off.

Required:

Prepare for *internal management purposes* the trading, profit and loss and appropriation account for Leisureplay Plc for the year ended 31 December 2001 and balance sheet as at 31 December 2001.

Q **9** You have the following summarised trial balance for Stock High Plc as at 31 March 2001. Further details are provided in the notes.

	£000	£000
Turnover		1,250
Cost of sales	400	
Administrative expenses	200	
Distribution expenses	150	
Patents	50	
Land and buildings at cost	800	
Motor vehicles at cost	400	
Land and buildings accumulated depreciation as at 1 April 2000		140
Motor vehicles accumulated depreciation as at 1 April 2000		150
Long-term loan		60
Profit and loss account as at 1 April 2000		36
Share premium account		25
Revaluation reserve		30
General reserve		25
Ordinary share capital		450
Preference share capital		100
Taxation paid	86	
Ordinary dividends paid	50	
Creditors		12
Stock as at 31 March 2001	20	
Debtors	100	
Cash	22	
	2,278	2,278

Notes (In £000s except for note 3):

1. At the balance sheet date £8 is owing for taxation, £15 for ordinary dividends and £10 for preference dividends.
2. Depreciation is to be charged at 2% on cost for land and buildings (used for administration) and 20% on cost for motor vehicles (used for selling and distribution).
3. Authorised share capital is 600,000 £1 ordinary shares and £150,000 £1 preference shares.

Required:

Prepare the profit and loss account for the year ended 31 March 2001 and the balance sheet as at 31 March 2001 as it would appear in the published accounts.

Chapter 8

Main Financial Statements: The Cash Flow Statement

'Cash is King. It is relatively easy to "manufacture" profits but creating cash is virtually impossible.'

UBS Phillips and Drew (January 1991). *Accounting for Growth*, p. 32

Learning Outcomes

After completing this chapter you should be able to:

- ✔ Explain the nature of cash and the cash flow statement.
- ✔ Demonstrate the importance of cash flow.
- ✔ Investigate the relationship between profit and cash flow.
- ✔ Outline the direct and indirect methods of cash flow statement preparation.

In a Nutshell

- Cash is key to business success.
- Cash flow is concerned with cash received and cash paid, unlike profit which deals with income earned and expenses incurred.
- Reconciling profit to cash flow means adjusting for movements in working capital and for non-cash items, such as depreciation.
- Large companies provide a cash flow statement as the third major financial statement.
- Sole traders, partnerships and small companies may, but are not required to, prepare a cash flow statement.
- The two ways of preparing a cash flow statement are the direct and the indirect methods.
- Most companies use the indirect method of cash flow statement preparation.

Introduction

Cash is king. It is the essential lubricant of business. Without cash, a business cannot pay its employees' wages or pay for goods or services. As Real-Life Nugget 8.1 shows, at its most extreme, this can lead to a business's failure. A business records cash in the bank account in the books of account. Small businesses may sometimes prepare a cash flow statement directly from the bank account. More usually, however, the cash flow statement is prepared indirectly by deducing the figures from the profit and loss account and balance sheet. The cash flow statement, at its simplest, records the cash inflows and cash outflows classified under certain headings such as cash flows from operating (i.e., trading) activities. All companies (except small ones) must prepare cash flow statements in line with financial reporting regulations. However, some sole traders, partnerships and smaller companies also provide them, often at the request of their bank. As Real-Life Nugget 8.2 shows, banks are well aware of the importance of cash.

As well as preparing cash flow statements on the basis of past cash flows, businesses will continually monitor their day-to-day cash inflows and outflows. As we shall see in Chapter 17, they also prepare cash budgets which look to the future. Cash management, therefore, concerns the past, present and future activities of a business.

Real-Life Nugget 8.1

Cash Bloodbath

With last week's collapse of Boo.com – the first big liquidation of a dot.com company in Europe – the internet gold rush has taken on the appearance of a bloodbath.

The company's principal failing – and there were many, it was burning cash at a rate of $1m a week – was to forget that in the new economy the old rules still apply.

Source: *Accountancy Age*, 25 May 2000, p. 26

Importance of Cash

Cash is the lifeblood of a business. Cash is needed to pay the wages, to pay the day-to-day running costs, to buy stock and to buy new fixed assets. The generation of cash is, therefore, essential to the survival and expansion of businesses. Money makes the world go round! In many ways, the concept of cash flow is easier to understand than that of profit. Most people are more familiar with cash than profit. Cash is, after all, what we use in our everyday lives.

At its most stark, if a business runs out of cash it will not be able to pay its creditors and it will cease trading. As Jack Welch, a successful US businessman, has said, 'There's one thing you can't cheat on and that's cash and Enron didn't have any cash for the last three years. Accounting is odd, but cash is real stuff. Follow the cash' (Guardian, 27 February 2002, p. 23).

Real-Life Nugget 8.2

Cash Flow

Bankers *do* know about cash flow. They have to live with it on Friday, every Friday, in any number of companies up and down the country. Where there is insufficient cash to pay the wages, really agonising decisions result. Should the company be closed, with all the personal anguish it will cause, or should it be allowed to limp on, perhaps to face exactly the same agonising dilemma in as little as a week's time?

Source: B. Warnes (1984), *The Genghis Khan Guide to Business*, Osmosis Publications, London, p. 6

It is far easier to manipulate profit than it is to manipulate cash flow. This is highlighted by Real-life Nugget 8.3. Phillips and Drew, a firm of city fund managers (now called UBS Global Management), basically state that cash is essential to business success.

Real-Life Nugget 8.3

Importance of Cash

'In the end, investment and accounting all come back to cash. Whereas "manufacturing" profits is relatively easy, cash flow is the most difficult parameter to adjust in a company's accounts. Indeed, tracing cash movements in a company can often lead to the identification of unusual accounting practices. The long term return of an equity investment is determined by the market's perception of the stream of dividends that the company will be able to pay. We believe that there should be less emphasis placed on the reported progression of earnings per share and more attention paid to balance sheet movements, dividend potential and, most important of all, cash.'

Source: UBS Phillips and Drew (January 1991), *Accounting for Growth*, p. 1

Context

The cash flow statement is the third of the key financial statements which medium and large companies provide. It summarises the company's cash transactions over time. At its simplest, the cash flow is related to the opening and closing cash balances.

Opening cash + Inflows – Outflows = Closing cash

Pause for Thought 8.1

Yes, But What Exactly is Cash?

Cash is cash! However, there are different types of cash, such as petty cash, cash at bank, bank deposit accounts, or deposits repayable on demand or with notice. How do they all differ?

The basic distinction is between cash and bank. However, the terms are often used loosely and interchangeably. Cash is the cash available. In other words, it physically exists, for example, a fifty pound note. Petty cash is money kept specifically for day-to-day small expenses, such as purchasing coffee. Cash at bank is normally kept either in a current account (which operates via a cheque book for normal day-to-day transactions) or in a deposit account (basically a store for surplus cash). Deposits repayable on demand are very short-term investments which can be repaid within one working day. Deposits requiring notice are accounts where the customer must give a period of notice for withdrawal (for example, 30 days).

Cash inflows are varied, but may, for example, be receipts from sales or interest from a bank deposit account. Cash outflows may be payments for goods or services, or for capital expenditure items such as motor vehicles.

All *large* companies are required to provide a cash flow statement. There are two methods of preparation. The first is the **direct method** which categorises cash flow by function, for example receipts from sales. A cash flow statement, using the direct method, can be prepared from the bank account and is the most readily understandable. The second method is the **indirect method.** This uses a 'detective' approach. It deduces cash flow from the existing balance sheets and profit and loss account and reconciles operating profit to operating cash flow. The cash flow statement, using the indirect method, is not so readily comprehensible. Unfortunately, this is the method most often used.

Cash and the Bank Account

As we saw in Chapter 4, cash is initially recorded in the bank account. In large businesses, a separate book is kept called the cash book. Debits are essentially good news for a company in that they increase cash in the bank account, whereas credits are bad news in that they decrease cash in the bank account. From the bank account it is possible to prepare a simple cash flow statement.

Pause for Thought 8.2

Cash Inflows and Outflows

What might be some examples of the main sources of cash inflow and outflow for a small business?

Cash Inflow	Cash Outflow
Cash from customers for goods	Payments to suppliers for goods
Interest received from bank deposit account	Payments for services, e.g., telephone, light and heat
Cash from sale of fixed assets	Repay bank loans
Cash introduced by owner	Payments for fixed assets, e.g., motor vehicles
Loan received	Interest paid on bank loan

Let us take the example once more of Gavin Stevens' bank account (see Figure 8.1). As Figure 8.1 shows, we have essentially summarised the figures from the bank account and reclassified them under certain headings.

Figure 8.1 Simple Cash Flow Statement for Gavin Stevens

Taking Gavin Stevens' bank account:

Bank

	£		£
1 Jan. Capital	200,000	2 Jan. Hotel	110,000
7 Jan. Ireton	1,965	2 Jan. Van	3,000
7 Jan. Hepworth	2,500	2 Jan. Purchases	2,000
		4 Jan. Electricity	300
		4 Jan. Wages	1,000
		7 Jan. Hogen	250
		7 Jan. Lewis	1,000
		7 Jan. Bal. c/f	86,915
	204,465		204,465
8 Jan. Bal. b/f	86,915		

From the bank account, we can summarise the main cash flows and record them in a cash flow statement, as follows:

Gavin Stevens
Cash Flow Statement up to 7 January

	£	£
Opening Cash Balance		
Add *Inflows*		
Capital invested (1)	200,000	
Trading (2)	4,465	204,465
Less *Outflows*		
Capital expenditure (3)	113,000	
Trading (4)	4,550	117,550
Closing Cash Balance		86,915

Notes:

(1) Represents the initial capital investment. Often termed a 'financing' cash flow.
(2) Represents money received from debtors (Ireton £1,965 and Hepworth £2,500). Often termed cash flow from a 'trading' or 'operating' activity.
(3) Represents the purchase of fixed assets (hotel £110,000 and van £3,000). Often termed cash flow from 'investing' activities.
(4) Represents the money paid for goods and services (purchases £2,000, electricity £300, wages £1,000, Hogen £250, Lewis £1,000). Often termed cash flow from a 'trading' or 'operating' activity.

For sole traders and partnerships, there is no regulatory requirement for a cash flow statement in the UK. Small companies are also exempt. Many organisations do, nevertheless, prepare one. Companies are regulated by an accounting standard, Financial Reporting Standard 1.

This standard lays down certain main headings for categorising cash flows (see Figure 8.2). From now on, we will use these categories for all organisations. Unfortunately, these headings are very cumbersome and often lack transparency.

Figure 8.2 Main Headings for Cash Flow Statements

	Simplified Meaning	*Examples of Inflows*	*Examples of Outflows*
Net Cash Flow from Operating Activities[1]	Cash flows from the normal trading activities of a business	i. Cash for sale of goods	i. Payment for purchase of goods ii. Expenses paid
Returns on Investments and Servicing of Finance	Cash received from investments or paid on loans	i. Interest received ii. Dividends received	i. Interest paid
Taxation	Cash paid to government for taxation	i. Taxation refunds	i. Taxation paid
Capital Expenditure and Financial Investment	Cash flows relating to the purchase and sale of i. fixed assets ii. investments	i. Receipts for sale of fixed assets, e.g., motor vehicles ii. Sale of investments	i. Payments for fixed assets, e.g., motor vehicles ii, Purchase of investments
Acquisitions and Disposals[2]	Payments for the purchase or sale of other companies	i. Cash paid to buy another company	i. Cash received for sale of another company
Equity Dividends Paid	Dividends companies pay to shareholders	None	Dividends paid
Financing	Cash flows relating to the issuing or buying back of shares or loan capital	Cash received from the issue of i. shares ii. loans	Cash paid to buy back i. shares ii. loans

1. Where cash flow is positive we use the term net cash inflow , where negative net cash outflow .
2. This item mainly applies to groups of companies. They are outside the scope of this chapter.

Relationship between Cash and Profit

Cash and profit are fundamentally different. In essence, cash flow and profit are based on different principles. Cash flow is based on cash received and cash paid (see Figure 8.3). By contrast, profit is concerned with income earned and expenses incurred.

Figure 8.3 Cash Flow and Profit

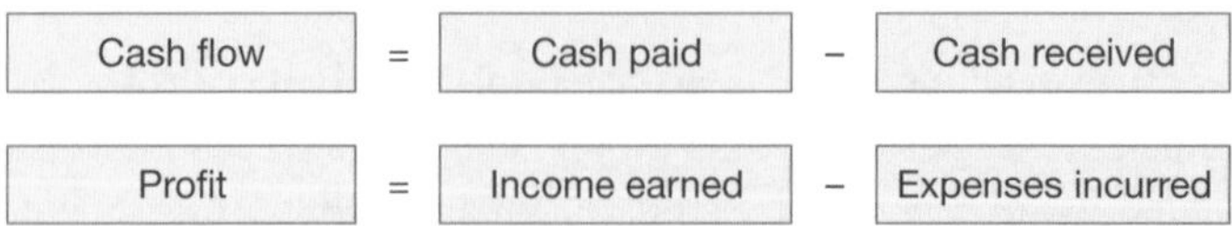

In a sense, the difference between the two merely results from the timing of the cash flows. For example, a telephone bill owing at the year end is included as an accrued expense in the profit and loss account, but is not counted as a cash payment. However, next year the situation will reverse and there will be a cash outflow, but no expense.

An important difference between profit and cash flow is depreciation. Depreciation is a non-cash flow item. Cash flows occur only when fixed assets are bought or sold. Real-Life Nugget 8.4 demonstrates this.

Real-Life Nugget 8.4

Cash Loss vs. Stated Loss

As always, there is the need to distinguish between a stated loss, per the profit and loss account, and a cash loss. One company the author handled was running at an apparently frightening loss of £25,000 per month, but on closer examination there was not too much to worry about. It had a £30,000 monthly depreciation provision. It was in reality producing a cash-positive profit of £5,000 per month. It had *years* of life before it. This gave all the time needed to get the operation right.

Source: B. Warnes (1984), *The Genghis Khan Guide to Business*, Osmosis Publications, London, p. 63

Figure 8.4 shows how some common items are treated in the profit and loss account and cash flow statement. Some items, such as sale of goods for cash, appear in both. However, amounts owing, such as a telephone bill, appear only in the profit and loss account. By contrast, money received from a loan only affects the cash flow statement.

Figure 8.4 Demonstration of How Some Items Affect the Profit and Loss Account and Some Affect the Cash Flow Statement

Transaction	*In Profit and Loss Account*	*In Cash Flow Statement*
i. Sale of goods for cash	Yes	Yes
ii. Sale of goods on credit	Yes	No
iii. Telephone bill for year owing	Yes	No
iv. Telephone bill for year paid	Yes	Yes
v. Cash purchase of fixed assets	No	Yes
vi. Profit on sale of fixed assets	Yes	No
vii. Cash from sale of fixed assets	No	Yes
viii. Money received from a loan	No	Yes
ix. Bank interest received for a year	Yes	Yes

Sometimes, a business may make a profit, but run out of cash. This is called overtrading and happens especially when a business starts trading.

Pause for Thought 8.3

Overtrading, Cash Flow vs. Profit

A company, Bigger is Better, doubles its sales every month.

Month	1	2	3	4
	£000	£000	£000	£000
Sales	10	20	40	80
Purchases	(8)	(16)	(32)	(64)
Profit	2	4	8	16

It pays its purchases at once, but has to wait two months for its customers to pay for the sales. The bank, which has loaned £10,000, will close down the business if it is owed £50,000. *What happens?*

Month	1	2	3	4
	£000	£000	£000	£000
Cash at bank	10	2	(14)	(36)
Cash in	–	–	10	20
Cash out	(8)	(16)	(32)	(64)
Cash at bank	2	(14)	(36)	(80)

The result: Bye-bye, Bigger is Better. Even though the business is trading profitably, it has run out of cash. This is because the first cash is received in month 3, but the cash outflows start at once.

Preparation of Cash Flow Statement

In this section, we present the two methods of preparing cash flow statements (the direct and indirect methods). In Figure 8.5 (on the next page) a cash flow statement is prepared for a sole trader using the **direct method**, which classifies *operating* cash flows by function or type of activity (e.g., receipts from customers). In essence, this resembles the cash flow statement for Gavin Stevens in Figure 8.1 on page 165. We assume a bank has requested a cash flow statement and that it is possible to extract the figures directly from the company's accounting records. We then present the cash flow statement for a company using the more conventional **indirect method** (see Figures 8.7 and 8.8 on pages 172 and 173). In this case, we derive the operating cash flow from the profit and loss account and balance sheets.

Direct Method

This method of preparing cash flow statements is relatively easy to understand. It is made of functional flows such as payments to suppliers or employees. These are usually extracted from the cash book or bank account. Figure 8.5 below demonstrates the direct method.

Figure 8.5 Preparation of a Sole Trader's Cash Flow Statement Using the Direct Method

You have extracted the following aggregated cash figures from the accounting records of Richard Hussey, who runs a book shop. The bank has requested a cash flow statement. Prepare Hussey's cash flow statement for year ended 31 December 2001

	£		£
Cash receipts from customers	150,000	Interest received	850
Cash payments to suppliers	60,000	Interest paid	400
Cash payments to employees	30,000	Cash from sale of motor car	3,000
Cash expenses	850		
Loan received and paid into the bank	8,150	Payment for new motor car	4,350

Richard Hussey
Cash Flow Statement Year Ended 31 December 2001

	£	£
Net Cash Inflow from Operating Activities		
Receipts from customers	150,000	
Payments to suppliers	(60,000)	
Payments to employees	(30,000)	
Expenses	(850)	59,150
Returns on Investment and Servicing of Financing		
Interest received	850	
Interest paid	(400)	450
Capital Expenditure and Financial Investment		
Sale of motor car	3,000	
Purchase of motor car	(4,350)	(1,350)
Financing		
Loan	8,150	8,150
Increase in Cash		66,400

We used the headings in Figure 8.2 on page 166. The *net cash inflow from operating activities* represents all the cash flows relating to trading activities (i.e., buying or selling goods). By contrast, *returns on investment and servicing of financing* deals with interest received or paid, resulting from money invested or money borrowed. *Capital expenditure and financial investment* are concerned with the cash spent on, or received from, buying or selling fixed assets. Finally, *financing* represents a loan paid into the bank.

From Richard Hussey's cash flow statement it is clear that cash has increased by £66,400. However, the statement also clearly shows the separate components such as a positive operating cash flow of £59,150. By looking at the cash flow statement, Richard Hussey can quickly gain an overview of where his cash has come from and where it has been spent. The principles underlying the direct method of preparation are similar to those used in the construction of a cash budget (see Chapter 17).

Pause for Thought 8.4

Profit and Positive Cash Flow

If a company makes a profit, does this mean that it will have a positive cash flow?

No! Not necessarily. A fundamental point to grasp is that if a company makes a profit this means that its *assets will increase*, but this increase in assets *will not necessarily be in the form of cash*. Assets other than cash may increase (e.g., fixed assets, stock or debtors) or liabilities may decrease. This can be shown by a quick example. Noreen O. Cash has two assets: stock £25,000 and cash £50,000. Noreen makes a profit of £25,000, but invests it all in stock. We can, therefore, compare the two balance sheets.

	Before	After		Before	After
	£	£		£	£
Stock	25,000	50,000	Capital	75,000	75,000
Cash	50,000	50,000	Profit	–	25,000
	75,000	100,000		75,000	100,000

There is a profit, but it does not affect cash. The increase in profit is reflected in the increase in stock.

Indirect Method

The most common method of cash flow statement preparation is the indirect method. This method, which can be more difficult to understand than the direct method, has three steps.

- First, **we must adjust profit before taxation to arrive at operating profit.**
- Second, **we must reconcile operating profit to operating cash flow by adjusting for changes in working capital and for other non-cash flow items such as depreciation.** By adjusting the operating profit to arrive at operating cash flow, we effectively bypass the bank account. Instead of directly totalling all the operating cash flows from the bank account, we work indirectly from the figures in the profit and loss account and the opening and closing balance sheets.
- Third, **we can prepare the cash flow statement.**

These steps are outlined in Figure 8.6 on the next page and the direct and indirect methods are compared.

We will now look in more detail at the first two steps. Figure 8.8 on page 173 illustrates them. We then work through a full example, Any Company plc, in Figure 8.9 on page 174.

1. Calculation of Operating Profit by Adjusting Profit before Taxation

In the indirect method, we need to calculate operating cash flow (i.e., net cash flow from operating activities). To do this we need to first calculate operating profit so that we can reconcile operating profit to operating cash flow. Operating profit is calculated by *adjusting profit before taxation to operat-*

Figure 8.6 **Comparison of Direct and Indirect Methods of Preparing Cash Flow Statements**

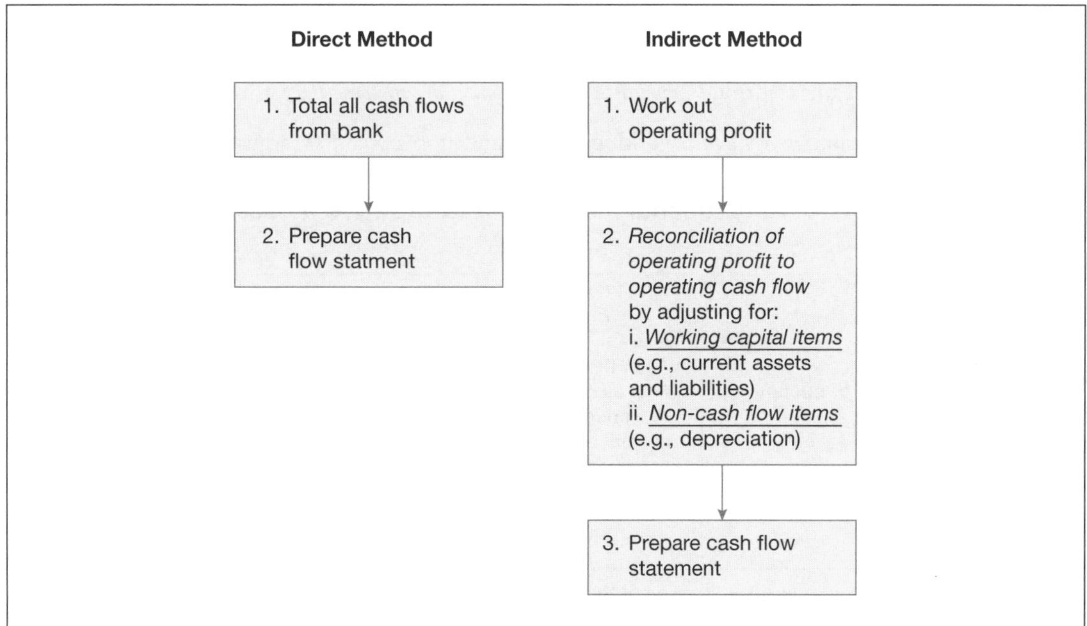

ing profit. The profit before tax figure must be adjusted by adding interest paid and deducting interest received, which are investment rather than operating items.

2 Reconciliation of Operating Profit to Operating Cash Flow

It is possible to identify two main types of adjustment needed to adjust operating profit to operating cash flow: (i) *working capital adjustments* and (ii) *non-cash flow items*, such as depreciation. It is important to emphasise that we need to consider operating cash flow and operating profit. The term 'operating' is used in accounting broadly to mean trading activities such as buying or selling goods or services.

(i) Working Capital Adjustments Effectively, working capital adjustments represent short-term timing adjustments between the profit and loss account and cash flow statement. They principally concern stock, debtors and creditors. Essentially, an increase in stock, debtors or prepayments (or a decrease in creditors or accruals) means less cash flowing into a business for the current year. For example, when debtors increase there is a delay in receiving the money. There is thus less money in the bank. By contrast, a decrease in stock, debtors or prepayments (or an increase in creditors or accruals) will mean more cash flowing into the business.

(ii) Non-Cash Flow Items Two major items are depreciation and profit or loss on the sale of fixed assets. These two items are recorded in the profit and loss account, but not in the cash flow statement. **Depreciation** (which has reduced profit) must be **added back to profit** to arrive at cash flow. By contrast, **profit on sale of fixed assets** (which has increased profit) must be **deducted from profit** to arrive at *operating cash flow*. Cash actually spent on purchasing fixed assets or received from selling fixed assets is included under *capital expenditure and financial investment* in the cash flow statement.

Figure 8.7 provides examples of both working capital and non-cash flow adjustments.

Figure 8.7 Summary of Adjustments Made to Profit to Arrive at Cash Flow

Item	*Effect on Cash Flow*	*Adjustment*
i. Working Capital (Source: **Comparison of opening and closing balance sheets**) Increase in stock Increase in debtors Increase in prepayments Decrease in creditors Decrease in accruals	All of these **reduce** cash flow as more cash is tied up in working capital (i.e., current assets less current liabilities)	**Deduct from profit** to arrive at cash flow
Decrease in stock Decrease in debtors Decrease in prepayments Increase in creditors Increase in accruals	All of these **increase** cash flow as less cash is tied up in working capital (i.e., current assets less current liabilities)	**Add to profit** to arrive at cash flow
ii. Non-Cash Flow Items (Source: **Profit and loss account**) Depreciation Loss on sale of fixed assets	Have **no effect** on cash flow, but were deducted from profit as expenses	**Add back to profit** to arrive at cash flow
Profit on sale of fixed assets	Has **no effect** on cash flow, but increased profit as other income	**Deduct from profit** to arrive at cash flow

Figure 8.8 on the next page now demonstrates the first two steps (calculation of operating profit and reconciliation of operating profit to operating cash flow). **The increases or decreases in working capital items are established by comparing the individual current assets and current liabilities in the opening and closing balance sheets.** By contrast, the **non-cash flow items are taken from the profit and loss account.**

Figure 8.8 **Illustration of Profit Adjustments**

Colette Ash has the following extracts from her business profit and loss account for year ending 31 December 2001 and the balance sheets as at 31 December 2000 and 31 December 2001. Reconcile her operating profit to her operating cash flow.

Profit and Loss Account	£	**Balance Sheets**	31.12.2000	31.12.2001
Profit before Taxation	95,000	**Current Assets**	£	£
After deducting:		Stock	4,000	5,300
Depreciation	3,000	Debtors	3,250	3,000
Interest paid	10,000	Prepayments	350	300
After adding:		Cash	6,300	10,500
Profit on sale	1,000	**Current Liabilities**		
of fixed assets		Creditors	(1,850)	(1,750)
Interest received	5,000	Accruals	(650)	(700)

i. Calculation of Operating Profit

Before reconciling operating profit to operating cash flow, we must adjust our profit before taxation for interest paid and interest received which are investment rather than operating items. These items will appear in our cash flow statement as *returns on investments and servicing of finance*. Interest paid has already been deducted from our profit before taxation and interest received has already been credited to profit. We must reverse these entries. We, therefore, have

	£
Profit before Taxation	95,000
Add Interest paid	10,000
Less Interest received	(5,000)
Operating Profit	100,000

ii. Reconciliation of Operating Profit to Operating Cash Flow

We are now in a position to adjust the operating cash flow for all changes in working capital and for all non-cash flow items (i.e., depreciation and profit on sale of fixed assets).

C. Ash
Reconciliation of Operating Profit to Operating Cash Flow

	£	£
Operating Profit		100,000
Add:		
Decrease in debtors	250	
Decrease in prepayments	50	
Increase in accruals	50	
Depreciation	3,000	3,350
Deduct:		
Increase in stock	(1,300)	
Decrease in creditors	(100)	
Profit on sale of fixed assets	(1,000)	
		(2,400)
Net Cash Inflow from Operating Activities		100,950

The three-stage process is now illustrated in Figure 8.9 which shows the calculation of a cash flow statement for Any Company Ltd. The profit and loss accounts and balance sheets are provided. Then steps 1–3 which follow show how a cash flow statement would be prepared using the indirect method.

Figure 8.9 Preparation of the Cash Flow Statement of any Company plc using the Indirect Method

Any Company Plc
Profit and Loss Account Year Ended 31 December 2001 (extracts)

	£000
Profit before Taxation (see note below)	150
Taxation	(30)
Profit after Taxation	120
Dividends	(20)
Retained Profit	100

Note:
This is after having added interest received of £15 to profit and having deducted interest paid of £8 from profit.

Balance Sheets

	31 December 2000		31 December 2001	
	£000	£000	£000	£000
Fixed Assets				
Intangible Assets				
Patents		30		50
Tangible Assets				
Plant and machinery				
Cost	300		500	
Accumulated depreciation	(50)	250	(60)	440
Total fixed assets		280		490
Current Assets				
Stock	50		40	
Debtors	20		45	
Prepayments	25		30	
Cash	45	140	40	155
Current Liabilities				
Creditors	(35)		(15)	
Accruals	(5)	(40)	(10)	(25)
Net current assets		100		130
Total assets less current liabilities		380		620
Long-term Creditors		(80)		(110)
Total net assets		300		510
Capital and Reserves		£000		£000
Ordinary share capital		250		360
Profit and loss		50		150
Total shareholders' funds		300		510

Notes:

1 There are no disposals of fixed assets. Therefore, the increases in fixed assets between 2000 and 2001 are purchases of fixed assets.
2 Taxation and dividends in the profit and loss account equal the amounts actual paid. This will not always be so.

Step 1: Calculation of Operating Profit

We need first to adjust net profit before taxation (£150) taken from profit and loss account, by adding back interest paid (£8) and deducting interest received (£15). These items are *investment* not operating flows. This is because we wish to determine operating or trading profit. Thus:

	£000
Net Profit before Taxation	150
Add Interest paid	8
Deduct Interest received	(15)
Operating Profit	143

Figure 8.9 Preparation of the Cash Flow Statement *(continued)*

Step 2: Reconciliation of Operating Profit to Operating Cash Flow

This involves taking the company's operating profit and then adjusting for:

(i) movements in working capital (e.g., increase or decreases in stock, debtors, prepayments, creditors and accruals).

(ii) non-cash flow items such as depreciation, and profit or loss on sale of fixed assets.

	£000	£000
Operating Profit		143
Add:		
Decrease in stock (1)	10	
Increase in accruals (1)	5	
Depreciation (2)	10	25
Deduct:		
Increase in debtors (1)	(25)	
Increase in prepayments (1)	(5)	
Decrease in creditors (1)	(20)	(50)
Net Cash Inflow from Operating Activities		118

(1) Represents increases, or decreases, in the current assets and current liabilities sections between the two balance sheets (i.e., movements in working capital).
(2) Difference in accumulated depreciation in the two balance sheets represents depreciation for year (i.e., represents a non-cash flow item).

Step 3: Cash Flow Statement Year Ended 31 December

This involves deducing the relevant figures in the cash flow statement by using the existing figures from the profit and loss account and the opening and closing balance sheets.

	£000	£000
Net Cash Inflow from Operating Activities (see above)		118
Returns on Investments and Servicing of Finance		
Interest received (1)	15	
Interest paid (1)	(8)	7
Taxation		
Taxation paid (1)	(30)	(30)
Capital Expenditure and Financial Investment		
Patents purchased (2)	(20)	
Plant and machinery purchased (2)	(200)	(220)
Equity Dividends Paid (1)	(20)	(20)
Financing		
Increase in long-term creditors (2)	30	
Increase in share capital (2)	110	140
Decrease in Cash (2)		(5)

	£000
Opening Cash	45
Decrease in cash	(5)
Closing Cash	40

Notes:
1. Figure from profit and loss account.
2. Represents increase or decrease from balance sheets.

All the adjusted figures, therefore, involved comparing the two balance sheets or taking figures direct from the profit and loss account. In Figure 8.9, the *dividends and taxation* in the profit and loss were *assumed to be the amounts paid.* This will not always be so. Where this is not the case, it is necessary to do some detective work to arrive at cash paid! This is illustrated for dividends paid and tax paid in Figure 8.10.

Figure 8.10 Deducing Cash Paid for Dividends or Tax – the Sherlock Holmes Approach

If we have only details of dividends payable or tax payable and the amount for the year, *we need to deduce dividends paid* or tax paid by a bit of detective work. For example, S. Holmes Ltd has the following information:

Profit and Loss Account (extracts)		**Balance Sheets (extracts)**		
			2001	2002
	£		£	£
Tax	9,000	Tax payable	6,500	7,500
Dividends	4,000	Dividends proposed	4,000	5,000

How much did S. Holmes pay for dividends and taxation?

Effectively, we know the opening and closing amounts owing (i.e., accruals) and the profit and loss charge. The amount paid is the balancing figure.

	Opening accrual	**+**	**Profit and loss**	**–**	**Amount paid**	**=**	**Closing accrual**
∴ Tax:	£6,500	+	£9,000	–	**£8,000**	=	£7,500
Dividends:	£4,000	+	£4,000	–	**£3,000**	=	£5,000

Or for those who like 'T' accounts.

Tax or dividends account

	£		£
Amount paid	**x**	Opening accrual	x
Closing accrual	x	Profit and Loss	x
	x		x

Tax

	£		£
Amount paid	*8,000*	Opening accrual	6,500
Closing accrual	7,500	Profit and Loss	9,000
	15,500		15,500

Dividends

	£		£
Amount paid	*3,000*	Opening accrual	4,000
Closing accrual	5,000	Profit and Loss	4,000
	8,000		8,000

Thus tax paid = £8,000 and dividends paid = £3,000

Essentially, we find the total liability by adding the amount owing at the start of the year to the amount incurred during the year recorded in the profit and loss account. If we then deduct the amount owing at the end of the year, we arrive at the amount paid.

Cash flow statements provide important insights into a business's inflows and outflows of cash. From J.D. Wetherspoon's cash flow statement (see The Company Camera 8.1), for example, we can see that in 2000 there was a cash inflow from operating activities of £76.2 million. There was also a net investment in pubs of £146.8 million (i.e., net cash outflow from capital expenditures). Finally, Wetherspoon financed its operations mainly by bank loans of £124 million. Overall, Wetherspoon's cash increased by £20.9 million. Wetherspoon also reports its free cash flow (£45.5 million). Essentially, this is a company's cash flow from ongoing activities excluding financing.

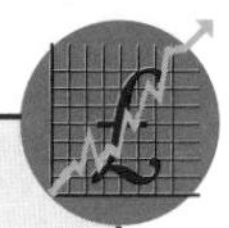

The Company Camera 8.1

Cash Flow Statement For the Year Ended 30 July 2000

	Notes	2000 £000	2000 £000	1999 £000	1999 £000
Net cash inflow from operating activities	10	**76,165**	**76,165**	60,863	60,863
Returns on investments and servicing of finance					
Interest received		**2,412**	**2,412**	782	782
Interest paid – existing pubs		**(13,710)**	**(13,710)**	(12,117)	(12,117)
Interest paid and capitalised into new pubs		**(3,921)**		(2,548)	
Net cash outflow from returns on investment and servicing of finance		**(15,219)**		(13,883)	
Taxation					
Advance corporation tax paid		**–**		(636)	
Corporation tax paid		**(1,100)**		–	
		(1,100)	**(1,100)**	(636)	(636)
Capital expenditure					
Purchase of tangible fixed assets for existing pubs		**(14,471)**	**(14,471)**	(8,804)	(8,804)
Proceeds of sale of tangible fixed assets		**4,277**		76,526	
Investment in new pubs and pub extensions		**(136,612)**		(106,390)	
Net cash outflow from capital expenditures		**(146,806)**		(38,668)	
Equity dividends paid		**(3,785)**	**(3,785)**	(3,037)	(3,037)
Net cash (outflow)/inflow before financing		**(90,745)**		4,639	
Financing					
Issue of ordinary shares		**46,566**		973	
Advances under bank loans		**124,353**		50,000	
Advances under US senior notes		**86,815**		–	
Repayments of secured bank loans		**(187,882)**		(5,784)	
Net cash inflow from financing		**69,852**		45,189	
(Decrease)/increase in cash	11	**(20,893)**		49,828	
Free cash flow	9		**45,511**		37,051
Cash flow per ordinary share	9		**22.3p**		18.8p

Source: J.D. Wetherspoon plc, 2000 Annual Report p. 25

Most companies comment on their cash flow in their annual reports. Rolls-Royce, for example, summarises its cash flow activities in The Company Camera 8.2.

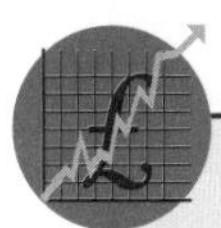

The Company Camera 8.2

Net Cash Flow from Operating Activities

Cash

The Group cash flow statement is shown on pages 42 and 43 of the financial statements. Year-end net funds, before the impact of acquisitions, amounted to £301 million (1998 £302 million). The average overdraft for 1999 at £573 million was £393 million higher than 1998, giving rise to the increased interest charge.

Net cash flow from operating activities was £359 million (1998 £285 million). Other sources of funds were £618 million from increased borrowings and £4 million from the issue of shares. Major outflows were acquisitions (£666 million), net capital expenditure (£199 million) and an increase in net working capital (£97 million).

Source: Rolls-Royce plc, Annual Report 1999, p. 22

Conclusion

Cash and cash flow are at the heart of all businesses. Cash flow is principally concerned with cash received and cash paid. It can thus be contrasted with profit which is income earned less expenses incurred. Cash is initially entered into the bank account or cash book. Companies usually derive the cash flow statement from the profit and loss account and balance sheets, not the cash book. This is known as the indirect method of cash flow statement. The cash flow, after the profit and loss account and the balance sheet, is the third major financial statement. As well as preparing cash flow statements based on past cash flows, managers will constantly monitor current cash flows and forecast future cash flows. Cash is much harder to mainipulate than profits. 'Accounting sleight of hand might shape profits whichever way a management team desires, but it is hard to deny that a cash balance is what it is. No more, no less.' (E. Warner, *Guardian*, 16 February 2002, p. 26).

Discussion *Questions*

Questions with numbers in blue have answers at the back of the book.

Q 1 At the start of this chapter, it was stated that 'cash is king' and that it is relatively easy 'to manufacture profits, but virtually impossible to create cash'. Discuss this statement.

Q 2 What is the relationship between profit and cash flow?

Q 3 The direct method of preparing the cash flow statement is the easiest to understand, but most companies use the indirect method. Why do you think this might be so?

Q 4 Preparing a cash flow statement using the indirect method is like being an accounting detective. Discuss this view.

Q 5 State whether the following are true or false. If false, explain why.

(a) Depreciation and profit from sale of fixed assets are both non-cash flow items and must be added back to operating profit to arrive at operating cash flow.
(b) Stock, debtors and fixed assets are all items of working capital.
(c) Decreases in current assets such as stock, debtors and prepayments must be added back to profit to arrive at cash flow.
(d) We need to adjust profit before taxation for non-operating items (such as interest paid or received) to arrive at operating profit.
(e) The indirect method of cash flow statement is seldom used by large companies.

Numerical *Questions*

These questions are designed to gradually increase in difficulty. Questions with numbers in blue have answers at the back of the book.

Q 1 Bingo has the following items in its accounts:

(a) Dividends payable
(b) Cash from loan
(c) Sale of goods on credit
(d) Purchase of goods for cash
(e) Cash purchase of fixed assets
(f) Cash on sale of motor car.
(g) Loan repaid.
(h) Taxation payable.
(i) Receipts from share capital issue.
(j) Bank interest paid.

Required:
Are the above items recorded in the profit and loss account, the cash flow statement, or both? If these items appear in the cash flow statement, state which heading would be most appropriate when using the direct method (e.g., net cash inflow from operating activities).

Q 2 The cash flows below were extracted from the accounts of Peter Piper, a music shop owner.

	£		£
Loan repaid	25,000	Purchase of office equipment	15,000
Sale of property	25,000	Interest paid	350
Interest received	1,150	Payments to suppliers	175,000
Payments to employees	55,000	Expenses paid	10,000
Receipts from customers	250,000		

Required:
Prepare a cash flow statement using the **direct** method for the year ended 31 December 2001.

Q3 The *cash flows* below were extracted from the accounts of Picasso and Partners, a painting and decorating business.

	£		£
Bank interest paid	1,000	Purchase of a building	88,000
Loan received	9,000	Sale of office furniture	2,300
Cash for sale of a motor car	4,000	Payment for a motor car	12,000
Interest received	300		

Required:
Prepare a cash flow statement under the **indirect** method for the year ended 31 December 2001. You know that the operating profit was £111,000 with £75,000 of working capital adjustments to be deducted and £15,000 of non-cash adjustments to be added back to arrive at operating cash flow. The operating profit has already been adjusted for the interest paid and received (so do not adjust again!).

Q4 Diana Rink Ltd, a chain of off-licences, has the following extracts from the accounts.

Profit and Loss Account

	£
Operating profit	95,000
Depreciation for year	8,000
Profit on sale of fixed assets	3,500

Balance Sheets as at 31 December

	2000 £	2001 £
Current Assets		
Stock	19,000	16,000
Debtors	10,000	11,150
Prepayments	5,000	3,500
Cash	10,000	3,250
Current Liabilities		
Creditors	1,700	2,000
Accruals	750	1,000

Required:
Prepare a statement which reconciles operating profit to operating cash flow

Q5 Brian Ridge Ltd, a construction company, has extracted the following *cash flows* from its books as at 30 November 2001.

	£		£
Operating profit	25,000	Interest paid	500
Increase in stock over year	3,500	Increase in long-term creditors	4,600
Increase in debtors over year	1,300	Purchase of plant and machinery	18,350
Increase in creditors over year	800		
Depreciation for year	6,000	Share capital issued	3,200
Tax paid	23,500	Dividends paid	550
Interest received	3,000	Purchase of patents	1,650

Required:
Prepare a cash flow statement using the indirect method. The operating profit has already been adjusted for the interest paid and received (so do not adjust again!).

*Q*6 You have the following extracts from the profit and loss account and balance sheets for Grow Hire Ltd, a transport company.

Grow Hire Ltd
Profit and Loss Account Year Ended 31 December 2001 (extracts)

	£000
Profit before Taxation (Note)	112,000
Taxation paid	(33,600)
Net Profit after Taxation	78,400
Dividends paid	(35,800)
Retained Profit	42,600

Note: After adding interest received £13,000 and deducting interest paid £6,500.

Grow Hire Ltd
Balance Sheets as at 31 December

	2000		2001	
	£000	£000	£000	£000
Fixed Assets				
Intangible Assets				
Patents		8,000		42,200
Tangible Assets				
Land and buildings:				
Cost	144,000		164,000	
Accumulated depreciation	(28,000)	116,000	(44,000)	120,000
Total fixed assets		124,000		162,200
Current Assets				
Stock	112,000		110,000	
Debtors	18,000		11,000	
Cash	7,000		10,000	
	137,000		131,000	
Current Liabilities				
Creditors	(45,000)		(20,000)	
Accruals	(4,000)		(5,000)	
	(49,000)		(25,000)	
Net current assets		88,000		106,000
Total assets less current liabilities		212,000		268,200
Long-term Creditors		(16,000)		(28,000)
Total net assets		196,000		240,200
Capital and Reserves		£000		£000
Share capital		177,000		178,600
Profit and loss		19,000		61,600
Total shareholders' funds		196,000		240,200

There were no sales of fixed assets during the year.

Required:

Prepare a cash flow statement using the indirect method for the year ended 31 December 2001.

Q7 You have the following information regarding dividends and taxation for Brain and Co., a software house.

Profit and Loss Account (Extract) (from Year to 31 December 2001)

	£
Profit before Taxation	106,508
Taxation	(51,638)
Profit after Taxation	54,870
Dividends	(27,329)
Retained Profit	27,541

Balance Sheets as at 31 December (Extracts)

	2000	2001
Current Liabilities	£	£
Tax payable	50,320	65,873
Dividends payable	23,100	29,400

Required:

Calculate tax paid and dividends paid.

Q8 A construction company, Expenso Ltd, has the following summaries from the profit and loss accounts and balance sheets for the year ended 30 September 2002.

	£000
Turnover	460,750
Cost of Sales	(328,123)
Gross Profit	132,627
Other Income	
Interest received	868
	133,495
Expenses includes interest paid £85,000	(123,478)
Profit before Taxation	10,017
Taxation	(3,005)
Profit after Taxation	7,012
Dividends	(4,105)
Retained Profit	2,907

*Q*8 Expenso Ltd (continued)

Expenso Ltd
Balance sheets as at 30 September

	2001		2002	
	£000	£000	£000	£000
Fixed Assets				
Intangible Assets				
Patents		4,000		4,500
Tangible Assets				
Land and buildings:				
Cost	20,000		26,000	
Accumulated depreciation	(7,000)		(8,000)	
Net book value	13,000		18,000	
Plant and machinery:				
Cost	25,000		30,000	
Accumulated depreciation	(8,500)		(10,000)	
Net book value	16,500	29,500	20,000	38,000
Total fixed assets		33,500		42,500
Current Assets				
Stock	2,800		6,400	
Debtors	3,200		4,500	
Cash	8,800		1,500	
	14,800		12,400	
Current Liabilities				
Creditors	(4,600)		(5,000)	
Accruals	(400)		(350)	
Dividends	(3,600)		(3,800)	
Taxation	(4,200)		(3,200)	
	(12,800)		(12,350)	
Net current assets		2,000		50
Total assets less current liabilities		35,500		42,550
Long-term Creditors		(12,100)		(12,505)
Total net assets		23,400		30,045
Capital and Reserves		£000		£000
Share capital		15,030		18,768
Profit and loss account		8,370		11,277
Total shareholders' funds		23,400		30,045

Note: There were no sales of fixed assets during the year.

Required: Prepare a cash flow statement using the indirect method for the year ended 30 September 2002.

Chapter 9

Interpretation of Accounts

'More money has been lost reaching for yield than at the point of a gun.'

Raymond Revoe Jr, *Fortune* 18 April 1994, *Wiley Book of Business Quotations* (1998), p. 19

Learning Outcomes

After completing this chapter you should be able to:

- ✔ Explain the nature of accounting ratios.
- ✔ Appreciate the importance of the main accounting ratios.
- ✔ Calculate the main accounting ratios and explain their significance.
- ✔ Understand the limitations of ratio analysis.

In a Nutshell

- Ratio analysis is a method of evaluating the financial information presented in accounts.
- Ratio analysis is performed after the book-keeping and preparation of final accounts.
- There are six main types of ratio: profitability, efficiency, liquidity, gearing, cash flow, and investment.
- Three important profitability ratios are return on capital employed (ROCE), gross profit ratio and net profit ratio.
- Four important efficiency ratios are debtors collection period, creditors collection period, stock turnover ratio and asset turnover ratio.
- Two important liquidity ratios are the current ratio and the quick ratio.
- Five important investment ratios are dividend yield, dividend cover, earnings per share (EPS), price earnings ratio, interest cover.
- Ratios can be viewed collectively using Z scores or pictics.
- For some, predominately non-profit oriented businesses, it is appropriate to use non-standard ratios, such as performance indicators.
- Four limitations of ratios are that they must be used in context, the absolute size of the business must be considered, ratios must be calculated on a consistent and comparable basis and international comparisons must be made with care.

Introduction

The interpretation of accounts is the key to any in-depth understanding of an organisation's performance. Interpretation is basically when users evaluate the financial information, principally from the profit and loss account and balance sheet, so as to make judgements about issues such as profitability, efficiency, liquidity, gearing (i.e., amount of indebtedness), cash flow, and success of financial investment. The analysis is usually performed by using certain 'ratios' which take the raw accounting figures and turn them into simple indices. The aim is to try to measure and capture an organisation's performance using these ratios. This is often easier said than done!

Context

The interpretation of accounts (or ratio analysis) is carried out after the initial bookkeeping and preparation of the accounts. In other words, the transactions have been recorded in the books of account using double-entry bookkeeping and then the financial statements have been drawn up (see Figure 9.1). For this reason, the interpretation of accounts is often known as financial statement analysis. The financial statements which form the basis for ratio analysis are principally the profit and loss account and balance sheet.

Figure 9.1 Main Stages in Accounting Process

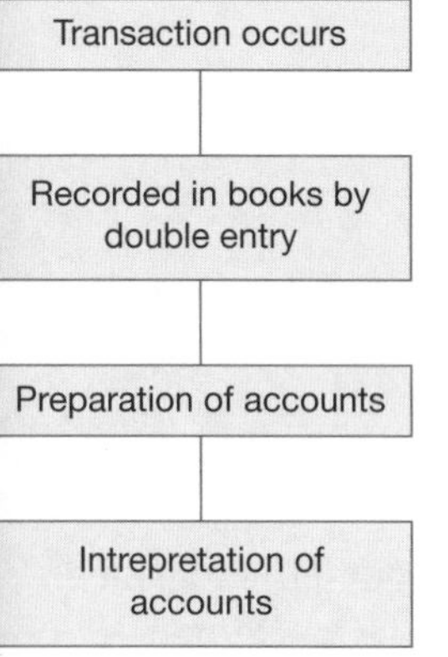

Overview

Two useful techniques, when interpreting a set of accounts, are (i) vertical and horizontal analysis, and (ii) ratio analysis. Vertical and horizontal analysis involve comparing key figures in the financial statements. In vertical analysis, key figures (such as sales in the profit and loss account and total net assets in the balance sheet) are set to 100%. Other items are then expressed as a percentage of 100. In horizontal analysis, the company's profit and loss account and balance sheet figures are compared across years. We return to vertical and horizontal analysis later in the chapter. For now, we focus on ratio analysis.

Broadly, ratio analysis can be divided into six major areas: profitability, efficiency, liquidity, gearing, cash flow and investment. The principal features are represented diagrammatically in Figure 9.3 on page 188, but set out in more detail in Figure 9.2 on the next page.

It is important to appreciate that there are potentially many different ratios. The actual ratios used will depend on the nature of the business and the individual preferences of users. The interpretation of accounts and the choice of ratios is thus inherently subjective. The ratios in Figure 9.3 have been chosen because generally they are appropriate for most businesses and are commonly used.

Figure 9.2 Principal Features of the Main Areas of the Interpretation of Accounts

Main Area	*Main Source of Ratios*	*Main Ratios*	*Overview Definition*
1. Profitability	Mainly derived from profit and loss account	1. Return on capital employed (ROCE) 2. Gross profit ratio 3. Net profit ratio	$\frac{\text{Profit before tax and loan interest}}{\text{Average capital employed}}$ $\frac{\text{Gross profit}}{\text{Sales}}$ $\frac{\text{Net profit before tax}}{\text{Sales}}$
2. Efficiency	Mixture of profit and loss account and balance sheet information	1. Debtors collection period 2. Creditors collection period 3. Stock turnover ratio 4. Asset turnover ratio	$\frac{\text{Average debtors}}{\text{Credit sales}}$ $\frac{\text{Average creditors}}{\text{Credit purchases}}$ $\frac{\text{Cost of sales}}{\text{Average stock}}$ $\frac{\text{Sales}}{\text{Average total assets}}$
3. Liquidity	Mainly from balance sheet	1. Current ratio 2. Quick ratio	$\frac{\text{Current assets}}{\text{Current liabilities}}$ $\frac{\text{Current assets} - \text{stock}}{\text{Current liabilities}}$
4. Gearing	Mainly from balance sheet	1. Gearing ratio	$\frac{\text{Long-term borrowing}}{\text{Total long-term capital}}$
5. Cash flow	Cash flow statement	1. Cash flow ratio	$\frac{\text{Total cash inflows}}{\text{Total cash outflows}}$
6. Investment	Mainly share price and profit and loss account information	1. Dividend yield 2. Dividend cover 3. Earnings per share 4. Price/earnings ratio 5. Interest cover	$\frac{\text{Dividend per ordinary share}}{\text{Share price}}$ $\frac{\text{Profit after tax and preference shares}}{\text{Ordinary dividends}}$ $\frac{\text{Profit after tax and preference dividends}}{\text{Number of ordinary shares}}$ $\frac{\text{Share price}}{\text{Earnings per share}}$ $\frac{\text{Profit before tax and loan interest}}{\text{Loan interest}}$

The 16 ratios, therefore, cover six main areas. Each of the above ratios can yield many more; for example, the gross profit ratio in Figure 9.2 is currently divided by sales. However, gross profit per employee (divide by number of employees) or gross profit per share (divide by number of shares) are also possible. The fun and frustration of ratio analysis is that there are no fixed rules. None of the ratios, except for earnings per share, is a regulatory requirement.

Figure 9.3 **Main Ratios**

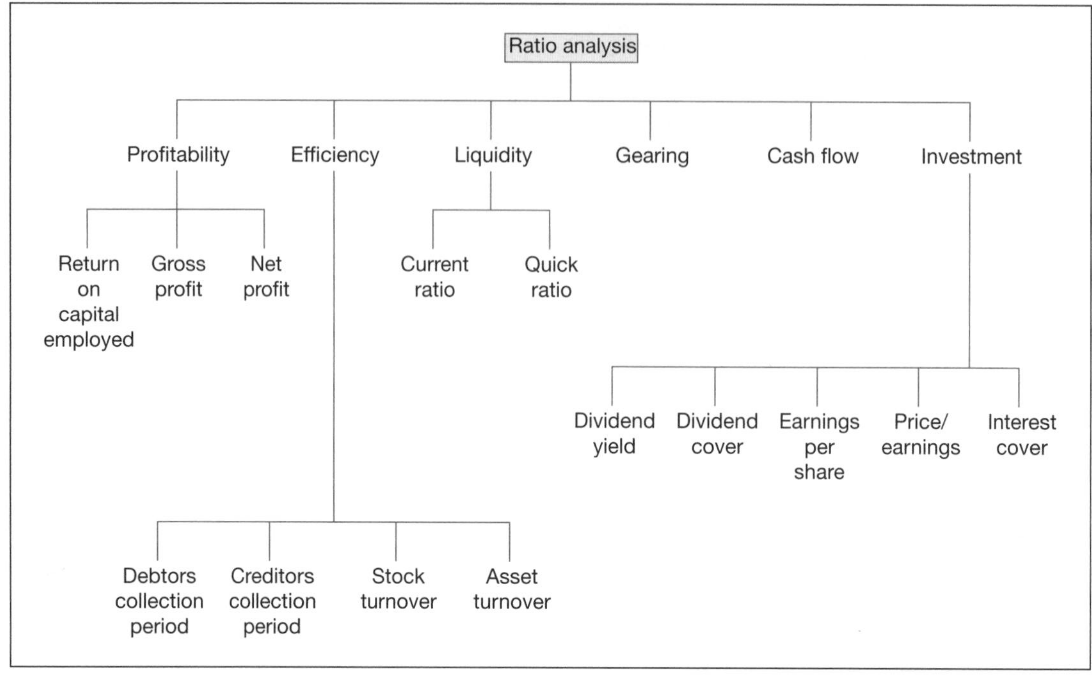

Importance of Ratios

Once the managers have prepared the accounts, then many other groups, such as investment analysts, will wish to comment on them. This is expressed in Soundbite 9.1. Different users will be interested in different ratios. For example, shareholders are primarily interested in investment ratios that measure the performance of their shares. By contrast, bankers and loan creditors may be interested primarily in liquidity (i.e., can the company repay its loan?). Ratios are important for three main reasons. First, they provide a quick and easily digestible snapshot of an organisation's achievements. It is much easier to glance at a set of ratios and draw conclusions from them than plough through the often quite complex financial statements. Second, ratios provide a good yardstick by which it is possible to compare one company with another (i.e., inter-firm comparisons) or to compare the same company over time (intra-firm comparisons). Third, ratio analysis takes account of size. One company may make more absolute profit than another. At first glance, it may, therefore, seem to be doing better than its competitor. However, if absolute size is taken into account it may in fact be performing less well.

Soundbite 9.1

Analysis

'Captains of industry change things. We merely comment on them. Somebody has to bring the news of the relief of Mafeking'.

Angus Phaure, August, *County Natwest Business* (October 1990)

Source: *The Book of Business Quotations* (1991), p. 13

Pause for Thought 9.1

Ratios and Size

Two companies, David and Goliath, have net profits of £1 million and £100 million. Is it obvious that Goliath is doing better than David?

No! We need to take into account the size of the two businesses. David may be doing worse, but then again . . . Imagine that David's sales were £5 million, while Goliath's sales were £5,000 million. Then, in £ millions,

	David			Goliath		
$\frac{\text{Net profit}}{\text{Sales}}$	$\frac{1}{5}$	=	20%	$\frac{100}{5{,}000}$	=	2%

Suddenly, Goliath's performance does not look so impressive. Size means everything when analysing ratios.

Closer Look at Main Ratios

It is now time to look in more depth at the main categories of ratio and at individual ratios. The ratios are mainly derived from the accounts of John Brown. Although John Brown is a limited company, many of the ratios are also potentially usable for partnerships or sole traders. John Brown's financial statements are given at the back of this chapter as Appendix 9.1.

Some ratios (return on capital employed, debtors and creditors collection period, stock turnover and asset turnover) use average figures from two years' accounts. In practice, two years' figures are not always available. In this case, as for John Brown, the closing figures are used on their own. When this is done, then any conclusions must be drawn cautiously.

In order to place the interpretation of key ratios in context, I have, wherever the information was available, referred to figures calculated from the UK's top 200 publicly quoted companies. This information was collected from Fame and Extel, two corporate databases, in November 2000.

Profitability Ratios

The profitability ratios seek to establish how profitably a business is operating. Profit is a key measure of business success and, therefore, these ratios are keenly watched by both internal users, such as management, and external users, such as shareholders. There are three main profitability ratios (return on capital employed, gross profit ratio, and net profit ratio). The figures used are from John Brown Plc (see Appendix 9.1).

(i) Return on Capital Employed

This ratio considers how effectively a company uses its capital employed. It compares net profit to capital employed. A problem with this ratio is that different companies often use different versions of capital employed. At its narrowest, a company's capital employed is ordinary share capital and reserves. At its widest, it might equal ordinary share capital and reserves, preference shares, long-term loans (i.e., debentures) and current liabilities. Different definitions of capital employed necessitate different definitions of profits.

The most common definition measures *profit before tax and loan interest* against *long-term capital* (i.e., ordinary share capital and reserves, preference share capital, long-term capital). Therefore, for John Brown we have:

$$\frac{\text{Profit before tax and loan interest}}{\text{Long-term capital (ordinary share capital and reserves, preference share capital and long-term loans)}} = \frac{50 + 10}{150 + 65 + 50 + 70} = \frac{60}{335} = 17.9\%$$

Essentially, the 17.9% indicates the return which the business earns on its capital. The key question is: could the capital be used anywhere else to gain a better return? In this example, with only one year's balance sheet we can only take one year's capital employed. If we have an opening and a closing balance sheet, we can take the average capital employed over the two balance sheets.

(ii) Gross Profit Ratio

The gross profit ratio (or gross profit divided by sales) is a very useful ratio. It calculates the profit earned through trading. It is particularly useful in a business where stock is purchased, marked up and then resold. For example, a retail business selling car batteries (see Figure 9.4) may well buy the batteries from the manufacturer and then add a fixed percentage as mark-up. In the case of pubs, it is traditional to mark up the purchase price of beer by 100% before reselling to customers.

Figure 9.4 Gross Profit Illustration

Snowfield batteries buys car batteries from a wholesaler for £40 each and resells them for £60 each. What will Snowfield's gross profit be?

It will simply be sales (£60) – purchases (£40) = gross profit (£20). Expressed as a percentage this is $\frac{20}{60} = 33.33\%$.

This is useful because Snowfield will know that its gross profit ratio should be 33.33%. If it is not, then there may be a problem, such as theft of stock.

In the case of John Brown, the gross profit is

$$\frac{\text{Gross profit}}{\text{Sales}} = \frac{100}{200} = 50\%$$

This is the *direct* return that John Brown makes from buying and selling goods.

(iii) Net Profit Ratio

The net profit ratio (or net profit divided by sales) is another key financial indicator. Whereas gross profit is calculated *before* taking expenses into account, the net profit ratio is calculated *after* expenses. This ratio may be calculated before or after taxation. For John Brown, the alternatives are:

$$\frac{\text{Net profit before taxation}}{\text{Sales}} = \frac{50}{200} = 25\%$$

$$\frac{\text{Net profit after taxation}}{\text{Sales}} = \frac{35}{200} = 17.5\%$$

The most popularly used alternative is net profit before taxation. This assumes that taxation is a factor that cannot be influenced by a business. This is the ratio which will be used from now on. As Real-Life Nugget 9.1 shows, most companies have traditionally, and still today, operate on net profit margins less than 10%. Across the top UK 200 public limited companies this ratio was 8.0% in November 2001.

> *Real-Life Nugget 9.1*
>
> **Net Profit Margins**
>
> However as we have already said profits are only likely to be a comparatively minor factor in cash flow anyway. *After-tax* profits in even the most spectacularly successful company will rarely run at more than about 10% of annual turnover and most companies will operate at well below this figure, say, 7% or 8% *before* tax.
>
> *Source*: B. Warnes (1984) *The Genghis Khan Guide to Business*, Osmosis Publications, p. 66

Efficiency Ratios

The efficiency ratios look at how effectively a business is operating. They are primarily concerned with the efficient use of assets. Four of the main efficiency ratios are explained below (debtors collection period, creditors collection period, stock turnover, and asset turnover). The first two are related in that they seek to establish how long debtors and creditors take to pay. The figures used are from John Brown (see Appendix 9.1).

(i) Debtors Collection Period

This ratio seeks to measure how long customers take to pay their debts. Obviously, the quicker a business collects and banks the money, the better it is for the company. This ratio can be worked out on a monthly, weekly or daily basis. This book prefers the daily basis as it is the most accurate method. The calculation for John Brown follows:

$$\text{Daily basis} = \frac{\text{Average debtors}}{\text{Credit sales per day}} = \frac{40}{200/365} = 73 \text{ days}$$

It, therefore, takes 73 days for John Brown to collect its debts. It is important to note that 'credit' sales (i.e., not cash sales) are needed for this ratio to be fully effective. This information, although available internally in most organisations, may not be readily ascertainable from the published accounts. Normally, the average of opening and closing debtors is used to approximate average debtors. When this figure is not available (as in this case), we just use closing debtors. Across the top 200 UK plcs it took 42 days to collect money from debtors in November 2001.

(ii) Creditors Collection Period

In many ways, this is the mirror image of the debtors collection period. It calculates how long it takes a business to pay its creditors. The slower a business is to pay the longer the business has the money in the bank! As with the debtors collection period, we can calculate this ratio either monthly, weekly or daily. Once more, we prefer the daily basis. This is calculated below for John Brown.

$$\text{Daily basis} = \frac{\text{Average creditors}}{\text{Credit purchases per day}} = \frac{50}{100/365} = 183 \text{ days}$$

It is usually not possible to establish accurately the figure for credit purchases from the published accounts. In John Brown, cost of sales is used as the nearest equivalent to credit purchases (remember that cost of sales is opening stock add purchases less closing stock). As with the debtors collection ratio, strictly we should use average creditors for the year (i.e. normally, the average of opening and closing creditors). If this is not available, as in this case, we use closing creditors.

It is often important to compare the debtors and creditors ratios. For John Brown, this is:

$$\frac{\text{Debtors collection period (in days)}}{\text{Creditors collection period (in days)}} = \frac{73 \text{ days}}{183 \text{ days}} = 0.40$$

In other words John Brown collects its cash from debtors in 40% of the time that it takes to pay its creditors. The management of working capital is effective.

Pause for Thought 9.2

Debtors and Creditors Collection Period

Businesses whose debtors collection periods are much less than their creditors collection periods are managing their working capital well. Can you think of any businesses which might be well placed to do this?

Businesses which sell direct to customers, generally for cash, would be prime examples. Pubs and supermarkets operate on a cash basis, or with short-term credit (cheques or credit cards). Their debtor collection period is very low. However, they may well take their time to pay their suppliers. If they have a high turnover of goods, they may collect the money for their goods from customers before they have even paid their suppliers.

(iii) Stock Turnover Ratio

This ratio effectively measures the speed with which stocks move through the business. This varies from business to business and product to product. For example, crisps and chocolate have a high stock turnover, while diamond rings have a low turnover. Strictly this ratio compares cost of sales to average stock. Where this figure is not available, we use the next best thing, closing stock. Thus for John Brown, we have:

$$\frac{\text{Cost of sales}}{\text{Average stock}} = \frac{100}{60} = 1.66 \text{ times}$$

(iv) Asset Turnover Ratio

This ratio compares sales to total assets employed (i.e., fixed assets and current assets). Businesses with a large asset infrastructure, perhaps a steel works, have lower ratios than businesses with minimal assets, such as management consultancy or dot.com businesses. Once more, where the information is available, it is best to use average total assets. For John Brown, average total assets are not available, we therefore use this year's total assets:

$$\frac{\text{Sales}}{\text{Average total assets}} = \frac{200}{395} = 0.51 \text{ times}$$

In other words, every year John Brown generates about half of its total assets in sales. This is very low. There are many other potential asset turnover ratios where sales are compared to, for example, fixed assets or total net assets.

Liquidity Ratios

Liquidity ratios are derived from the balance sheet and seek to test how easily a firm can pay its debts. Loan creditors, such as bankers, who have loaned money to a business are particularly interested in these ratios. There are two main ratios (the current ratio and quick ratio). Once more we use John Brown (Appendix 9.1).

(i) Current Ratio

This ratio tests whether the short-term assets cover the short-term liabilities. If they do not, then there will be insufficient liquid funds immediately to pay the creditors. For John Brown this ratio is:

$$\frac{\text{Current assets}}{\text{Current liabilities}} = \frac{120}{60} = 2$$

In other words, the short-term assets are double the short-term liabilities. John Brown is well covered. Across the top 200 UK plcs, this ratio was 1.31 in November 2000. In other words, current assets just covered current liabilities.

(ii) Quick Ratio

This is sometimes called the 'acid test' ratio. It is a measure of extreme short-term liquidity. Basically, stock is sold, turning into debtors. When debtors pay, the business gains cash. The quick ratio excludes stock, the least liquid (i.e., the least cash-like) of the current assets, to arrive at an immediate test of a company's liquidity. If the creditors come knocking on the door for their money, can the business survive? For John Brown we have:

$$\frac{\text{Current assets} - \text{stock}}{\text{Current liabilities}} = \frac{120 - 60}{60} = 1.0$$

For John Brown, the answer is yes. John Brown has just enough debtors and cash to cover its immediate liabilities. Across the top 200 UK plcs in November 2000 this ratio was 0.91. Thus, immediate assets just failed to cover immediate liabilities.

Gearing

Like liquidity ratios, gearing ratios are derived from the balance sheet. Gearing effectively represents the relationship between the ordinary shareholders' funds and the debt capital of a company. Essentially, ordinary shareholders' funds represent the capital owned by the ordinary shareholders. By contrast, debt capital is that supplied by external parties (normally preference shareholders and loan holders).

So far, so good. However, the role of preference share capital and short-term liabilities is worth discussing. First, preference share capital is technically part of shareholders' funds, but preference shareholders **do not own** the company and usually receive a fixed dividend. We therefore treat them as debt. Second, current liabilities and short-term loans, to some extent, do finance the company. However, generally gearing is concerned with *long-term* borrowing. Figure 9.5 now summarises shareholders' funds and long-term borrowings.

Figure 9.5 Main Elements of the Gearing Ratio

Ordinary Shareholders' Funds	*Long-Term Borrowings*
Ordinary share capital Share premium account Revaluation reserve General reserve Profit and loss account Other reserves	Preference share capital Long-term loans (i.e., long-term creditors, also known as debentures)

We can now calculate the gearing ratio for John Brown. The preferred method used in this book is to compare long-term borrowings to total long-term capital employed (i.e., ordinary shareholders' fund plus long-term borrowings). Thus we have for John Brown:

$$\frac{\text{Long-term borrowings}}{\text{Total long-term capital}} = \frac{\text{Preference share capital and debentures}}{\text{Ordinary share capital, profit and loss account and preference share capital and long-term loans (i.e., long-term creditors)}}$$

$$= \frac{50 + 70}{150 + 65 + 50 + 70} = \frac{120}{335} = 36\%$$

In other words, 36% (or 36 pence in every £1) of John Brown is financed by long-term non-ownership capital. Across the top 200 UK plcs in November 2000 the gearing ratio was about 22%. Essentially, the more highly geared a company, the more risky the situation for the owners when profitability is poor. This is because interest on long-term borrowings will be paid first. Thus, if profits are poor, there may be little, if anything, left to pay the dividends of ordinary shareholders. Conversely, if profits are booming there will be relatively more profits left for the ordinary shareholders since the return to the 'borrowers' is fixed. When judging the gearing ratio, it is thus important to bear in mind the overall profitability of the business.

Cash Flow

The cash flow ratio, unlike the other ratios we have considered so far, is prepared from the cash flow statement, not the profit and loss account or balance sheet. There are many possible ratios, but the one shown here simply measures total cash inflows to total cash outflows. This is illustrated in Figure 9.6 (it is based on Figure 8.9 on page 175, from Chapter 8, Any Company Plc).

Figure 9.6 The Cash Flow Ratio

Any Company Plc has the following cash inflows and outflows (in £000s).

	Inflows	**Outflows**
Net Cash Inflow from Operating Activities	118	
Returns on Investments and Servicing of Finance	15	8
Taxation		30
Capital Expenditure and Financial Investment		220
Equity Dividends Paid		20
Financing	140	
Totals	273	278

Therefore, our cash flow ratio is:

$$\frac{\text{Total cash inflows}}{\text{Total cash outflows}} = \frac{273}{278} = 0.98$$

To all intents and purposes, our total cash inflows thus match our total cash outflows.

Investment Ratios

The investment ratios differ from the other ratios, as they focus specifically on returns to the shareholder (dividend yield, earnings per share and price/earnings ratio) or the ability of a company to sustain its dividend or interest payments (dividend cover and interest cover). The ratios once more are calculated from John Brown (see Appendix 9.1). The first four ratios covered below are mainly of concern to the shareholders. The fifth, interest cover, is of more interest to the holders of long-term loans. Many companies give details of investment ratios in their annual reports. The Company Camera 9.1 shows the earnings per share, dividends per share and dividend cover for Manchester United from 1995 to 1999.

The Company Camera 9.1

Investment Ratios

	1999	1998	1997	1996	1995
Earnings per share (pence)	5.9	7.6	7.4	4.6	5.9
Dividends per share (pence)	1.800	1.700	1.550	1.300	1.125
Dividend cover (times)	3.3	4.4	4.7	3.5	5.2

Source: Manchester United plc, 1999 Annual Report, p. 47

Dividend Yield

This ratio shows how much dividend the ordinary shares earn as a proportion of their market price. The market price for the shares of leading public companies is shown daily in many newspapers, such as the *Financial Times*, the *Guardian*, the *Telegraph* or *The Times*. Dividend yield can be shown as net or gross of tax (dividends are paid net after deduction of tax; gross is inclusive of tax). The calculation of gross dividend varies according to the tax rate and tax rules. For simplicity, we just show the *net* dividend yield.

For John Brown, the dividend yield is:

$$\frac{\text{Dividend per ordinary share}}{\text{Share price}} = \frac{\pounds 10\text{m} \div \pounds 150\text{m}}{\pounds 0.67} = \frac{0.067}{\pounds 0.67} = 10\%$$

The dividend yield is perhaps comparable to the interest at the bank or building society. However, the increase or decrease in the share price over the year should also be borne in mind. The return from the dividend combined with the movement in share price is often known as the total shareholders' return. Across the top 200 UK plcs in November 2000, the dividend yield ratio was 4.1%.

(ii) Dividend Cover

This represents the 'safety net' for ordinary shareholders. It shows how many times profit available to pay ordinary shareholders' dividends covers the actual dividends. In other words, can the current dividend level be maintained easily. For John Brown we have:

$$\frac{\text{Profit after tax and preference dividends}}{\text{Ordinary dividends}} = \frac{30}{10} = 3.0$$

Thus, dividends are covered three times by current profits. As The Company Camera 9.2 shows, J.D. Wetherspoon's dividend is well covered by profit available. Across the top 200 UK plcs in November 2000, dividend cover was 2.5.

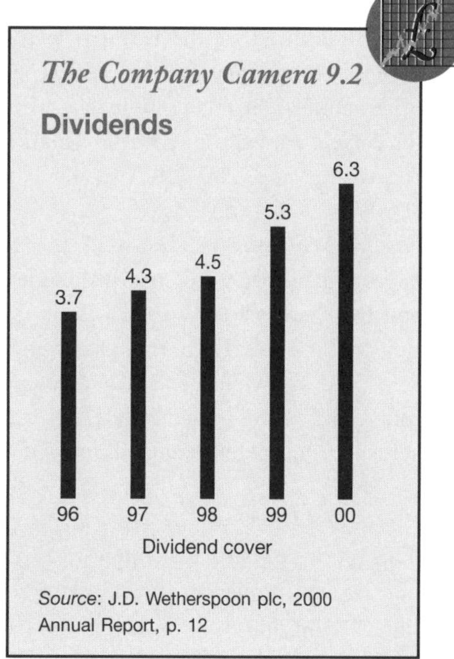

Source: J.D. Wetherspoon plc, 2000 Annual Report, p. 12

(iii) Earnings per Share (EPS)

Earnings per share (EPS) is a key measure by which investors measure the performance of a company. Its importance is shown by the fact that it is required to be shown in the published accounts of listed companies (unlike the other ratios). It measures the earnings attributable to a particular ordinary share. For John Brown it is:

$$\frac{\text{Profit after tax and preference dividends}}{\text{Number of ordinary shares}} = \frac{30}{150}$$

$$= 20\text{p}$$

Each share thus earns 20 pence. Manchester United's EPS was 7.6p in 1998. As The Company Camera 9.3 shows, the number of ordinary shares is adjusted for the fact that some share options may be taken up to create new shares. This is called diluted EPS. Across the top 200 UK plcs in November 2000, EPS was 27.8 pence.

> ***The Company Camera 9.3***
>
> **Earnings per Share**
>
> **10 Earnings per ordinary share**
>
> The calculation of earnings per share is based on earnings of £11,950,000 (1999 – £15,388,000) being the profit for the year after taxation and the weighted average number of ordinary shares in issue for the year of 259,768,040 (1999 – 259,768,040). The weighted average number of ordinary shares used in the calculation of diluted earnings per share is 260,790,097 (1999 – 260,104,590). This has been adjusted for the effect of potentially dilutive share options under the Group's share option schemes.
>
> *Source*: Manchester United plc, 2000 Annual Report, p. 37

(iv) Price/Earnings (P/E) Ratio

This is another key stock market measure. It uses EPS and relates it to the share price. A high ratio means a high price in relation to earnings and indicates a fast-growing, popular company. A low ratio usually indicates a slower-growing, more established company. If we look at John Brown, we have:

$$\frac{\text{Share price}}{\text{Earnings per share}} = \frac{67}{20} = 3.35$$

This indicates that the earnings per share is covered three times by the market price. In other words, it will take more than three years for current earnings to cover the market price. Across the top 200 UK plcs in November 2000, the P/E ratio was 36.

The P/E ratio is shown in the financial pages of newspapers along with dividend yield and the share price. In Real-Life Nugget 9.2, we show details from the *Guardian* for the aerospace and defence and automobiles sectors. This shows that the P/E ratio for aerospace and defence ranged from 5.5 to 29.7.

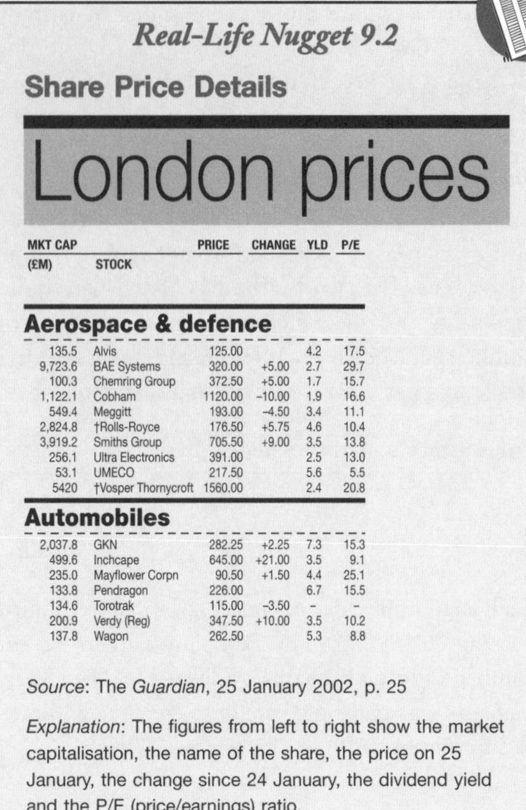

Real-Life Nugget 9.2

Share Price Details

London prices

MKT CAP (£M)	STOCK	PRICE	CHANGE	YLD	P/E
Aerospace & defence					
135.5	Alvis	125.00		4.2	17.5
9,723.6	BAE Systems	320.00	+5.00	2.7	29.7
100.3	Chemring Group	372.50	+5.00	1.7	15.7
1,122.1	Cobham	1120.00	−10.00	1.9	16.6
549.4	Meggitt	193.00	−4.50	3.4	11.1
2,824.8	†Rolls-Royce	176.50	+5.75	4.6	10.4
3,919.2	Smiths Group	705.50	+9.00	3.5	13.8
256.1	Ultra Electronics	391.00		2.5	13.0
53.1	UMECO	217.50		5.6	5.5
5420	†Vosper Thornycroft	1560.00		2.4	20.8
Automobiles					
2,037.8	GKN	282.25	+2.25	7.3	15.3
499.6	Inchcape	645.00	+21.00	3.5	9.1
235.0	Mayflower Corpn	90.50	+1.50	4.4	25.1
133.8	Pendragon	226.00		6.7	15.5
134.6	Torotrak	115.00	−3.50	–	–
200.9	Verdy (Reg)	347.50	+10.00	3.5	10.2
137.8	Wagon	262.50		5.3	8.8

Source: The *Guardian*, 25 January 2002, p. 25

Explanation: The figures from left to right show the market capitalisation, the name of the share, the price on 25 January, the change since 24 January, the dividend yield and the P/E (price/earnings) ratio.

(v) Interest Cover

This ratio is of particular interest to those who have loaned money to the company. It shows the amount of profit available to cover the interest payable on long-term borrowings. Long-term borrowings can be defined as either preference shares and long-term loans or simply long-term loans. We will use only *long-term loans* here. This ratio is similar to dividend cover. It represents a safety net for borrowers. How much could profits fall before they failed to cover interest? However, it is worth pointing out that interest is paid out of cash, not profit. For John Brown, we have:

$$\frac{\text{Profit before tax and loan interest}}{\text{Loan interest}} = \frac{50 + 10}{10} = 6$$

Loan interest is thus covered six times (i.e., well-covered). Profits would have to fall dramatically before interest was not covered. Over the top 200 UK public limited companies in November 2000, this ratio was 14.12.

Worked Example

Having explained the 16 ratios, it is now time to work through a full example. In order to do this, we use the summarised accounts of Stevens, Turner Plc (last seen in Chapter 7). Although adapted slightly, these are essentially the accounts in Figure 7.14 (see page 149). The main change is that there are now two years' figures and percentages.

Figure 9.7 **Illustrative Example on Interpretation of Accounts**

Summarised figures for Stevens, Turner Plc 20X1 and 20X2					
	20X1	20X1	20X2	20X2	(Increase/decrease)
	£000	%	£000	%	%
Profit and Loss Accounts					
Sales (all credit)	350	100	450	100	+29
Cost of Sales (all credit)	(135)	(39)	(150)	(33)	+11
Gross Profit	215	61	300	67	+40
Loan interest	(8)	(2)	(10)	(2)	+25
Administrative expenses	(77)	(22)	(95)	(22)	+23
Profit before Taxation	130	37	195	43	+50
Taxation	(26)	(7)	(39)	(8)	+50
Profit after Taxation	104	30	156	35	+50
Ordinary dividends	(6)	(2)	(12)	(3)	+100
Preference dividends	(3)	(1)	(3)	(1)	-
Retained Profit	95	27	141	31	+48
Balance Sheets					
Fixed Assets					
Intangible Assets	50	7	250	31	+400
Tangible Assets	600	90	611	75	+2
Total fixed assets	650	97	861	106	+32
Current Assets					
Stock	25	4	42	5	+68
Debtors	80	12	38	5	–52
Bank	75	11	30	3	–60
	180	27	110	13	–39
Creditors: Amounts Falling Due within One Year	(80)	(12)	(60)	(7)	–25
Net current assets	100	15	50	6	–50
Total assets less current liabilities	750	112	911	112	+21
Creditors: Amounts Falling Due after More than One Year	(80)	(12)	(100)	(12)	+25
Total net assets	670	100	811	100	+21
Capital and Reserves	£000	%	£000	%	
Share Capital					
Ordinary share capital (£1 each)	300	45	300	37	-
Preference share capital (£1 each)	150	22	150	18	-
	450	67	450	55	-
Reserves					
Share premium account	25	4	25	3	-
Revaluation reserve	30	4	30	4	-
General reserve	20	3	20	3	-
Profit and loss account	145	22	286	35	+97
Total shareholders' funds	670	100	811	100	+21
Market Price	£1		£1.50		

Vertical and Horizontal Analysis

Before calculating the ratios it is useful to perform vertical and horizontal analysis.

Vertical Analysis

Vertical analysis is where key figures in the accounts (such as sales, balance sheet totals) are set to 100%. The other figures are then expressed as a percentage of 100%. For example, cost of sales for 20X1 is 135, it is thus 39% of sales (i.e., 135 of 350). Vertical analysis is a useful way to see if any figures have changed markedly during the year. Real-Life Nugget 9.3 presents a graph using vertical analysis for Tesco's 1999 results. In this case, total assets are shown as 100%. In addition, Tesco's liquidity ratio (probably current ratio) and gearing are compared to that of other retailers.

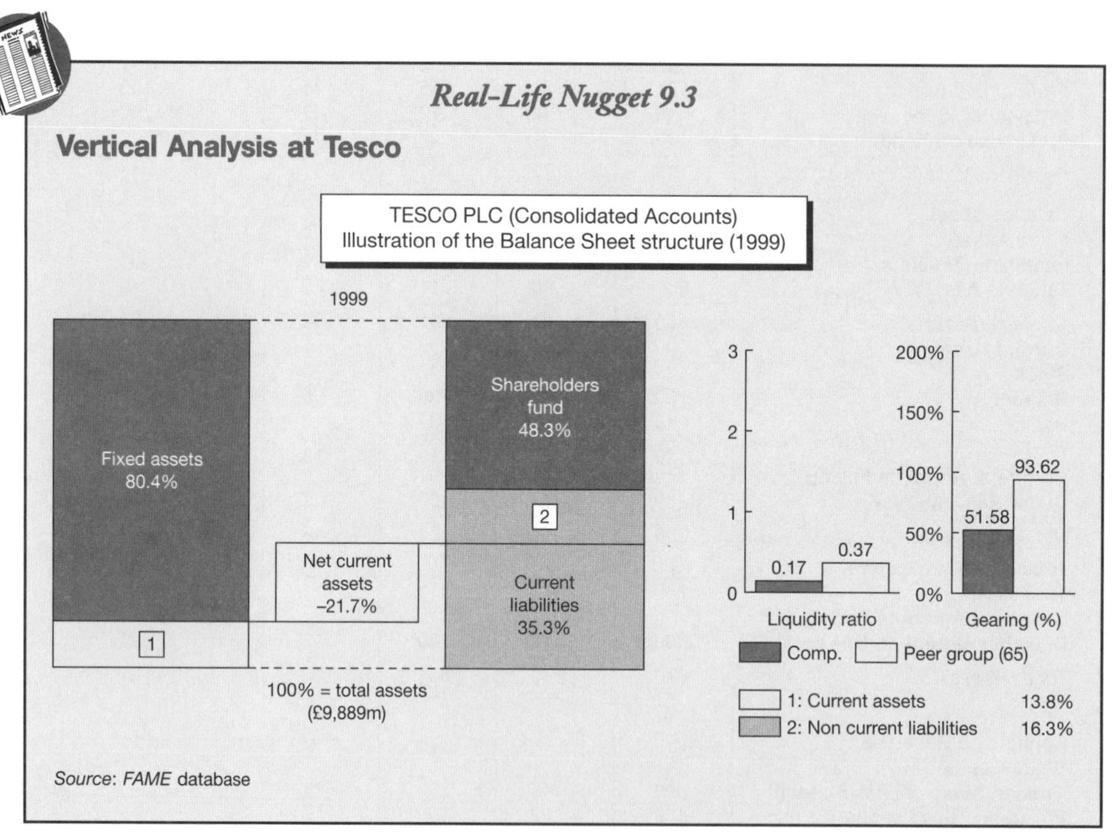

Real-Life Nugget 9.3

Vertical Analysis at Tesco

Source: *FAME* database

In Stevens, Turner Ltd, in 20X2, we can see from the profit and loss account that loan interest and administrative expenses represent 2% and 22% of sales, whereas in the balance sheet, in 20X2, tangible assets represent 75% of total net assets. We need to assess whether or not these figures appear reasonable.

Horizontal Analysis

Whereas vertical analysis compares the figures within the same year, horizontal analysis compares the figures across time. Thus, for example, we see that sales have increased from £350,000 in 20X1 to £450,000 in 20X2, a 29% increase. We need to investigate any major changes which look out of line. For example, why have there been so many changes in current assets: stock, debtors, and cash have all changed markedly (i.e., stock up 68%, debtors down 52%, and bank down 60%)? These may represent normal trading changes, or then again . . . .

Interpretation

We will now work through the various categories of ratio and make some observations. When reading these it needs to be borne in mind that normally these observations would be set in the context of the industry in which the company operates and in the economic context. They should, therefore, be taken as illustrative not definitive. We will use the available data. This is most comprehensive for 20X2 (as we can use the 20X1 comparative data).

(i) Profitability Ratios

Figure 9.8 Profitability Ratios for Stevens, Turner Plc 20X1 and 20X2

Ratios	20X1	20X2
1. Return on Capital Employed		
$\frac{\text{Profit before tax and loan interest}}{\text{Average capital employed*}}$	$\frac{130 + 8}{750^{*}} = 18.4\%$	$\frac{195 + 10}{(911^{*} + 750^{**}) \div 2} = 24.7\%$
*(i.e., ordinary share capital and reserves, preference share capital and long-term capital, i.e. long-term creditors)	*(i.e., 300 + 25 +30 + 20 + 145 + 150 + 80)	*(i.e., 300 + 25 + 30 + 20 + 20 + 286 + 150 + 100)
In 20X1, only one year-end figure is available.		**(i.e., 300 + 25 + 30 +20 + 145 + 150 + 80)
2. Gross Profit Ratio		
$\frac{\text{Gross profit}}{\text{Sales}}$	$\frac{215}{350} = 61.4\%$	$\frac{300}{450} = 66.7\%$
3. Net Profit Ratio		
$\frac{\text{Net profit before tax}}{\text{Sales}}$	$\frac{130}{350} = 37.1\%$	$\frac{195}{450} = 43.3\%$

Brief Discussion

Essentially, these profitability ratios tell us that Stevens, Turner Plc's return on capital employed is running at between 18% and 25%, having increased over the year. This represents the return from the net assets of the company. Meanwhile, the business is operating on a high gross profit margin. This has also increased over the year. Finally, the net profit ratio has also increased, perhaps because the relative cost of sales has reduced.

(ii) Efficiency Ratios

Figure 9.9 Efficiency Ratios for Stevens, Turner Plc 20X1 and 20X2

Ratios	20X1	20X2
1. Debtors Collection Period		
$\frac{\text{Average debtors}}{\text{Credit sales per day}}$	$\frac{80^*}{350 \div 365} = 83$ days	$\frac{(80 + 38)^* \div 2}{450 \div 365} = 48$ days
	*only year-end figure available	*average of two year-end figures
2. Creditors Collection Period		
$\frac{\text{Average creditors}}{\text{Credit purchases per day}^*}$	$\frac{80^*}{135 \div 365} = 216$ days	$\frac{(80 + 60)^* \div 2}{150 \div 365} = 170$ days
*in this case cost of sales	*only year-end figure available	*average of two year-end figures
3. Stock Turnover Ratio		
$\frac{\text{Cost of sales}}{\text{Average stock}}$	$\frac{135}{25^*} = 5.4$ times	$\frac{150}{(25 + 42)^* \div 2} = 4.48$ times
	*only year-end figure available.	*average of two year-end figures
4. Asset Turnover Ratio		
$\frac{\text{Sales}}{\text{Average total assets}}$	$\frac{350}{50 + 600 + 180^*} = 0.42$ times	$\frac{450}{(50 + 250) + (600 + 611) + (180 + 110)/2^*} = 0.50$ times
(i.e., intangible, tangible and current)	*the intangible and tangible fixed assets and current assets figures for 20X1	*the intangible and tangible fixed assets and current assets figures for 20X1 and 20X2 averaged (i.e. ÷ 2)

Brief Discussion

There have been substantial reductions in the debtors and creditors collection periods. Debtors now pay in 48 rather than 83 days. By contrast, Stevens, Turner pays its creditors in 170 days not 216 days. By receiving money more quickly than paying it, Stevens, Turner is benefiting as its overall bank balance is healthier. However, it must be careful not to antagonise its suppliers as 170 days is a long time to withhold payment. Stock is moving more slowly this year than last. However, each item of stock is still replaced $4\frac{1}{2}$ times each year. Finally, the asset turnover ratio is disappointing. Sales are considerably lower than total assets, even though there is some improvement over the year.

(iii) Liquidity Ratios

Figure 9.10 Liquidity Ratios for Stevens, Turner Plc 20X1 and 20X2

Ratios	20X1	20X2
1. Current Ratio		
$\frac{\text{Current assets}}{\text{Current liabilities}}$	$\frac{180}{80} = 2.2$	$\frac{110}{60} = 1.8$
2. Quick Ratio		
$\frac{\text{Current assets} - \text{stock}}{\text{Current liabilities}}$	$\frac{180 - 25}{80} = 1.9$	$\frac{110 - 42}{60} = 1.1$

Brief Discussion

There is a noted deterioration in both liquidity ratios. The current ratio has fallen from 2.2 to 1.8. Meanwhile, the quick ratio has declined from 1.9 to 1.1. While not immediately worrying, Stevens, Turner needs to pay attention to this.

(iv) Gearing Ratio

Figure 9.11 Gearing Ratio for Stevens, Turner Plc 20X1 and 20X2

Ratio	20X1	20X2
$\frac{\text{Long term borrowings*}}{\text{Total long-term capital**}}$	$\frac{230^{*}}{750^{**}} = 30.7\%$	$\frac{250^{*}}{911^{**}} = 27.4\%$
*Preference shares and long-term loans (i.e., long-term creditors)	*150 + 80 = 230	*150 + 100 = 250
**Preference shares, long-term loans (i.e. long-term creditors), ordinary shares, share premium account, revaluation reserve, general reserve, profit and loss account.	**150 + 80 + 300 + 25 + 30 + 20 + 145 = 750	**150 + 100 + 300 + 25 + 30 + 20 + 286 = 911

Brief Discussion

Gearing has declined over the year from 30.7% to 27.4%. In 20X2, 27.4 pence in the £ of the long-term capital employed is from borrowed money, rather than 30.7 pence last year. This improvement is due to the increase in retained profit during the year.

(v) Cash Flow Ratio

Figure 9.12 Cash Flow Ratio for Stevens, Turner Plc 20X1 and 20X2

From the balance sheets and profit and loss accounts in Figure 9.7, we can determine the cash flow statement. From this cash flow statement we can work out the cash inflows and cash outflows.

Stevens, Turner Plc
Reconciliation of Operating Profit to Operating Cash Flow Year Ended 31.12.20X1

	£000	£000
Profit before Tax and Loan Interest*		205
Add:		
Decrease in debtors	42	42
Deduct:		
Increase in stock	(17)	
Decrease in creditors	(20)	(37)
Net Cash Inflow from Operating Activities		210

*For simplicity, we assume no depreciation.

Stevens, Turner Plc
Cash Flow Statement Year Ended 31.12.20X2

	£000	£000
Net Cash Inflow from Operating Activities		210
Returns on Investments and Servicing of Finance		
Interest paid	(10)	(10)
Taxation		
Taxation paid	(39)	(39)
Capital Expenditure and Financial Investment		
Purchase of intangible assets	(200)	
Purchase of tangible assets	(11)	(211)
Equity Dividends Paid	(15)	(15)
Financing		
Increase in loan capital	20	20
Decrease in Cash		(45)

	£000
Opening Cash	75
Decrease in cash	(45)
Closing Cash	30

Therefore our cash flow ratio is:

	Cash Inflows £000	Cash Outflows £000
Net Cash Inflow from Operating Activities	210	
Returns on Investment and Servicing of Finance		10
Taxation		39
Capital Expenditure and Financial Investment		211
Equity Dividends		15
Financing	20	
	230	275

$$\frac{\text{Total cash inflows}}{\text{Total cash outflows}} = \frac{230}{275} = 0.84.$$

More cash is flowing out than is flowing in. The main reason for this is the purchase of fixed assets.

(vi) Investment Ratios

Figure 9.13 Investment Flow Ratios for Stevens, Turner Plc 20X1 and 20X2

Ratios	20X1	20X2
1. Dividend Yield		
$\frac{\text{Dividend per ordinary share*}}{\text{Share price}}$	$\frac{2^*}{100p} = 2\%$	$\frac{4^*}{150p} = 2.67\%$
*Net ordinary dividend divided by number shares	*6 ÷ 300 = 2p	*12 ÷ 300 = 4p
2. Dividend Cover		
$\frac{\text{Profit after tax and preference dividends}}{\text{Ordinary dividends}}$	$\frac{101}{6}$ = 16.8 times	$\frac{153}{12}$ = 12.8 times
3. Earnings per Share		
$\frac{\text{Profit after tax and preference dividends}}{\text{Number of ordinary shares}}$	$\frac{101}{300}$ = 33.7p	$\frac{153}{300}$ = 51p
4. Price/Earnings Ratio		
$\frac{\text{Share price}}{\text{Earnings per share}}$	$\frac{100}{33.7}$ = 3.0	$\frac{150}{51}$ = 2.9
5. Interest Cover		
$\frac{\text{Profit before tax and loan interest}}{\text{Loan interest}}$	$\frac{130 + 8}{8}$ = 17.2 times	$\frac{195 + 10}{10}$ = 20.5 times

Brief Discussion

The dividend yield is quite low at around 2% to 3%. However, it must be remembered that the share price has increased rapidly by 50p, and it is unusual to have strong capital growth and high dividends at the same time. Both dividend cover and interest cover are high. If necessary the company has the potential to increase dividends and interest. Earnings per share (EPS) has increased over the year and is now running at an improved 51 pence. It is this rise in EPS which may have fuelled the share price increase. The P/E ratio, however, is still very modest at 2.9.

Report Format

Students are often required to write a report on the performance of a company using ratio analysis. A report is not an essay! It has a pre-set style, usually including the following features:

- Terms of reference
- Heading
- Introduction
- Major sections
- Recommendations
- Appendix

As an illustration, Figure 9.14 presents a *concise* overall report on Stevens, Turner Plc for 20X1 and 20X2. A real report would be longer than this, but this report gives a good insight into the use of report format.

Figure 9.14 **Illustrative Report on Financial Performance of Stevens, Turner Plc for the Year Ended 20X2**

Report on the Financial Performance of Stevens, Turner Ltd Year Ended 20X2

1.0 Terms of Reference
The Managing Director requested a report on the financial performance of Stevens, Turner Plc for the year ended 20X2 using appropriate ratio analysis.

2.0 Introduction
The profit and loss account, balance sheet and cash flow data were used to prepare 16 ratios to assess the company's performance for 20X2. The 20X2 financial results were compared to those in 20X1. The underpinning ratios with their calculations are presented in the appendix. This report briefly covers the profitability, efficiency, liquidity, gearing, cash flow, and investment ratios.

3.0 Profitability
The company has traded quite profitably over the year. The return on capital employed and gross profit ratios have increased over the year from 18.4% to 24.7%, and from 61.4% to 66.7%, respectively. The net profit ratio improved from 37.1% to 43.3%. This ratio is extremely good.

4.0 Efficiency
The collection of money from debtors is still quicker than the payment of suppliers (48 days vs 170 days), which is good for cash flow. Both collection periods have declined over the year. The stock is turned over 4.48 times per year which is usual for this type of business. Finally, the asset turnover ratio appears quite low. However, once more this reflects the nature of the business.

5.0 Liquidity
Liquidity is an area to watch for the future. Both the current ratio and quick ratio have declined markedly over the year (from 2.2 to 1.8, and 1.9 to 1.1, respectively). While this is not immediately worrying, this ratio should not be allowed to slip any further.

6.0 Cash flow
The cash flow ratio is 0.84. More cash is flowing out than is coming in. The main reason appears to be the purchase of fixed assets.

7.0 Gearing
The dependence on outside borrowing has declined during the year from 30.7% to 27.4%. This is good news.

8.0 Investment
Both dividends and loan interest remain well-covered (respectively 12.8 times and 20.5 times). Overall, shareholders are receiving a good return for their investment. Share price has increased by 50 pence, which compensates for the low dividend yield of 2.67%. The earnings per share remains a healthy 51p (up from 33.7p). Finally, the P/E ratio has remained steady at 2.9.

9.0 Conclusions
Overall, Stevens, Turner Ltd has had a good year in terms of profitability, investment performance, gearing, and efficiency ratios. The one area we really need to pay attention to is cash flow and liquidity. While not immediately worrying, this area should be carefully monitored.

Appendix 1 (Extract)

1. Return on capital employed	20X1		20X2	
$\frac{\text{Net profit before tax and loan interest}}{\text{Average capital employed}}$	$\frac{138}{750}$ =	18.4%	$\frac{205}{830}$ =	24.7%

Note: All the ratios are calculated in Figures 9.8 to 9.13.

Holistic View of Ratios

So far we have looked at individual ratios. However, although useful, one ratio on its own may potentially be misleading or may even be manipulated through creative accounting. Therefore, there have been attempts to look at ratios collectively. Two main approaches are briefly discussed here.

1. The Z Score Model

The idea behind this model, which was first developed in the US, is to select ratios which when combined have a high predictive power. In the UK, an academic, Richard Taffler, developed the model using two groups of failed and non-failed companies. After a comprehensive study of accounts, he produced a model for listed industrial companies. This model proved successful in distinguishing between those companies which would go bankrupt and those companies which would not.

2. Pictics

Pictics are an ingenious way of presenting ratios. Essentially, each pictic is a face. The different elements of the face are represented by different ratios. Real-Life Nugget 9.4 demonstrates two pictics. The face on the left represents a successful business while that on the right is an unsuccessful business. The size of the smile represents profitability while the length of the nose represents liquidity. Pictics are an easy way of presenting multi-dimensional information. Although it is easy to dismiss pictics as a joke, they have proved remarkably successful in controlled research studies.

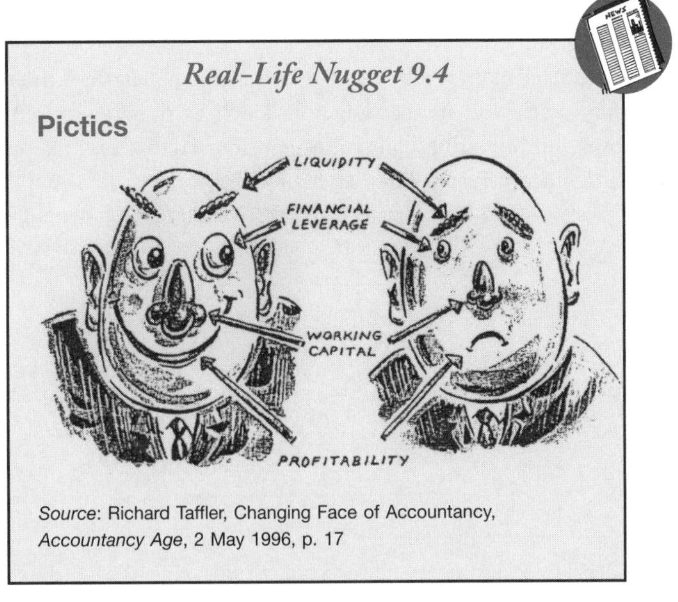

Real-Life Nugget 9.4

Pictics

Source: Richard Taffler, Changing Face of Accountancy, *Accountancy Age*, 2 May 1996, p. 17

Performance Indicators

The conventional mix of ratios may be unsuitable for some businesses, in particular those where non-financial performance is very important. Examples of such organisations include the National Health Service and the railways. Such businesses use customised performance measures often called performance indicators. For the National Health Service, indicators such as number of operations, or bed occupancy rate, may be more important than net profit.

Pause for Thought 9.3

Performance Indicators

Which performance indicators do you think would be useful when assessing the performance of individual railway operating companies?

Potentially, there are many performance indicators. For example:

- Percentage of trains late
- Miles per passenger
- Volume of freight moved
- Passengers per train
- Number of complaints
- Number of accidents

Organisations like the rail companies need to balance financial considerations (such as making profits for shareholders) with non-financial factors (such as punctuality). As Real-Life Nugget 9.5 on the next page shows, rail operating companies consider factors such as punctuality, passenger and train numbers as well as income from fares and subsidies.

Real-Life Nugget 9.5

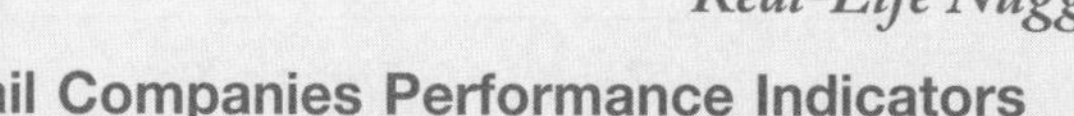

Rail Companies Performance Indicators

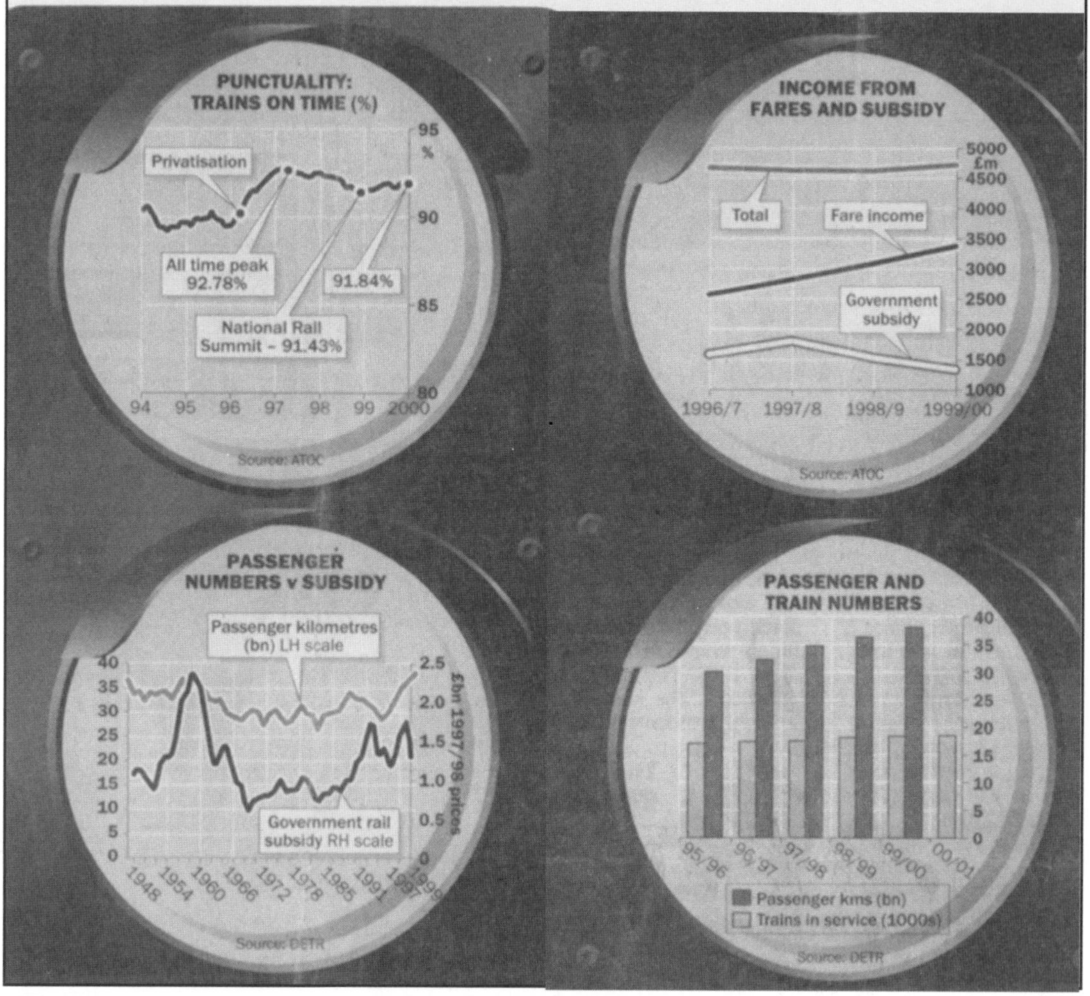

Source: Rail Companies in line for huge public handout, Arthur Leathley, *The Times*, May 25 2000, p. 4

Limitations

Ratio analysis can be a useful financial tool. However, certain problems associated with ratio analysis must be appreciated.

1. Context

Ratios must be used in context. They cannot be used in isolation, but must be compared with past results or industry norms. Unless such a comparative approach is adopted ratio analysis is fraught with danger.

2. Absolute Size

Ratios give no indication of the relative size of the result. If net profit is 10%, we do not know if this is 10% of £100 or 10% of £1 million. Both the ratio *and* the size of the organisation need to be taken into account.

3. Like with Like

We must ensure that we are comparing like with like. The accounting policies of different companies differ and this needs to be appreciated. If companies A and B, for example, use a different rate of depreciation then a net profit of 10% for company A may equal 8% for company B.

4. International Comparison

The comparison of companies in different countries is even more problematic than same-country comparisons. The economic and business infrastructure in Japan, for example, is very different from that in the US. Traditionally, this has led to the current ratio in the US being much higher than in Japan.

Real-Life Nugget 9.6

Financial Ratios

Those of us with a grounding in accounting already know that financial information, although often (and naively) assumed to be precise, is necessarily based on significant assumptions and varying underlying principles. This means that accurate financial comparisons between companies (even those within the UK) cannot be made without a considerable amount of additional research and even restatement. In a cross-border analysis situation, the problem is further compounded by important accounting differences.

Despite this, traditional performance indicators such as profit margin, return on capital employed (ROCE), earnings per share (EPS) and the price earnings (P/E) ratio are widely published and used in decision-making, often one suspects without any great attention being paid to what lies behind them. For example, banks, credit rating agencies, auditors, investment analysts, merger and acquisition teams and the financial press all use such financial ratios in their daily work.

Source: M. Gardiner and K. Bagshaw (1997) Financial Ratios: Can You Trust Them, *Management Accounting*, September, p. 30

Despite the above limitations, ratios are widely used. As Real-Life Nugget 9.6 shows, even though these accounting ratios are based on significant assumptions and varying underlying principles, they are still commonly employed by banks, credit rating agencies and other users.

Conclusion

Ratio analysis is a good way to gain an overview of an organisation's activities. There are a whole range of ratios on profitability, efficiency, liquidity, gearing, cash flow and investment. Taken together these ratios provide a comprehensive view of a company's financial activities. They are used to compare a company's performance over time as well as to compare different companies' financial performance. For certain businesses, particularly those not so profit-orientated, performance indicators provide a useful alternative. When calculating ratios care is necessary to ensure that the underlying figures have been drawn up in a consistent and comparable way. However, when used carefully, ratios are undoubtedly very useful.

Discussion *Questions*

Questions with numbers in blue have answers at the back of the book.

Q **1** What do you understand by ratio analysis? Distinguish between the main types of ratio analysis.

Q **2** Do the advantages of ratio analysis outweigh the disadvantages? Discuss.

Q **3** 'Each of the main financial statements provides a distinct set of financial ratios.' Discuss this statement.

Q **4** Devise a set of non-financial performance indicators which might be appropriate for monitoring:

(a) The Police
(b) The Post Office (now known as Royal Mail)

Q **5** State whether the following are true or false? If false, explain why.

(a) Gross profit ratio $= \dfrac{\text{Gross profit}}{\text{Sales}}$

(b) Net profit ratio $= \dfrac{\text{Net profit after taxation}}{\text{Average capital employed}}$

(c) Current asset ratio $= \dfrac{\text{Current assets} - \text{stock}}{\text{Current liabilities}}$

(d) Debtors collection period $= \dfrac{\text{Debtors}}{\text{Credit sales per day}}$

(e) Asset turnover ratio $= \dfrac{\text{Sales}}{\text{Fixed assets}}$

(f) Dividend yield $= \dfrac{\text{Dividend per ordinary share}}{\text{Sales}}$

(g) Earnings per share $= \dfrac{\text{Profit after tax and preference dividends}}{\text{Number of ordinary shares}}$

Numerical *Questions*

Questions with numbers in blue have answers at the back of the book.

Q 1 The information below is from the accounts of John Parry.

Trading and Profit and Loss Account Year Ended 31 December 2001

	£	£
Sales		150,000
Less *Cost of Sales*		
Opening stock	25,000	
Add Purchases	75,000	
	100,000	
Less Closing stock	30,000	70,000
Gross Profit		80,000
Less Expenses		30,000
Net Profit		50,000

Other information	31.12.2000	31.12.2001
	£	£
Total assets	50,000	60,000
Closing capital	300,000	500,000
Debtors	18,000	19,000
Creditors	9,000	10,000

Note: All of John Parry's sales and purchases are on credit.

Required:

Calculate the following profitability and efficiency ratios:

(a) Return on capital employed
(b) Gross profit ratio
(c) Net profit ratio
(d) Debtors collection period
(e) Creditors collection period
(f) Stock turnover ratio
(g) Asset turnover ratio,

Q 2 Henry Mellett has the following extracts from his balance sheet as at 31 March 2001.

	£
Current Assets	
Stock	18,213
Debtors	12,407
Cash	1,283
Current Liabilities	
Creditors	14,836
Long-term Creditors	30,000
Total net assets	150,000

Required:

Calculate the following liquidity and gearing ratios:

(a) Current ratio
(b) Quick ratio
(c) Gearing ratio.

Q 3 Jane Edwards Ltd has prepared her cash flow statement under the direct method. She has the following main cash flows.

	£		£
Cash from customers	125,000	Dividends paid	8,000
Cash paid to employees	18,300	Taxation paid	16,000
Cash paid to suppliers	9,250	Purchase of fixed assets	80,000
Issue of shares	29,000	Sale of fixed assets	35,000
Buy back loan	8,000		

Required:

Calculate the cash flow ratio.

Q 4 From the following information for Clatworthy Plc calculate the investment ratios as indicated.

	£000
Turnover	1,000
Profit before Taxation	750
(after charging loan interest of £40,000)	
Taxation	(150)
Profit after Taxation	600
Preference dividends	(20)
Ordinary dividends	(40)
Retained Profit	540

Market price ordinary shares £1.25. Number of ordinary shares in issue are 500,000.

Required:

(a) Dividend yield
(b) Dividend cover
(c) Earnings per share
(d) Price/earnings ratio
(e) Interest cover.

Q5 The abridged accounts for N.O. Hope Plc are given below.

Profit and Loss Accounts	2000	2001
	£000	£000
Sales	400	440
Cost of Sales	(300)	(330)
Gross Profit	100	110
Administrative expenses	(15)	(25)
Distribution expenses	(5)	(10)
Profit before Taxation	80	75
Taxation	(16)	(15)
Profit after Taxation	64	60
Dividends	(46)	(45)
Retained Profit	18	15

Balance Sheets		
Fixed Assets	£000	£000
Intangible Assets	20	20
Tangible Assets	120	235
	140	255
Current Assets		
Stock	80	40
Debtors	40	20
Bank	20	10
	140	70
Creditors: Amounts Falling Due within One Year	(70)	(75)
Net current assets (liabilities)	70	(5)
Total assets less current liabilities	210	250
Creditors: Amounts Falling Due after More Than One Year	(20)	(40)
Total net assets	190	210

Q5 N.O. Hope Plc (continued)

	£000	£000
Capital and Reserves		
Share Capital		
Ordinary share capital (£1 each)	120	125
Preference share capital (£1 each)	17	17
	137	142
Reserves		
Capital reserves		
Share premium account	10	10
Revaluation reserve	10	10
Other reserves		
General reserve	8	8
Profit and loss account	25	40
	53	68
Total shareholders' funds	190	210

Required:
Prepare a horizontal and vertical analysis. Highlight three figures that may need further enquiry.

Q6 The following two companies Alpha Industries and Beta Industries operate in the same industrial sector. You have extracted the following ratios from their accounts.

	Alpha	Beta
Return on capital employed	9%	20%
Gross profit ratio	25%	25%
Net profit ratio	7%	14%
Current ratio	2.1	1.7
Quick ratio	1.7	1.3
Price/Earnings ratio	4	8
Dividend cover	2	4

Required:
Compare the financial performance of the two companies. All other things being equal, which company would you expect to have the higher market price?

*Q*7 Anteater Plc has produced the following summary accounts.

Profit and Loss Account Year Ended 31 December 2001

	£000
Sales	1,000
Cost of sales	(750)
Gross Profit	250
Administrative expenses (includes debenture interest £3,000)	(120)
Distribution expenses	(30)
Profit before Taxation	100
Taxation	(20)
Profit after Taxation	80
Preference dividends	(10)
Ordinary dividends	(40)
Retained Profit	30

Balance Sheet as at 31 December 2001

	£000	£000
Fixed Assets		420
Current Assets		
Stock	40	
Debtors	50	
Cash	30	
	120	
Creditors: Amounts Falling Due within One Year	(40)	
Net current assets		80
Total assets less current liabilities		500
Creditors: Amounts Falling Due after More Than One Year		(100)
Total net assets		400
Capital and Reserves		
Share Capital		£000
Ordinary share capital (£1 each)		300
Preference share capital (£1 each)		20
		320
Reserves		
Capital reserves		
Share premium account		10
Other reserves		
Profit and loss account		70
Total shareholders' funds		400

Share price £2.00.

*Q*7 Anteater Plc (continued)

Required:
From the above accounts prepare the following ratios:

(a) Profitability ratios
(b) Efficiency ratios
(c) Liquidity ratios
(d) Gearing ratio
(e) Investment ratios.

*Q*8 You are an employee of a medium-sized, light engineering company. Your managing director, Sara Potter, asks you to analyse the accounts of your company, Turn-a-Screw Ltd, with a competitor, Fix-it-Quick.

Profit and Loss Accounts Year Ended 31 December 2001

	Turn-a-Screw	Fix-it-Quick
	£000	£000
Sales	2,500	2,800
Cost of Sales	(1,000)	(1,200)
Gross Profit	1,500	1,600
Administrative expenses (includes loan interest)	(900)	(1,170)
Distribution expenses	(250)	(200)
Profit before Taxation	350	230
Taxation	(76)	(46)
Profit after Taxation	274	184
Preference dividends	(30)	(20)
Ordinary dividends	(124)	(94)
Retained Profit	120	70

Q **8** Turn-a-Screw Ltd (continued)

Balance Sheets as at 31 December 2001

	Turn-a-Screw		Fix-it-Quick	
	£000	£000	£000	£000
Fixed Assets				
Tangible assets		1,820		1,765
Current Assets				
Stock	120		115	
Debtors	100		115	
Cash	10		25	
	230		255	
Creditors: Amounts Falling Due within One Year	(190)		(225)	
Net current assets		40		30
Total assets less current liabilities		1,860		1,795
Creditors: Amounts Falling Due after More than One Year (10% interest)		(250)		(400)
Total net assets		1,610		1,395
Capital and Reserves		£000		£000
Share Capital				
Ordinary share capital (£1 each)		850		860
Preference share capital (£0.50 each)		300		200
		1,150		1,060
		£000		£000
Reserves				
Capital reserves				
Share premium account		125		–
Other reserves				
Profit and loss account		335		335
Total shareholders' funds		1,610		1,395
Share price		£1.44		£1.00

Required:

Using the accounts of the two companies calculate the appropriate:

(a) Profitability ratios
(b) Efficiency ratios
(c) Liquidity ratios
(d) Gearing ratio
(e) Investment ratios.

Briefly comment on your main findings for each category.

Q9 You have been employed temporarily by a rich local businessman, Mr Long Pocket, as his assistant. He has been told at the golf club that Sunbright Enterprises Plc, a locally based company, would be a good return for his money. The last five years' results are set out below.

Profit and Loss Accounts Year Ended 31 December

	1997	1998	1999	2000	2001
	£000	£000	£000	£000	£000
Sales	1,986	2,001	2,008	2,010	2,012
Cost of Sales	(1,192)	(1,221)	(1,406)	(1,306)	(1,509)
Gross Profit	794	780	602	704	503
Expenses (including loan interest)	(633)	(648)	(487)	(606)	(437)
Profit before Taxation	161	132	115	98	66
Taxation	(32)	(26)	(23)	(19)	(13)
Profit after Taxation	129	106	92	79	53
Preference dividends	(8)	(8)	(8)	(9)	(9)
Ordinary dividends	(25)	(26)	(27)	(32)	(35)
Retained Profit	96	72	57	38	9

Balance Sheets as at 31 December

	1997	1998	1999	2000	2001
	£000	£000	£000	£000	£000
Fixed Assets					
Tangible assets	500	580	660	780	878
Current Assets					
Stock	24	26	27	45	68
Debtors	112	120	121	130	134
Cash	25	24	30	21	9
	161	170	178	196	211
Creditors: Amounts Falling Due within One Year	(83)	(90)	(111)	(126)	(210)
Net current assets	78	80	67	70	1
Total assets less current liabilities	578	660	727	850	879
Creditors: Amounts Falling Due after More than One Year (10% interest)	(100)	(110)	(120)	(130)	(150)
Total net assets	478	550	607	720	729

Q 9 Sunbright Enterprises Plc (continued)

Capital and Reserves	£000	£000	£000	£000	£000
Share Capital					
Ordinary share capital (£1 each)	250	250	250	300	300
Preference share capital (£1 each)	88	88	88	100	100
	338	338	338	400	400
Reserves					
Capital reserves					
Share premium account	12	12	12	25	25
Other reserves					
Profit and loss account	128	200	257	295	304
Total shareholders' funds	478	550	607	720	729
Share price	£1.10	£1.08	£1.07	£1.05	£0.95

Required:

Analyse the last five years' financial results for the company and calculate the appropriate ratios. Present your advice as a short report.

Note: Horizontal analysis, vertical analysis and a calculation of the cash flow ratio are not required.

Appendix 9.1: John Brown Plc

John Brown Plc has the following abridged results for the year ending 31 December 2001

John Brown Plc
Trading and Profit and Loss Account Year Ended 31 December 2001

	£m	£m
Sales		200
Cost of sales		(100)
Gross Profit		100
Less *Expenses*		
General	40	
Loan interest	10	50
Profit before Taxation		50
Taxation		(15)
Profit after Taxation		35
Preference dividends (10%)	(5)	
Ordinary dividends	(10)	(15)
Retained Profit		20

Appendix 9.1: John Brown Plc (continued)

Balance Sheet as at 31 December 2001

	£m	£m	£m
Fixed Assets			**275**
Current Assets			
Stock	60		
Debtors	40		
Cash	20	120	
Current Liabilities			
Creditors	(50)		
Proposed dividends and tax	(10)	(60)	
Net current assets			60
Total assets less current liabilities			335
Long-term Creditors			(70)
Total net assets			265

	£m	£m
Capital Employed		
Capital and Reserves		
Share Capital		
Ordinary share capital (150m £1)		150
Preference share capital (50m £1)		50
		200
Reserves		
Other reserves		
Opening profit and loss account	50	
Profit for year	15	
Closing profit and loss account		65
Total shareholders' funds		265

The market price of the ordinary shares was 67p.

Section B

Financial Accounting: The Context

In Section A, we looked at the accounting techniques which underpin the preparation and interpretation of the financial statements of sole traders, partnerships and limited companies. These techniques do not exist in a vacuum. In this section, we examine five crucial aspects of the context in which these accounting techniques are applied.

Chapter 10 investigates the regulatory and conceptual frameworks within which accounting operates. The regulatory framework provides a set of rules and regulations which govern accounting. The conceptual theory is broader and seeks to set out a theoretical framework to underpin accounting. Then, in Chapter 11, the main potential alternative measurement systems which can underpin the preparation of accounts are laid out. This chapter shows how using different measurement systems can yield different profits and different balance sheet valuations.

The annual report, the main way in which public limited companies communicate financial information to their shareholders, is discussed in Chapter 12. This chapter outlines the nature, context and function of the annual report. Both the content and the presentation of the annual report are examined.

Finally, Chapters 13 and 14 investigate two interesting aspects of financial accounting: creative accounting and international accounting. Creative accounting explores the flexibility within accounting and shows how managers may manipulate financial information out of self-interest. Finally, Chapter 14 provides a broad international view of accounting. It shows that different countries have different accounting environments. Moreover, this chapter also shows the progress which has been made towards the harmonisation of accounting practices both in the UK and worldwide. In particular, the role of the International Accounting Standards Board is investigated.

Chapter 10

Regulatory and Conceptual Frameworks

'Regulation is like salt in cooking. It's an essential ingredient – you don't want a great deal of it, but my goodness you'd better get the right amount. If you get too much or too little you'll soon know.'

Sir Kenneth Berrill, *Financial Times* (6 March 1985)

Source: *The Book of Business Quotations* (1991), p. 47

With no regulations, then accountants could use their professional judgement and receive high fees.

On the other hand, with a regulatory framework, accountants can interpret the regulations and still receive high fees.

Nothing like a win-win situation!

©MMI Mike Jones

Learning Outcomes

After completing this chapter you should be able to:

- ✔ **Outline the traditional corporate model.**
- ✔ **Understand the regulatory framework.**
- ✔ **Explain corporate governance.**
- ✔ **Understand the conceptual framework.**

In a Nutshell

- Directors, auditors and shareholders are the main parties in traditional corporate model.
- The regulatory framework provides a set of rules and regulations for accounting.
- At the international level, the International Accounting Standards Board provides a broad regulatory framework of International Accounting Standards.
- In the UK, the two main sources of regulation are the Companies Acts and accounting standards.
- Financial statements must give a true and fair view.
- The UK accounting standard-setting regime is the Financial Reporting Council, the Accounting Standards Board, the Urgent Issues Task Force and the Financial Reporting Review Panel.
- Corporate governance is the system by which companies are directed and controlled.
- A conceptual framework is a coherent and consistent set of accounting principles which will help in standard setting.
- Some major elements in a conceptual theory are the objectives of accounting, users, user needs, information characteristics and measurement models.
- The most widely agreed objective is to provide information for decision making.
- Users include shareholders and analysts, lenders, creditors, customers and employees.
- Key information characteristics are relevance, reliability, comparability and understandability.

Introduction

So far, we have looked at accounting practice – focusing on the preparation and interpretation of the financial statements of sole traders, partnerships and limited companies. In particular, we considered practical aspects of accounting such as double-entry bookkeeping, the trial balance, the profit and loss account, the balance sheet, the cash flow statement and ratio analysis. Accounting practice does not, however, take place in a vacuum. It is bounded both by a regulatory framework and a conceptual framework. These frameworks have grown up over time to bring order and fairness into accounting practice. They have been devised principally in relation to limited companies, but are also relevant to some extent to sole traders and partnerships.

The regulatory framework is essentially the set of rules and regulations which govern corporate accounting practice. At the international level the regulatory framework is provided by the International Accounting Standards Board. In the UK, regulations are set down mainly by government in Companies Acts and by independent private sector regulation in accounting standards. The conceptual framework seeks to set out a theoretical and consistent set of accounting principles by which financial statements can be prepared.

Traditional Corporate Model: Directors, Auditors and Shareholders

In the traditional corporate model there are three main groups. As Figure 10.1 shows, these three groups interact.

Figure 10.1 The Traditional Corporate Model

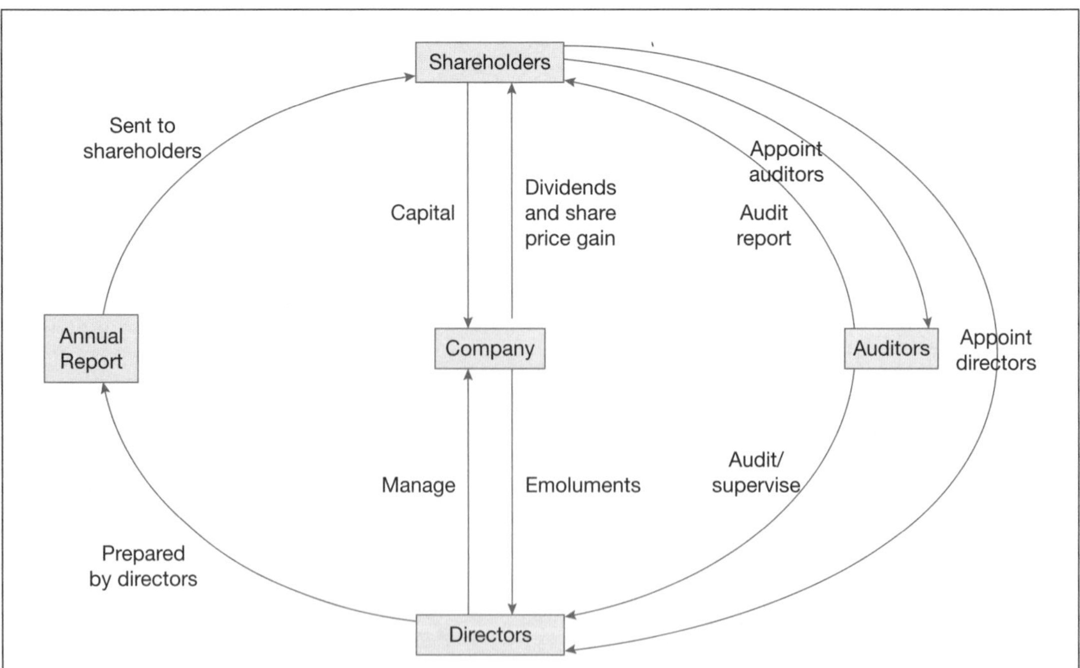

The directors are responsible for preparing the accounts – in practice this is usually delegated to the accounting managers. These accounts are then checked by professionally qualified accountants, the auditors. Finally, the accounts are sent to the shareholders.

1. Directors

The directors are those responsible for running the business. They are accountable to the shareholders, who in theory appoint and dismiss them. The relationship between the directors and shareholders is sometimes uneasy. The shareholders own the company, but it is the directors who run it. This relationship is often termed a 'principal–agent' relationship. The shareholders are the principals and the directors are the agents. The principals delegate the management of the company to directors. However, the directors are still responsible to the shareholders.

The directors are responsible for preparing the accounts which are sent to the shareholders. These accounts, prepared annually, allow the shareholders to assess the performance of the company and of the directors. They also provide information to shareholders to enable them to make share trading decisions (i.e., to hold their shares, to buy more shares or to sell their shares). As a reward for running the company, the directors receive emoluments. These may take the form of a salary, profit bonuses or other benefits in kind such as share options or company cars.

2. Auditors

Unfortunately, human nature being human nature, there is a problem with such arms-length transactions. In a nutshell, how can the shareholders trust accounts prepared by the directors? For example, how can they be sure that the directors are not adopting creative accounting in order to inflate profits and thus pay themselves inflated profit-related bonuses? One way is by auditing.

The auditors are a team of professionally qualified accountants who are *independent* of the company. They are appointed by the shareholders on the recommendation of the directors. It is their job to check and report on the accounts. This checking and reporting involves extensive work verifying that the transactions have actually occurred, that they are recorded properly and that the monetary amounts in the accounts do indeed provide a true and fair view of the company's annual accounts. An audit (see Definition 10.1) is thus an independent examination and report on the accounts of a company.

Definition 10.1

The Audit

Working definition
An independent examination and report on the accounts.

Formal definition
The systematic independent examination of a business's, normally a company's, accounting systems, accounting records and financial statements in order to provide a report on whether or not they provide a true and fair view of the business's activities.

For their time and effort the auditors are paid often quite considerable sums. The auditors prepare a formal report for shareholders. This is part of the annual report. In The Company Camera 10.1 we attach an auditors' report prepared by Deloitte and Touche, the auditors of Vodafone, on Vodafone's 2001 accounts.

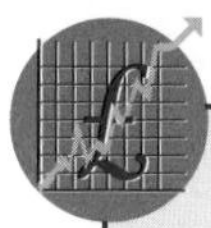

The Company Camera 10.1

Auditors' Report

Auditors' report to the members of Vodafone Group Plc

We have audited the financial statements on pages 25 to 60, which have been prepared under the accounting policies set out on pages 29 and 30, and the detailed information disclosed in respect of directors' remuneration and share options set out in the Board's Report to Shareholders on Directors' Remuneration on pages 14 to 22. We have also audited the financial information prepared in accordance with accounting principles generally accepted in the United States set out on pages 61 to 63.

We have reviewed the pro forma proportionate financial information on page 64 which has been prepared in accordance with the bases set out on page 64.

Respective responsibilities of directors and auditors

The directors are responsible for preparing the Annual Report & Accounts, including, as described above, preparation of the financial statements, which are required to be prepared in accordance with applicable United Kingdom law and accounting standards. The directors are also responsible for the preparation of the financial information prepared in accordance with accounting principles generally accepted in the United States and the pro forma proportionate financial information prepared in accordance with the bases referred to above. Our responsibilities, as independent auditors, are established by statute, the Auditing Practices Board, the UK Listing Authority, and by our profession's ethical guidance.

We report to you our opinion as to whether the financial statements give a true and fair view and are properly prepared in accordance with the Companies Act 1985. We also report to you if, in our opinion, the directors' report is not consistent with the financial statements, if the Company has not kept proper accounting records, if we have not received all the information and explanations we require for our audit, or if information specified by law or the Listing Rules regarding directors' remuneration and transactions with the Company and other members of the Group is not disclosed.

We review whether the Corporate Governance statement on pages 11 to 13 reflects the Company's compliance with the seven provisions of the Combined Code specified for our review by the UK Listing Authority, and we report if it does not. We are not required to consider whether the Board's statements on internal control cover all risks and controls, or form an opinion on the effectiveness of the Group's Corporate Governance procedures or its risk and control procedures.

We read the other information contained in the Annual Report & Accounts, including the directors' report on Corporate Governance, and consider whether it is consistent with the audited financial statements. We consider the implications for our report if we become aware of any apparent misstatements or material inconsistencies with the financial statements.

The Company Camera 10.1 (continued)

Bases of opinions

We conducted our audit in accordance with United Kingdom auditing standards issued by the Auditing Practices Board which are similar to auditing standards in the United States. An audit includes examination, on a test basis, of evidence relevant to the amounts and disclosures in the financial statements. It also includes an assessment of the significant estimates and judgements made by the directors in the preparation of the financial statements, and of whether the accounting policies are appropriate to the circumstances of the Company and the Group, consistently applied and adequately disclosed.

We planned and performed our audit so as to obtain all the information and explanations which we considered necessary in order to provide us with sufficient evidence to give reasonable assurance that the financial statements are free from material misstatement, whether caused by fraud or other irregularity or error. In forming our opinion we also evaluated the overall adequacy of the presentation of information in the financial statements.

Our review of the pro forma financial information, which was substantially less in scope than an audit performed in accordance with Auditing Standards, consisted primarily of considering the nature of the adjustments made and discussing the resulting pro forma financial information with management.

Opinions

In our opinion:

- the financial statements give a true and fair view of the state of affairs of the Company and the Group as at 31 March 2001 and of the loss of the Group for the year then ended and have been properly prepared in accordance with the Companies Act 1985;
- the financial information set out on page 61 has been properly prepared in accordance with accounting principles generally accepted in the United States; and
- the pro forma proportionate financial information has been properly prepared in accordance with the bases set out on page 64, which are consistent with the accounting policies of the Group.

Deloitte & Touche
Chartered Accountants and Registered Auditors
Hill House
1 Little New Street
London EC4A 3TR
29 May 2001

Source: Vodafone, 2001 Annual Report and Accounts, p. 24

This auditors' report thus confirms that the directors of Vodafone have prepared a set of financial statements which have given a true and fair view of the company's accounts as at 31 March 2001. This is known as a clean or unqualified audit report. The shareholders of Vodafone can thus draw comfort from the fact that the auditors believe the accounts do give a true and fair view and faithfully reflect the economic performance of the company over the year.

3. Shareholders

The shareholders (in the US known as the stockholders) own the company. They have provided the share capital by way of shares. Their reward is twofold. First, they may receive an annual dividend which is simply a cash payment from the company based on profits. Second, they may benefit from any increase in the share price over the year. However, companies may make losses and share prices can go down as well as up, so this reward is not guaranteed. In the developed world, more and more companies are owned by large institutions (such as investment trusts or pension funds) rather than private shareholders.

The shareholders of the company receive an annual audited statement of the company's performance. This is called the annual report. It comprises the financial statements and also a narrative explanation of corporate performance. Included in this annual report is an auditors' report.

It is important to realise that shareholders are only liable for the capital which they contribute to a company. This capital is known as *share capital* (i.e., the capital of a company is divided into many shares). These shares limit the liability of shareholders and so we have limited liability companies. Shares, once issued, are bought or sold by shareholders on the stock market. This enables people, who are not involved in the day-to-day running of the business, to own shares. This division between owners and managers is often known as the divorce of ownership and control. It is a fundamental underpinning of a capitalist society.

Pause for Thought 10.1

Risk and Reward

In the corporate model each of the three groups are rewarded for their contributions. This is called the 'risk and reward model'. Can you work out each group's risk and reward?

	Contribution (risk)	*Reward*
Shareholders	Share capital	Dividends and increase in share price
Directors	Time and effort	Salaries, bonuses, benefits-in-kind such as cars or share options.
Auditors	Time and effort	Auditors' fees

Regulatory Framework

The corporate model of directors, shareholders and auditors is one of checks and balances. The directors manage the company, receive directors' emoluments and recommend the appointment of the auditors to the shareholders. The shareholders own the company, but do not run it, and rely upon the auditors to check the accounts. Finally, the auditors are appointed by shareholders on the recommendation of the directors. They receive an auditors' fee for the work they undertake when they check the financial statements prepared by managers.

Pause for Thought 10.2

Checks and Balances

Is auditing enough to stop company directors pursuing their own interests at the expense of the shareholders?

Auditing is a powerful check on directors' self-interest. The directors prepare the accounts and the auditors check that the directors have correctly prepared them and that they give a 'true and fair' view. However, there are problems. The auditors, although technically appointed by the shareholders at the company's *annual general meeting* (i.e., a meeting called once a year to discuss a company's accounts), are recommended by directors. Auditors are also paid, often huge fees, by the company. The auditors do not wish to upset the directors and lose those fees. Given the flexibility within accounts, there are a whole range of possible accounting policies which the directors can choose. The regulatory framework helps to narrow this range of potential accounting policies and gives guidance to both directors and auditors. The auditors can, therefore, point to the rules and regulations if they feel that the directors' accounting policies are inappropriate. The regulatory framework is, therefore, a powerful ally of the auditor.

This system of checks and balances is fine, in principle. However, it is rather like having two football teams and a referee with no rules. The regulatory framework, in effect, provides a set of rules and regulations to ensure fair play. As Definition 10.2 shows, at the national level, these rules and regulations may originate from the government, the accounting standard setters or, more rarely, for listed companies, the stock exchange. The principal aim of the regulatory framework is to ensure that the financial statements present a true and fair view of the financial performance and position of the organisation.

Definition 10.2

The National Regulatory Framework

Working definition
The set of rules and regulations which govern accounting practice, mainly prescribed by government and the accounting standard-setting bodies.

Formal definition
The set of legal and professional requirements with which the financial statements of a company must comply. Company reporting is influenced by the requirements of law, of the accounting profession and of the Stock Exchange (for listed companies).

Chartered Institute of Management Accountants (2000), *Official Terminology*

In most countries, including the UK, the main sources of authority for the regulatory framework are either via the government, through companies legislation, or via accounting standard-setting bodies through accounting standards. As Soundbite 10.1 suggests, there is a need not to overregulate. At the international level, there is now a set of International Accounting Standards (IAS) issued by the International Accounting Standards Board (IASB). The IASB is steadily growing in importance. Its standards are aimed primarily at large international companies (see Chapter 14 for a fuller discussion of IAS). An increasing number of companies now use IAS.

Soundbite 10.1

Regulations

'If you destroy a free market you create a black market. If you have ten thousand regulations, you destroy all respect for the law.'

Winston S. Churchill

Source: *The Book of Unusual Quotations* (1959), pp. 240–241

Regulatory Framework in the UK

In the UK, there are two main sources of authority for regulation: the Companies Acts and accounting standards. There are some additional requirements from the Stock Exchange for listed companies, but given their relative unimportance, they are not discussed further here.

In the UK, as in most countries, the regulatory framework has evolved over time. As accounting has grown more complex, so has the regulatory framework which governs it. At first, the only requirements that companies followed were those of the Companies Acts. However, in 1970 the first accounting standards set by the Accounting Standards Steering Committee were issued. Today, UK companies must adhere both to the requirements of Company Acts and to accounting standards. The overall aim of this regulatory framework is to protect the interests of all those involved in the corporate model. Specifically, there is a need to provide a 'true and fair view' of a company's affairs.

True and Fair View

Section 226[2] of the 1985 Companies Act requires that 'the balance sheet should give a true and fair view of the state of affairs of the company as at the end of the financial year; and the profit and loss account shall give a true and fair view of the profit or loss of the company for the financial year.' The true and fair concept is thus of overriding importance. Unfortunately, it is a particularly nebulous concept which has no easy definition. A working definition is, however, suggested in the box below. In essence, there is a presumption that the accounts will reflect the underpinning economic reality. Generations of accountants have struggled unsuccessfully to pin down the exact meaning of the phrase. In general, to achieve a true and fair view accounts should comply with the Companies Acts and accounting standards.

Soundbite 10.2

Statutes

'The greater the number of statutes, the greater the number of thieves and brigands.'

Lao-Tse

Source: *The Book of Unusual Quotations* (1959), p. 148

Definition 10.3

Working Definition of a True and Fair View

A set of financial statements which faithfully, accurately and truly reflect the underlying economic transactions of an organisation.

Occasionally, however, where compliance with the law would not give a true and fair view, a company may override the legal requirements. However, the company would have to demonstrate clearly why this was necessary.

Companies Acts

Companies Acts are Acts of Parliament which lay down the legal requirements for companies including regulations for accounting. There have been a succession of Companies Acts which have gradually increased the reporting requirements placed on UK companies. Initially, the Companies Acts provided only a broad legislative framework. However, later Companies Acts (CAs), especially the CA 1981, have imposed a significant regulatory burden on UK companies. The CA 1981 introduced the European Fourth Directive into UK law. Effectively, this Directive was the result of a deal between the United Kingdom and other European Union members. The United Kingdom exported the true and fair view concept, but imported substantial detailed legislation and standardised formats for the profit and loss account and balance sheets. The CA 1981, therefore, introduced a much more prescriptive 'European' accounting regulatory framework into the UK.

Accounting Standards

Whereas Companies Acts are governmental in origin, accounting standards are set by non-governmental bodies. At the international level, International Accounting Standards (IAS) are set by the International Accounting Standards Board. These are voluntary for UK companies and are designed essentially for internationally-oriented companies which wish to list on foreign stocks markets. The US market does not accept IAS at present without reconciliation to US GAAP. However, the European Union has recently proposed that all listed European companies should comply with IAS. For domestic purposes, UK companies must follow UK accounting standards.

There are, as Figure 10.2 on the next page shows, four main constituents of the UK's accounting standards regulatory framework: the Financial Reporting Council (FRC), the Accounting Standards Board (ASB), the Urgent Issues Task Force (UITF) and the Financial Reporting Review Panel (FRRP). The current UK accounting standards regulatory framework was set up in 1990.

1. Financial Reporting Council (FRC)

The Financial Reporting Council is a supervisory body which ensures that the overall system is working.

2. The Accounting Standards Board (ASB)

The Accounting Standards Board is the engine of the accounting standards process. The ASB has a full-time chairman and a full-time technical director plus eight part-time members.

Figure 10.2 UK's Accounting Standards' Regulatory Framework

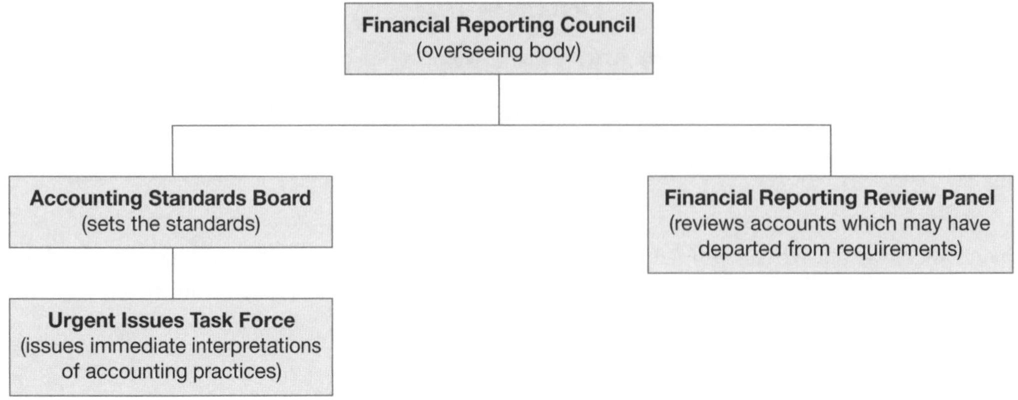

In the UK, accounting standards are called Financial Reporting Standards (FRS). The ASB issues FRS which are applicable to the accounts of all organisations and are intended to give a true and fair view. These Financial Reporting Standards have replaced most of the Standard Statements of Accounting Practice which were issued from 1970 to 1990 by the ASB's predecessor, the Accounting Standards Committee. As Definition 10.4 shows, in essence, accounting standards are pronouncements which must normally be followed in order to give a true and fair view.

Definition 10.4

Accounting Standards

Working definition

Accounting pronouncements which must be followed in order to give a true and fair view within the regulations.

Formal definition

'Accounting standards are authoritative statements of how particular types of transaction and other events should be reflected in financial statements and accordingly compliance with accounting standards will normally be necessary for financial statements to give a true and fair view.'

Source: Foreword to *Accounting Standards*, Accounting Standards Board, 1993, para.16.

Accounting standards are mandatory in that accountants are expected to observe them. They cover specific technical accounting issues such as stock, depreciation, and research and development. These standards essentially aim to improve the quality of accounting in the UK. They narrow the areas of difference and variety in accounting practice, set out minimum disclosure standards and disclose the accounting principles upon which the accounts are based. Overall, accounting standards provide a comprehensive set of guidelines which preparers and auditors can use when drawing up the financial statements.

3. Urgent Issues Task Force (UITF)

The Urgent Issues Task Force is the 'Flying Squad' of the standard-setting process. The development of standards takes time. The UITF, therefore, was set up to react quickly to new situations. Recommendations are made to curb undesirable interpretations of accounting standards or to prevent accounting practices which the ASB considers undesirable.

4. The Financial Reporting Review Panel (FRRP)

The FRRP investigates contentious departures from accounting standards. It is the 'detective' arm of the ASB. Possible departures from accounting standards and from the Companies Acts are referred to the FRRP by interested parties (for example, shareholders). The FRRP is thus reactive rather than proactive, responding to rather than seeking out complaints. The FRRP questions the directors of the companies investigated. The last resort of the FRRP is to take miscreant companies to court to force them to revise their accounts. However, so far the threat of court action has been enough.

Trafalgar House plc is one company which fell foul of the FRRP. As Real-Life Nugget 10.1 shows, by using a particular accounting policy (i.e., classifying its properties as fixed assets *not* current assets), it increased its accounting profit by £102.7 million. In this case, the FRRP intervened and Trafalgar House adjusted its accounts.

Real-Life Nugget 10.1

FRRP and Trafalgar House Plc

'. . . the group suffered a loss of £102.7 million on the valuation of properties previously classified as current assets. Had their classification remained unchanged, this loss would have been charged in the profit and loss account. It would have reduced the 1991 pre-tax profit from £122.4 million to £19.7 million. The classification of assets as "fixed" or "current" depends on whether they are intended for use on a continuing basis in the company's activities. If not intended for such use, they are current assets. Trafalgar House presumably concluded that a change of intention warranted reclassification of the property concerned.'

Source: Trevor Pijper (1993) *Creative Accounting: The Effect of Financial Reporting in the UK* (Basingstoke: Macmillan, p. 27)

In another more recent case the FRRP review panel forced London Underground to transfer £4,896m from 'capital employed' to 'deferred income'. As Real-Life Nugget 10.2 on the next page indicates, this is a major £5bn adjustment to the balance sheet.

Real-Life Nugget 10.2

FRRP and London Underground

London Underground forced to make £5bn accounts change

by Gavin Hicks

A change in accounting procedures pressed on London Underground by the financial reporting watchdog has forced the tube operator to make a £5bn adjustment to its balance sheet.

LUL has been told by the Financial Reporting Review Panel that it did not accept some of its accounting practices and must change the way it records government grants.

Following an FRRP review of the 1998/99 accounts, LUL has had to move £4,896m from 'capital employed' to 'deferred income'.

LUL has dealt with the grants using the same procedures as the Department of Environment, Transport and Regions, which has special accounting dispensations.

Wrapping LUL's knuckles, the financial reporting watchdog said it did not have the same privileges.

A statement from LUL said: 'This is a technical accounting change and has no impact on the finanial substance of LUL.'

It goes on to say the grants were fully disclosed in the 1998/99 accounts which it says were supported by the DETR and KPMG, the external auditor, who declined to comment.

LUL said the change would not effect investment and had no effect on its assets.

The FRRP ruling follows a complaint reported in *Accountancy Age* in March when Smith Williamson partner John Newman claimed LUL had overstated its assets. That complaint was rejected.

An FRRP statement said: 'The directors now concur with the Panels and, in March 2000 accounts recently published, reclassify all grants recieved as deferred income and accordingly take account of such amounts in determining net assets.'

Source: *Accountancy Age*, 19 October 2000, p. 1

Corporate Governance

From the 1990s, corporate governance has grown in importance. Effectively, corporate governance is the system by which companies are directed and controlled (see Real-Life Nugget 10.3 below).

Real-Life Nugget 10.3

Corporate Governance

'Corporate governance is the system by which companies are directed and controlled. Boards of directors are responsible for the governance of their companies. The shareholders' role in governance is to appoint the directors and the auditors and to satisfy themselves that an appropriate governance structure is in place. The responsibilities of the board include setting the company's strategic aims, providing the leadership to put them into effect, supervising the management of the business and reporting to shareholders on their stewardship. The board's actions are subject to laws, regulations and the shareholders in general meeting.

Within that overall framework, the specifically financial aspects of corporate governance (the committee's remit) are the way in which boards set financial policy and oversee its implementation, including the use of financial controls, and the process whereby they report on the activities and progress of the company to the shareholders.'

Source: *Report of the Committee on the Financial Assets of Corporate Governance* (1992), Gee and Go., p. 15

The financial aspects of corporate governance relate principally to internal auditing, the way in which the board of directors set financial policy and the process by which the directors report to the shareholders on the activities and progress of the company. These aspects have been investigated in the UK by several committees including the Cadbury Committee (1992), the Greenbury Committee (1995), the Hampel Committee (1998) and the Turnbull Committee (1999).

The continuing interest in corporate governance arose in part for two reasons. First, there were some unexpected failures of major companies such as BCCI, Polly Peck and Maxwell Communications. Second, there were extensive criticisms in the press of 'fat-cat' directors. These directors, often of newly privatised companies (i.e., companies which were previously state-owned and -run), were generally perceived to be paying themselves huge and unwarranted salaries.

As a result of the Cadbury Committee and other subsequent committees, there were attempts to tighten up corporate governance. In particular, there was a concern with the amount of information companies disclosed, with the role of non-executive directors (i.e., directors appointed from outside the company), with directors' remuneration, with audit committees (committees ideally controlled by non-executive directors which oversee the appointment of external auditors and deal with their reports), with relations with institutional investors and with systems of internal financial control set up by management.

In the annual report, companies now set out extensive details of directors' remuneration and disclose information about corporate governance. The auditors review these corporate governance elements to check that they comply with the principles of good governance and code of best practice as set out in the London Stock Exchange's rules. The Company Camera 10.2 above presents part of J.D. Wetherspoon's statement on internal control. The directors acknowledge their responsibility to establish controls such as those to protect against the unauthorised use of assets.

> *The Company Camera 10.2*
>
> **Internal Control**
>
> The company has adopted the transitional approach to implementation of the Turnbull guidance on the review of internal controls.
>
> The company has established the procedures necessary to implement the guidance under 'Internal control: Guidance for Directors on the Combined Code', from the start of the financial year 2000/2001.
>
> The directors acknowledge their responsibility for the company's system of internal financial control, which can be defined as the controls established in order to provide reasonable assurance that the assets have been protected against unauthorised use, that proper accounting records have been maintained and that the financial information which is produced is reliable. Such a system can, however, provide only reasonable and not absolute assurance against material misstatement or loss.
>
> *Source*: J.D. Wetherspoon, 2000 Annual Report, p. 22

Conceptual Framework

Since the 1960s, standard-setting bodies (such as the Financial Accounting Standards Board (FASB), in the USA, the International Accounting Standards Board (IASB) and the Accounting Standards Board in the UK) have sought to develop a conceptual framework or statement of principles which will underpin accounting practice. As Definition 10.5 shows, the basic idea of a conceptual framework is to create a set of fundamental accounting principles which will help in standard setting.

> *Definition 10.5*
>
> **Conceptual Framework**
>
> The development of a coherent and consistent set of accounting principles which underpin the preparation and presentation of financial statements.

A major achievement of the search for a conceptual theory has been the emergence of the decision-making model. As Real-Life Nugget 10.4 sets out, there is a need to provide decision-useful information to investors. The five essential elements of a conceptual framework are broadly agreed by all three major standard-setting bodies: objectives, users, user needs, information characteristics and measurement rules. These elements are briefly discussed below.

> *Real-Life Nugget 10.4*
>
> **Conceptual Framework**
>
> First – and of fundamental importance – all involved in global financial reporting must have a common mission or objective. At the heart of that mission is a conceptual framework which must focus on the investor, provide decision-useful information, and assume that capital is allocated in a manner that achieves the lowest cost in our world markets. I believe we all have an understanding and acceptance of providing decision-useful information for investors, but not all standard setters and not all standards yet reflect that mission.
>
> *Source*: International Accounting Standards Board, IASC Insight, p. 12

1. Objectives

Both the US Financial Accounting Standards Board (FASB) and the International Accounting Standards Board (IASB) broadly agree that 'The objective of financial statements is to provide information about the financial position, performance and changes in financial position of an enterprise that is useful to a wide range of users in making decisions' (*Framework for the Preparation and Presentation of Financial Statements*, IASB, 1999, para. 12). This is widely known as the decision-making model (see Figure 10.3). In other words, the basic idea of accounting is to provide accounting information to users which fulfils their needs, thus enabling them to make decisions. Encompassed within this broad definition is the idea that financial statements show how the managers have accounted for the resources entrusted to them by the shareholders. This accountability is often called stewardship. To enable stewardship and decision making, the information must have certain information characteristics and use a consistent measurement model.

Figure 10.3 Decision-Making Model

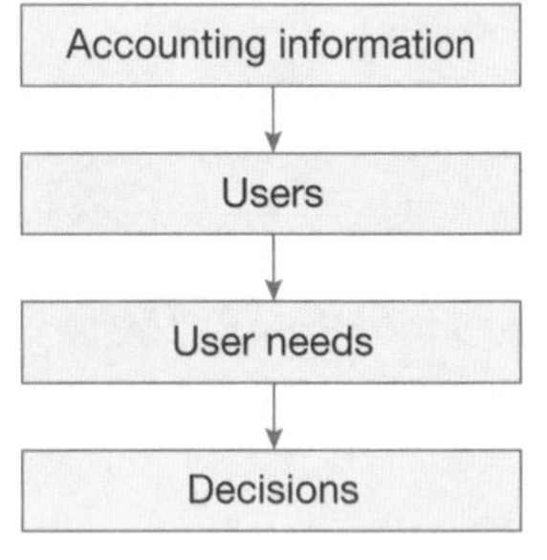

In the UK, the ASB has developed a Statement of Principles. In its latest version, the Statement takes a broader definition of the objectives of financial reporting than either the FASB or the IASB. 'The objective of financial statements is to provide information about the reporting entity's financial position, performance and changes in financial position that is useful to a wide range of users for assessing the stewardship of the entity's management and for making economic decisions' (Accounting Standards Board, *Statement of Principles*, 1999). Thus, the ASB sees the *objective of financial reporting* to be (i) *stewardship of management*, and (ii) *making economic decisions*.

Stewardship and decision making are discussed in more depth in Chapter 12. However, at this stage it is important to introduce them. Stewardship is all about accountability. It seeks to make the directors accountable to the shareholders for their stewardship or management of the company. Corporate governance is one modern aspect of stewardship. Decision making, by contrast, focuses on the needs of shareholders to make economic decisions, such as to buy or sell their shares. As performance measurement and decision making have grown in importance so has the profit and loss account. In a sense, decision making and stewardship are linked, as information is provided to shareholders so that they can make decisions about the directors' stewardship of the company.

Essentially, stewardship and decision making are user-driven and take the view that accounting should give a 'true and fair' view of a company's accounts. By contrast, the public relations view suggests that there are behavioural reasons why managers might seek to prepare accounts that favour their own self-interest. Self-interest and 'true and fair' may well conflict. In this section, we focus only on the officially recognised roles of accounting (stewardship and decision making). Discussion of the public relations role and the conflicting multiple accounting objectives is covered in Chapter 12.

Pause for Thought 10.3

Stewardship and Decision Making

Why are assets and liabilities most important for stewardship, but profits most important for decision making?

Stewardship is about making individuals accountable for assets and liabilities. In particular, stewardship focuses on the physical existence of assets and seeks to prevent their loss and/or fraud. Stewardship is, therefore, about keeping track of assets rather than evaluating how efficiently they are used.

Decision making is primarily concerned with monitoring performance. Therefore, it is primarily concerned with whether or not a business has made a profit. It is less concerned with tracking assets.

2. Users

The main users are usually considered to be the present and future shareholders. Indeed, shareholders are the only group required by law to be sent an annual report. Shareholders comprise individual and institutional shareholders. Besides shareholders, there are a number of other users. Those identified by the International Accounting Standards Board include:

- lenders, such as banks or loan creditors
- suppliers and other trade creditors
- employees
- customers
- governments and their agencies, and the
- general public.

In addition to this list we can add:

- academics
- management

- analysts and advisers, and
- pressure groups such as Friends of the Earth.

Broadly, we can see that this list is the same as that discussed in Chapter 1. In Chapter 1, however, we distinguished between internal users (management and employees) and external users (the rest)).

Generally the accounts are pitched at the shareholders. Satisfying their interests is generally thought to cover the main concerns of the other groups. The annual report adopts a general purpose reporting model. This provides a comprehensive set of information targeted at all users. It does not, therefore, specifically target the needs of one user group.

3. User Needs

User needs vary. However, commonly users will want answers to questions such as:

- How profitable is the organisation?
- How much cash does it have in the bank?
- Is it likely to keep trading?

In order to answer these questions, users will need information on the profitability, liquidity, efficiency and gearing of the company. This is normally provided in the three key financial statements: the profit and loss account, the balance sheet and the cash flow statement. Users will also be interested in the softer, qualitative information provided, for example, in accounting narratives such as the chairman's statement.

4. Information Characteristics

In order to be useful to users, the financial information needs to possess certain characteristics. Both the International Accounting Standards Board and the UK's Accounting Standards Board focus on four principal characteristics: relevance, reliability, comparability and understandability. It is helpful to classify these characteristics into those relating to content (relevance and reliability) and those relating to presentation (comparability and understandability) (see Figure 10.4).

Figure 10.4 Overview of Information Characteristics

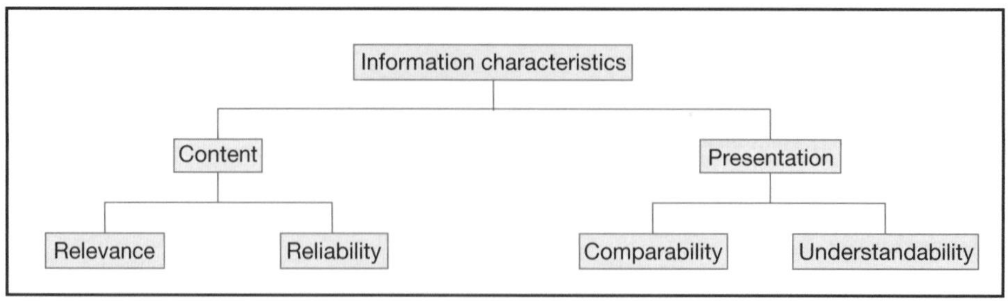

A. Content

Relevance. Relevant information affects users' economic decisions. Relevance is a prerequisite of usefulness. Examples of relevance are information that helps to predict future events or to confirm or correct past events.

Reliability. Reliable information is free from material error and is unbiased. Like relevance, reliability is a predeterminant of usefulness. Five characteristics underpin reliability. First, reliable information is representationally faithful (i.e., it validly describes the underlying events). Reliable information thus reflects the substance of economic reality not just its legal form. For example, a leased asset is included on a balance sheet even though the company leases, but does not own the asset. Second, reliable information must be neutral, i.e., not biased. Third, the information must be free from material error. Fourth, reliable information must be prudent and not deliberately overstate assets or income, or understate liabilities or expenses. And, finally, information must be complete.

B. Presentation

Comparability. Accounts should be prepared on a consistent basis and should disclose accounting policies. This will then allow users to make inter-company comparisons and intra-company comparisons over time.

Understandability. Information that is not understandable is useless. Therefore, information must be presented in a readily understandable way. In practice, this means conveying complex information as simply as possible rather than 'dumbing down' information. Whether financial information is understandable will depend on the way in which it is characterised, aggregated, classified and presented.

The inclusion of financial information in the financial statements crucially depends on its materiality. If information is immaterial (i.e., will not affect users' decisions), then, in practice, it does not need to be reported. In addition to materiality, the ASB recognises that trade-offs are inevitable where conflicts arise between relevance, reliability, comparability and understandability. For example, out-of-date information may be useless. Therefore, some detail (reliability) may be sacrificed to speed of reporting (relevance).

5. Measurement Model

The objectives, users, user needs and information characteristics have proved relatively easy to agree upon. However, the choice of an appropriate measurement model for profit measurement and asset determination has caused much controversy. The measurement basis that underpins financial statements remains a modified form of historical cost. In other words, income, expenses, assets and liabilities are recorded at the date of their original monetary transaction. Unfortunately, although historical cost is relatively well understood and easy to understand, it understates assets and overstates profits, especially in times of inflation. Most commentators agree that historical cost, therefore, is flawed. However, there is no consensus on a suitable replacement. The alternative measurement models are more fully discussed in Chapter 11.

Pause for Thought 10.4

Critics of the Conceptual Framework

The conceptual framework has been criticised for not achieving very much and being a social document rather than a theory document. What do you think these criticisms mean and are they fair?

The conceptual framework has brought into being the decision-making model. However, there is little agreement on the appropriate measurement model for accounting. In other words, should we continue to use historical cost or should we move towards some alternative measurement model, that perhaps accounts for the effects of inflation? Critics have seen this failure to agree on a measurement model as a severe blow to the authority of the conceptual framework. In addition, there is concern that the conceptual framework has not really developed a theoretically coherent and consistent set of accounting principles at all. These critics argue that the conceptual framework is primarily descriptive – just describing what already exists. A descriptive framework is not a theoretical framework. Finally, some critics argue that the real reason for the search for a conceptual framework is to legitimise and support the notion of a standard-setting regime independent of government. The conceptual framework should thus be seen as a social document which supports the existence of an independent accounting profession.

Conclusion

In order to appreciate accounting practice properly, we need to understand the regulatory and conceptual frameworks within which it operates. These frameworks were primarily devised for published financial statements, such as those in the corporate annual report. The regulatory framework is the set of rules and regulations which governs corporate accounting practice. The International Accounting Standards Board sets International Accounting Standards. The two major strands of the UK's regulatory framework are the Companies Acts and accounting standards. The UK's accounting standards regulatory framework consists of four elements: the Financial Reporting Council, the Accounting Standards Board, the Urgent Issues Task Force and the Financial Reporting Review Panel. Corporate governance is the system by which companies are governed.

A conceptual framework is an attempt to create a set of fundamental accounting principles which will help standard-setting. A major achievement of the search for a conceptual framework has been the emergence of the decision-making model. The essence of this is that the objective of financial statements is to provide financial information useful to a wide range of users for making economic decisions. A second objective is to provide financial information for assessing the stewardship of managers. In order to be useful, this information must be relevant, reliable, comparable and understandable. Although there is general agreement on the essentials of a decision-making model, there is little consensus on which measurement model should underpin the decision-making process.

Selected Reading

The references below will give you further background. They are roughly divided into those on the regulatory framework and those on the conceptual framework.

Regulatory Framework

Bartlett, S.A. and M.J. Jones (1997), Annual Reporting Disclosures 1970–90: An Exemplification, *Accounting, Business and Financial History*, Vol. 7, No. 1, pp. 61–80.
This article looks at how the accounts of one firm, H.P. Bulmers (Holdings) plc, the cider makers, were affected by changes in the regulations from 1970 to 1990.

Blake, J. and H. Lunt (2001), *Accounting Standards*, Financial Times/Prentice Hall.
A good, easy-to-digest, student-orientated guide to the main standards.

Pijper, T. (1993), *Creative Accounting: The Effect of Financial Reporting in the U.K.* (Basingstoke, Macmillan).
Provides a good, easy-to-follow overview of the UK's regulatory system.

The Combined Code (1998), London Stock Exchange, June.
This provides a comprehensive set of recommendations arising from the various corporate governance reports (i.e., Cadbury, Greenbury, Hampel). UK-listed companies now follow these.

The Financial Aspects of Corporate Governance (The Cadbury Committee Report) (1992), Gee and Co.
The first, and arguably the most influential, report into corporate governance. Authoritative.

UK GAAP (2000), Ernst and Young, Butterworth.
Provides a very comprehensive guide to UK and IAS standards as well as to the ASB's *Statement of Principles*. This guide is updated annually.

Conceptual Framework

Outlines of Potential Conceptual Frameworks by Professional Bodies

Accounting Standards Board (1999), Statement of Principles, *Accountancy*.
This synopsis offers a good, quick insight into current thinking in the UK about a conceptual theory.

Accounting Standards Setting Committee (ASSC) (1975), *The Corporate Report* (London).
A benchmark report which outlined the Committee's, at the time groundbreaking, thoughts about the theory of accounting. Easy-to-read.

Financial Accounting Standards Board (FASB) (1978), *Statement of Financial Accounting Concepts No. 1, Objectives of Financial Reporting by Business Enterprises* (Stanford, FASB).
Offers an insight into the US view of a conceptual theory.

International Accounting Standards Board (IASB) (2000), *Framework for the Preparation and Presentation of Financial Statements* in International Accounting Standards, 2000.
A very influential document outlining the IASB's views.

Critical Theorists

As well as the conventional view of the conceptual theory, many observers are critical of the whole process. Two are given below.

Archer, S. (1993), 'On the Methodology of a Conceptual Framework for Financial Accounting. Part 1: An Historical and Jurisprudential Analysis', *Accounting, Business and Financial History*, Vol. 3, No. 1, pp. 199–227. Easier to read than Hines, below. It sees the conceptual framework as legitimating the accounting standard process rather than being a primarily theory-driven document.

Hines, R.D. (1991), 'The FASB's Conceptual Framework, Financial Accounting and the Maintenance of the Social World', *Accounting, Organizations and Society*, Vol. 16, No. 4, pp. 313–331.
A very challenging read. It presents an unorthodox outlook on the whole conceptual framework project.

Discussion *Questions*

Questions with numbers in blue have answers at the back of the book.

Q1 What is the role of directors, shareholders and auditors in the corporate model?

Q2 What is the decision-making model? Assess its reasonableness.

Q3 Discuss the view that if a regulatory framework did not exist it would have to be invented.

Q4 Companies often disclose 'voluntary' information over and above that which they are required to do. Why do you think they do this?

Q5 What is a conceptual framework and why do you think so much effort has been expended to try to find one?

Chapter 11

Measurement Systems

'What is a Cynic? A man who knows the price of everything and the value of nothing.'

Oscar Wilde, *Lady Windermere's Fan* (1892)

Wiley Book of Business Quotations (1998), p. 349

Learning Outcomes

After completing this chapter you should be able to:

- ✔ Explore the importance of accounting measurement systems.
- ✔ Critically evaluate historical costing.
- ✔ Investigate the alternatives to historic costing.

In a Nutshell

- Measurement systems determine asset valuation and profit measurement.
- The capital maintenance concept is concerned with maintaining the capital of a company.
- Historical cost, which records items at their original cost, is the most widely used measurement system.
- Current purchasing power adjusts historical cost for changes in the purchasing power of money (e.g., inflation).
- Replacement cost is based on the cost of replacing assets.
- Net realisable value is based on the orderly sale value of the assets.
- Present value is based on the present value of the discounted net cash inflows of an asset.
- Exceptions to historical cost are the valuing of stock at the lower of cost or net realisable value or, in the UK, the revaluation of fixed assets.

Soundbite 11.1

Measurement

'What you measure is what you get.'

Robert S. Kaplan and David P. Norton, *Harvard Business Review*, January–February, 1992.

Source: *The Wiley Book of Business Quotations* (1998), p. 295

Introduction

Measurement systems underpin not only profit, but also asset valuation. Essentially, a measurement system is the way in which the elements in the accounts are valued. Traditionally, historical cost has been the accepted measurement system. Incomes, expenses, assets and liabilities have been recorded in the accounting system at cost at the time that they were first recognised. Unfortunately, historical cost, although easy to use, has several severe limitations; for example, it does not take inflation into account. However, although the limitations of historical cost accounting are well known, accountants have been unable to agree on any of the main alternatives such as current purchasing power, replacement cost, net realisable value or present value.

Overview

Measurement systems are the processes by which the monetary amounts of items in the financial statement are determined. These systems are fundamental to the determination of profit and to the measurement of net assets. In essence, the measurement system determines the values obtained. Potentially, there are five major measurement systems: historical cost, current purchasing power, replacement cost, net realisable value and present value. The essentials of these five measurement systems are set out in Figure 11.1 below.

Figure 11.1 The Alternative Measurement Systems

Measurement System	*Explanation*	*Capital Maintenance System*
Historical Cost Systems		
i. Historical cost	Monetary amounts recorded at the date of original transaction.	Financial capital maintenance.
ii. Current purchasing power	Historical cost adjusted by general changes in purchasing power of money (e.g., inflation), often measured using the retail price index (RPI).	Financial capital maintenance.
Current Value Systems		
i. Replacement cost	Assets valued at the amounts needed to replace them with an equivalent asset.	Physical capital maintenance.
ii. Net realisable value	Assets valued at the amount they would fetch in an orderly sale.	Physical capital maintenance.
iii. Present value	Assets valued at the discounted present values of future cash inflows.	Physical capital maintenance.

Measurement systems are underpinned by the idea of capital maintenance. Capital maintenance determines that a profit is made only after capital is maintained. This capital can be monetary (monetary capital maintenance) or physical (physical capital maintenance). These terms are explained in more detail in Figure 11.2.

Figure 11.2 Capital Maintenance Concepts

What exactly is a capital maintenance concept and why is it important?

A capital maintenance concept is essentially a way of determining whether the 'capital' of a business has improved, deteriorated or stayed the same over a period of time. There are two main capital maintenance concepts (*financial capital maintenance* and *physical capital maintenance*). **Under financial capital maintenance we are primarily concerned with monetary measurement, in particular, the measurement of the net assets.** This is true using both historical cost and current purchasing power. For example, under *historical cost* the capital maintenance unit is based on actual monetary units (i.e., actual pounds in the UK). Under *current purchasing power*, it is actual pounds adjusted by the rate of inflation. In both cases, if our closing net assets (as measured in £s) are higher than our opening net assets we make a profit. **Under physical capital maintenance, the physical productive capacity (i.e., operating capacity of the business) must be maintained.** For example, can we still produce the same amount of goods or services at the end of a period as we could at the start? We maintain the operating capacity in terms of *replacement costs*, *realisable values* or *present values* (i.e., discounted future cash inflows). For example, under replacement costs, we are concerned with valuing the operating capacity of the business at the replacement cost of individual assets and liabilities.

Historical cost and current purchasing power both stem from the normal bookkeeping practice of recording transactions at the date they occur in monetary amounts. For current purchasing power, these amounts are then adjusted by the general changes in the purchasing power of money. Under both measurement systems the concern is to maintain the monetary amount of the enterprise's net assets. In both cases, we are therefore concerned with financial capital maintenance.

Replacement cost, net realisable value and present value are sometimes known as current value systems. They seek to maintain the physical (or operating) capital of the enterprise. The three systems differ in how they seek to do this. Replacement cost measures assets at the amount it would cost to *replace* them with an equivalent asset. Net realisable value measures assets at their sale value. Present value measures the business at the present discounted values of future net cash inflows. These three current value systems can be combined into a 'value to the business' model. A detailed description of this model, sometimes called current value accounting, is beyond the scope of this book.

It is, however, important to realise two fundamental points. First, historical cost still remains the most common measurement basis adopted by enterprises. Second, although historical cost is much criticised, there is no consensus about which measurement system, if any, should replace it. Agreement on a measurement system is where attempts to arrive at a consensual conceptual theory have all floundered.

Measurement Systems

In this section, we discuss two of the most important measurement systems: historical cost and replacement cost. Readers interested in the other three measurement systems are referred to more advanced texts such as Geoffrey Whittington's *Inflation Accounting: An Introduction to the Debate.*

Historical Cost

Historical cost has always been the most widely used measurement system. Essentially, transactions are recorded in the books of account at the date the transaction occurred. This original cost is maintained in the books of account and not updated for any future changes in value that might occur. To illustrate, if we paid £5,000 for a building in 1980, this will be the cost that is shown in the balance sheet when we prepare our accounts in 2002. This is even when the building has increased in value to say £20,000 through inflation. The depreciation will be based on the original value of the asset (i.e., £5,000 not £20,000).

The main strength of historical cost is that it is objective. In other words, you can objectively verify the original cost of the asset. You only need to refer to the original invoice. In addition, historical cost is very easy to use and to understand. Finally, historical cost enables businesses to keep track of their assets.

There is, however, one crucial problem with historical cost. It uses a fixed monetary capital maintenance system which does not take inflation into account. This failure to take into account changing prices can cause severe problems.

Pause for Thought 11.1

Historical Cost and Asset-Rich Companies

The balance sheets of asset-rich companies, such as banks, may not reflect their true asset values, if prepared under historical cost accounting. Why do you think this might be?

If we take banks and building societies as examples of asset-rich companies, these businesses have substantial amounts of prime location fixed assets. Almost in every town, banks occupy key properties in central locations. These properties were also often acquired many years ago, indeed possibly centuries ago. Using strict historical cost, these buildings would be recorded in the balance sheet at very low amounts. This is because over time money values have changed. If a prime site was purchased for £1,000 in 1700 that might have been worth a lot then. Today, it might be worth say £400 million. Thus, fixed assets will be radically understated, unless revalued.

Replacement Cost

Replacement cost attempts to place a realistic value on the assets of a company. It is concerned with maintaining the operating capacity of a business. Essentially, replacement cost asks the question: what would it cost to replace the existing business assets with identical, equivalent assets at today's prices?

Replacement cost is an alternative method of measuring the assets and profits of a business rather than principally a method of tackling inflation. In the Netherlands, replacement costing has been successfully used by many businesses, such as Heineken. As The Company Camera 11.1 shows, Heineken values its tangible fixed assets at replacement cost based on expert valuation. Indeed, the problem for the Dutch is not so much the difficulties of using replacement cost, but of convincing the rest of the world that it is a worthwhile system.

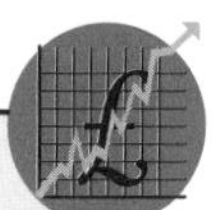

The Company Camera 11.1

Replacement Cost and Heineken

Fixed assets

Tangible fixed assets are valued at replacement cost and, with the exception of land, less accumulated depreciation. The following table of average useful life is used to determine depreciation:

Plants	30–40 years
Machinery and installations	10–30 years
Other fixed operating assets	5–10 years

The replacement cost is based on valuations made by internal and external experts, taking technical economic developments into account and supported by the experience gained in the construction of breweries throughout the world. Projects under construction are stated at historical cost.

Source: Heineken, 2000 Annual Report, p. 58

Replacement cost accounting in the UK is still practised by a few companies. BG, formerly British Gas, for example, uses a variant of replacement costing called current cost accounting. This can have a dramatic effect upon profits and balance sheet values. For example, British Gas's historical profit for 1995 was £986 million while its current cost profit was £607 million. Its tangible assets were £12,202 million under historical cost, but £23,659 million under current costs. It has been suggested that British Gas has deliberately adopted current cost accounting (see Real-Life Nugget 11.1 on the next page).

Real-Life Nugget 11.1

British Gas and Current Cost

Implications

The impact of applying current cost to British Gas's accounts is twofold. First, reported profit is markedly reduced. Second, total net assets are markedly strengthened. These outcomes, one can speculate, are unlikely to upset the firm's management. Public opinion has been extremely hostile to British Gas, and the lower reported profits created by current cost could have an important psychological impact. The company can more easily ward off claims of excessive profits as return on capital employed is depressed.

Source: M.J. Jones and H. Mellet (1997), 'The Accounting Messages of British Gas', *Management Accounting*, p. 53.

The main problem with replacement cost is that, in practice, it is often difficult to arrive at an objective value for the replacement assets. However, in many cases specific indices are available for certain classes of assets, allowing more accurate valuations.

Deficiencies of Historical Cost Accounting

Figure 11.3 on on the next page shows how historical cost accounting can give a misleading impression of the profit for the year and of the value of assets in the balance sheet. In particular, strictly following historical cost will have the effect of:

(i) encouraging companies to pay out more dividends to shareholders than is wise,
(ii) making companies appear more profitable than they really are, and,
(iii) impairing the ability of companies to replace their assets.

Figure 11.3 **The Deficiencies of Historical Cost Accounting**

A company's only asset is a building, purchased 10 years ago for £20,000. The replacement cost for an equivalent building is now £200,000. The company, which deals only in cash, has profits of £10,000 per annum, it distributes 50% of its profits as dividends. The asset is written off over 20 years.

(i) Historical Cost Accounts in Year 10

Profit and Loss Account	£	Balance Sheet	£
Profit before depreciation	10,000	**Fixed Assets**	20,000
Depreciation	(1,000)	Accumulated depreciation	(10,000)
	9,000	*Total fixed assets*	10,000
Dividends	(4,500)	Cash	55,000
Retained Profit	4,500	*Net assets*	65,000

(a) Over the first ten years, the company's profits, or as it only deals in cash, cash flow is £100,000. It has paid out £45,000 in dividends leaving £55,000 cash in the company. This looks healthy.
(b) The shareholders are happy receiving an annual dividend.
(c) Return on capital employed (taking closing net assets) is:

$$\frac{£9,000}{£65,000} = 13.8\%$$

Everything, therefore, seems pretty good. *Unfortunately, the company has only £65,000 in net assets which is not enough to replace the fixed assets which will cost £200,000!*

(ii) Replacement Cost Accounts in Year 10

Profit and Loss Account	£	Balance Sheet	£
Profit before depreciation	10,000	**Fixed Assets**	200,000
Depreciation	(10,000)	Accumulated depreciation	(100,000)
	–	*Total fixed assets*	100,000
		Cash	100,000
		Net assets	200,000

(a) In this case, the company makes no profit because the increased depreciation has wiped out all the profits. There is no profit out of which to pay dividends. If the company had paid out dividends during the 10 years it would have no money left to replace the fixed assets.
(b) The net worth has risen considerably. This is a plus for the company. However, not paying out dividends is a considerable minus.
(c) There is no return on capital employed!
Superficially, everything appears less rosy. *However, the firm can continue in business because it can just about replace its fixed assets* (in actual fact, its net assets equals the amount needed to replace the fixed assets). This assumes that the building could be sold for £100,000!

In practice, many UK companies now use a modified form of historical cost accounting. This involves revaluing fixed assets, often every five years. Depreciation is then based on the revised valuation. However, in other countries, such as the US, Germany and France, there is still rigid adherence to historical cost.

Illustrative Example of Different Measurement Systems

In Figure 11.4, we pull together some of the threads and show how the valuation of an individual asset can vary considerably depending upon the chosen measurement system.

Figure 11.4 Example of Different Measurement Systems

JoJo bought a van two years ago for £10,000. She expects to keep the van for five years. The used van guide states the van is now worth £2,500. Replacement cost for a van in a similar condition is £4,000. The future net cash flows will be £4,000 for the next three years (assume the cash flows occur at the end of the year) and she can borrow money at 10%. The retail price index was 100 when the van was bought and it is 120 now.

	Appropriate value £
Historical Cost We base our calculations on the original historical cost of £10,000. Using straight-line depreciation (£10,000 ÷ 5) = £2,000 p.a. *Thus, £10,000 – £4,000 (two years' depreciation)*	6,000*
Current Purchasing Power We base our calculation on the original historical cost less depreciation. In the calculation above, this was £10,000 – £4,000 = £6,000*. We then adjust this for inflation. This is measured using the retail price index, which has increased from 100 to 120. $\text{£}6{,}000 \times \frac{\text{Closing RPI}}{\text{Opening RPI}} \left(\frac{120}{100}\right)$	7,200
Net Realisable Value In this case, our calculations are based upon the amount of money we would receive for the van if we sold it. *Used van guide*	2,500
Replacement Cost In this case, we base our calculations on the amount it would cost to replace the van with a similar asset in a similar condition. *Similar value asset*	4,000
Present Value Here, we are interested in looking at the future cash flows generated by the asset. We then discount them back to today's value (see Chapter 21 for further information about discounting).	

£	Discount Factor*	£	
4,000	0.9091	3,636	
4,000	0.8264	3,306	
4,000	0.7513	3,005	
		9,947	9,947

*10% interest discounted back, assumes cash flow is on the last day of each year.

We can, therefore, see that different measurement systems give different asset valuations. There are thus five different valuations ranging from £2,500 to £9,947.

	£
• Historical cost	6,000
• Current purchasing power	7,200
• Net realisable value	2,500
• Replacement cost	4,000
• Present value	9,947

It is important to note that, in practice, each measurement system itself could potentially yield many different asset valuations, depending on the underlying assumptions and estimations. For example, present value is crucially dependent on the estimated discount rate (10%), the estimated future cash flows (£4,000), and the timing of those cash flows.

Real Life

The merits of historical cost accounting and the advantages and disadvantages of the competing alternative measurement systems have been debated vigorously for at least 40 years. However, with some rare exceptions, most companies worldwide still use historical cost.

This is not to say that experimentation has not occurred. In the Netherlands, for example, Philips, one of the world's leading companies, used replacement cost accounting for over a generation. Finally, Philips abandoned replacement cost, not because of replacement cost's inadequacies, but because of the failure of international financial analysts to understand Philips accounts. Today, there are still companies in the Netherlands, such as Heineken, which use replacement cost. In the UK too, there are a few companies, usually ex-privatised utilities with extensive infrastructure assets, such as British Gas, which use replacement costs. The Company Camera 11.2 shows that BG carries some of its tangible assets at depreciated replacement cost, unless they have lost value in which case they are carried at estimated value in use.

The Company Camera 11.2

Replacement Cost and BG

Tangible fixed assets

Transco's regulatory tangible fixed assets are included in the balance sheet at depreciated replacement cost or, where lower, the estimated value in use . . . Investment properties are carried at valuation. All other categories of tangible fixed assets are carried at depreciated historical cost.

Source: BG, 1999 Annual Report, p. 52

In both the UK and the US in the 1970s, there were serious attempts to replace historical cost accounting initially with current purchasing, but later with current value accounting (a mixture of the three current value systems). These methods were thought to be superior to historical cost accounting when dealing with inflation, which was at that time quite high. They were also believed to provide a more realistic valuation of company assets. In the end these attempts failed. The reasons for their failure were quite complex. However, in general, accountants preferred the objectivity of a tried-and-tested, if somewhat flawed, historical cost system to the subjectivity of the new systems. In addition, rates of inflation fell.

Although the backbone of the accounts is historical cost, there is some limited use of alternative measurement systems (see Figure 11.5). In addition, in the UK, many companies revalue their fixed assets every five years. This is particularly common in companies that have many fixed assets, such as hotel chains. However, this periodic revaluation of fixed assets means that UK accounts are prepared on a different basis to those in countries, such as France or the US, where periodic revaluations are not permitted.

Figure 11.5 Use of Alternative Measurement Systems

'The measurement basis most commonly adopted by enterprises in preparing their financial statements is historical cost. This is usually combined with other measurement bases. For example, inventories [i.e. stocks] are usually carried at the lower of cost and net realisable value, marketable securities may be carried at market value and pension liabilities are carried at their present value. Furthermore, some enterprises use the current cost basis as a response to the inability of the historical cost accounting model to deal with the effects of changing prices of non-monetary assets.'

Source: International Accounting Standards Board (2000), *Framework for the Preparation and Presentation of Financial Statements*, para. 101

Conclusion

Different measurement systems will give different figures in the accounts for profit and net assets. The mostly widely used measurement system, historical cost, records and carries transactions in the accounts at their original amounts. Historical cost, however, does not deal well with changes in asset values resulting from, for example, inflation. There are four other main measurement systems (current purchasing power, replacement cost, net realisable value, present value). Current purchasing power adjusts historical cost for general changes in the purchasing power of money. Replacement cost records assets at the amounts needed to replace them with equivalent assets. Net realisable value records assets at the amounts they would fetch in an orderly sale. Finally, present value discounts future cash inflows to today's monetary values. Although historical cost is the backbone of the accounting measurement systems, there are departures from it, such as the valuation of stock at the lower of cost or net realisable value. In particular, in the UK, many companies revalue their fixed assets.

Selected Reading

The topic of accounting measurement systems can be extremely complex. The first three readings below have been deliberately selected because they are quite accessible to students. Students wishing for a fuller insight into the debate are referred to the book by Geoffrey Whittington.

1. Accounting Standards Steering Committee (1975), *The Corporate Report*, Section 7, pp. 61–73.
   Although now 25 years old, this report provides a very good, easy-to-read, introduction to the topic.
2. International Accounting Standards Board (IASB) (2000) 'Framework for the Preparation and Presentation of Financial Statements', in *International Accounting Standards* (2000), paras. 99–110.
   It presents more modern thinking on the topics and is reasonably easy-to-follow.
3. Jones, M.J. and Mellett, H. (January 1997), 'The Accounting Messages of British Gas', *Management Accounting*, pp. 52–53.
   This article looks at the impact which using current cost has on British Gas's profit.

For the Enthusiast:

Whittington, G. (1983), *Inflation Accounting: An Introduction to the Debate* (Cambridge University Press).
For students who enjoy a challenge. Gives a thorough grounding in the inflation debate, which is at the heart of choosing different measurement systems.

Discussion *Questions*

Questions with numbers in blue have answers at the back of the book.

Q 1 'Accounting measurement systems are the skeleton of the accounting body.' Critically evaluate this statement.

Q 2 Why is historical cost still so widely used, if it is so deeply flawed?

Q 3 What is the difference between a financial capital maintenance concept and a physical capital maintenance concept?

Q 4 Why do many UK companies revalue their fixed assets? How might this affect profit? Why is this practice unusual internationally?

Chapter 12

The Annual Report

'It is a yearly struggle: the conflict between public relations experts determined to put a sunny face on somewhat drearier figures, and those determined to tell it like it is, no matter how many "warts" there are on the year's story. The annual report is a vital instrument designed – ideally – to tell the story of a company, its objectives, where the company succeeded or failed, and what the company intends to do next year.'

Kirsty Simpson (1997), 'Glossy, expensive and useless', *Australian Accountant*, September, pp.16–18

Learning Outcomes

After completing this chapter you should be able to:

- ✔ Explain the nature of the annual report.
- ✔ Outline the multiple, conflicting objectives of the annual report.
- ✔ Discuss the main contents of the annual report.
- ✔ Evaluate how the annual report is used for impression management.

In a Nutshell

- The annual report is a key corporate financial communication document.
- It is an essential part of corporate governance.
- It serves multiple, and sometimes conflicting, roles of stewardship/accountability, decision making and public relations.
- It comprises key audited financial statements: profit and loss account, balance sheet, and cash flow statement.
- It normally includes at least 20 identifiable sections.
- It also includes important non-audited sections such as the chairman's statement and an operating and financial review.
- Most important companies provide group acounts.
- Goodwill is an important intangible asset in many group accounts.
- Managers use the annual report for impression management.

Introduction

The annual report is well-entrenched as a core feature of corporate life. This yearly-produced document is the main channel by which directors report corporate annual performance to their shareholders. All leading companies worldwide will produce an annual report. In the UK, both listed and unlisted companies produce one. Many other organisations, such as the British Broadcasting Corporation, now also produce their own versions of the annual report. Indeed, the Labour Government produced the first governmental annual report in 1998. Traditionally, the annual report was a purely statutory document. The modern annual report, however, now has multiple functions, including a public relations role. Modern reports comprise a mixture of voluntary and statutory, audited and unaudited, narrative and non-narrative, financial and non-financial information. They are also governed by a regulatory framework which includes Companies Acts and accounting standards. The modern annual report has become a complex and sophisticated business document.

Definition

In essence, an annual report is a document produced to fulfil the duty of the directors to report to shareholders. It is produced annually and is a mixture of financial and non-financial information. As Definition 12.1 shows, it is a report containing both audited financial information and unaudited, non-financial information.

Definition 12.1

Annual Report

Working definition
A report produced annually by companies comprising both financial and non-financial information.

Formal definition
A document produced annually by companies designed to portray a 'true and fair' view of the company's annual performance, with audited financial statements prepared in accordance with company legislation and other regulatory requirements, and also containing other non-financial information.

Context

The annual report has evolved over time into an important communications document, especially for large publicly listed companies. The earliest annual reports arose out of the need to make directors accountable to their shareholders. The main financial statement was the balance sheet. In order to ensure that the financial statements fairly represented corporate performance, the annual report was audited. The annual report, therefore, has always played a key role in the control of the directors by the shareholders. The central role of the annual report in external reporting can be seen in Figure 12.1.

Figure 12.1 The Annual Report

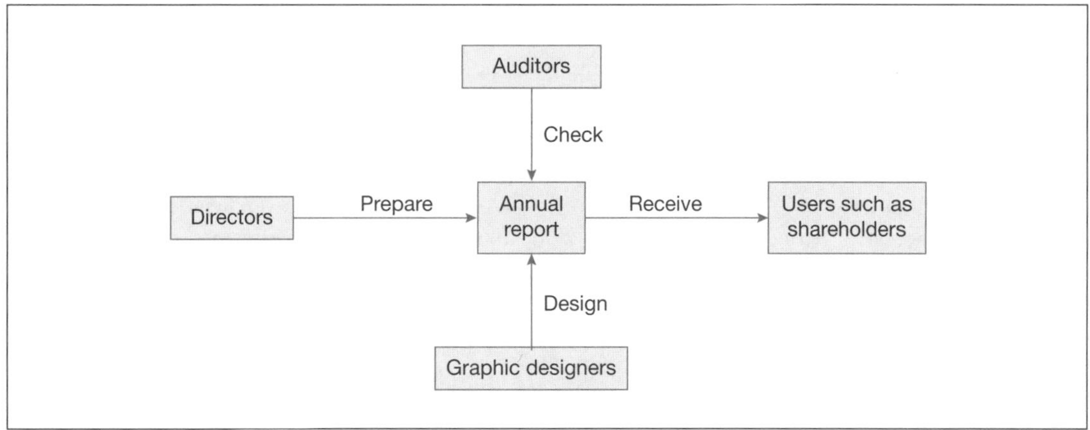

In essence, the directors are responsible for the preparation of the financial statements from the accounting records. The actual preparation is normally carried out by accounting staff. An annual report is then compiled, often with the help of a company's public relations department and graphic designers. These graphic designers are responsible for the layout and design of the annual report (providing, for example, colourful graphs and photographs). The annual report's financial content is then audited and disseminated to the main users, principally the shareholders. As discussed in Chapter 10, much of the annual report is mandated by a regulatory framework consisting principally of the requirements of the Companies Acts and accounting standards.

Multiple Roles

The annual report is a social as well as a financial document. Therefore, as society evolves, so does the annual report. The earliest annual reports were stewardship documents. Today's annual report is much more complex, being an amalgam of stewardship and accountability, decision making and public relations. These concepts are discussed below. The first two roles are those traditionally recognised by standard setters. However, the public relations role is more driven by preparer self-interest.

(i) Stewardship and Accountability

Effectively, stewardship involves the directors reporting their actions to the shareholders. This reflects the origins of financial reporting. In the middle ages, the stewards who managed the estates used to render an annual account of the master's assets (for example, livestock and cereals) to the lord of the manor. The main aim of the accounts, or annual statement, was thus for the lord of the manor to keep a check on the steward's activities. In particular, there was a concern that the steward should not defraud the lord of the manor. An important aspect of stewardship is this accountability. Accountability is traditionally seen as referring to the control and safeguarding of the assets of a company.

Gradually, as the economy became more sophisticated so did the accountability mechanisms. At first, there was a rudimentary statement of assets and liabilities, showing how much the organisation owned and was owed. This gradually evolved into the modern balance sheet. However, the fundamental aim was still to account for the assets and liabilities of the organisation. Accountability tended to diminish in importance with the rise of decision making.

An important modern aspect of stewardship is corporate governance. Essentially, in the 1990s, several well-publicised corporate financial scandals (e.g., Polly Peck, Maxwell) led to a growing concern with monitoring the activities of directors. In addition, the privatisation of the utilities created considerable concern over the salaries of so-called 'fat-cat directors'. The committees which looked at corporate governance in the 1990s all stressed the role that corporate financial communication could play in increasing directors' accountability to shareholders. The accounting scandals such as Enron and WorldCom in the US have reawakened interest in corporate governance. Real-Life Nugget 12.1 discusses this issue.

Real-Life Nugget 12.1

Directors' pay

Alex Brummer
Financial Editor

The call by Stephen Byers for world class salaries for world class performance is a terrific slogan. The trouble is that it fits only a handful of the executives and companies in the Guardian's pay survey. Of the 35 or so directors in the million-pounds-plus pay club only a handful – such as those at the drugs companies SmithKline Beecham and Glaxo Wellcome – deliver a world class product. Others such as Royal & Sun Alliance, headed by Bob Mendelsohn, who took home £2.4m last year, just about register in Britain, certainly not on the global stage. As for the remuneration committees which sanctioned increases of 26% in directors' wage packets – in a year when trading profits went up just 6.9% – they should (if world class performance really counted) all be fired.

But the timid Byers, clearly fearful of upsetting pals in the boardroom, could not bring himself to say anything which might smack of controversy.

Nevertheless, there are some areas of merit in what he has to say, if only he would show a sense of determination. It is ludicrous that some four years after Sir Richard Greenbury first exposed the excesses and unfairness in the system, ministers still creep through the undergrowth for fear of disturbing the big beasts. The disclosure principle has been around for quite a time but annual reports tend to obscure rather than clarify. There is a lack of clear tables showing the comparators used, there is no description of differences of opinion within remuneration committees. If it is possible for members of the Bank of England monetary policy committee to express their views publicly on something as critical as interest rates, why not some better disclosure of remuneration committee voting patterns?

Source: The *Guardian*, 20 July, 1999

Pause for Thought 12.1

'Fat Cat' Directors

There has been a great furore about the salaries of 'fat-cat' directors. What justification do you think they tend to give for their salaries? What do their critics argue?

The directors' view

Conventionally, directors argue that they are doing a complex and difficult job. They are running world-class businesses and they, therefore, need to be paid world-class salaries. They also create and add shareholder value because of increased share prices and, therefore, they deserve to be well paid.

The critics' view

Yes, but if the directors are paid on the basis of performance, then we would expect them *not* to get big bonuses when their organisations are doing less well. However, generally this does not happen. Also, much of the increase in share prices that directors ascribe to themselves is caused by a general rise in the stock market.

(ii) Decision Making

In the twentieth century, decision making has increasingly replaced stewardship as the main role of accounting. This reflects wider developments in society, business and accounting. In particular, decision making is associated with the rise of the modern industrial company. Industrialisation led to increasingly sophisticated businesses and to the creation of the limited liability company with its divorce of ownership and control. Shareholders were no longer involved in the day-to-day running of the business. They were primarily interested in increases in the value of their share price and in any dividends they received. These dividends were based on profits. Consequently, the profit and loss became more important relative to the balance sheet. The primary interest of shareholders shifted from cash and assets to profit. Thus, performance measurement and decision making replaced asset management and stewardship as the prime objective of financial information.

Pause for Thought 12.2

Engines of Capitalism

Limited liability companies have been called the engines of capitalism. Why do you think this is so?

Effectively, limited liability companies are very good at allowing capital to be allocated throughout an economy. There are several advantages to investors. First, they can invest in many companies not just one. Second, they can sell their shares very easily, assuming a buyer can be found. Third, they stand to lose only the amount of capital they have originally invested. Their personal assets are thus safe.

These new shareholder concerns were officially recognised by two reports in the 1960s and 1970s in the US and the UK. Both reports, *A Statement of Basic Accounting Theory* (American Accounting Association, 1966) in the US and *The Corporate Report* (Accounting Standards Steering Committee, 1975) in the UK, proved turning points in the development of accounting. Before then stewardship had been the generally acknowledged role of accounting. After them, decision-usefulness was generally recognised as the prime criterion. In a sense, decision making and stewardship are linked. For shareholders need to make decisions about how well the directors have managed the company.

In a nutshell, the purpose of the annual report was recognised to be:

> 'to communicate economic measurements of and information about the resources and performance of the reporting entity useful to those having reasonable rights to such information'. (*The Corporate Report*, 1975, para. 3.2)

The decision-making model had been born!

As Definition 12.2 shows, the modern objective of accounting is still recognised as providing users with information so that they can make decisions.

Definition 12.2

Decision-Making Objective of Annual Report

Working definition

Providing users, especially shareholders, with financial information so that they can make decisions such as buying or selling their shares.

Formal definition

'The objective of financial statements is to provide information about the financial position, performance and changes in financial position of an enterprise that is useful to a wide range of users in making economic decisions.'

Source: International Accounting Standards Board (2000), *Framework for the Preparation and Presentation of Financial Statements*.

For the shareholder, these economic decisions might involve the purchase or sale of shares. Other users will have be different concerns. For example, banks might be principally interested in whether or not to lend a company more money.

(iii) Public Relations Role

The public relations role reflects the annual report's development over the last 20 years as a major marketing tool. Company management has come to realise that the annual report represents an unrivalled opportunity to 'sell' the corporate image. In part, this only reflects human nature. We all wish to look good. It is a rare person who never attempts to massage the truth, for example, at a job interview. The public relations role of annual reports does, however, provoke strong reactions by some commentators (see Real-Life Nugget 12.2 on the next page).

Real-Life Nugget 12.2

Public Relations and the Annual Report

'Queen Isabella was said to have washed only three times in her life, and only once voluntarily. That was when she was married. The other two times were at her birth and death. No wonder Columbus left to discover a new world. Why this olfactory analysis of history? Because this is the time of year when we are inundated with corporate annual reports, and in most of them the letter to the shareholders smells as wretched as Queen Isabella must have.

One of the sad truths about malodorous things is that people tend to get used to them in time. But I'll never become accustomed to the public-relations pap I read in most annual reports. Every year tens of thousands of stale, vapid, and uninspired letters to shareholders appear in elaborate annual reports. They are printed on expensive paper whose gloss and sheen are exceeded only by the glitzy words of the professional PR writer who ghosted the message. They will be read by shareholders who don't understand them – or believe them. Quite often they are hype. Sometimes they are dull. Some are boastful, others apologetic. And they are generally ambiguous.'

Source: Sal Marino, *Industry Week*, 5 May, 1997, p. 12

Conflicting Objectives

The standard-setting organisations generally only recognise the first two objectives of financial statements (also by implication of annual reports): stewardship and decision making. In effect, these two objectives clash with the public relations role. This is because the stewardship and decision-making roles rely upon the notion of providing a neutral and objective view of the company. However, the public relations view is where managers seek to present a favourable, not a neutral, view of a company's activities. This causes stress, particularly if a company did not perform as well as market analysts had predicted. In these cases, as we see in Chapter 13, there is great pressure for the company management to indulge in such impression management.

Main Contents of the Annual Report

Every annual report is unique. Ranging usually from about 30 to 60 pages, a company's report presents a variety of corporate financial and non-financial information. The traditional financial statements (e.g., balance sheet, profit and loss account, cash flow statement) and accompanying financial information (e.g., notes to accounts) are normally audited. Other parts, such as the chairman's statement and operating and financial review, are not. However, most auditors are reluctant to let directors make statements that are blatantly inconsistent with the audited accounts. In addition, the report is a mixture of voluntary and mandatory (i.e., prescribed by regulation) information, and narrative and non-narrative information. In Figure 12.2 on the next page, the main sections of a typical annual report are outlined.

Figure 12.2 **Main Sections of a Typical Annual Report**

Section	*Audited Formally*	*Narrative (N) Non-Narrative (NN)*	*Mandatory (M) Voluntary (V)*
1. Profit and loss account	Yes	NN	M
2. Balance sheet	Yes	NN	M
3. Cash flow statement	Yes	NN	M
4. Statement of total recognised gains and losses	Yes	NN	M
5. Reconciliation of movements in shareholders funds	Yes	NN	M
6. Note on historical cost profits and losses	Yes	NN	M
7. Accounting policies	Yes	N	M
8. Notes to the accounts	Yes	N	M
9. Principal subsidiaries	Yes	N	M
10. Chairman s statement	No	N	V
11. Directors report	No	N	M
12. Operating and financial review	No	N	V
13. Review of operations	No	N	V
14. Statement of corporate governance	No	N	M
15. Auditors report	Not applicable	N	M
16. Statement of directors responsibilities for the financial statements	No	N	M
17. Shareholder information	No	N	V
18. Highlights	No	NN	V
19. Historical summary	No	NN	V
20. Shareholder analysis	No	NN	V

The main sections of the annual report can be divided into audited and non-audited statements. These are shown in Figure 12.3 on the next page and discussed below.

The Audited Statements

Nine audited financial sections are normally included by most UK companies in their annual reports. The first three (the profit and loss account (see Chapter 4), the balance sheet (see Chapter 5) and the cash flow statement (see Chapter 8)) have already been covered in depth earlier. They are, therefore, only lightly touched on here. The remaining audited statements are all comparatively recent. They can be divided into subsidiary statements and explanatory material. All the sections are presented in a relatively standard way, following guidance laid down in the Companies Acts and accounting standards.

Figure 12.3 Overview of the Annual Report

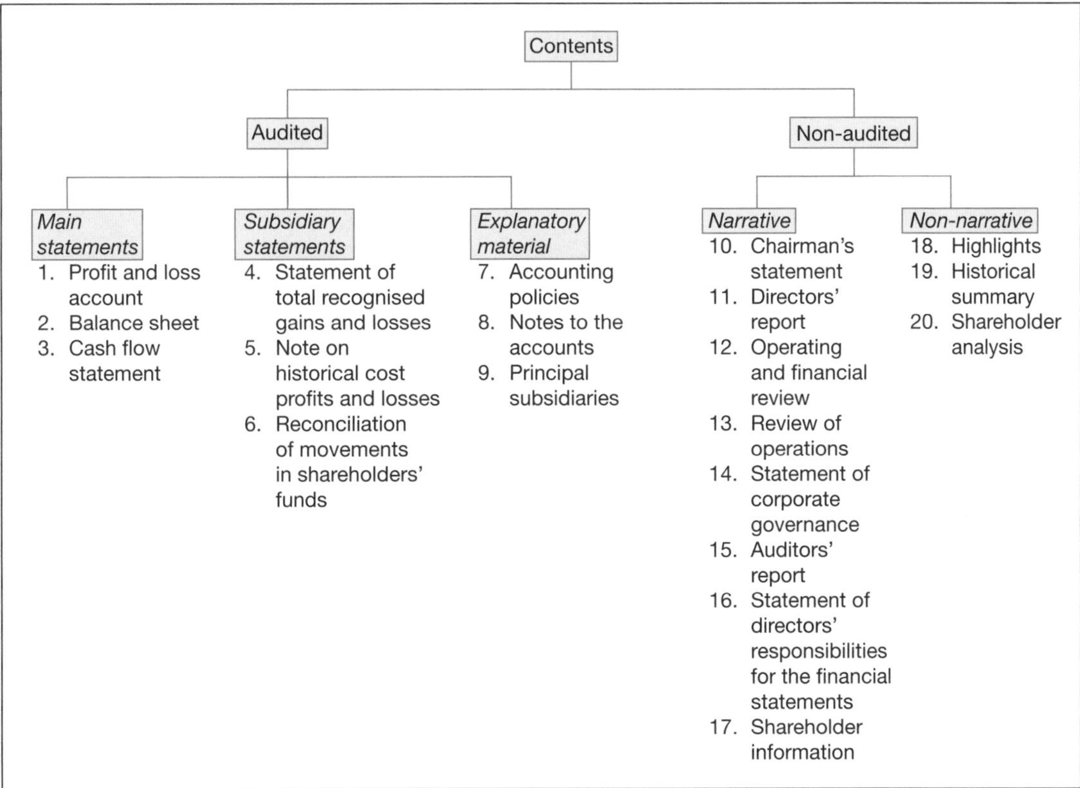

Main Statements

1. Profit and Loss Account. The profit and loss account is widely recognised as the primary financial statement. It focuses on the revenue earned and expenses incurred by the business during the accounting period. Importantly, this is not the same as cash received and cash paid. Most companies follow a format for the profit and loss account determined by UK Financial Reporting Standard 3, *Reporting Financial Performance*. An illustration of the profit and loss account for J. Sainsbury plc is given in The Company Camera 7.3 on page 147 in Chapter 7.

2. Balance Sheet. Nowadays, the balance sheet is often seen as of secondary importance to the profit and loss account as a decision-making document. It focuses on assets, liabilities and shareholders funds (i.e., capital employed) at a particular point in time (the balance sheet date). The balance sheet, along with the profit and loss account, is prepared from the trial balance. The balance sheet is commonly used to assess the liquidity of a company, whereas the profit and loss account focuses on profit. An illustration of the balance sheet for J. Sainsbury plc is given in the Company Camera 7.4 on page 148 in Chapter 7.

3. Cash Flow Statement. Unlike the previous two statements, which use the matching basis and are prepared from the trial balance, the cash flow statement is usually prepared by deduction from the profit and loss account and balance sheets. It is a relatively new statement, introduced in 1991. The objective of the cash flow statement is to report and categorise cash inflows and outflows during a particular period.

Subsidiary Statements

4. Statement of Total Recognised Gains and Losses (STRGL). This is another comparatively recent statement. The STRGL begins with the profit for the year taken from the profit and loss account. It then deals with *non-trading gains and losses* traditionally taken to reserves not to the profit and loss account. The STRGL attempts to highlight all shareholder gains and losses and not just those from trading. These gains and losses might, for example, be surpluses on property revaluation. Alternatively, as in the case of Tesco in 2001 (see The Company Camera 12.1), they may be a loss on foreign currency net investments.

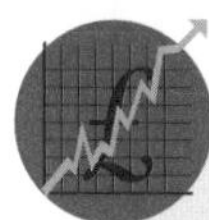

The Company Camera 12.1

Statement of Total Recognised Gains and Losses

52 weeks ended 24 February 2001

	Group		Company	
	2001	2000	2001	2000
	£m	£m	£m	£m
Profit for the financial year	767	674	66	42
Loss on foreign currency net investments	(2)	(36)	(4)	(3)
Total recognised gains and losses relating to the financial year	765	638	62	39

Source: Tesco plc, 2001 Annual Report, p. 19

5. Note on Historical Cost Profits and Losses. If the accounts are prepared under the historical cost convention then the original cost of assets is recorded in the accounts. However, sometimes assets, particularly fixed assets, will be revalued. These assets are not then included in the accounts at their original purchase price. Depreciation on the revalued fixed assets will then be more than on the original cost. This note records any such differences caused by departures from the historical cost convention.

6. Reconciliation of Movements in Shareholders' Funds. This statement highlights major changes to the wealth of shareholders. These include profit (or loss) for the year, annual dividends and new share capital. As Tesco's reconciliation shows (The Company Camera 12.2 on next page), they can also include loss on foreign currency net investments.

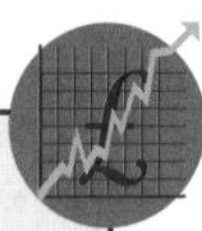

The Company Camera 12.2

Reconciliation of Movements in Shareholders' Funds

52 weeks ended 24 February 2001

	Group		Company	
	2001	2000	2001	2000
	£m	£m	£m	£m
Profit for the financial year	767	674	66	42
Dividends	(340)	(302)	(340)	(302)
	427	372	(274)	(260)
Loss on foreign currency net investments	(2)	(36)	(4)	(3)
New share capital subscribed less expenses	110	30	174	54
Payment of dividends by shares in lieu of cash	52	21	52	21
Net addition/(reduction) to shareholders' funds	587	387	(52)	(188)
Opening shareholders' funds	4,769	4,382	2,511	2,699
Closing shareholders' funds	5,356	4,769	2,459	2,511

Source: Tesco plc, 2001 Annual Report, p.19

Explanatory Material

7. *Accounting Policies.* Companies must describe the accounting policies they use to prepare the financial statements. The flexibility inherent within accounting means that companies have a choice of accounting policies in areas such as foreign currencies, goodwill, pensions, sales and stocks. Different accounting policies will result in different accounting figures. Tesco's policy on stocks is given as an illustration (see The Company Camera 12.3). It is clearly stated that stocks are valued at the lower of cost and net realisable value.

The Company Camera 12.3

Policy on Stocks

Stocks comprise goods held for resale and properties held for, or in the course of, development and are valued at the lower of cost and net realisable value. Stocks in stores are calculated at retail prices and reduced by appropriate margins to the lower of cost and net realisable value.'

Source: Tesco plc, 2001 Annual Report, p .22

8. Notes to the Accounts. These notes provide additional information about items in the accounts. They are often quite extensive. For example, in Tesco's 2001 report the main three financial statements take up three pages, but the 32 notes take up a further 16 pages. These notes flesh out the detail of the three main financial statements. They cover a variety of topics. For example, the first five Tesco notes cover segmental analysis, operating profit, employee profit-sharing, profit on ordinary activities before taxation, and employment costs. The notes to accounts can be crucial. 'The numbers are just part of the story. The balance sheet is just a snap shot. It captures some of the picture but not all of it so that's why the notes to the accounts are important' (Jill Treanor, *The Guardian*, March 6, 2000, p. 28).

9. Principal Subsidiaries. Most large companies consist of many individual companies arranged as the parent (or holding company) and its subsidiaries. Collectively, they are known as groups (see later in this chapter for a fuller explanation of groups). In the case of a group, there will be a listing of the parent (i.e., main) company's subsidiary companies (i.e., normally those companies where over 50% of the shares are owned by the parent company) and associate companies (normally where between 20% and 50% of shares are owned by the parent company).

The Non-Audited Sections

The amount of non-audited information in annual reports has mushroomed over the last 20 years. It has caught even experienced observers by surprise (see, for example, Real-Life Nugget 12.3).

Real-Life Nugget 12.3

The Mushrooming Importance of Non-Audited Information

'In this survey we are interested in the pages outwith the statutory financial statements . . . What surprised us was that the "narrative" pages exceeded or equalled the number in the statutory financial statements in half of our survey companies' annual reports.'

Source: Arthur Andersen, (1996), *What's the Story*, p. 5

The non-audited information is extremely varied, but can be broadly divided into narrative and non-narrative information. The *narrative* information consists mainly of the chairman's statement, the directors' report, the operating and financial review, and the auditors' report. By contrast, the *non-narrative* information mainly comprises the highlights and the historical summary. Although these sections are not audited, the auditor is required to review the non-narrative information to see if there are material misstatements or material inconsistencies with the financial statements. If there are, the auditors consider whether any information needs to be amended. Unfortunately, all this is very subjective and, in reality, little guidance is given to auditors.

Narrative Sections

10. Chairman's Statement. This is the longest-established accounting narrative. It is provided voluntarily by nearly all companies. The chairman's statement provides a personalised overview of the company's performance over the past year. Most chairman's statements cover strategy, the financial performance and future prospects. It is also traditional for the chairman to thank the employees and retiring directors.

Pause for Thought 12.3

Auditing the Accounting Narratives

What difficulties do you think an auditor might have if called upon to audit the narrative sections of the annual report, such as the chairman's statement?

The main difficulty is deciding how to audit the written word. Usually, auditors audit figures. They can thus objectively trace these back to originating documentation. The problem with accounting narratives is that they are very subjective. How do you audit phrases such as 'We have had a good year' or 'Profit has increased substantially'?

11. Directors' Report. The directors' report is prescribed by law. Its principal objective is to supplement the financial information with information that is considered vital for a full appreciation of the company's activities. Items presented here (or elsewhere in the accounts – an increasingly common practice) might include any changes in the company's activities, proposed dividends, and charitable and political gifts.

12. Operating and Financial Review. The operating and financial review (OFR) represents a major innovation in UK financial reporting. For the first time, regulators formally recognised the importance of qualitative, non-financial information. It enables companies to provide a formalised, structured and narrative explanation of financial performance. The ASB introduced the OFR in 1993. The OFR is voluntary not mandatory. It has two parts: first, the operating review which discusses items such as a company's operating results, profit and dividends; second, the financial review which covers items such as capital structure and treasury policy. The OFR can be treated as a stand-alone document or may be split up and incorporated into other narrative sections.

13. Review of Operations. This section forms a natural complement to the chairman's statement. Whereas the chairman provides the overview, the chief executive reviews the individual business operations, often quite extensively. Normally, the chief executive discusses, in turn, each individual business or geographical segment.

14. Statement of Corporate Governance. This statement arise out of the drive to make directors more accountable to their shareholders. The corporate governance statement is governed by stock market requirements. The issues usually covered are risk management, treasury management, internal controls, going concern and auditors. A major objective is to present a full and frank discussion of the directors' remuneration.

15. Auditors' Report. The audit is an independent examination of the financial statements. An example of a typical auditors' report is given for Tesco's in The Company Camera 12.4 on the next page.

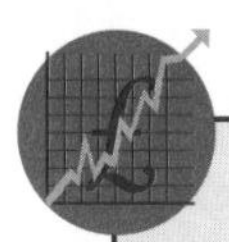

The Company Camera 12.4

Independent Auditors' Report to the Members of Tesco plc

We have audited the financial statements which comprise the profit and loss account, the balance sheet, the cash flow statement, the statement of total recognised gains and losses and the related notes, including the information on Directors' emoluments and share details included within tables one to five, in the remuneration report on pages 13 to 16, which have been prepared under the historical cost convention and the accounting policies set out in the statement of accounting policies on pages 22 and 23.

Respective responsibilities of Directors and auditors

The Directors' responsibilities for preparing the annual report and the financial statements, in accordance with applicable United Kingdom law and accounting standards, are set out in the statement of Directors' responsibilities.

Our responsibility is to audit the financial statements in accordance with relevant legal and regulatory requirements, United Kingdom Auditing Standards issued by the Auditing Practices Board and the Listing Rules of the Financial Services Authority.

We report to you our opinion as to whether the financial statements give a true and fair view and are properly prepared in accordance with the United Kingdom Companies Act 1985. We also report to you if, in our opinion, the Directors' report is not consistent with the financial statements, if the company has not kept proper accounting records, if we have not received all the information and explanations we require for our audit or if information specified by law or the Listing Rules regarding Directors' remuneration and transactions is not disclosed.

We read the other information contained in the annual report and consider the implications for our report if we become aware of any apparent misstatements or material inconsistencies with the financial statements. The other information comprises only the Directors' report, the Chairman's statement, the operating and financial review, the corporate governance statement and the report of the Directors on remuneration.

We review whether the corporate governance statement reflects the company's compliance with the seven provisions of the Combined Code specified for our review by the Listing Rules, and we report if it does not. We are not required to consider whether the Board's statements on internal control cover all risks and controls, or to form an opinion on the effectiveness of the company's or Group's corporate governance procedures or its risk and control procedures.

Basis of audit opinion

We conducted our audit in accordance with Auditing Standards issued by the Auditing Practices Board. An audit includes examination, on a test basis, of evidence relevant to the amounts and disclosures in the financial statements. It also includes an assessment of the significant estimates and judgements made by the Directors in the preparation of the financial statements, and of whether the accounting policies are appropriate to the company's circumstances, consistently applied and adequately disclosed.

We planned and performed our audit so as to obtain all the information and explanations which we considered necessary in order to provide us with sufficient evidence to give reasonable assurance that the financial statements are free from material misstatement, whether caused by fraud or other irregularity or error. In forming our opinion we also evaluated the overall adequacy of the presentation of information in the financial statements.

Opinion

In our opinion the financial statements give a true and fair view of the state of affairs of the company and the Group at 24 February 2001 and of the profit and cash flows of the Group for the year then ended and have been properly prepared in accordance with the Companies Act 1985.

PricewaterhouseCoopers
Chartered Accountants and Registered Auditors
London 9 April 2001

Source: Tesco plc, 2001 Annual Report, p. 17

Companies are legally required to publish the auditors' report. In essence, the report states whether the financial statements present a 'true and fair view' of the company's activities over the previous financial year. It sets out the respective responsibilities of directors and auditors as well as spelling out the work carried out to arrive at the auditors' opinion.

The auditors' report thus outlines the respective responsibilities of directors and auditors, the basis of the audit opinion and how the auditors arrived at their opinion. PricewaterhouseCoopers, the auditors in this case, are one of the UK's leading auditing partnerships.

16. Statement of Directors' Responsibilities for the Financial Statements. This statement (see The Company Camera 12.5) was introduced because of a general misconception by the general public of the purpose of an audit compared with the actual nature of an audit as understood by auditors. The directors must spell out their responsibilities which include (i) keeping proper accounting records; (ii) preparing financial statements in accordance with the Companies Act 1985; (iii) applying appropriate accounting policies; and (iv) following all applicable accounting standards.

The Company Camera 12.5

Directors' Responsibilities for the Preparation of the Financial Statements

The directors are required by the Companies Act 1985 to prepare financial statements for each financial year which give a true and fair view of the state of affairs of the company and the Group as at the end of the financial year and of the profit or loss for the financial year.

The directors consider that in preparing the financial statements on pages 18 to 39 the company has used appropriate accounting policies, consistently applied and supported by reasonable and prudent judgements and estimates, and that all accounting standards which they consider to be applicable have been followed.

The directors have responsibility for ensuring that the company keeps accounting records which disclose, with reasonable accuracy, the financial position of the company and which enable them to ensure that the financial statements comply with the Companies Act 1985.

The directors have general responsibility for taking such steps as are reasonably open to them to safeguard the assets of the Group and to prevent and detect fraud and other irregularities.

Source: Tesco plc, 2001 Annual Report, p. 17

17. Shareholder Information. Companies increasingly include a variety of shareholder information. This might, for example, include a financial calendar, share price details or shareholder analysis (see item 20). The information may be narrative or non-narrative in nature.

Non-Narrative Sections

18. Highlights. This very popular feature normally occurs at the start of the annual report, often accompanied by graphs of selected figures. This section provides an at-a-glance summary of selected figures and ratios. In Tesco's report (The Company Camera 12.6), for example, group sales, group profit, profit before tax, earnings per share and dividends per share are the financial ratios highlighted. The financial highlights may be seen as an abridged version of the historical summary.

The Company Camera 12.6

Financial Highlights

Group Sales	UP **11.9%**
Group profit before tax	UP **12.0%**
Earnings per share	UP **11.1%**
Dividend per share	UP **11.2%**

Source: Tesco plc, 2001 Annual Report, p. 1

19. Historical Summary. The historical summary is a voluntary recommendation of the stock exchange. Indeed, it is one of the very few regulations set out by the stock exchange. Usually, companies choose to present five years of selected data from both the balance sheet and profit and loss account.

20. Shareholder Analysis. In many ways this item supplements item 17, shareholder information. It provides detailed analysis of the shareholders, for example, by size of shareholding.

These 20 items are by no means exhaustive, for example, H.P. Bulmer (Holdings) plc, the cider makers, provides an annual statement of corporate objectives.

Presentation

The style of the annual report is becoming much more important. Companies are increasingly presenting key financial information as graphs rather than as tables. This information is voluntary and generally often supplements the mandatory information. Many companies use graphs to provide oases of colour and interest in otherwise dry statutory documents. In many cases, the graphs are presented at the front along with the highlights. Tesco, for example, in 1999 provides five-year graphs of group sales, group operating profit, earnings per share, and operating cash flow and capital expenditure (see The Company Camera 12.7).

The Company Camera 12.7

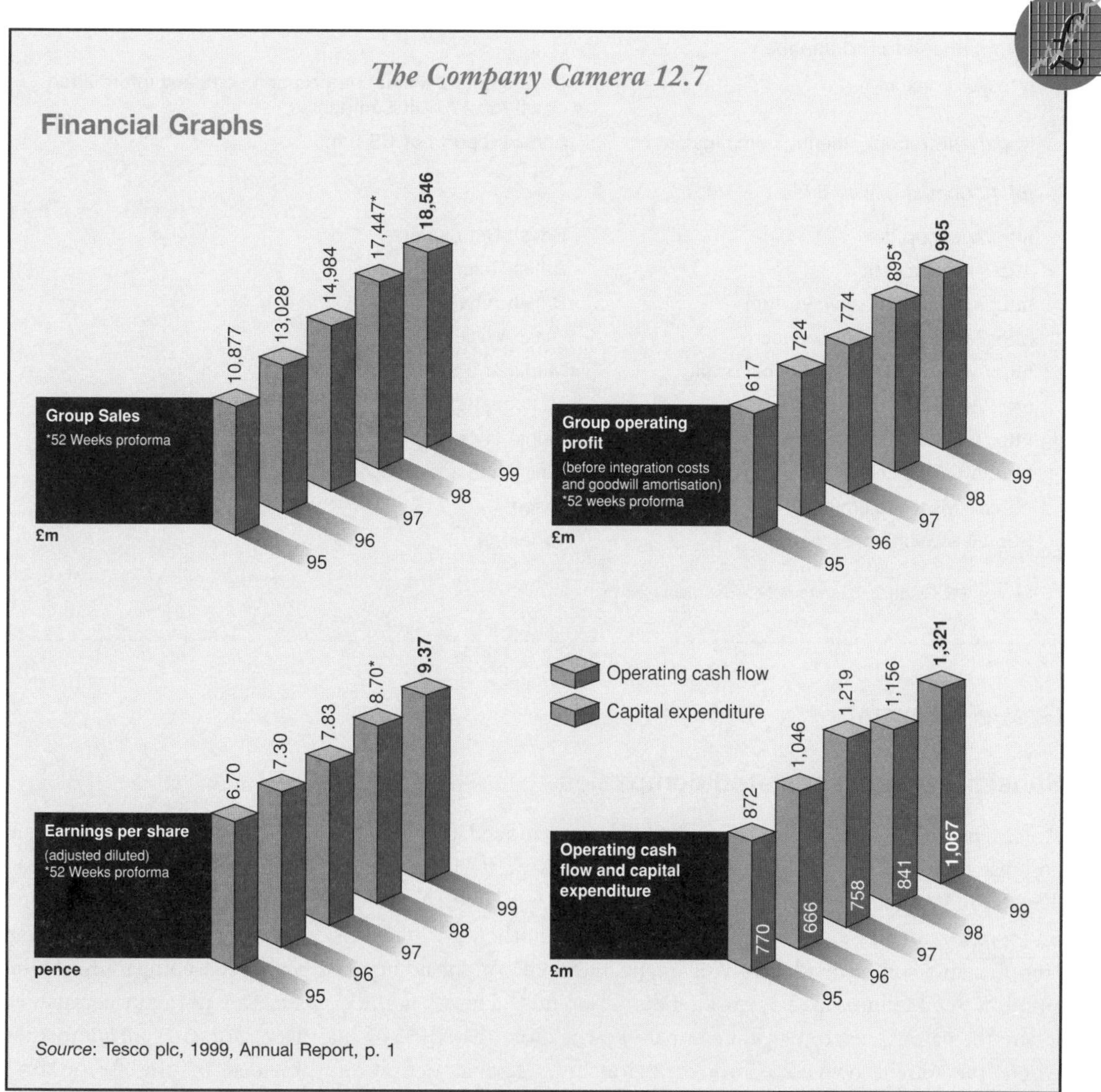

Source: Tesco plc, 1999, Annual Report, p. 1

Photographs are also common in annual reports. These may be of employees or products. However, often they act as 'mood music' with no obvious relationship to the actual content of the report.

The increased use of the Internet provides many opportunities for companies to present their annual reports on their web sites. Many companies are now experimenting with this new presentational format, often using varied presentational methods. Figure 12.4 gives some well-known company Web sites and students are encouraged to visit them.

Figure 12.4 Well-Known Company Web Sites

Internet Address	*Description*
(a) Information on Companies	
http://bized.ac.uk	Case studies, annual reports and company information on all top 100 UK companies
http://www.reportgathering.com.bigaz.htm	Annual reports of US firms
***(b)** 10 Company Web Sites*	
http://www.bp.com	British Petroleum
http://www.bt.com	British Telecom
http://www.british-airways.com	British Airways
http://www.glaxowellcome.com	Glaxo Wellcome
http://www.marks-and-spencer.co.uk	Marks and Spencer
http://www.manutd.com	Manchester United
http://www.sainsburys.co.uk	Sainsbury's
http://www.sb.com	Smithkline and Beecham
http://www.tesco.co.uk	Tesco
http://www.vodafone.co.uk	Vodafone

Source: PC Guide 2: *The Internet Guide*, Mark Goode

Group Accounts

Subsidiary and associated companies

An interesting feature of most of the UK's, and indeed the world's, largest companies is that they are structured as groups. It is the annual reports of these groups of companies such as Vodafone, Toyota or Walmart that are most often publicly available. A group of companies is one where one or more companies is owned or controlled by another. In many cases, these groups are extremely complex and complicated involving many hundreds of subsidiary and associated companies. At its simplest (see Definition 12.3, working definition on the next page), *subsidiaries* are normally companies where the parent (i.e., top group company) owns more than 50% of shares and *associates* are companies where the parent owns 20–50% of shares. The formal definitions provided by the Accounting Standards Board (ASB) and the Companies Act are expressed in considerably more complex language.

Definition 12.3

1. Subsidiary company

Working definition

A company where more than half the shares are owned by another company or which is effectively controlled by another company, or is a subsidiary of a subsidiary.

Formal definition

A subsidiary undertaking is one where the parent undertaking fulfils any of the following:

(a) Holds a majority of the voting rights . . .
(b) Is a member of the undertaking and has the right to appoint or remove directors . . .
(c) Has the right to execute a dominant influence over the undertaking . . .
(d) Controls . . . a majority of the voting rights.
(e) Has a participating interest . . . and . . . actually exercises a dominant influence.
(f) Is the parent undertaking of the subsidiary undertakings of its subsidiary undertakings.

Source: Accounting Standards Board, *Financial Reporting Standard* 2, Subsidiary Undertakings, para.14, adapted and abridged

2. Associated company

Working definition

A company in which 20–50% of the shares are owned by another company or one in which another company has a significant influence.

Formal definition

An 'associated undertaking' means an undertaking in which an undertaking included in the consolidation has a participating interest and over whose operating and financial policy it exercises a significant influence, and which is not:

(a) a subsidiary undertaking of the parent company, or
(b) a joint venture . . .

Where an undertaking holds 20 per cent or more of the voting rights, it should be presumed to exercise such an influence . . .

Source: Companies Act 1985, para. 20, Schedule A, adapted and abridged

Group accounts are prepared using special accounting procedures, which are beyond the scope of this particular book (interested readers could try Elliot and Elliot, *Advanced Financial Accounting and Reporting*, or Alexander and Britton, *Financial Reporting*). In essence, the group profit and loss account and group balance sheet attempt to portray the whole group's performance and financial position rather than that of individual companies.

Thus, in Figure 12.5, the group comprises seven companies. Company A is the parent company and owns over 50% of companies B and C, making them subsidiaries. Company B also owns more than 50% of the shares of companies B1 and B2. Companies B1 and B2 thus become sub-subsidiaries of Company A. These four companies (B, C, B1 and B2) are therefore consolidated as subsidiaries using normal accounting procedures. In addition, an appropriate proportion of companies D and E are taken into the group accounts since these two companies are associates as between 20% and 50% of the shares are held. Overall, therefore we have the aggregate financial performance of the whole group. One group set of financial statements is prepared. It is these group accounts that are normally published in the annual report.

Figure 12.5 Example of Group Structure

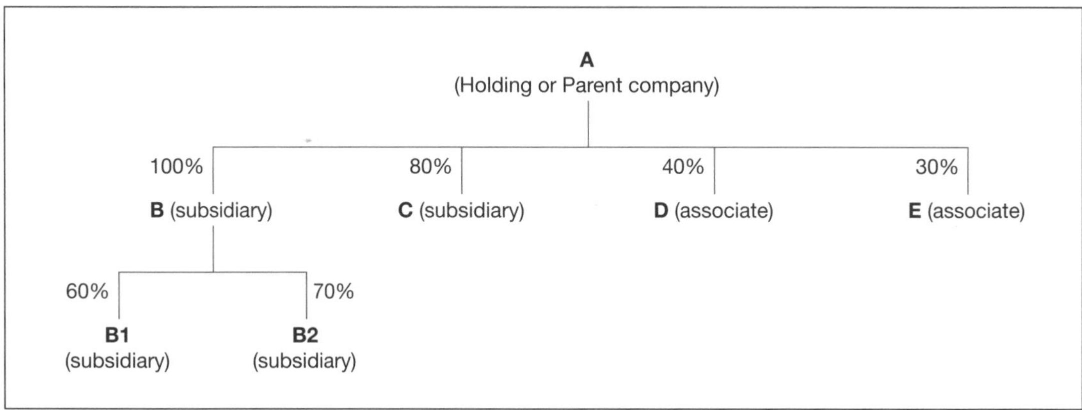

Pause for Thought 12.4

Group Accounts

Nowadays, most leading companies must prepare group accounts. Can you think of any problems they might encounter?

There are many! Many companies will have hundreds of subsidiaries. All the information must be supplied to head office. At head office, it must all be collected and collated. Different subsidiaries may have different year-end accounting dates, operate in different countries using different accounting policies and different currencies. Some of the subsidiaries will have been acquired or sold, or the shareholdings of the parent company will have changed, during the year. All these are potential problems.

Goodwill

Goodwill is a particular feature of group companies. Goodwill, in accounting terms, is only recognised when one company takes over another. It represents the purchase price less the amount paid for the net assets (i.e., the value placed on the earnings power of a business over and above its net assets value). This can often be quite considerable. In other words, when the market values of the real world meet the historical costs of the accounting world there is a valuation gap. That gap is termed goodwill. Financial Reporting Standard 10 recommends that all organisations wishing to give a true and fair view of their financial performance should write off goodwill to the profit and loss account annually over a period of up to 20 years. The effect of this is that goodwill is initially recorded in the balance sheet and then written off to the profit and loss account (see Figure 12.6).

Figure 12.6 **Illustrative Example on Goodwill**

Superactive plc pays £100,000 for £80,000 net assets of a company, Takenover Ltd. Goodwill is written off over 20 years through a deduction of £1,000 annually in the income statement. The financial statements after one year would look like this.

Superactive plc	
Profit and Loss Account	£
Expenses	
Goodwill	1,000
Balance Sheet	
Fixed Assets	
Intangible Assets	
Goodwill (£20,000 – £1,000)	19,000

This process of writing off goodwill is called amortisation. It is similar to depreciation. Organisations are also permitted, if they can make a case, to write off goodwill over a period of greater than 20 years or even not at all.

Impression Management

Managers have significant incentives to try to influence the financial reporting process in their own favour. These incentives may be financial and non-financial. Financially, managers may be keen, for example, to maximise their own remuneration. If remuneration is based on profits, they may seek to adopt accounting policies that will increase rather than decrease profits. In non-financial terms, managers, like all human beings, will try to portray themselves in a good light. This may result, for example, in managers selectively disclosing only positive features of the annual performance.

Pause for Thought 12.5

Impression Management

Can you think of any non-accounting situations where human beings indulge in impression management?

There are many, just to take two: interviews and dating. At an interview, normal, sane candidates try to give a good impression of themselves in order to get the job. This may involve trying to stress their good points and downplay their bad points. When dating, you try to look good to impress your partner. Once more, most normal people will try to present themselves in a favourable light. You want to impress your dates not repel them.

In this section, three illustrative examples of impression management are discussed: creative accounting, narrative enhancement and use of graphs.

Creative Accounting

Creative accounting will be dealt with more fully in Chapter 13. Put simply, creative accounting is the name given to the process whereby managers use the flexibility inherent within the accounting process to manipulate the accounting numbers. Flexibility within the accounting system is abundant. By itself, flexibility allows managers to choose those accounting policies that will give a true and fair view of the company's activities. However, there are also opportunities for managers to choose policies which portray themselves in a good light. The worst excesses are covered by the regulatory framework. To see how the flexibility within accounting can alter profit, we take stock and depreciation as examples.

Stock

In accounting terms

Assets – **Liabilities** = **Capital**

In other words, if assets increase so will capital. As capital, at its simplest, is accumulated profits, if we increase stock we will increase profits. Stock is an easy asset to manipulate, if we wish to increase our profits. We could, for example, do an extremely thorough stocktake at the end of one year, recording and valuing items which normally would have been overlooked.

Pause for Thought 12.6

Stock and Creative Accounting

A company has one asset stock. Its abridged balance sheet is set out below.

	£		£
Stock	*50,000*	*Capital*	*30,000*
		Profit	*20,000*
	50,000		*50,000*

If the company revalues its stock to £60,000, what will happen to profit?

The answer is that profit increases by £10,000, as the new balance sheet shows.

	£		£
Stock	60,000	Capital	30,000
		Profit	30,000
	50,000		60,000

Depreciation

Depreciation is the expense incurred when fixed assets such as plant and machinery are written down in value over their useful lives. Unfortunately, estimates of useful lives vary. For example, if an asset has an estimated useful life of five years then depreciation, using a straightline basis, would be 20% per year. If the asset's estimated useful life was ten years, depreciation would be 10% per year. In other words, by extending the useful life, we halve the depreciation rate and halve the amount that is treated as an expense in the profit and loss account. Managers can thus alter profit by choosing a particular rate of depreciation. They would argue that they are more fairly reflecting the useful life of the asset.

Narrative Enhancement

Narrative enhancement occurs when managements use the narrative parts of the annual report to convey a more favourable impression of performance than is actually warranted. They may do this by omitting key data or stressing certain elements. Many companies stress, for example, their 'good' environmental performance. Indeed, social and environmental disclosures are nowadays exceedingly common. Since such disclosures are voluntary and reviewed rather than audited, there is great potential for companies to indulge in narrative enhancement. This can be seen from Real-Life Nugget 12.4.

Real-Life Nugget 12.4

Environmental Accounting

An interesting example of narrative measurement is given by Craig Deegan and Ben Gordon. They studied the environmental disclosure practices of Australian corporations. The number of positive and negative words of environmental disclosure in annual reports from 1980 to 1991 are recorded for 25 companies. They find:

	1980	1985	1988	1991
Mean positive disclosure	12	14	20	105
Mean negative disclosure	0	0	0	7

They conclude:

> 'The environmental disclosures are typically self-laudatory, with little or no negative disclosures being made by all firms in the study.'

Source: Craig Deegan and Ben Gordon, (1996), *Accounting and Business Research*, p. 198

Graphs

Graphs are a voluntary presentational medium. Used well they are exceedingly effective. However, they also present managers with significant opportunities to manage the presentation of the annual report. For example, a variety of research studies show that managers are exceedingly selective in their use of graphs. They tend to display time-series trend graphs when performance is good, with these graphs presenting a rising trend of corporate performance. By contrast, when the results are poor, graphs are omitted.

Even when included, there is a potential for graphical misuse. For example, graphs may be (and often are) drawn with non-zero axes that enhance the perception of growth. Or graphs may simply be drawn inaccurately. Currently, graphs are not regulated, therefore companies are free to use them creatively.

An interesting example is shown in The Company Camera 12.8 on the next page. These four graphs represent the financial performance of Polly Peck just before the company collapsed into bankruptcy. Although not inaccurately drawn, they do present a very effective, and misleading, display. Who could guess that this company was about to fail?

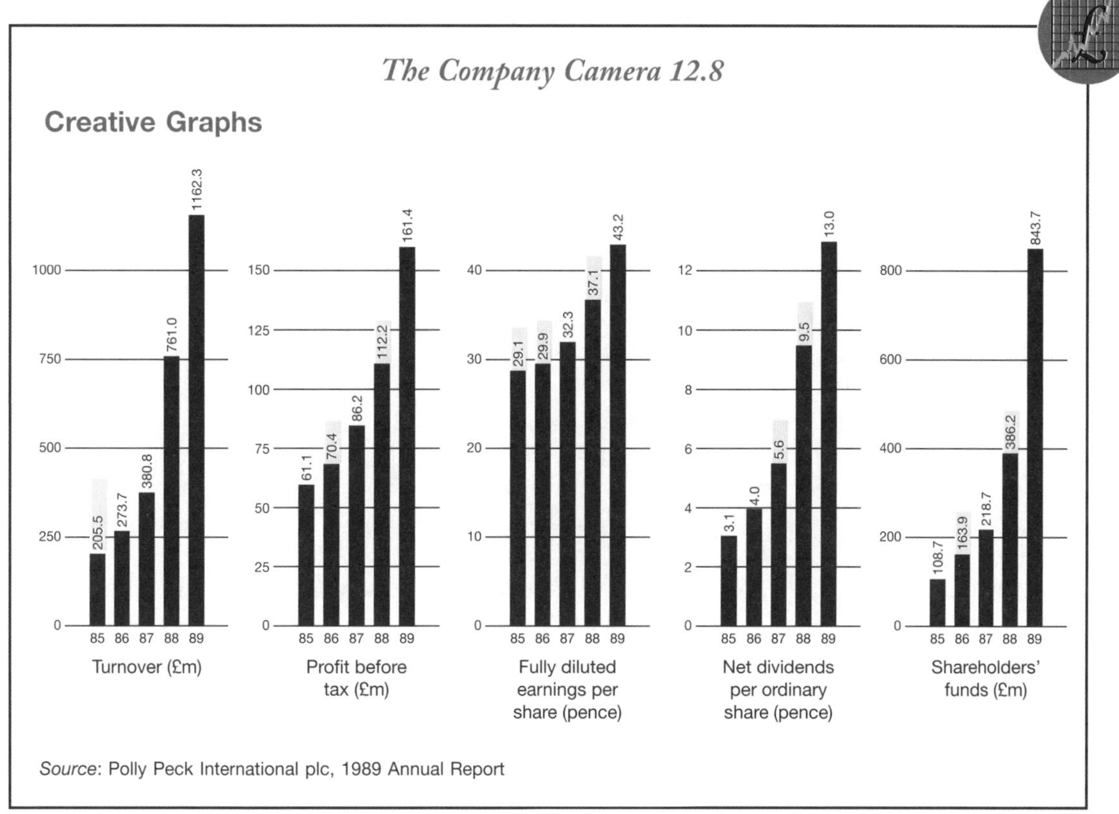

The Company Camera 12.8

Source: Polly Peck International plc, 1989 Annual Report

Conclusion

The annual report is a key part of the process by which managers report to their shareholders. It is central to the corporate governance process. There are three overlapping objectives: stewardship and accountability, decision making and public relations. The annual report is built around three core audited financial statements: the profit and loss account, the balance sheet and the cash flow statement. However, another six important audited financial statements are normally present: statement of total recognised gains and losses; reconciliation of movements in shareholders' funds; note on historical cost profits and losses; accounting policies; notes to the accounts; and principal subsidiaries. Most large companies comprise groups consisting of subsidiary and associated companies. The main accounts are therefore group accounts.

The modern annual report also consists of non-audited sections. These can be divided into narratives and non-narratives. The eight narratives comprise the chairman's statement, the directors' report, the operating and financial review, the review of operations, the statements of corporate governance, the auditors' report, the statement of directors' responsibilities for the financial statements and shareholder information. The three non-narratives are the highlights, the historical summary and the shareholder analysis. As well as these sections, the modern annual report commonly uses graphs and photographs to enhance its presentation. Managements face many incentives to influence the financial reporting process in their favour. This can be done, for example, through creative accounting, narrative enhancement or the use of graphs.

Selected Reading

Unfortunately, there is no one book or article which really covers the modern annual report. Readers are referred to the following three sources which cover some valuable material.

1. Arthur Andersen (1996), '*What's the Story: A Survey of Narrative Reporting in Annual Reports*', London. This publication takes a good look at the rapid growth of the narrative parts of the annual report.
2. Deegan, C. and Gordon, B. (1996), 'A study of environmental disclosure practices of Australian companies', *Accounting and Business Research*, Vol. 26, No. 5, pp. 187–99.
   This article provides an interesting insight into how individual companies accentuate the good news and downplay bad news of their environmental activities.
3. Coopers and Lybrand, *Form and Content of Company Accounts* (1986)
   A good overview of legislation which governs the preparation of the financial statements. A handy reference item. The last important Companies Act was in 1985.
4. McKinstry, S. (1996) 'Designing the annual reports of Burton plc from 1930 to 1994', *Accounting, Organizations and Society*, Vol. 21, No. 1, pp. 89–111.
   This rather heavyweight article looks at how one company's annual reports have changed over time, concentrating particularly on the public relations aspects.

Discussion *Questions*

Questions with numbers in blue have answers at the back of the book.

Q **1** Explain the role that the annual report plays in the corporate governance process.

Q **2** Evaluate the stewardship/accountability, decision-making and public relations roles of the annual report and identify any possible conflicts.

Q **3** In your opinion what are the six most important sections of the annual report. Why have you chosen these sections?

Q **4** Why do companies prepare group accounts?

Q **5** What do you understand by the term 'impression management'? Why do you think that managers might use the annual report for impression management?

Chapter 13

Creative Accounting

'Every company in the country is fiddling its profits. Every set of published accounts is based on books which have been gently cooked or completely roasted. The figures which are fed twice a year to the investing public have all been changed to protect the guilty. It is the biggest con trick since the Trojan Horse.'

Ian Griffiths (1986), *Creative Accounting*, Sidgwick and Jackson, p. 1

Learning Outcomes

After completing this chapter you should be able to:

- ✔ Explain the nature of creative accounting.
- ✔ Outline the managerial incentives for creative accounting.
- ✔ Demonstrate some common methods of creative accounting.
- ✔ Understand the real-life relevance of creative accounting.

In a Nutshell

- Creative accounting involves managers using the flexibility within accounting to serve their own interests.
- The regulatory framework tries to ensure that accounts correspond to economic reality.
- Management will indulge in creative accounting, *inter alia*, to flatter profits, smooth profits or manage gearing.
- Income, stock, depreciation, interest payable, goodwill and brands can all be managed creatively.
- In extreme cases, such as Polly Peck, WorldCom or Enron creative accounting can contribute to bankruptcy.
- Several well-known publications, in particular, *Accounting for Growth*, have documented actual cases of creative accounting.
- Companies can also creatively manage the published version of their accounts through, for example, creative graphics.
- Regulators, such as the Accounting Standards Board, try to curb creative accounting. This creates a creative accounting 'arms race'.

Introduction

Creative accounting became a hot topic in the late 1980s. Attention was drawn, by commentators such as Ian Griffiths (author of *Creative Accounting*), to how businesses use the flexibility inherent in accounting to manage their results. In itself, flexibility is good because it allows companies to choose accounting policies that present a 'true and fair view'. However, by the judicious choice of accounting policies and by exercising judgement, accounts can serve the interests of the preparers rather than the users. Creative accounting is not illegal but effectively, through creative compliance with the regulations, seeks to undermine a 'true and fair view' of accounting. Creative accounting can involve manipulating income, expenses, assets or liabilities through simple or exceedingly complex schemes. The current regulatory framework can partially be seen as a response to creative accounting. It attempts to ensure that accounting represents economic reality and presents a true and fair view of the company's activities. However, new regulations bring new opportunities for creative compliance and thus creative accounting. As Real-Life Nugget 13.1 shows, even well-known companies are accused of questionable accounting. Enron, once the seventh biggest US company, which went into liquidation in 2001 is believed to have indulged in creative accounting. Other US companies such as WorldCom have been involved in accounting scandals in which creative accounting has played a contributory role.

Real-Life Nugget 13.1

Microsoft and Cookie Jar Accounting

The Securities and Exchange Commission (SEC), which has been cracking down on so-called 'cookie-jar' accounting, has mounted a probe of Microsoft's accounting for financial reserves . . .

The SEC customarily does not comment on its investigations. Microsoft, however, in an apparent attempt to prevent bad publicity and any negative effect on its stock price, recently revealed the existence of the probe in a conference call with analysts and reporters . . .

. . . Cookie-jar accounting is the practice of hiding assets in reserves when times are good so that they can be used as a fallback when times are bad. It is not illegal as such but the SEC is, nevertheless, adamant that there are limits beyond which a company cannot go. Specifically, the SEC has been targeting questionable accounting for restructuring charges and restructuring reserves. "Some companies like the idea so much that they establish restructuring reserves every year", said Walter Schuetze, Chief of the SEC's enforcement division, in a recent speech.

Source: J.R. Peterson, *The Accountant*, July 1999, p. 1, 'Microsoft faced with SEC accounting probe'.

Helpnote: Cookie jar accounting is so called because you are hiding assets away when times are good, ready to use them when times are bad. It is thus 'saving up for a rainy day'.

Definition

Creative accounting is a slippery concept, which evades easy definition. As Definition 13.1 shows, there are perhaps three key elements in creative accounting: flexibility, management of the accounts and serving the interests of managers.

> ### *Definition 13.1*
>
> **Creative Accounting**
>
> **Working definition**
> Using the flexibility within accounting to manage the measurement and presentation of the accounts so that they serve the interests of the preparers.
>
> **Formal definition**
> 'A form of accounting which, while complying with all regulations, nevertheless gives a biased impression (generally favourable) of the company's performance.'
>
> *Source*: Chartered Institute of Management Accounting (2000), *Official Terminology*

(i) Flexibility

Accounting is very flexible. There are numerous choices, for example, for measuring depreciation, valuing stock or recording sales. This flexibility underpins the idea that the financial statements should give a 'true and fair view' of the state of affairs of the company and of the profit. Accounting policies should thus, in theory, be chosen to support a true and fair view. In many cases they are, but the flexibility within accounting does sometimes enable managers to present a more favourable impression of the company's performance than is perhaps warranted. Indeed, within accounting there is a continuum (see Figure 13.1).

Figure 13.1 Flexibility within Accounting

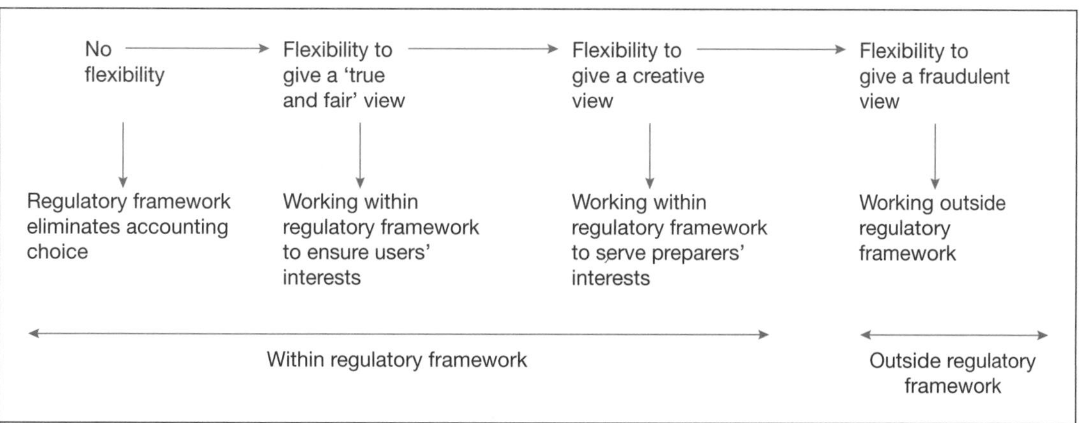

This continuum starts with a completely standardised accounting system. This gives way to flexibility so as to present a true and fair view. Next we have flexibility to account creatively. Finally, there is fraud, which involves non-compliance with the regulations rather than 'bending' them. In Soundbite 13.1, Zero Mostel seems to be urging Bloom to adopt practices that appear more fraudulent than creative. In the US, for example, WorldCom has been accused of fraudulently overstating profits by $3.8 bn (*Financial Times*, June 26, 2002, p. 1).

Soundbite 13.1

Creative Accounting or Fraud

'Bloom, do me a favor move a few decimal points around. You can do it. You're an accountant. You're a noble profession. The word 'count' is part of your title.'

Zero Mostel in *The Producers* by Mel Brooks.

Source: *The Executive's Book of Quotations* (1994), p. 3

(ii) Management of the accounts

Unfortunately, in practice, the directors may choose accounting policies more to fulfil managerial objectives than to satisfy the requirements of users for a 'true and fair' view. Accounting thus becomes a variable to be managed rather than an instrument for providing true and fair information.

(iii) Interests of managers

Accounting theory suggests that the aim of the accounts is to provide financial information to users so that they can make decisions. However, creative accounting privileges the interests of managers. Users *may*, indeed, benefit from creative accounting, but managers *will* definitely benefit.

Pause for Thought 13.1

Users' and Preparers' Interests

How do you think the interests of regulators, users and preparers might conflict?

The aim of the regulators is that the financial statements should provide a 'true and fair view' of the accounts. This involves concepts of neutrality, lack of bias and faithfully representing externally reality. In theory, users, such as shareholders, are likely to support this aim. Preparers, by contrast, are likely to wish to manage the accounts in their own interests. Preparers may thus indulge in reporting strategies, such as profit smoothing or flattering profits, which keep share prices high. Interestingly, existing shareholders may support these strategies since they benefit from them. However, other users and potential shareholders will be less happy.

Managerial Motivation

Managers have incentives to adopt creative accounting. Essentially, managers are judged, and rewarded, on the performance of their companies. It is, therefore, in managers' interests that their companies meet expectations. For example, if managers have profit-related bonuses, it makes sense for them to maximise their profits so they get bigger bonuses. Other managerial incentives might be the ownership of shares and share options, the need to smooth profits or to manage gearing.

Shares and Share Options

Managers may own shares. They also may have share options which allow them to buy shares today at a set price and then sell them for a higher price at a future date. If the stock market expects a certain amount of profit from a company, then managers may wish to adopt creative accounting to deliver that profit. Otherwise, the share price will fall and the managers will lose out.

Profit Smoothing

The stock market prefers a steady progression in earnings to an erratic earnings pattern. Companies with erratic earnings typically have lower share prices than those with steadier performances. Their lower share price makes these companies more vulnerable to takeover than companies with smoother profit trends. Managers of companies that are taken over may lose their jobs. Therefore, managers have incentives to smooth profits. Current shareholders are also likely to benefit from profit smoothing as the share price remains high.

Pause for Thought 13.2

Profit Smoothing

Two firms (A and B) in the same industry have the following profit trends. Which do you think might be favoured by the stock market?

Years	*1*	*2*	*3*	*4*
	£m	*£m*	*£m*	*£m*
A	*1*	*2*	*4*	*8*
B	*4*	*(1)*	*15*	*(3)*

At first glance, A looks the better bet. Its profits steadily rise, doubling each year. However, company B, whose profit are irregular, actually makes the same cumulative profits as A (i.e., £15m). Overall, the stock market will probably favour A with its steady growth. Indeed, in year 4, company A might even be able to make a successful bid for B! Company A's share price at that date will probably be much higher than company B's.

Manage Gearing

As well as managing profits, companies have incentives to manage gearing or conceal debt. Companies may wish to borrow money. However, existing borrowers may put restrictions on the amount of any new debt that can be raised. Managers may attempt to circumvent these restrictions using creative accounting. Enron, for instance, attempted to conceal the amount of debt it possessed.

The managerial incentives vary from firm to firm. For example, in some regulated industries (such as gas or water), managers may actually wish to reduce sales and profits so as to stop the government putting price restrictions on them. However, the key point about managerial incentives is that they encourage managers to serve their own or the company's interests rather than present a true and fair view of the company's performance.

Soundbite 13.2

Restating your Figures

'The easiest way to do a snow job on investors (or on yourself) is to change one factor in the accounting each month. Then you can say, "It's not comparable with last month or last year. And we can't really draw any conclusion from the figures".'

Robert Townsend, *Up the Organization* (Knopf, 1970).

Source: *The Wiley Book of Business Quotation* ((1998), p. 89

Methods of Creative Accounting

There are innumerable methods of creative accounting. These arise mainly from the flexibility of accounting and the existence of so many acceptable accounting policies. If you change your accounting policies, you will change your results. Indeed, as Robert Townsend suggests in Soundbite 13.2, one easy way of creative accounting is continually to change your accounting policies. In the US, WorldCom, which collapsed in 2002, was accused of repeatedly restating its accounts. However, the consistency concept does to some extent limit companies' abilities to do this.

Many creative accounting procedures are complex and often undetectable to the analyst. In this section, we will look at six of the more straightforward techniques. The aim is to give some illustrative examples rather than to present a comprehensive list. The worst excesses of creative accounting have been curbed by accounting rules and regulations that have been developed to try to ensure that the accounts correspond to economic reality. In many cases, some of the techniques listed will be used by management taking advantage of accounting's flexibility to give a true and fair view of the company's activities. In other cases, these techniques will be used creatively. To the onlooker it is generally difficult to distinguish these two contrasting uses. Motivation is the key distinguishing feature. Where managers attempt to serve their own interests rather than present a true and fair view, creative accounting is occurring.

Those readers who wish a more in-depth look at the subject are referred to Ian Griffiths, *New Creative Accounting*, Terry Smith's *Accounting for Growth*, or Kamal Naser's *Creative Financial Accounting*.

(i) Inflating Income

When should we recognise a sale as a sale? This may, at first, seem a silly question. However, the date of sale is not always obvious. For example, is it when (i) we dispatch goods to a customer, (ii) we invoice the customer, or (iii) we receive the money? Normally, it is when we invoice the customer. However, in complex businesses there is often a fair degree of latitude about sales recognition. The problem is that sales recognition is not as precise as cash flow (see Real-Life Nugget 13.2). If you take a big construction project (like John Laing's building of the Cardiff Millennium Rugby Stadium), for example, when should you recognise sales and take profits? There are rules to help in profit determination, but these rules still permit a good deal of flexibility.

Another troublesome area associated with income recognition is warranty provision (i.e. setting aside money to deal with customer returns). You can deal with this, in advance, and estimate a provision or deal with it on an actual return-by-return basis. Moreover, warranties can be treated as a reduction in sales or as an expense. The differing treatments can result in differing profits.

Real-Life Nugget 13.2

Income Recognition

'Once it is accepted that actual cash flows do not present a true and fair view of the company's performance, then the door marked creativity is pushed wide open. As long as a company can justify with a degree of reasonableness that its income recognition policy is soundly based then it has *carte blanche* to do pretty much what it likes.'

Source: Ian Griffiths (1995), *New Creative Accounting*, p. 16

Pause for Thought 13.3

Income Recognition

I was once involved in auditing a company selling agricultural machinery like combine harvesters. The company's year end was 30 June. Farmers wanted the machinery invoiced in March, but would pay in August when the machinery was delivered. Why do you think the farmers wished to do this and what income recognition policy was best for the company?

Essentially, this arrangement benefited both the company and the farmers. The company would take its sales in March, arguing this was the invoice date. The sales thus appeared before the June year end. The farmer would treat the purchase in March so that they could set off the machinery against taxation. As the tax year runs from April 6 to April 5, the farmers would hope to receive the capital allowances in one tax year, and pay for the machinery in the next tax year. Everybody was happy.

(ii) Stock

Stock provides a rich area for the creative accountant. The key feature about stock is that if you increase your stock you increase your profit (see Figure 13.2 below). The beauty of stock is also that, in many businesses, stock is valued once a year at an annual stock-take. When carrying out these stock-takes, it is relatively easy to take an optimistic or a pessimistic view of the value of stock.

Figure 13.2 Stock as an Example of Creative Accounting

If there is only one asset, stock, worth say £10 million, then if capital is £5 million and this year's profit is £5 million, then we have balance sheet A:

Balance Sheet A

	£m		£m
Capital	5	Stock	10
Profit	5		
	10		10

The company could:
(1) adopt a more generous stock valuation policy, perhaps by lowering the provisions for obsolete stock (increases stock by £0.5 million),
(2) do a particularly rigorous stock-take (increases stock by £1.0 million undiscovered stock). The balance sheet now looks like balance sheet B.

Balance Sheet B

	£m		£m
Capital	5.0	Stock	11.5
Profit	6.5		
	11.5		11.5

Hey presto! We have increased our profits by £1.5 million.

(iii) Depreciation

Depreciation is the allocation of the cost of fixed assets over time. It is an expense recorded in the profit and loss account. If the amount of depreciation changes then so will profit. The depreciation process is subject to many estimations, such as the life of the asset, which may alter the depreciation charge. A simple example is given in Figure 13.3 below. In essence, lengthening expected lives boosts profits while reducing them reduces profits. If management lengthens the assets' lives because it judges that the assets will last longer, this is fair enough. If the motivation, however, is to boost profits then this is creative accounting.

Figure 13.3 Depreciation

A business has a profit of £10,000. It has £100,000 worth of fixed assets. Currently it depreciates them straight line over 10 years. However, the company is thinking of changing its depreciation policy to 20 years straight line. Will this affect profit?

	Original policy	New policy
	£	£
Profit before depreciation	10,000	10,000
Depreciation	(10,000)	(5,000)
Profit after depreciation	–	5,000

The answer is yes. By changing the depreciation policy, profit has increased by £5,000. In fact, the company looks much healthier.

The pace of technological change creates shorter asset lives. It would be assumed, therefore, that most companies would reduce their expected asset lives. However, as UBS Phillips and Drew point out below, in Real-Life Nugget 13.3, this is not necessarily so.

Real-Life Nugget 13.3

Change of Depreciation Lives

'Most changes in depreciation policies tend to be a lengthening of the expected lives rather than a shortening. Indeed it is difficult to identify any UK company which has recently reduced the depreciation life of any of its assets which has been of significance in terms of depressing reported profits. This is despite a general perception that many mechanical assets' true life expectations are shortening.'

Source: UBS Phillips and Drew (1991), *Accounting for Growth*, p. 11

Finally, Smith (1992) draws attention to British Airports Authority's (BAA) decision to lengthen its terminal and runway lives. Runway lives lengthened from 23.5 years in 1990 to 100 years in 1998. Annual depreciation was thus reduced.

(iv) Capitalisation of Costs such as Interest Payable

The capitalisation of costs involves the simple idea that a debit balance in the accounts can either be an expense or an asset. Expenses are deducted from sales and reduce profit. Assets are capitalised. Fixed assets are particularly important in this context. Only the depreciation charged on fixed assets is treated as an expense and reduces profit. Therefore, it will often benefit companies to treat certain expenses as fixed assets.

An example of this is interest costs. Where companies borrow money to construct fixed assets, they can argue, and often do, that interest on borrowing should be capitalised. However, some commentators such as Phillips and Drew find this a dubious practice (see Real-Life Nugget 13.4). Indeed, Phillips and Drew point out that some UK companies would actually make a loss, not a profit, if they did not capitalise their interest.

> *Real-Life Nugget 13.4*
>
> **Capitalisation of Costs**
>
> 'Virtually every UK listed property company, with the notable exception of Land Securities, makes use of capitalised interest (and often other costs as well) to defer the P&L impact of developments. While commercial property prices were rising rapidly, investors and banks did not worry about the amount of interest being capitalised. However, in more difficult property market conditions, such as those at present, the substantial difference between profits and cash flow caused by the significant capitalisation of costs can become critical in investment terms.
>
> Often, there is clear justification for capitalising interest when it relates to a [sic] asset being constructed for use in the company's line of business, such as a supermarket, an aircraft or a new factory. The justification becomes much less convincing when the asset is being built for sale, such as property development and house building, for instance.'
>
> *Source*: UBS Phillips and Drew (1991), *Accounting for Growth*, p. 10

(v) Goodwill

When one company acquires another, goodwill is created. Goodwill is the surplus of the amount paid for the acquired company over the fair value of the net assets acquired. One of the key problems with goodwill is its accounting treatment. Companies face a choice. They can either argue that goodwill is an asset whose value will not decline in the future and capitalise it on the balance sheet without amortisation (amortisation is an accounting term, rather like depreciation, meaning writing an amount off each year to the profit and loss account). Or, alternatively, they can amortise it over a period of up to 20 years to the profit and loss account. The amounts of goodwill can often be enormous; therefore, the choice of accounting treatment can often have a significant effect on profit. This is demonstrated in Real-Life Nugget 13.5. This takes the accounts of The Great Universal Stores, a UK company, and demonstrates how by choosing different goodwill policies one can arrive at different balance sheets and profit and loss accounts.

Real-Life Nugget 13.5

Goodwill – The Case of The Great Universal Stores (1997)

	Option 1 Write-off reserves Old UK (pre-1998)/Germany/ Netherlands)	*Option 2 Not write off (New UK)*	*Option 3 40 years (US)*	*Option 4 20 years amortise (New UK/IASB)*	*Option 5 5 years (Japan/Ne-therlands)*	*Option 6 4 years (Germany)*
Balance Sheet	£m	£m	£m	£m	£m	£m
Net Assets	2,859	2,859	2,859	2,859	2,859	2,859
Goodwill	–	982	957	933	786	736
Total net assets	2,859	3,841	3,816	3,792	3,645	3,595
Index	100	134	133	133	127	126
Profit and Loss Account	£m	£m	£m	£m	£m	£m
Profit for financial year	379	379	379	379	379	379
Goodwill amortisation	–	–	(25)	(49)	(196)	(246)
Net Profit	379	379	354	330	183	133
Index	100	100	93	87	48	35
Return on capital employed	379 / 2,859 = 13.3%	379 / 3,841 = 9.9%	354 / 3,816 = 9.3%	330 / 3,792 = 8.7%	183 / 3,645 = 5.0%	133 / 3,595 = 3.7%
EPS	37.6p	37.6p	35.0p	32.7p	18.0p	13.2p
Index	100	100	93	87	48	35

Basic facts for year ended 1997:		
	Net Assets	£2,859m
	Goodwill	£982m
	Profit for Financial Year	£279m
	EPS	37.6p

Notes:
1. Countries are always changing their accounting regulations. This example was correct for 1997.
2. Where alternative policies are permitted countries are listed under multiple columns. Calculations use normal maximum length (in years) specified in national legislation. In some cases, countries permit longer periods.
3. Remember total net assets equals capital employed.

Source: Adapted from Michael Jones (1999), 'Goodwill an International Bugbear', *Accountancy International*, March, p. 77

For example, by using option 1 you can show a net profit of £379 million and obtain a return on capital employed of 13.3%. However, option 6 would give a profit of only £133 million and a return on capital employed of only 3.7%. All the options shown are allowed somewhere in the world. In the UK, options 2, 4, 5 and 6 are permitted. Again if the policy is specifically chosen because the management believe it reflects economic reality, this is fine. However, there is the possibility that companies will choose their goodwill policy with one eye on delivering an appropriate profit figure.

(vi) Brands

Brand valuation is a contentious issue within accounting. Some companies argue that it is appropriate to value brands so as better to reflect economic reality. By contrast, other observers believe that valuing brands is too subjective and judgemental and that the real motive behind brand valuation is for companies to boost asset values on the balance sheet. Brand accounting is a relatively new phenomenon in the UK. The idea is that, in many cases, brands are worth incredible amounts of money (think, for example, of Guinness or Kit Kat). Indeed, it is estimated that Coca Cola's brand name adds £2bn additional value to the company per year (Fiona Gilmore, *Accountancy Age*, May 2001).

Traditionally, brands have not formally been recognised as assets. However, from the mid-1980s, UK companies such as Grand Metropolitan and Cadburys started to include brands in their balance sheets. These brands are an asset and help to boost the assets in the balance sheet. They are *not* amortised (i.e. written off).

Traditional accountants are still suspicious of brands. This is because they are difficult to measure and, in essence, they are subjective. The situation in the UK is, therefore, something of an uneasy compromise. Acquired brands can be capitalised (i.e., included in the balance sheets). However, companies are not permitted to capitalise internally-generated brands. Overall, some companies capitalise their brands and some do not. There is also a great variety of ways in which brands are valued. In other words, there is great potential for creativity.

Soundbite 13.3

Grasping Reality

'Some have suggested that Rolls-Royce accounts are fine because they meet with UK GAAP accounting rules, but let's remember that Enron complied with US rules. The question is whether you get a grasp of reality from the accounts, and I don't know that you do.'

Source: *The Guardian*, 12 February 2002, p. 23, Terry Macalister quoting Terry Smith of Collins Stewart brokers.

Example

In order to demonstrate that creative accounting can make a difference, an example, Creato plc, is presented in Figure 13.4 on the next page. Adjustments are made for income, stock, depreciation, capitalisation of interest and goodwill.

Figure 13.4 Example of Creative Accounting

Creato plc
Profit and Loss Account for Year Ended 31 December 2001

	Notes	£000	£000
Sales	1		100
Less *Cost of Sales*			
Opening stock		10	
Add Purchases		40	
		50	
Less Closing stock	2	15	35
Gross Profit			65
Less *Expenses*			
Depreciation	3	12	
Interest payable	4	15	
Goodwill	5	8	
Other expenses		35	70
Retained Loss			(5)

Notes
1. Creato has a prudent income recognition policy, a less conservative one would create an additional £10,000 sales.
2. Closing stock could be valued, less prudently, at £18,000.
3. Depreciation is charged over five years; a few competitors charge depreciation over ten years even though this is longer than the realistic expected life.
4. £10,000 of the interest payable is interest on borrowings used to finance a new factory.
5. The company has amortised the goodwill over ten years. This period is realistic. It has definitely decided it cannot argue for the goodwill having an indefinite life. The goodwill arose on a purchase this year.

If we indulge in a spot of creative accounting we can transform Creato plc's profit and loss account.

Creato plc
Profit and Loss Account for Year Ended 31 December 2001

		£000	£000
Sales	1		110
Less *Cost of Sales*			
Opening stock		10	
Add Purchases		40	
		50	
Less Closing stock	2	18	32
Gross Profit			78
Less *Expenses*			
Depreciation	3	6	
Interest payable	4	5	
Depreciation on capitalised interest payable	4	1	
Goodwill	5	4	
Other expenses		35	51
Retained Profit			27

Notes
1. We can simply boost sales by £10,000 and be less conservative.
2. If we value closing stock at £18,000, this will reduce cost of sales, thus boosting gross profit.
3. By doubling the life of our fixed assets, we can halve the depreciation charge.
4. If we have borrowed the money to finance fixed assets, then we can capitalise some of the interest payable. Interest payable thus reduces from £15,000 to £5,000. We assume here that we will then depreciate this capitalised interest over ten years (this company's new policy for fixed assets). Thus, we are charging £1,000 depreciation on the capitalised interest payable. We thus boost profit by £9,000 (i.e., £10,000 saved less £1,000 extra depreciation).
5. We have chosen to amortise our goodwill not capitalise it. If we chose to capitalise it we would charge no goodwill at all to the profit and loss account. However, why not adopt a twenty-year life for goodwill, thus halving our goodwill charge.

Hey presto! We have transformed a loss of £5,000 into a profit of £27,000.

Real Life

It should be stressed that creative accounting is very much a real-life phenomenon. Extensive research has demonstrated its existence. Of particular interest are two empirical studies: *Accounting for Growth* and *Company Pathology*. Although dating from the 1990s, these studies, which have not been repeated more recently, demonstrate quite clearly the existence of creative accounting. *Accounting for Growth* was published twice: first, by fund managers, UBS Phillips and Drew in 1991 as a report and second by Terry Smith in 1992 as a book. It caused considerable controversy – in fact, it resulted in Terry Smith, one of the analysts responsible for the research, leaving UBS Phillips and Drew. However, given the active attention paid by the Accounting Standards Board to curbing creative accounting since the 1990s, many of the worst abuses have been curtailed. Essentially, as Real-Life Nugget 13.6 shows, UBS Phillips and Drew wished to draw attention to the recent growth in creative accounting.

Real-Life Nugget 13.6

Growth in Innovative Accounting

'The last ten years has been a time of major innovation as far as accounting techniques are concerned. Complex adjustments relating to acquisitions, disposals and many other transactions have become commonplace with the ultimate goal being to report continuous growth in earnings per share. The task now facing fund managers is to cut through the accounting camouflage in order to interpret the underlying trends. The penalty for getting the analysis wrong is the risk of substantial share price underperformance and in extreme cases, total loss.'

Source: UBS Phillips and Drew (1991), *Accounting for Growth*, p. 1

Pause for Thought 13.4

Accounting for Growth

In their report, UBS Phillips and Drew identified the innovative accounting practices used by 185 UK companies. Why do you think this caused such a storm?

Before *Accounting for Growth* was published, there was much speculation about creative accounting. However, there was little systematic evidence. The Phillips and Drew report identified 165 leading companies (out of 185 they investigated) which had used at least one innovative accounting practice. It named names! The speculation turned into reality. The companies named were unhappy. As some of them were clients of UBS Phillips and Drew, some of the companies felt let down. The result of the storm was that Terry Smith left UBS Phillips and Drew and published his book, *Accounting for Growth*, on his own.

UBS Phillips and Drew analysed 185 UK listed companies. They identified 11 innovative (i.e., creative) accounting practices and drew up an accounting health check. They found that 165 companies used at least one innovative accounting practice, 17 used five or more and three used seven (see Real-Life Nugget 13.7). Interestingly, two of the high-scoring companies subsequently went bankrupt: Maxwell Communications and Tiphook.

Real-Life Nugget 13.7

High Scores in Health Check

Companies using the most accounting techniques in Phillips and Drew's 'Health Check'

Company	*Sector*	*Frequency*
British-Aerospace	Engineering	7
Maxwell	Media	7
Burton Group	Stores	7
Dixons	Stores	6
Cable and Wireless	Telephone networks	6
Blue Circle	Building materials	5
TI Group	Engineering	5
Bookers	Food manufacturing	5
Asda	Food retailer	5
Granada	Leisure	5
Next	Stores	5
Sears	Stores	5
LEP	Business services	5
Laporte	Chemicals	5
British Airways	Transport	5
Tiphook	Transport	5
Ultramar	Oils	5

Source: M.J. Jones (1992), Accounting for Growth: Surviving the Accounting Jungle, *Management Accounting*, February, p. 22

In *Company Pathology*, County Natwest Woodmac studied 45 'deceased' companies from 1989–90. They drew attention to questionable accounting practices, such as the capitalisation of interest. In only three out of the 45 cases did the audit report warn of the impending disaster. In only two out of the 45 cases did the pre-collapse turnover fall. Finally, in only six out of 45 cases did the last reported accounts show a loss. As Real-Life Nugget 13.8 on the next page shows, the report was often quite scathing.

Real-Life Nugget 13.8

Analysis of Failed Companies

'A downturn in earnings per share is a lagging rather than a leading indicator of trouble. Accounting Standards give companies far too much scope for creative accounting. One set of accounts were described by an experienced and well qualifed fund manager as "*a complete joke*". Auditors' reports seldom give warning of impending disaster.'

Source: County NatWest Woodmac (1991), *Company Pathology*, p. 4

Case Studies

(i) Polly Peck

An interesting example of a spectacular company collapse where creative accounting was present is Polly Peck. Polly Peck's demise is well-documented not only in Terry Smith's *Accounting for Growth*, but also in Trevor Pijper's *Creative Accounting: The Effect of Financial Reporting in the UK.*

Essentially, Polly Peck was one of the fastest expanding UK firms in the 1990s. It was headed by a charismatic chairman, Asil Nadir. It started off in food and electronics, and expanded rapidly. County NatWest Wood Mackenzie calculated that in the year to November 1989, Polly Peck's shares had grown faster than any other UK company. The 1989 results were full of optimism. For example, profit before tax had increased from £112 million to £161 million.

Indeed, the optimism was continued in the 1990 interim results. Eleven days after their publication, and six day before the collapse, Kitcat and Aitken (city analysts) reported that profits would increase substantially.

The collapse of Polly Peck cannot be attributed solely to creative accounting. There was fraud and deception and, in addition, there was the blind faith of bankers and shareholders. The signs were there for those who wished to look. For example, debt rose from £65.9 million in 1985 to £1,106 million in 1989. However, creative accounting did play its part in two key areas. First, Polly Peck capitalised the acquired brands, such as Del Monte, and thus strengthened its balance sheet. Second, and more seriously, the company indulged in currency mismatching. As explained in Figure 13.5 this flattered Polly Peck's profit and loss account at the expense of seriously damaging its balance sheet. The brand valuation, in effect, helped to boost a seriously depleted balance sheet.

Figure 13.5 Polly Peck's Currency Mismatching

Polly Peck borrowed in Swiss francs, a strong currency, and paid back a low rate of interest. These borrowings were invested in Turkey, which had a weaker currency. Polly Peck was paid a high rate of interest. Unfortunately, however, the Turkish dinar depreciated against the Swiss franc. This meant that Polly Peck made capital losses of £44.7 million in 1989 on its borrowings. Meanwhile, its profit and loss had been flattered by £12.5 million (£68.1 million received less £55.6 million paid) because of the high rate of interest on the dinar deposits as compared with the low interest on the matched Swiss franc borrowings.

(ii) Enron

The spectacular collapse of Enron in December 2001 has brought creative accounting once more back on the agenda, centre stage. Enron, at one time, was the seventh biggest US company. However, from August 2000 its share price began to fall as a result of doubts about the strength of its balance sheet and significant sales of shares by managers. Enron's main business was to supply and make markets in oil and gas throughout the world. Enron first made gains on investments in technology and energy businesses followed by losses. Following these losses, Enron built up huge debts which have been estimated at $80 billion. From the accounts it was not obvious that these liabilities existed. They were buried in rather complex legal jargon (see The Company Camera 13.1).

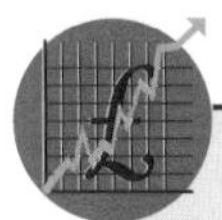

The Company Camera 13.1

Extract from Enron's Notes to the Accounts

Enron is a guarantor on certain liabilities of unconsolidated equity affiliates and other companies totalling approximately $1,863 million at December 31, 2001, including $538 million related to EOTT trade obligations [EOTT Energy Partners]. The EOTT letters of credit and guarantees of trade obligation are secured by the assets of EOTT. Enron has also guaranteed $386 million in lease obligations for which it has been indemnified by an "Investment Grade" company. Management does not consider it likely that Enron would be required to perform or otherwise incur any losses associated with the above guarantees . In addition, certain commitments have been made related to capital expenditures and equity investments planned in 2001.

Source: Enron, Annual Report (2000), Notes to the Accounts, p. 48

In order to manipulate income, to avoid reporting losses and keep its debts off the group's balance sheet Enron set up special purpose entities (SPEs). Under US regulations if the SPEs were not controlled by Enron and if outside equity capital controlled at least 3% of total assets then Enron would not have to bring the SPEs into its group accounts. It would thus not disclose its debts. Investors would not therefore realise the net indebtedness of the company. Many SPEs appear not to have been incorporated into the group accounts quite legally. However, in other cases it is alleged that control was held by Enron not by third parties and that Enron had provided third parties with funds so that the 3% was not truly held independently. If this is the case, there was prima facie false accounting and questions began to be raised about the role of Arthur Andersen, the company's auditors. Arthur Andersen have also been accused of shredding documents relating to Enron. The role of the auditors as independent safeguards of the investors appears to have broken down. Finally, it has been reported in the press that Enron was receiving up-front payments for the future sales of natural gas or crude oil and effectively treating loans as sales.

Overall, therefore, Enron demonstrates that even well-known companies still indulge in creative accounting. In Enron's case, however, like that of Polly Peck, the borderline between creative accounting and fraud became blurred.

Creative Presentation

As well as creative accounting, companies can present their accounts in a flattering way. One of the ways they can do this is by using graphs. For example, they may use graphs only in years when the company has made a profit (the graph will show a rising trend). Alternatively, graphs may be deliberately drawn so as to exaggerate a rising trend. This may, for example, be done by using a non-zero axis. An example of creative graphical presentation is shown in Real-Life Nugget 13.10. In particular, it should be noted that the earnings per share graph is inconsistent with the other three. It is for three years not four and has a non-zero axis. The overall result is that it perhaps presents a more favourable view of the company's results than would otherwise be warranted.

Real-Life Nugget 13.10

Creative Graphs: T.I.P. Europe plc, 1989 Annual Report

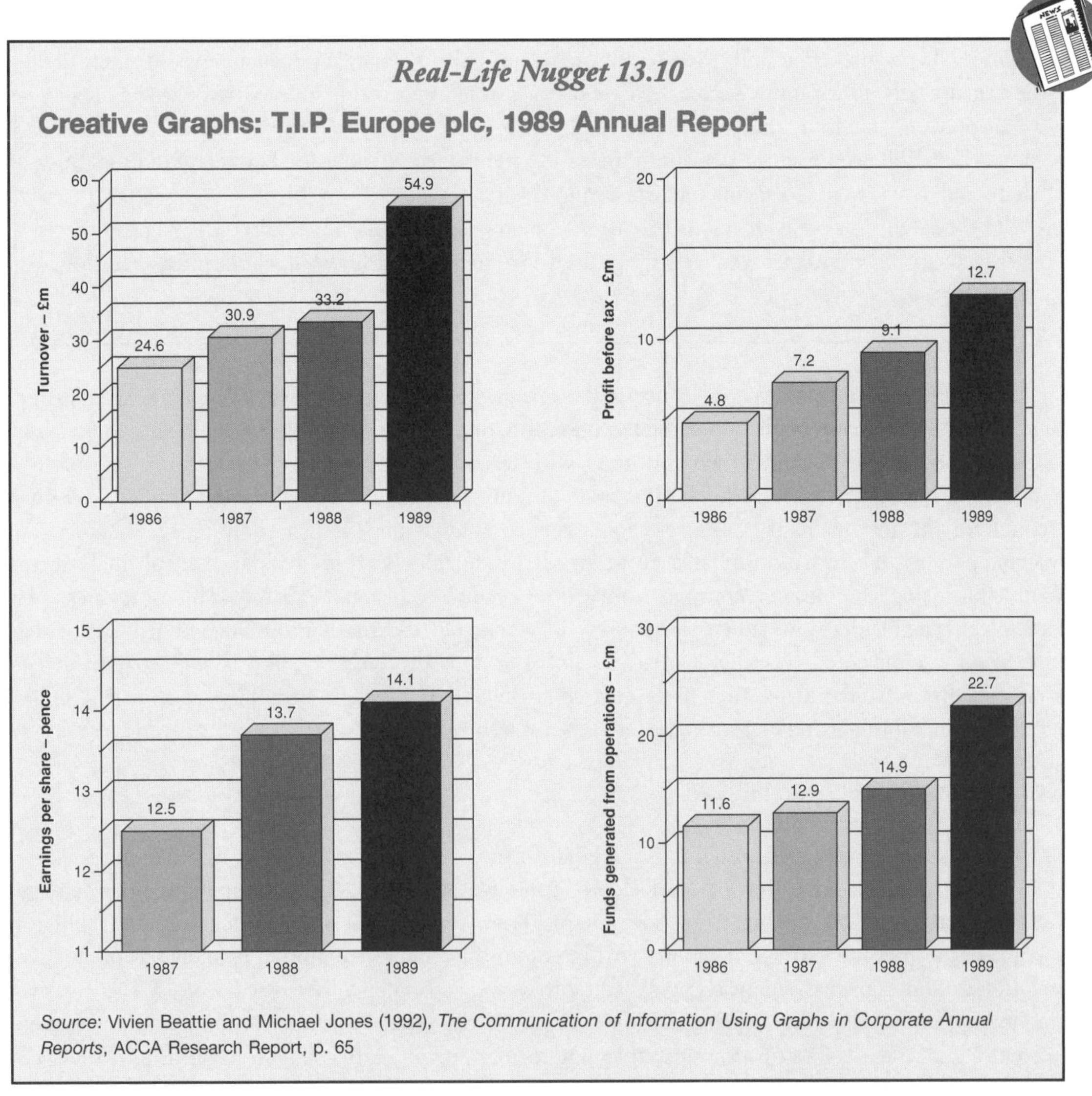

Source: Vivien Beattie and Michael Jones (1992), *The Communication of Information Using Graphs in Corporate Annual Reports*, ACCA Research Report, p. 65

Controlling Creative Accounting

One of the key objectives of the Accounting Standards Board, which was set up in the 1990, was to control creative accounting. This is shown in the Financial Reporting Council's *The State of Financial Reporting – a Review* (1991) (paras 2.1 and 2.2) – see Real-Life Nugget 13.11.

Real-Life Nugget 13.11

Regulatory Response to Creative Accounting

'The corporate confidence developed during the 1980s boom, the associated readiness by banks to lend and by companies to borrow, the growth of innovative accounting practices (e.g. off balance sheet financing and the development of hybrid financial instruments) sometimes designed solely to avoid an increase in reported company gearing, coupled with a framework of accounting standards that was being outpaced by such developments, have made the recession which followed a correspondingly more chastening experience for bankers, creditors and shareholders, as well as for financial reporting itself.

The existence of weaknesses in the arrangements for formulating and securing compliance with accounting standards was well recognised long before the boom declined into recession.'

Source: Trevor Pijper (1993), *Creative Accounting: The Effect of Financial Reporting in the UK*, p. 15

Since 1990, there has been a concerted attempt by regulators to curb creative accounting. Undoubtedly, they have made substantial progress in many areas. Companies now have much less scope for creativity, for example, when dealing with leased assets. However, it is often a case of two steps forward, one step back. Indeed, there is a continuing battle between the regulators and creative accountants. Some merchant banks actively advise companies on 'creative compliance'. There is an evolving pattern of creative compliance from avoidance, to rules, back to avoidance. Atul Shah documents this using the case of complex convertible securities issued by listed UK companies. He concludes: '[r]egulators were slow to respond, and when they did make pronouncements, companies once again circumvented the rules with the help of various professionals. A "dialectic of creativity" is created, from avoidance to rules to avoidance again' (Shah, 1998, p. 36). This continual struggle between companies and regulators causes a creative accounting 'arms race'.

Conclusion

Creative accounting emerged into the limelight in the 1980s. However, it is still alive and kicking today. As the collapse of Enron in 2001 clearly demonstrated. Managers have incentives to manage their accounting profits so that they serve managerial, rather than shareholder, interests. This is possible because of the extreme flexibility within accounting. There are numerous methods of creative accounting: some are extremely complex and others very simple. When used excessively, creative accounting can be positively dangerous; for example, it has contributed to corporate collapses. The Accounting Standards Board attempts, through regulation, to curb creative accounting. However, there is an ongoing battle as companies seek ways around the regulations.

Selected Reading

1. *Accounting for Growth*
   This report was a real accounting bombshell. It documents the use by 185 UK companies of 11 innovative accounting practices. The original report (see (a) below) was issued by UBS Phillips and Drew. Then Terry Smith published (b) below after he left UBS Phillips and Drew. Those who can't get hold of the original report/book, or who want a quick summary, could try Jones (c).
   (a) UBS Phillips and Drew (1991), *Accounting for Growth.*
   (b) Smith, T. (1992), *Accounting for Growth* (Century Business).
   (c) Jones, M.J. (1992), 'Accounting for growth: Surviving the accounting jungle', *Management Accounting*, February pp. 20–22.
2. County NatWest WoodMac (1991), *Company Pathology.*
   This report provides an interesting study into creative accounting by 45 companies.
3. Griffiths, I. (1986) *Creative Accounting* (Sidgwick and Jackson) (1995) and *New Creative Accounting* (Macmillan). Both very good reads which provide a journalist's view of the debate.
4. Naser, K. (1993) *Creative Financial Accounting: Its Nature and Use.*
   Written by an academic, this book provides perhaps a more balanced view of creative accounting than those by Griffiths and Smith.
5. Pijper, T. (1993) *Creative Accounting: The Effect of Financial Reporting in the UK* (Macmillan).
   Another good overview of creative accounting.
6. Shah, A.K. (1998) 'Exploring the influences and constraints on creative accounting in the United Kingdom', *The European Accounting Review*, Vol. 7, No.1 , pp. 83–104.
   This article provides a good insight into evolving struggle between regulators and creative accountants.

Discussion *Questions*

Questions with numbers in blue have answers at the back of the book.

Q 1 What is creative accounting? And why do you think that it might clash with the idea that the financial statements should give a 'true and fair view' of the accounts?

Q 2 Does creative accounting represent the unacceptable face of accounting flexibility?

Q **3** Why do you think that there are those strongly in favour and those strongly against creative accounting?

Q **4** What incentives do managers have to indulge in creative accounting?

Q **5** Will creative accounting ever be stopped?

Q6 You are the financial accountant of Twister Plc. The managing director has the following draft accounts. She is not happy.

Twister Plc: Draft Profit and Loss Account Year Ended 30 June 2001

	Notes	£000	£000
Sales	1		750
Less: *Cost of Sales*			
Opening stock		80	
Add Purchases		320	
		400	
Less Closing stock	2	60	340
Gross Profit			410
Less *Expenses*			
Depreciation	3	60	
Interest payable	4	30	
Goodwill	5	20	
Other expenses		312	422
Net Loss			(12)

Notes

1. The company's sales policy is to record sales prudently, one month after invoicing the customer so as to allow for any sales returns. If the company recorded sales when invoiced, this would increase sales by £150,000.
2. This is a prudent valuation, a more optimistic valuation gives £65,000.
3. Depreciation is currently charged on fixed assets over ten years. This is a realistic expected life, but a competitor charges depreciation over 15 years.
4. Half the interest payable relates to the borrowing of money to finance the construction of fixed assets.
5. Goodwill represents one-fifth of the goodwill on a subsidiary purchased this year. Most companies in the same industrial sector adopt a 20-year amortisation period even though this appears realistically rather too long.

Required:

Using the accounts, and the notes above, present as flattering a profit as you can.

Chapter 14

International Accounting

'Whether we are ready or not, mankind now has a completely integrated financial and informational market place capable of moving money and ideas to any place on this planet in minutes.'

W. Wriston, *Risk and Other Four-letter Words*, p. 132. *The Executive's Book of Quotations* (1990), p. 150

Learning Outcomes

After completing this chapter you should be able to:

- ✔ **Explain and discuss the main divergent forces.**
- ✔ **Understand the macro and micro approaches to the classification of international accounting practices.**
- ✔ **Understand the accounting systems and environments in France, Germany, the UK and the US.**
- ✔ **Examine the convergent forces upon accounting, especially harmonisation in the European Union and standardisation through the International Accounting Standards Board.**

In a Nutshell

- Global trade and investment make national accounting very constrictive.
- Divergent forces are those factors that make accounting different in different countries.
- The main divergent forces are: objectives, users, sources of finance, regulation, taxation, the accounting profession, spheres of influence and culture.
- Countries can be classified into those with macro and micro accounting systems.
- France and Germany are macro countries with tight legal regulation, creditor orientation and weak accounting professions.
- The UK and the US are micro countries, guided by the idea of fair presentation, with influential accounting standards, an investor-oriented approach and a strong accounting profession.
- Internationalisation causes pressure on countries to depart from national standards.
- The three main potential sources of convergence internationally are the European Union, the International Accounting Standards Board and US standards.

Introduction

Increasingly, we live in a global world where multinational companies dominate world trade. The world's stock exchanges are active day and night. Accounting is not immune from this globalisation. There is an increasing need to move away from a narrow national view of accounting and see accounting in an international context. The purpose of this chapter is to provide an insight into these wider aspects of accounting. In particular, we explore the factors that cause accounting to be different in different countries, such as objectives, users, regulation, taxation and the accounting profession. Accounting in several important countries (France, Germany, the UK and the US) is also explored. Finally, this chapter looks at the international pressures for convergence towards one universal world accounting system. In essence, therefore, the aim of this chapter is to provide a brief overview of the international dimension to accounting.

Context

Perhaps surprisingly, given the variety of peoples and cultures throughout the world, the fundamental techniques of accounting are fairly similar in most countries. In other words, in most countries businesses use double-entry bookkeeping, prepare a trial balance and then a profit and loss account and balance sheet. This system of accounting techniques was developed in Italy in the fifteenth century and spread around the world with trade.

However, the context of accounting in different countries is very different and causes transnational differences in the measurement of profit and net assets. The differences between countries are caused by so-called 'divergent forces', such as objectives, users, sources of finance, regulation, taxation, accounting profession, spheres of influence and culture. These divergent forces, which are examined in more detail in the next section, cause accounting in the UK to be different to that in, say, France or the US.

These international accounting differences caused few problems until the globalisation of international trade. However, with the rise of global trade and the erosion of national borders (see Soundbite 14.1), the variety of world accounting practices has been seen by many as a significant problem, particularly for multinational companies.

For large multinational companies, it is cheaper and easier to have just one set of world accounting standards. This is the aim of the International Accounting Standards Board (IASB). Meanwhile, within Europe there is pressure to harmonise accounting to create one set of Europe-wide standards. These pressures for the harmonisation and standardisation of accounting are called 'convergent forces'. In essence, therefore, these divergent and convergent forces pull accounting internationally in different directions. Large global corporations, such as Glaxo-Wellcome, Microsoft and Toyota, dominate world trade. As Real-Life Nugget 14.1 shows (see next page), the sales of many of these corporations are greater than the gross national product of many countries. The sales of Mitsubishi and Mitsui in 1995 were greater than the GDP of countries such as Turkey, Thailand, Denmark, Hong Kong or Norway. In fact, only 23 countries in the world had a gross domestic product which exceeded the sales of either of these two companies.

Soundbite 14.1

National Boundaries

'National borders are no longer defensible against the invasion of knowledge, ideas, or financial data.'

Walter Wriston, *Risk and Other Four-Letter Words*, p. 133.

Source: *The Executive's Book of Quotations* (1994), p. 202

Real-Life Nugget 14.1

The World's Top Economic Entities

Country/company by GDP/sales

		(Dollars Bn.)
1.	United States	7100
2.	Japan	4964
3.	Germany	2252
4.	France	1451
5.	United Kingdom	1095
24.	Mitsubishi (Japan)	184
25.	Mitsui (Japan)	181
26.	Itochu (Japan)	169
27.	Turkey	169
28.	General Motors (US)	169
29.	Sumitomo (Japan)	168
30.	Marubeni (Japan)	161
31.	Thailand	160
32.	Denmark	156
33.	Hong Kong	142
34.	Ford Motors (US)	137
35.	Norway	136

Source: FT Profile, Copyright Financial Times Information. Simon Caulkin, 'It's bigger than Turkey, Thailand or Denmark – but it's not a country', The Management Column (*Observer*, March 1998, p. 9)

Divergent Forces

Divergent forces are those factors that cause accounting to be different in different countries. These may be internal to a country (such as taxation system) or external (such as sphere of influence). There is much debate about the nature and identity of these divergent forces. For example, some writers exclude objectives and users. Each country has a distinct set of divergent forces and the relative importance of each divergent force varies between countries. These divergent forces are interrelated. The main divergent forces discussed in this chapter are shown in Figure 14.1 below.

(i) Objectives

The objectives of an accounting system are a key divergent force. There are two major objectives: economic reality, and planning and control. The basic idea behind economic reality is that accounts should provide a 'true

Figure 14.1 The Divergent Forces that Determine National Accounting Systems and Environments

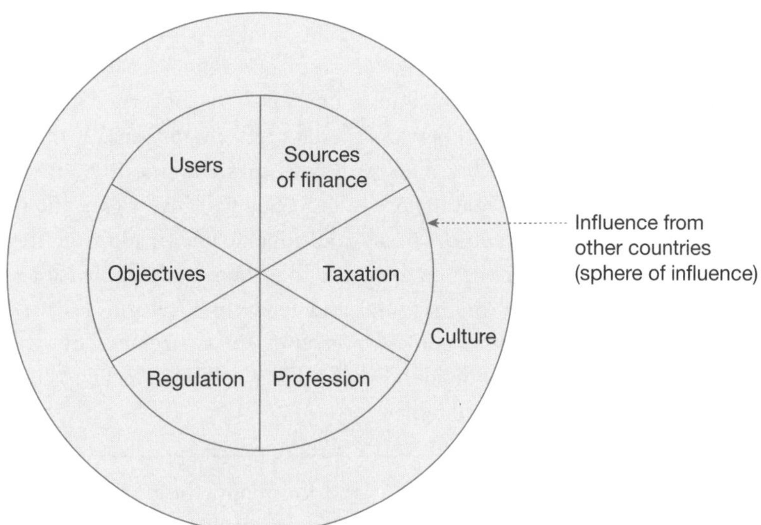

and fair' view of the financial activities of the company. The UK holds this view. The US takes a similar view: 'present fairly . . . in accordance with generally accepted accounting principles (GAAP)'. However, in the US there is a stronger presumption than in the UK that following GAAP will present a fair view of the company's activities. The 'true and fair' concept is basically linked to the underlying idea that the purpose of accounting is to provide users with financial information so that they can make economic decisions.

In contrast to economic reality, planning and control places more emphasis on the provision of financial information to government. Accounting in countries such as France and Germany is more concerned with collecting statistical and regularised information for comparability and planning. For example, in France accounting is seen as promoting national economic and fiscal planning. In France and Germany, the concept of a true and fair view is incorporated in national law. Unlike in the UK, this is narrowly interpreted as compliance with the regulations.

(ii) Users

Users and objectives are closely interrelated. Users are also linked closely to another divergent force, sources of finance. In some countries, such as the UK and the US, shareholders are the main users of accounts. In the UK, institutional investors (such as insurance companies and pension firms) dominate, while in the US the role of the private individual is comparatively more important. When shareholders dominate, the accounting system focuses on profitability and the profit and loss account becomes the main financial statement. However, recently standard setters have moved towards a balance sheet approach.

Pause for Thought 14.1

Users

If the main users of accounts are banks rather than shareholders, how will that affect the items in the financial statements which they scrutinise?

Bankers essentially lend companies money and are concerned about two main things: loan security and loan repayments. They are, therefore, likely to focus on the balance sheet, particularly on assets, gearing and liquidity. They will wish to make sure that any assets on which the loans are secured have not been sold or lost value. They will investigate the gearing ratios to see that the company has not taken on too many additional loans. In addition, they will scrutinise the amount of cash and the liquidity ratios to ensure that there is enough cash to pay the loan interest.

Shareholders invest their capital and look for a return. They will, therefore principally be concerned with the profit and loss account. In particular, they will look at earnings, earnings per share, dividends and dividend cover. They will wish to be assured that the company is profitable and will remain profitable.

In other countries, such as France and Germany, the power of the shareholder is less. In France, for example, the users of accounting are more diverse, ranging from the government for economic planning, to the tax authorities for fiscal planning, to the banks. In Germany, the main users are the tax authorities and the banks. In France and Germany, the users are generally relatively more concerned about liquidity and the balance sheet than in the UK and the US. However, in both France and Germany as more multinational companies are listed, the power of the shareholder is increasing.

(iii) Sources of Finance

A key question for any company is how to finance its operations. Apart from internally generated profits, there are two main sources: equity (or shareholder) finance and debt (or loan) finance. In some countries, such as the UK and the US, equity finance dominates. In other countries, for example, France and Germany, there is a much greater dependence on debt finance. The relative worldwide importance of the stock exchange (where equity is traded and raised) can be seen in Figure 14.2.

Figure 14.2 Relative Strength of Selected Major Stock Exchanges as at 31 December 1997

Country	*Exchange*	*Domestic Companies*	*International Companies*	*Market Capitalisation*
		No.	No.	£m
Australia	Australia	1,159	60	178,853
Canada	Toronto	1,362	58	358,126
France	Paris	744	193	410,851
Germany	Frankfurt	700	N/A	501,555
Hong Kong	Hong Kong	638	20	250,544
Japan	Tokyo	1,805	60	1,287,476
UK	London	2,465	526	1,251,425
US	New York	2,691	356	5,463,413
	NASDAQ	5,033	454	1,033,183

N/A = Not available

Source: London Stock Exchange (1998), *Factfile*, p. 41

The presence of a strong equity stock market leads countries to have an investor and an economic reality orientated accounting system. There is a focus on profitability and the importance of auditing increases. By contrast, a strong debt market is associated with the increased importance of banks.

(iv) Regulation

By regulation, we mean the rules that govern accounting. These rules are primarily set out by government as statute or by the private sector as accounting standards (see Chapter 10 for more detail on the UK). The balance and interaction between statutory law and standards within a particular country is often subtle. The regulatory burden is usually significantly greater for public listed companies than for smaller non-listed companies.

In the UK and the US, for example, accounting standards are very important. However, whereas in the UK there are Companies Acts applicable to all companies, in the US, the Securities Exchange Commission, an independent regulatory organisation with quasi-judicial powers, regulates the listed US companies. Other US companies are regulated by state legislation, which is often minimal.

In some countries, such as France and Germany, accounting standards are not very important. Indeed, France has no formal standard-setting body equivalent to the UK's Accounting Standards Board, while Germany only set up a standard-setting organisation in 1999. In France and Germany accounting regulation originates from the government, in Germany by way of a Commercial Code and in France by way of a comprehensive government manual for accounting.

(v) Taxation

Strictly, the influence of taxation is part of a country's regulatory framework. However, as the relationship between accounting and taxation varies significantly between countries, it is discussed separately here. Essentially, national taxation systems may be classed as '*independent*' of, or '*dependent*' on, accounting. In independent countries, such as the UK, the taxation rules and regulations do not determine the accounting profit. There are, in effect, two profits (calculated quite legally!) – one for tax and the other for accounting. The US uses a broadly independent tax system. However, there is a curious anomaly. If US companies wish to lower their taxable profits, they can use the Last-in-First-out (LIFO) method of stock valuation (which usually values stock lower than alternative stock valuation methods, and thus lower profits). (See Chapter 16 for a fuller discussion of LIFO.) However, the companies *must* also use this method in their shareholder accounts.

In France and Germany, however, accounting profit is dependent upon taxable profit. In effect, the rules used for taxation determine the rules used for accounting. Accounting, therefore, becomes a process of minimising tax paid rather than showing a 'true and fair view' of the company's financial activities.

(vi) Accounting Profession

Accounting professions worldwide vary considerably in their size and influence. In the UK and US accountants play an important interpretive and judgemental role. Although accounting standards are independently set, qualified accounting professionals generally constitute the majority of the members of the accounting standards boards. In France and Germany, professionally qualified accountants have far less influence on accounting. There is less flexibility in the accounting rules and regulations and less need for judgement.

Figure 14.3 (see next page) shows the age and size of some professional accountancy bodies. In the UK and the US, the accounting profession is very strong and influential. In the US, for example, in 1997 there were about 328,000 Certified Public Accountants (CPA). By contrast, in Germany, there were only 8,000 *Wirtschaftprüfers*, the German equivalent of the CPA.

Figure 14.3 Age and Size of Selected Professional Accountancy Bodies

Country	*Body*	*Founding Date of Original Body*	*Approximate Membership (000) in 1997*
Australia	Australian Society of Certified Practising Accountants	1952 (1886)	83
	Institute of Chartered Accountants in Australia	1928 (1885)	29
Canada	Canadian Institute of Chartered Accountants	1902 (1880)	60
France	Ordre des Experts Comptables	1942	15
Germany	Institut der Wirtschaftsprüfer	1932	8
Japan	Japanese Institute of Certified Public Accountants	1948 (1927)	12
Netherlands	Nederlands Instituut van Register Accountants	1895	10
New Zealand	New Zealand Society of Accountants	1909 (1894)	23
United Kingdom and Ireland	Institute of Chartered Accountants in England and Wales	1880 (1870)	112
	Institute of Chartered Accountants of Scotland	1951 (1854)	14
	Association of Chartered Certified Accountants	1939 (1891)	53
	Institute of Chartered Accountants in Ireland	1888	10
United States	American Institute of Certified Public Accountants	1887	328

Note: Dates of earliest predecessor bodies in brackets.

Source: Nobes, C. and Parker, R. (2000), *Comparative International Accounting*, Table 1.2, (abridged), p.5.

(vii) Spheres of Influence

Accounting is not immune from the wider forces of economics and politics. Many countries' accounting systems have been heavily influenced by other countries. The UK, for example, imported double-entry bookkeeping from Italy in the fifteenth century. It then exported this to countries such as Hong Kong, India, Kenya, Malaysia, Nigeria and Singapore. These countries still follow a model inherited from the UK. The UK also originally exported its accounting system to the US. However, from the early 1900s, the US has influenced the UK rather than vice versa. Overall, it is perhaps possible to identify three broad spheres of influence: the UK, the US and the continental European. Currently, the UK is being influenced by both the US and the more prescriptive continental European model.

(viii) Culture

Culture is perhaps the most elusive of the divergent forces. Hofstede, in *Culture's Consequences* (Sage, 1980), defines culture as 'the collective programming of the mind which distinguishes the members of one human group from another'. Building on the work of Hofstede, accounting researchers have sought to establish whether culture influences accounting. For example, do the French or the British have a collective culture which influences accounting? Using such concepts as professionalism, uniformity, conservatism and secrecy, they have found some, if not overwhelming, support.

Classification

Accounting researchers have sought to classify national accounting systems using, for example, divergent forces, cultural characteristics, and accounting measurement and disclosure practices. A useful classification system is described by Nobes (1983). Broadly, his system classifies countries as either having macro or micro accounting systems. Figure 14.4 below uses the divergent forces to summarise the major national characteristics of macro and micro countries. It can thus be seen that macro countries, such as France and Germany, are typified by governmentally-orientated systems where the influence of tax is strong, but that of the accounting profession is weak. Meanwhile, micro countries, such as the UK and the US, are typified by investor-orientated systems where tax influence is weak, but the influence of the accounting profession is strong.

Figure 14.4 Main National Characteristics of the Macro and Micro Classification

Divergent Force	*Macro*	*Micro*
Objectives	Planning and control	Economic reality
Users	Government, banks, tax	Shareholders
Sources of Finance	Banks	Stock exchange
Regulation	Government, relatively prescriptive	Private sector, relatively flexible
Taxation	Dominant, dependent system	Subordinate, independent system
Accounting Profession	Weak, uninfluential	Strong, influential
Spheres of Influence	Continental	UK, US
Examples	France, Germany	UK, US

Country Snapshots

Every country's accounting system and environment is a unique mixture of divergent forces. In this sector, we look at the accounting systems and environments of four important developed countries: France, Germany, UK and the US. The aim is to provide a quick overview of the distinctive natural characteristics of each country's accounting systems.

France

The French accounting system is often cited as a good example of a standardised accounting system. It thus contrasts with the UK and US systems. Essentially, the French accounting system was introduced in order to provide the government with economic information for planning and controlling the French economy. The French system centres around *Le Plan Comptable Générale* [General

Accounting Plan]. This sets out a uniform system of accounting to be used throughout France. Of special interest is a chart of accounts where each item in the accounts is given a number. These numbers can then be used to aggregate items from different companies. Overall, *Le Plan Comptable Générale* resembles a comprehensive accounting manual which French companies follow.

In addition to the *Le Plan Comptable Générale*, there are various detailed accounting and taxation laws. The role of taxation is particularly important. In essence, tax law drives accounting law. Standish (in Nobes and Parker, 2000, p. 205) comments:

> 'An enterprise wishing to take advantage of various tax concessions must accordingly enter the relevant tax assessable income or tax deductible charges in its accounts as if valid for tax purposes, even if the effect is to generate assets or liabilities that do not conform to the accounting criteria for asset and liability recognition and measurement'.

The accounting profession in France is much smaller than in the UK and US. There were 15,000 professional accountants in 1997 (Nobes and Parker, 2000, p. 5). Compared with other countries, the profession is not influential. There are, for example, no accounting standards in France comparable to those in the UK and US.

Soundbite 14.2

France

'We're [France] somewhere between the United States and Germany on transparency, probably closer to US practices and certainly less able than German business to hide systematic abuses. We're going toward the US model, but it's going to take several years more.'

Alan Minc quoted in *International Herald Tribune* (October 10, 1996)

Source: *The Wiley Book of Business Quotations* (1998), p. 158

There are, however, signs that accounting in France is changing (see Soundbite 14.2 above). In their group accounts, French companies are now permitted to use non-French accounting principles. Many large French groups, therefore, use US accounting principles or International Accounting Standards (IAS). For example, according to their respective Web sites: 18 French companies were using US standards and one French company was using IAS in June 2001 for their group accounts. However, individual company accounts must still be prepared using French principles.

Germany

Germany has always had a relatively distinctive accounting system (see Real-Life Nugget 14.2). Essentially, it is very prescriptive, government controlled and creditor orientated. Currently, there is no comprehensive body of accounting standards. The main regulations are contained in the Companies Acts, Commercial Code and Publication Law. These laws set out very detailed regulations.

Real-Life Nugget 14.2

German Accounting System

'Traditionally, German accounting has followed a relatively distinct path. Germany, along with France, has always represented a continental accounting tradition based on a creditor rather than a shareholder approach. The Franco-German approach is in direct contrast to the Anglo-Saxon view of accounting. This latter view, typified by UK/US accounting, espouses an economic reality, user-based approach, with the main user being the shareholder.'

Source: Jones, M.J. (1999), 'Germany: An Accounting System in Change', *Accountancy International*, August, pp. 64–65

The measurement systems underpinning German accounting have traditionally been seen as very conservative. This is especially so when compared with UK accounting policies. For instance, as Real-Life Nugget 14.3 shows, in 1997 Rover, the UK car maker, would have made a profit of £147m under UK rules. Under German rules it reported a loss of £363m. Taxation law also dominates accounting, for example, expenses are only allowable for tax if they are deducted in the financial accounts.

Real-Life Nugget 14.3

German Accounting Rules

By Gavin Hinks

Rover accounts drawn up according to UK accounting rules show the car manufacturer lost considerably less in the lead up to last week's controversial sale than the figure reported in BMW's accounts.

The figures shows that for the year ending December 1998, Rover lost around £160m less than that stated in accounts drawn up under harsher German accounting standards.

According to UK accounts, Rover made losses of £509m in 1998. But BMW accounts show the company lost £670m.

As morale plummeted at Rover's Longbridge works following the threat of thousands of job losses as a result of the sale to venture capital company Alchemy, a spokesman said the trend of lower losses in the company's UK accounts continued into 1999.

Last year, when controversy over Rover first appeared, *Accountancy Age* revealed the discrepancy between the two sets of accounts. An analysis of four years' of figures, from 1994 to 1997, revealed Rover had actually made a profit of £147m according to UK rules while the German books stated a loss of £363m.

Under German accounting policies investments are depreciated faster, there are more possibilities for making provisions and there are different rules for the valuation of stocks.

BMW was expected to move to International Accounting Standards which may have improved the position of Rover, but BMW this week refused to comment on which accounting standards are now being used.

Reading University Professor Chris Nobes, a member of the International Accounting Standards committee, said the problem was not unique to Rover. 'Normal German practice is more conservative than Britain's,' he added.

Chemical companies like Bayer and Hoechst, and Deutsche Bank, have moved away from German standards but the number switching is small. Lufthansa and Volkswagen use German standards.

Alchemy, the buyer of Rover, is certain to use UK accounting standards when dealing with its new purchase.

Source: *Accountancy Age*, 23 March 2000, p. 9

Soundbite 14.3

Germany

'Their people would come here and put down our people, our work ethic. I had a little problem with that. I finally slammed my door shut and told my German counterpart that I didn't need him telling us how good he was and how weak we were. We never had any problems after that.'

Time (October 9, 1989)

Source: *The Wiley Book of Business Quotations* (1998), p. 294

The German accounting profession is very small. In 1997, there were only 8,000 German professionally qualified accountants (Nobes and Parker, 2000, p. 5). Although small, the German profession is very well qualified. However, in Germany most accounting developments originate from the government rather than from the accounting profession.

By international standards, the German stock market is comparatively small. Its total market capitalisation is only about half that of the UK's. German industry is financed principally by banks. The focus of German accounting, therefore, has traditionally been on assets rather than profit.

However, German accounting is changing. Listed German companies, like French companies, are now allowed to prepare group accounts using foreign accounting principles. At the start of the 1990s, no German companies were listed in the US. However, Daimler Benz caused a minor sensation in Germany by listing on the US exchange in 1993. By June 2001, 14 Germany companies were listed on the New York Street Exchange according to its Web site. Similarly, by June 2001, according to the IASB website, 66 German companies were using IAS.

The UK

The UK is still an important player in accounting worldwide. However, it is no longer the world leader. The current UK system contains elements of US and Continental European accounting. The generally recognised objective of accounting in the UK is to give a 'true and fair view' of a company's financial activities to the users, most notably the shareholders. There are two sources of regulation: company law and accounting standards. Company law, which now includes the European Fourth and Seventh Directives, sets out an increasingly, prescriptive accounting framework. The financial reporting standards, set by the Accounting Standards Board, provide guidance on particular accounting issues. Uniquely, the UK has a Financial Reporting and Review Panel, which investigates companies suspected of non-compliance with the standards. The UK, therefore, has an unusual mix of regulations.

The accounting profession in the UK is very influential and accounting in the UK is a relatively high status profession. It is also very fragmented, comprising the Institute of Chartered Accountants in England and Wales (ICAEW), the Institute of Chartered Accountants of Scotland, the Institute of Chartered Accountants in Ireland, the Association of Chartered Certified Accountants, the Chartered Institute of Management Accountants and the Chartered Institute of Public Finance and Accountancy. The ICAEW is the largest body numbering 112,000 professionally qualified accountants in 1997 (see Figure 14.3 on page 309).

Pause for Thought 14.2

Unique Features of UK Accounting System

Every country's accounting system and environment is unique. Can you think of three features of the UK's accounting and environment which contribute, in combination, to the uniqueness of the UK's accounting system?

There are many special features in the UK's system. Some of the main ones are listed below.

- Combination of a 'true and fair view' with Companies Acts and accounting standards.
- The fragmentation of the accounting profession into six institutes.
- The dominance of the institutional investor in the stock market.
- The Financial Reporting Review Panel's role as guardian of good accounting practice.
- The ability of companies to depart from historical cost by revaluing their fixed assets.
- The ability of companies to capitalise their goodwill in the balance sheet and not necessarily write it off to the profit and loss account.

The main users of accounts in the UK are the shareholders. In particular, in the UK, institutional investors dominate. The stock market is very active and its market capitalisation is very high. Unlike most other countries, there is a separation between accounting profit and taxable profit. In essence, there are two distinct bodies of law. Certainly, taxable profit does not determine accounting profit.

The US

The US dominates world accounting. Most important developments in accounting originate in the US. A particular feature of the US environment is the position of the Securities Exchange Commission (SEC). The SEC was set up in the 1930s after the Wall Street crash of 1929. It is an independent regulatory institution with quasi-judicial powers. Listed US companies have to file a detailed annual form, called the 10-K with the SEC. The SEC also supervises the operation of the US standard-setting body, the Financial Accounting Standards Board (FASB). This body has been designated by the American Institute of Certified Public Accountants (AICPA) as the US's standard-setting body.

FASB is the most active of the world's national standard-setting bodies. The members of FASB have to sever their prior business and professional links. FASB has published over 130 standards. It is supported by the Emerging Issues Task Force, which examines new and emerging accounting issues. The US accounting standard-setting model proved the blueprint for the current UK system.

US listed companies are thus well regulated. However, curiously for the mass of US companies, which are not listed, there may be very few accounting or auditing requirements. Company legislation is a state not a federal matter. Each state, therefore, sets its own laws.

The accounting profession in the US is both numerous and influential. There were 328,000 members of AICPA in 1997 (see Figure 14.3 on page 309). Most of the world's leading audit firms have their head offices in the US.

The objective of accounting in the US is generally recognised as being to provide users with information for decision making. Most of the finance for US industry is provided by shareholders. Unlike in the UK, the majority of shareholders in the US are still private shareholders. The market capitalisation of the US stock market is the greatest in the world. Many of the world's largest companies such as Microsoft or The General Electric Company are US. The high-quality financial information produced by US financial reporting standards is often argued to help the efficiency of the US stock market (see Real-Life Nugget 14.4). However, in 2001 and 2002 a series of high-profile accounting scandals involving companies such as Enron, WorldCom and Xerox have led to a questioning of the quality of US accounting and auditing standards.

Real-Life Nugget 14.4

US Financial Reporting

Financial Reporting and Capital Markets

There is a clear connection between the efficient and effective US capital markets and the high quality of US financial reporting standards. US reporting standards provide complete and and transparent information to investors and creditors. The cost of capital is directly affected by the availability of credible and relevant information. The uncertainty associated with a lack of information creates an additional layer of cost that affects all companies not providing the information regardless of whether the information, if provided, would prove to be favorable or unfavorable to individual companies. More information always equates to less uncertainty, and it is clear that people pay more for better certainty.

Source: Edmund Jenkins, Global Reporting Standards, IASC *Insight*, June 1999, p. 11

In the US, as the aim of the accounts is broadly to reflect economic reality: the tax and accounting systems are generally independent. There is, however, one curious exception: last-in first-out (LIFO) stock valuation. Many US companies adopt LIFO in their taxation accounts as it lowers their taxable profit. However, if they do so, federal tax laws state they must also use LIFO in their financial accounts. However, since LIFO stock values are generally old and out of date, this fails to reflect economic validity (see Chapter 16 for more on LIFO).

Convergent Forces

Convergent forces are pressures upon countries to depart from their current national standards and adopt more internationally-based standards. Convergent forces thus oppose divergent forces. The advantage of national standards is that they may reflect a particular country's circumstances. Unfortunately, the disadvantage is that they impair comparability between countries and are a potential barrier to international trade and investment.

Pause for Thought 14.3

Pressures for Convergence

How might national accounting standards prove a barrier for a multinational company, such as Glaxo-Wellcome, or a large institutional investor, such as the Mercury Asset Management Fund?

The world of trade and investment is now global. Companies such as Glaxo-Wellcome will, therefore, trade all over the world and have subsidiaries in numerous countries. They will sometimes have to deal with accounting requirements in literally hundreds of countries. To do this takes time, effort and, more importantly, money. If there was one set of agreed international accounting standards, this would make the life of many multinationals easier and cheaper.

Large institutional investors will also be active in many countries. For them, there is a need to compare the financial statements of companies from different countries. For example, an institutional investor, such as the Mercury Asset Management Fund, may wish to achieve a balanced portfolio and invest funds in the motor car industry. It would wish to compare companies such as BMW in Germany, Fiat in Italy, General Motors in the US and Toyota in Japan. To do this effectively, there is a need for comparable information. A common set of international accounting standards would provide the comparable information.

There are three main potential sources of convergence for European companies: the European Union (EU), the International Accounting Standards Board (IASB) and the United States. The US is very much the default option, if convergence is not achieved through the EU or the IASB. The process of convergence through the EU is normally termed *harmonisation*, while that through the IASB is called *standardisation*.

European Union (EU)

The European Union is concerned with harmonising the economic and social policies of its member states. Accounting represents part of the economic harmonisation within Europe. In countries such as France and Germany, accounting regulation has always been seen as a subset of a more general legal regulation. Differences in accounting between member states are seen as a barrier to the harmonisation of trade.

The main legal directives affecting accounts are the Fourth Directive and the Seventh Directive. The Fourth Directive is particularly important to the UK. It is based on the German Company Law of 1965. The Fourth Directive ended up as a compromise between the traditional UK and the Franco-German approaches to accounting. The UK approach was premised on flexibility and individual judgement. By contrast, the continental approach set out a prescriptive and detailed legal framework. In the end, the Fourth Directive married the two approaches. The UK for the first time accepted a standardised format for presenting company accounts and much more detailed legal regulation. However, France and Germany agreed to incorporate into law the British concept of the presentation of a true and fair view.

International Accounting Standards Board (IASB)

The International Accounting Standards Committee (IASC) was founded in 1973 by Sir Henry Benson to work for the improvement and harmonisation of accounting standards worldwide. The IASC was renamed the International Accounting Standards Board (IASB) in 2001 (to simplify matters we use IASB for both IASC and IASB throughout this book). Originally, there were nine members: Australia, Canada, France, Germany, Japan, Mexico, the Netherlands, the UK and Ireland, and the US. Member bodies are the national professional bodies of different countries. The IASB has grown rapidly and, in 2000, 140 countries were members. The member bodies use their best endeavours to ensure that their countries follow International Accounting Standards (IAS). Standards are thus not mandatory, but persuasive.

At first, the IASB merely codified the world's standards. After this initial step, the IASB began to work towards the improvement of standards. The IASB set out a restricted number of options within an accounting standard from which companies could then choose. Up until the mid-1990s, it is fair to say that the IASB made only limited progress. A threefold differentiation in the IASB's impact is possible: lesser developed countries, European countries and capital market countries. Lesser developed countries, such as Malaysia, Nigeria and Singapore, adopted IAS because doing so was cheaper than developing their own standards. In Continental Europe, the IAS were seen both as a problem and a solution. They were a problem in that generally IAS were seen to adopt a primarily investor-orientated approach to accounting which conflicted with the traditional Continental European tax-driven, creditor-based model. They were a solution in that IAS were preferable to US standards. Increasingly, in the early 1990s, French and German companies adopted US standards. Karl Van Hulle, Head of the EU's Accounting Unit commented in 1995, 'It would be crazy for Europe to apply American standards, as it would be crazy for the Americans to apply European standards. We ought to develop those standards which we believe are the best for us or for our companies'. Finally, for capital market countries such as the UK and the US, the IAS were generally already similar to the national standards. Even so, there was a great reluctance, particularly by the US, to accept IAS.

A breakthrough agreement came in 1995. IOSCO (The International Organisation of Securities Commissions), the body which represents the world's stock exchanges, agreed that when the IASB had developed a set of core standards it would consider them for endorsement and would recommend them to national stock exchanges as an alternative to national standards. The advantage to IOSCO was that there would be a common currency of standards which could be used internationally. In particular, there was the hope that non-US companies could trade on the New York Stock Exchange without having to use US Generally Accepted Accounting Principles (GAAP) or provide a reconciliation to US GAAP.

The IASB subsequently experienced severe problems compiling a set of core standards. It missed a couple of important deadlines. However, in 2000 IOSCO did endorse the IAS core standards but this does not affect the rights of individual stock exchanges (see Real-Life Nugget 14.5).

Real-Life Nugget 14.5

The International Accounting Standards Committee

IASC sets the standards

IASC has reached a major landmark in its drive to develop global accounting standards.

Its core standards have been endorsed by IOSCO, the international organisation of securities regulators. At its annual meeting in Sydney, IOSCO resolved that it recommends its members to permit incoming multinational issuers to use the IASC standards for cross-border offerings and listings. The core standards are all of IASC's standards except for a few that have specialised application.

What does the endorsement decision mean for the future? All of IOSCO's major members, except the US and Canada, already accept use of IASC's standards in cross-border listings without any conditions. This is not expected to change.

SIR BRYAN CARSBERG

The IOSCO decision does not directly affect the position in the national jurisdictions of its members. Action by the SEC is needed before IASC's standards can be given higher recognition in the US.

The SEC published a concept release about use of international standards in February, and when the comments have been analysed, we shall learn more about the way forward.

In the meantime, we can draw certain encouraging inferences. We know the SEC supported the IOSCO resolution. The resolution refers to the possibility that some administrations will require 'reconciliation of certain items'. This seems to offer the prospect of a major step forward from the present position, in which a virtually complete reconciliation is required.

However, the IASC Board has proposed a restructuring of IASC that is the key to future prospects. It will achieve the highest levels of independence in decision-making and create the possibility of an effective partnership with national standard setters to work on elimination remaining differences between national and international standards.

The final stage is a vote of members that took place on 24 May. If the decision on endorsement by IOSCO is followed by final approval of the restructuring by members, we shall be securely on the path to global accounting standards.

● *Sir Bryan Carsberg is secretary general of the International Accounting Standards Committee*

Source: Bryan Carsberg, 'IASC sets the standards', *Accountancy Age*, 25 May 2000, p. 26

The US Securities Exchange Commission (SEC), which controls the world's most important stock market, thus still has to endorse IAS. However, interestingly, in the spring of 2000 the European Union proposed that by 2005 all listed European companies should comply with IAS.

One particular problem is that the US appears reluctant to adopt IAS. Although in principle, the US favours world standards, it has several concerns about IAS. It feels they are not as rigorous as US standards and is also worried about their enforcement. However, the shortcomings of US accounting standards revealed by recent US accounting scandals make IAS potentially more attractive to US regulators. In 2001 the IASB was restructured and a new chairman, Sir David Tweedie, was appointed.

Pause for Thought 14.4

US Acceptance of IAS Standards

Cynics argue that it is in the US's interests deliberately to delay accepting International Accounting Standards. Why do you think this might be?

US standards are probably the most advanced of any country in the world. The US is also the richest country in the world and a source of potential capital for companies from other countries. To gain access to US finance, many overseas companies list on the US Stock Exchange. However, to do this they must adopt US standards or provide reconciliations to US standards. As time passes, more foreign companies adopt US standards. Cynics, therefore, believe that the US may be playing a waiting game. The longer the US delays approving IAS, the more foreign companies will list on the US exchange. These cynics argue, therefore, that it is in the US's interests to delay accepting IAS.

From a UK perspective, the Accounting Standards Board appears to have accepted the need for eventual international harmonisation. The current policy is to depart from an international consensus only when there are particular legal or tax difficulties or when the UK believes the international approach is wrong. Recently, the ASB has made important efforts to harmonise UK and IAS standards in key areas such as goodwill, taxation and pensions. Jones (1998, p. 32), for example, comments, 'as we, therefore, enter the next millennium, it is likely that UK standards and IAS will probably be compatible in all substantial respects'. The UK is also harmonising its accounting practices with those currently used in the US (see Real-Life Nugget 14.6).

Real-Life Nugget 14.6

UK and US Harmonisation

Global harmonisation of accounting standards has taken two transatlantic steps forward with the news that both the UK and US are prepared, in two specific cases, to give up business-friendly domestic rules in the interests of international convergence and more transparent accounting.

Both issues – accounting for tax in the UK, and for mergers and acquisitions in the US – are fundamental building blocks of financial reporting. The removal of a transatlantic divide on both offers investors comparability and perhaps reminds them how uneven the present system is.

Sir David Tweedie, chairman of the UK's Accounting Standards Board, was brutally honest when he announced proposals that would see some leading companies face a sharp drop in reported earnings as they are required to account fully for tax liabilities.

Source: Jim Kelly, 'Removing the transatlantic divide', *Accountancy*, 16 September 1999

US Standards

If the SEC refuses to endorse the IAS, it is possible that US standards will become a substitute for worldwide accounting standards. More and more companies worldwide are using US standards. The advantage is that this enables them to list on the New York Stock Exchange. This is the world's largest stock exchange and a ready source of capital.

Conclusion

Different countries have different accounting environments. These accounting environments are determined by divergent forces, such as objectives, users, sources of finance, regulation, taxation, the accounting profession, spheres of influence and culture. These divergent forces are interrelated and the mix of divergent forces is unique for each country. Countries can be classified as having macro or micro accounting systems. In macro countries, such as France or Germany, the emphasis is on tight legal regulation, with a creditor orientation and a weak accounting profession. By contrast, in micro countries such as the UK and the US, there is a focus on regulation via standards rather than the law and an overall investor-orientated approach with a strong accounting profession.

The great variety of accounting systems worldwide impedes the growth in world trade. Consequently, there are pressures for countries to depart from purely nationally-based accounting. The pressures result from the growth in multinational companies and in cross-border trade and investment. These convergent forces thus counteract the divergent forces.

At the European level, the European Union works towards the harmonisation of accounting practices. On the global level, the International Accounting Standards Board sets International Accounting Standards. Having made a slow start, the IASB is now gathering momentum. In particular, in 1995, an agreement was made with the International Organisation of Securities Commissions that the IASB would draw up a set of core accounting standards. The idea is that these will then be used for all cross-border listings. So far, this has not yet happened. The US Securities and Exchange Commission, in particular, has reservations. In addition, the European Union now supports IAS.

Selected Reading

1. Books

International Accounting is blessed with some very good, comprehensive books. As International Accounting is continually changing, it is always necessary to check that you have the latest edition.

Nobes, C.W., and R.H. Parker (2000), *Comparative International Accounting* (Fifth Edition), (Prentice Hall International: London).

This is the longest standing of the books. Nobes and Parker write some of the chapters themselves, but there are also useful chapters by Peter Standish on France and Klaus Langer on Germany.

Roberts, C., P. Weetman and P. Gordon (2002), *International Financial Accounting: A Comparative Approach* (Financial Times Management: London).

This book provides a comprehensive coverage of the issues in this chapter.

Walton, P., A. Haller, and B. Raffournier (1998), *International Accounting*, (International Thompson Business Press: London).

Once more this is a useful book. Each chapter is written by a specialist.

2. Articles

Students may find the following two articles provide a reasonable coverage of the International Accounting Standards Board and Germany, respectively.

Jones, M.J. (1998), 'The IASB: Twenty-five years old this year', *Management Accounting*, May, pp. 30–32.

Jones, M.J. (1999), 'Germany: An accounting system in change', *Accountancy International*, August, pp. 64–5.

The key article on the micro–macro classification of accounting systems is listed below. Also there is a useful chapter in Nobes and Parker (2000).

Nobes, C.W. (1983), 'A judgmental international classification of financial reporting practices', *Journal of Business Finance and Accounting*, Spring, pp. 1–19.

Discussion *Questions*

Questions with numbers in blue have answers at the back of the book.

Q **1** Why is the study of international accounting important?

Q **2** What are the main divergent forces and which are the most important?

Q **3** Compare and contrast the main features of the Anglo-American and continental accounting systems.

Q **4** Taking any one country, rank the divergent forces in order of importance.

Q **5** Has the UK more in common with the US than with France?

Q **6** Will the convergent forces outweigh the divergent forces?

Section C

Management Accounting

So far, we have looked at financial accounting. Financial accounting is much more publicly visible than management accounting. Management accounting takes place within businesses and is essential to the running of a business. It can be usefully divided into (1) cost accounting, which includes costing (Chapter 16), and planning and control (Chapters 17 and 18), and (2) decision making (Chapters 19–22).

This section investigates some major issues involved in management accounting. In Chapter 15, a brief introduction into the nature of management accounting is provided. Much of the terminology is introduced and an overview of the rest of the book is provided. Chapters 16, 17 and 18 look at cost accounting, one of the two main streams of management accounting. In Chapter 16, the role of costing in stock valuation and pricing is investigated. The cost allocation process is then outlined together with the use of different costing methods in different industries. Both traditional absorption costing and the more modern activity-based costing are discussed. Chapters 17 and 18 look at planning and control. Chapter 17 sets out the cost control technique of budgeting. The different types of budget are also briefly discussed. Then in Chapter 18 standard costing is explored. The individual variances are explained and then a worked example is shown.

Chapters 19–22 look at the decision-making aspects of management accounting. Chapter 19 investigates the managerial techniques (such as break-even analysis and contribution analysis) used for short-term operational decision making. Then Chapter 20 looks at the relatively new management accounting topic of strategic management accounting. This shows how management accounting influences strategic business decisions. In Chapter 21, we investigate capital budgeting. This chapter investigates the different techniques (such as payback and discounted cash flow) used to evaluate long-term investment decisions. Finally, Chapter 22 investigates two aspects of both internal and external sources of finance: short-term financing via the efficient management of working capital and long-term financing such as share capital and loan capital.

Chapter 15

Introduction to Management Accounting

'Jones was loyal to his staff, and diplomatic, but clearly had been bewildered by what he found when he arrived. "Well, we've only just got an audited balance sheet for 1998", he said, "We just don't have any contemporary operating or financial data on which to make management decisions in this airline at the moment. It's the old garbage-in, garbage-out syndrome."'

Graham Jones commenting on the Greek airline company *Olympic Airlines*

Source: Matthew Gwyther, 'Icarus Descending', *Management Today* January 2000, pp. 52–53

Learning Outcomes

After completing this chapter you should be able to:

- ✔ Explain the nature and importance of management accounting.
- ✔ Outline the relationship between financial accounting and management accounting.
- ✔ Explain the main branches of cost accounting and decision making.
- ✔ Discuss cost minimisation and revenue maximisation.
- ✔ Understand some of the major terms used in management accounting.

In a Nutshell

- Management accounting is the provision of accounting information to management to help with costing, planning and control, and decision making.
- Whereas the main focus of financial accounting is external, management accounting is internally focused.
- Management accounting has its origins in costing; however, nowadays costing is less important as new areas such as strategic management accounting develop.
- Management accounting can be broadly divided into cost accounting (costing, and planning and control) and decision making (short-term and long-term).
- Costing consists of recovering costs for pricing and for the valuation of stock.
- Planning and control consists of planning and controlling future costs using budgeting and standard costing.
- Decision making involves short-term decision making, strategic management accounting, capital budgeting and sources of finance.
- Traditionally, management accounting has been criticised for focusing on minimising costs rather than maximising revenue.

Introduction

Management accounting is concerned with the internal accounting within a business. Essentially, it is the provision of both financial and non-financial information to managers so that they can manage costs and make decisions. In many ways, therefore, management accounting is less straightforward than financial accounting. Management accounting varies markedly from business to business and management accountants, in effect, carry around a toolkit of techniques, their 'tools of the trade'. The purpose of this section is to try to explain these techniques and to fit them into an overall picture of management accounting.

Pause for Thought 15.1

Management Accounting

Why do you think management accounting is so called?

Management accounting is a relatively new term and has been in widespread use only since the 1950s. The term combines management and accounting. It suggests that accountants have a managerial role within the company. They are, in effect, more than just a functional specialist group. The term management accounting has come, therefore, to represent all the management and accounting activities carried out by accountants within a business. This involves not only costing, and planning and control, but also managerial decision making.

Context

The management accountant works within a business. The focus of management accounting is thus internal rather than external. In this book, we simplify the official definition (see Definition 15.1) and take the purpose of management accounting as being to provide managers with accounting information in order to help with costing, planning and control, and decision making. As the formal, official definition shows, management accounting concerns business strategy, planning and control, decision making, efficient resource usage, performance improvement, safeguarding assets, corporate governance and internal control.

Definition 15.1

Management Accounting

Working definition

The provision of financial and non-financial information to management for costing, planning and control, and decision making.

Formal definition

'The application of the principles of accounting and financial management to create, protect, preserve and increase value so as to deliver that value to the stakeholders of profit and not-for-profit enterprises, both public and private. Management accounting is an integral part of management, requiring the identification, generation, presentation, interpretation and use of information relevant to:

- formulating business strategy;
- planning and controlling activities;
- decision making;
- efficient resource usage;
- performance improvement and value enhancement;
- safeguarding tangible and intangible assets;
- corporate governance and internal control.'

Source: Chartered Institute of Management Accounting (2000), *Official Terminology*

In essence, costing concerns (1) setting a price for a product or service so that a profit is made and (2) arriving at a correct valuation for stock. Planning and control involves planning and controlling future costs using budgeting and standard costing. Decision making involves managers solving problems using various problem-solving techniques.

Pause for Thought 15.2

Problem Solving

Can you think of five problems the management accountant may need to solve?

The list is potentially endless. However, here are ten.

1. What products or services should be sold?
2. How much should be sold?
3. Should a product be made in-house or bought in?
4. Should the firm continue manufacturing the product?
5. At what level of production will a profit be made?
6. Which products or services are most profitable?
7. How can the firm minimise its working capital?
8. How can the firm maximise revenue?
9. Which areas should the firm diversify into?
10. Should new finance be raised via debt or equity?

Relationship with Financial Accounting

The orientation of management accounting is completely different to that of financial accounting. As Figure 15.1 shows, financial accounting is concerned with providing information (such as the balance sheet and profit and loss account) to shareholders about past events. Management accountants will also be interested in such information, often on a monthly basis. However, management accountants will also require a broader range of internal management information for costing, planning and control, and decision making. Importantly, whereas financial accounting works within a statutory context, management accounting does not. Management accounting is thus much more varied and customised than financial accounting.

Figure 15.1 Relationship between Financial and Management Accounting

Financial Accounting	*Management Accounting*
1. Aims to provide information principally for external users, such as shareholders.	1. Aims to provide information for internal users such as management.
2. Concerned with recording information using double-entry bookkeeping.	2. Not so concerned with recording. Information is needed for costing, planning and control, decision making etc.
3. Works within a statutory context.	3. There is no statutory context.
4. Main statements are balance sheet and profit and loss account.	4. Not so concerned with preparing these financial statements.
5. Basically looks back to the past.	5. Looks to the future.
6. The end product is the annual reporting package in a standardised format.	6. Different businesses produce very different management information.

Financial accounting, in fact, has influenced the development of management accounting. Management accounting emerged much later than financial accounting. Indeed, early management accountants were primarily cost accountants (see Figure 15.2).

Armstrong and Jones (1992) argue that management accountants have been involved in a 'collective mobility' project where cost accountants have redefined themselves as management accountants dealing with wider management accounting and strategic management issues. In part, this has been an attempt to move away from costing, which was perceived as low status, and to emulate the Institute of Chartered Accountants in England and Wales, which was considered as high status.

Figure 15.2 The Cost Accountant

'The cost accountant is essentially a practical man and it is important for the registered student to have first-hand experience of works and factory practice and routine, including, for example, the following: operation of various machines; shop floor organisation; purchase; storage and control of materials; design, planning and progress of work; inspection; work study; maintenance; warehousing and distribution.'

Source: *The Cost Accountant*, December 1957, p. 280. As cited in P. Armstrong and C. Jones (1992), *Management Accounting Research*, pp. 53–75

Perhaps because of its comparatively humble origins, management accounting has often been seen as subservient to financial accounting. Johnson and Kaplan (1987) argued in *Relevance Lost* that most management accounting practices had been developed by 1925 and that since then there has been comparatively little innovation. Under this view, management accounting has major problems. In particular, product costing is distorted, decision-making information becomes unreliable and management accounting reflects external reporting requirements rather than modern management needs.

Opinions differ on the current relevance of management accounting. However, it is true that management accounting has struggled to adapt to changes in the business environment. In particular, it has been relatively slow to adapt to the decline of manufacturing industry and the rise of service organisations (see Figure 15.3). These difficulties are likely to be exacerbated by the rise of knowledge-based companies. Management accounting is also struggling to adapt to globalisation and technological change. As we will see, however, new management techniques have arisen (such as activity-based costing, strategic management accounting and just-in-time stock valuation) which seek to address the criticisms of management accounting.

Figure 15.3 Service Industries

'There are many service organisations which do not have finished good stocks or work-in-progress. They require management accounting information to ascertain the cost of each service and its contribution to total company profits. These organisations do not have to conform to any financial accounting requirements for the purpose of tracing costs to various services. Nevertheless, most service organisations adopted traditional product cost accounting techniques based on arbitrary overhead allocation, to trace costs to the different business segments.'

Source: Drury, C. (1996), *Cost and Management Acounting*, pp. 833–34

Overview

The diversity of management accounting makes if difficult to separate out individual strands. However, this book broadly splits management accounting into two.

1. **Cost Accounting**. This involves:
   (i) *Costing* (recovering costs as a basis for pricing and for stock valuation), and
   (ii) *Planning and control* (i.e., planning and controlling future costs using budgeting and standard costing).
2. **Decision Making**. This involves:
   (i) *Short-term decision making* (such as break-even analysis, contribution analysis and the efficient management of working capital), and
   (ii) *Long-term decision making* (i.e., strategic management accounting, capital budgeting, and management of sources of long-term finance).

These two major areas are shown in Figure 15.4 on the next page.

Figure 15.4 Overview of Management Accounting

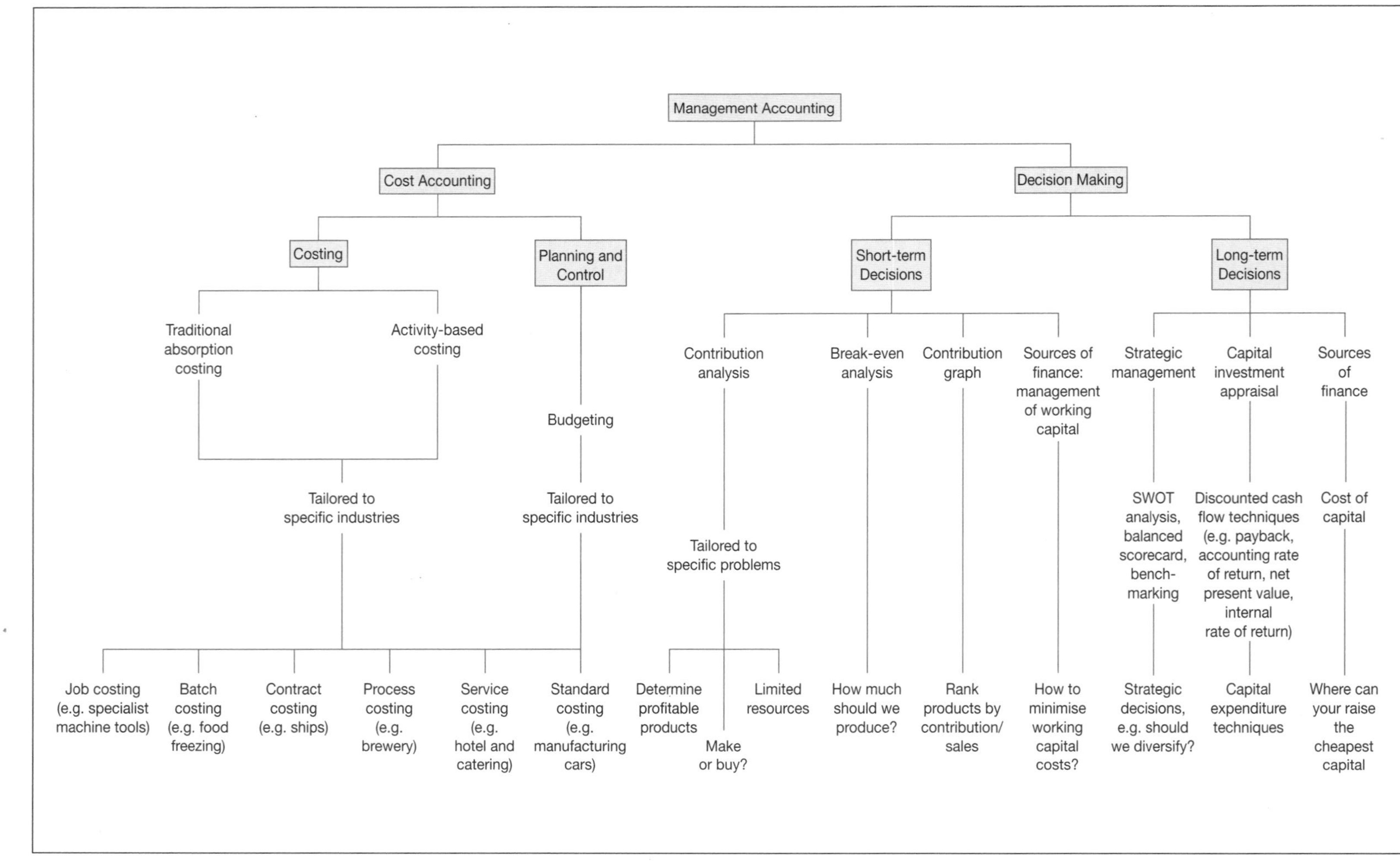

The major areas shown in figure 15.4 are now discussed briefly below and then more fully in the chapters which follow. Key terms that underpin management accounting are introduced. These new terms are highlighted in **bold** in the text and a full definition is then provided in Definitions 15.2 on page 330 and 15.3 on page 331 which follow the same order as in the text. These terms are more fully discussed later in the book.

Cost Accounting

Costing (Chapter 16)

Costing has its origins in manufacturing industry. The basic idea underpinning costing is cost recovering (i.e. the need to recover all the **costs** of making a product into the price of the final product). Broadly, cost recovery is achieved using a technique called **total absorption costing.** Total absorption costing seeks to recover both **direct costs** and **indirect costs** into a product or service. This form of costing thus seeks to establish the total costs of a product so that they can be recovered in the final selling price. As well as costing for pricing, costing is used for stock valuation. In this case, either **absorption costing** or **marginal costing** is used.

Different industries will have different costing systems. For example, medicinal tablets will necessitate batch processing, while shipbuilding uses contract costing. A more modern technique called activity-based costing is often used.

Planning and Control (Chapters 17 and 18)

Planning and control aims to control current costs and plan for the control of future costs. The major techniques that underpin cost control are budgeting and standard costing.

- *Budgeting* (Chapter 17). Budgeting involves setting future targets. Actual results are then compared with these budgeted results. Any differences will be investigated.
- *Standard costing* (Chapter 18) is a standardised version of budgeting. Standard costing involves using preset costs for direct labour, direct materials and overheads. Actual costs are then compared with the **standard costs**. Any **variances** are then investigated.

Decision Making

Decision making involves choosing between alternatives. When making a choice it is essential to consider only those costs that are relevant to the actual decision, i.e., *relevant costs.*

Short-Term Decisions (Chapter 19)

In many ways, distinguishing between short-term, operational decisions and long-term strategic decisions is somewhat arbitrary. However, it proves a useful basic distinction. Short-term decisions are dealt with in Chapter 19 and the first part of Chapter 22. Long-term decisions are covered in Chapters 20–21 and in the second half of Chapter 22.

Businesses face many short-term decisions. Essentially, as Soundbite 15.1 shows, decision making is about choice. Businesses need to make numerous decisions.

Soundbite 15.1

Choice or Deciding

'The art of management is about choice.'

G. Richard Thoman, *Business Week* (December 9, 1996).

Source: *The Wiley Book of Business Quotations* (1998), p. 295

Definition 15.2

The Basic Building Blocks of Management Accounting: Some Key Cost Accounting Terms

i Costing

Cost. A cost is simply an item of expenditure (planned or actually incurred).

Total absorption costing. This form of costing is used to recover all the costs (both direct and indirect) incurred by a company into the price of the final product or service.

Direct costs. Direct costs are those costs that can be directly identified and attributed to a product or service, for example, the amount of direct labour that is incurred making a product. Sometimes these costs are called *product costs.*

Indirect costs. Indirect costs are those costs that cannot be directly identified and attributed to a product or service, examples are administrative, selling and distribution costs. These costs are totalled and then recovered into the product or service in an indirect way. For example, if there were £10,000 administrative costs and 10,000 products, each product might be allocated £1 of administrative costs. Indirect costs are often called *overheads* or *period costs.*

Absorption costing. Absorption costing is the form of costing used for valuing stock for external financial reporting. It recovers the costs of all the overheads that can be directly attributed to a product or service. However, unlike marginal costing, which can sometimes be used for stock valuation, it includes both fixed and variable production overheads.

Marginal costing. Marginal costing excludes fixed costs from the costing process. It focuses on sales, variable costs and contribution. Fixed costs are written off against contribution. It can be used for decision making or for valuing stock. When valuing stock only variable production overheads are included in the stock valuation. (See Definition 15.3 for explanations of *fixed costs*, *variable costs* and *contribution.*) Marginal costing is called contribution analysis when used for decision making.

ii Planning and Control

Standard costs. Standard costs are individual cost elements (such as direct materials, direct labour and variable overheads) which are estimated in advance. Normally, the quantity and the price of each cost element is estimated separately. Actual costs are then compared with standard costs to determine variances.

Variances. Variances are the difference between the budgeted costs and the actual costs in both budgeting and standard costing. These variances are then investigated by management.

For example, should a business make a product itself or buy in the product? Should a business continue making or providing a product or service? At what price will a product break even (i.e., make neither a profit nor a loss)? These decisions are solved using various problem-solving techniques such as break-even analysis and contribution analysis. In such short-term decisions, it is essential to distinguish between **fixed costs** and **variable costs.**

A key aspect of short-term decision making is contribution. Contribution is sales less variable costs. In other words, fixed costs are excluded. Contribution is at the heart of two common management accounting techniques: **break-even analysis** and **contribution analysis.** This area of management accounting is sometimes called **cost-volume-profit analysis.**

Definition 15.3

The Basic Building Blocks of Management Accounting: Some Key Decision-Making Terms

Short-Term Decisions

Fixed costs. Fixed costs are those costs that *will not vary* with production or sales (for example, insurance) in an accounting period. They will not, therefore, be affected by short-term decisions such as whether or not production is increased.

Variable costs. Variable costs, however, *will vary* with production and sales (for example, the metered cost of electricity). Short-term decision making is primarily concerned with variable costs. Sales less variable costs gives *contribution*, a useful concept in short-term decision making.

Break-even analysis. Break-even analysis involves calculating the point at which a product makes neither a profit nor a loss. Fixed costs are divided by the contribution per unit (i.e. sales less variable costs divided by number of products). This gives the break-even point.

Contribution analysis. When there is more than one product, this technique is useful in determining which product is the most profitable. It compares the relative contribution of each product. This can also be called cost-volume-profit analysis.

Long-Term Decisions

Discounted cash flow. Discounted cash flow discounts the future expected cash inflows and outflows of a potential project back to their present value today. Decisions can then be made on whether or not to go ahead with the project.

Strategic Management Accounting (Chapter 20)

Whereas costing represents the oldest branch of management accounting, strategic management accounting represents the newest. Strategic management accounting represents a response to criticisms that management accounting is outdated and lacking in innovation. Using strategic management accounting, the management accountant relates the activities of the firm to the wider external environment. Strategic management is thus concerned with the long-term strategic direction of the firm.

Pause for Thought 15.3

Cost Accounting and Management Accounting

A distinction is sometimes made between cost accounting and management accounting. What do you see as the essential difference?

Management accounting has its origins in cost accounting. However, gradually cost accounting and management accounting have been seen as distinct. Cost accounting is seen, at its narrowest, as focusing on cost collection and cost recovery and more widely as cost recovery and control. However, management accounting is viewed as a much broader term which not only encompasses cost accounting, but also involves decision making and, more recently, strategic management accounting. Cost accounting is thus typically portrayed as routine and low-level, whereas management accounting is seen as a higher-level activity.

Capital Investment Appraisal (Chapter 21)

Capital budgeting involves the financial evaluation of future projects. This evaluation is carried out by using various techniques such as payback, the accounting rate of return, net present value and the internal rate of return. Under net present value and the internal rate of return, the technique of **discounted cash flow** analysis is used.

Sources of Finance (Chapter 22)

The topic of business finance is immense. The aspects covered in this book are the financing of the business in the short and the long term. This covers both the internal generation of funds and the raising of external finance. In the short term, the management of working capital involves the management accountant in a series of short-term decisions about the optimal level of stock and debtors. Meanwhile, the long-term financing of the business involves the choice between internal financing through retained profits and external financing through either leasing, loans or share capital.

Pause for Thought 15.4

Short-Term vs Long-Term Decisions

What do you think are the essential differences between short-term and long-term decisions?

Short-term decisions are operational, day-to-day decisions which typically involve the firm's internal environment. For example, what quantity of a particular product should we make or what should be the price of a particular product.

By contrast, long-term decisions are strategic, non-operational decisions. These typically involve a firm's external environment. So, for example, they may involve the need to diversify or the need to make acquisitions and disposals. Alternatively, they might evaluate whether a long-term capital investment is worthwhile.

Cost Minimisation and Revenue Maximisation

The management accountant can make a business more efficient through cost minimisation or revenue maximisation. Cost minimisation attempts to reduce costs. This may be achieved by tight budgetary control or cutting back on expenditure.

Some authors argue strongly that management accounting needs to refocus on revenue maximisation. In part, revenue maximisation is achieved through the new focus on strategic management accounting which looks outwards to the external environment rather than inwards. Substantial opportunities arise to adopt new techniques such as customer database mining. This latter technique looks at customer databases and seeks to extract from them customer data which will expand the business's revenue.

Management accountants have often been criticised for being overly concerned with cost cutting (see, for example, Lesley Jackson, H.P. Bulmer Holding plc's UK Finance Director, in Real-Life Nugget 15.1).

Soundbite 15.2

Sales Maximisation

'Legend tells of the traveller who went into a county store and found the shelves lined with bags of salt. "You must sell a lot of salt", said the traveller, "Nah", said the storekeeper. "I can't sell no salt at all. But the feller who sells me salt – boy, can *he* sell salt".'

Martin Mayer, *The Bankers*, 1974.

Source: *The Executive's Book of Quotations* (1994), p. 255

Real-Life Nugget 15.1

Cost minimisation

Jackson's time as general commercial manager gave her the opportunity to look at a business from a different angle. She says it made her a better accountant.

'A lot of accountants tend to look at cost minimalisation and low risk. They tend to have a more conservative mindset. I became slightly more maverick in this sense,' she says.

It was this broader outlook on business and varied skills-sets that gave Jackson the edge over other candidates for the role at Bulmers.

Source: Michelle Perry, 'Brewing up a profit', *Accountancy Age*, 3 May, 2001, p.20

Use of Computers

The theory and practice of management accounting is shown in the next seven chapters. For most of the techniques shown, in practice, a dedicated computer program would be used. Alternatively, a spreadsheet could be set up so as to handle the calculations. These programs enable complicated and often complex real-life situations to be modelled. However, it is essential to be able to appreciate which figures should be input into the computer. As Peter Williams states.

> 'While it may be possible for any company with a PC to produce a set of management information using relatively low-cost accounting software, there is no guarantee that the output will be true and fair.'
>
> (*Accountancy Age*, 2 March, 2000, p. 23)

Conclusion

Management accounting is the internal accounting function of a firm. It can be divided into cost accounting (costing, and planning and control), and decision making (short-term and long-term). In costing, the two main aspects are pricing and stock valuation. In planning and control, budgeting and standard costing are used. Decision making has four main strands: short-term decision making; strategic management accounting; capital budgeting; and sources of finance. Management accountants are increasingly moving away from costing towards new areas such as strategic management accounting. In effect, this reflects the decline of manufacturing industry in developed countries.

Selected Reading

Armstrong, P. and C. Jones (1992), 'The decline of operational expertise in the knowledge-base of management accounting: An examination of some post-war trends in the qualifying requirements of the Chartered Institute of Management Accountants', *Management Accounting Research*, Vol. 3, pp. 53–75.
An interesting look at how the management accounting profession has gradually changed over time. Originally concerned with costing, it now has a much wider focus.

Drury, C. (1996), *Management and Cost Accounting* (International Thomson Business Press: London).
This is a comprehensive text on management and cost accounting. Once students have mastered the basics, this represents a good book for future reading.

Johnson, T. and R.S. Kaplan (1987), *Relevance Lost: The Rise and Fall of Management Accounting* (Harvard University Press).
A benchmarking book which triggered a relook at management accounting. After this book a new management accounting emerged consisting of topics such as activity-based costing and strategic management accounting.

Kaplan, R.S. (1984), 'Yesterday's accounting undermines production', *Harvard Business Review*, July/August, pp. 95–101.
This provides a good overview of the problems with traditional management accounting.

Discussion *Questions*

Questions with numbers in blue have answers at the back of the book.

Q 1 What are the main branches of management accounting and what are their main functions?

Q 2 Why do you think that management accounting has been so keen to lose its costing image?

Q 3 What are the following types of cost and why are they important?

- (a) Direct cost
- (b) Indirect cost
- (c) Fixed cost
- (d) Variable cost
- (e) Standard cost

Q4 The management accountant has been described as a professional with a toolkit of techniques. How fair a description do you think this is?

Q5 Why have management accountants been criticised for being cost minimisers and how might they be revenue maximisers?

Q6 State whether the following statements are true or false. If false explain why.

(a) The two main branches of management accounting are cost accounting and decision making.

(b) Total absorption costing is where all the overheads incurred by a company are recovered in the valuation of stock.

(c) The difference between absorption costing and marginal costing as a form of costing for stock valuation is that absorption costing includes direct materials, direct labour and all production overheads. By contrast, marginal costing only includes direct materials, direct labour and all *variable* production overheads. Marginal costing, therefore, excludes fixed production overheads.

(d) Strategic management accounting is concerned principally with short-term operational decisions.

(e) Discounted cash flow discounts the future expected cash flows of a project back to their present-day monetary values.

Chapter 16

Costing

'Watch the costs and the profits will take care of themselves.'

Andrew Carnegie, quoted in R. Sobel and D.B. Silicia, *The Entrepreneurs – An American Adventure* (Houghton Mifflin, 1986)

Source: *Wiley Book of Business Quotations* (1998), p. 89

Learning Outcomes

After completing this chapter you should be able to:

- ✔ **Explain the nature and importance of costing.**
- ✔ **Discuss the process of traditional costing.**
- ✔ **Understand the nature of activity-based costing.**
- ✔ **Distinguish between different costing systems.**
- ✔ **Discuss target costing.**

In a Nutshell

- Costing is a subset of cost accounting and is used as a basis for stock valuation and for cost-plus pricing.
- There are direct costs and indirect costs or overheads.
- The six stages in traditional cost recovery are (1) recording costs, (2) classifying costs, (3) allocating indirect costs to departments, (4) reapportioning costs from service to productive departments, (5) calculating an overhead recovery rate, and (6) absorbing costs into products and services.
- Activity-based costing is a sophisticated version of cost recovery which uses cost allocation drivers based on activities to allocate costs.
- Different industries have different costing methods such as job costing, batch costing, standard costing, contract costing, process costing and service costing.
- Non-production overheads are included in cost-plus pricing, but not in stock valuation.
- In stock valuation, either all production costs (absorption costing) or only variable production costs (marginal costing) can be allocated.
- Target costing, developed by the Japanese, is based on market prices and set at the pre-production stage.
- Companies in trouble often cut costs, such as labour costs, in order to improve their profitability.

Introduction

Management accounting can be broadly divided into cost accounting and decision making. In turn, cost accounting has two major strands: costing, and planning and control. Costing involves ascertaining all the costs of a product or service so as to form the basis for pricing and for stock valuation. The first management accountants were essentially cost accountants. Their job was to make sure that manufactured products were priced so as fully to recover all the costs incurred in making them. This process of recording, classifying, allocating the costs and then absorbing those costs into individual products and services still remains the basis of cost recovery. However, the traditional methods of cost recovery were geared up for manufacturing industries and assumed that overhead costs were relatively small. The decline of manufacturing industry, the increasing importance of overhead costs, and the rise of service industries has caused a need to rethink some of the basics of cost recovery. In particular, the technique of activity-based costing has gained in popularity.

Importance of Cost Accounting

Cost accounting is a common term used to embrace both costing, and planning and control. Definition 16.1 shows two formal definitions of cost accounting. The first is by the Institute of Cost and Management Accountants (ICMA), the predecessor body of the author of the second definition, the Chartered Institute of Management Accountants (CIMA). It is interesting to see that the definition has widened considerably. In particular, the more recent CIMA definition now explicitly mentions budgeting and standard costing.

Definition 16.1

Cost Accounting

Working definition

The determination of actual and standard costs, budgeting and standard costing.

Formal definition

1. 'The classification, recording and appropriate allocation of expenditure in order to determine the total cost of products or services.'

This earlier definition by ICMA is a good description of cost recovery.

2. 'The establishment of budgets, standard costs and actual costs of operations, processes, activities or products; and the analysis of variances, profitability or the social use of funds. The use of the term 'costing' is not recommended except with a qualifying adjective, e.g. standard costing.'

This CIMA (2000) definition embraces both cost accounting, and planning and control.

In this book, we take cost accounting to involve two main parts. The first is costing, which is the process of determining the actual costs of products and service. This usually looks to the past. And, the second is costing for planning and control where expected costs are determined for future periods either through budgeting or standard costing.

Costs are the essential building blocks for both financial and management accounting. In costing, costs represent *actual* items of expenditure. In budgeting and standard costing, costs represent *future* (expected) items of expenditure. Costs represent a major building block of management accounting. In financial accounting, it is important to match costs against revenue to determine profit.

An important part of this matching process in manufacturing industry is the allocation of production costs to stock. Either all production costs (absorption costing) or variable production costs (marginal costing) can be allocated. Sometimes marginal costing is termed variable costing. In Figure 16.1, we use absorption costing, in which all the production costs are allocated to stock. These costs will be included in the cost of opening stock in the next accounting period and then matched against sales to determine profit. Later on, in Figure 16.12 on page 352, both absorption and marginal costing are shown for comparative purposes.

Figure 16.1 Allocation of Costs to Stock Using Absorption Costing

Stockco manufactures only toys. The direct costs of manufacturing 1,000 toys is £1,000. Total production overheads for the toys (for example, factory light and heat) are £500. At the end of the year 200 toys are in stock. What is the cost of the closing stock?

	1,000 units	Per unit
	£	£
Direct costs	1,000	1.00
Production overheads	500	0.50
	1,500	1.50

The closing stock is thus 200 @ 1.50 = £300

Another extremely important function of costing is as a basis for pricing. The selling price of a product is key to making a profit. In many businesses, the fundamental economic law is that the cheaper the price of the goods, the more goods will be sold. This sentiment is expressed in Real-Life Nugget 16.1.

Real-Life Nugget 16.1

Pricing Policy

Retailing used to be so simple. If shopkeepers wanted to increase sales, they cut their prices, putting their goods within reach of people who previously could not afford them.

In developed countries, as incomes have risen, higher prices have become more affordable. But if retailers thought the need for discounting would fade, they were wrong.

Source: Richard Tomkins, 'Marketing Value for Money', *Financial Times*, 14 May 1999

In practice, there are two methods of pricing: market pricing and cost-plus pricing. In market pricing, the focus is external. The prices charged by competitors are examined together with the amount that customers are willing to pay. In cost-plus pricing, by contrast, the focus is internal. The total costs of making the product are established (i.e., all the direct costs and all the indirect costs or overheads) and then a profit percentage or profit mark-up is added.

Pause for Thought 16.1

Cost-Plus Pricing

What is the purpose of cost-plus pricing and how can it prove dangerous in a competitive market?

Cost-plus pricing seeks to recover all the overheads into a product or service. However, it is only as good as the method chosen to recover the overheads, and the estimates made. The problem is that prices are often set by the market rather than by the company. A company may determine the cost of a product and then its selling price. However, in a competitive market, this price may be higher than competitors' prices or be more than customers wish to pay. When adopting cost-plus pricing it is always important to perform a reality check and ask: can we really sell the product or service at this price? In practice, therefore, it is advisable to use both cost-plus pricing and market pricing together.

In general, companies strive to keep their costs as low as possible. Those companies that can do this, such as Wal-Mart, the American retailer that purchased Asda, have a key competitive advantage (see Real-Life Nugget 16.2).

Real-Life Nugget 16.2

Costing as Competitive Advantage

But Wal-Mart is not just another Aldi, Lidl or Netto. A typical Aldi store stocks only about 700 product lines. A Tesco superstore sells far more: about 36,000, including food and general merchandise. But even Tesco is left in the shade by the typical Wal-Mart store, which sells about 100,000 lines. So how can European retailers compete with Wal-Mart? Certainly not on price, says Elliott Ettenberg, chairman and chief executive of Customer Strategies Worldwide, a New York retail consultancy.

'Wal-Mart, as its core competency, has built a low-cost distribution system. That's their competitive advantage,' Mr. Ettenberg says. 'The fact that they're in groceries, that they're in pharmaceuticals, that they're in general merchandise – all of that is a by-product of their ability to construct a process that has all the costs taken out from it right from the very beginning.'

Source: Richard Tomkins, 'Marketing Value for Money', *Financial Times*, 14 May 1999

Types of Cost

A cost is simply an amount of expenditure which can be attributed to a product or service. It is possible to distinguish between two main types of cost: **direct costs** and **indirect costs**. Direct costs are simply those that **can** be directly attributed to a product or service. Indirect costs are those that **cannot** be directly attributed. Indirect costs are also known as overheads. In manufacturing industry, direct costs have declined over time while indirect costs have risen. This makes costing more difficult. Figures 16.2 and 16.3 (see next page) demonstrate two cost structures, for a manufactured product and for a service product, respectively. In both cases, we are totalling all our costs so that we can recover them into the final selling price. This is known as **total absorption costing**. This should be distinguished from absorption costing and marginal costing which, as we have just discussed, are used for stock valuation not for pricing.

Figure 16.2 Cost Structure for a Manufactured Product (for Example, a Computer)

	£	*Examples*
Direct materials	50*	Plastics, steel
Direct labour	120*	Manufacturing labour
Direct expenses	10*	Royalties
Prime Cost	180	
Production overheads	60†	Supervisors' wages
Production Cost	240	
Administrative overheads	50†	Staff costs, office expenses
Selling and distribution overheads	60†	Advertising
Total Cost	350	
Profit	50	
Selling Price	400	

*Direct costs, i.e. those that can be directly attributed to the product
†Indirect costs, i.e. those that cannot be directly attributed to the product

Essentially, a proportion of the cost of the manufactured product can be directly attributed. The direct materials are those such as the plastic, steel and glass that are actually used to make the product. The direct labour is the labour cost actually incurred in making the product. The direct expenses are royalties paid per product manufactured. These three costs (direct materials, direct labour and direct expenses) are known as the **prime cost**. **The remaining costs are all overheads**. The production overheads are those associated with the production process. Examples are supervisors' wages or the electricity used in the factory area. By contrast, administrative overheads (for example, staff costs, office expenses, accountants' fees) and selling and distribution overheads (for example, transport and advertising) are incurred outside the production area. An essential element of costing is that as we move further away from the direct provision of a product or service it becomes more difficult to allocate the costs fairly. Thus,

Soundbite 16.1

'I haven't heard of anybody who wants to stop living on account of the cost.'

Ken Hubbard

Source: R. Flesch (1959) *The Book of Unusual Quotations*, p. 51

it is easiest to allocate the direct costs, but most difficult to allocate the administrative costs. The Company Camera 16.1 provides an example of administrative overheads for the Woolwich Building Society.

The Company Camera 16.1

Administrative Expenses – Staff costs

	2000 £m	*1999* £m
Salaries and accrued incentive payments	155.7	160.8
Social security costs	17.2	17.1
Pension costs	9.8	9.9
Post-retirement benefits	0.7	0.6
Other staff costs	2.4	9.9
	185.8	198.3

Other staff costs reported above for 1999 represent exceptional costs relating to staff reductions.

Source: Woolwich plc, Annual Report 2000, p. 22

In Figure 16.3, we use the example of a computer helpline which callers phone to receive advice. In this case, the only direct costs will probably be direct labour. The administrative, selling and distribution costs will be proportionately higher. The key problem in cost recovery is thus how to recover the overheads. Direct costs can be directly allocated to a cost unit (i.e., an individual product or service unit, such as a customer's phone call). However, indirect costs have to be totalled and then divided up amongst all the cost units. This is much more problematic.

Figure 16.3 Cost Structure for a Service Product (for Example, the Cost per Customer Call for a Computer Helpline)

	£	*Examples*
Direct labour	4*	Telephone operator
Prime Cost	4	
Administrative expenses	5†	Management, office expenses
Selling and distribution costs	4†	Advertising, marketing
Total Cost	13	
Profit	2	
Selling Price	15	

*Direct costs, i.e. those that can be directly attributed to the service
†Indirect costs, i.e. those that cannot be directly attributed to the service

Pause for Thought 16.2

Manufacturing a Television

From the following information can you work out the cost and selling price of each television using total absorption costing?

(i) Direct materials £25; direct labour £50; direct expenses £3 per television.
(ii) £18,000 production overheads; £14,000 administrative expenses; and £15,000 selling and distribution costs.
(iii) 1,000 televisions produced. Profit mark-up of 20% on total cost.

We would thus have:	£
Direct materials	25
Direct labour	50
Direct expenses	3
Prime Cost	78
Production overheads (N1)	18
Production Cost	96
Administrative expenses (N1)	14
Selling and distribution costs (N1)	15
Total Cost	125
Profit (20% × £125, i.e. 20% mark-up on cost)	25
Selling Price	150

N1: To work out overheads per television, divide the total overheads by number of televisions produced.

$$\text{Production overheads} = \frac{£18{,}000}{1{,}000} = £18 \text{ per television}$$

$$\text{Administrative expenses} = \frac{£14{,}000}{1{,}000} = £14 \text{ per television}$$

$$\text{Selling and distribution costs} = \frac{£15{,}000}{1{,}000} = £15 \text{ per television}$$

In this case, we are thus absorbing all the overheads into the product whether they represent direct overheads or indirect overheads (such as production, administrative or selling and distribution expenses).

Traditional Costing

The aim of costing is simply to recover (also known as 'to absorb') the costs into an identifiable product or service so as to form the basis for pricing and stock valuation. In pricing all the overheads are recovered. For stock valuation, only those overheads directly related to the production of stock can be recovered (i.e., direct costs and attributable production overheads). In the next two sections, we look at total absorption costing for pricing, using traditional costing and activity-based costing. In traditional cost recovery or cost absorption, there are six major steps. We will then look at activity-based costing, which is a more modern way of tackling cost recovery. Both traditional total absorption costing and activity-based costing can be used in manufacturing or non-manufacturing industries.

In manufacturing industry, total absorption costing means recovering all the costs from those departments in which products are manufactured (*production departments*) and from those that supply support activities such as catering, administration, or selling and distribution (*service support departments*) into the cost of the end product. In service industries we recover all the costs from those departments that deliver the final service to customers (*service delivery departments*) and from the *service support departments*. We set out below the six steps for recovering costs into products (see Figure 16.4). Diagrammatically, this process is illustrated in Figure 16.5 on the next page. **However, exactly the same process would be used for recovering costs into services.**

Figure 16.4 Six Steps in Traditional Total Absorption Costing

1. Record all the costs.
2. Classify all the costs.
3. Allocate all the indirect costs to the departments of a business.
4. Reallocate costs from service support departments to production departments.
5. Calculate an overhead recovery rate.
6. Absorb both the direct costs and the indirect costs (or overheads) into individual products.

In essence, businesses need to make a profit. Revenue from sales, therefore, needs to exceed the total costs incurred in making a product. Businesses are thus very keen to ascertain their costs so that they can set a reasonable price for their products. Sometimes, especially for long-term contracts or one-off jobs, this can prove very difficult. For example, Cardiff's Millennium stadium, although a triumph of construction, was a financial disaster for John Laing (see Real-Life Nugget 16.3 on next page).

Step 1: Recording

The first step in cost recovery is to record all the costs. It is important to appreciate that many businesses have integrated financial accounting and management accounting systems. There is thus no need to keep separate records for both financial accounting and management accounting.

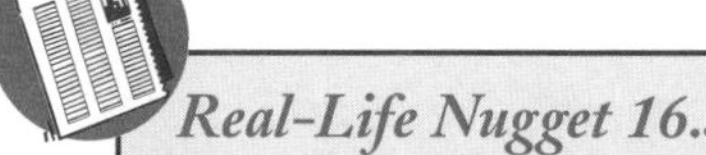

Real-Life Nugget 16.3

Pricing Long-Term Contracts

'Cardiff is the cause,' said Weir. 'The race to finish for the October kick-off has caused further unexpected costs.'

Weir blamed the tender process: 'It was a guaranteed maximum price job and we got the bidding wrong. We have delivered a super project, but a financial disaster.'

Last year John Laing had exceptional losses of £26m on the stadium. There may be more to come, 'but not on the same scale', said Weir.

'We have a four-cylinder business firing on three. We are taking action to get construction right.'

Source: Roger Nuttall, 'Cardiff Stadium a Penalty for Laing', *The Express*, 10 September 1999

Direct costs, such as direct materials and direct labour, can be directly traced to individual products or services. For example, in manufacturing industries, a job may have a job card on which the direct material and direct labour incurred is recorded.

Indirect costs, such as rent or administrative costs, are much more difficult to record. There are two main problems: estimation and total amount. The first problem is that it is unlikely that the actual costs will be known until the end of a period. Transport costs, for example, may be known only after the journeys have been made. They will, therefore, have to be estimated. The second problem is that the total costs are recorded centrally. They must then be allocated to products.

Step 2: Classification

This involves categorising and grouping the costs before allocating them to departments. We may, for example, need to calculate the total rent or total costs for light and heat.

Figure 16.5 Diagrammatic Representation of Traditional Total Absorption Costing

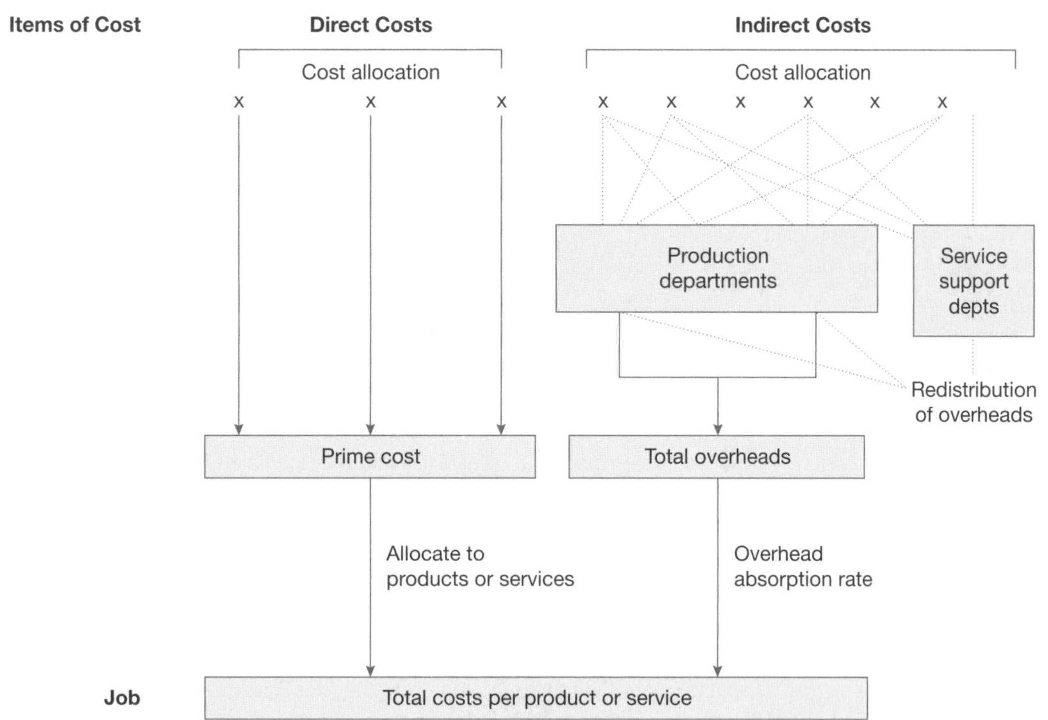

Step 3: Allocate indirect costs to departments

Direct costs can be directly traced to goods or services. For overheads, the process is indirect. The key to overhead allocation is to find an appropriate basis for allocation. These are called allocation bases or cost drivers. Figure 16.6 shows some possible allocation bases for various types of cost.

Figure 16.6 Departmental Allocation Bases

Types of Cost	*Possible Allocation Bases*
Power	Number of machine hours
Depreciation, insurance	Value of fixed assets
Rent, business rates, light and heat	Relative area of floor space
Canteen, office expenses	Number of employees

Step 4: Reallocate service support department costs to production departments

Once all the costs are allocated, we must reallocate them to the production departments. This is because we need to recover these costs into *specific products* which are *made* only in the *production departments.*

Step 5: Overhead recovery rate

Once we have allocated our costs to production departments, we need to absorb these costs into the final product. We must choose a suitable recovery rate. The choice of rate reflects the nature of the activity and could be any of the following (or even others):

- Rate per unit
- Rate per direct labour hour
- Rate per machine hour
- Rate per £ materials
- Rate per £ direct labour
- Rate per £ prime cost

Figure 16.7 illustrates this process

Figure 16.7 Overhead Recovery Rate

A firm has £12,000 indirect costs in department F. It also has the following data available:

	£	
Direct materials	30,000	240,000 units produced
Direct labour	70,000	6,000 direct labour hours are used
Prime Cost	100,000	50,000 machine hours are used

Calculate the various overhead recovery rates discussed above in step 5.

Figure 16.7 Overhead Recovery Rate (continued)

The various overhead recovery rates are:

1.	Per unit	$= \frac{£12,000}{240,000}$	=	£0.05 per unit
2.	Per direct labour hour	$= \frac{£12,000}{6,000}$	=	£2.00 per labour hour
3.	Per machine hour	$= \frac{£12,000}{50,000}$	=	£0.24 per machine hour
4.	Per £ material	$= \frac{£12,000}{30,000}$	=	£0.40 per £ material
5.	Per £ labour	$= \frac{£12,000}{70,000}$	=	£0.17 per £ labour
6.	Per £ prime cost	$= \frac{£12,000}{100,000}$	=	£0.12 per £ prime cost

In practice, only one of these six overhead recovery rates would be used, most likely direct labour hours.

Step 6: Absorption of costs into products

Once we have calculated an absorption rate we can absorb our costs. In practice, there will be many products of varying size and complexity. Figure 16.8 illustrates how we would now recover our costs into products.

Figure 16.8 Recovery into Specific Jobs

If, for example, the following Job X007 was one of the 200,000 units produced and we recover our overheads *per direct labour hour* then we might have the following situation.

Job X007		£
Direct materials	(10 kilos at £2)	20
Direct labour	(12 hours at £12)	144
Prime Cost		164

The direct materials are directly recovered into the job; we must now recover the indirect costs. We recover into our product 12 hours at £2.00 per hour (as calculated in Figure 16.7) = £24.00.

Our total cost is therefore	£
Prime Cost	164
Overheads	24
Total Cost	188
Profit: 25% mark-up on cost	47
Selling Price	235

Comprehensive Example

In Figure 16.9 on the next two pages a comprehensive example of cost recovery looks at all six steps in the cost recovery process.

Figure 16.9 Comprehensive Cost Recovery Example

Millennium Plc has the following three departments:

A. Production
B. Production
C. Service support

Type of Cost	*Proposed Basis of Apportionment*	£
Rent and business rates	Floor area	4,000
Repairs and maintenance	Amount actually spent	1,000
Canteen	Number of employees	500
Depreciation	Cost of fixed assets	2,100
		7,600

It is estimated that:

(i) Department A has a floor area of 4,000 sq.ft., Department B has a floor area of 3,000 sq.ft., and Department C has a floor area of 1,000 sq.ft.
(ii) The following direct labour hours will be used: A (2,000 hours), B (3,000 hours), and C (300 hours)
(iii) Department A has 50 employees, Department B 30 employees and Department C 20 employees
(iv) Department A has machinery costing £15,000, Department B has £25,000 machinery and Department C has £30,000 machinery
(v) The wage rates are £11 per hour for A, £12 per hour for B, and £10 per hour for C
(vi) Repairs are to be A £200; B £300; C £500.

It is estimated that 70% of Department C's facilities are used by Department A, and 30% by Department B, and that C's direct material and direct labour for the year will be £3,000 and £1,500. The overhead will be recovered by reference to the amount of direct labour hours used. Profit is to be at 25% on cost.

Millennium Plc wishes to prepare a quotation for the following job C206.

Direct material	£400
Direct labour	Department A 20 hours
	Department B 10 hours

We have already recorded (step 1), and classified (step 2) our information so the next stage (step 3) is to allocate our indirect costs (i.e. overheads) to departments.

Step 3: Overhead Allocation to Departments

Type of Cost	*Allocation*	*Total* £	*A* £	*B* £	*C* £
C's direct labour	Traced directly	3,000			3,000
C's direct material	Traced directly	1,500			1,500
Rent and business rates (calculation shown below in Helpnote point 1)	Area (4,000: 3,000: 1,000)	4,000	2,000	1,500	500
Repairs and maintenance	Actual	1,000	200	300	500
Canteen	No. of employees (50: 30: 20)	500	250	150	100
Depreciation	Cost of fixed assets (15,000: 25,000: 30,000)	2,100	450	750	900
Total		12,100	2,900	2,700	6,500

Figure 16.9 **Comprehensive Cost Recovery Example (*continued*)**

Helpnotes:

1. The allocations for each type of cost are made by allocating each cost across the departments in proportion to the total cost. For example, the rent and business rates total area is 8,000 sq.ft. Thus:

   A (4,000/8,000) × £4,000 = £2,000.

   B (3,000/8,000) × £4,000 = £1,500, and

   C (1,000/8,000) × £4,000 = £500

2. We must include our direct labour and direct materials for Department C in our calculations of the overhead absorption rate because although direct for Department C, they are indirect for Departments A and B, which are the production departments.

Step 4: Reallocate Service Support Department Costs to Production Departments

The next stage is to reallocate the service support department costs to the production departments. Since 70% of C is used by A, and 30% by B, it is only fair to reallocate them in this proportion.

	Total	A	B	C
	£	£	£	£
Total overheads	12,100	2,900	2,700	6,500
Reallocation %		70%	30%	(100%)
		4,550	1,950	(6,500)
New total	12,100	7,450	4,650	

If there were more service support departments, we would have to continue to reallocate the costs until they were all absorbed into production departments. In this case the allocation is finished.

Step 5: Calculate an Overhead Recovery Rate

It is now time to work out an overhead recovery rate.

	Total	A	B
	£	£	£
Total overheads	12,100	7,450	4,650
Direct labour hours		2,000	3,000
Recovery rate		3.73	1.55

Step 6: Absorption of Costs into Products

Finally, we must absorb both our direct and indirect costs into our products as a basis for pricing.

Job C206

		£
Direct labour	A. 20 hours at £11	220.00
	B. 10 hours at £12	120.00
Direct material		400.00
Prime Cost		740.00
Overheads	A. 20 hours at £3.73	74.60
	B. 10 hours at £1.55	15.50
Total Cost		830.10
25% mark-up on cost		207.52
Selling Price		1037.62

Total absorption costing (as we have just demonstrated in Figure 16.9) is where we try to recover all our overheads (direct and indirect) into a product or service. It forms the basis of job, contract, batch, process and service costing.

Pause for Thought 16.3

Traditional Product Costing

Traditional product costing usually uses either direct machine hours or direct labour hours to determine its overhead recovery rate. Can you see any problems with this in a service and knowledge-based economy?

In manufacturing industry, products are made intensively using machines and direct labour. It makes sense, therefore, to recover overheads using those measures. However, in service industries or knowledge-based industries, products or services may have very little direct labour input and do not use machines. The financial services industry or Internet companies, for example, have very few industrial machines or direct labour. In this case machine hours and labour hours become, at best, irrelevant and, at worst, very dangerous when pricing goods or services. In these and other industries new methods of allocating cost are necessary. The stimulus to find these new methods has led to the development of activity-based costing. This seeks to establish activities as a basis for allocating overheads.

Activity-Based Costing

Traditional total absorption costing was developed in manufacturing industries. In these industries, there are usually substantial quantities of machine hours or direct labour hours. These volume-related allocation bases are used to allocate overheads. However, recently traditional absorption costing has been criticised for failing to respond to the new post-manufacturing industrial environment and for lacking sophistication. Allocation using direct cost bases often, therefore, fails to reflect the true distribution of the costs. It is argued that this leads to inaccurate pricing.

Activity-based costing aims to remedy these defects. It is based on the premise that activities that occur within a firm cause overhead costs. By identifying these activities, a firm can achieve a range of benefits, including the ability to cost products more accurately. Essentially, activity-based costing is a more sophisticated version of the traditional product costing system. There is a six-stage process.

Figure 16.10 Six Steps in Activity-Based Costing

1. Record all the costs.
2. Classify all the costs.
3. Identify activities.
4. Identify cost drivers and allocate overheads to them.
5. Calculate activity-cost driver rates.
6. Absorb both the direct costs and indirect costs into a product or service.

We will now work through these six steps. The process is portrayed graphically in Figure 16.11 at the bottom of the page.

Steps 1 and 2

The first two steps are the same as for traditional costing.

Step 3: Identify activities

The firm identifies those activities that determine overhead costs. Production is one such activity.

Step 4: Identify cost drivers and allocate appropriate overheads to them

Activity-based accounting seeks to link cost recovery to cost behaviour. Activity-cost drivers determine cost behaviour. Activity-cost drivers for a production department, for example, might be the number of purchase orders, the number of set-ups, maintenance hours, the number of inspections, the number of despatches, machine hours or direct labour hours.

Step 5: Calculate activity-cost driver rates

Here, for each activity driver we calculate an appropriate cost driver rate. This is the total costs for a cost driver divided by the number of activities. An example might be £2,000 total costs for material inspections divided by 250 material inspections.

Step 6: Absorb both direct and indirect costs into a product or service

This process is similar to traditional product costing. The total costs are allocated to products using activity cost drivers; they are then divided by the number of products.

Figure 16.11 Diagrammatic Representation of Activity-Based Costing

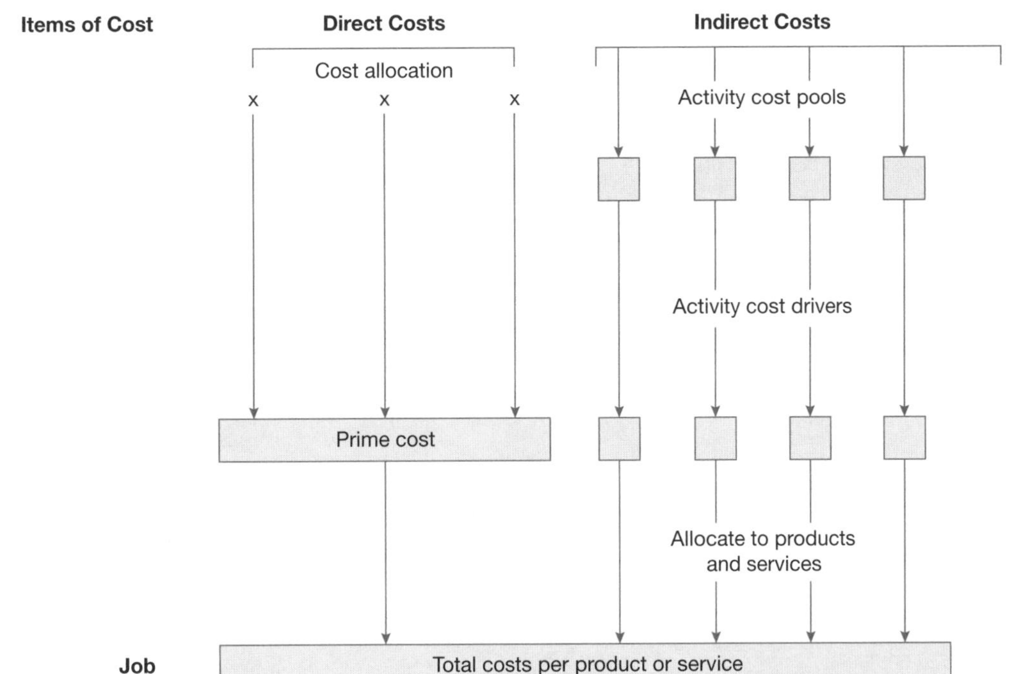

In Figure 16.12 an activity-based costing example is given for Fireco. This has two products (A and B) and four activity-cost drivers have been identified (set-ups, purchase orders, inspections and sales invoices.

Figure 16.12 Activity-Based Costing Example

Fireco has the following costs: for setting up machinery £8,000, for ordering material £3,000, for inspecting the material £2,000 and sales ledger expenses £8,000. Selling price is cost plus 25%. There is also the following information:

	A	B
	£	£
Direct materials	8,000	7,000
Direct labour	10,000	15,000
Number of set-ups	20	30
Number of purchase orders	4,000	1,000
Number of inspections	100	150
Number of sales invoices processed	7,000	3,000
Number of products	16,000	10,000

What would be the activity-cost driver rates, total cost and selling price?

As we have already completed steps 1–3 (recording, classifying, and identifying cost drivers with associated overheads), we can proceed to step 4, calculating the activity cost-driver rates.

Step 4: Calculation of Activity-Cost Driver Rates

Activity	*Set-ups*	*Purchase orders*	*Inspections*	*Sales invoices*
Cost	£8,000	£3,000	£2,000	£8,000
Cost driver	50 set-ups	5,000 purchase orders	250 inspections	10,000 sales invoices
Cost per unit of cost driver	£160	£0.60	£8.00	£0.80

Here we simply divided the total costs of each activity by the total number of the activities. There are, for example, 20 set-ups for A and 30 set-ups for B making 50 set-ups in total. We divide this into the total cost of £8,000 to arrive at £160 per set-up.

Step 5: Absorb both the Direct and Indirect Costs into the Products

Activity	*Set-ups*	*Purchase orders*	*Inspections*	*Sales*	*Total costs**
A	20 × £160 = £3,200	4,000 × £0.60 = £2,400	100 × £8.00 = £800	7,000 × £0.80 = £5,600	£12,000
B	30 × £160 = £4,800	1,000 × £0.60 = £600	150 × £8.00 = £1,200	3,000 × £0.80 = £2,400	£9,000
Total	£8,000	£3,000	£2,000	£8,000	£21,000

Here, we multiplied the cost per unit of cost driver by the number of cost drivers. For example, we have 20 set-ups (i.e., number of cost drivers) for A and 30 set-ups (i.e., number of cost drivers) for B. Each set-up costs £160 (i.e., cost per unit of cost driver).

	A	B
	£	£
Total overhead costs*	12,000	9,000
Number of products	16,000	10,000
Overhead per product	£0.75	£0.90

Figure 16.12 Activity-Based Costing Example (*continued*)

Here, we took the total overhead costs for A (£12,000) and for B (£9,000) and divided by the number of products to arrive at the overhead per product. We can now prepare the product cost statements for products A and B. Direct materials and direct labour are calculated by dividing the total costs for each product by the total number of products (i.e., for A's direct materials, we divide £8,000 by 16,000 products).

	A Product cost £	B Product cost £
Direct materials	0.50	0.70
Direct labour	0.62	1.50
Prime Cost	1.12	2.20
Overheads	0.75	0.90
Total Cost	1.87	3.10
Profit (25% mark-up on cost)	0.47	0.78
Sales Price	2.34	3.88

Activity-based costing is thus a more sophisticated version of traditional product costing. It can give more accurate and reliable information. It also significantly increases the company's ability to manage costs and is suitable for both manufacturing and service companies. However, importantly, it involves much more time and effort to set up than traditional product costing and may not be suitable for all companies.

Costing for Stock Valuation

Non-Production Overheads

It is important to distinguish between costing as a basis of pricing, which we have just looked at, and costing as the basis for stock valuation. The key difference is in the treatment of non-production overheads.

Pause for Thought 16.4

Non-Production Overheads

Why are administrative, sales and marketing costs not included as overheads in stock valuation?

The problem of non-production overheads is a tricky one. The administrative, sales and marketing costs, for example, are required when calculating the total cost of a product as a basis for pricing. However, they should not be included when calculating a stock valuation. The reason is that these costs are not incurred in actually making the product. It would, therefore, be incorrect to include them in the cost of the product. These non-production costs are often called period costs. This is because they are allocated to the **period** *not* the **product**.

When valuing stock it is permissible to include production overheads, but not non-production overheads. However, for cost-plus pricing we need to take into account all overheads. These include non-production overheads such as sales, administrative and marketing expenses. As Figure 16.13 shows, there can be quite a difference.

Figure 16.13 Costing for Stock Valuation and for Cost-Plus Pricing

A product has the following cost structure. 100,000 products are manufactured.

	Stock valuation (Marginal costing)	Stock valuation (Absorption costing)	Cost-plus pricing (Total absorption costing)
	£000	£000	£000
Direct materials	10	10	10
Direct labour	25	25	25
Prime Cost	35	35	35
Variable production overheads	10	10	10
Fixed production overheads	–	5	5
Total Production Costs	45	50	50
Administrative costs			10
Marketing costs			8
Sales costs			7
Total Cost			75
Profit: 33.3% mark-up on cost			25
Selling Price			100

At what value should each item be recorded in stock and what is the selling price per item?

i. *Stock* using marginal costing $\frac{£45,000}{100,000} = 0.45p$

ii. *Stock* using absorption costing $\frac{£50,000}{100,000} = 0.50p$

iii. *Selling price* using total absorption costing $\frac{£100,000}{100,000} = £1.00$

Helpnote: Using **marginal costing,** only the **variable production overheads** that can be attributed are included in the stock valuation. In **absorption costing,** we include **all the production overheads.** It is absorption costing that is used for stock valuation in *financial reporting*. In **total absorption costing,** we recover **all the costs** into the product's final selling price.

Different stock valuation measures: FIFO, LIFO, AVCO

The inclusion of production overheads in finished goods stock is one key problem in stock valuation. Another difficulty, which primarily concerns raw material stock, is the choice of stock valuation methods. There are three main methods: FIFO, LIFO and AVCO. All three are permitted in management accounting and for stock valuation in the financial accounts in the US. However, in the UK only FIFO and AVCO are permitted for stock valuation in financial accounting.

Stock valuation is not quite as easy as it may at first seem. It depends on which assumptions you make about the stock sold. Is the stock you buy in first, the first to be sold (first-in-first out (FIFO))? Or is the stock you buy in last, sold first (last-in-first-out (LIFO))? If the purchase price of stock changes then this assumption matters.

In Figure 16.14, we investigate an example of the impact that using FIFO or LIFO has upon the valuation of raw materials stock. We also include a third method, AVCO (average cost), which takes the average purchase price of the goods as their cost of sale.

Figure 16.14 FIFO, LIFO and AVCO Stock Valuation Methods

Stockco purchases its stock on the first day of the month. It starts with no stock. Its purchases and sales over the first three months are as follows:

		Kilos	*Cost per kilo*	*Total cost*
January 1	*Purchases*	*10,000*	*£1.00*	*£10,000*
February 1	*Purchases*	*15,000*	*£1.50*	*£22,500*
March 1	*Purchases*	*20,000*	*£2.00*	*£40,000*
		45,000		*£72,500*
March 30	*Sales*	*(35,000)*		
March 30	*Closing stock*	*10,000*		

What is the closing stock valuation using FIFO, LIFO and AVCO?

(i) FIFO: Here, the first stock purchased is the first sold. The 35,000 kilos sold therefore, used up all the January stock (10,000 kilos), and the February stock (15,000 kilos), and (10,000 kilos) of the March stock. We, therefore, have left in stock 10,000 kilos of material valued at the March purchase price of £2.00 per kilo. Closing stock is, therefore, 10,000 × £2 = £20,000.

(ii) LIFO: Here the last stock purchased is assumed to be the first to be sold. The 35,000 kilos sold, therefore, used up:

20,000	kilos from March
15,000	kilos from February
35,000	

We, therefore, have remaining 10,000 kilos from January at £1 per kilo = £10,000.

(iii) AVCO: Here the stock value is pooled and the average cost of purchase is taken as the cost of the goods sold. We purchased 45,000 kilos for £72,500 (i.e. £1.611 per kilo). The cost of our stock, is therefore, 10,000 kilos × average cost £1.611 = £16,110.

We can, therefore, see that our *stock valuations* vary considerably.

	£
FIFO	20,000
LIFO	10,000
AVCO	16,110

It should be appreciated that stock valuation is distinct from physical stock management. In most businesses, good business practice dictates that you usually physically issue the oldest stock first (i.e., adopt FIFO). However, in stock valuation you are allowed to choose what is acceptable under the regulations.

Pause for Thought 16.5

FIFO, LIFO and AVCO and Cost of Sales

Do you think that using a different stock valuation method in Figure 16.14 would affect cost of sales or profit?

Yes!! Whichever valuation method is used, both cost of sales and profit are affected. Essentially, the cost of purchases will be split between stock and cost of sales.

	Total cost	*Cost of sales*	*Stock*
	£	£	£
FIFO	72,500	52,500 (N1)	20,000
LIFO	72,500	62,400 (N2)	10,000
AVCO	72,500	56,390 (N3)	16,110

			£
(N1) FIFO represents:	January	10,000 kilos at £1.00	10,000
	February	15,000 kilos at £1.50	22,500
	March	10,000 kilos at £2.00	20,000
			52,500

			£
(N2) LIFO represents:	February	15,000 kilos at £1.50	22,500
	March	20,000 kilos at £2.00	40,000
			62,500

		£
(N3) AVCO represents:	35,000 kilos at £1.611	56,390

Profit is affected because if cost of sales is less, stock, and thus profit, is higher and vice versa. In this case, using FIFO will show the greatest profit as its cost of sales is lowest. LIFO will show the lowest profit. AVCO is in the middle!

Different Costing Methods for Different Industries

So far we have focused mainly on allocating costs to individual jobs in manufacturing. However, job costing is not appropriate in many situations. Different industries have different costing problems. To solve these problems different types of costing have evolved. In order to give a flavour of this, we look below at four industries (see Figure 16.15).

Figure 16.15 Overview of Different Industries' Costing Methods

	Industry	**Costing Method**
1	Manufacturing in batches	Batch costing
2	Shipbuilding	Contract costing
3	Brewery	Process costing
4	Hotel and catering	Service costing

1. Batch Costing

Batch costing is where a number of items of a similar nature are processed together. There is thus not one discrete job. Batch costing can be used in a variety of situations, such as manufacture of medicinal tablets.

Figure 16.16 Batch Costing

A run of 50,000 tablets are made for batch number x1.11. There are 2,000 defective tablets and 50 tablets per box. Overheads are recovered on the basis of £6 per labour hour. The material was 3 kg at £5.00 per kg, and labour rate A 6 hours at £9.00, and rate B 7 hours at £10.00.

The costing statement might look as follows:	£
Direct materials (3 kg at £5.00)	15.00
Direct labour (13 hours: A 6 at £9	54.00
B 7 at £10)	70.00
Prime Cost	139.00
Overheads (13 hours at £6)	78.00
Total Cost	217.00
Boxes (48,000 ÷ 50)	960
Cost per box	22.6p

2. Contract Costing

A contract can be looked at as a very long job, or a job lasting more than one year. It occurs in big construction industries such as shipbuilding and aircraft building. A long-term contract extends over more than one year and creates the problem of when to take profit. If we have a three-year contract do we take all our profit at the start of our contract, at the end, or equally throughout the three years?

The answer has gradually evolved over the years, and there are now certain recognised guidelines for taking a profit or loss. Generally, losses should be taken as soon as it is realised they will occur. By contrast, profits should be taken so as to reflect the proportion of the work carried out. This is shown in The Company Camera 16.2 which shows John Laing's (until recently a construction company) accounting policy for long-term contracts.

The Company Camera 16.2

Long-Term Contracts

Profits on long-term contracts are calculated in accordance with industry standard accounting practice and do not therefore relate directly to turnover. Profit on current contracts is only taken at a stage near enough to completion for that profit to be reasonably certain. Provision is made for all losses incurred to the accounting date together with any further losses that are foreseen in bringing contracts to completion.

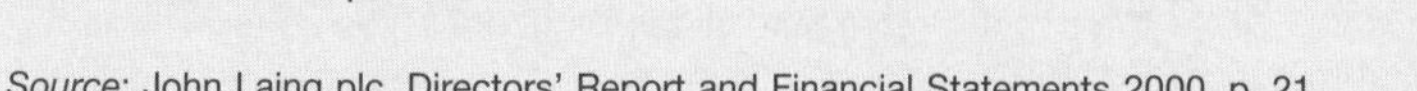

Source: John Laing plc, Directors' Report and Financial Statements 2000, p. 21

In practice, allocating profits can be very complex. We will use a simplified formula: **Profit to be taken = % contract complete × total estimated contract profit × 2/3**. Usually, the customer and the supplier will negotiate and set a price for a contract in advance. Figure 16.17 provides an illustrative example.

Figure 16.17 Contract Costing

O&P Ferries requires a new ferry to be built. It asks Londonside Shipbuilders for a quotation. Londonside looks at its costs and decides that the ship will cost £4 million direct materials, £4 million direct labour and £5 million indirect overheads. They then quote O&P Ferries £19 million for the job. The aim to complete the job in 3 years. Londonside therefore estimate a £6 million profit (£19 million less £13 million). When should they take the profit?

The answer is that the situation is reviewed as the contract progresses, and profit taken according to the formula given earlier. For example, if we have the following costs:

	Year 1		**Year 2**		**Year 3**	
In £ millions	£	£	£	£	£	£
Contract Price (Remains fixed)		19		19		19
Costs Incurred to Date						
Direct materials	2		3		5	
Direct labour	2		2		5	
Indirect overheads	2		3		6	
Total costs incurred to date	**6**		**8**		**16**	
Estimated future costs	8		5		–	
Total Costs (estimated and incurred)		**14**		**13**		**16**
Estimated Profit		5		6		3
% Contract complete		6/14ths		8/13ths		Complete

At the end of year 3 the contract is finished, therefore, the estimated profit will be the actual profit.

So if we apply our formula we can see how it works:

	% contract complete	×	**Total estimated contract profit**	×	$\frac{2}{3}$	**This year's profit** £	**Total estimated profit to date** £
Year 1	$\frac{6}{14}$	×	£5m	×	$\frac{2}{3}$ = £1.43m	£1.43m	£1.43m
Year 2	$\frac{8}{13}$	×	£6m	×	$\frac{2}{3}$ = £2.46m	£1.03m	£2.46m

We take an additional £1.03 million in year 2 (i.e., the profit to date, £2.46m, less year 1's profit, £1.43m).

	This year's profit	Total estimated profit to date
Year 3	£0.54m	£3.0m

In year 3, we know the actual profit is £3 million. We can therefore take all the profit not taken so far. This is £3 million – £2.46 (already taken) = £0.54 million.

3. Process Costing

Process costing is used in industries with a continuous production process, e.g., beer brewing. Products are passed from one department to another, and then processed further. At any one point in time, therefore, many of the products will only be partially complete. To deal with this problem of partially completed products, the concept of equivalent units has developed. At the end of a process if we partially finish units then we take the percentage of completion and convert to fully completed equivalent units. If, for example, we have 1,000 litres of beer that are half way through the beer-making process this would equal 500 litres fully complete (i.e., there would be the equivalent of 500 litres).

Figure 16.18 Process Costing

A firm manufactures beer. There are two processes A and B. There are 1,000 litres of opening stock of beer for process B, the fermentation stage (the product has already been through process A). They are 50% complete. During the year 650 litres are finished and transferred (i.e., completed). The closing work in progress consists of 800 litres, 75% completed. The costs incurred in that process during that year are £15,000.

To find out the cost per litre, we must first find out how many litres are effectively produced during the period. We can do this by deduction. First of all we can establish the total stock that was completed by the end of the year. This will equal the stock we have finished and transferred (650 litres) plus the closing stock (800 litres, 75% complete, equals 600 litres). From this 1,250 litres we need to take the opening stock (1,000 litres, 50% complete, equals 500 litres). We have, therefore, our effective production of 750 litres.* The cost is then £15,000 divided by the 750 equivalent litres, equals £20 per litre.

	Equivalent litres
Finished and transferred	650
Closing stock (800 × 75%)	600
Total completed	1,250
Opening stock (1,000 × 50%)	(500)
**Effective production*	750

Therefore, using 'equivalent' litres our cost per litre for process B is:

$$\frac{£15{,}000}{750} = £20$$

4. Service Costing

Service costing concerns services such as canteens. In large businesses, for example, canteens might be run as independent operating units that have to make a profit. The cost of a particular service is simply the total costs for the service divided by the number of services provided. Service costing uses the same principles as job costing.

Figure 16.19 Service Costing for Canteens

A canteen serves 10,000 meals: 6,000 are roast dinners and 4,000 salads. The roast dinners are bought in for £2.50 each and the salads for £1.25. The canteen costs are £1,000. They are apportioned across the number of meals served. What is the cost of each meal?

	Roast Dinners	Salads
	£	£
Bought-in price	2.50	1.25
Overheads per meal $\left(\frac{£1,000}{10,000} = 0.10p\right)$	0.10	0.10
	2.60	1.35

Target Costing

So far we have focused principally on cost-plus pricing. However, the Japanese have introduced a concept called target costing, which focuses on market prices. Essentially, a price is set with reference to market conditions and customer purchasing patterns. A target profit is then deducted to arrive at a target cost. This target cost is set in order to allow a company to achieve a certain market share and a certain profit. The target profit is set before the produce is manufactured.

The costs are then examined and re-examined in order to make the target cost. Often this is done by breaking down the product into many individual components and costing them separately. The product may be divided into many functions using 'functional analysis'. Functions may include attractiveness, durability, reliability, and style among other things. Each function is priced. Target costing may also be used in conjunction with life cycle costing. Life cycle costing (see Chapter 20) involves tracing all the costs of a product over their entire life cycle. Target costing will be the first stage in this process.

Cost-Cutting

A final key reason why it is essential for a business to have a good knowledge of its costs is for cost-cutting. As we mentioned in Chapter 15, the two major ways to improve profitability are by improving sales or by cutting costs. Whereas improving sales is a long-term solution, cost-cutting is a short-term solution. It is particularly useful when a business is in trouble. By announcing cost-cutting measures, the business signals to the City and to its investors that it is serious about improving its profitability. Unfortunately, one of the most important elements of most companies costs is labour. Therefore, companies often shed labour. When doing this companies may run into trouble with trade unions or with politicians. Centrica's reduction of the number of staff in its gas showrooms is a good example of this (see Real-Life Nugget 16.4 on the next page).

Real-Life Nugget 16.4

Cutting Labour Costs

Centrica, Britain's largest gas retailer, plans to axe almost 1,500 jobs in its underperforming high street energy stores, drawing the wrath of unions who claimed the act was a 'betrayal of staff'.

The company said it would close its 243 British Gas Energy Centres and incur £60m in costs to be charged this financial year.

Centrica chief executive Roy Gardner revealed the operating losses for those stores had grown to an estimated £25m in the June half, compared with £33m for the whole 1998 year.

Source: Anne Hyland, 'Centrica Faces Anger over Store Job Cuts', *Daily Telegraph*, 20 July 1999

As an alternative to cutting labour costs, companies sometimes attempt to cut other costs such as training or research and development. By doing this, companies may sacrifice long-term profitability for short-term profitability.

Conclusion

Costing is an important part of management accounting. It involves recording, classifying, allocating and absorbing costs into individual products and services. Cost recovery is used to value stock by including production overheads and for cost-plus pricing by determining the total costs of a product or service. Traditionally, overheads have been allocated to products or services primarily using volume-based measures such as direct labour hours or machine hours. However, more recently, activity-based measures such as number of purchase orders processed have been used. Different industries use different costing systems such as batch costing, contract costing, process costing and service costing. Target costing, based on a product's market price, has been developed in Japan. Businesses in trouble often cut costs, such as labour, in order to try to improve their profitability.

Discussion *Questions*

Questions with numbers in blue have answers at the back of the book.

Q **1** What is costing and why is it important?

Q **2** Compare and contrast the traditional and the more modern activity-based approaches to costing.

Q **3** Overhead recovery is the most difficult part of cost recovery. Discuss.

Q 4 Target costing combines the advantages of both market pricing and cost-plus pricing. Discuss.

Q 5 State whether the following statements are true or false. If false, explain why.

(a) A cost is an actual past expenditure.

(b) When recovering costs for pricing we use total absorption costing. However, for stock valuation we use absorption costing or marginal costing.

(c) When using traditional total absorption costing it is important to identify activity cost drivers.

(d) In batch costing if we had 100 units that were 50% complete this would equal 50 equivalent units.

(e) In stock valuation, marginal costing is normally used in valuing stock in financial reporting.

Numerical *Questions*

Questions with numbers in blue have answers at the back of the book.

Q 1 Sorter has the following costs:

(a) Machine workers' wages
(b) Cost clerks' wages
(c) Purchase of raw materials
(d) Machine repairs
(e) Finance director's salary
(f) Office cleaners
(g) Delivery van staff's wages
(h) Managing director's car expenses
(i) Depreciation on office furniture
(j) Computer running expenses for office
(k) Loan interest
(l) Auditors' fees
(m) Depreciation on machinery
(n) Advertising costs
(o) Electricity for machines
(p) Bank charges

Required:

An analysis of Sorter's costs between:

(i) Direct materials
(ii) Direct labour
(iii) Production overheads
(iv) Administrative expenses
(v) Selling and distribution costs

Q 2 Costa has the following costs:

	£
Salaries of administrative employees	90,800
Wages of factory supervisors	120,000
Computer overhead expenses	
(⅔ in factory, ⅓ in administration)	9,000
Interest on loans	3,000
Wages: selling and distribution	18,300
Salaries: marketing	25,000

Q2 Costa (continued)

		£
Royalties		3,600
Raw materials used in production		320,000
Depreciation:	Machinery used for production	8,000
	Office fixtures and fittings	4,200
	Delivery vans	3,500
	Buildings (½ factory; ¼ office; ¼ sales)	10,000
Labour costs directly connected with production		200,000
Other production overheads		70,000
Commission paid to sales force		1,200

Required:

A determination of Costa's:

(i) prime cost
(ii) production cost
(iii) total cost

Q3 Makemore has three departments: Departments A and B are production and Department C is a service support department. It apportions its overheads as follows (see brackets):

		£
Supervisors' salaries	(number of employees)	25,000
Computer advisory	(number of employees)	18,000
Rent and business rates	(floor area)	20,000
Depreciation on machinery	(cost)	21,000
Repairs	(actual spend)	4,000
		84,000

You have the following information:

(a) Department A 1,000 employees, B 2,000 employees, C 500 employees.
(b) Department A 10,000 sq. feet, B 6,000 sq. feet, and C 4,000 sq. feet.
(c) Department A machinery costs £30,000, B machinery costs £15,000.
(d) Repairs are £2,800, £1,100 and £100, respectively, for departments A, B and C.
(e) Department C's facilities will be reallocated 60% for A and 40% for B.
(f)

	Department A	Department B
Direct labour hours	80,000	40,000
Machine hours	100,000	200,000

Required:

An apportionment of the overheads to products A and B

(i) using direct labour hours, and
(ii) using machine hours.

Q **4** Flight has two products, the 'Takeoff' and the 'Landing'. They go through two departments and incur the following costs:

		Takeoff		*Landing*	
Dept. A –	Direct labour	10 hours	£10 per hour	9 hours	£7 per hour
	Direct materials	10 kilos	£5 per kilo	7 kilos	£12 per kilo
Dept. B –	Direct labour	8 hours	£12 per hour	5 hours	£8 per hour
	Direct materials	5 kilos	£15 per kilo	6 kilos	£10 per kilo

Takeoff's indirect overheads are absorbed at £7 per labour hour for Dept. A and £5 per hour Dept. B.

Landing's indirect overheads are absorbed at £6 per labour hour for Dept. A and £4 per hour for Dept. B.

Selling price is to be 20% on cost.

Required:

The prices for which Flight should sell 'Takeoff' and 'Landing'.

Q **5** An up-market catering company, Spicemeals, organises banquets and high-class catering functions. A particular function for the Blue Devils university drinking club has the following costs for 100 guests.

100 starters	at	£1.00 each
50 main meals	at	£3.00 each
50 main meals	at	£2.50 each
200 sweets	at	£1.50 each
200 bottles of wine	at	£5.00 each
100 coffees	at	£0.20 each

Direct labour:	Supervisory	8 hours	at	£15.00
	Food preparation	30 hours	at	£8.00
	Waitressing	200 hours	at	£5.00

Overheads relating to this job based on total hours are recovered at £1.10 per hour. There are £180,000 general overheads within the business and about 300 functions. Profit is to be 15% on cost.

Required:

The price Spicemeals should charge for this function and the amount for each guest.

Q6 A medicinal product 'Supertab' is produced in batches. Batch number x308 has the following costs:

Direct labour: Grade	1	200 hours	at	£5.00
	2	50 hours	at	£8.00
	3	25 hours	at	£15.00
Direct materials: Type	A	10 kilos	at	£8.00
	B	5 kilos	at	£10.00
	C	3 kilos	at	£15.00

Production overheads are based on labour hours at £3 per hour. 20,000 tablets are produced, but there is a wastage of 10%. Non-production overheads are £15,000 for the month. There are usually 250 batches produced per month. There are 50 tablets in a container. The selling price will be 25% on cost.

Required:
What is the selling price of each container?

Q7 Dodo Airways has agreed a tender for a new aircraft at £20m. The company making the product has the following cost structure for this long-term contract.

	Tender		Year 1		Year 2		Year 3	
	£m	£m	£m	£m	£m	£m	£m	£m
Sales price		20		20		20		20
Direct materials	3		1		2		4.5	
Direct labour	8		2		5		8.0	
Overheads	4	15	1	4	2	9	4.5	17
Profit		5		16		11		3
Estimated costs to complete				12		7		–

Required:
What is the profit Dodo should take every year? Use the formula given in this chapter (see page 358).

Q8 Serveco is a home-service computer company. There are two levels of computer service offered: basic and enhanced service. You have the following details on each.

	Basic	*Enhanced*
Total basic call out time	25,000 hours	37,500 hours
Total travelling time	25,000 hours	5,000 hours
Parts serviced/replaced	50,000	100,000
Technical support (mins)	75,000	100,000
Service documentation (units)	100,000	25,000
Number of call outs	50,000	10,000

Q8 Serveco (continued)

You have the following costs:

	Basic	*Enhanced*	*Total*
Basic computer operatives' labour	£20 per hour	£25 per hour	
Spare parts installed			£100,000
Technical support cost			£125,000
Service documentation cost			£300,000

Serveco requires a profit mark-up of cost plus 25%.

Required:
Calculate an appropriate standard call out charge for the basic and enhanced services using activity-based costing.

Q9 A company, Rugger, manufactures two products: the Try and the Conversion. The company has traditionally allocated its production overhead costs on the basis of the 100,000 direct hours used in the manufacturing department. Direct labour costs £5 per hour. The company is now considering using activity-based costing. Details of the overheads and cost drivers are as follows:

Production Overheads	*Total Cost (£)*	*Cost Driver*	*Total*
(a) Manufacturing	10,000	Assembly-line hours	100,000 hours
(b) Materials handling	60,000	Number of stores notes	1,500 notes
(c) Inspection	40,000	Number of inspections	600 inspections
(d) Set-ups	5,000	Number of set-ups	500 set-ups

You have the following information about the products.

	Try	*Conversion*
Number of units	15,000	2,500
Assembly-line hours (direct labour) per unit	6 hours	4 hours
Direct materials per unit	£8	£100
Number of stores notes	600	900
Number of inspections	257	343
Number of set-ups	200	300

Required:
Calculate a product cost using

(i) traditional total absorption costing, recovering overheads using direct labour hours;
(ii) activity-based costing, and then
(iii) comment on any differences.

'The budget is God'.

Slogan at Japanese Company, Topcom (*Economist*, January 13, 1996)

Source: *The Wiley Book of Business Quotations* (1998), p. 90

Chapter 17

Planning and Control: Budgeting

Learning Outcomes

After completing this chapter you should be able to:

- ✔ Explain the nature and importance of budgeting.
- ✔ Outline the most important budgets.
- ✔ Prepare the major budgets and a master budget.
- ✔ Discuss the behavioural implications of budgets.

In a Nutshell

- The two major branches of cost accounting are costing, and planning and control.
- Budgeting is a key element of planning and control.
- Budgets are ways of turning a firm's strategic objectives into practical reality.
- Most businesses prepare, at the minimum, a cash budget.
- Large businesses may also prepare a sales, a debtors and a creditors budget.
- Manufacturing businesses may prepare a raw materials, a production cost, and a finished goods budget.
- Individual budgets fit into a budgeted trading and profit and loss account and a budgeted balance sheet.
- Budgeting has behavioural implications for the motivation of employees.
- Some behavioural aspects of budgets are spending to budget, padding the budget and creative budgeting.

Introduction

Cost accounting can be divided into costing, and planning and control. Budgeting and standard costing are the major parts of planning and control. Budgeting, or budgetary control, is a key part of businesses' planning for the future. A budget is essentially a plan for the future. Budgets are thus set in advance as a way to quantify a firm's objectives. Actual performance is then monitored against budgeted performance. For small businesses, the cash budget is often the only budget. Larger businesses, by contrast, are likely to have a complex set of interrelating budgets. These build up into a budgeted trading and profit and loss account, and budgeted balance sheet. Although usually set for a year, budgets are also linked to the longer-term strategic objectives of an organisation.

Nature of Budgeting

In many ways, it would be surprising if businesses did not budget. For budgeting is part of our normal everyday lives. Whether it is a shopping trip, a university term or a night out on the town, we all generally have an informal budget of the amount we wish to spend. Businesses merely have a more formal version of this 'informal' personal budget.

Pause for Thought 17.1

Personal Budgets

You are planning to jet off for an Easter break in the Mediterranean sun. What sort of items would you include in your holiday budget?

There would be a range of items, for example,

- transport costs to and from the airport
- cost of flight to Mediterranean
- cost of hotel
- cost of meals
- spending money
- entertainment money
- money for gifts

All these together would contribute to your holiday budget.

Planning and control is one of the two major branches of cost accounting. This is shown in Figure 17.1. Standard costing is, in reality, a more tightly controlled and specialised type of budget. Although often associated with manufacturing industry, standard costs can, in fact, be used in a wide range of businesses.

Figure 17.1 Main Branches of Cost Accounting

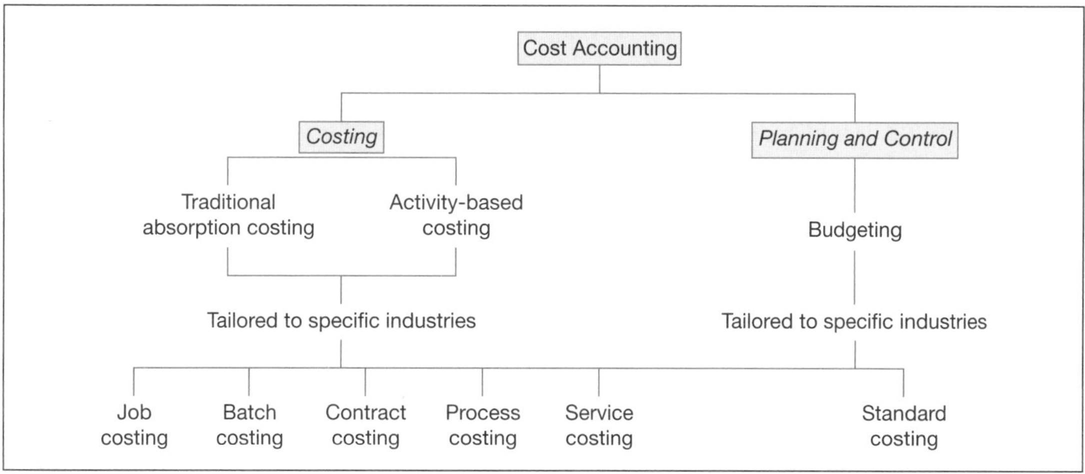

Budgeting can be viewed as a way of turning a firm's long-term strategic objectives into reality. As Figure 17.2 shows, a business's objectives are turned into forecasts and plans. These plans are then compared with the actual results and performance is evaluated.

Figure 17.2 The Budgeting Process

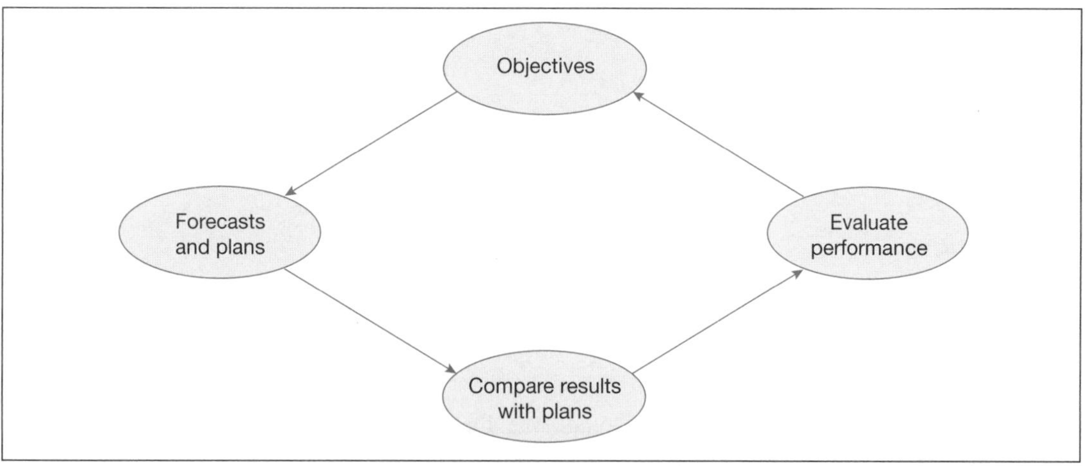

In small businesses, this process may be relatively informal. However, for large businesses there will be a complex budgeting process. The period of a budget varies. Often there is an annual budget broken down into smaller periods such as months or even weeks.

As Definition 17.1 shows, a budget is a quantitative financial plan which sets out a business's targets.

Definition 17.1

Budget

Working definition
A future plan which sets out a business's financial targets.

Formal definition
'A quantitative statement, for a defined period of time, which may include planned revenues, expenses, assets, liabilities and cash flows. A budget provides a focus for the organisation, aids the co-ordination of activities, and facilitates control.'

Source: Chartered Institute of Management Accountants (2000), *Official Terminology*.

Four major aspects of budgets are planning, coordinating, motivation and control. Budgets, therefore, combine both technical and behavioural features. These behavioural aspects are discussed more fully later in the chapter.

(i) Planning

This involves setting out a comprehensive plan appropriate to the business. For small businesses, this may mean a cash budget. For larger businesses, there will probably be a formalised and sophisticated budgeting system.

(ii) Coordinating

A key aspect of the budgetary process is that it relates the various activities of a company to each other. Therefore, sales is related to purchases, and purchases to production. The business can be viewed as an interlocking whole.

(iii) Motivation

By setting targets the budget has important motivational aspects. If the targets are too hard, they can be demotivating. If too easy, they will not provide any motivation at all.

(iv) Control

This is achieved through a system of making individual managers responsible for individual budgets. When actual results are compared against target results, individual managers will be asked to explain any differences (or variances). The manager's performance is then evaluated. Budgets, as Real-Life Nugget 17.1 shows (see next page), are an indispensable form of administrative control.

Real-Life Nugget 17.1

Budgeting

Budgeting now occupies a central position in the design and operation of most management accounting systems. Almost regardless of the type of organisation, the nature of its problems and the other means for influencing behaviour, the preparation of a quantitative statement of expectations regarding the allocation of the organisation's resources tends to be seen as an essential, indeed indispensable, feature of the battery of administrative controls. Nevertheless, despite its wide acceptance, budgeting remains one of the most intriguing and perplexing of management accounting procedures.

Source: A. Hopwood (1974), *Accounting and Human Behaviour*, p. 39

In large businesses, budgets are generally set by a budgetary committee. This will normally involve managers from many different departments. The sales manager, for example, may provide information on likely future sales, which will form the basis of the sales budget. However, it is important that individual budgets are meshed together to provide a coordinated and coherent plan. As Soundbite 17.1 shows, the starting point for this year's budget is usually last year's budget. A form of budgeting called *zero based budgeting* assumes the activities are being incurred for the first time. The budgetary process is normally ongoing, however, with meetings during the year to review progress against the current budget and to set future budgets.

Although budgets are set within the business, external factors will often constrain them. A key external constraint is demand. It is futile for a company to plan to make ten million motorised skateboards if there is demand for only five million. Indeed, potential sales are the principal factor that limits the expansion of many businesses. Usually, the sales budget is determined first. Some businesses employ a *materials requirement planning* (MRP) system. Based on sales demand, MRP coordinates production and purchasing to ensure the optimal flow of raw materials and optimal level of raw material stocks. Another important budgetary constraint in manufacturing industry is production capacity. It is useless planning to sell ten million motorised skateboards if the production capacity is only five million. A key element of the budgetary process is thus harmonising demand and supply.

Nowadays, it is rare for businesses not to have budgets. As Figure 17.3 shows, budgets have many advantages. The chief disadvantages are that budgets can be inflexible and create behavioural problems. Budgets can be inflexible if set for a

Soundbite 17.1

Last Year's Budget

'The largest determining factor of the size and content of this year's budget is last year's budget.'

Aaron Wildavsky, The Politics of the Budgetary Process, 1964.

Source: *The Executive's Book of Quotations* (1994), p. 39

Figure 17.3 The Benefits of Budgeting

- Strategic planning can be more easily linked to management decisions
- Standards can be set to aid performance evaluation
- Plans can be set in financial terms
- Managers can be made responsible for budgets
- Budgets encourage cooperation and coordination

year. It is common to revise budgets regularly to take account of new circumstances. This is easier when they are prepared using spreadsheets. The behavioural problems may be created, for example, when individuals attempt to manipulate the budgeting process in their own interest. This is discussed later.

Pause for Thought 17.2

Budgets

Budgets are often said to create inflexibility as they are typically set for a year in advance. Can you think of any ways to overcome this inflexibility?

This inflexibility can be dealt with in two ways: rolling budgets and flexible budgets. With rolling budgets the budget is updated every month. There is then a new twelve-month budget. The problem with rolling budgets is that it takes a lot of time and effort to update budgets regularly. Flexible budgets attempt to deal with inflexibility by setting a range of budgets with different activity levels. So a company might budget for servicing 100,000, 150,000 or 200,000 customers. In this case, there would, in effect, be three budgets for three different levels of business activity.

Cash Budget

The cash budget is probably the most important of all budgets. Almost all companies prepare one. Indeed, banks will often insist on a cash budget before they lend money to small businesses. For a sole trader, the cash budget is often as important as the trading and profit and loss account and balance sheet. It reflects the need to balance profitability and liquidity. There are similarities, but also differences, between the cash budget and the cash flow statement that we saw in Chapter 8. They are similar in that both the cash budget and cash flow statement chart the flows of cash within a business. The differences arise in that the cash flow statement is normally prepared in a standardised format in accordance with accounting regulations and looks backwards in time. By contrast, the cash budget is not in a standardised format and looks forward in time.

Essentially, the cash budget looks into the future. Figure 17.4 gives the format of the cash budget. We start with the opening cash balance. Receipts are then recorded, for example, cash received from cash sales, from debtors or from the sale of fixed assets. Cash payments are then listed, for example, cash purchases of goods, payments to creditors or expenses paid. Receipts less payments provide the monthly cash flow. Opening cash and cash flow determine the closing cash balance.

Figure 17.4 The Format of a Cash Budget

	Jan.	Feb.	March	April	May	June	Total
	£	£	£	£	£	£	£
Opening cash	X	X	X	X	X	X	X
Add *Receipts*							
Debtors	X	X	X	X	X	X	X
	X	X	X	X	X	X	X
Less *Payments*							
Payments for goods	X	X	X	X	X	X	X
Expenses	X	X	X	X	X	X	X
Other payments	X	X	X	X	X	X	X
	X	X	X	X	X	X	X
Cash flow	Y	Y	Y	Y	Y	Y	Y
Closing cash	X	X	X	X	X	X	X

In Figure 17.5, we show an actual example of how in practice a cash budget is constructed.

Figure 17.5 Illustrative Example of a Cash Budget

Jason Chan has £12,000 in a business bank account. His projections for the first six months trading are as follows.

(i) Credit sales will be: January £8,800, February £8,800, March £9,000, April £12,500, May £20,200, June £30,000. Debtors pay in the month following sale.

(ii) Goods supplied on credit will be: January £7,000, February £10,000, March £9,800, April £10,400, May £7,000, June £8,000. Creditors are paid one month in arrears.

(iii) Loan receivable 1 March £4,000 to be repaid in full plus £400 interest on 1 June.

(iv) Drawing £500 per month.

Prepare the cash budget.

	Jan.	Feb.	March	April	May	June	Total
	£	£	£	£	£	£	£
Opening cash	12,000	11,500	12,800	15,100	13,800	15,400	12,000
Add *Receipts*							
Debtors		8,800	8,800	9,000	12,500	20,200	59,300
Loan received			4,000				4,000
	–	8,800	12,800	9,000	12,500	20,200	63,300
Less *Payments*							
Payments for goods		7,000	10,000	9,800	10,400	7,000	44,200
Loan repayment						4,400	4,400
Drawings	500	500	500	500	500	500	3,000
	500	7,500	10,500	10,300	10,900	11,900	51,600
Cash flow	*(500)*	*1,300*	*2,300*	*(1,300)*	*1,600*	*8,300*	*11,700*
Closing cash	11,500	12,800	15,100	13,800	15,400	23,700	23,700

Helpnotes:

(i) The receipts from debtors and payments by creditors are thus running one month behind actual sales and actual purchases, respectively. Thus, for example, the debtors will pay the £30,000 of sales made in June in July and they are, therefore, not recorded in this budget.

(ii) The cash flow column is simply total receipts less total payments. Cash outflows (where receipts are less than payments) are recorded in brackets.

Other Budgets

A business may have numerous other budgets. Indeed, each department of a large business is likely to have a budget. In this section, we will look at three key budgets common to many businesses: sales budget; debtors budget and creditors budget

In the next section, we will look at three additional budgets that are commonly found in manufacturing businesses: raw materials budget, finished goods budget and production cost budget. Finally, we will bring the budgets together in a comprehensive example.

i Sales Budget

The sales budget is determined by examining how much the business is likely to sell during the forthcoming period. Figure 17.6 provides an example. Sales budgets like many other budgets can be initially expressed in units before being converted to £s. In many businesses, the sales budget is the key budget as it determines the other budgets. The sales budget is, therefore, often set first.

Figure 17.6 Example of a Sales Budget

A business has two products, Alpha and Omega. It is anticipated that sales of the Alpha will run at 500 units throughout January to June. However, the Omega will start at 1,000 units and rise by 100 units per month. Each Alpha sells at £35, each Omega sells at £50. Prepare the sales budget.

	Jan.	Feb.	March	April	May	June	Total
	£	£	£	£	£	£	£
Alpha	17,500	17,500	17,500	17,500	17,500	17,500	105,000
Omega	50,000	55,000	60,000	65,000	70,000	75,000	375,000
Total sales	67,500	72,500	77,500	82,500	87,500	92,500	480,000

ii Debtors Budget

The debtors budget begins with opening debtors (often taken from the opening balance sheet) to which are added credit sales (often taken from the sales budget). Cash receipts are then deducted, leaving closing debtors. An example of the format for a debtors budget is provided in Figure 17.7, while Figure 17.8 provides an illustrative example.

Figure 17.7 Format of a Debtors Budget

	Jan.	Feb.	March	April	May	June	Total
	£	£	£	£	£	£	£
Opening debtors	X	X	X	X	X	X	X
Credit sales	X	X	X	X	X	X	X
	X	X	X	X	X	X	X
Cash received from debtors	(X)	(X)	(X)	(X)	(X)	(X)	(X)
Closing debtors	X	X	X	X	X	X	X

Figure 17.8 Example of a Debtors Budget

Sara Peters has opening debtors of £800. Debtors pay one month in arrears. Sales are forecast to be £900 in January rising by £200 per month. Prepare the debtors budget.

	Jan.	Feb.	March	April	May	June	Total
	£	£	£	£	£	£	£
Opening debtors	800	900	1,100	1,300	1,500	1,700	800
Credit sales	900	1,100	1,300	1,500	1,700	1,900	8,400
	1,700	2,000	2,400	2,800	3,200	3,600	9,200
Cash received	(800)	(900)	(1,100)	(1,300)	(1,500)	(1,700)	(7,300)
Closing debtors	900	1,100	1,300	1,500	1,700	1,900	1,900

iii Creditors Budget

In many ways, the creditors budget is the mirror image of the debtors budget. It starts with opening creditors (often taken from the opening balance sheet), adds credit purchases and then deducts cash paid. The result is closing creditors. The format for the creditors budget is given in Figure 17.9, while Figure 17.10 provides an illustrative example.

Figure 17.9 Format of a Creditors Budget

	Jan.	Feb.	March	April	May	June	Total
	£	£	£	£	£	£	£
Opening creditors	X	X	X	X	X	X	X
Credit purchases	X	X	X	X	X	X	X
	X	X	X	X	X	X	X
Cash paid to creditors	(X)	(X)	(X)	(X)	(X)	(X)	(X)
Closing creditors	X	X	X	X	X	X	X

Figure 17.10 Example of a Creditors Budget

Jon Matthews has opening creditors of £1,200. Creditors are expected to pay one month in arrears. In January, purchases are forecast to be £9,000 rising by £100 per month. Prepare the creditors budget.

	Jan.	Feb.	March	April	May	June	Total
	£	£	£	£	£	£	£
Opening creditors	1,200	9,000	9,100	9,200	9,300	9,400	1,200
Credit purchases	9,000	9,100	9,200	9,300	9,400	9,500	55,500
	10,200	18,100	18,300	18,500	18,700	18,900	56,700
Cash paid	(1,200)	(9,000)	(9,100)	(9,200)	(9,300)	(9,400)	(47,200)
Closing creditors	9,000	9,100	9,200	9,300	9,400	9,500	9,500

Manufacturing Budgets

Manufacturing companies normally hold more stock than other businesses. It is, therefore, common to find three additional budgets: a production cost budget; a raw materials budget; and a finished goods budget. Often these budgets are expressed in units, which are then converted into £s. For ease of understanding, we express them here only in financial terms.

(i) Production Cost Budget

The production cost budget, as its name suggests, estimates the cost of production. This involves direct labour, direct materials and production overheads. There may often be sub-budgets for each of these items. The production cost format is shown in Figure 17.11, while Figure 17.12 shows an example. Once the production cost is determined, the finished goods budget can be prepared. It is important to realise that the budgeted production levels are generally determined by the amount the business can sell.

Figure 17.11 Format of a Production Cost Budget

	Jan.	Feb.	March	April	May	June	Total
	£	£	£	£	£	£	£
Direct materials	X	X	X	X	X	X	X
Direct labour	X	X	X	X	X	X	X
Production overheads	X	X	X	X	X	X	X
	X	X	X	X	X	X	X

Figure 17.12 Example of a Production Cost Budget

Ray Anderson has the following forecast details from his production department. Direct materials are £6 per unit, direct labour is £8 per unit and production overheads are £4 per unit. 1,000 units will be made in January rising by 100 units per month. Prepare the production cost budget.

	Jan.	Feb.	March	April	May	June	Total
	£	£	£	£	£	£	£
Direct materials	6,000	6,600	7,200	7,800	8,400	9,000	45,000
Direct labour	8,000	8,800	9,600	10,400	11,200	12,000	60,000
Production overheads	4,000	4,400	4,800	5,200	5,600	6,000	30,000
	18,000	19,800	21,600	23,400	25,200	27,000	135,000
Units	1,000	1,100	1,200	1,300	1,400	1,500	7,500

(ii) Raw Materials Budget

The raw materials budget is particularly useful as it provides a forecast of how much raw material the company needs to buy. This can supply the purchases figure for the creditors budget. The raw materials budget format is shown in Figure 17.13. It starts with opening stock of raw materials (often taken from the opening balance sheet); purchases are then added. The amount used in production is then subtracted, arriving at closing stock. An example is shown in Figure 17.14.

Figure 17.13 Format of a Raw Materials Budget

	Jan.	Feb.	March	April	May	June	Total
	£	£	£	£	£	£	£
Opening stock of raw materials	X	X	X	X	X	X	X
Purchases	X	X	X	X	X	X	X
	X	X	X	X	X	X	X
Used in production	(X)	(X)	(X)	(X)	(X)	(X)	(X)
Closing stock of raw materials	X	X	X	X	X	X	X

Figure 17.14 Example of a Raw Materials Budget

Dai Jones has opening raw materials stock of £1,200. Purchases of raw materials will be £600 in January, increasing by £75 per month. Production will be 400 units per month using £2 raw material per unit. Prepare the raw materials budget.

	Jan.	Feb.	March	April	May	June	Total
	£	£	£	£	£	£	£
Opening stock of raw materials	1,200	1,000	875	825	850	950	1,200
Purchases	600	675	750	825	900	975	4,725
	1,800	1,675	1,625	1,650	1,750	1,925	5,925
Used in production	(800)	(800)	(800)	(800)	(800)	(800)	(4,800)
Closing stock of raw materials	1,000	875	825	850	950	1,125	1,125

(iii) Finished Goods Budget

The finished goods budget is similar to the raw materials budget except it deals with finished goods. As Figure 17.15 shows, it starts with the opening stock of finished goods (often taken from the balance sheet), the amount produced is then added (from the production cost budget). The cost of sales (i.e., cost of the goods sold) is then deducted. Finally, it finishes with the closing stock of finished goods. The finished goods budget is useful for keeping a check on whether the business is producing sufficient goods to meet demand. Figure 17.16 gives an example of a finished goods budget.

Figure 17.15 Format of Finished Goods Budget

	Jan. £	Feb. £	March £	April £	May £	June £	Total £
Opening stock of finished goods	X	X	X	X	X	X	X
Produced	X	X	X	X	X	X	X
	X	X	X	X	X	X	X
Cost of sales	(X)	(X)	(X)	(X)	(X)	(X)	(X)
Closing stock of finished goods	X	X	X	X	X	X	X

Figure 17.16 Example of a Finished Goods Budget

Ranjit Patel has £8,000 of finished good stocks in January. 1,000 units per month will be produced at a product cost of £10 each. Sales will be £10,000 in January rising by £1,000 per month. Gross profit is 25% of sales. Prepare the finished goods budget.

	Jan. £	Feb. £	March £	April £	May £	June £	Total £
Opening stock of finished goods	8,000	10,500	12,250	13,250	13,500	13,000	8,000
Produced	10,000	10,000	10,000	10,000	10,000	10,000	60,000
	18,000	20,500	22,250	23,250	23,500	23,000	68,000
Cost of sales*	(7,500)	(8,250)	(9,000)	(9,750)	(10,500)	(11,250)	(56,250)
Closing stock of finished goods	10,500	12,250	13,250	13,500	13,000	11,750	11,750
*Cost of sales is 75% of sales	10,000	1,000	12,000	13,000	14,000	15,000	75,000

Comprehensive Budgeting Example

Once all the budget have been prepared, it is important to gain an overview of how the business is expected to perform. Normally, a budgeted trading and profit and loss account and budgeted balance sheet are drawn up. This is often called a master budget. The full process is now illustrated using the example of Jacobs Engineering (see Figure 17.17), a manufacturing company. A manufacturing company is used so as to illustrate the full range of budgets discussed in this chapter. An overview of the whole process is presented in Figure 17.18 on page 383. It must be appreciated that Figure 17.18 does not include all possible budgets (for example, many businesses have labour, selling and administration budgets). However, Figure 17.18 does give a good appreciation of the basic budgetary flows.

Figure 17.17 Comprehensive Budgeting Example

Jacobs Engineering is a small manufacturing company. There are the following forecast summarised details.

Jacobs Engineering Ltd
Balance Sheet (Abridged) as at 31 December 2001

	£	£	£
Fixed Assets			100,000
Current Assets			
Debtors (Nov. £13,000, Dec £14,000)	27,000		
Stock of raw materials	10,000		
Stock of finished goods	15,000	52,000	
Current Liabilities			
Creditors (Nov. £12,000, Dec. £13,000)	(25,000)		
Bank overdraft	(7,000)	(32,000)	
Net current assets			20,000
Total net assets			120,000

	£
Share Capital and Reserves	
Share Capital	
Ordinary share capital	90,000
Reserves	
Profit and loss account	30,000
Total shareholders' funds	120,000

Notes:

1. Depreciation is 10% straight line basis per year.
2. Purchases will be £10,000 in January, increasing by £300 per month. They are payable two months after purchase.
3. Sales will be 450 units of product A at average production cost plus 25% and 500 units of product B at average production cost plus 20%. Debtors will pay two months in arrears.
4. Average production cost *per unit* remains the same as last year and is the same for product A and product B. It consists of direct materials £10, direct labour £7, production overheads £3. Both direct labour and production overheads will be paid in the month used. Production 1,000 units per month.
5. Non-production expenses are £4,000 per month, and will be paid in the month incurred.

Required

Prepare the sales budget, cash budget, debtors budget, creditors budget, production cost budget, finished goods budget, raw materials budget, budgeted trading, profit and loss account and budgeted balance sheet for six months ending 30 June, 2002.

Figure 17.17 Comprehensive Budgeting Example (*continued*)

Sales Budget

	Jan.	Feb.	March	April	May	June	Total
	£	£	£	£	£	£	£
Product A*	11,250	11,250	11,250	11,250	11,250	11,250	67,500
Product B**	12,000	12,000	12,000	12,000	12,000	12,000	72,000
	23,250	23,250	23,250	23,250	23,250	23,250	139,500

* (450 units × (£20 average production cost, i.e. direct materials £10, direct labour £7 and production overheads £3) plus 25%

** (500 units × (£20 average production cost, i.e. direct materials £10, direct labour £7 and production overheads £3) plus 20%

Cash Budget

	Jan.	Feb.	March	April	May	June	Total
	£	£	£	£	£	£	£
Opening cash	(7,000)	(20,000)	(33,000)	(33,750)	(34,800)	(36,150)	(7,000)
Add Receipts							
Debtors	13,000	14,000	23,250	23,250	23,250	23,250	120,000
	13,000	14,000	23,250	23,250	23,250	23,250	120,000
Less Payments							
Payment for goods	12,000	13,000	10,000	10,300	10,600	10,900	66,800
Non-production expenses	4,000	4,000	4,000	4,000	4,000	4,000	24,000
Direct labour	7,000	7,000	7,000	7,000	7,000	7,000	42,000
Production overheads	3,000	3,000	3,000	3,000	3,000	3,000	18,000
	26,000	27,000	24,000	24,300	24,600	24,900	150,800
Cash flow	*(13,000)*	*(13,000)*	*(750)*	*(1,050)*	*(1,350)*	*(1,650)*	*(30,800)*
Closing cash	(20,000)	(33,000)	(33,750)	(34,800)	(36,150)	(37,800)	(37,800)

Debtors Budget

	Jan.	Feb.	March	April	May	June	Total
	£	£	£	£	£	£	£
Opening debtors	27,000	37,250	46,500	46,500	46,500	46,500	27,000
Credit sales	23,250	23,250	23,250	23,250	23,250	23,250	139,500
	50,250	60,500	69,750	69,750	69,750	69,750	166,500
Cash received	(13,000)	(14,000)	(23,250)	(23,250)	(23,250)	(23,250)	(120,000)
Closing debtors	37,250	46,500	46,500	46,500	46,500	46,500	46,500

Figure 17.17 Comprehensive Budgeting Example (*continued*)

Creditors Budget

	Jan.	Feb.	March	April	May	June	Total
	£	£	£	£	£	£	£
Opening creditors	25,000	23,000	20,300	20,900	21,500	22,100	25,000
Credit purchases	10,000	10,300	10,600	10,900	11,200	11,500	64,500
	35,000	33,300	30,900	31,800	32,700	33,600	89,500
Cash paid	(12,000)	(13,000)	(10,000)	(10,300)	(10,600)	(10,900)	(66,800)
Closing creditors	23,000	20,300	20,900	21,500	22,100	22,700	22,700

Production Cost Budget

	Jan.	Feb.	March	April	May	June	Total
	£	£	£	£	£	£	£
Direct materials	10,000	10,000	10,000	10,000	10,000	10,000	60,000
Direct labour	7,000	7,000	7,000	7,000	7,000	7,000	42,000
Production overheads	3,000	3,000	3,000	3,000	3,000	3,000	18,000
	20,000	20,000	20,000	20,000	20,000	20,000	120,000

Finished Goods Budget

	Jan.	Feb.	March	April	May	June	Total
	£	£	£	£	£	£	£
Opening stock of finished goods	15,000	16,000	17,000	18,000	19,000	20,000	15,000
Produced	20,000	20,000	20,000	20,000	20,000	20,000	120,000
	35,000	36,000	37,000	38,000	39,000	40,000	135,000
Cost of sales*	(19,000)	(19,000)	(19,000)	(19,000)	(19,000)	(19,000)	(114,000)
Closing stock of finished goods	16,000	17,000	18,000	19,000	20,000	21,000	21,000

*(950 units × £20)

Raw Materials Budget

	Jan.	Feb.	March	April	May	June	Total
	£	£	£	£	£	£	£
Opening stock of raw materials	10,000	10,000	10,300	10,900	11,800	13,000	10,000
Purchases	10,000	10,300	10,600	10,900	11,200	11,500	64,500
	20,000	20,300	20,900	21,800	23,000	24,500	74,500
Used in product'n*	(10,000)	(10,000)	(10,000)	(10,000)	(10,000)	(10,000)	(60,000)
Closing stock of raw materials	10,000	10,300	10,900	11,800	13,000	14,500	14,500

*1,000 units per month × £10 direct materials.

Figure 17.17 **Comprehensive Budgeting Example (*continued*)**

Jacobs Ltd Engineering

Budgeted Trading and Profit and Loss Account for Year Ending 30 June 2002

	£	£	Source budget
Sales		139,500	**Sales**
Less *Cost of Sales*		114,000	**Finished goods**
Gross Profit		25,500	
Less *Expenses*			
Depreciation	10,000		–
Expenses	24,000	34,000	
Net Loss		(8,500)	–

Jacobs Ltd Engineering

Budgeted Balance Sheet as at 30 June 2002

	£ *Cost*	£ *Accumulated depreciation*	£ *Net book value*	Source Budget
Fixed Assets	100,000	(10,000)	90,000	
Current Assets				
Debtors	46,500			**Debtors**
Stock of raw materials	14,500			**Raw materials**
Stock of finished goods	21,000	82,000		**Finished goods**
Current Liabilities				
Creditors	(22,700)			**Creditors**
Bank overdraft	(37,800)	(60,500)		**Cash**
Net current assets			21,500	
Total net assets			111,500	
Share Capital and Reserves				
Share Capital				**Opening balance sheet**
Ordinary share capital			90,000	
Reserves				
Opening profit and loss account		30,000		**Opening balance sheet**
Less: Loss for year		(8,500)	21,500	**Trading and profit and loss account**
Total shareholders' funds			111,500	

After the budgetary period, the actual results are compared against the budgeted results. Any differences between the two sets of results will be investigated and action taken, if appropriate. This basic principle of investigating why any variances have occurred is common in both budgeting and standard costing.

Figure 17.18 Budgeting Overview

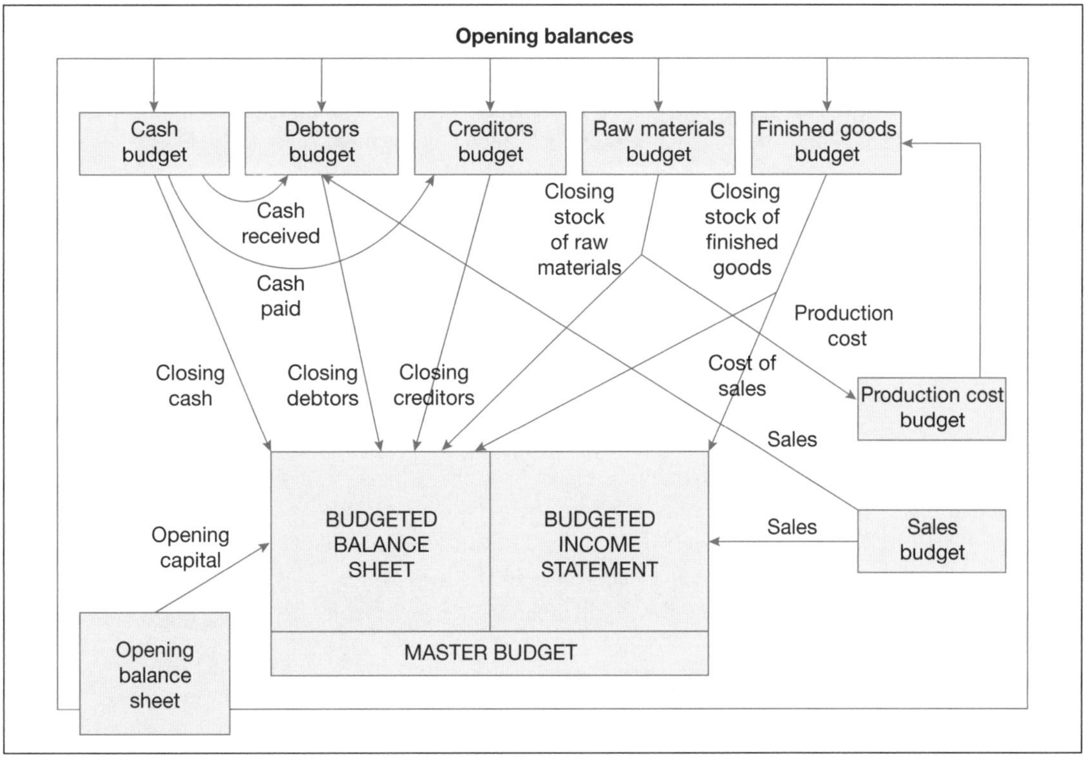

Behavioural Aspects of Budgeting

It is important to realise that there are several, sometimes competing, functions of budgets, for example, planning, coordinating, motivation and control. Budgeting is thus a mixture of technical (planning and coordinating) and behavioural (motivation and control) aspects. In short, budgets affect the behaviour of individuals within firms.

The budgetary process is all about setting targets and individuals meeting those targets. The problem is that an optimal target for the business is not necessarily optimal for the individual employee. Conflicts of interest, therefore, arise between the individual and the company. It may, for example, be in a company's interests to set a very demanding budget. However, it is not necessarily in the employee's interests. Budgets can, therefore, have powerful motivational or demotivational effects. For employees, in particular, budgets are often treated with great suspicion. Managers often use budgets deliberately as a motivational tool (see Real-Life Nugget 17.2 on the next page).

Real-Life Nugget 17.2

Budgetary Optimism

Where senior managers and accountants use a single budget and are reluctant to relinquish it as a motivational tool, which they often are, the problem gives rise to the 'bottom drawer phenomenon'. The main budget is designed to include some reflection of aspired performance levels, but the senior accountants keep another version in the bottom drawer of their desks which tries to make some allowance for the prevailing corporate optimism. Sometimes this merely takes the form of an overall 5 per cent or 10 per cent, but in other organisations, it is done on a departmental basis, with 5 per cent off Fred's budget and possibly 15 per cent off Joe's, depending on estimates of the aspirational element in each case. This is obviously done because senior management required a more realistic statement of expected performance for decision and planning purposes.

Source: A. Hopwood (1974), *Accounting and Human Behaviour*, p. 63

To demonstrate the behavioural impact of budgeting, we look below at three behavioural practices associated with budgeting: spending to budget; padding the budget; and creative budgeting.

Soundbite 17.2

Spending to Budget

'People spend what's in a budget, whether they need it or not.'

George W. Sztykiel, *Fortune* (December 28, 1992)

Source: *The Wiley Book of Business Quotations* (1998), p. 90

(i) Spending to Budget

Many companies allocate their expense budget for the year. If the budget is not spent, then the department loses the money. This is a double whammy as often next year's budget is based on this year's. So an underspend this year will also result in less to spend next year. It is in managers' individual interests (but not necessarily in the firm's interests) to avoid this. Managers, therefore, will spend money at the last-minute on items such as recarpeting of offices. This is the idea behind the cartoon at the start of this chapter!

Pause for Thought 17.3

Spending to Budget

Jane Morris has a department and is allocated £120,000 to spend for year ended 31 December 2002. In 2003, the budget is based on the 2002 budget plus 10% inflation. She spends prudently so that by 1 December she has spent £90,000. Normal expenditure would be a further £10,000. Thus, she would have spent £100,000 in total. What might she do?

Well, it is possible she might give the £20,000 unspent back to head office. However, this will result in only £110,000 (£100,000 + 10%) next year. More likely she will try to spend the surplus. For example, buy

- a new computer
- new office furniture
- new carpets

(ii) Padding the Budget

Budgets are set by people. Therefore, there is a great temptation for individuals to try to create slack in the system to give themselves some leeway. For example, *you might think* that your department's sales next year will be £120,000 and the expenses will be £90,000. Therefore, you think you will make £30,000 profit. However, at the budgetary committee *you might argue* that your sales will be only £110,000 and your expenses will be £100,000. You are, therefore, attempting to set the budget so that your profit is £10,000 (i.e., £110,000 – £100,000). You have, therefore, built in £20,000 budgetary slack (£30,000 profit [true position] less £10,000 profit [argued position]). Real-Life Nugget 17.3 demonstrates the negotiating process involved in padding the budget.

Real Life Nugget 17.3

Padding the Budget

Managers seek out and receive facts and opinions in order to arrive at an estimate of what they should ask for in the light of what they can expect to get and then, with due *'padding'* made to allow for anticipated cuts, they seek to market their budgetary demands. As in the initial example, the support of other parties can be actively canvassed and demands packaged in the most appealing form. Tangible results can be given undue weight and complex activities described in either the simplest or the most complex of terms. . . . Emphasis can be placed on the qualitative rather than measurable advantages, the forthcoming rather than current results, and the procedures rather than the outcome.

Source: A. Hopwood (1974), *Accounting and Human Behaviour*, p. 56

(iii) Creative Budgeting

If departmental managers are rewarded on the basis of the profit their department makes, then they may indulge in creative budgeting. Creative budgeting may, for example, involve deferring expenditure planned for this year until next year (see, for example, Figure 17.19 on the next page).

Figure 17.19 Creative Budgeting

Jasper Grant gets a bonus of £1,000 if he meets his budgeted profit of £100,000 for the year ending in December. At the end of November, he has the following information.

	£
Sales commission to date	200,000
Extra sales commission for December	25,000
Expenses to date	97,000
Expenses to be incurred	
: Weekly press advertising	3,000
: Repairs	8,000
: Necessary expenses	25,000

How can Jasper meet budget?

At the moment, Jasper's situation is as follows:

	£
Sales commission (£200,000 + £25,000)	225,000
Less: *Expenses* (£97,000 incurred + £36,000 anticipated)	133,000
Actual Profit	92,000
Budgeted Profit	100,000
Budgetary Shortfall	(8,000)

Jasper would, therefore, fail to meet his budgeted profit.

However, if the advertising and repairs were deferred until the next period, Jasper would make the budget. So he might be tempted to defer them.

	£
Sales commission	225,000
Less: Expenses to date	97,000
Necessary expenses	25,000
Actual Profit	103,000
Budgeted Profit	100,000
Surplus over Budget	3,000

Jasper would, therefore, gain his bonus as he would meet his target. However, what is good for Jasper is not necessarily good for his firm. By not advertising or having repairs done quickly it is likely that the long-term productivity of the firm will suffer.

Conclusion

Budgeting is planning for the future. This is important as a business needs to compare its actual performance against its targets. Most businesses prepare a cash budget and large businesses often prepare a complex set of interlocking budgets which culminate in a budgeted trading and profit and loss account and a budgeted balance sheet. Budgets have a human as well as a technical side. As well as being useful for planning and coordination, budgets are used to motivate and monitor individuals. Sometimes, therefore, the interests of individuals and businesses may conflict.

Discussion *Questions*

Questions with numbers in blue have answers at the back of the book.

Q 1 What the advantages and disadvantages of budgeting?

Q 2 Why do some people think that the cash budget is the most important budget?

Q 3 The behavioural aspects of budgeting are often overlooked, but are extremely important. Do you agree?

Q 4 State whether the following statements are true or false? If false, explain why.

(a) The four main aspects of budgets are planning, coordinating, control and motivation.
(b) The commonest limiting factor on the budgeting process is production.
(c) A master budget is formed by feeding in the results from all the other budgets.
(d) Depreciation is commonly found in a cash budget.
(e) Spending to budget, padding the budget and creative budgeting are all common behavioural responses to budgeting.

Numerical *Questions*

Questions with numbers in blue have answers at the back of the book.

Q 1 Jill Lee starts her business on 1 July with £15,000 in the bank. Her plans for the first six months are as follows.

(a) Payments for goods will be made one month after purchase:

January	February	March	April	May	June
£21,000	£19,500	£18,500	£23,400	£25,900	£31,100

(b) All sales will be cash sales:

January	February	March	April	May	June
£25,200	£27,100	£21,200	£20,250	£48,300	£37,500

(c) Expenses will be £12,000 in January and will rise by 10% per month. They will be paid in the month incurred.

Required:
Prepare Jill Lee's cash budget from 1 July to 31 December.

Q 2 John Rees has the following information for the six months 1 July to 31 December.

(a) Opening cash balance 1 July £8,600

(b) Sales at £25 per unit:

	April	May	June	July	Aug.	Sept.	Oct.	Nov.	Dec.
Units	100	130	150	180	200	210	220	240	280

Debtors will pay two months after they have bought the goods.

Q2 John Rees (continued)

(c) Production in units:

	April	May	June	July	Aug.	Sept.	Oct.	Nov.	Dec.	Jan.
Units	140	140	140	180	200	190	200	210	260	200

(d) Raw materials costing £10 per unit are delivered in the month of production and will be paid for three months after the goods are used in production.

(e) Direct labour of £6 per unit will be payable in the same month as production.

(f) Other variable production expenses will be £6 per unit. Two-thirds of this cost will be paid for in the same month as production and one-third in the month following production.

(g) Other expenses of £200 per month will be paid one month in arrears. These expenses have been at this rate for the past two years.

(h) A machine will be bought and paid for in September for £8,000.

(i) John Rees plans to borrow £4,500 from a relative in December. This will be banked immediately.

Required:
Prepare John Rees's cash budget from 1 July to 31 December.

Q3 Fly-by-Night plc has the following sales forecasts for two products: the Moon and the Star.

(a) The Moon will sell at 1,000 units in January, rising by 50 units per month. From January to March each Moon will sell at £20, with the price rising to £25 from April to June.

(b) The Star will sell 2,000 units in January rising by 100 units per month. Each Star will sell at £10.

Required:
Prepare the sales budget for Fly-by-Night from January to June.

Q4 David Ingo has opening debtors of £2,400 (November £1,400, December £1,000). The debtors will pay two months in arrears. Credit sales in January will be £1,000 rising by 10% per month.

Required:
Prepare D. Ingo's debtors budget for January to June.

Q5 Thomas Iger has opening creditors of £2,900 (£400 October, £1,200 November, £1,300 December). Creditors will be paid three months in arrears. Credit purchases in January will be £2,000 rising by £200 per month until March and then suffering a 10% decline in April and remaining constant.

Required:
Prepare T. Iger's creditors budget for January to June.

Q 6 Brenda Ear will have production costs per unit of £5 raw materials, £5.50 direct labour and £2 variable overheads. Production will be 700 units in January rising by 50 units per month.

Required:
Prepare B. Ear's production cost budget for January to June.

Q 7 Roger Abbit has £1,000 opening stocks of raw materials. Purchases will be £900 in January rising by £200 per month. Production will be 240 units from January to March at £4 raw materials per unit, rising to 250 units per month at £5 raw materials per unit from April to June.

Required:
Prepare R. Abbit's raw materials budget for January to June.

Q 8 Freddie Ox has £9,000 of opening finished goods stocks. In July, 1,500 units will be produced at a production cost of £10 each. Production will increase at 100 units per month; production cost remains steady. Sales will be £15,000 in July rising by £1,500 per month. Gross profit is 20% of sales.

Required:
Prepare F. Ox's finished goods budget for July–December.

Q 9 Asia is a small manufacturer. There are the following details.

Asia plc
Abridged Balance Sheet as at 31 December 2001

	£	£	£
Fixed Assets			99,500
Current Assets			
Debtors (Nov. £10,000, Dec. £11,000)	21,000		
Stock (raw materials)	10,000		
Stock (finished goods)	9,500	40,500	
Current Liabilities			
Creditors (Nov. £3,500, Dec. £3,500)	(7,000)		
Bank	(8,000)	(15,000)	
Net current assets			25,500
Total net assets			125,000
Share Capital and Reserves			£
Share Capital			
Ordinary share capital			103,000
Reserves			
Profit and loss account			22,000
Total shareholders' funds			125,000

Q **9** Asia (continued)

Notes

1. Fixed assets are at cost. Depreciation is at 10% straight line basis per year.
2. Purchases will be £4,800 in January increasing by £200 per month. They will be paid two months after purchase.
3. Sales will be £15,000 in January increasing by £400 per month. Debtors pay two months in arrears. They will be based on market price with no formal mark-up from gross profit.
4. Production cost per unit will be: direct materials £12; direct labour £10; production overheads £2 (direct labour and production overheads will be paid in the month incurred). Production 400 units per month. Sales 380 units per month.
5. Expenses will run at £6,000 per month. They will be paid in the month incurred.

Required:

Prepare sales budget, cash budget, debtors budget, creditors budget, production cost budget, raw materials budget, finished goods budget, trading and profit and loss account and balance sheet for six months ending 30 June 2002.

Q **10** Peter Jenkins manages a department and has the following budget for the year.

	£	£
Sales		100,000
Discretionary costs:		
Purchases	(20,000)	
Advertising	(10,000)	
Training	(8,000)	
Repairs	(19,000)	(57,000)
		43,000
Non-discretionary costs:		
Labour (split equally throughout the year)		(18,000)
Profit		25,000

Peter receives a budget of 10% of profit for any quarter in which he makes a minimum profit of £8,000. If he makes less than £8,000 profit, he receives no bonus. Any quarter in which he makes a loss he will earn no profit, but will not incur a penalty.

Required:

Calculate the maximum and minimum bonuses Peter could expect. Assume Peter has *complete discretion* about when the sales will be earned and when the discretionary costs will be incurred.

Chapter 18

Planning and Control: Standard Costing

'No amount of planning will ever replace dumb luck'

Anonymous.

Source: *The Executive's Book of Quotations* (1994), p. 217

Learning Outcomes

After completing this chapter you should be able to:

- ✔ Explain the nature and importance of standard costing.
- ✔ Outline the most important variances.
- ✔ Calculate variances and prepare a standard costing operating statement.
- ✔ Interpret the variances.

In a Nutshell

- Costing, and planning and control are the two main branches of cost accounting.
- Standard costing, along with budgeting, is one of the key aspects of planning and control.
- Standard costing is a sophisticated form of budgeting based on predetermined costs for cost elements such as direct labour or direct materials.
- There are sales and cost variances.
- Variances are deviations of the actual results from the standard results.
- Standard cost variances can be divided into quantity variances and price variances.
- There are direct materials, direct labour, variable overheads and fixed overheads cost variances.
- Standard cost variances are investigated to see why they have occurred.

Introduction

Cost accounting has two main branches: costing, and planning and control. Standard costing, along with budgeting, is one of the two parts of planning and control. Standard costing can be seen as a more specialised and formal type of budgeting. A particular feature of standard costing is the breakdown of costs into various cost components such as direct labour, direct materials, variable overheads and fixed overheads. In standard costing, **expected** (or **standard**) **sales or costs are compared against actual sales or costs**. Any differences between them (called **variances**) are then investigated. Variances are divided into those based on quantity and those based on prices. Standard costing, therefore, links the future (i.e., expected costs) to the past (i.e., costs that were actually incurred). Standard costing has proved a useful planning and control tool in many industries.

Nature of Standard Costing

Standard costing is a sophisticated form of budgeting. Standard costing was originally developed in manufacturing industries in order to control costs. Essentially, standard costing involves investigating, often in some detail, the distinct costing elements which make up a product such as direct materials, direct labour, variable overheads and fixed overheads. After this investigation the costs which should be incurred in making a product or service are determined. For example, a table might be expected to be made using eight hours of direct labour and three metres of wood. These become the standard quantities for making a table.

Definition 18.1

Standard Cost

Working definition

A predetermined calculation of the costs that *should* be incurred in making a product.

Formal definition

'The planned unit cost of the products, components or services produced in a period. The standard cost may be determined on a number of bases. The main uses of standard costs are in performance measurement, control, stock valuation and in the establishment of selling prices.'

Source: Chartered Institute of Management Accounting (2000), *Official Terminology*

After a table is made the direct labour and direct materials actually used are compared with the standard. For example, if nine hours are taken rather than eight, this is one hour worse than standard. However, if two metres of wood rather than three metres are used, this is one metre better than standard.

These differences from standard are called variances, which may be favourable (i.e., we have performed better than expected) or unfavourable (i.e., we have performed worse than expected). Unfavourable variances are sometimes called adverse variances. The essence of a good standard costing system is to establish the reason for any variances. Although standard costing is associated with manufacturing industry, it can be adopted in other industries, such as hotel and catering.

Setting standards involves making a decision about whether you are aiming for ideal, attainable or normal standards. They are all different. As Real-Life Nugget 18.1 shows, there is no such thing as a perfect standard. Ideal standards are those which would be attained in an ideal world. Unfortunately, the real world differs from the ideal world. Attainable standards are more realistic and can be reached with effort. Normal standards are those which a business usually attains. Attainable standards are the standards most often used by organisations.

Standards are usually set by industrial engineers in conjunction with accountants. This often causes friction with the workers who are actually monitored by the standards. Real-Life Nugget 18.2 gives a flavour of these tensions.

Real-Life Nugget 18.1

A Manager under Pressure

'We had unrealistic production allowances in the budget last year, but I got them adjusted a bit. It's done by the industrial engineers and the accountants, but I have a bit of a say in it. Some are too slack now, but they are very workable. There is never a perfect standard. The industrial engineers are back in the caveman era. Well, if there is going to be some imperfections, why not have them in my favour? We can modify them and we have done quite well. Why should I beat myself'

Source: A. Hopwood, *Accounting and Human Behaviour* (1974), p. 82

Real-Life Nugget 18.2

Setting the Standards

'Remember those bastards are out to screw you, and that's all they got to think about. They'll stay up half the night figuring out how to beat you out of a dime. They figure you're going to try to fool them, so they make allowances for that. They set [rates] low enough to allow for what you do. It's up to you to figure out how to fool them more than they allow for . . . .'

Source: W.P. Whyte, *Money and Motivation* (1955), pp. 15–16, as quoted in A. Hopwood, *Accounting and Human Behaviour* (1974), p. 6

Standard Cost Variances

In order to explain how standard costing works, the illustration of manufacturing a table is continued. In a succession of examples, more detail is gradually introduced. Figure 18.1 on the next page introduces the basic overview information for Alan Carpenter who is setting up a furniture business. In May, he makes a prototype table. To simplify the presentation in the figures *favourable* variances are shorted to 'Fav'. and *unfavourable* variances to 'Unfav'.

Figure 18.1 Overview of Standard Costs for Alan Carpenter for May

	Standard	Actual	Variance	
	£	£	£	
Sales	30	33	3	Fav.
Costs	20	22	(2)	Unfav.
Profit	10	11	1	Fav.

Before the prototype table was actually made, Alan Carpenter thought it would cost £20 and that he could sell it for £30. He, therefore, anticipated a profit of £10. In actual fact, the table was sold for £33. Carpenter thus made £3 more sales than anticipated. However, the table cost £22 (£2 more than anticipated). Overall, therefore, actual profit is £11 rather than the predicted standard profit of £10. Carpenter benefits £3 from selling well, but loses £2 through excessive costs. There is an overall favourable sales variance of £3, and an overall unfavourable cost variance of £2 which gives a £1 favourable profit variance.

We now expand the example and look at Alan Carpenter's operations for June. In June he anticipates making and selling 10 tables. In actual fact, he makes and sells only 9 tables (see Figure 18.2).

Figure 18.2 Standard Costs for Alan Carpenter for June

	Budget 10 Tables	*Actual* 9 Tables
	£	£
Sales	300	290
Costs	200	170
Profit	100	120

Calculation of variances

i. Flex the Budget	*Budget* 10 Tables	*Flexed Budget* 9 Tables	*Actual* 9 Tables	*Variance*	
	£	£	£	£	
Sales	300	270	290	20	Fav.
Costs	200	180	170	10	Fav.
Profit	100	90	120	30	Fav.

ii. Calculate Variances

	£	
Budgeted Profit	100	
Sales quantity variance (1 table at £10 profit)	(10)	Unfav.
Budgeted profit for actual production	90	
Sales price variance	20	Fav.
Cost variance	10	Fav.
Actual Profit	120	

Before calculating the variances, we must flex the budget. Flexing the budget simply means adjusting the budget to take into account the *actual quantity produced*. We need to do this because we need to calculate the costs *which would have been incurred if we actually made* nine tables. If we made only nine tables, we would logically expect to incur costs for nine rather than ten tables. In addition, we would expect to sell nine not ten tables. The budget must, therefore, be calculated on the basis of the nine actual tables made rather than the ten predicted.

Pause for Thought 18.1

Flexing the Budget

Why would it be so misleading if we failed to flex our budget?

If we failed to flex our budget then we would not compare like with like. For example, we would be trying to compare costs based on an output of ten tables with costs incurred actually making nine tables. We need to adjust for this, otherwise the budget will be distorted.

By making nine rather than ten tables, the profit for one table is lost. This was £10 (as there was £100 profit for ten tables). There is thus a £10 drop in profits. This is called a **sales quantity variance**. (Note that we are *assuming*, at this stage, that all the *costs will vary in direct relation to the number of tables.*)

We can now compare the sales and costs actually earned and incurred for nine tables with those that were expected to be earned and incurred. As Figure 18.2 shows, the nine tables were expected to sell for £270, but were actually sold for £290. In other words, we received £20 more than we had anticipated for our tables. This was the **sales price variance**. Note that the sales quantity variance is caused by lost profit. By contrast, the sales price variance is caused by the fact we are selling the tables for more than we anticipated. There is thus a **favourable sales price variance** of £20.

The budgeted cost for nine tables was £180. However, the actual cost incurred is only £170. There is thus a favourable cost variance of £10. The difference between the budgeted profit of £100 and the actual profit of £120 can be explained by these three variances (unfavourable sales quantity variance (£10), favourable sales price variance (£20), and favourable cost variance (£10)).

So far, we have seen that it is important to flex the budget, and looked at the sales volume variance and the sales price variance. It is now time to look in more detail at the cost variances. When looking at cost variances, it is important to distinguish between variable and fixed costs. As we saw in Chapter 15, variable costs are those costs that directly vary with a product or service (for example, direct materials, direct labour or production overheads). The more products made, the more the variable costs. Thus, if it costs £20 to make one table, it will cost £200 to make ten tables. Fixed costs, however, do not vary. You will pay the building's insurance of £10 per month whether you make one table or 100 tables. In continuing the Alan Carpenter example, the costs are now divided into fixed and variable.

Essentially, *all the variable costs* (direct materials, direct labour, and variable overheads) *have two elements – price and quantity*. For example, when making a table we might predict using the following standard prices and standard quantities:

	Price		*Quantity*
Direct labour	£5 per hour	for	2 hours
Direct materials	£1 per metre	for	5 metres of wood
Variable overheads (incurred on basis of direct labour hours)	£2.0 per hour	for	2 hours

When our actual price and actual quantity vary from standard, we will have price and quantity variances. In other words, the overall cost variances for direct labour, direct materials and variable overheads can be divided into a quantity and a price variance. **Price variances** are caused when the **standard price** of a product *differs* from the **actual price. Quantity variances** are caused when the **standard quantity** *differs* from the **actual quantity**. These differences between the standard and the actual quantity when multiplied by standard price will give us the overall quantity variance. Favourable variances are where we have done better than expected; unfavourable variances are where we have done worse than expected. For fixed overheads, which do not vary with production, we compare the actual fixed overheads incurred with the standard. We call this difference, the fixed overheads quantity variance.

An overview of all these variances is given in Figure 18.3 and in Figure 18.4 (on the opposite page) the main elements of all the variances is summarised. For simplification, this book uses the terms price and quantity variances throughout. Often, however, more technical terminology is used (these alternative technical terms are shown in the second column of Figure 18.4).

Figure 18.3 Diagram of Main Variances

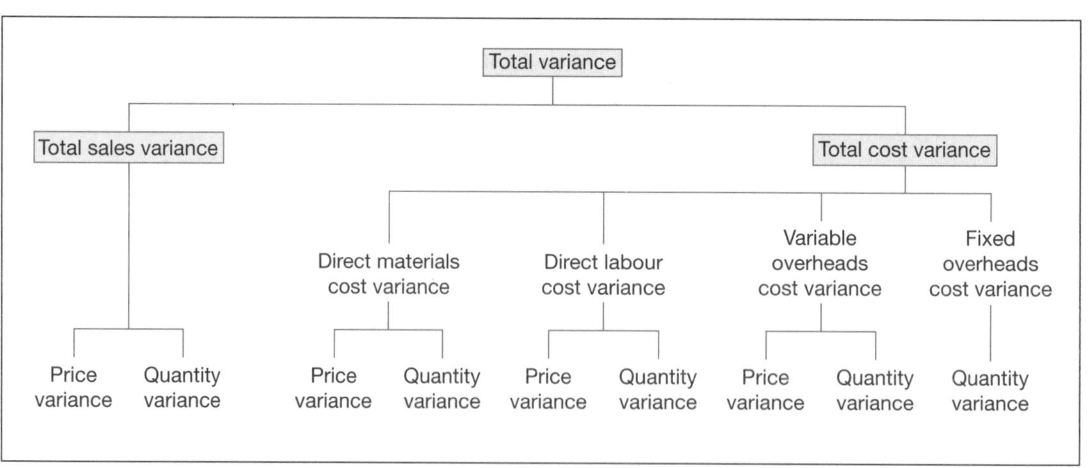

Figure 18.4 Calculation of the Main Variances

Variances	*Technical Name*	*Calculation*
i. SALES		
(a) Overall Variance	Sales	Not usual to calculate an overall sales variance
(b) Price	Price	(Standard price per unit – actual price per unit) × actual quantity of units sold
(c) Quantity	Volume	(Standard quantity of units sold – actual quantity of units sold) × standard contribution per unit
ii. COSTS		
Direct Materials		
(a) Overall Variance	Cost	Standard cost of materials for actual production less actual cost of materials used in production
(b) Price	Price	(Standard price per unit of material – actual price per unit of material) × actual quantity of materials used.
(c) Quantity	Usage	(Standard quantity of materials for actual production – actual quantity of materials used) × standard material price per unit of materials
Direct Labour		
(a) Overall Variance	Cost	Standard cost of labour for actual production less actual cost of labour used in production
(b) Price	Rate	(Standard price per hour – actual price per hour) × actual quantity of labour hours used
(c) Quantity	Efficiency	(Standard quantity of labour hours for actual production – actual quantity of labour hours used) × standard labour price per hour
Variable Overheads		
(a) Overall Variance	Cost	Standard cost of variable overheads for actual production less actual cost of variable overheads for production
(b) Price	Efficiency	(Standard variable overhead price per hour – actual variable overheads prices per hour) × actual quantity of labour hours used
(c) Quantity	Expenditure	(Standard quantity of labour hours for actual production – actual quantity of labour hours used) × standard variable overheads price per hour
Fixed Overheads		
(a) Quantity	Spending	(Standard fixed overheads – actual fixed overheads)

Figure 18.4 Calculation of the Main Variances (*continued*)

Helpnote
At first sight this table looks daunting. However, in essence the way we calculate all the overall variances, price variances and quantity variances is remarkably similar. The key essentials are explained below. *Note that in standard costing, standard refers to the budget.*

i. Overall Variances
Here we are interested in comparing the budgeted cost of the actual items we produce **(standard cost of actual production)** with the actual cost of items we produce **(actual cost of production).** The only variation is whether we are talking about direct materials, direct labour or variable overheads. The overall variance can be split into **price and quantity variances.** There is no overall variance for sales or fixed overheads. It is important to realise that, in practice, the **overall variance is normally calculated just by flexing the budget.** The difference between the flexed budget and the actual results gives the overall variance.

ii. Price Variances
Here we are interested in comparing the **standard price for the actual quantity used or sold** with the **actual price for the actual quantity used or sold.** The only variation is whether we are relating this to sales, direct materials, direct labour, or variable overheads. For variable overheads we use the actual quantity of labour hours used, as we are recovering variable overheads using labour hours. There is no price variance for fixed overheads. The standard formula is:

(Standard price – actual price) × actual quantity used or sold

iii. Quantity Variances
In this case, we are interested in comparing the budgeted cost of the actual items produced or sold **(standard cost of actual production or standard quantity for sales)** with the actual quantity produced or sold **(actual quantity used or sold).** This gives us the quantity difference. This is then multiplied by the standard contribution per item for the sales variance and by the standard price per unit **(normally, metres or hours)** for the other variances. The only variation is whether we are considering the standard price of direct materials, direct labour or variable overheads, or the standard selling price. For variable overheads, we use the standard quantity of labour hours and the actual quantity of labour hours used as we are recovering variable overheads using labour hours. The standard formula is:

Standard quantity of actual production (or, for sales, standard quantity for sales) less actual quantity used (or, for sales, actual quantity sold) × standard price (or, for sales, standard contribution per item)

For fixed overheads, the quantity variance is simply standard fixed overheads less actual fixed overheads.

In order to see how these variances all interlock, we now expand the Alan Carpenter example (see Figure 18.5).

Figure 18.5 **Worked Example: Alan Carpenter's Standard Costs for June**

i. Standard Cost per Table	£	£
Selling Price		30
Direct materials: 5 metres of wood at £1 per metre	(5)	
Direct labour: 2 hours at £5 per hour	(10)	
Variable overheads (recovered on labour hours):		
2 hours at £2 per hour	(4)	
Fixed overheads: £10 in total divided by 10 tables	(1)	(20)
Budgeted Profit		10

ii. Budgeted (i.e. standard) Results for June: 10 Tables	£	£
Selling Price		300
Direct materials: 50 metres of wood at £1 per metre	(50)	
Direct labour: 20 hours at £5 per hour	(100)	
Variable overheads: 20 hours at £2 per hour	(40)	
Fixed overheads	(10)	(200)
Budgeted Profit		100

iii. Actual Results for June: 9 Tables	£	£
Selling Price		290
Direct materials: 44 metres of wood at £1.20 per metre	(52.80)	
Direct labour: 16 hours at £4.50	(72.00)	
Variable overheads	(33.00)	
Fixed overheads	(12.20)	(170)
Actual Profit		120

Using figure 18.5, we now calculate the variances in two steps: flexing the budget and calculating the individual variances.

1. Flexing the Budget

Note that fixed overheads are not flexed! This is because by their very nature they do not vary with production. The flexed budget is used in the calculation of variances so that the levels of activity are the same and can be used as a basis for meaningful comparisons.

	Budget (i.e., standard)		*Flexed Budget* (i.e., standard quantity of actual production)		*Actual Results*		*Overall Variances*	
	10 Tables		9 Tables		9 Tables			
	£	£	£	£	£	£	£	
Sales		300		270		290.00	20.00	Fav.
Direct materials	(50)		(45)		(52.80)		(7.80)	Unfav.
Direct labour	(100)		(90)		(72.00)		18.00	Fav.
Variable overheads	(40)	(190)	(36)	(171)	(33.00)	(157.80)	3.00	Fav.
*Contribution**		110		99		132.20		
Fixed overheads		(10)		(10)		(12.20)	(2.20)	Unfav.
Profit		100		89		120.00		

*This is sometimes known as profit before fixed overheads.

By flexing the budget, *we have automatically calculated (1) the sales price variance, and (2) the overall cost variances* for direct materials, direct labour and variable overheads. It is important to note that the *flexed budget gives us the standard quantity of the actual production.* The cost variances will then be broken down into the individual price and quantity variances (see Figure 18.6 on the next page). However, *first* we need to calculate the sales quantity variance.

2. Calculating Individual Variances

(a) Sales Quantity Variance

Standard quantity of units sold	−	actual quantity of units sold	×	contribution*	
10	−	9	×	£11**	= £11 Unfav.

Notes:

* This is budgeted profit before fixed overheads (£110) divided by the budgeted number of tables (10). Therefore, we have, £110 ÷ 10 = £11. Note that this is calculated *before* fixed overheads. This is because fixed overheads will remain the same whatever the value of sales. They must be excluded from the calculation and their variance calculated separately.

** This represents the difference between the budgeted profit (£100) less the flexed budget (£89). *In practice, this is the easiest way to calculate this variance.*

(b) Individual Price and Quantity Variances

When we calculate the price and quantity variances based on our flexed budget on page 399, we can use the standard price and quantity formulas given below (see Figure 18.6 on the next page). We then need to remember that the standard price for sales is the standard selling price per unit, for direct materials it is the price per unit of direct materials used and so on. For quantity we need to remember that we are concerned with the standard quantity of materials, labour, etc. Figure 18.6 is based on Figure 18.5 and our flexed budget both on page 399.

Figure 18.6 Calculation of Individual Price and Quantity Variances for Alan Carpenter

Overall Variance =	Price Variance +	Quantity Variance
	(standard price – actual price sold or used) × actual quantity sold or used	*(standard quantity of actual production – actual quantity used) × standard price*
	Price calculations **Notes**	**Quantity calculations** **Notes**
(a) *Sales* *i. Sales price* *Note the sales price variance is automatically given by flexing the budget*	(£30 – £32.22*) × 9 tables = £20 Fav. **(1)** *£290 sold ÷ 9 tables sold = (£32.22)	ii. Sales quantity See calculation above (2a) for sales quantity variance **(2)**
(b) *Direct Materials* **Overall (£7.80) Unfav.**	(£1– £1.20 × 44 metres = *(£8.80) Unfav.* **(3)** **= (£8.80) Unfav.** +	(45 metres – 44 metres) × £1 = £1 Fav. **(4)** **£1 Fav.**
(c) *Direct Labour* **Overall £18 Fav**	(£5 – £4.50) × 16 hours = £8 Fav. **(5)** **= £8 Fav.** +	(18 hours – 16 hours) × £5 = £10 Fav. **(6)** **£10 Fav.**
(d) *Variable Overheads* **Overall £3 Fav.**	(£2.00 – £2.06*) × 16 hours = *(£1) Unfav.* **(7)** *£33 actual at 16 hours = £2.06 per hour **= (£1) Unfav.** +	(18 hours – 16 hours) × £2 = £4 Fav. **(8)** **£4 Fav.**
(e) *Fixed Overheads* **Overall (£2.20) Unfav.**	=	Standard less actual £10 – £12.20 = (£2.20) Unfav. **(9)** **(£2.20) Unfav.**

Notes:

1. **Sales Price Variance:** The standard selling price is £30 per unit. However, we sell at £32.22 per table (£290 ÷ 9). Therefore, overall, across the nine tables, we make £20 (9 × £2.22) more than we expected.
2. **Sales Quantity Variance:** The standard quantity expected to be sold was ten tables. We actually sold nine tables. Therefore, we lose the contribution on one table of £11. This was the overall contribution expected of £110 divided by the standard quantity of ten tables.
3. **Direct Materials Price Variance:** The standard price (£1.00) per metre is less than the actual price (£1.20) paid per metre by £0.20 pence per metre. As 44 metres were used this results in an unfavourable variance of £8.80.
4. **Direct Materials Quantity Variance:** The standard quantity to make nine tables is 45 metres. However, only 44 were actually used. As the standard price was £1 per metre, Alan Carpenter gains £1, which is a favourable variance.
5. **Direct Labour Price Variance:** The standard price (£5.00) per labour hour is greater than the actual price paid (£4.50). As 16 labour hours were used, the result is an £8 favourable variance.
6. **Direct Labour Quantity Variance:** The standard quantity to make nine tables is 18 hours. Alan Carpenter has taken 16 hours. As the standard price per labour hour is £5, there is a £10 favourable variance.
7. **Variable Overhead price Variance:** The standard recovery rate is £2.00 per hour. However, Alan Carpenter has recovered £33 in the 16 hours actually worked (i.e., £2.0625 per hour). Given the 16 hours worked, this results in an overall over-recovery of £1 (i.e. 16 × £0.0625). In other words, we expected to recover £32, but we actually recovered £33 (16 hours × £2). We thus recovered £1 more than expected, which is unfavourable.
8. **Variable Overheads Quantity Variance:** It was anticipated that 18 hours would be used to recover variable overheads. In actual fact, 16 hours were used. Since £2 per hour is recovered, there is an under-recovery of £4. This is favourable.
9. **Fixed Overhead Volume Variance:** The fixed overhead was expected to be £10. In fact, it was £12.20. This is £2.20 more than expected. Therefore, we have a £2.20 unfavourable variance.

We can now draw up a standard cost reconciliation statement. This statement reconciles the budgeted profit of £100 to the actual profit of £120.

Alan Carpenter: Standard Cost Reconciliation Statement for June				
	£	£	£	
Budgeted Profit			100	
Sales quantity variance			(11)	Unfav.
Budgeted Profit at Actual Sales			89	
	Fav.	Unfav.		
Variances	£	£		
Sales price	20			
Direct materials price		8.80		
Direct materials quantity	1			
Direct labour price	8			
Direct labour quantity	10			
Variable overhead price		1.00		
Variable overhead quantity	4			
Fixed overhead variance	—	2.20		
	43	12.00	31	Fav.
Actual Profit			120	

Interpretation of Variances

A key aspect of both budgeting and standard costing is the investigation of variances. In Figure 18.7 we look at some possible reasons for variances.

Figure 18.7 Possible Causes of Variances

Variance	*Favourable*	*Unfavourable*
1. Sales Quantity	We sell more than we expect because of • good market conditions • good marketing.	We sell less than we expect because of • poor market conditions • bad marketing.
2. Sales Price	We sell at a higher price than expected because of • unexpected market demand • good economy.	We sell at a lower price than expected because of • tough competition • poor economy.
3. Direct Materials Quantity	We use more material than expected because of • poor-quality material • sloppy production.	We use less material than expected because of • high-quality material • efficient production.
4. Direct Materials Price	Our material costs are more than expected because of • a price rise.	Our material costs are less than we expect because • we find an alternative cheaper supplier.
5. Direct Labour Quantity	We use more labour than we expected because of • inefficient cheaper workers who take longer • labour problems.	We use less labour than expected because of • efficient workers who are quicker.
6. Fixed Overheads	The overall costs are more than expected because of • inflation.	The overhead costs are less than expected because • we change to a cheaper source.

These variances represent differences in what actually happened to what we expected to happen. It must be remembered that variances are only as good as the original budgets or standards that are set. Therefore, if unrealistic, incorrect or unattainable standards are set, this will create misleading variances. As we saw earlier, setting the original standards can often cause severe tensions between the standard setters and the employees. In standard costing a more sophisticated, systematic approach is taken to the investigation of variances than in budgeting.

The nature of variable overheads means that these variances are tied into the base for absorption (for example, direct labour hours). The causes for the variable overhead variances will, therefore, tend to reflect those for the direct labour hours. It is not, therefore, particularly meaningful to investigate the reasons for these variances separately. Variable overheads have, therefore, been excluded from Figure 18.7.

Soundbite 18.1

Behavioural Aspects of Standards

'All persons concerned with setting standards are striving to gain some measure of personal control over factors which are important in their organisational lives.'

Source: A. Hopwood, *Accounting and Human Behaviour* (1974), pp. 6/7

Pause for Thought 18.2

Compensating Variances

Why might there be compensating variances? For example, an unfavourable material quantity variance, but a favourable material price variance.

Sometimes the quantity and price variances are interconnected. For example, we plan to make a product and we budget for a high quality material. If we then substitute a low quality material, it will be likely to cost less. However, we will need to use more. Therefore, we would have an unfavourable material quantity variance, but a favourable price variance.

Conclusion

Standard costing is a sophisticated form of budgeting. It was originally used in manufacturing industries, but is now much more widespread. Standard costing involves predetermining the cost of the various elements of a product or service (such as selling price, direct materials, direct labour and variable overheads). These standard costs are then compared with the actual costs and any differences, called variances, investigated. These variances are split into price and quantity variances. Standard costing allows costs to be monitored very closely. In many businesses, it is thus a very useful form of planning and control.

Discussion *Questions*

Questions with numbers in blue have answers at the back of the book.

Q 1 'Setting the standards is the most difficult part of standard costing.'
What considerations should be taken into account when setting standards?

Q 2 'Standard costing is good for planning and control, but unless great care is taken can often be very demotivational.' Discuss.

Q 3 Standard costing is more about control than motivation. Do you agree with this statement?

Q 4 'The key to standard setting is providing a good, fair initial set of standards.' Discuss.

Q 5 State whether the following statements are true or false. If false, explain why.

(a) Favourable cost variances are where actual costs are less than standard costs.
(b) Flexing the budget means adjusting the budget to take into account the actual prices incurred.
(c) Price and quantity variances are the main constituents of the overall cost variances.
(d) The direct materials price variance is: (standard price per unit of material – actual price per unit of material) × actual quantity of materials used.
(e) The direct labour quantity variance is: (standard price of labour hours for actual production – actual price of labour hours used) × standard labour price per hour.

Numerical *Questions*

Questions with numbers in blue have answers at the back of the book.

Q 1 Stuffed restaurant has the following results for May 2002

	Budget	*Actual*
Number of meals	10,000	12,000
	£	£
Price per meal	10.00	10.60
Food cost	30,000	37,200
Labour cost	35,000	36,000
Variable overheads	5,000	6,000
Fixed overheads	3,000	3,100

Required:

(i) Calculate the flexed budget.
(ii) Calculate the sales price variance.
(iii) Calculate the overall cost variances for materials, labour, variable overheads and fixed overheads (note: you do not have enough information to calculate the more detailed price and quantity variances).
(iv) Calculate the sales quantity variance.
(v) Discuss the variances. In particular, highlight what extra information might be needed.

Q2 Engines Incorporated, a small engineering company, has the following results for April 2002 for its product, the Widget. It budgeted to sell 12,500 widgets at £9.00 each. However, 16,000 widgets were actually sold at £8.80 each. The budgeted and actual costs are given below.

	Budget	*Actual*
Number of widgets	12,500	16,000
	£	£
Price per widget	9.00	8.80
Direct materials	40,000	42,000
Labour cost	32,000	29,000
Variable overheads	6,000	8,000
Fixed overheads	8,000	10,000

Required:

(i) Calculate the flexed budget.

(ii) Calculate the sales price and sales quantity variances.

(iii) Calculate the overall cost variances for materials, labour, variable overheads and fixed overheads (note: you do not have enough information to calculate the more detailed price and quantity variances).

(iv) Discuss the variances. In particular, highlight what extra information might be needed.

Q3 Birch Manufacturing makes bookcases. The company has the following details of its June production.

	Estimated	*Actual*
Number of bookcases	10,000	11,000
Metres of wood	100,000	120,000
Price per metre	0.50p	0.49p

Required:

Calculate the:

(i) overall direct materials cost variance

(ii) direct materials price variance

(iii) direct materials quantity variance.

Q 4 Sweatshop has the following details of direct labour used to make tracksuits for July.

(a) Standard: 550 sweatshirts at 2 hours at £3.50 per hour.
(b) Actual production: 500 sweatshirts at 1,050 hours for £3,780.

Required:

Calculate the:

(i) overall direct labour cost variance
(ii) direct labour price variance
(iii) direct labour quantity variance.

Q 5 Toycare manufactures puzzlegames. It has the following details of its March production.

	Estimated	*Actual*
Number of puzzlegames	11,000	12,000
Kilos of raw materials	5,500	4,800
Price per kilo	0.45p	0.46p
Direct labour (hours)	6,600	4,800
Direct labour price per hour	£5.50	£5.30

Required:

Calculate the:

(i) overall direct materials cost variance
(ii) direct materials price variance
(iii) direct materials quantity variance
(iv) overall direct labour cost variance
(v) direct labour price variance
(vi) direct labour quantity variance.

Q 6 Wonderworld has the following details for its variable overheads for August on its Teleporter. Each Teleporter is expected to take two labour hours and variable overheads are expected to be £2.50 per labour hour. Wonderworld expects to make 100,000 teleporters. In actual fact, it makes 110,000 teleporters using 230,000 labour hours. Its budgeted fixed overheads were £10,000. However, it actually spends £9,800 on fixed overheads. Actual variable overheads are £517,500.

Required:

Calculate the:

(i) overall variable overheads cost variance
(ii) variable overheads price variance
(iii) variable overheads quantity variance
(iv) fixed overheads variance.

Q7 Special Manufacturers has the following details for its variable overheads for July on its Startrek product. Each Startrek is expected to take three labour hours and variable overheads are expected to be £4.25 per labour hour. The firm expects to make 200,000 Startreks. In actual fact, it makes 180,000 Startreks using 630,000 labour hours. Its budgeted fixed overheads were £25,000. However, actually it spends £23,000 on fixed overheads. Actual variable overheads are £2,800,000.

Required:
Calculate the:
(i) overall variable overheads cost variance
(ii) variable overheads price variance
(iii) variable overheads quantity variance
(iv) fixed overheads variance.

Q8 Peter Peacock plc manufactures a subcomponent for the car industry. There are the following details for August.

(a) *Budgeted data*
Sales: 200,000 subcomponents at £2.80 each
Direct labour: 40,000 hours at £7.25 per hour
Direct materials: 100,000 sheets of metal at £1.25 each
Variable overheads: £40,000 recovered at £1.00 per direct labour hour
Fixed overheads: £68,000

(b) *Actual data*
220,000 subcomponents were actually sold and produced
Sales: 220,000 subcomponents at £2.78 each
Direct labour: 43,500 hours at £7.30 per hour
Direct material: 125,000 sheets of metal at £1.20
Variable overheads: £42,500
Fixed overheads: £67,000

Required:
(i) Calculate the flexed budget and overall variances.
(ii) Calculate the individual price and quantity variances.
(iii) Calculate a standard cost reconciliation statement for August.
(iv) Comment on the results.

Q9 Supersonic plc manufactures an assembly mounting for the aircraft industry. In July, it was expected that 80,000 assembly mountings would be sold at £18.80 each. In actual fact, 76,000 assembly mountings were sold for £20.00 each. The cost data is provided below.

	Budgeted data	*Actual data*
Direct materials	200,000 sheets of metal at £2.10 each	150,000 sheets of metal at £2.20 each
Direct labour	50,000 hours at £8.00 per hour	51,000 hours at £7.50 per hour
Variable overheads	£45,000 recovered at £0.90 per direct labour hour	£41,000
Fixed overheads	£78,000	£80,000

Required:

(i) Calculate the flexed budget and overall variances.

(ii) Calculate the individual price and quantity variances.

(iii) Calculate a standard cost reconciliation statement for July.

(iv) Comment on the results.

Chapter 19

Short-Term Decision Making

'So there I was, fresh from the annual meeting of the Society for Judgement and Decision Making, and behaving like Buridan's Ass – the imaginary creature which starved midway between two troughs of hay because it couldn't decide which to go for.'

Peter Aytan, 'Ditherer's Dilemma', *New Scientist*, 12 February 2000, p. 47

Learning Outcomes

After completing this chapter you should be able to:

- ✔ Explain the nature of short-term business decisions.
- ✔ Understand the concept of contribution analysis.
- ✔ Investigate some of the decisions for which contribution analysis is useful.
- ✔ Draw up break-even charts and contribution graphs

In a Nutshell

- In business, decision making involves choosing between activities and involves looking forward, using relevant information and financial evaluation.
- Businesses face a range of short-term decisions such as how to maximise limited resources.
- It is useful to distinguish between costs that vary with production or sales (variable costs) and costs that do not (fixed costs).
- Sales less variable costs equals contribution.
- Contribution less fixed costs equals net profit.
- Contribution and contribution per unit are useful when making short-term business decisions.
- Contribution analysis can help determine which products or services are most profitable, which are making losses, whether to buy externally rather than make internally and how to maximise the use of a limited resource.
- Throughput accounting attempts to remove bottlenecks from a production system; it treats direct labour and variable overheads as fixed.
- Break-even analysis shows the point at which a product makes neither a profit nor a loss.
- Both break-even charts and contribution graphs are useful ways of portraying business information.

Introduction

Businesses, like individuals, are continually involved in decision making. A decision is simply a choice between alternatives. It is forward looking. Business decisions may be long-term strategic ones about raising long-term finance or capital expenditure. Alternatively, the decisions may be short-term, day-to-day, operational ones, such as whether to continue making a particular product. This particular chapter looks at short-term decisions. When making short-term decisions, it is important to consider only factors relevant to the decision. Depreciation, for example, will not generally change whatever the short-term decision. It is not, therefore, included in the short-term decision-making calculations.

Decision Making

Individuals make decisions all the time. These may be short-term decisions: Shall we go out or stay in? If we go out, do we go for a meal, to the cinema or to a pub? Or long-term decisions: Shall we get married? Shall we buy a house? These decisions involve choosing between various competing alternatives.

Managers also make continual decisions about the short-term and long-term future. Shall we make a new product or not? Which product shall we make: A or B? Figure 19.1 gives some examples of short-term business decisions.

Figure 19.1 Some Short-Term Managerial Decisions

- Which products should the business continue to make this year?
- Which departments should the business close down this year?
- How should the business maximise limited resources this year?
- At what level of production does the business currently break even?
- How can the business maximise current profits?

Whatever the nature of the decision, informed business decision making will share certain characteristics, such as being forward-looking, using relevant information and involving financial evaluation.

(a) Forward-looking

Decisions look to the future and, therefore, require forward-looking information. Past costs that have no ongoing implications for the future are irrelevant. These costs are sometimes known as **sunk costs** and should be *excluded from decision making.*

(b) Relevant information

When choosing between alternatives, we are concerned only with information which is relevant to the particular decision. For example, after arriving in the centre of town by taxi, you are trying to choose between going to the cinema or pub. The taxi fare is a sunk cost which is not relevant to your decision. You cannot alter the past. *Relevant costs and revenues are, therefore, those costs and revenues that will affect a decision. Sunk costs are non-relevant costs.*

To make an informed decision only relevant information is needed. This information may be financial or non-financial. The non-financial information will normally, however, have indirect financial consequences. For example, a drop in the birthrate may have financial consequences for suppliers of baby products.

(c) Financial evaluation

In business, effective decision making will involve financial evaluation. This means gathering the relevant facts and then working out the financial benefits and costs of the various alternatives. There are a variety of techniques available to do this, which we will be discussing in subsequent chapters.

Decisions may have **opportunity costs**. *Opportunity costs are the potential benefit lost by rejecting the best alternative course of action.* If you work in the union bar at £5 per hour and the next best alternative is the university bookshop at £3 per hour, then the opportunity cost is £3 per hour. This is because you forgo the chance of working in the bookshop at £3 per hour.

However, from the above, it should not be assumed that business decision making is always wholly rational. As we all know, many factors, not all of them rational, enter into real-life decision making (see Real-Life Nugget 19.1).

Real-Life Nugget 19.1

Decision Making

The upshot? Although you might think that asking what you want should be the flip-side of asking what you don't want, it isn't. When you reject something you focus more on the negative features; when you select something you're focusing on the positive. So whether it's an ice cream or a prospective employee, deciding what (or who) you want by a process of rational elimination may not actually give you what you want. There is no simple panacea to ease the pain of choice other than passing the buck, of course (hence the irksome cliché: "No, you decide").

As for you sadistic purveyors of all this choice, don't get too smug. When Sheena Sethi-lyengar, from the Massachusetts Institute of Technology, set up a tasting counter for exotic jams in a grocery, she found that too much choice can be bad for business. More customers stopped to sample from a 24-jam counter than from a 6-jam counter. But only 3 per cent bought any jam when 24 were on offer, compared with 30 per cent when there were 6 to choose from.

Source: Peter Aytan, 'Ditherer's Dilemma', *New Scientist*, 12 February, 2000, p. 471

Contribution Analysis

When making short-term decisions, a technique called contribution analysis has evolved. This technique has several distinctive features (see Figure 19.2).

Figure 19.2 Key Features of Contribution Analysis

- Distinction between those costs that are the same whatever the level of production or service (*fixed costs*) and those that vary with the level of production or service (*variable costs*).
- Profit is no longer the main criterion by which decisions are judged. The key criterion becomes *contribution to fixed costs.*
- Calculations are often performed on the basis of *unit costs* rather than in total.
- Contribution analysis focuses on the *extra cost of making an extra product* or providing an extra unit of service.

It is now important to look more closely at the key elements of contribution analysis: fixed costs, variable costs and contribution. The analysis and interaction of these elements is sometimes called cost-profit-volume analysis. In this book, the term contribution analysis is used because it is considered easier to understand and more informative.

(i) Fixed costs

Fixed costs *do not change* if we sell more or less products or services. **They are thus irrelevant for short-term decisions**. For a business, fixed costs might be business rates, depreciation, insurance or rent. On a normal household telephone bill, for example, the fixed cost is the amount of the rental. It should be stressed that fixed costs will not change over the short term, but over the long term all costs will change (see Pause for Thought 19.1).

Pause for Thought 19.1

Fixed Costs

In the long run all fixed costs are variable. Why do you think this is so?

This is because at some stage the underlying conditions will change. For example, at a certain production level it will be necessary to buy extra machines, this will necessitate extra depreciation and insurance both normally fixed costs. Alternatively, if a factory closes then even the fixed costs will no longer be incurred.

(ii) Variable costs

These costs *do vary* with the level of production or service provided. **They are thus relevant for short-term decisions**. If we take a business, variable costs might be direct labour, direct materials, or overheads directly linked to service or production.

In practice, it may be difficult to ascertain which costs are fixed and which are variable. In addition, some costs will have elements of both a fixed and variable nature. For example, an electricity bill has a fixed standing charge and then an amount per unit of electricity used. However, to simplify matters, we shall treat costs as either fixed or variable.

(iii) Contribution

Contribution to fixed overheads, or contribution in short, is simply sales less variable costs. If sales are greater than variable costs, it means that for each product made or service provided the business contributes to its fixed overheads. Once a business's fixed overheads are covered, a profit will be made.

Contribution, as we shall see, is a very useful technique which enables businesses to choose the most profitable goods and services. Contribution analysis is sometimes called **marginal costing**. This comes from economics where a marginal cost is the *extra cost or 'incremental' cost needed to produce one more good or service*. Figure 19.3 demonstrates how contribution analysis works.

Figure 19.3 Demonstration of Contribution Analysis

Clueless has two products (X and Y) and the following abridged trading and profit and loss account.

	£	£
Sales		100,000
Less: *Costs*		
Direct materials	25,000	
Direct labour	35,000	
Overheads	20,000	80,000
Net Profit		20,000

Which of the two products is the most profitable?

To answer this question, we need additional information about X and Y and about which costs are fixed and which are variable. We can then use contribution analysis.

Additional information	**X**	**Y**
Sales units	1,000	2,000
	£	£
Sales	50,000	50,000
Direct materials	15,000	10,000
Direct labour	20,000	15,000
Variable overheads	7,000	3,000
Fixed overheads	10,000	

Contribution Analysis

	X (1,000 units)				**Y (2,000 units)**			
	Per unit		Total		Per unit		Total	
	£	£	£	£	£	£	£	£
Sales		50		50,000		25		50,000
Less: *Costs*								
Direct materials	15		15,000		5.0		10,000	
Direct labour	20		20,000		7.5		15,000	
Variable overheads	7	42	7,000	42,000	1.5	14	3,000	28,000
Contribution		8		8,000		11		22,000
X's contribution								8,000
Total Contribution								30,000
Fixed overheads								(10,000)
Net Profit								20,000

We can thus see that:

1. X make a contribution of £8 per unit, while Y makes a contribution of £11 per unit.
2. X contributes £8,000 to fixed overheads, while Y contributes £22,000 to fixed overheads.

The contribution data in Figure 19.3 can be used to answer a series of 'what if' questions, varying the levels of sales for X and Y. Contribution analysis is thus very versatile (as Figure 19.4 shows).

Figure 19.4 'What if' Questions for Products X and Y

(i) What if sales of X and Y double?
(ii) What if sales of X and Y halve?

(i) If sales of X and Y double, the contribution would double. Thus,

	£
X (Existing contribution £8,000)	16,000
Y (Existing contribution £22,000)	44,000
	60,000
Fixed overheads	(10,000)
Net Profit	50,000

(ii) If sales of X and Y halve the contribution will halve. Thus

	£
X (Existing contribution £8,000)	4,000
Y (Existing contribution £22,000)	11,000
	15,000
Fixed overheads	(10,000)
Net Profit	5,000

Helpnote:
The net profit (originally £20,000) increases and decreases by more than the direct increase or decrease in sales units. The contribution varies in line with sales, but fixed overheads do not vary. The overall net profit, in turn, therefore does not alter in direct proportion to the change in sales or contribution.

Decisions, Decisions

Contribution analysis can be used in a range of possible situations. All these involve the basic business questions:

- Are we maximising the firm's contribution by producing the most profitable products?
- Is the product making a positive contribution to the firm? If not, cease production.
- Should we make the products in house?
- Are we making the most of limited resources?

Although the decisions are different, the basic approach is the same (see Helpnote 19.1).

Helpnote 19.1

Basic Contribution Approach to Decision Making

1. Separate the costs into fixed and variable.
2. Allocate sales and costs to different products.
3. Calculate contribution (sales less variable cost) for each product:
   (a) in total
   (b) where appropriate, per unit or per unit of limiting factor.

It is important to realise that contribution analysis provides a rational approach to decision making. However, it should not be seen as providing a definitive answer. In the end, making the right decision will also involve an element of judgement. This is the sentiment expressed in Soundbite 19.1.

Soundbite 19.1

Decision Making as an Art

'Management is more art than science. No one can say with certainty which decisions will bring the most profit, any more than they can create instructions over how to sculpt a masterpiece. You just have to feel it as it goes.'

Richard D'Aveni, *Financial Times* (September 1, 1992)

Source: *The Wiley Book of Business Quotations* (1998), p. 311

(i) Determining the Most Profitable Products

If a company makes a range of products or services, we can use contribution analysis to see which are the most profitable (see Figure 19.5).

Figure 19.5 Determining the Most Profitable Products or Services

A garage provides its customers with three services: the basic service, the deluxe service and the superdeluxe service. It has the following details.

	Selling price	*Direct labour*	*Direct materials*
	£	£	£
Basic	75	35	10
Deluxe	95	45	12
Superdeluxe	120	65	14

Variable overheads are 50% direct labour. Fixed overheads are £1,000 per month.
Which services are the most profitable?

We need to calculate *per unit*	Basic		Deluxe		Superdeluxe	
	£	£	£	£	£	£
Sales Price		75.00		95.00		120.00
Less: *Variable costs*						
Direct materials	10.00		12.00		14.00	
Direct labour	35.00		45.00		65.00	
Variable overheads	17.50		22.50		32.50	
Total variable costs		62.50		79.50		111.50
Contribution		12.50		15.50		8.50
Ranked by profitability		2		1		3

The deluxe service is the most profitable (£15.50 contribution per unit), followed by the basic service (£12.50 contribution per unit) and the superdeluxe (£8.50 contribution per unit). Note that we do not take fixed costs into account. This is because they are irrelevant to the decision.

(ii) Should We Cease Production of Any Products?

The key here is to see whether or not any products or services are making a negative contribution.

Pause for Thought 19.2

Negative Contribution

Why should we discontinue any product or service with a negative contribution?

Products and services with negative contribution are bad news. This means that every extra product or service makes no contribution to our fixed costs. In actual fact, the more products or services we provide the greater our loss. This is because our variable costs per unit are greater than the selling price per unit.

If we take the example in Figure 19.6.

Figure 19.6 Dropping Loss-Making Products

We have the following information for Dolly, which makes three sweets: the mixtures, the sweeteners and the gobsuckers.

	£	£
Current Sales		
Mixtures (100,000 at 50p)		50,000
Sweeteners (200,000 at 20p)		40,000
Gobsuckers (400,000 at 10p)		40,000
		130,000
Less: *Costs*		
Direct materials	50,000	
Direct labour	40,000	
Variable overheads	20,000	
Fixed overheads	10,000	
Total costs		120,000
Net Profit		10,000

The variable costs are split between the products: 50% to mixtures, 25% to sweeteners and 25% to gobsuckers.
Are all these products profitable? If not, what is the effect on profit of dropping the unprofitable one?

Figure 19.6 Dropping Loss-Making Products (continued)

We need to rearrange our information to identify contribution per sweet.

	Mixtures		Sweeteners		Gobsuckers	
	£	£	£	£	£	£
Sales		50,000		40,000		40,000
Less: *Variable costs*:						
Direct materials	25,000		12,500		12,500	
Direct labour	20,000		10,000		10,000	
Variable overheads	10,000	55,000	5,000	27,500	5,000	27,500
Contribution		(5,000)		12,500		12,500
Total Contribution ((£5,000) + £12,500 + £12,500)						20,000
Fixed overheads						(10,000)
Net Profit						10,000

Sweeteners and gobsuckers thus make positive contributions of £12,500 each and are therefore profitable. By contrast, mixtures makes a negative contribution of £5,000. If we drop mixtures, profit increases by £5,000, we can see this below.

	£
Sweeteners	12,500
Gobsuckers	12,500
Total Contribution	25,000
Less: Fixed overheads	(10,000)
Net Profit	15,000

(iii) The Make or Buy Decision

Here we need to compare the cost of providing goods or services internally with the cost of buying in the goods or services. We compare the variable costs of making them internally with the external costs. Businesses often outsource (i.e., buy in) their non-essential activities. In Real-Life Nugget 19.2 British Airways has outsourced its engineering and IT services.

Real-Life Nugget 19.2

British Airways and Outsourcing

At the height of its prosperity in the mid-1990s, BA brought in an ambitious plan to cut £1 billion off its £8 billion annual costs. So far it has found savings of about £700m; it hopes to hack out the other £300m by March. A cull of 1,000 middle managers should lop a further £225m off costs. This was proclaimed as a precautionary measure to prepare BA for the next downturn, when price competition was bound to intensify. BA also started to outsource as much as it could, getting rid of such services as engineering, IT and catering. Its bosses talked of 'a virtual airline', one that concentrated only on selling seats and operating flights. Analysts noted approvingly that BA had been the first international airline to emerge smiling from the 1990–91 slump, thanks to similar prompt action.

Source: 'Diving for Cover', *The Economist*, 16 October, 1999

This sort of decision is also often faced by local governments in tendering or contracting out services such as cleaning (see Figure 19.7).

Figure 19.7 The Make or Buy Decision

A local government is looking at a particular cleaning contract. One of the existing local government departments has bid for a particular contract, with the following costs.

	£
Direct materials	28
Direct labour, 25 hours at £5.10	
Variable overheads	12

Speedyclean, a private company, has offered to do the contract for £165, should we accept?

	£
Internally	
Direct materials	28.00
Direct labour (25 hours at £5.10)	127.50
Variable overheads	12.00
	167.50
Externally	165.00

Yes, on pure cost grounds it should be awarded externally to Speedyclean.

Pause for Thought 19.3

Other Factors in Make or Buy Decisions

What other factors, apart from the costs, should you take into account in make and buy decisions?

In practice, make or buy decisions can involve many other factors. For example:

- Have the internal employees other work?
- Is the external price sustainable over time?
- Do we want to be dependent on an external provider?
- Will we be able to take action against the external provider if there is a deficient service?
- What consequences will awarding the contract externally have for morale, staff turnover?

(iv) Maximising a Limiting Factor

Businesses often face a situation where one of the key resource inputs is a limiting factor on production. For example, the quantity of direct materials may be limited or labour hours may be restricted. The basic idea, in this case, is to **maximise the contribution of the limiting factor**. Figure 19.8 on the next page demonstrates this concept for a hotel which has three restaurants, but a limited amount of direct labour hours.

Figure 19.8 **Maximising Contribution per Limiting Factor**

A hotel has three restaurants (Snack, Bistro and Formal). There are only 1,150 labour hours available at £10 per hour. If the restaurants opened normally for the coming week then 1,300 hours would be used: 400 hours for Snack, 500 hours for Bistro and 400 hours for Formal. Under normal opening you expect the following:

	Snack		*Bistro*		*Formal*	
Customers	2,000		3,000		2,000	
	£	£	£	£	£	£
Sales		10,000		12,000		16,000
Less: *Costs*						
Direct materials	2,000		2,500		7,500	
Direct labour	4,000		5,000		4,000	
Variable overheads	1,000		1,250		2,000	
Fixed overheads*	2,000	9,000	3,000	11,750	2,000	15,500
Net Profit		1,000		250		500

*Allocated by number of customers

You are required to maximise profit.

To maximise profit, we need to **maximise our contribution per unit of limiting resource,** i.e., labour hours. We need to maximise this as *labour is limited to 1,150 hours.* We must, therefore, first determine our overall contribution and the contribution per labour hour.

	Snack		*Bistro*		*Formal*	
Labour hours	400		500		400	
	£	£	£	£	£	£
Sales		10,000		12,000		16,000
Less: *Variable costs*						
Direct materials	2,000		2,500		7,500	
Direct labour	4,000		5,000		4,000	
Variable overheads	1,000	7,000	1,250	8,750	2,000	13,500
Contribution		3,000		3,250		2,500

Contribution per labour hour:

Contribution / Hours	£3,000 / 400 hours	= £7.50	£3,250 / 500 hours	= £6.50	£2,500 / 400 hours	= £6.25

Above, we have divided the coming week's contribution by the number of labour hours available. We ought, therefore, to use our labour first in the Snack, then in the Bistro, and only, lastly, on the Formal restaurant. This is because our contribution is greatest for the Snack at £7.50 per hour, next for the Bistro at £6.50 per hour and least for the Formal at £6.25 per hour. Our maximum profit (to the nearest £) is therefore:

			£
Snack	400 hours (i.e. the Snack's capacity)	× £7.50	3,000
Bistro	500 hours (i.e. the Bistro's capacity)	× £6.50	3,250
Formal	250 hours (i.e. the balance)	× £6.25	1,563
1150 hours			7,813
Less: Fixed costs			(7,000)
Net Profit			813

Any other allocation would result in less profit. For example, if we allocated the hours according to *maximum contribution per customer* (i.e., in the order Snack, Formal, Bistro).

			£
Snack	400 hours (i.e. Snack's capacity)	× £7.50	3,000
Formal	400 hours (i.e. Formal's capacity)	× £6.25	2,500
Bistro	350 hours (i.e. Balance)	× £6.50	2,275
	1150 hours		7,775
Less: Fixed costs			(7,000)
Net Profit			775

Throughput Accounting

Throughput accounting is a relatively new approach to production management, and uses a variant of contribution per limiting factor. This approach looks at a production system from the perspective of bottlenecks. It essentially asks what are a system's main bottlenecks? For instance, is it shortage of machine hours in a certain department? Every effort is then made to eliminate the bottlenecks. Goldratt and Cox in *The Goal* (1992) look at throughput contribution (defined as sales less direct materials) as a key measure. Interestingly, therefore, all other costs are treated as fixed. Thus, direct labour and variable overheads are seen as fixed overheads in this system. Figure 19.9 provides an example of a throughput operating statement. The throughput contribution is £40,000 (i.e., sales of £105,000 less direct materials at £65,000).

Figure 19.9 **Throughput Accounting**

Sanderson Engineering has the following details from its accounting records for May 2002.

	£000		£000
Sales	105,000	Production overheads	10,000
Direct materials	65,000	Administrative expenses	6,000
Direct labour	18,000	Selling and distribution expenses	5,000

Prepare a throughput accounting statement.

Sanderson Engineering Throughput Accounting Statement for May 2002

	£000	£000
Sales		105,000
Direct materials		(65,000)
Throughput Contribution		40,000
Direct labour	(18,000)	
Production overheads	(10,000)	
Administrative expenses	(6,000)	
Selling and distribution expenses	(5,000)	(39,000)
Net Profit		1,000

Break-Even Analysis

The contribution concept is particularly useful when determining the break-even point of a firm. The break-even point is simply that point at which a firm makes neither a profit nor a loss. A firm's break-even point can be expressed as follows:

Sales − Variable costs − Fixed costs = 0

Figure 19.10 The Essentials of Break-Even Analysis

Break-Even Point	Contribution	Break-Even Point in Units
The point at which a business makes neither a profit or a loss	Sales – Variable costs	$\frac{\text{Fixed costs}}{\text{Contribution per unit}}$

In other words, the break-even point is the point where contribution equals fixed costs. We can find the break-even point in units by dividing fixed costs by contribution per unit. Figure 19.11 shows how the break-even point works.

Figure 19.11 An Example of Break-Even Analysis

A restaurateur, William Bunter, has expected sales of 15,000 meals at £20 each. His variable costs are £8 per meal. If the fixed costs are £120,000, what is the break-even point? What is sales revenue at break-even profit?

We, therefore, have

$$\frac{\text{Fixed costs}}{\text{Contribution per unit (i.e. meal)}} = \frac{£120{,}000}{£20 - £8} = \frac{£120{,}000}{£12} = 10{,}000 \text{ meals}$$

(sales – variable cost per unit (i.e. meal))

Sales revenue at break-even is thus 10,000 × £20 = £200,000.

Assumptions of Break-Even Analysis

The beauty of break-even analysis is that it is comparatively straightforward. However, break-even analysis is underpinned by several key assumptions. Perhaps the main one is **linearity**. Linearity assumes that the behaviour of the sales and costs will remain constant despite increases in the level of sales. Sales and variable costs are assumed always to be strictly variable and fixed costs are assumed to be strictly fixed. For instance, in Figure 19.11 it is assumed that sales will remain at £20 per meal, variable costs will remain at £8 per meal and fixed costs will remain at £120,000, whether we sell 1,000 meals, 10,000 meals or 100,000 meals. In practice, it is more likely that these costs will be fixed or variable within a particular range of activity (often called the **relevant range**). The break-even point also implies a precision which is perhaps unwarranted. A better description might be the break-even area.

Other Uses of Break-Even Analysis

Break-even analysis can also form the basis of more sophisticated analyses such as (i) calculating the margin of safety, (ii) the basis for 'what-if' analysis, or (iii) the basis for graphical analysis.

(i) Margin of Safety

Bunter may wish to calculate how much he has sold over and above the break-even point. This is called the margin of safety. Bunter's margin of safety is calculated using a general formula:

$$\frac{\text{Actual units sold} - \text{units at break-even point}}{\text{Actual units}}$$

The break-even point can be calculated in either (a) units (i.e., in this case, meals) or (b) in money. Therefore,

$$\text{(a) Bunter's margin of safety (units)} = \frac{15{,}000 - 10{,}000}{10{,}000} = 50\%$$

$$\text{(b) Bunter's margin of safety (£s)} = \frac{£300{,}000 - £200{,}000}{£200{,}000} = 50\%$$

(ii) What-if Analysis

The break-even point can also be used as a basis for 'what-if' analysis. For instance, we know that each unit sold in excess of the break-even point adds one unit's contribution to profit. Similarly, each unit less than the break-even point creates a loss of one unit's contribution. So, we can easily answer questions such as 'what is the profit or loss if Bunter sells (a) 8,000 meals or (b) 13,000 meals?'

(a) **8,000 meals**. This is 2,000 meals less than the break-even point of 10,000 meals. The loss is, therefore, 2,000 meals × contribution per meal. Thus,

2,000 meals × £12 contribution per meal = £24,000 loss

(b) **13,000 meals**. This is 3,000 meals more than the break-even point of 10,000 meals. The profit is thus.

3,000 meals × £12 contribution per meal = £36,000 profit

The break-even point is a very flexible concept and provides potentially rewarding insights into business.

(iii) Graphical Break-Even Point

Another benefit of break-even analysis is that it can be shown on a graph (see Figure 19.12).

Figure 19.12 Graphical Break-Even Point

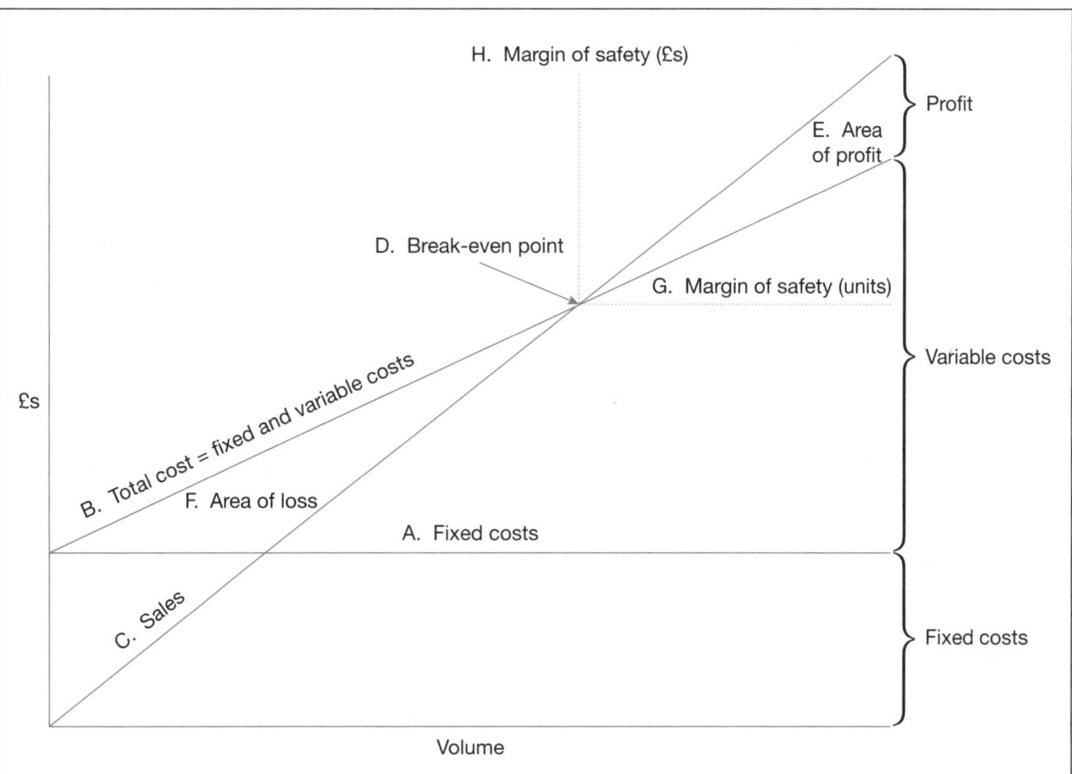

On the graph, it is important to note the following points:

A. **Fixed costs.** This is a straight horizontal line.
B. **Variable costs (total cost).** This is a straight line which starts on the vertical Y axis and 'piggybacks' the fixed cost line.
Effectively, the variable cost line also represents the **total cost (fixed cost plus variable costs).**
C. **Sales.** The sales line starts at the origin and then climbs steadily.
D. **Break-even point.** This is the point where the sales line (C) and the total cost line (B) (i.e., variable costs line) cross.
E. **Area of profit.** This is where the sales line (C) is higher than the total cost line (B). A profit is thus being made.
F. **Area of loss.** This is where the sales line (C) is lower than the total cost line (B). There is thus a loss.
G. **Margin of safety (units).** This is the difference between current sales and the sales needed to break even in units.
H. **Margin of safety (£s).** This is the difference between current sales and the sales needed to break even in £s.

Figure 19.13 is the graph for William Bunter. Sales are set at four levels: 0 meals; 5,000 meals; 10,000 meals; and 15,000 meals. Break-even point is 10,000 meals.

Figure 19.13 Bunter's Break-Even Chart

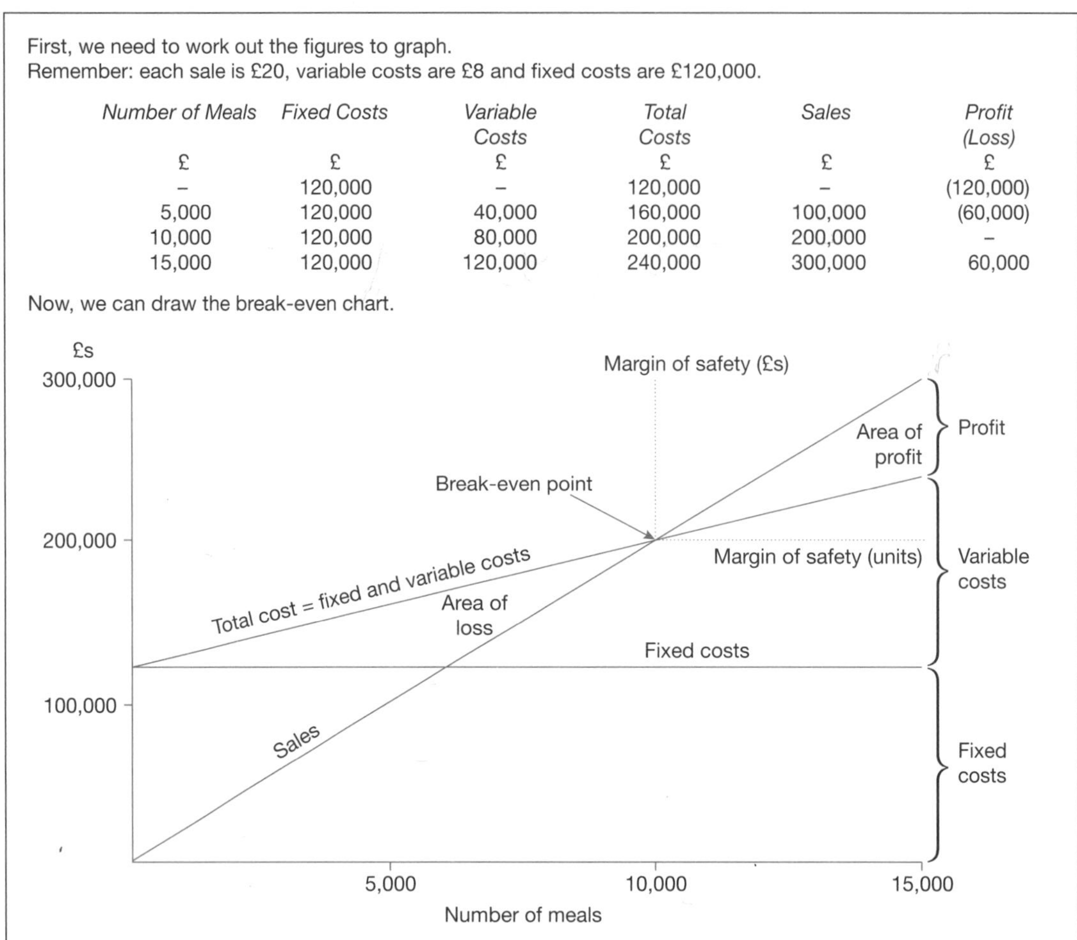

First, we need to work out the figures to graph.
Remember: each sale is £20, variable costs are £8 and fixed costs are £120,000.

Number of Meals	*Fixed Costs*	*Variable Costs*	*Total Costs*	*Sales*	*Profit (Loss)*
£	£	£	£	£	£
–	120,000	–	120,000	–	(120,000)
5,000	120,000	40,000	160,000	100,000	(60,000)
10,000	120,000	80,000	200,000	200,000	–
15,000	120,000	120,000	240,000	300,000	60,000

Now, we can draw the break-even chart.

Contribution Graph

A further limitation of the break-even chart is that it can be used for only one product. This disadvantage is overcome by using a contribution graph (see Figure 19.14 below). This is sometimes called a *profit/volume chart*. However, in this book we use contribution graph as it is easier to understand. A contribution graph looks a bit like a set of rugby posts! It is based on the idea that each unit sold generates one unit's contribution. Initially, this contribution covers fixed costs and then generates a profit. The horizontal line represents level of sales (either in units or £s). Above the horizontal line is profit, while below the line is loss. The diagonal line represents contribution. In effect, it is the cumulative profit or loss plotted against cumulative sales. So when sales are zero there is a loss (point A). This loss is, in effect, the total fixed costs. The company then breaks even at point B. At this point contribution equals fixed costs. Above point B each unit sold adds one unit of contribution to the company's profit. Finally, point C represents maximum cumulative sales and maximum contribution.

Figure 19.14 Contribution Graph

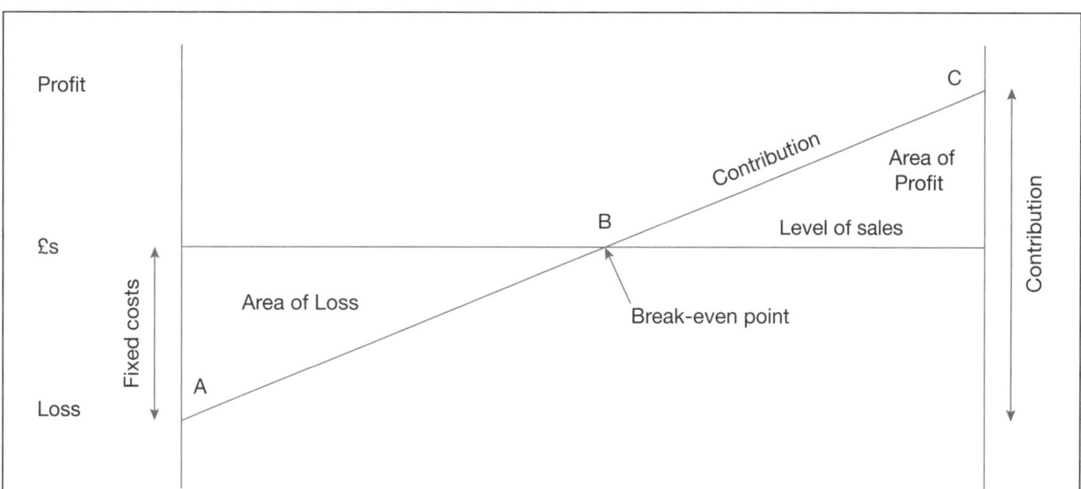

The relationship between contribution and sales is defined as $\frac{\text{Contribution}}{\text{Sales}}$. Sometimes this is known as the profit/volume ratio.

This ratio provides an easy way of comparing the contributions of different products. We take, as an example, a department store which has three departments: toys, clothes and records (see Figure 19.15).

Figure 19.15 Contribution Graph

Department	*Sales*	*Variable Costs*	*Contribution*	*Contribution/Sales Ratio*		*Ranking*
	£	£	£	%		
Toys	20,000	10,000	10,000	50	(£10,000/£20,000)	1
Clothes	40,000	30,000	10,000	25	(£10,000/£40,000)	3
Records	60,000	40,000	20,000	33 1/3	(£20,000/£60,000)	2
Total	120,000	80,000	40,000	33 1/3	(£40,000/£120,000)	
Fixed costs			(20,000)			
Net Profit			20,000			

Using this information draw a contribution graph.

First we draw up a cumulative profit/loss table ranked by the highest contribution/sales ratio. The cumulative profit/loss is simply cumulative contribution less fixed costs.

	Cumulative Sales	*Cumulative Contribution*	*Cumulative Profit/(Loss)*
	£	£	£
Fixed costs			(20,000)
Toys	20,000	10,000	(10,000)
Records	80,000	30,000	10,000
Clothes	120,000	40,000	20,000

We are now in a position to draw up a graph.

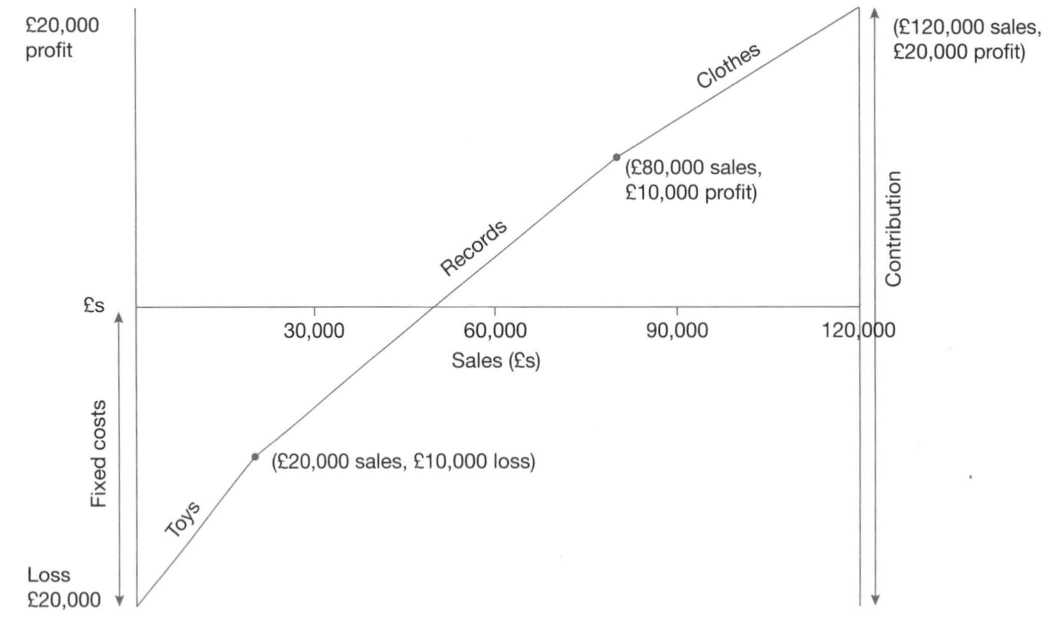

In many industries, there are substantial fixed costs. In the retail industry, for example, as Real-Life Nugget 19.3 shows, the concept of break-even becomes very important.

Real-Life Nugget 19.3

Fixed Costs and Supermarkets

The question is of particular interest in Britain, where the competition authorities are investigating newspaper claims that the big four supermarket chains Tesco, J. Sainsbury, Asda and Safeway are making excessive profits.

In recent years these four chains have tried to avoid price wars because price-slashing makes little sense in markets where a few big operators with similar cost structures operate.

The reason is that each has substantial fixed costs, making a profit by driving more than enough volume through its stores to cover its overheads. If one operator cuts prices, the others are compelled to follow for fear of seeing their own volumes fall below the break-even point. So nobody gains market share, and everybody's profits fall.

Source: Richard Tomkins, 'Marketing Value for Money', *Financial Times*, 14 May 1999

Conclusion

Businesses constantly face short-term decisions such as how to maximise a limited resource. When making these decisions it is useful to distinguish between fixed and variable costs. Fixed costs do not, in the short run, change with either production or sales (for example, insurance or depreciation). Variable costs, by contrast, do change when the production volume or sales volume changes (examples are direct materials and direct labour). Sales less variable costs gives contribution. Contribution is a useful accounting concept. By calculating contribution we can, for example, determine which products are the most profitable. Break-even analysis is another useful concept that builds on contribution analysis. The break-even point is the point at which a business makes neither a profit or a loss. It is determined by dividing fixed costs by the contribution per unit. The break-even chart shows the break-even point graphically. The contribution graph is a useful way of graphing the profit or loss of one or more products.

Discussion *Questions*

Questions with numbers in blue have answers at the back of the book.

Q 1 Distinguish between fixed and variable costs. Why are fixed costs irrelevant when making a choice between certain alternatives such as whether to produce more of product A or product B?

Q **2** What is contribution per unit and why is it so useful in short-term decision making?

Q **3** What are the strengths and weaknesses of break-even analysis?

*Q*4 State whether the following statements are true or false? If false, explain why.

(a) Fixed costs are those that do not vary with long-term changes in the level of sales or production.

(b) Contribution is sales less variable costs.

(c) Break-even point is $\dfrac{\text{Variable costs}}{\text{Contribution per unit}}$

(d) Contribution/sales ratio is $\dfrac{\text{Profit}}{\text{Sales}}$

(e) Non-financial items are not important in decision making.

Numerical *Questions*

Questions with numbers in blue have answers at the back of the book.

*Q*1 Jungle Animals makes 10 model animals with the following cost structure.

	Selling Price	*Direct Labour*	*Direct Materials*	*Variable Overheads*
	£	£	£	£
Alligators	1.00	0.50	0.20	0.35
Bears	1.20	0.60	0.10	0.30
Cougars	1.10	0.66	0.15	0.33
Donkeys	1.15	0.60	0.18	0.30
Eagles	1.20	0.56	0.12	0.28
Foxes	0.90	0.50	0.10	0.25
Giraffes	1.05	0.40	0.25	0.20
Hyenas	1.25	0.56	0.10	0.28
Iguanas	0.95	0.40	0.12	0.20
Jackals	0.80	0.40	0.13	0.20

Required:

(i) Calculate the contribution per toy.

(ii) Calculate the contribution/sales ratio per toy.

(iii) Which two toys bring in the greatest contribution?

(iv) Which three toys have the highest contribution/sales ratio?

(v) Which two toys would you not manufacture at all?

*Q*2 An insurance company, Riskmore, has four divisions: car, home, personal and miscellaneous. These four divisions account, respectively, for 40%, 30%, 20% and 10% of sales and 25% each of the variable costs. Riskmore has the following summary profit and loss account.

	£
Sales	200,000
Less: *Costs*	
Variable costs	100,000
Fixed costs	50,000
Net Profit	50,000

Required:

Calculate the profitability of the divisions.

Q3 Scrooge Ltd is looking to outsource its accounts department. Ghost Ltd, has approached Scrooge and offered to provide the service for £160,000. Scrooge Ltd ascertains the following costs are involved internally.

Clerical labour	10,000 hours at £5
Supervisory labour	6,000 hours at £10
Direct materials	£25,000
Variable overheads	£3 per clerical labour hour
Fixed overheads	£8,000

Required:

Calculate whether or not Scrooge Ltd should accept Ghost's bid. State any assumptions you have made and other factors you might take into account.

Q4 A large hotel, The Open Umbrella, has two kiosks. One sells sweets and is open 35 hours per week. The second sells newspapers and magazines and is open 55 hours per week. Unfortunately, next week labour is restricted to 70 hours. Labour is £5 per hour. Last week's results when both were fully open and 90 labour hours were available are set out below.

	Kiosk 1 (Sweets)		Kiosk 2 (Newspapers)	
	£	£	£	£
Sales		900		1,200
Less: *Costs*				
Direct labour	175		275	
Direct materials	600		700	
Variable overheads	70	845	110	1,085
Contribution		55		115

Required:

How would you maximise the profits using the 70 labour hours available?

Q5 Globeco makes four geographical board quiz games: France, Germany, UK and US. Globeco has the following recent results.

	France		*Germany*		*UK*		*US*	
Units sold	1,000		1,500		4,000		6,000	
	£	£	£	£	£	£	£	£
Sales		2,000		3,000		12,000		24,000
Less: *Variable Costs*								
Direct labour	800		1,350		6,000		15,600	
Direct materials	200		300		1,300		1,400	
Variable overheads	80		135		600		1,560	
Fixed overheads (equal allocation)	1,000		1,000		1,000		1,000	
		2,080		2,785		8,900		19,560
Net Profit (loss)		(80)		215		3,100		4,440

Q5 Globeco (continued)

For next year, there are only 3,000 direct labour hours available. Last year's results used 4,750 direct labour hours. Direct labour is paid at £5 per hour. The maximum sales (in units) are predicted to be 2,000 France, 3,500 Germany, 6,000 UK and 8,000 US.

Required:
Calculate the most profitable production schedule, given that direct labour hours, the limiting factor, are restricted to 3,000 hours.

Q6 Freya manufactures heavy-duty hammers. They each cost £4 in direct materials and £3 in variable expenses. They sell for £10 each. Fixed costs are £30,000. Currently 20,000 hammers are sold.

Required:
(i) What is the break-even point?
(ii) What is the profit if the number of hammers sold is:
(a) 4,000 (b) 14,000
(iii) What is the current margin of safety in (a) units and (b) £s?
(iv) Draw a break-even chart.

Q7 Colin Xiao runs a restaurant which serves 10,000 customers a month. Each customer spends £20, variable costs per customer are £15. Fixed costs are £10,000.

Required:
(i) What are the current break-even point and margin of safety in £s?
(ii) Calculate the new break-even point and margin of safety in £s if the average spend per customer is:
(a) £17 (b) £19 (c) £25
Assume all other factors remain the same.

Q8 A computer hardware distributor, Modem, has three branches in Cardiff, Edinburgh and London. It has the following financial details.

	Sales	*Direct Labour*	*Other Variable Overheads*
	£	£	£
Cardiff	200,000	60,000	105,000
Edinburgh	300,000	80,000	150,000
London	1,000,000	350,000	520,000

Head office fixed overheads are £150,000.

Required:
(i) Calculate the contribution/sales ratios.
(ii) Draw the contribution graph.

Chapter 20

Strategic Management Accounting

'Today's management accounting information, driven by the procedures and cycle of the organisation's financial reporting system, is too late, too aggregated, and too distorted to be relevant for managers' planning and control decisions.'

H.T. Johnson and R.S. Kaplan (1987), *Relevance Lost: The Rise and Fall of Management Accounting*, p. 1

Learning Outcomes

After completing this chapter you should be able to:

- ✔ **Explain the nature and importance of strategic management accounting.**
- ✔ **Understand and explain techniques used to assess the current position of the business.**
- ✔ **Appreciate the techniques of SWOT analysis, balanced scorecard and benchmarking.**
- ✔ **Discuss the strategic choices facing companies.**

In a Nutshell

- Strategic management accounting is externally orientated and concerns a business's future long-term strategy.
- Strategic management accounting is a relatively new topic.
- The three stages of strategic management accounting are (i) *assessment* of the current position of the business, (ii) *appraisal* of the current position of the business, and (iii) strategic *choice* of future direction of the business.
- The assessment of the current position of the business is concerned with both the external and internal environment and may use techniques such as value chain analysis, life cycle analysis and the product portfolio matrix.
- The appraisal of the current position of the business may use SWOT (strengths, weaknesses, opportunities and threats) analysis, the balanced scorecard and benchmarking.
- Strategic choice may involve exploiting inherent strengths such as the business's products or customer base and/or external diversification through acquisition or merger.

Introduction

Strategic management accounting attempts to involve management accountants in wider business strategy. It concerns a business's future, long-term direction. As Soundbite 20.1 shows, essentially strategy is concerned with where a business is now and where it wants to be in the future. Strategic management accounting is thus an attempt by management accountants to move away from a narrow, functional specialism towards full participation in the long-term strategic planning of businesses. In a sense, this continues a long historical process whereby management accountants have consistently widened the scope of their activities, for example, from costing to management accounting. It also meets the criticism that management accounting has failed to respond to changing environmental circumstances and focuses too much on a business's internal activities and too little on a business's external environment. In short, traditional management accounting is criticised for being too narrow. Strategic management accounting is a very contentious topic. For some, it represents the next stage in the evolution of the management accounting profession. For others, it is seen as a step too far, an unsuitable activity for management accountants.

Soundbite 20.1

Strategy

'There's no rocket science to strategy . . . you're supposed to know where you are, where your competition is, what your cost position is, and where you want to go. Strategies are intellectually simple, their execution is not.'

Lawrence A. Bossidy, *Harvard Business Review*, March–April, 1995.

Source: *The Wiley Book of Business Quotations* (1998), p. 88

Nature of Strategic Management Accounting

Strategic management accounting is relatively new. It dates from the 1980s. At this time, considerable concern was expressed that management accounting had lost its way. Johnson and Kaplan capture this concern in a book entitled *Relevance Lost: The Rise and Fall of Management Accounting*. They argue that conventional management accounting has failed to adapt to a changing industrial environment, that traditional product costing systems are increasingly inappropriate, that financial accounting dominates management accounting and that management accounting needs to reflect the external environment (see Real-Life Nugget 20.1).

Real-Life Nugget 20.1

Criticisms of Traditional Management Accounting

'Most [of today's] accounting and control systems have major problems: they distort product costs; they do not produce the key financial data required for effective and efficient operations; and the data they do produce reflect external reporting requirements far more than they do the reality of the new manufacturing environment.'

Source: R.S. Kaplan (1984), 'Yesterday's accounting undermines production', *Harvard Business Review*, p. 95

Strategic management accounting is one response to these criticisms. It attempts to involve management accountants in strategic business decisions and in the planning of a business's future long-term direction. The exact nature and extent of strategic management accounting is still, however, somewhat vague. In Definition 20.1, we provide a working definition and the Chartered Institute of Management Accountants' formal definition.

Definition 20.1

Strategic Management Accounting

Working definition

A form of management accounting which considers both an organisation's internal and external environments.

Formal definition

'A form of management accounting in which emphasis is placed on information which relates to factors external to the firm as well as non-financial information and internally-generated information.'

Source: Chartered Institute of Management Accountants (2000), *Official Terminology*

This book takes strategic accounting to be a form of management accounting which emphasises the external and future environment of the business, but also takes into account the internal environment of a business. In essence, it is the way in which a business seeks to implement its long-term strategic objectives, such as to be the world's leading supplier of a particular good or service. Figure 20.1 shows the three main stages of strategic management accounting. These stages are outlined briefly here and then discussed in more detail in the following section.

Figure 20.1 Overview of Strategic Management Accounting

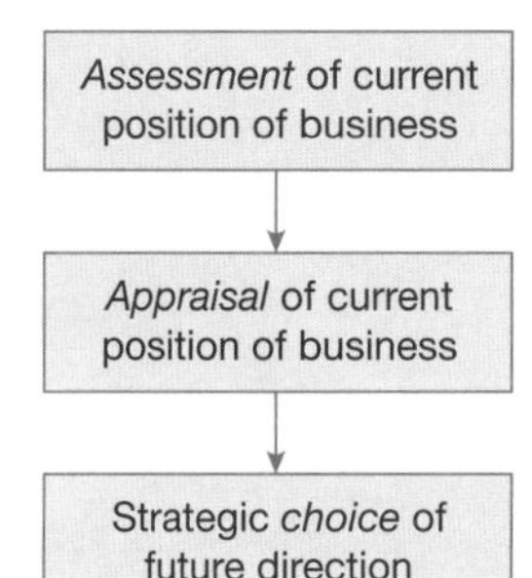

(i) Assessment of current position of the business

This is essentially an information gathering stage. Information is gathered on the current internal and external environment of the business.

(ii) Appraisal of current position of the business

This involves a hard look at the current position of the business investigating its strengths, weaknesses and competitiveness.

(iii) Strategic choice of future direction of the business

Once the current position of the business has been identified and appraised, it is time to choose the future direction of the business. This may involve diversification into new areas or the fuller exploitation of old areas.

It is important to appreciate that these strategic management activities are ongoing, not one-off. A well-run business will continually be assessing and appraising its current position and looking at its strategic choices. Assessment and appraisal are not distinct activities, but overlap. However, it is useful here to deal with them separately. As Soundbite 20.2 shows, it is harder to implement than plan business strategy.

Soundbite 20.2

Implementation of Strategy

'Strategy is easy, implementation is hard . . . [O]nly the superb companies actually find a way to do what they say they're going to do.'

Raymond Smith, Speech (November 1, 1995).

Source: *The Wiley Book of Business Quotations* (1998) p. 88

Assessment of Current Position of the Business

The first step in strategic management accounting is an assessment of the current position of the business. In essence, this is an information-gathering exercise. Information is gathered about the national and international external environment and the business's internal environment. Figure 20.2 outlines some of the issues that might be covered.

Figure 20.2 Information Useful in Assessment of Current Position

External Environment	*Internal Environment*
Political and legal environment Economic environment Social and cultural environment Technological change Interest and pressure groups Environmental issues Competitors	Resources Current products Operating systems Internal organisation Customers Sources of finance

External Environment

The seven factors listed in Figure 20.2, and discussed briefly below, serve as a basis for assessing both the current national and international environment. The importance of each factor varies from business to business.

Political and Legal Environment

This may involve the legal framework, national and international laws, and government policy. For example, privatised utilities in the UK, such as the water companies, are subject to a complex regulatory system.

Economic Environment

The economy is a key consideration for any business. Indeed, correctly assessing the current and future state of the economy is probably the most important external factor in business success.

Pause for Thought 20.1

Economic Environment

What are some of the considerations which a business might take into account when assessing the economy?

These are varied, but they include:

- current economic indicators (such as economic growth, inflation, interest rates, taxation and unemployment) at the regional, national and international level
- stage in economic cycle (boom or slump)
- long-term trends (such as towards a knowledge-based economy)
- government economic policy (such as spending, subsidy, privatisation)
- international trade (different economic conditions in different countries, exchange rates).

Social and Cultural Environment

Society continually evolves. For example, in developed countries there are increasing numbers of retired people and single-person households. These long-term demographic trends need to be matched to a company's products and services.

Technological Change

A business needs to look closely at technological change. Well-positioned businesses are best-placed to exploit social trends such as the use of mobile phones or lap-top computers.

Interest and Pressure Groups

It is important to realise that a business operates within society not outside society. Businesses must, therefore, keep an eye on public opinion, which is often manifested in pressure groups. For example, mutual building societies often face pressure from groups wishing to abolish their mutual status and turn them into publicly listed companies.

Environmental Issues

Nowadays, environmental issues are critically important for businesses. Companies must increasingly take on board environmental concerns. Indeed, many businesses have separate environmental departments. Some companies, such as Body Shop, have built their businesses upon a strong environmental ethic.

Competitors

A business needs to be aware of its competitors and the competitive environment in which it operates. Both customers and suppliers have potential power over a business. Businesses must also be aware of potential entrants to an industry as well as potential substitute products. For example, the entry of the US supermarket chain Walmart into the UK caused repercussions for UK companies such as Sainsbury and Tesco.

Internal Environment

A business's internal environment, like its external environment, varies from business to business. The six factors listed in Figure 20.2 (resources, current products, operating systems, internal organisation, customers and sources of finance) are reviewed below. We will then look at three techniques useful in assessing the internal environment: value chain analysis, product life cycle and the product portfolio matrix.

Resources

The resources potentially important in an internal assessment include materials, management, fixed assets, working capital, human resources, brands and other intangibles, and intellectual capital. Nowadays, in an increasingly knowledge-driven society, the last three resources are becoming ever more important. An important element of the resource assessment is the determination of any limiting factors or bottlenecks which will stop a business fulfilling its plans.

Current Products

It is important for a company to review its current product mix. This may include its current competitiveness, a product life cycle analysis, a product profitability review and an analysis of its customers.

Operating Systems

Operating systems are those information systems that form the backbone of a company, such as the sales invoicing system or the payroll system. Their effectiveness needs to be carefully assessed.

Internal Organisation

The effectiveness of the organisational structure of a business, such as the relationship between the head office and the department, needs to be carefully examined.

Customers

A key element of any internal assessment is the customer base. There is a need to gather data and profile customers, for example, by age, ethnicity, sex and social class.

Sources of Finance

It is useful to assess the sources of short-term and long-term capital and the mix between debt and equity. This important topic will be covered in more detail in Chapter 22.

Three techniques potentially useful in the internal assessment are now discussed: the value chain, the product life cycle and the product portfolio matrix.

(i) Value chain analysis

Value chain analysis was devised by Porter (1985). It represents a systematic way of looking at a business. Porter's idea is that each business has a value chain consisting of primary activities and support activities. The primary activities represent a set of value-creating activities from the handling of the raw materials (receiving goods) to after-sales service (service). These primary activities are supported by a set of support activities such as the personnel department (human resource management) or computer department (technology department) (see Figure 20.3).

Figure 20.3 Value Chain Analysis

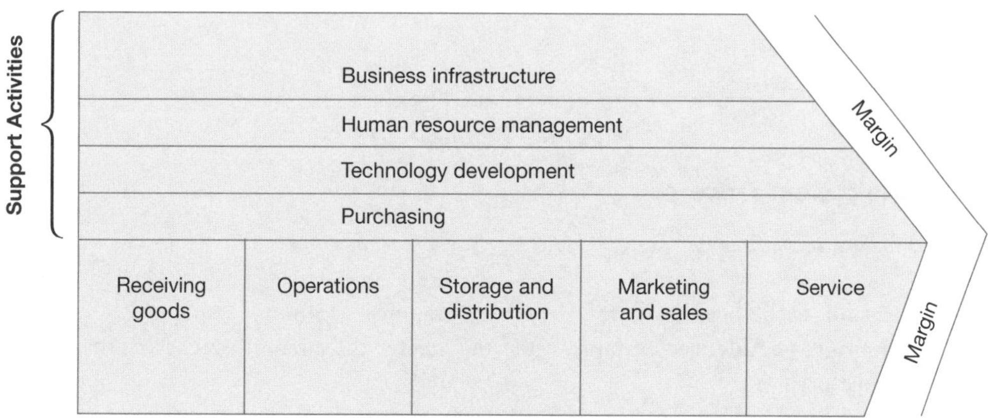

Source: Adapted from M.E. Porter (1985), *Competitive Advantage: Creating and Sustaining Superior Performance*, p. 37. Some of the terminology has been changed to aid understanding.

In Figure 20.4 we set out a value chain for a retail computer superstore.

Figure 20.4 Value Chain for a Retail Computer Superstore

Activity	*Definition*	*Example*
A. Primary Activities		
1. Receiving goods	Activities of receiving, handling inputs	Warehousing, transport, stock control
2. Operations	Conversion of inputs to outputs	Preparing computers for sale, service staff
3. Storage and distribution	Activities concerned with storing and distributing the product	Packaging, warehousing, delivery
4. Marketing and sales	Activities which inform and persuade customers	T.V. advertising campaigns
5. Service	After sales	Computer helplines
B. Support Activities		
1. Purchasing	Acquisition of resources	Computers for resale
2. Technology development	Techniques and work organisation	New computer system
3. Human resources management	Recruiting, training, developing and rewarding people	Induction course, computer training course
4. Business infrastructure	Information and planning systems	Sales invoicing system, budgetary system

The cost of these activities is determined by certain cost drivers (i.e., factors which determine the cost of each activity) such as location. Once these cost drivers have been established, then the management accountant can work out the organisation's sustainable competitive advantage. This involves reducing the costs of each activity while increasing the final sales.

Pause for Thought 20.2

Value Chain Cost Drivers

Porter identified 10 possible cost drivers. How many can you identify?

- **Location.** Is the business in the correct geographical location?
- **Economies of scale.** For example, often the greater the amount purchased, the cheaper the cost per unit.
- **Learning curve.** How experienced is the company at delivering its product or service?
- **Capacity utilisation.** Is the company fully utilising its production capacity?
- **Linkages.** How well are business linkages utilised, such as relationships with suppliers?
- **Interrelationships.** How good are the relations with other units within the group?
- **Timing.** Does the business buy and sell (assets, for example) at the right time?
- **Integration.** How well are the business's individual departments integrated?
- **Discretionary policies.** For example, has the business chosen the optimal computer operating systems?
- **Institutional factors.** Are the business's organisational structures (for example, its management structures) optimal?

(ii) Product life cycle analysis

Most products have a life cycle. Just as with human beings, this involves birth (known as introduction), growth, maturity, decline and senility (known as withdrawal). The key point is that each of these stages is associated with a certain level of sales and profit. Essentially, a product will make a profit once it becomes established. A period of growth will follow in both sales and profits. A period of maturity will then exist and, in this period, profitability will be maximised. The product's sales and profits will then decline and, finally, the product will be withdrawn.

A life cycle analysis graph is shown in Figure 20.5. It is important to realise that for some products, such as the latest children's toy, the product life cycle is short. On the other hand, for certain products, such as a prestige car, the life cycle is very long.

Figure 20.5 Life Cycle Analysis

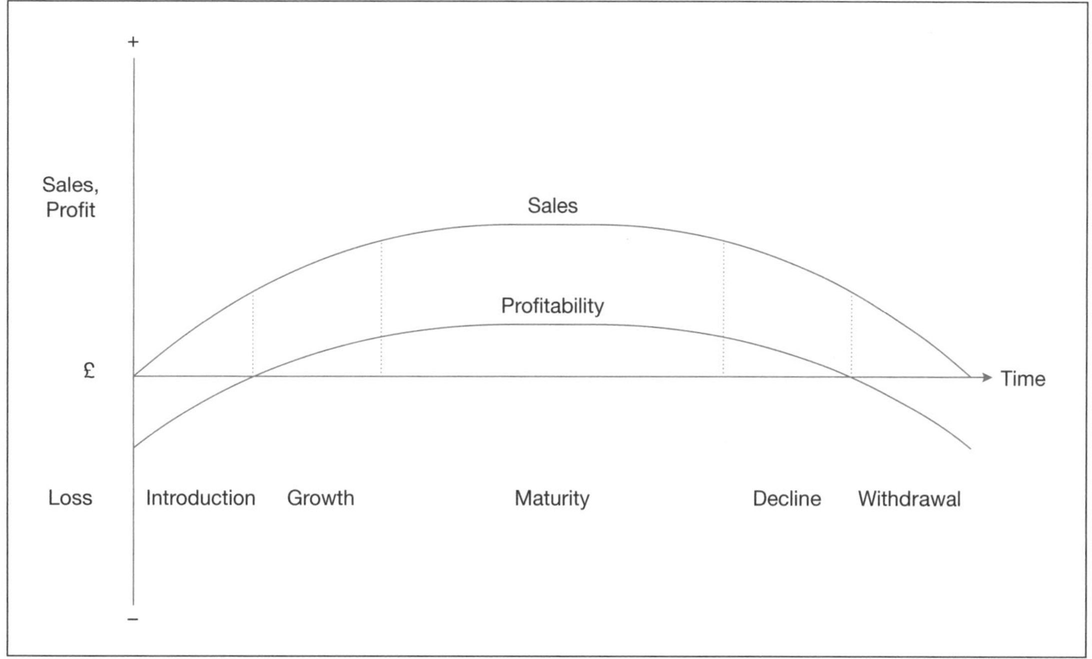

Pause for Thought 20.3

Product Life Cycles

Are product life cycles today longer or shorter than in the past? If so, what are the consequences for a business?

It is difficult to give a definitive answer. However, the expectation would be shorter. Technology and society are changing at an ever-faster pace, with new technology flourishing. Consumers also, in general, become rich with greater spending power. You would, therefore, expect fashion, styles and products to change more rapidly. Most products thus have shorter life cycles.

The consequences are that businesses need to invest more resources into product development and associated activities such as research and development. If they don't they will not keep pace with the changing market. In addition, it is likely that they will have to relaunch existing products and spend more on advertising and marketing.

Businesses will obviously try, through advertising and other means, to prolong the length of their product life cycle, especially in its mature phase. However, whatever the length of the product's life cycle, a business needs to develop new products for the eventual replacement of the existing product. Just as products have life cycles, it is also true that industries have life cycles (see, for example, Real-Life Nugget 20.2).

Real-Life Nugget 20.2

Industry Life Cycle

Industries are born, grow, mature and die, and for investors the most spectacular gains are to be had when an industry breaks through from immaturity to growth. This may seem a simple-minded truism to rank alongside "buy low, sell high" but, according to research by the investment bank Schroder Salomon Smith Barney, plotting the position of each sector on the growth/decline curve is fiendishly tricky.

The reason? Industries simply do not behave as they should.

On paper it looks straightforward, with the life-cycle of an industry resembling a motorway flyover. There is an up ramp (immaturity and growth), a flat stretch (maturity) and a down ramp (decline). Salomon is too delicate to include the final stage, but, as Edward G Robinson said in Double Indemnity: "The last stop's the cemetery."

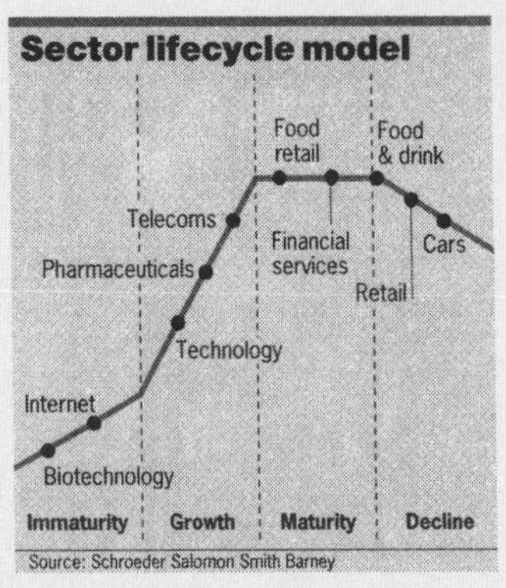

Source: Don Atkinson, 'Dead but refusing to lie down', The *Guardian*, June 7, 2000, p. 27

(iii) The product portfolio matrix

The Boston Consulting Group (BCG) developed a product portfolio matrix based partially on the ideas behind the product life cycle. There are four major categories of products: stars, cash cows, question marks and dogs.

(a) **Stars.** Stars are characterised by high market growth and high market share. Perhaps the mobile phone in its developmental stage. Stars may be cash earners or cash drains, requiring heavy capital expenditure.

(b) **Cash cows.** Cash cows are a company's dream product. They are well-established products which require little capital expenditure, but generate high returns.

(c) **Question marks.** They have low market share, but are in high market growth industries. So the potentially difficult problem for businesses is should they invest in marketing, advertising or capital expenditure in order to gain market share? Alternatively, should they completely withdraw from the market?

(d) **Dogs.** Dogs are cash traps with low market growth and low market share. Dogs tie up capital and often should be 'withdrawn'.

The product portfolio matrix is set out in Figure 20.6.

Figure 20.6 Boston Consulting Group's Product Portfolio Matrix

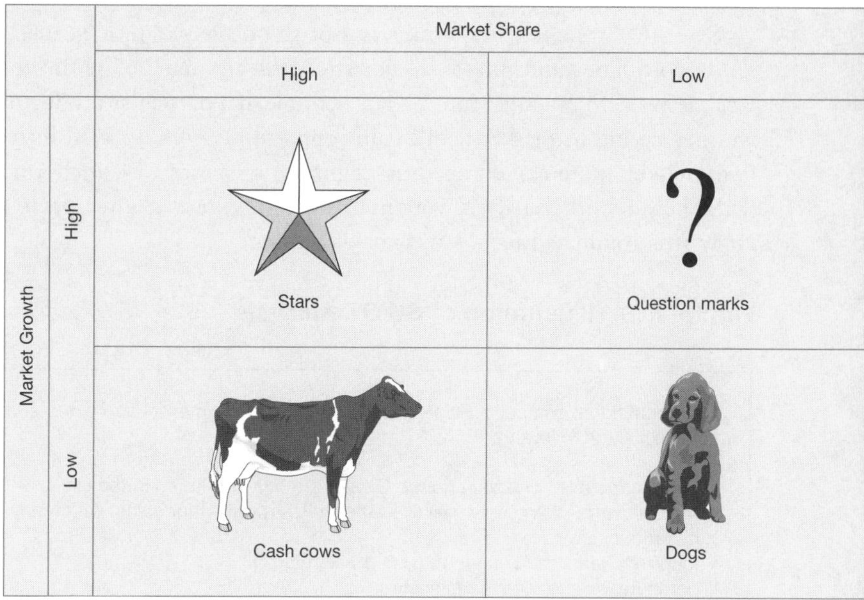

Appraisal of Current Position of Business

Once the current position of the business has been assessed, it needs to be appraised. We focus here on three major appraisal methods: SWOT analysis; the balanced scorecard; and benchmarking. These appraisal methods take both financial and non-financial factors into account. The appraisal of a business also includes ratio analysis (see Chapter 9).

(i) SWOT Analysis

SWOT analysis is a way of critically assessing a business's strengths and weaknesses, opportunities and threats. SWOT analysis is thus wide-ranging and looks at internal and external factors. As Figure 20.7 shows, SWOT analysis embraces the whole business and might include marketing, products, finance, infrastructure, management, organisational structure and resources. This list is

Figure 20.7 Typical Factors included in SWOT Analysis

(i) Marketing
- What is our market share?
- Are advertising campaigns effective?
- Are customers happy?

(ii) Products
- Is the branding strong?
- Do we have cash cows or dogs?
- Where are the products in their life cycle?

(iii) Finance
- Are the accounting ratios healthy?
- What is the contribution per product?
- Are the sources of finance secure?
- What is the product cost structure?
- Are we profitable and liquid?
- Do we have optimal gearing?

(iv) Infrastructure
- What is the age of our fixed assets?
- What is the market value of the assets?
- Have we sufficient space for expansion?

(v) Management
- What is the age spread?
- Have we training programmes in place?
- Do we have an appropriate skills base?

(vi) Organisational Structure
- Is our organisational structure appropriate?
- Is there a clear command structure?
- Is the management style appropriate?

(vii) Resources
- Are our sources of supply secure?
- Have we any resource bottlenecks?
- Have we good stock control?

illustrative rather than exhaustive. SWOT analysis involves the management accountant in detailed financial analysis using quantitative and qualitative data.

Once the strengths and weaknesses of the business are identified, we can investigate the potential opportunities and threats. Major strengths are matched with profitable opportunities and major weaknesses with potential threats. For example, if a particular product is selling well in Italy [strength], we may decide to market it in Spain [opportunity]. Or if an advertising campaign has gone wrong [weakness], a potential competitor might see a chance to launch a new competing product [threat].

So how does this analysis work in practice? Figure 20.8 attempts to answer this question. A SWOT analysis is conducted for a sweet manufacturer.

Figure 20.8 Illustration of SWOT Analysis

A major national company, Sweetco, has conducted an internal and external assessment of its products. It establishes the following:

- two products the Guzzler and the Geezer are only marketed in the UK
- the Guzzler is selling very well; the Geezer is struggling for sales, with many complaints, and is at the end of its product life
- long-term loan: repayment date is approaching
- computer system needs replacing
- management training is poor
- supplies of basic materials for products are running out.

Prepare a SWOT analysis

SWOT Analysis

Strengths	**Weaknesses**
Guzzler is a cash cow	Geezer is a dog Long-term loan Poor management training Lack of raw materials
Opportunities	**Threats**
Expand Guzzler into Europe Bring out a similar product to Guzzler Renegotiate a cheaper long-term finance package Introduce effective training programmes	Supplies of raw materials will run out Without loan, cash flow position is unhealthy. Geezer will tie up resources.

In this case, therefore, the Guzzler is clearly a strength, but the Geezer is a weakness. Sweetco, therefore, needs to focus on exploiting the Guzzler and minimising the problems caused by the Geezer.

(ii) Balanced Scorecard

The balanced scorecard attempts to look at a business from multiple perspectives. Definition 20.2. shows a working definition and the formal CIMA definition.

Definition 20.2

The Balanced Scorecard

Working definition

A system of corporate appraisal which looks at financial and non-financial elements from a financial, customer, internal business and innovation perspective.

Formal definition

'An approach to the provision of information to management to assist strategic policy formulation and achievement. It emphasises the need to provide the user with a set of information which addresses all relevant areas of performance in an objective and unbiased fashion. The information provided may include both financial and non-financial elements, and cover areas such as profitability, customer satisfaction, internal efficiency and innovation.'

Source: Chartered Institute of Management Accountants (2000), *Official Terminology*

The balanced scorecard takes a strategic, holistic view of an organisation. It combines both financial and non-financial information to provide a multi-perspective view of the organisation's activities. The balanced scorecard was adapted by Norton and Kaplan (1992) from the practice of some US companies. To Kaplan's credit, it seeks to address Kaplan's earlier criticisms of management accounting (such as lack of external focus) which were expressed in Real-Life Nugget 20.1 (see page 432). The balanced scorecard is a way of appraising a business's performance from four perspectives: *a financial perspective* (e.g., how profitable is it?), *a customer perspective* (e.g., how good is the after-sales service?), *an internal business perspective* (e.g., how efficient is our manufacturing process?), and *an innovation and learning perspective* (e.g., how many new products do we produce?). In essence, each perspective is set goals and then a set of performance measures devised. Thus, from the financial perspective the goals may be survival and profitability. These might be measured using cash flow and return on capital employed. As Soundbite 20.3 shows, a balanced scorecard is concerned with implementing rather than formulating strategy.

Soundbite 20.3

The Balanced Scorecard

'The [balanced] scorecard is not a way of formulating strategy. It's a way of understanding and checking what you have to do throughout the organization to make your strategy work.'

David Norton and Robert Kaplan, quoted in *Financial Times* (April 1, 1997)

Source: *The Wiley Book of Business Quotations* (1998), p. 295

Businesses such as Apple Computers have successfully used the balance scorecard. Figure 20.9 gives an example of a balanced scorecard.

Figure 20.9 Illustration of the Balanced Scorecard

Financial Perspective	
Goals	*Measures*
Survival	Cash flow
Profitability	Return on capital employed

Customer Perspective	
Goals	*Measures*
Quality product	Defect rate
Good after-sales service	Response time

Internal Business Perspective	
Goals	*Measures*
Managerial efficiency	Throughput of products
Advertising success	Increase in customer spend

Innovation and Learning Perspective	
Goals	*Measures*
Continual improvement	Number of employee suggestions
Develop new products	Number of new product launches

The balanced scorecard thus looks at a business in a comprehensive way. It devises goals and then seeks to measure them. Finally, it combines financial and non-financial measures. One problem with the balanced scorecard is that by taking a multiple-perspective view, no single goal is overriding. It is, therefore, difficult to prioritise the goals. In addition, the appropriate selection of a balanced set of measures may prove difficult, in practice.

(iii) Benchmarking

Benchmarking measures a business against its competitors. The concept is easy – find the best and compare oneself against them across a series of performance indicators. These might be, for example, customer service, number of complaints, or debtors collection period.

Definition 20.3

Benchmarking

Working definition

The comparison of a business with its direct competitors or industry norms.

Formal definition

'The establishment, through data gathering, of targets and comparators, through whose use relative levels of performance (and particularly areas of underperformance) can be identified. By the adoption of identified best practices it is hoped that performance will improve.'

Chartered Institute of Management Accountants (2000), *Official Terminology*

By benchmarking, companies aim to improve their own performance by comparing themselves against their competitors. The problem is that the most effective comparison is to competitors in the same industry. These competitors will be reluctant to provide the data. Benchmarking is thus more usually conducted against industry norms. A practical example of benchmarking is given in Real-Life Nugget 20.3.

Real-Life Nugget 20.3

Benchmarking in the Dairy Industry

Farmers generally supply their milk to local, regionalised dairies. These dairies may take the milk of, say, 20–30 farmers. Every month the dairy provides these farmers with a printout of how they are doing in comparison with the other farmers. This printout might contain, for example, milk yield per cow in litres, cows per hectare, yield from grazing, litres per cow. From this printout, the individual farmers can judge their relative productivity. The printouts are anonymous so that the farmers can only identify themselves. An extract from a printout is given below.

ANNUAL ROLLING RESULTS TO: August 2000

STOCK		**MILK**		**FEED**			**FORAGE**
Cows in herd	Cows/ hectare	Milk price pence	Yield litres/ cow	All concs tonnes/ cow	Av concs price £/tonne	All bought feed costs pence per litre	Yield from grazing litres/cow
133	2.75	16.19	7022	1.35	95	1.88	—
87	3.46	17.33	5836	1.06	108	1.97	3065
99	1.91	18.57	8005	1.35	106	2.04	2742
175	2.43	18.46	6707	1.59	95	2.24	—
99	2.49	15.42	7870	1.57	111	2.31	2181

Source: By kind permission of *Promar International*

Strategic Choice of Future Direction

Once the current position has been assessed and appraised, there is a need to plan for future strategy. In essence, there are two main directions. A business can either exploit its inherent strengths, for example, by expanding its product range or by more effectively utilising its customer base. Or alternatively, a business can diversify externally, for example, by acquiring another business. We explore these two directions briefly below.

(i) Exploit Inherent Strengths

If a particular product or service is successful, a business may consider expanding its range to similar, but as yet untapped, markets. For instance, if the product is selling well in Brazil, the company might expand its sales to other South American countries.

Pause for Thought 20.4

Exploiting a Product's Strength

Can you think of any ways other than exploiting new geographical markets in which you could use an existing product's strength?

It might be possible to do the following:

- target existing buyers to buy more
- target a new market segment, such as a new age range (e.g., market a child's book to adults, such as the Harry Potter series) or social group
- sell other products in association with the main product (e.g., the range of Barbie products with the Barbie doll).

An alternative to expanding its geographical range is for the business to exploit its customer base more fully. In order to gain relevant information, market research may be carried out. This may seek to ascertain the size of the customer base, the contribution of each sales item, product market share, sales growth and product demand.

Another technique is customer profitability analysis. In essence, this involves a detailed analysis of revenue looking at customer profitability and customer mix. To do this the customers are analysed by characteristics such as age, sex, spending power or social class. The aim is to focus on the most profitable customers.

Various techniques are now available which facilitate customer profitability analysis. Database mining, for example, enables a sales database to be critically examined for trends and other useful information. Interestingly, the findings from using these techniques are not always obvious, as Real-Life Nugget 20.4 shows.

Real-Life Nugget 20.4

Database Mining

A telecommunications company analysed its database using database mining. The customers were rated as low, medium and high spenders. The initial reason for the investigation was to maximise sales by encouraging the low spenders to spend more. In actual fact, the company found that the way to maximise sales was quite the reverse. The big spenders could easily be encouraged to spend still more.

(ii) External Diversification

A business may grow internally, acquire or merge with another business or cooperate with other businesses through a joint venture. This section concentrates on the acquisition/merger alternative. An acquisition is where one business takes over another (i.e., acquires more than 50% of another company's shares), whereas a merger is where two businesses combine their resources to form a new company. From an accounting perspective, mergers have the advantage that a business's profits are pooled and available for distribution to shareholders. In acquisitions not only are profits not available for distribution, but goodwill is created (see Chapter 12 for a more detailed discussion of goodwill).

Businesses seek mergers/acquisitions because they are a quick way of growing and also there may be certain operating advantages such as economies of scale. Vodafone, for example, acquired Mannesmann, a German company, so that it could extend its mobile phones into the European, and, in particular, the German, market. There are also other more specific reasons. For example, a manufacturing company's SWOT analysis might indicate a shortage of raw materials. To safeguard its supplies, the company may then purchase a key supplier.

Conclusion

Strategic management accounting attempts to overcome the inward-looking nature of management accounting. It is concerned with the long-term strategy of a business. Strategic management accounting is still evolving and is an elusive concept. However, three stages are identifiable. First, an assessment of the current position of a business. This involves an audit of a company's external environment (for example, political, legal and economic environment) and internal environment (for example, resources and current products). The techniques of value chain analysis, life cycle analysis and the product portfolio matrix are sometimes used. The second stage involves an appraisal of the business's current position. This may use SWOT analysis (strengths, weaknesses, opportunities and threats), the balanced scorecard (using financial and non-financial performance measures in a multi-perspective context) and benchmarking against competitors. And, lastly, there is the strategic choice of future direction. This may involve focusing internally on finding new product markets or more fully exploiting the customer database. Alternatively, a business may choose to diversify externally through a merger or acquisition.

Selected Reading

Johnson, H.T. and R.S. Kaplan (1987), *Relevance Lost: The Rise and Fall of Management Accounting*, Harvard University Press.
A ground-breaking book which re-examined the history of management accounting and argued it was out of touch with contemporary business needs.

Kaplan, R.S. (1984), 'Yesterday's accounting undermines production', *Harvard Business Review*, July/August, pp. 95–101.
In the same vein as *Relevance Lost*. It argues management accounting has not kept sufficiently up-to-date.

Kaplan, R.S. and D.P. Norton (1992), 'The Balanced Scorecard: measures that drive performance', *Harvard Business Review*, January–February, pp. 71–9.
Offers an alternative to the perceived inadequacies of management accounting.

Porter, M.E. (1985), *Competitive Advantage: Creating and Sustaining Superior Performance*, New York, Free Press.
A well-respected book which looks at how businesses can gain competitive advantage through techniques such as value chain analysis.

Discussion *Questions*

Questions with numbers in blue have answers at the back of the book.

Q1 What is strategic management accounting and how does it address some of the limitations of traditional management accounting?

Q2 Management accountants and strategy do not mix very well. Management accountants should, therefore, not involve themselves in strategic management accounting. Discuss the main arguments for and against this view.

Q3 Briefly outline the nature of the following techniques and then discuss their strengths and weaknesses:

(a) value chain analysis
(b) life cycle analysis
(c) the product portfolio matrix
(d) SWOT analysis
(e) the balanced scorecard
(f) benchmarking

Q4 Is growth through exploiting the internal resources of a company better than growth through acquisition/merger?

Q5 State whether the following statements are true or false. If false, explain why.

(a) Strategic management accounting looks at both the internal and external environments of a business.
(b) The cost of activities in value chain analysis is determined by key cost drivers.
(c) The product portfolio matrix consists of dogs, cats, cows and horses.
(d) The balanced scorecard uses only financial information.
(e) SWOT analysis stands for strengths, weaknesses, openings and threats.

Numerical *Questions*

Questions with numbers in blue have answers at the back of the book.

Q 1 You have the following details for three products (the maxi, mini and midi) for five years.

	Maxi		*Mini*		*Midi*	
Year	Sales	Profits (losses)	Sales	Profits (losses)	Sales	Profit-ability
	£	£	£	£	£	£
1	8,000	(3,000)	100,000	33,000	–	–
2	25,000	4,000	60,000	18,000	1,000	(3,000)
3	100,000	25,000	30,000	6,000	25,000	8,000
4	110,000	26,000	10,000	(3,000)	50,000	18,000
5	60,000	13,000	1,000	(4,000)	55,000	26,000

Required:
In Year 3, state at which stage of their life cycle you would anticipate the products to be (introduction, growth, maturity, decline or withdrawal)?

Q 2 Four products (the apple, orange, pear and banana) have the following financial profiles.

	Market Share	*Market Growth*
Apple	70%	80%
Orange	15%	70%
Pear	80%	10%
Banana	20%	15%

Required:
Classify the products into either stars, cash cows, question marks or dogs. Explain if you would expect them to be profitable.

Q **3** You have the following details about Computeco, a computer games manufacturer and distributor. An internal and external audit establishes the following details of two products, the Kung and the Fu.

	Market Share	*Market Growth*	*Sales* (000's)	*Contribution* (000's)
Kung	70%	80%	£10,000	£10,000
Fu	18%	20%	£3,000	£500

The Kung is currently marketed only in the UK; the Fu is marketed throughout the world. The product life cycles are five years. The Kung is 18 months old; the Fu is 4-years-old. The top developmental programmer has just left. Computeco has just installed a state-of-the-art computer system. The company is small, friendly and well-connected. A Japanese company has just approached Computeco asking it to market their product.

Required:
Prepare a SWOT analysis stating your recommendations to management for the future.

Q4 You have gathered the following information for a regional railway company.

(a) The mission statement of the company is to survive, be profitable, be safe, give customers good service, run an efficient service, maintain a good infrastructure, and constantly improve.

(b) The main financial indicators are cash flow, return on capital employed and return on investment.

(c) The company has recently offered full-ticket refunds to customers who are unhappy with the service provided or where trains are more than 10 minutes late.

(d) The company publishes monthly details of accidents.

(e) The company aims to use staff more efficiently and maximise the train operating times.

(f) The company operates a continual improvement programme based on staff suggestions.

(g) A new capital expenditure programme has expanded considerably the rolling stock.

Required:

From this information, prepare a balanced scorecard.

Q5 You sell the Feelgood, a health cushion, to 100,000 customers. Each cushion costs £25. You obtain the following breakdown of sales by customer category.

Sales Matrix

Age	*Sales*	*Geographical Location*	*Sales*	*Sex*	*Sales*	*TV Viewing*	*Sales*	*News-paper*	*Sales*
	£		£		£		£		£
0–20	10,000	South	150,000	Male	50,000	BBC	200,000	*Mail*	150,000
21–40	50,000	Midland	20,000	Female	200,000	ITV	20,000	*Sun*	10,000
41–60	80,000	North	70,000			Sky	30,000	*Telegraph*	20,000
60+	110,000	Overseas	10,000					*Times*	10,000
								Express	60,000
	250,000		250,000		250,000		250,000		250,000

Required:

(i) Using this customer matrix, state which customers you might target.

(ii) State whether you need any further information.

(iii) State where you might target your advertising campaign.

'There are no maps to the future.'

A.J.P. Taylor

Chapter 21

Long-Term Decision Making: Capital Investment Appraisal

Learning Outcomes

After completing this chapter you should be able to:

- ✔ Introduce and explain the nature of capital investment.
- ✔ Outline the main capital investment appraisal techniques.
- ✔ Appreciate the time value of money.
- ✔ Explain the use of discounting.

In a Nutshell

- Capital investment decisions are long-term, strategic decisions, such as building a new factory.
- Capital investment decisions involve initial cash outflows and then subsequent cash inflows.
- Many assumptions underpin these cash inflows and outflows.
- There are four main capital investment techniques. Two (payback and accounting rate of return) *do not* take into account the time value of money. Two *do* (net present value and internal rate of return).
- Payback is the simplest method. It measures how long it takes for a company to recover its initial investment.
- The accounting rate of return uses profit not cash flow and measures the annual profit over the initial capital investment.
- Net present value discounts estimated future cash flows back to today's values.
- Internal rate of return establishes the discount rate at which the project breaks even.
- Sensitivity analysis is often used to model future possible alternative situations.

Introduction

Management accounting can be divided into cost recovery and control, and decision making. In turn, decision making consists of short-term and long-term decisions. Whereas strategic management sets the overall framework within which the long-term decisions are made, capital investment appraisal involve long-term choices about specific investments in future projects. These projects may include, for example, investment in new products or new infrastructure assets. Without this investment in the future, firms would not survive in the long term. However, capital investment decisions are extremely difficult as they include a considerable amount of crystal ball gazing. This chapter looks at four techniques (the payback period, the accounting rate of return, net present value and the internal rate of return) which management accountants use to help them peer into the future.

Nature of Capital Investment

Capital investment is essential for the long-term survival of a business. Existing fixed assets, for example, will wear out and need replacing. The capital investment decision operationalises the strategic, long-term plans of a business. As Figure 21.1 shows, long-term capital expenditure decisions can be distinguished from short-term decisions by their time span, topic, nature and level of expenditure, by the external factors taken into account and by the techniques used.

Figure 21.1 Comparison of Short-Term Decisions and Long-Term Capital Investment Decisions

Characteristic	*Short-Term*	*Long-Term*
1. Time Span	Maximum 1 to 2 years, mostly present situation	Upwards from 2 years
2. Topic	Usually concerned with current operating decision, e.g., discontinue present product	Concerned with future expenditure decisions, e.g., build new factory
3. Nature	Operational	Strategic
4. Level of Expenditure	Small to medium	Medium to great
5. External Factors	Generally not so important	Very important, especially interest rate, inflation rate
6. Sample Techniques	Contribution analysis, break-even analysis	Payback, accounting rate of return, net present value, and internal rate of return

Capital investment decisions are thus usually long-term decisions which may sometimes look 10 or 20 years into the future. They are often the biggest expenditure decisions which a business faces. Some examples of capital investment decisions might be deciding whether or not to build a new factory or whether to expand into a new product range. A football club, such as Manchester United, for example, might have to decide whether or not to build a new stadium. A common capital expenditure decision will involve choosing between alternatives. For example, which of three particular stadiums should we build? Or which products should we currently develop for the future? These decisions are particularly important given the fast-changing world (see for example, Soundbite 21.1).

A key problem with any capital investment decision is taking into account all the external factors and correctly forecasting future conditions. As the quotation by A.J.P. Taylor at the start of the chapter stated: 'There are no maps to the future'. In Real-Life Nugget 21.1 Peter Aytan wonders why everything takes longer to finish and costs more than originally budgeted.

Soundbite 21.1

Fast-Changing Corporate World

'Fully one third of our more than $13 billion in worldwide revenues comes from products that simply did not exist five years ago.'

Ralph S. Larsen, *Johnson and Johnson* (October 22, 1992)

Source: *The Wiley Book of Business Quotations* (1998), p. 83

Real-Life Nugget 21.1

Forecasting the Future

Trouble ahead

WHY does everything take longer to finish and cost more than we think it will?

The Channel Tunnel was supposed to cost £2.6 billion. In fact, the final bill came to £15 billion. The Jubilee Line extension to the London Underground cost £3.5 billion, about four times the original estimate. There are many other examples: the London Eye, the Channel Tunnel rail link.

This is not an exclusively British disease. In 1957, engineers forecast that the Sydney Opera House would be finished in 1963 at a cost of A$7 million. A scaled-down version costing $102 million finally opened in 1973. In 1969, the mayor of Montreal announced that the 1976 Olympics would cost C$120 million and "can no more have a deficit than a man can have a baby". Yet the stadium roof alone – which was not finished until 13 years after the games – cost C$120 million.

Source: Pater Aytan, *New Scientist*, 29 April 2000, p. 43

The basic decision is simply whether or not a particular capital investment decision is worthwhile. In business, this decision is usually made by comparing the initial cash outflows associated with the capital investment with the later cash inflows. We can distinguish between the initial investment, net cash operating flows for succeeding years and other cash flows.

Pause for Thought 21.1

A Football Club's New Stadium

A football club is contemplating building a new stadium. What external factors should it take into account?

There are countless factors. Below are some that might be considered.

- How much will the stadium cost?
- How many extra spectators can the new stadium hold?
- How much can be charged per spectator?
- How many years will the stadium last?
- What is the net financial effect when compared with the present stadium?
- How confident is the club about future attendance at matches?

(i) Initial investment

This is our initial capital expenditure. It will usually involve capital outflows on infrastructure assets (for example, buildings or new plant and machinery) or on working capital. This initial expenditure is needed so that the business can expand.

(ii) Net cash flows

These represent the operating cash flows expected from the project once the infrastructure assets are in place. Normally, we talk about annual net cash flow. This is simply the cash inflows less the cash outflows calculated over a year. For convenience, *cash flows are usually assumed to occur at the end of the year.*

(iii) Other flows

These involve other non-operating cash flows. For example, the taxation benefits from the initial capital expenditure or the cash inflow from scrap.

Pause for Thought 21.2

Assumptions in Capital Investment Decisions

A big problem in any capital investment decision is the assumptions that underpin it. Can you think of any of these?

- Costs of initial outlay
- Tax effects
- Cost of capital
- Inflation
- Cash inflows and outflows over period of project. These, in turn, may depend on pricing policy, external demand, value of production etc.

Capital Investment Appraisal Techniques

The four main techniques used in capital investment decisions are payback period, accounting rate of return, net present value and internal rate of return. An overview of these four techniques is presented in Figure 21.2. These techniques are then discussed below.

Figure 21.2 Four Main Types of Investment Appraisal Techniques

Feature	*Payback*	*Accounting Rate of Return*	*Net Present Value*	*Internal Rate of Return*
Nature	Measures time period in which *cumulative cash inflows* overtake *cumulative cash outflows*	Assesses *profitability* of initial investment	*Discounts* future cash flows to present	Determines the *rate of return* at which a project *breaks even*
Ease of use	Very easy	Easy	May be difficult	Quite difficult
Takes time value of money into account	No	No	Yes	Yes
Main assumptions	Value and volume of cash flows	Reliability of annual profits	Value and volume of cash flows, cost of capital	Value and volume of cash flows, cost of capital
Focus	Cash flows	Profits	Cash flows	Cash flows

It is important to realise that the payback period and the accounting rate of return take into account only the *actual* cash inflows and outflows. However, net present value and accounting rate of return take into account the *time value of money*.

Pause for Thought 21.3

Time Value of Money

Why do you think it is important to take into account the time value of money?

There is an old saying that time is money! In the case of long-term capital investment decisions, it certainly is. Would you prefer £100 now or £100 in 10 years time? That one is easy! But what about £100 now or £150 in five years time? There is a need to standardise money in today's terms. To do this, we need to attribute a time value to money. In practice, we take this time value to be the rate at which a company could borrow money. This is called the cost of capital. If our cost of capital is 10%, we say that £100 today equals £110 in one year's time, £121 in 2 years time and so on. If we know the cost of capital of future cash flows we can, therefore, discount them back to today's cash flows.

As Real-Life Nugget 21.2 shows, amounts spent yesterday can mean huge sums today. Compounding is the opposite of discounting! If we discounted the $136,000,000,000 dollars back from 1876 to 1607 using a 10% discount rate we should arrive at one dollar.

Real-Life Nugget 21.2

Compound Growth

Compound Interest

From a speech in Congress more than 100 years ago:

It has been supposed here that had America been purchased in 1607 for $1, and payment secured by bond, payable, with interest annually compounded, in 1876 at ten percent, the amount would be – I have not verified the calculation – the very snug little sum of $136,000,000,000; five times as much as the country will sell for today. It is very much like supposing that if Adam and Eve have continued to multiply and replenish once in two years until the present time, and all their descendants had lived and had been equally prolific, then, saying nothing about twins and triplets, there would now be actually alive upon the earth, a quantity of human beings in solid measure more than thirteen and one-fourth times the bulk of the entire planet.

Source: Peter Hay (1988), *The Book of Business Anecdotes*, Harrap Ltd, London, p. 10

Each of the four capital investment appraisal techniques is examined using the information in Figure 21.3.

Figure 21.3 Illustrative Example of a Financial Service Company Wishing to Invest in a New On-Line Banking Service

The Everfriendly Building Society is contemplating launching a new on-line banking service, called the Falcon. There are three alternative approaches, each involving £20,000 initial outlay. In this case, cash inflows can be taken to be the same as profit. Cost of capital is 10%.

		Projects	
Year	**A**	**B**	**C**
Cash flows	£	£	£
0 (i.e. now)	(20,000)	(20,000)	(20,000)
1	4,000	8,000	8,000
2	4,000	6,000	8,000
3	8,000	6,000	6,000
4	6,000	3,000	6,000
5	6,000	2,000	3,000

Helpnote: The cash outflow is traditionally recorded as being in year 0 (i.e., today) in brackets. The cash inflows occur from year 1 onwards and are conventionally taken at the end of the year. To simplify matters, in this example, cash inflows have been taken to be the same as net profit. Finally, cost of capital can be taken as the amount that it cost the Everfriendly Building Society to borrow money.

Payback Period

The payback period is a relatively straightforward method of investment appraisal. It simply measures the cumulative cash inflows against the cumulative cash outflows until the project recovers its initial investment. The payback method is useful for screening projects for an early return on the investment. Ideally, it should be complemented by another method such as net present value.

Definition 21.1

Payback Period

The payback period simply measures the cumulative cash inflows against the cumulative cash outflows. The point at which they coincide is the payback point.

Specific advantages

1 Easy to use and understand.
2 Conservative.

Specific disadvantages

1 Fails to take into account cash flows after payback.
2 Does not take into account the time value of money.

Taking Everfriendly Building Society's Falcon project, when do we recover the £20,000? Figure 21.4 shows this is after 3.67 years for project A, 3 years for project B and 2.67 years for project C.

Figure 21.4 Payback Using Everfriendly Building Society

	Projects		
	A	**B**	**C**
Year	£	£	£
0 (i.e. now) Cash outflows	(20,000)	(20,000)	(20,000)
Cumulative cash inflows			
1	4,000	8,000	8,000
2	8,000	14,000	16,000
3	16,000	**20,000**	**22,000**
4	**22,000**	23,000	28,000
5	28,000	25,000	31,000
Payback year	3.67 years*	3 years	2.67 years*

*For these two projects, payback will be two-thirds of the way through a year. This is because (taking project A to illustrate) after three years our cumulative inflows are £16,000 and after four years they are £22,000. Assuming, and it is a big assumption, a steady cash flow, we reach payback point after two-thirds of a year (i.e., £4,000 needed for payback, divided by £6,000 cash inflows).

Payback is a relatively straightforward investment technique. It is simple to understand and apply and promotes a policy of caution in the investment decision. The business always chooses the investment which pays off the initial investment the most quickly. However, although useful, payback has certain crucial limitations.

Pause for Thought 21.4

Limitations of Payback

Can you think of any limitations of payback?

Two of the most important limitations of payback are that it ignores both cash flows after the payback and the time value of money. Thus, a project may have a slow payback period, but have substantial cash flows once it is established. These will not be taken into account. In addition, cash flows in later years are treated as being the same value as cash flows in early years. This ignores the time value of money and may distort the capital investment decision.

Accounting Rate of Return

This method, unlike the other three methods, focuses on the profitability of the project, rather than its cash flow. Thus it is distinctly different in orientation from the other methods. The basic definition of the accounting rate of return is:

$$\text{Accounting rate of return} = \frac{\text{Average annual profit}}{\text{Capital investment}}$$

However, once we look more closely we run into potential problems. What exactly do we mean by 'profit' and 'capital investment'? For profit, do we take into account interest, taxation and depreciation? For capital investment, do we take the initial capital investment or the average capital employed over its life? Different firms will use different versions of this ratio. In this book, profit before interest and taxation, and initial capital investment are preferred (see Definition 21.2 on the next page). This is because it is similar to conventional accounting ratios and seems logical! The accounting rate of return is easy to understand and use. Its main disadvantages are that profit and capital investment have many possible definitions and, like payback, the accounting rate of return does not take into account the time value of money.

Definition 21.2

Accounting Rate of Return

Accounting rate of return is a capital investment appraisal method which assesses the viability of a project using annual profit and initial capital invested. We can define it as:

$$\frac{\text{Average annual profit before interest and taxation}}{\text{Initial capital investment}}$$

Specific advantages

1 Takes the whole life of a project.
2 Similar to normal accounting ratios.

Specific disadvantages

1 Many definitions of profit and capital investment possible.
2 Does not consider the time value of money.

The accounting rate of return is applied to the Everfriendly Building Society in Figure 21.5.

Figure 21.5 Accounting Rate of Return Using Everfriendly Building Society

	Project Cash Flows		
	A	**B**	**C**
Year	£	£	£
0 (i.e. now) Cash outflows	(20,000)	(20,000)	(20,000)
Net cash inflows (note)			
1	4,000	8,000	8,000
2	4,000	6,000	8,000
3	8,000	6,000	6,000
4	6,000	3,000	6,000
5	6,000	2,000	3,000
Total	28,000	25,000	31,000
Average profit	£28,000 / 5 years	£25,000 / 5 years	£31,000 / 5 years
	= 5,600	= 5,000	= 6,200
Our accounting rate of return is therefore:	£5,600 / £20,000	£5,000 / £20,000	£6,200 / £20,000
	= 28%	= 25%	= 31%

We would, therefore, choose project C because its accounting rate of return is the highest.

Note: In this case, cash inflow equals net profit. This will not always be the case.

Returns on investment are treated very seriously by businesses. In Real-Life Nugget 21.3, Vodafone has a projected rate of return of at least 15% on its capital investment. However, it is not absolutely clear from the press extract whether it has used the accounting rate of return.

Real-Life Nugget 21.3

Rates of Return

Vodafone AirTouch yesterday gave the first indication of how much money it expects to make from third generation mobile networks during the release of better-than-expected full-year results.

Key Hydon, finance director, said that the projected rate of return on the large capital investment required for licences and infrastructure was at least 15 per cent.

Although the £6bn cost of Vodafone's new UK licence will be spread over its 20-year lifespan, most of the estimated £4bn network spending and handset subsidy will be required in the first four or five years.

This implies that the company is projecting potential operating profits of about £75m above its existing performance.

Source: Dan Roberts, 'Telecommunications group releases internal targets to counter criticism it overpaid for licence', *Financial Times*, 31 May 2000

Net Present Value

The net present value and internal rate of return can be distinguished from the payback period and accounting rate of return because they take into account the time value of money. Essentially, time is money. If you invest £100 in a bank or building society and the interest rate is 10%, the £100 is worth £110 in one year's time (100×1.10), and £121 in two year's time (100×1.10^2, or 110×1.10).

We can use the same principle to work backwards. If we have £110 in the bank in a year's time with a 10% interest rate, it will be worth £100 today (£110×0.9091, i.e. $100 \div 110$). Similarly, £121 in two years' time would be worth £100 today (£121×0.8264, i.e. $100 \div 121$).

Fortunately, we do not have to calculate discount rates all the time! We use discount tables (see Appendix 21.1 at the end of this chapter). So to obtain a 10% interest rate in two year's time, we look up the number of years (two) and the discount rate (10%). We then find a discount factor of 0.8264!

As Definition 21.3 shows, net present value uses the discounting principle to work out the value in today's money of future expected cash flows. It uses the cost of capital as the discount rate.

Pause for Thought 21.5

Discounting

You are approached by your best friend, who asks you to lend her £1,000. She promises to give you back £1,200 in two years' time. Your money is in a building society and earns 6%. Putting friendship aside, do you lend her the money?

Pause for Thought 21.5 (continued)

To work this out, we need to compare like with like. We could either work forwards (i.e., multiplying £1,000 by the 6% earned over two years (1.06^2) = £1,123.60). £1,000 today is worth £1,123.60 in two years' time. Or more conventionally, we work back from the future using the time value of money, in this case 6%.

Thus, £1,200 in one year's time	=	£1,200 × (100/106, i.e., 0.9434)
	=	£1,132.08 today
While £1,200 in two years' time	=	£1,200 × (100/112.36, i.e., 106 × 1.06)
	=	£1,200 × 0.8900
	=	£1,068 today

Therefore, as £1,200 in two year's time is the equivalent of £1,068 today you should accept your friend's offer as this is £68 more than you currently have.

Definition 21.3

Net Present Value

Net present value is a capital investment appraisal technique which discounts future expected cash flows to today's monetary values using an appropriate cost of capital.

Specific advantages

1 Looks at all the cash flows.
2 Takes into account the value of money.

Specific disadvantages

1 Estimation of cost of capital may be difficult.
2 Assumes all cash flows occur at end of year.
3 Can be complex.

Conventionally, the cost of capital is taken as the rate at which the business can borrow money. However, the estimation of the cost of capital can be quite difficult. In Chapter 22 we look in more detail at how companies can derive their cost of capital. Cost of capital is a key element of capital investment appraisal. Essentially, a business is seeking to earn a higher return from new projects than its cost of capital. If this is achieved, the projects will be viable. If not, they are unviable.

Behavioural factors can also play their part in making a decision. This is shown in Real-Life Nugget 21.4.

Real-Life Nugget 21.4

Behavioural Implications of Cost of Capital

If a company decides it will only go ahead with projects that show a discounted return on investment of better than 15 per cent, anyone putting forward a project will ensure that the accompanying figures indicate a better than 15 per cent return. If the criterion is raised to 20 per cent, the figures will be improved accordingly. And there is no way in which the person analysing the project can contest those figures – unless he can find a flaw in the performance of the projection rituals.

Source: Graham Cleverly (1971), *Managers and Magic*, Longman Group Ltd, London, p. 86

Figure 21.6 applies net present value to the Everfriendly Building Society.

Figure 21.6 Net Present Value as Applied to the Everfriendly Building Society

	Project Cash Flows			*Discount Rate*	*Discounted Cash Flows*		
	A	**B**	**C**	**10%**	**A**	**B**	**C**
Year	£	£	£		£	£	£
0 (i.e. now)	(20,000)	(20,000)	(20,000)	1	(20,000)	(20,000)	(20,000)
1	4,000	8,000	8,000	0.9091	3,636	7,273	7,273
2	4,000	6,000	8,000	0.8264	3,306	4,958	6,611
3	8,000	6,000	6,000	0.7513	6,010	4,508	4,508
4	6,000	3,000	6,000	0.6830	4,098	2,049	4,098
5	6,000	2,000	3,000	0.6209	3,725	1,242	1,863
Total discounted cash flows					20,775	20,030	24,353
Net Present Value (NPV)					775	30	4,353

All three projects have a positive net present value and, therefore, are worth carrying out. However, project C has the highest NPV and thus should be chosen if funds are limited.

When using the net present value, it is useful to follow the steps laid out in Helpnote 21.1.

Helpnote 21.1

Calculating Net Present Value

1. Calculate initial cash flows.
2. Choose a discount rate (usually given), normally based on the company's cost of capital.
3. Discount original cash flows using discount rate.
4. Match discounted cash flows against initial investment to arrive at net present value.
5. Positive net present values are good investments. Negative ones are poor investments.
6. Choose the highest net present value. This is the project that will most increase the shareholders' wealth.

Pause for Thought 21.6

Discount Tables

Using the discount tables given in Appendix 21.1, what are the following discount factors:

(i) 8% at 5 years, (ii) 12% at 8 years, (iii) 15% at 10 years, (iv) 20% at 6 years, and (v) 9% at 9 years.

From Appendix 21.1, we have

(i) 0.6806; (ii) 0.4039; (iii) 0.2472; (iv) 0.3349; (v) 0.4604.

In practice, calculating net present values becomes very complicated. For example, you have to take into account taxation, inflation, etc. However, net present value is a very versatile technique, allowing you to determine which projects are worth investing in and to choose between competing projects.

Internal Rate of Return (IRR)

The internal rate of return is a more sophisticated discounting technique than net present value. As Definition 21.4 shows, it can be defined as the rate of discount required to give a net present value of zero. Another way of looking at this is the maximum rate of interest that a company can afford to pay without suffering a loss on the project. Projects are accepted if the company's cost of capital is less than the IRR. Similarly, projects are rejected if the company's cost of capital is higher than the IRR. The advantages of the internal rate of return are that it takes into account the time value of money and calculates a break-even rate of return. However, it is complex and difficult to understand.

Definition 21.4

The Internal Rate of Return

The internal rate of return represents the discount rate required to give a net present value of zero. It pays a company to invest in a project if it can borrow money for less than the IRR.

Specific advantages

1 Uses time value of money.
2 Determines the break-even rate of return.
3 Looks at all the cash flows.

Specific disadvantages

1 Difficult to understand.
2 No need to select a specific discount rate.
3 Complex, often needing a computer.
4 In certain situations, gives misleading results (for example, where there are unconventional cash flows).

The relationship of the internal rate of return to net present value is shown in Figure 21.7 and the internal rate of return is then applied to the Everfriendly Building Society in Figure 21.8 on the next page.

Figure 21.7 Relationship Between the Net Present Value and Internal Rate of Return

A project has the following net present values (NPVs).

£2,000 NPV at 10% discount rate.
(£1,000) NPV at 15% discount rate.
What is the internal rate of return?

We can find the internal rate of return either (i) *diagrammatically* or (ii) *mathematically* by a process called interpolation.

(i) Diagrammatically

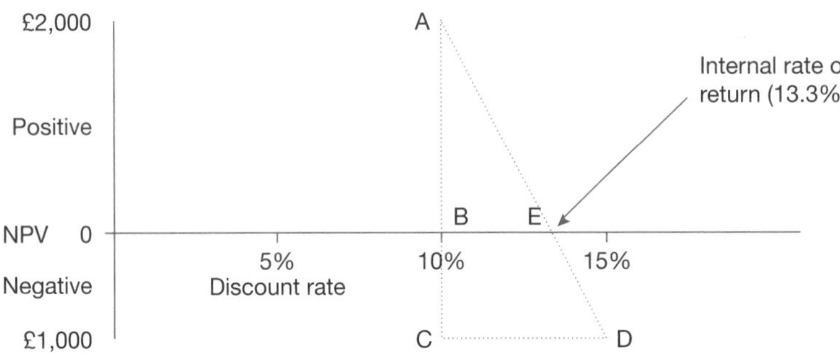

A = Discount rate 10%, £2,000 NPV
D = Discount rate 15%, (£1,000) NPV
E = Internal rate of return (13.3%)

(ii) Mathematically

We need to find point E. To do this, we can find the distance from B to E using the principles of similar triangles from mathematics: $\frac{BE}{CD} = \frac{AB}{AC}$

$$\therefore \frac{BE}{5} = \frac{2,000}{3,000}$$

$$\therefore \frac{BE}{5} = \frac{10,000}{3,000} = 3.3\%$$

$\therefore$ IRR = 10% + 3.33% = 13.3%

This can be re-written as follows:

IRR = Lowest discount rate + difference in discount rates $\times$ $\frac{\textbf{lowest discount rate NPV}}{\textbf{difference in NPVs}}$

$$13.3\% = 10\% + 5\% \times \frac{2,000}{3,000}$$

It is this formula that we will use from now on.

Figure 21.8 Internal Rate of Return as Applied to the Everfriendly Building Society

From Figure 21.6, we know that at 10% our net present values are all positive, A £775, B £30 and C £4,353. To solve the problem mathematically, we need to ascertain a negative present value for each project. So, first of all, we try 15%.

Year	Project Cash Flows A £	Project Cash Flows B £	Project Cash Flows C £	Discount Rate 15%	Discounted Cash Flows (DCF) A £	Discounted Cash Flows (DCF) B £	Discounted Cash Flows (DCF) C £	Discount Rate 20%	DCF C £
0	(20,000)	(20,000)	(20,000)	1	(20,000)	(20,000)	(20,000)	1	(20,000)
1	4,000	8,000	8,000	0.8696	3,478	6,957	6,957	0.8333	6,666
2	4,000	6,000	8,000	0.7561	3,024	4,537	6,049	0,6944	5,555
3	8,000	6,000	6,000	0.6575	5,260	3,945	3,945	0.5787	3,472
4	6,000	3,000	6,000	0.5718	3,431	1,715	3,431	0.4823	2,894
5	6,000	2,000	3,000	0.4972	2,983	994	1,492	0.4019	1,206
Total discounted cash inflows					18,176	18,148	21,874		19,793
Net Present Value (NPV)					(1,824)	(1,852)	1,874		(207)

15% produced A (£1,824), B (£1,852) and C £1,874 in net present values. We still need a negative value for C. So above we also ran a higher discount rate 20% for C. This gives us a negative NPV of (£207).

To solve this we can now use our formula.

$$\text{IRR} = \text{Lowest discount rate} + \text{difference in discount rates} \times \frac{\text{lowest discount rate NPV}}{\text{difference in NPVs}}$$

$$\text{A. } 10\% + (15 - 10\%) \times \frac{£775}{£775 + £1{,}824} = 10\% + (5\% \times 0.2982) = 11.5\%$$

$$\text{B. } 10\% + (15 - 10\%) \times \frac{£30}{£30 + £1{,}852} = 10\% + (5\% \times 0.01594) = 10.1\%$$

$$\text{C. } 10\% + (20 - 10\%) \times \frac{£4{,}353}{£4{,}353 + £207} = 10\% + (10\% \times 0.9546) = 19.6\%$$

Our ranking for the IRR method is thus project C, then A and, finally, B. As all projects have IRRs which exceed our cost of capital of 10% all three projects are potentially viable. However, for project B the IRR is only marginally greater than the cost of capital.

When calculating the IRR, there are four main steps. These are shown in Helpnote 21.2.

The IRR can be a very useful technique. However, the complexity is often off-putting. In practice, students will be pleased to learn that the calculations to arrive at the IRR are generally done using a computer program. Normally, the results from NPV and IRR will be consistent. However, in some cases, such as projects with unconventional cash flows (e.g., alternating cash inflows and outflows), the NPV and IRR may give different results. In these cases, NPV is probably the most reliable.

If there was a clash between the results from the different methods then it would be difficult to choose. Different businesses will prefer different methods depending on their priorities. I would probably choose net present value as the superior method. This is because it takes into account the time value of money, is more reliable, and is easier to understand and use than the internal rate of return.

Helpnote 21.2

Calculating the Internal Rate of Return (IRR)

1. Calculate a positive NPV for all projects.
2. Calculate a negative NPV for all projects. Use trial and error. The higher the discount rate, the lower the NPV.
3. Calculate out the IRR using the formula: IRR = Lowest discount rate + difference in discount rates

$$\times \frac{\text{lowest discount rate NPV}}{\text{difference in NPVs}}$$

4. Choose the project with the highest IRR. When evaluating a single project, the IRR will be chosen when it is higher than the company's cost of capital.

Now that we have calculated the results for the Everfriendly Building Society using all four methods, we can compare them (see Figure 21.9). Project C is clearly the superior method.

Figure 21.9 Comparison of the Projects for Everfriendly Building Society Using the Four Capital Appraisal Methods

	Projects		
	A	**B**	**C**
Payback Period	3.67 years	3 years	2.67 years
Ranking	3	2	1
Accounting Rate of Return	28%	25%	31%
Ranking	2	3	1
Net Present Value	£775	£30	£4,353
Ranking	2	3	1
Internal Rate of Return	11.5%	10.1%	19.6%
Ranking	2	3	1

Project C is the superior project using all methods, so we could choose project C.

Before we leave the capital investment appraisal techniques, it is important to reiterate that the results achieved will only be as good as the assumptions that underpin them. This is powerfully expressed in Real-Life Nugget 21.5.

Real-Life Nugget 21.5

Investment Appraisal Techniques

> But where the very existence of angels is in doubt, debating how many of them can dance on a pin seems a sterile exercise. Where the validity of one's original information is suspect, performing ever more sophisticated calculations seems pointless. Nonetheless, there can hardly be a company in which new projects are not required to be subjected to a ritual calculation of the 'internal rate of return', 'net present value', 'payback period' or some other criterion of profitability. And yet most of the time the performers themselves are not convinced of the validity of the material information they are manipulating.

Source: Graham Cleverly (1971), *Managers and Magic*, Longman Group Ltd, London, p. 86

Other Factors

There are many other factors that affect capital investment appraisal. Below we discuss three of the most important: sensitivity analysis, inflation and taxation.

1. Sensitivity analysis

The assumptions underpinning the capital investment decision mean that it is often sensible to undertake some form of **sensitivity analysis**. Sensitivity analysis involves modelling the future to see if alternative scenarios will change the investment decision. For example, a project's estimated cash outflow might be £10 million, its estimated inflows might be £6 million and its cost of capital might be 10%. All three of these parameters would be altered to assess any impact upon the overall results.

2. Inflation

The effects of inflation should also be included in any future capital investment model. Inflation means that a pound in a year's time will not buy as much as a pound today, as the price of goods will have risen. Therefore, we need to adjust for inflation. This can be done in two ways.

1. *By adjusting future cash flow*. Under this method, we increase the future cash flows by the expected rate of inflation. If inflation was 3% over a year we would, therefore, increase an expected cash flow of £100 million by 3% to £103 million. We would then discount these adjusted flows as normal.
2. *By adjusting the discount rate*. If we were using a discount rate of 10%, we would deduct the inflation rate. If inflation was 3%, we would then discount the future cash flows at 7%.

3. Taxation

As we saw in Chapter 8 cash flow and profit are distinctly different. Discounted cash flows are based on cash flows not on accounting profit. In order to arrive at forecast cash flows from forecast profit we must, therefore, adjust the profit for non cash-flow items. In particular, we need to add back depreciation (disallowed by taxation authorities) and deduct any capital allowances. Capital allowances are allowed by the taxation authorities as a replacement for depreciation. They can be set off against taxable profits. Capital allowances, thus, reduce cash outflows. Normally, companies will pay their taxation nine months after their year end.

Conclusion

Capital investment appraisal methods are used to assess the viability of long-term investment decisions. Capital investment projects might be a football club building a new stadium or a company building a new factory. Capital investment decisions are long-term and often very expensive. It is, therefore, very important to test their viability. Four capital investment appraisal methods are commonly used: payback period; accounting rate of return; net present value; and internal rate of return. Unlike the last two, the first two do not take into account the time value of money. The payback period simply assesses how long it takes a company's cumulative cash inflows to outstrip its initial investment. The accounting rate of return assesses the profitability of a project using average annual profit over initial capital investment. Net present value discounts back estimated future cash flows to today's values using cost of capital as a discount rate. Finally, internal rate of return establishes the discount rate at which a project breaks even. All four appraisal methods are based on many assumptions about future cash flows. These assumptions are often tested using sensitivity analysis.

Discussion *Questions*

Questions with numbers in blue have answers at the back of the book.

Q 1 (a) Why do you think capital investment is necessary for companies?
(b) What sort of possible capital investment might there be in:
(i) the shipping industry?
(ii) the hotel and catering industry?
(iii) manufacturing industry?

Q **2** Discuss the general assumptions that underpin capital investment appraisal.

Q **3** Briefly outline the four main capital investment appraisal techniques and then discuss the specific advantages and disadvantages of each technique.

Q **4** Why is time money?

Q 5 State whether the following statements are true or false. If false, explain why.

(a) The four main capital investment appraisal techniques are payback period, accounting rate of return, net present value and internal rate of return.

(b) The payback period and net present value techniques generally use discounted cash flows.

(c) The accounting rate of return is the only investment appraisal technique that focuses on profits not cash flows.

(d) The discount rate normally used to discount cash flows is the interest rate charged by the Bank of England.

(e) Sensitivity analysis involves modelling future alternative scenarios and assessing their impact upon the results of capital investment appraisal techniques.

Numerical *Questions*

Questions with numbers in blue have answers at the back of the book.

These questions gradually increase in difficulty. Students may find question 6, in particular, testing.

Q 1 What are the appropriate discount factors for the following:

(i) 5 years at 10% cost of capital
(ii) 6 years at 9% cost of capital
(iii) 8 years at 13% cost of capital
(iv) 4 years at 12% cost of capital
(v) 3 years at 14% cost of capital
(vi) 10 years at 20% cost of capital

Q 2 A company, Fairground, has a choice between investing in one of three projects: the Rocket, the Carousel or the Dipper. The cost of capital is 8%. There are the following cash flows.

	Rocket	*Carousel*	*Dipper*
	£	£	£
Initial outlay	(18,000)	(18,000)	(18,000)
Cash inflow			
Year 1	8,000	6,000	10,000
Year 2	8,000	4,000	6,000
Year 3	8,000	14,000	5,000

Required:

An evaluation of Fairground using:

(i) the payback period
(ii) the accounting rate of return (assume cash flows are equivalent to profits)
(iii) net present value
(iv) the internal rate of return.

Q3 Wetday is evaluating three projects: the Storm, the Cloud, and the Downpour. The company's cost of capital is 12%. These projects have the following cash flows.

	Storm	*Cloud*	*Downpour*
Year	£	£	£
0 (Initial outlay)	(18,000)	(12,000)	(13,000)
Cash inflow			
1	4,000	5,000	4,000
2	5,000	2,000	4,000
3	6,000	3,000	4,000
4	7,000	2,500	4,000
5	8,000	3,000	4,000

Required:
Calculate

(i) the payback period
(ii) the accounting rate of return (assume cash flows equal profits)
(iii) the net present value
(iv) the internal rate of return.

Q4 A company, Choosewell, has £30,000 to spend on capital investment projects. It is currently evaluating three projects. The initial capital outlay is on a piece of machinery that has a four-year life. Its cost of capital is 9%.

	Ready		*Steady*		*Go*	
		£		£		£
Initial capital outlay		(30,000)		(15,000)		(15,000)
	Inflows	Outflows	Inflows	Outflows	Inflows	Outflows
Year	£	£	£	£	£	£
1	36,000	24,000	25,000	16,000	16,000	8,000
2	36,000	14,000	18,000	11,000	13,000	6,500
3	32,000	26,000	17,000	12,000	12,000	6,000
4	4,000	5,000	3,000	4,000	6,000	6,000

Required:
Calculate

(i) the payback period
(ii) the accounting rate of return
(iii) the net present value
(iv) the internal rate of return.

Q5 A football club, Manpool, is considering investing in a new stadium. There are the following expected capital outlays and cash inflows for two prospective stadiums.

Year	*Bowl* £000	*Superbowl* £000
Outlays		
0	(1,000)	(1,000)
1	(1,000)	(1,000)
Net inflows		
1	50	300
2	100	300
3	150	300
4	200	300
5	250	300
6	300	300
7	350	300
8	400	300
9	450	300
10	500	300

Assume the football club can borrow money at respectively
(a) 5% (b) 8% (c) 10%

Required:

(i) Calculate which stadium should be built and at which rate using net present value?
(ii) What is the internal rate of return for the two stadiums?

***Q*6** A company, Myopia, has the following details for its new potential product, the Telescope.

Year	*Capital Outlay*	*Capital Inflow*	*Sales*	*Interest*	*Expenses (Excludes Depreciation)*	*Taxation*
	£	£	£	£	£	£
0	(700,000)	–	–	–	–	–
1			340,000	20,000	(40,000)	–
2			270,000	20,000	(130,000)	(4,500)
3			320,000	20,000	(160,000)	(18,375)
4			345,000	20,000	(185,000)	(24,281)
5			430,500	20,000	(205,000)	(48,361)
6			330,300	20,000	(160,300)	(35,033)
7			200,600	20,000	(100,100)	(16,825)
8			145,300	20,000	(46,500)	(18,034)
9			85,200	20,000	(28,100)	(6,925)
10		5,000	38,600	20,000	(8,300)	–

Myopia's cost of capital is 10%. The capital outlay is for the Jodrell machine which will last 10 years. The capital inflow of £5,000 is the scrap value after 10 years.

Required:
Calculate

(i) the payback period
(ii) the accounting rate of return
(iii) the net present value
(iv) the internal rate of return.

Helpnote. In this question, which has a different format from those encountered so far, it is first necessary to calculate *profit before interest and tax* (i.e., sales less expenses and depreciation). After this we need to adjust for interest, taxation, depreciation (based on the initial capital outlay) and other cash flows to arrive at a final cash flow figure.

Appendix 21.1 Present Value of £1 at Compound Interest Rate (1 + r)

Years	Interest rates (r)										
(n)	1%	2%	3%	4%	5%	6%	7%	8%	9%	10%	
1	0.9901	0.9804	0.9709	0.9615	0.9524	0.9434	0.9346	0.9259	0.9174	0.9091	1
2	0.9803	0.9612	0.9426	0.9246	0.9070	0.8900	0.8734	0.8573	0.8417	0.8264	2
3	0.9706	0.9423	0.9151	0.8990	0.8638	0.8396	0.8163	0.7938	0.7722	0.7513	3
4	0.9610	0.9238	0.8885	0.8548	0.8227	0.7921	0.7629	0.7350	0.7084	0.6830	4
5	0.9515	0.9057	0.8626	0.8219	0.7835	0.7473	0.7130	0.6806	0.6499	0.6209	5
6	0.9420	0.8880	0.8375	0.7903	0.7462	0.7050	0.6663	0.6302	0.5963	0.5645	6
7	0.9327	0.8706	0.8131	0.7599	0.7107	0.6651	0.6227	0.5835	0.5470	0.5132	7
8	0.9235	0.8535	0.7894	0.7307	0.6768	0.6274	0.5820	0.5403	0.5019	0.4665	8
9	0.9143	0.8368	0.7664	0.7026	0.6446	0.5919	0.5439	0.5002	0.4604	0.4241	9
10	0.9053	0.8203	0.7441	0.6756	0.6139	0.5584	0.5083	0.4632	0.4224	0.3855	10

Years	Interest rates (r)										
(n)	11%	12%	13%	14%	15%	16%	17%	18%	19%	20%	
1	0.9009	0.8929	0.8850	0.8772	0.8696	0.8621	0.8547	0.8475	0.8403	0.8333	1
2	0.8116	0.7972	0.7831	0.7695	0.7561	0.7432	0.7305	0.7182	0.7062	0.6944	2
3	0.7312	0.7118	0.6931	0.6750	0.6575	0.6407	0.6244	0.6086	0.5934	0.5787	3
4	0.6587	0.6355	0.6133	0.5921	0.5718	0.5523	0.5337	0.5158	0.4987	0.4823	4
5	0.5935	0.5674	0.5428	0.5194	0.4972	0.4761	0.4561	0.4371	0.4190	0.4019	5
6	0.5346	0.5066	0.4803	0.4556	0.4323	0.4104	0.3898	0.3704	0.3521	0.3349	6
7	0.4817	0.4523	0.4251	0.3996	0.3759	0.3538	0.3332	0.3139	0.2959	0.2791	7
8	0.4339	0.4039	0.3762	0.3506	0.3269	0.3050	0.2848	0.2660	0.2487	0.2326	8
9	0.3909	0.3606	0.3329	0.3075	0.2843	0.2630	0.2434	0.2255	0.2090	0.1938	9
10	0.3522	0.3220	0.2946	0.2697	0.2472	0.2267	0.2080	0.1911	0.1756	0.1615	10

Helpnote:
This table shows what the value of £1 today will be in the future, assuming different interest rates. Therefore, £1 today will be worth 90.91 pence in one year's time if the interest rate is 10%. If you are given cash flows at a future time you need to use the interest rates in the table to adjust them to today's monetary value. Thus, if we have £5,000 in five years' time and our interest rate is 10%, then, using the table, this will be worth £5,000 × 0.6209 or £3,104 in today's money.

Chapter 22

The Management of Working Capital and Sources of Finance

'It doesn't take very long to screw up a company. Two, three months should do it. All it takes is some excess inventory [stock], some negligence in collecting, and some ignorance about where you are.'

Mary Baechler in *Inc.*, October (1994) quoted in *The Wiley Book of Business Quotations* (1998), p. 86

Learning Outcomes

After completing this chapter you should be able to:

✔ Explain the nature and importance of sources of finance.
✔ Discuss the nature of short-term financing.
✔ Analyse the ways in which the long-term finance of a company may be provided.
✔ Understand the concept of the cost of capital.

In a Nutshell

- Sources of finance are vital to the survival and growth of a business.
- There are internally and externally generated sources of finance.
- Sources of finance can be short-term or long-term.
- Short-term and long-term sources of finance are normally matched with current assets and long-term, infrastructure assets, respectively.
- Short-term internal sources of finance concern the more efficient use of cash, debtors and stock.
- Techniques for the internal management of working capital involve the debtors collection model, the economic order quantity and just-in-time stock management.
- Short-term external sources of finance include a bank overdraft, a bank loan, debt factoring, invoice discounting, and the sale and buy back of stock.
- Retained profits are where a company finances itself from internal funds.
- Three major sources of external long-term financing are leasing, share capital and long-term loans.
- Leasing involves a company using, but not owning, an asset.
- Share capital is provided by shareholders who own the company and receive dividends.
- Loan capital providers do not own the company and receive interest.
- Cost of capital is the effective rate at which a company can raise finance.

Introduction

Sources of finance are vital to a business. they allow it to survive and grow. Sources of finance may be raised internally and externally. The efficient use of working capital is an important internal, short-term source of funds, while retained profits can be an important internal, long-term source. External funds may also be short-term or long-term. Bank loans and debt factoring are examples of external short-term finance. External long-term sources allow businesses to carry out their long-term strategic aims. Share capital and loan capital are examples of long-term finance. Every business aims to optimise its use of funds. This involves minimising short-term borrowings and financing long-term infrastructure investments as cheaply as possible.

Nature of Sources of Finance

Sources of finance may be generated internally or raised externally. These sources of finance may be short-term or long-term. If short-term, they will typically be associated with operational activities involving the financing of working capital (such as stock, debtors or cash). If long-term, they will typically be associated with funding longer-term infrastructure assets, such as the purchase of land and buildings.

Normally, it is considered a mistake to borrow short and use long (e.g., use bank overdrafts for long-term purposes). For instance, if a bank overdraft was used to finance a new building, then the company would be in trouble if the bank suddenly withdrew the overdraft facility.

As Figure 22.1 shows, *internal* sources of finance may be generated over the *short term* through the efficient management of working capital, or over the *long term* by reinvesting retained profits.

Figure 22.1 Overview of Sources of Finance

	Short-term (less than 2 years)	*Long-term (over 2 years)*
Internal (within firm)	(a) Efficient cash management (b) Efficient debtors management (c) Efficient stock management	(a) Retained profits
External (outside firm)	(a) Bank overdraft (b) Bank loans (c) Invoice discounting (d) Debt factoring (e) Sale and leaseback	(a) Leasing (b) Share capital (c) Long-term borrowing

External sources of capital may be raised in the *short term* by cash overdrafts, bank loans, invoice discounting, debt factoring or sale and leaseback. Over the *longer term* the three most important external sources of finance are leasing, share capital and long-term borrowing. Although 'short-term' is defined as less than two years, this is, in fact, very arbitrary. For example, some leases may be for less than two years while bank overdrafts and bank loans will sometimes be for more than two years. However, this division provides a useful and convenient simplification.

Short-Term Financing

Short-term financing is often called the management of working capital (current assets less current liabilities). One of the main aims is to reduce the amount of short-term finance that companies need to borrow for their day-to-day operations. The more money that is tied up in current assets, the more capital is needed to finance those current assets. Essentially, as well as using its working capital as efficiently as possible, a company may use short-term borrowings to finance its stock, debtors or cash needs.

Figure 22.2 provides an overview of a company's short-term sources of finance. This includes *internal* management techniques for the efficient management of working capital (for example, debtor collection model and economic order quantity) and the main *external* sources of funds.

Figure 22.2 Overview of Short-Term Sources of Finance

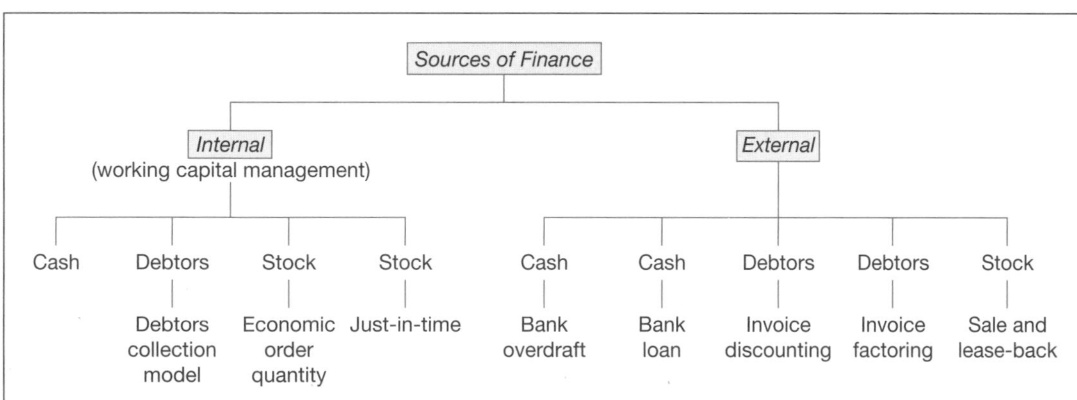

Internal Financing

Companies will try to minimise their levels of working capital so as to avoid short-term borrowings. In this section, we look at the main elements of efficient working capital management and three important techniques used to control working capital levels. In Chapter 5, we looked at the working capital cycle.

(a) Cash

As we saw in Chapter 8, cash is the lifeblood of a business. Companies need cash to survive. Businesses will try to keep enough cash to manage their day-to-day business operations (e.g., purchase stock and pay creditors), but not to maintain excessive amounts of cash. Businesses prepare cash budgets (see Chapter 17), which enable them to forecast the levels of cash that they will need to finance their operations. They may also use the liquidity and quick ratio to assess their level of cash (see Chapter 9). If despite careful cash management the business's short-term cash requirements are insufficient then the business will have to borrow.

(b) Debtors

Debtors management is a key activity within a firm. It is often called credit control. Debtors result from the sale of goods on credit. There is, therefore, the need to carefully monitor the receipts from debtors to see that they are in full and on time.

There will often be a separate department of a business concerned with credit control. The credit control department will, for example, establish credit limits for new customers, monitor the age of debts and chase up bad debts. In particular, they may draw up a debtors age schedule. This will profile the age of the debts and allow old debts to be quickly identified.

Debtors Collection Model. A useful technique designed to maintain the most efficient level of debtors for a company is the debtors collection model. A debtors collection model balances the extra revenue generated by increased sales with the increased costs associated with extra sales (i.e., credit control costs, bad debts and the delay in receiving money). The model assumes that the more credit granted the greater the sales. However, these extra sales are offset by increased bad debts as the business sells to less trustworthy customers. Whether the delay in receipts means that the business receives less interest or pays out more interest depends on whether or not the bank account is overdrawn. Usually, the cost of capital (i.e., effectively, the company's borrowing rate: see later in this chapter) is used to calculate the financial costs of the delayed receipts. Figure 22.3 on the next page illustrates the debtors collection model.

(c) Stock

For many businesses, especially for manufacturing businesses, stock is often an extremely important asset. Stock is needed to create a buffer against excess demand, to protect against rising prices or against a potential shortage of raw materials and to balance sales and production.

Stock control is concerned not primarily with valuing stock (see Chapter 16), but with protecting the stock physically and ensuring that the optimal level is held. Stock may be stolen or may deteriorate. For many businesses such as supermarkets the battle against theft and deterioration is never-ending. A week-old lettuce in a supermarket is not a pleasant asset!! For supermarkets, stock can also be a competitive advantage (see Real-Life Nugget 22.1).

Real-Life Nugget 22.1

Stock

A Way to Look at It

When F. W. Woolworth opened his first store, a merchant on the same street tried to fight the new competition. He hung out a big sign: "Doing business in this same spot for over fifty years." The next day Woolworth also put out a sign. It read: "Established a week ago; no old stock."

Source: Peter Hay (1988), *The Book of Business Anecdotes*, Harrap Ltd, London, p. 275

Figure 22.3 Debtors Collection Model

Bruce Bowhill is the finance director of a business with current sales of £240,000. If credit rises, so do bad debts. The contribution is 20%, the cost of capital is 15%, the credit control costs are £10,000 per annum at all levels of sales.

Credit	*Annual sales*	*Bad Debts*
Nil	£240,000	–
1 month	£320,000	1%
2 months	£500,000	5%
3 months	£650,000	10%

What is the most favourable level of sales?

We need to balance the increased contribution earned by increased sales with the increased costs of easier credit (i.e., credit control, bad debts and cost of capital).

	Nil credit	*1 month*	*2 months*	*3 months*
	£	£	£	£
Sales	240,000	320,000	500,000	650,000
Contribution 20%	48,000	64,000	100,000	130,000
Cost of credit control		(10,000)	(10,000)	(10,000)
Bad debts 1% sales		(3,200)		
5% sales			(25,000)	
10% sales				(65,000)
Cost of capital relating to delay in payment: 15% of average debtors*		(4,000)	(12,500)	(24,375)
Revised Contribution	48,000	46,800	52,500 **	30,625

*Average debtors per month
= £26,667 (£320,000 ÷ 12) (1 month)
= £83,333 (£500,000 ÷ 6) (2 months)
= £162,500 (£650,000 ÷ 4) (3 months)

**The optimal level is thus two months as it has the highest revised contribution.

Two common techniques associated with efficient stock control are the economic order quantity model and just-in-time stock management.

(i) Economic Order Quantity (EOQ). The EOQ model seeks to determine the optimal order quantity needed to minimise the costs of ordering and holding stock. These costs are the costs of placing the order and the carrying costs. Carrying costs are those costs incurred in keeping an item in stock, such as insurance, obsolescence, interest on borrowed money or clerical/security costs. The costs of ordering stock are mainly the clerical costs. The EOQ can be determined either graphically or algebraically. Figure 22.4 demonstrates both methods. A key assumption underpinning the EOQ model is that the stock is used in production at a steady rate.

Figure 22.4 Economic Order Quantity

Tree plc has the following information about the Twig, one of its stock items. The Twig has an average yearly use in production of 5,000 items. Each Twig costs £1 and the company's cost of capital is 10% per annum. For each Twig, insurance costs are 2p per annum, storage costs are 2p per annum and the cost of obsolescence is 1p. The ordering costs are £60 per order.

i. Graphical Solution

We must first compile a table of costs at various order levels. For example, at an order quantity of 500, Tree needs 10 orders per annum (i.e., to buy 5,000 Twigs) which would give a total order cost of £600. The average quantity in stock (250) will be half the order quantity (500). If it costs 15 pence to carry an item, the total carrying cost will be £37.50 (i.e., 250 × 15p). Finally, the total cost will be £637.50 (total order cost £600 plus total carrying cost £37.50).

Order quantity (Q)	*Average number or orders per annum*	*Order cost*	*Total order cost*	*Average quantity in stock*	*Carrying cost per item of stock*	*Total carrying cost*	*Total cost*
			(1)	(2)	(3)	(4)	(5)
		£	£		£	£	£
500	10	60	600	250	0.15	37.50	637.50
1,000	5	60	300	500	0.15	75.00	375.00
2,000	2.5	60	150	1,000	0.15	150.00	300.00
2,500	2	60	120	1,250	0.15	187.50	307.50
5,000	1	60	60	2,500	0.15	375.00	435.00

Notes from table:

(1) Number of orders per annum multiplied by order cost.

(2) This represents half the order quantity. It assumes steady usage and instant delivery of stock.

(3) Cost of capital 10p (10% × £1), insurance 2p, storage 2p, obsolescence 1p. All per item.

(4) Average quantity in stock (2) multiplied by the carrying cost of £0.15 (3).

(5) Total order cost (1) and total carrying cost (4).

The graph can now be drawn with order quantity on the horizontal axis and annual costs along the vertical axis. The intersection of carrying cost and ordering cost represents the optimal re-order level. The optimal order quantity is thus close to 2,000.

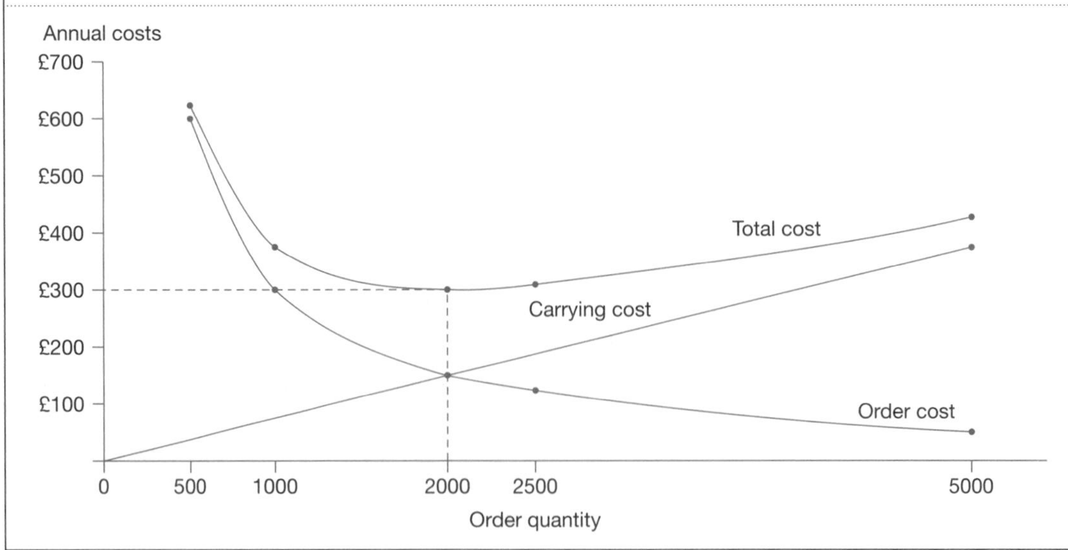

Figure 22.4 **Economic Order Quantity (*continued*)**

ii. Algebraic Solution
Fortunately, there is no need to know how to derive this formula, just apply it.

$$Q = \sqrt{\frac{2AC}{i}}$$ where Q = economic order quantity
A = average annual usage
C = cost of each order being placed
i = carrying cost per unit per annum

$$\therefore Q = \sqrt{\frac{2 \times 5{,}000 \times 60}{0.15}} = 2{,}000 \text{ Twigs}$$

(ii) Just-in-Time. Just-in-time was developed in Japan, where it has proved an effective method of stock control. It seeks to minimise stock holding costs by the careful timing of deliveries and efficient organisation of production schedules. At its best, just-in-time works by delivering stock just before it is used. The amount of stock is thus kept to a minimum and stock holding costs are also minimised. In order to do this, there is a need for a very streamlined and efficient production and delivery service. The concept behind just-in-time has been borrowed by many UK and US firms. Taken to its logical extreme, just-in-time means that no stocks of raw materials are needed at all. One potential problem with just-in-time is that if stock levels are kept at a minimum there is no stock buffer to deal with unexpected emergencies. For example, in the fuel blockade of Autumn 2000 in the UK, many supermarkets ran out of food because they had kept low levels of stock in their stores.

External Financing

(a) Cash

If a company is not generating enough cash from trading it may need to borrow. The most common method of borrowing is via a bank overdraft or a bank loan. Bank overdrafts are very flexible. Most major banks will set up an overdraft facility for a business as long as they are sure the business is viable. A bank overdraft is a good way to tackle the fluctuating cash flows experienced by many businesses.

Pause for Thought 22.1

Bank Overdrafts

A bank overdraft represents a flexible way for a business to raise money. Can you think of drawbacks?

Overdrafts usually carry relatively high rates of interest. Overdrafts often carry variable rates of interest and are subject to a limit, which should not be exceeded without authorisation from the bank. They can be withdrawn at very short notice and are normally repayable on demand. Generally, interest is determined by time period and security. The shorter the time period, the higher the interest rate typically paid. In addition, if the loan is not secured on an asset (i.e., is unsecured) the rate of interest charged will once again be higher. Small businesses without a track record may often find it difficult to get a bank overdraft. Even when an overdraft is granted, the bank may insist that it is secured against the company's assets.

An alternative to a bank overdraft is a bank loan. The exact terms of individual bank loans will vary. However, essentially a loan is for a set period of years and this may well be more than two years. The rate of interest on a loan will normally be lower than on a bank overdraft. Loans may be secured on business assets, for example specific assets, such as stock or motor vehicles.

> *Soundbite 22.1*
>
> **Debt Factoring**
>
> **'Handing over your sales ledger to a factor was once viewed in the same league as Dr Faustus flogging his soul off to the devil – a path of illusory rides that would lead only to inevitable business ruin and damnation.'**
>
> Jerry Frank, *No Longer a Deal with the Devil*
>
> Source: *Accountancy Age*, 18 May, 2000, p. 27

(b) Debtors

Since debtors are an asset, it is possible to raise money against them. This is done by debt factoring or invoice discounting.

(i) Debt factoring. Debt factoring is, in effect, the subcontracting of debtors. Many department stores, for example, find it convenient to subcontract their credit sales to debt factoring companies. The advantage to the business is twofold. First, it does not have to employ staff to chase up the debtors. Second, it receives an advance of money from the factoring organisation. There are, however, potential problems with factoring. The debt factoring company is not a charity and will charge a fee, for example 4% of sales, for its services. In addition, the debt factoring company will charge interest on any cash advances to the company. Finally, the company will lose the management of its customer database to an external party. As Soundbite 22.1 shows, debt factoring has traditionally been viewed with some suspicion.

(ii) Invoice discounting. Invoice discounting, in effect, is a loan secured on debtors. The financial institution will grant an advance (for example, 75%) on outstanding sales invoices (i.e., debtors). Invoice discounting can be a one-off, or a continuing, arrangement. An important advantage of invoice discounting over debt factoring is that the credit control function is not contracted out. The company, therefore, keeps control over its records of debtors. Figure 22.5 compares debt factoring with invoice discounting.

Figure 22.5 Comparison of Debt Factoring and Invoice Discounting

Element	*Debt Factoring*	*Invoice Discounting*
Loan from financial institution	Yes	Yes
Sales ledger (i.e., keeping records of debtors)	Management by financial institution	Managed by company
Time period	Continuing	Usually one-off, but can be continuing

The aim of debtors management is simply to collect money from debtors as soon as possible. For an optimal cash balance, with no considerations of fairness, a business will benefit if it can accelerate its receipts and delay its payments. Receipts from customers and payments to suppliers are measured using the debtors/creditors collection period ratio (see Chapter 9).

(c) Stock

As with debtors, it is sometimes possible to borrow against stock. However, the time period is longer. Stock needs to be sold, then the debtors need to pay. Stock is not, therefore, such an attractive basis for lending for the financial institutions. However, in certain circumstances, financial institutions may be prepared to buy the stock now and then sell it back to the company at a later date.

Pause for Thought 22.2

Sale and Buy Back of Stock

Can you think of any businesses where it may take such a long time for stock to convert to cash that businesses may sell their stock to third parties?

The classic example of sale and buy back occurs in the wine and spirit business. It takes a long time for a good whisky to mature. A finance company may, therefore, be prepared to buy the stock from the whisky distillery and then sell it back at a higher price at a future time. In effect, there is a loan secured against the whisky.

Another example might be in the construction industry. Here the financial insitution may be prepared to loan the construction company money in advance. The money is secured on the work-in-progress which the construction company has already completed. The construction company repays the loan when it receives money from the customers.

Long-Term Financing

There are potentially four main sources of long-term finance: retained profits; leasing; share capital; and loan capital (see Figure 22.6 below and Figure 22.7 on the next page).

Figure 22.6 Overview of Sources of Long-Term Finance

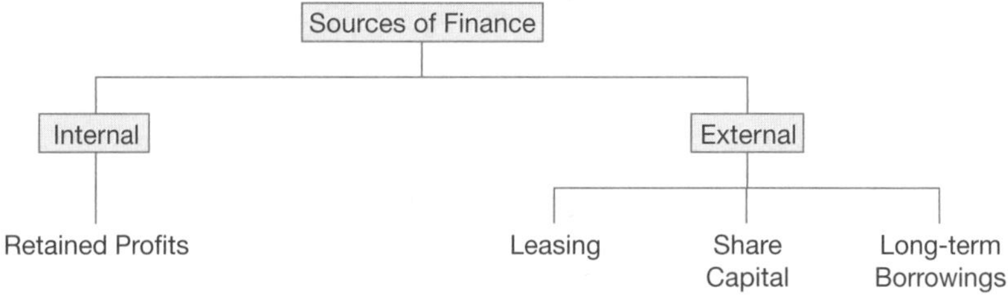

Figure 22.7 Sources of Long-Term Finance

(i) Internal Sources	*Main Features*
Retained profits	The company reinvests the profits it has made back into the business. It does not have to pay interest.
(ii) External Sources	
Leasing	A lessor leases specific assets to a company, such as a train or a plane. The company makes lease payments to the lessor.
Share capital	Money is raised from shareholders. Company pays dividends.
Long-term loan	Money is borrowed from banks or debt holder. Company repays interest.

Soundbite 22.2

Finance

'What is high finance? It's knowing the difference between one and ten, multiplying, subtracting and adding. You just add noughts. It's no more than that.'

John Bentley, *Sunday Mirror*, 8 April, 1973.

Source: *The Book of Business Quotations* (1991), p. 156

For long-term finance, there is a need to raise money as cheaply and effectively as possible. Long-term finance is usually used to fund long-term infrastructure projects and can often be daunting, given the huge sums involved (see Soundbite 22.2).

Internal Sources

Retained Profits

Retained profits (or revenue reserves) represent an alternative to external financing. In effect, the business is financing itself from its past successes. Instead of distributing its profits as dividends, the company invests them for the future. Shareholders thus lose out today, but hope to gain tomorrow. As the company grows using its retained profits, it will in the future make more profits, distribute more dividends and then have a higher share price. In theory, that is!

In many businesses, retained profits represent the main source of long-term finance. It is important to realise that retained profits, themselves, are not directly equivalent to cash. However, indirectly they represent the cash dividends that the company could have paid out to shareholders. Retained profits are particularly useful in times when interest rates are high.

Pause for Thought 22.3

Retained Profits

What might be a limitation of using only retained funds as a source of long-term finance?

Retained profits are an easy resource for a company to draw upon. There is, however, one major limitation: a company can only grow by its own efforts and when funds are available. Therefore, growth may be very slow. The situation is similar to buying a house, you could save up for 25 years and then buy a house. Or you could buy one immediately by taking out a mortgage. Most people, understandably, prefer not to wait.

External Sources

(a) Leasing

In principle, the leasing of a company's assets is no different to an individual hiring a car or a television for personal usage. The property, initially at least, remains the property of the lessor. The lessee pays for the lease over a period of time. Depending on the nature of the property, the asset may eventually become the property of the lessee or remain, forever, the property of the lessor.

The main advantage of leasing is that the company leasing the asset does not immediately need to find the capital investment. The company can use the asset without buying it. This is particularly useful where the leased asset will directly generate revenue which is then used to pay the leasing company. Many of the tax benefits of leasing assets have now been curtailed. However, leasing remains an attractive way of financing assets such as cars, buses, trains or planes. Normally, leasing is tied to specific assets.

In the end, the company leasing the assets will pay more for them than buying them outright. The additional payments are how the leasing companies make their money. Leases are usually used for medium or long-term asset financing. Hire purchase or credit sales, which share many of the general principles of leases, are sometimes used for medium to short-term financing.

Strictly, leasing is a source of assets rather than a source of finance. It does not result in an inflow of money. However, it is treated as a source of finance because if the asset had been bought outright, not leased, the business would have needed to fund the original purchase. Long-term leasing can be considered equivalent to debt. The asset is purchased by the lessor, but used and paid for by the lessee, including an interest payment. Long-term leases must be disclosed on the balance sheet. There is a recent trend where businesses that own property portfolios sell them for cash. They then lease back the original properties. This is the strategy being considered by Marks and Spencer in Real-Life Nugget 22.2.

Real-Life Nugget 22.2

Sale and Lease Back

Marks & Spencer, the troubled retail group, is poised to sell up to 40 of its high street stores across the country in a bid to raise up to £250m.

M&S, whose chief executive Peter Salsbury is conducting a strategy review to reverse the group's slump in profits, is said to be in talks with a select band of property investors. Under the proposals being considered, M&S would lease back the stores which are all understood to be prime sites and could include Oxford Circus, one of the retailer's larger London flagship stores.

Source: Doug Morrison, 'M&S £250m Stores Sale', *The Sunday Telegraph*, 21 March 1999

(b) Share Capital

Apart from reserves, there are two main types of corporate long-term capital: share capital and loan capital. Share capital is divided into ordinary shares and preference shares. Both sets of shareholders are paid dividends. However, while ordinary shareholders are owners; preference shareholders are not (Chapter 9 provides a fuller discussion of these issues).

When a company wishes to raise substantial sums of new money, the normal choice is either raising share capital or loan capital. Freeserve, the Internet service provider, for example, chose a stock market flotation raising an anticipated £1.9 billion from shares (see Real-Life Nugget 22.3).

Real-Life Nugget 22.3

Stock Market Flotations

Freeserve, the free internet access service launched last September by Dixons Group, the electrical retailer, is likely to be valued at £1.9 billion or £1,500 per subscriber, when it comes to the stock market. A flotation is likely to take place next month.

Details of the float, disclosed yesterday, emerged at a turbulent time for Internet stocks in the US, and are likely to cause controversy in the City.

Freeserve's value on a 'per subscriber' basis will be similar to that of BSkyB, the satellite broadcaster, 40 per cent owned by News International.

Source: Chris Ayres, Freeserve plans £1.9bn float, *The Times*, 16 June 1999, p. 25

There are three main methods by which a company may raise share capital: rights issue, public issue and placing. The main features of these alternatives are listed in Figure 22.8.

Figure 22.8 Types of Share Issue

Type	*Main Feature*
Rights Issue	Existing shareholders can buy more shares in proportion to their existing holdings.
Public Issue	Company itself directly offers shares to the public. A prospectus containing the company's details is issued. The shares may be issued at a fixed price or open to tender (i.e., bidding). A variation on the public issue is the offer for sale. The company sells the new issue of shares to a financial institution which then issues them on the company's behalf.
Placing	A company coming to the market for the first time may allow underwriters to 'place' shares with certain financial institutions. These institutions will then hold them or sell them.

A **rights issue** is therefore an issue to existing shareholders. In June 2001, for example, British Telecom asked its *existing* shareholders for £5.9 billion extra cash. By contrast, **public issues** and **placings** involve new shareholders. The public issue is distinguished from the placing chiefly by the fact that the shares are 'open' to public purchase rather than being privately allocated. In all three types of share issue, the company receives the amount that the shareholders pay for the shares. This money can then be used to finance expansion or on any other corporate activity. Stock market issues can make entrepreneurs potentially very rich, as happened to the creators of Lastminute.com (see Real-Life Nugget 22.4), although this wealth is dependent upon them selling their shares!

> ***Real-Life Nugget 22.4***
>
> **Getting Rich Quick**
>
> Yesterday's frenzied debut for Lastminute.com once again raises a question that may never be properly answered: how much is an internet company really worth?
>
> To many people a share is worth what the financial markets think it is, no more and no less. That would make Lastminute "worth" every penny of the £800m at which the total value of its shares were trading at one point yesterday. And no one can say investors were hoodwinked. They flocked to buy the shares despite the descant sung by analysts that they were overvalued.
>
> *Source*: Victor Keegan, 'Watch out for those last minute lemmings', The *Guardian*, 15 March 2000, p. 26.

(c) Loan Capital

Loan capital is long-term borrowing. The holders of the loans are paid loan interest. Loan capital can chiefly be distinguished from ordinary share capital by four features. First, loan interest is a *deduction* from profits *not*, like dividends, a *distribution* of profits. Second, loan capital, unlike ordinary share capital, is commonly repaid. Third, unlike ordinary shareholders, the holders of loan capital do not own the company. And, finally, most loans will be secured either on the general assets of the company or on specific assets, such as land and buildings. In other words, if the company fails, the loan holders have first call on the company's assets. They will be able to sell the assets of the company to recover their loans. Sometimes convertible loans are issued which may be converted into ordinary shares at a specified future date. Loans and debt have always attracted humourists, as Real-Life Nuggets 22.5 and 22.6 show below.

> ***Real-Life Nugget 22.5***
>
> **Loans**
>
> **The Importance of Being Seen**
>
> Around the turn of the century a speculator by the name of Charles Flint got into financial difficulties. He had a slight acquaintance with J.P. Morgan, Sr., and decided to touch him for a loan. Morgan asked him to come for a stroll around the Battery in lower Manhattan. The two men discussed the weather in some detail, and other pressing matters, when finally, after about an hour or so, the exasperated Flint burst out: 'But Mr. Morgan, how about the million dollars I need to borrow?'
>
> Morgan held out his hand to say goodbye: 'Oh, I don't think you'll have any trouble getting it now that we have been seen together.'
>
> Peter Hay (1988), *The Book of Business Anecdotes*, Harrap Ltd, London, p. 5

Real-Life Nugget 22.6

Debt

Stratagem

A moneylender complained to Baron Rothschild that he had lent 10,000 francs to a man who had gone off to Constantinople without a written acknowledgement of his debt.

'Write to him and demand back 50,000 francs,' advised the baron.

'But he only owes me 10,000,' said the moneylender.

'Exactly, and he will write and tell you so in a hurry. And that's how you will have acknowledgement of his debt.'

Peter Hay (1988), *The Book of Business Anecdotes*, Harrap Ltd, London, p. 5

As we saw in Chapter 9, the relationship between ordinary share capital and a company's fixed interest funds (long-term loans and preference shares) is known as gearing. Gearing is particularly important when assessing the viability of a company's capital structure.

Cost of Capital

From the details of a company's different sources of finance, it is possible to determine its cost of capital. In essence, the cost of capital is simply the cost at which a business raises funds. The two main sources of funds are debt (from long-term loans) and equity (from share capital). Normally, debt finance will be cheaper than equity. As Soundbite 22.3 below shows, it is an important business concept. Each source of finance will have an associated cost of capital. However, it is important to calculate the business's overall cost of capital (known as the **weighted average cost of capital** or WACC). WACC is the cost of capital normally used as the discount rate in capital investment appraisal (see Chapter 21).

The same principles can be adopted for personal finance and for business. We take the examples of Deborah Ebt, a student, in Figure 22.9 and Costco plc in Figure 22.10 (see next page).

Soundbite 22.3

Cost of Capital

'Business is the most important engine for social change in our society . . . And you're not going to transform society until you transform business. But you're not going to transform business by pretending it's not a business. Business means profit. You've got to earn your cost of capital. They didn't do it in the former Soviet Union, that's why it's former.'

Lawrence Perlman, *Twin Cities Business Monthly* (November 1994).

Source: *The Wiley Book of Business Quotations* (1998), p. 69

Figure 22.9 Calculation of Cost of Capital for a Student, D. Ebt

D. Ebt finances her college course as follows:

General expenses, by credit card	£3,000 at 25% interest per annum
Accommodation, by bank loan	£4,000 at 10% interest per annum
Car, by loan from uncle	£3,000 at 5% interest per annum

What is D. Ebt's overall weighted average cost of capital?

Source of Finance		*Proportion*	*Cost of Capital*	*Weighted Average Cost of Capital (WACC)*
	£	%	%	%
Credit card	3,000	30	25	7.5
Bank loan	4,000	40	10	4.0
Personal loan	3,000	30	5	1.5
	10,000	100		13.0

Helpnote

The weighted average cost of capital is the proportion of total debt financed by loan source multiplied by interest rate. Therefore, for credit cards (£3,000/£10,000) × 25% = 7.5%. The cost of capital for each source of finance is simply the interest rate.

Figure 22.9 thus shows that overall the weighted average cost of capital is 13% for D. Ebt. The most expensive source of finance was the credit card and the cheapest was the personal loan.

Figure 22.10 Example of a Company's Cost of Capital

Costco plc has 800,000 £1 ordinary shares currently quoted on the stock market at £2.50 each. It pays a dividend of 20p per share. Costco also has £200,000 worth of debt capital currently worth £1,000,000 on the stock market. The loan interest payable is 60,000.

Calculate Costco plc's weighted average cost of capital.

Source of Finance	*Market Value*	*Proportion %*	*Cost of capital*	*Weighted Average Cost of Capital (WACC)*
Equity	£2,000,000	66.67	8% (£0.20 ÷ £2.50)	5.3%
Debt	£1,000,000	33.33	6% (£60,000 ÷ £1,000,000)	2.0%
	£3,000,000	100.00		7.3%

The cost of capital is thus 7.3%.

Helpnote
We need to take the *market value* of the capital, not the original nominal value of the capital, as this is the value that the capital is *currently worth*. The cost of equity capital is simply the dividend divided by the share price and the cost of debt capital is the interest payable divided by the market price of the debt.

Figure 22.10 thus indicates that overall the weighted average cost of capital is 7.3%. The equity capital constitutes most of this (i.e. 5.3%).

The cost of capital is a potentially complex and difficult subject. However, the basic idea is simple, the company is trying to find the optimal mix of debt and equity which will enable it to fund its business at the lowest possible cost. Investors show great interest in a company's cost of capital (see Real-Life Nugget 22.7).

Real-Life Nugget 22.7

Railtrack's Cost of Capital

As for cost of capital, it's a matter of guesswork – or the capital-asset pricing model – to calculate the figure in the first place. However, the rail regulator reckons that Railtrack's cost of capital is close to 7.5 per cent before both tax and inflation. For investors, that figure looks pretty generous. Here's why: Railtrack will pay little tax in the coming years so, assuming inflation averages 2.5 per cent a year, the regulator is allowing Railtrack a post-tax return on its capital of nearly 10 per cent. That's approaching growth stock status and – using another branch of the capital-asset pricing model – means that investors could expect a total return from owning Railtrack shares of 10 per cent plus their current dividend yield, which is 2.8 per cent.

Source: Bearbull, 'Playing Monopoly', *The Investors Chronicle*, 4 August, 2000, p. 18

Conclusion

Sources of finance are essential to a business if it is going to grow and survive. We can distinguish between short-term and long-term sources of finance. Short-term financing concerns the management of working capital: cash, debtors and stock. This may involve the more efficient use of working capital using various techniques such as the debtors collection model and the economic order quantity model. Alternatively, it may involve raising loans from the bank by an overdraft, or loans secured on debtors by debt factoring, or by the sale and lease-back of stock. Long-term sources of capital are used to fund the long-term activities of a business. The four main sources are an internal source (retained profits) and three external sources (leasing, share capital and loan capital). Retained profits are not borrowings but, in effect, represent undistributed profits ploughed back into the business. Leasing involves a company using an asset, but not owning it. Shareholders own the company and are paid dividends. Loan capital providers do not own the company and are paid loan interest. A key function of managing long-term finance is to minimise a company's cost of capital which is the effective rate at which a company can raise funds.

Discussion *Questions*

Questions with numbers in blue have answers at the back of the book.

Q 1 Why are the sources of finance available to a firm so important? What are the main sources of finance and which activities of a business might they finance?

***Q* 2** What is working capital and how might a company try to manage it?

***Q* 3** What are the advantages and disadvantages of retained funds, debt and equity as methods of funding a business?

***Q* 4** What is the weighted average cost of capital and why is it an important concept in business finance?

Q 5 State whether the following statements are true or false. If false, explain why.

(a) Long-term sources of finance are usually used to finance working capital.
(b) A debtors collection model seeks the efficient management of debtors by balancing the benefits of extra credit sales against the extra costs of those sales.
(c) Two common techniques of stock control are the economic order quantity model and the just-in-time approach.
(d) There are four external long-term sources of finance: retained profits, leasing, share capital and loan capital.
(e) A rights issue, a public issue and a placing are three major ways in which a company can raise share capital.

Numerical *Questions*

Questions with numbers in blue have answers at the back of the book.

Q 1 Lathe plc is a small manufacturing firm. It buys in a subcomponent, the Tweak, for £1.50. The cost of insurance and storage per Tweak are £0.40, and £0.50 per item per year, respectively. It costs £25 for each order. The annual quantity purchased is 20,000.

Required:

(i) What is the economic order quantity? Solve this (a) graphically and (b) algebraically.
(ii) What are the total costs per annum at this level?

***Q* 2** A bookshop, Bookworm, buys 2,500 copies of the book, *Deep Heat*, per year. It costs the bookshop £0.80 per annum to carry the book in stock and £20 to prepare a new order.

Required:

(i) The total costs if orders are 1, 2, 5, 10 and 20 times per year.
(ii) The economic order quantity. An algebraic solution only is required.

***Q* 3** Winter Brollies wishes to revise its credit collection policy. Currently it has £500,000 sales, a credit policy of 25 days and an average collection period of 20 days. 1% of debtors default. Credit control costs are £5,000 for 25 days and 30 days; £6,000 for 60 days and 90 days. There is the following potential sales forecast.

Credit Period (Days)	*Average Collection Period (Days)*	*New Annual Sales (£)*	*Bad Debts (%)*
30	25	600,000	2
60	55	700,000	4
90	85	800,000	8

The cost of capital is 10% and the average contribution is 20% on selling price. Assume 360 days in year.

Required:

Advise Winter Brollies whether or not it should revise its credit policy.

Q 4 Albatross plc has the following information about its capital.

Source of Finance	*Current Market Value* £million	*Present Cost of Capital* %
Ordinary shares	4	12
Preference shares	1	10
Long-term loan	3	8

Required:

What is Albatross's weighted average cost of capital?

***Q*5** Nebula plc is looking at its sources of finance. It collects the following details. Currently, there are 500,000 ordinary shares in issue, with a market price of £1.50 and a current dividend of £0.30. The number of preference shares in issue is 300,000, with a market price of £1.00 and a dividend of 15p. There are £200,000 of long-term debentures carrying 12% interest. These are currently trading at £220,000.

Required:
Calculate Nebula's weighted average cost of capital.

Further Reading

In this book, I have tried to give non-specialist readers a comprehensive introduction to accounting. Most accounting textbooks are designed for accounting students and, therefore, often focus on demonstrating and explaining technical issues. These books must, therefore, be used by non-specialists with care.

In Section B, I have listed a considerable number of books and articles to supplement the theoretical chapters which deal with the conceptual and regulatory frameworks, measurement systems, the annual report, creative accounting and international accounting. Readers are referred back to these individual chapters.

Below, I detail some further reading which students may find useful for Section A on Financial Accounting and Section C on Management Accounting. There are separate references at the end of the chapter on Strategic Management Accounting in Section C. There is, in fact a much greater choice for Section A: Financial Accounting than for Section C on Management Accounting.

Section A: Financial Accounting

Alexander D. and A. Britton (2001) *Financial Reporting*, Thomson Business Press.

This book provides an in-depth look at financial accounting for those who have mastered the basics.

Edwards J.R. and J.R. Mellett (2002) *Introduction to Accounting*, Paul Chapman Publishing.

This book is pitched at a reasonable introductory level, but aimed at students specialising in accounting. Provides a good background supplemental text.

Elliott, B. and J. Elliott (2001) *Financial Accounting and Reporting*, Financial Times/Prentice Hall.

A book to be used with care. Provides in-depth coverage for those who have mastered the basics.

Gillespie, I., R. Lewis and K. Hamilton (2000) *Principles of Financial Accounting*, Prentice Hall.

This introductory text is aimed at first year students. It provides a good supplemental text. The spreadsheet approach may take some getting used to.

Horngren C.T., G.L. Sundem and J.A. Elliot (1999) *Introduction to Financial Accounting*, Prentice Hall.

This is a heavy-weight US text. Its in-depth approach and use of US terminology may make this book less accessible to UK students.

Wood F. and A. Sangster (1999) *Business Accounting I*, Financial Times/Pitman.

This well-established book is better on the practice than on context. It has numerous well-tried and well-tested examples.

Section C: Management Accounting

Arnold, J. and S. Turley (1996) *Accounting for Management Decisions*, Prentice Hall.

Provides a reasonable coverage of many of the key issues. At times may be too complex for non-specialist, first year students.

Drury C. (2000) *Management and Cost Accounting*, Thomson Business Press.

Provides a comprehensive coverage of management accounting for those who have grasped the basics.

Horngren C.T., G.L. Sundem and W.O. Stratton (1999) *Introduction to Management Accounting*, Prentice Hall.

A heavy-weight US text which mirrors the similar text on financial accounting. The in-depth approach and use of US terminology may make this book less accessible to UK students.

Weetman P., *Management Accounting: An Introduction*, (1999) Financial Times/Prentice Hall.

An accessible text for students who wish to rehearse their knowledge of the basics.

Glossary of Key Accounting Terms

This glossary contains most of the key accounting terms that students are likely to encounter. Words highlighted in bold are explained elsewhere in the glossary.

Absorption costing

The form of costing used for valuing stock for external financial reporting. All the overheads which can be attributed to a product are recovered. Unlike **marginal costing** which can sometimes be used for stock valuation, absorption costing includes both fixed and variable production overheads.

Acid test ratio

See **quick ratio**

Accounting

The provision of information to managers and owners so they can make business decisions.

Accounting concept

A principle underpinning the preparation of accounting information.

Accounting equation

The basic premise that **assets** equals **liabilities**.

Accounting period

The time period for which the accounts are prepared. Audited financial statements are usually prepared for a year.

Accounting policies

The specific accounting methods selected and followed by a company in areas such as sales, foreign currencies, stocks, goodwill and pensions.

Accounting rate of return

A method of **capital investment appraisal** which assesses the viability of a project using annual profit and initial capital invested.

Accounting standards
Accounting pronouncements which set out the disclosure and measurement rules businesses must follow to give a **true and fair view** when drawing up accounts.

Acounting Standards Board (ASB)
The Accounting Standards Board sets the UK's **accounting standards**.

Accruals
The amounts owed to the suppliers of services at the balance sheet date, for expenses such as telephone or light and heat.

Accruals concept
See **matching concept**

Accumulated depreciation
The total depreciation on **tangible fixed assets** including this year's and prior years' depreciation.

Activity-based costing
A cost recovery technique which identifies key activities and key activity **cost drivers**

Annual report
A report produced annually by **companies** comprising both financial and non-financial information.

Appropriation account
The sharing out of partners' profit after net profit has been calculated in the **profit and loss account**.

Asset turnover ratio
A ratio which compares sales to total assets employed.

Assets
Essentially, items owned or leased by a business. Assets may be tangible or intangible, current or fixed. Assets bring economic benefits through either sale (for example, stock) or use (for example, a car).

Associated company
A company in which 20–50% of the shares are owned by another company or in which another company has a significant influence.

Attainable standard cost
A **standard cost** which can be reached with effort.

Auditors
A team of professionally qualified accountants *independent* of a company. Appointed by the **shareholders** on the recommendation of the **directors**, the auditors check and report on the accounts prepared by the directors.

Auditors' report
A statement in a company's **annual report** which states whether the financial statements present a '**true and fair view**' of the company's activities over the previous financial year.

Authorised share capital
The amount of **share capital** that a company is *allowed* to issue to its shareholders.

Average cost (AVCO)
A method of stock valuation where stock is valued at the average purchase price (see also **first-in-first-out** and **last-in-first-out**).

Bad debts
Those debts that will definitely not be paid. They are an **expense** in the **profit and loss account** and are written off **debtors** in the **balance sheet**.

Balance off
In **double-entry bookkeeping**, the accounts are balanced off and, the figures for **assets** and **liabilities** are carried forward to the next period. In effect, this signals the end of an **accounting period.**

Balance sheet
A financial statement which is a snapshot of a business at a particular point in time. It records the **assets, liabilities** and **capital** of a business. Assets less liabilities equals capital. Capital is the owners' interest in the business.

Balanced scorecard
In **management accounting**, the balanced scorecard looks at a business from multiple perspectives such as financial, customer, internal business, and innovation and learning perspectives.

Bank overdraft
A business or individual owes the bank money.

Batch costing
A number of items of a similar nature are processed and costed together (e.g., baking bread).

Benchmarking
In **management accounting**, benchmarking measures a business against its competitors across a series of performance indicators (e.g., customer service or number of complaints).

Bookkeeping
The preparation of the basic accounts. Monetary transactions are entered into the books of account. A **trial balance** is then extracted, and a **profit and loss account** and a **balance sheet** are prepared.

Break-even analysis
Break-even analysis involves calculating the point at which a product or service makes neither a profit nor a loss. **Fixed costs** are divided by the contribution per unit giving the **break-even point**.

Break-even point
The point at which a firm makes neither a profit nor a loss. A firm's break-even point can be expressed as: Sales – variable costs – fixed costs = 0.

Budget
A future plan which sets out a business's financial targets.

Budgeting
Budgeting involves setting future targets. Actual results are then compared with budgeted results. Any **variances** are then investigated.

Called-up share capital
The amount of **issued share capital** that is fully paid up by **shareholders**. For example, a share may be issued for £1.50 and paid in three equal instalments. After two instalments the called-up share capital is £1.

Capital
Capital represents the owner's interest in the business. In effect, capital is a liability as it is owed by the business to the owner (e.g., **sole trader**, partner or **shareholder**). Capital is the assets of a business less its liabilities to third parties. Capital is accumulated wealth and is increased by profit, but reduced by losses.

Capital expenditure
A payment to purchase an **asset** with a long life such as a **fixed asset**.

Capital expenditure and financial investment
In a **cash flow statement**, cash flows relating to the purchase and sale of **fixed assets** and **investments**.

Capital investment appraisal
A method of evaluating long-term **capital expenditure decisions**.

Capital investment decisions
Usually long-term decisions (such as building a new factory).

Capital maintenance concept
A way of determining whether the '**capital**' of a business has improved, deteriorated or stayed the same over a period of time. There is both **financial** and **physical capital maintenance**.

Capital reserves
Reserves not distributable to shareholders as dividends (e.g., the **share premium account** or **revaluation reserve**).

Carriage inwards
The cost of delivering raw materials. Refers to the days when goods were delivered by horse-and-carriage.

Carriage outwards
The cost of delivering the finished goods. Refers to the days when goods were delivered by horse-and-carriage.

Carrying costs
Costs such as insurance, obsolescence, interest on borrowed money or clerical/security costs incurred in holding stock.

Cash and bank
The actual money held by the business either at the business as cash or at the bank.

Cash at bank
Money deposited with a bank.

Cash book
In large businesses, a separate book which records cash and cheque transactions.

Cash budget
This budget records the projected inflows and outflows of cash.

Cash cows
An element in the **product portfolio matrix**. Cash cows are a company's dream product. They are well-established, require little capital expenditure, but generate high returns.

Cash flow statement
A financial statement which shows the cash inflows and outflows of a business.

Chairman's statement
A statement in a company's **annual report** which provides a personalised overview of the company's performance over the past year. It generally covers strategy, financial performance and future prospects.

Companies Acts
Acts of Parliament which lay down the legal requirements for companies including accounting regulations.

Company
A business enterprise where the **shareholders** have **limited liability**.

Conceptual framework
A coherent and consistent set of accounting principles which underpin the preparation and presentation of financial statements.

Consistency concept
An accounting principle which states similar items should be treated similarly from year to year.

Contract costing

A form of **costing** in which costs are allocated to contracts (i.e., usually big jobs which occur in construction industries such as shipbuilding). A long-term contract extends over more than one year and creates the problem of when to take profit.

Contribution

Contribution to fixed overheads, or contribution in short, is **sales** less **variable costs**.

Contribution analysis

A technique for short-term decision making where **fixed costs, variable costs** and **contribution** are analysed. The objective is to maximise a company's contribution (and thus profit) when choosing between different operating decisions.

Contribution graph

A graph which plots cumulative contribution against cumulative sales. Also called a profit/volume graph.

Corporate governance

The system by which companies are directed and controlled. The financial aspects of corporate governance relate principally to internal auditing and the way in which the board of directors both set financial policy and report to the shareholders on the activities and progress of the company.

Cost

An item of expenditure (planned or actually incurred).

Cost accounting

Essentially, **costing** and **planning and control**.

Cost allocation

The process by which **indirect costs** are recovered into total cost or into stock.

Cost driver

In **activity-based costing**, a factor causing a change in an activity's costs.

Cost minimisation

Minimising cost either by tight budgetary control or cutting back on expenditure.

Cost recovery

The process by which costs are recovered into a product or service to form the basis of pricing or stock valuation.

Cost of capital

The rate at which a business raises funds.

Cost of sales
Essentially the cost of directly providing the sales.

Cost-volume-profit analysis
A **management accounting** technique which looks at the effect of changes in fixed costs, variable costs and sales on profit. Also called **contribution analysis**.

Costing
Recovering **costs** as a basis for pricing and stock valuation.

Creative accounting
Using the flexibility within accounting to manage the measurement and presentation of the accounts so that they serve the interests of the preparers.

Credit
An entry on the right-hand side of a **'T' account**. Records principally increases in **liabilities**, **capital** or **income**. May also record decreases in **assets** or **expenses**.

Credit control
Controlling **debtors** by establishing credit limits for new customers, monitoring the age of debts and chasing up **bad debts**.

Creditors
Amounts owed to trade suppliers for goods supplied on credit, but not yet paid.

Creditors budget
This **budget** forecasts the level of future creditors. It keeps a running balance of creditors by adding purchases and deducting cash payments.

Creditors collection period
Measures how long a business takes to pay its **creditors** by relating creditors to cost of sales.

Current assets
Those **assets** (e.g., **stocks**, **debtors** and cash) that a company uses in its day-to-day operations.

Current liabilities
The liabilities that a business uses in its day-to-day operations (e.g., **creditors**).

Current purchasing power
A **measurement system** where historical cost is adjusted by general changes in the purchasing power of money (e.g., inflation), often measured using the retail price index (RPI).

Current ratio
A short-term test of liquidity which determines whether short-term **assets** cover short-term **liabilities**.

Debenture
Another name for a **long-term loan**. Debentures may be **secured** or **unsecured loans**.

Debit
An entry on the left-hand side of a **'T' account**. Records principally increases in either **assets** or **expenses**. May also record decreases in **liabilities, capital** or **income**.

Debt factoring
Where the **debtors** are subcontracted to a third party who are paid to manage them.

Debtors
When sales are made on credit, but the customers have not yet paid.

Debtors age schedule
A credit control technique which profiles the age of the debts and allow old debts to be quickly identified.

Debtors budget
This **budget** forecasts the level of future debtors. It keeps a running balance of debtors by adding sales and deducting cash received.

Debtors collection model
A technique for managing **working capital** which seeks to maintain the most efficient level of debtors for a company. It balances the extra revenue generated by increased sales with the increased costs associated with extra sales (e.g., credit control costs, bad debts and the delay in receiving money).

Debtors collection period
A ratio which measures how long customers take to pay their debts by relating debtors to sales.

Decision making in management accounting
The choice between alternatives. Only **relevant costs and revenues** should be considered.

Decision-making objective of financial reporting
Providing users, especially shareholders, with financial information so that they can make decisions such as whether to buy or sell shares.

Depreciation
Depreciation attempts to match a proportion of the original cost of the **fixed assets** to the **accounting period** in which the fixed assets were used up as an annual expense.

Direct costs
Costs directly identifiable and attributable to a product or service (e.g., the amount of direct labour or direct materials incurred). Sometimes called *product costs.*

Direct labour overall variance
Standard cost of labour for actual production less actual cost of labour used in production.

Direct labour price variance
(Standard price per hour – actual price per hour) × actual quantity of labour used.

Direct labour quantity variance
(Standard quantity of labour hours for actual production – actual quantity of labour hours used) × standard labour price per hour.

Direct materials overall variance
Standard cost of materials for actual production less actual cost of materials used in production.

Direct materials price variance
(Standard price per unit of material – actual price per unit of material) × actual quantity of materials used.

Direct materials quantity variance
(Standard price per unit of material – actual price per unit of material) × actual quantity of materials used.

Direct method of preparing cash flow statement
Classifies *operating* cash flows by function or type of activity (e.g., receipts from customers).

Directors
Those responsible for running the business. Accountable to the **shareholders** who, in theory, appoint and dismiss them.

Directors' report
A narrative statement in an **annual report**. It supplements the financial information with information considered important for a full appreciation of a company's activities.

Discount allowed
A reduction in the selling price of a good or service allowed by the business to customers for prompt payment. Treated as an **expense**.

Discount factor
A factor by which future cash flows are discounted to arrive at today's monetary value.

Discount received
A reduction in the purchase price of a good or service granted to a business from the supplier for paying promptly. Treated as an **income**.

Discounted cash flow
The future expected cash inflows and outflows of a potential project discounted back to their present value today to see whether or not proposed projects are viable.

Dividend cover
A ratio showing how many times profit available to pay ordinary dividends covers actual dividends.

Dividend yield
A ratio showing how much dividend ordinary shares earn as a proportion of market price.

Dividends
A cash payment to shareholders rewarding them for investing money in a company.

Dogs
An element in the **product portfolio matrix**. Dogs are cash traps with low market growth and low market share.

Double-entry bookkeeping
A way of systematically recording the financial transactions of a company so that each transaction is recorded twice.

Doubtful debts
Debts which may or may not be paid. Usually, businesses estimate a certain proportion of their debts as doubtful.

Drawings
Money which a **sole trader** or partner takes out of a business as living expenses. It is, in effect, the owner's salary and is really a withdrawal of capital.

Earnings per share (EPS)
A key ratio by which investors measure the performance of a company.

Economic order quantity (EOQ)
A technique for managing **working capital**. The optimal EOQ is calculated so as to minimise the costs of ordering and holding stock.

Efficiency ratios
Ratios which show how efficiently a business uses its assets.

Entity concept
A business has a distinct and separate identity from its owner. This is obvious in the case of a large limited company where **shareholders** own the company and managers manage the company. However, there is also a distinction between a **sole trader's** or **partnership's** personal and business assets.

Equivalent units

In **process costing**, partially finished units are converted to fully completed equivalent units by estimating the percentage of completion.

Expenses

The day-to-day **costs** incurred in running a business, e.g., telephone, business rates and wages. Expenses are expenses even if goods and services are consumed, but not yet paid. Expenses are, therefore, different from cash paid.

Financial accounting

The provision of financial information on a business's financial performance targeted at external users, such as shareholders. It includes not only **double-entry bookkeeping**, but also the preparation and interpretation of the financial accounts.

Financial capital maintenance concept

This concept is primarily concerned with *monetary measurement*, in particular, the measurement of the net assets.

Financial Reporting Council (FRC)

A UK supervisory body which ensures that the overall accounting standard-setting system is working.

Financial Reporting Review Panel (FRRP)

The FRRP investigates contentious departures from accounting standards and is part of the UK's standard-setting regime.

Financing cash flows

In a **cash flow statement**, they relate to the issuing or buying back of shares or loan capital.

Finished goods stock

The final stock after the manufacturing process is completed, for example, finished tables. The cost includes materials and other manufacturing costs (e.g., labour and manufacturing overheads).

First-in-first-out (FIFO)

A method of stock valuation where the stock bought first is the first to be sold. See also **average cost** and **last-in-first-out**.

Fixed assets

Infrastructure assets used to run the business long-term and *not* used in day-to-day production. Includes **tangible fixed assets** (e.g., motor vehicles, land and buildings, fixtures and fittings, plant and machinery) and **intangible fixed assets** (e.g., goodwill).

Fixed costs

Costs that *do not vary* with production or sales (for example, insurance) in an **accounting period** and are not affected by short-term decisions. Often called fixed overheads.

Fixed overheads
See **fixed costs.**

Fixed overheads quantity variance
(Standard fixed overheads – actual fixed overheads).

Flexing the budget
Adjusting the **budget** to account for the *actual quantity produced.*

Gearing
The relationship between a company's ordinary shareholders' funds and the debt capital.

General reserve
A **revenue reserve** created to deal with general, unspecified contingencies such as inflation.

Going concern concept
The business will continue into the foreseeable future. **Assets, liabilities, incomes** and **expenses** are measured on this basis.

Goodwill
In takeovers, the purchase price less the amount paid for the net assets. It represents the value placed on the earning power of a business over and above its **net asset** value.

Gross profit
Sales less **cost of sales.**

Gross profit ratio
This ratio relates the profit earned through trading to sales.

Historical cost
A **measurement system** where monetary amounts are recorded at the date of original transaction.

Historical cost convention
The amount recorded in the accounts will be the *original* amount paid for a good or service.

Horizontal analysis
A form of ratio analysis which compares the figures in the accounts across time. It is used to investigate trends in the data.

Ideal standards
In **standard costing,** standards attained in an ideal world.

Impression management
Managers try to influence the financial reporting process in their own favour. Includes both **creative accounting** and **narrative enhancement.**

Income
The revenue earned by a business, e.g., **sales**. Income is income, even if goods and services have been delivered but customers have yet to pay. Income thus differs from cash received.

Income receivable
Received by the business from a third party, e.g., **dividends** receivable (from companies) or interest receivable (from the bank). The reverse of **interest payable**.

Indirect costs
Those costs *not* directly identifiable *nor* attributable to a product or service, e.g., administrative, and selling and distribution costs. These costs are totalled and then recovered indirectly into the product or service. Also called *indirect overheads* or *period costs*.

Indirect method of preparing cash flow statement
Operating cash flow is derived from the **profit and loss account** and **balance sheet** and not classified directly by function (such as receipts from sales).

Indirect overheads
See **indirect costs**.

Intangible assets
Fixed assets one cannot touch, unlike **tangible fixed assets** (such as land and buildings). Most common in **companies**.

Interest cover
A ratio showing the amount of profit available to cover the **interest payable** on long-term borrowings.

Interest payable
An expense related especially to bank loans. The reverse of **income receivable**.

Internal rate of return (IRR)
A **capital investment appraisal** technique. The internal rate of return represents the discount rate required to give a **net present value** of zero. It pays a company to invest in a project if it can borrow money for less than the IRR.

International Accounting Standards Board (IASB)
An international body founded in 1973 to work for the improvement and harmonisation of accounting standards worldwide. Originally called the International Accounting Standards Committee.

Interpretation of accounts
The evaluation of financial information, principally from the **profit and loss account** and **balance sheet**, so as to make judgements about profitability, efficiency, liquidity, gearing, cash flow, and the success of a financial investment. Sometimes called **ratio analysis**.

Investment ratios

Measures the returns to the shareholder (**dividend yield, earnings per share** and **price/earnings ratio**) or the ability of a company to sustain its dividend or interest payments (**dividend cover** and **interest cover**).

Investments

Assets such as stocks and **shares**.

Invoice discounting

The sale of debts to a third party for immediate cash.

Issued share capital

The share capital *actually* issued by a **company**.

Job costing

The recovery of costs into a specific product or service.

Just-in-time

A method of stock control developed in Japan. It seeks to minimise stock holding costs by the careful timing of deliveries and efficient organisation of production schedules. At its best, just-in-time delivers stock just before it is used.

Last-in-first-out (LIFO)

A method of stock valuation where the last stock purchased is the first sold. See also **average cost** and **first-in-first-out**.

Leasing

Where the **assets** are owned by a third party which the business pays to use them.

Liabilities

Amounts the business owes (e.g., creditors, bank loan). They can be short-term or long-term, third party liabilities or capital (i.e., liability owed by the business to the owner).

Limiting factor

Where production is constrained by a particular shortage of a key element, e.g., a restricted number of labour hours.

Limited liability

Shareholders are only liable to lose the amount of money they initially invested.

Liquidity ratios

Ratios derived from the **balance sheet** that measure how easily a firm can pay its debts.

Listed company
A **company** quoted on a stock exchange.

Loan capital
Money loaned to a company by third parties who do not own the company and are entitled to interest *not* dividends.

Loans
Amounts borrowed from third parties, such as a bank.

Long-term creditors
Amounts borrowed from third parties and repayable after a year. The most common are **long-term loans.**

Long-term loan
A loan, such as a bank loan, not repayable within a year. Sometimes called a **debenture**.

Management accounting
The provision of both financial and non-financial information to managers for **cost accounting, planning and control**, and **decision making**. It is thus concerned with the internal accounting of a business.

Margin of safety
In **break-even analysis**, the difference between current sales and break-even sales.

Marginal costing
Marginal costing excludes fixed overheads from the costing process. It focuses on **sales, variable costs** and **contribution. Fixed costs** are written off against contribution. It can be used for decision making or for valuing stock. When valuing stock only variable production overheads are included in the stock valuation.

Market value
The value shares fetch on the open market, i.e., their trading value. This may differ significantly from their **nominal value**.

Master budget
The overall budgeted **balance sheet** and **profit and loss account** prepared from the individual **budgets**.

Matching concept
Recognises **income** and **expenses** when accrued (i.e., earned or incurred) rather than when money is received or paid. Income is matched with any associated expenses to determine the appropriate profit or loss. Also known as the accruals concept.

Materials requirement planning (MRP) system
An MRP system is based on sales demand and coordinates production and purchasing to ensure the optimal flow of raw materials and optimal levels of **raw material stocks**.

Measurement systems

The processes by which the monetary amounts of items in the financial statements are determined. Such systems are fundamental to the determination of **profit** and to the measurement of **net assets.** There are five major measurement systems: **historical cost, current purchasing power, replacement cost, net realisable value** and **present value.**

Monetary measurement convention

Only items measurable in financial terms (for example, pounds or dollars) are included in the accounts. Atmospheric pollution is thus excluded, as it has no measurable financial value.

Narrative enhancement

Managers use the narrative parts of the **annual report** to convey a more favourable impression of performance than is actually warranted, e.g., by omitting key data or stressing certain elements.

Net assets

Total assets less **long-term loans** and **current liabilities.**

Net book value

The cost of **tangible fixed assets** less accumulated depreciation.

Net cash flow from operating activities

In a **cash flow statement,** cash flows from the normal trading activities of a business.

Net present value

A **capital investment appraisal** technique which discounts future expected cash flows to today's monetary values using an appropriate cost of capital.

Net profit

Sales less **cost of sales** less **expenses.**

Net profit ratio

A ratio which relates **profit** after **expenses** (i.e., **net profit**) to **sales.**

Net realisable value

A **measurement system** where assets are valued at what they would fetch in an orderly sale.

Nominal value

The face value of the shares when originally issued.

Normal standards

In **standard costing,** standards which a business usually attains.

Note on historical cost profits and losses

A statement in the **annual report** which records any differences caused by departures from the **historical cost convention** (e.g., revaluation and subsequent depreciation of fixed assets).

Notes to the accounts

In a company's **annual report**, they provide additional information about items in the accounts.

Objective of financial statements

To provide information about the financial position, performance and changes in financial position of an enterprise useful to a wide range of **users** in making decisions.

Operating and financial review

A statement in a company's **annual report** which enables companies to provide a formalised, structured and narrative explanation of financial performance. It has two parts. First, the operating review covers items such as a company's operating results, profit and dividends. Second, the financial review discusses items such as capital structure and treasury policy.

Operating cash flow

In a **cash flow statement**, operating profit adjusted for movements in **working capital** and non-cash flow items such as **depreciation**.

Operating profit

Net profit before taxation adjusted for interest paid and interest received.

Opportunity cost

The potential benefit lost by rejecting the best alternative course of action.

Ordinary (equity) share capital

Share capital issued to the **shareholders**, who own the company and are entitled to ordinary **dividends**.

Overall variances

In **standard costing**, the budgeted cost of the actual items produced (*standard cost of actual production*) is compared with the actual cost of items produced (*actual cost of production*).

Overheads

See **indirect costs**.

Padding the budget

In **budget** setting, where individuals try to create slack to give themselves some room for manoeuvre.

Partnership

Business enterprises run by more than one person, whose liability is normally unlimited.

Partnership capital accounts

The long-term capital invested into a partnership by the individual partners.

Partnership current accounts
The partners' share of the profits of the business. The main elements are the opening balances, salaries, profit for year, drawings and closing balances.

Patents
An **intangible asset** resulting from expenditure to protect rights to an invention.

Payback method
A method of **capital investment appraisal** which measures the cumulative cash inflows against the cumulative cash outflows to determine when a project will pay for itself.

Period costs
See **indirect costs**.

Periodicity convention
Accounts are prepared for a set period of time, i.e., an accounting period.

Physical capital maintenance concept
This concept is concerned with maintaining the physical productive capacity (i.e., operating capacity) of the business.

Planning and control
The planning and control of future costs using **budgeting** and **standard costing**.

Preference share capital
Share capital issued to **shareholders** who are *not* owners of the company and who are entitled to fixed dividends.

Prepayment
The amount paid in advance to the suppliers of services, e.g., prepaid insurance.

Present value
A **measurement system** where future cash inflows are discounted back to present-day values.

Price/earnings ratio
A ratio which measures **earnings per share** against share price.

Price variances
In **standard costing**, the *standard* price for the *actual* quantity used or sold is compared with the *actual* price for the *actual* quantity used or sold.

Prime cost
Direct materials, direct labour and direct expenses totalled.

Private limited company
A company where trading in shares is restricted.

Process costing
Used in industries with a continuous production process (e.g., beer brewing) where products progress from one department to another.

Product costs
See **direct costs**.

Product life cycle analysis
In **management accounting**, the analysis of product life cycles. The five stages of birth (known as introduction), growth, maturity, decline and senility (known as withdrawal) are associated with a certain level of sales and profit.

Product portfolio matrix
A strategic way of looking at a company's products and dividing them into **cash cows**, **dogs**, **question marks** and **stars**.

Production cost budget
This **budget** estimates the future cost of production, incorporating direct labour, direct materials and production overheads.

Production departments
Where products are manufactured.

Profit
Sales less purchases and **expenses**.

Profit and loss account
A financial statement which records the **income** and **expenses** of a business over the **accounting period**, normally a year. **Income** less **expenses** equals **profit**. By contrast, where expenses are greater than income, losses will occur. The balance from the profit and loss account is transferred annually to the **balance sheet** where it becomes part of **revenue reserves**.

Profitability ratios
They establish how profitably a business is performing.

Profit/volume chart
See **contribution graph**.

Provision for doubtful debts
Those debts a business is dubious of collecting. Deducted from **debtors** in the **balance sheet**. Only *increases* or *decreases* in the provision are entered in the **profit and loss account**.

Prudence concept
Income and **profit** should only be recorded in the books when an inflow of cash is certain. By contrast, any **liabilities** should be provided as soon as they are recognised even though the amount may be uncertain. Introduces an element of caution into accounting.

Public limited company
A **company** where shares are bought and sold by the general public.

Quantity variances
The budgeted cost of the actual items produced or sold (*standard cost of actual production* or *standard quantity sold*) is compared with the actual quantity produced or sold (*actual quantity used or sold*).

Question marks
An element of the **product portfolio matrix.** Question marks have low market share, but high market growth.

Quick ratio
Measures extreme short-term liquidity, i.e., **current assets** (excluding stock) against **current liabilities.** Sometimes called the 'acid test ratio'.

Ratio analysis
See **interpretation of accounts.**

Raw material stock
Stock purchased and ready for use, e.g., a carpenter with wood awaiting manufacture into tables.

Raw materials budget
This **budget** forecasts the future quantities of raw materials required. May supply the purchases figure for the **creditors budget.**

Reconciliation of movements in shareholders' funds
A financial statement in the **annual report** which highlights major changes to the wealth of shareholders such as profit (or loss) for the year, annual dividends and new share capital.

Reducing balance method of depreciation
A set percentage of **depreciation** is written off the **net book value** of **tangible fixed assets** every year.

Regulatory framework
The set of rules and regulations which govern accounting practice, mainly prescribed by government and the accounting standard-setting bodies.

Relevance
Relevant information affects **users'** economic decisions. Relevance is a prerequisite of usefulness and, for example, helps to predict future events or to confirm or correct past events.

Relevant costs and revenues

Costs that will affect a decision (as opposed to non-relevant costs, which will not).

Reliability

Reliable information is free from material error and is unbiased.

Replacement cost

A **measurement system** where assets are valued at the amounts needed to replace them with an equivalent asset.

Reserves

The accumulated profits (**revenue reserves**) or capital gains (**capital reserves**) to shareholders.

Retained profits

The **profit** a company has not distributed via **dividends**. An alternative to external financing. In effect, the business finances itself from its past successes.

Return on capital employed

A ratio looking at how effectively a company uses its capital. It compares **net profit** to capital employed. The most common definition measures **profit** before tax and **debenture** interest against long-term capital (i.e., **ordinary share capital** and **reserves, preference share capital, long-term loans**).

Returns on investments and servicing of finance

In a **cash flow statement**, cash received from investments or paid on loans.

Revaluation reserve

A **capital reserve** created when **fixed assets** are revalued to more than the original amount for which they were purchased. The revaluation is a gain to the shareholders.

Revenue expenditure

A payment for a current year's good or service such as purchases for resale or telephone expenses.

Revenue reserves

Reserves potentially distributable to shareholders as **dividends**, e.g., the **profit and loss account, general reserve**.

Review of operations

In a company's **annual report**, a narrative where the chief executive reviews the individual business operations.

Rights issue

Current **shareholders** are given the right to subscribe to new shares in proportion to their current holdings.

Sale and lease-back
Companies sell their **tangible fixed assets** to a third party and then lease them back.

Sales
Income earned from selling goods.

Sales budget
This **budget** estimates the future quantity of sales.

Sales price variance
(Standard price per unit – actual price per unit) × actual quantity of units sold.

Sales quantity variance
(Standard quantity of units sold – actual quantity of units sold) × standard contribution per unit.

Secured loans
Loans guaranteed (i.e., secured) by the **assets** of the company.

Securities Exchange Commission (SEC)
An independent regulatory institution in the US with quasi-judicial powers. US **listed companies** must file a detailed annual form, called the 10–K, with the SEC.

Sensitivity analysis
Involves modelling the future to see if alternative scenarios will change an investment decision.

Service costing
Service costing concerns specific services such as canteens run as independent operations. The cost of a particular service is the total costs for the service divided by the number of service units.

Service delivery departments
Departments in service industries that deliver the final service to customers.

Service support departments
Departments that supply support activities such as catering, administration, or selling and distribution.

Share capital
The total capital of the business is divided into shares. Literally a 'share' in the capital of the business.

Share options
Directors or employees are allowed to buy shares at a set price. They can then sell them for a higher price at a future date if the share price rises.

Share premium account

A **capital reserve** created when new shares are issued for more than their **nominal value**. For example, for shares issued for £150,000 with a nominal value of £100,000, the share premium account is £50,000.

Shareholders

The owners of the company who provide share capital by way of shares.

Sole trader

A business enterprise run by a sole owner whose liability is unlimited.

Spending to budget

Departments ensure they spend their allocated **budgets**. If they do not the department may lose the money.

Standard cost

A standard cost is the individual cost elements of a product or service (such as direct materials, direct labour and variable overheads) that are estimated in advance. Normally, the quantity and the price of each cost element are estimated separately. Actual costs are then compared with standard costs to determine **variances**.

Standard costing

A standardised version of **budgeting**. Standard costing uses preset costs for direct labour, direct materials and overheads. Actual costs are then compared with the standard costs. Any **variances** are then investigated.

Stars

An element of the **product portfolio matrix**. Stars are characterised by high market growth and high market share. Stars may be cash earners or cash drains requiring heavy capital expenditure.

Statement of directors' responsibilities for the financial statements

A statement in a company's **annual report** where directors spell out their responsibilities including (i) keeping proper accounting records; (ii) preparing financial statements in accordance with the Companies Act 1985; (iii) applying appropriate accounting policies; and (iv) following all applicable accounting standards.

Statement of total recognised gains and losses (STRGL)

A financial statement in the **annual report** which attempts to highlight all shareholder gains and losses and not just those from trading. The STRGL begins with the **profit** from the **profit and loss account** and then adjusts for *non-trading gains and losses*.

Stewardship

Making individuals accountable for **assets** and **liabilities**. Stewardship focuses on the physical monitoring of assets and the prevention of loss and fraud rather than evaluating how efficiently the assets are used.

Stock
Goods purchased and awaiting use (**raw materials**) or produced and awaiting sale (**finished goods**).

Stock turnover ratio
Measures the time taken for stock to move through a business.

Straight-line method of depreciation
The same amount of **depreciation** is written off the **tangible fixed assets** every year.

Strategic management accounting
A form of **management accounting** which considers an organisation's internal and external environments.

Subsidiary company
A **company** where more than half of the shares are owned by another company or which is effectively controlled by another company, or is a subsidiary of a subsidiary.

Sunk cost
A past cost with no ongoing implications for the future. It should thus be *excluded from decision making* as it is a non-relevant cost.

SWOT analysis
A strategic way of critically assessing a business's strengths and weaknesses, opportunities and threats.

'T' account (ledger account)
Each page of each book of account has a **debit** side (left-hand side) and a **credit** side (right-hand side). This division of the page is called a 'T' account.

'T' Account (ledger account)	
Assets and expenses on the left-hand side	Incomes, liabilities and capital on the right-hand side
DEBIT	CREDIT

Tangible fixed assets
Fixed assets one can touch (e.g., land and buildings, plant and machinery, motor vehicles, fixtures and fittings).

Target costing
A price is set with reference to market conditions and customer purchasing patterns. A target profit is then deducted to arrive at a target cost.

Third party liabilities

Amounts owing to third parties. They can be short-term (e.g., **creditors**, bank overdraft) or long-term (e.g., a bank loan).

Throughput accounting

This uses a variant of contribution per limiting factor to determine a production system's main bottlenecks (e.g., shortage of machine hours in a certain department).

Total absorption costing

Both **direct costs** and **indirect costs** are absorbed into a product or service so as to recover the total costs in the final selling price.

Total shareholders' funds

The **share capital** and **reserves** owned by both the ordinary and preference shareholders.

Trading and profit and loss account

The formal name for the full profit and loss account prepared by a **sole trader**.

Trading and profit and loss and appropriation account

The formal name for the profit and loss account prepared by a **company** or a **partnership**.

Trial balance

A listing of debit and credit balances to check the correctness of the **double-entry bookkeeping** system.

True and fair view

Difficult to define but, essentially, a set of financial statements which faithfully, accurately and truly reflect the underlying economic transactions of the organisation.

Unsecured loans

Loans which are not guaranteed (i.e., secured) by a company's **assets**.

Urgent Issues Task Force (UITF)

The UITF is part of the UK's standard-setting process. It makes recommendations to curb undesirable interpretations of existing accounting standards or prevent accounting practices which the **Accounting Standards Board** considers undesirable.

Users

Those with an interest in using accounting information, such as shareholders, lenders, suppliers and other trade creditors, customers, government, the public, management and employees.

Value chain analysis

In **management accounting**, a strategic way of determining a value chain for a business consisting of primary activities (e.g., receiving goods) and support activities (e.g., technology development).

Variable costs
These costs vary with production and sales (for example, the metered cost of electricity). Short-term decision making is primarily concerned with variable costs.

Variable overheads overall variance
Standard cost of variable overheads for actual production less actual cost of variable overheads for production.

Variable overheads price variance
(Standard variable overheads price per hour – actual variable overheads price per hour) × actual quantity of labour hours used.

Variable overheads quantity variance
(Standard quantity of labour hours for actual production – actual quantity of labour hours used) × standard variable overheads price per hour.

Variance
The difference between the budgeted costs and the actual costs in both **budgeting** and **standard costing**.

Vertical analysis
In **ratio analysis**, vertical analysis is where key figures in the accounts (such as sales, balance sheet totals) are set to 100%.

Work-in-progress stock
Partially completed stock (sometimes called stock in process) which is neither **raw materials** nor **finished goods**.

Working capital
Current assets less **current liabilities** (in effect, the operating capital of a business).

Zero-based budgeting
A **budget** based on the premise that the activities are being incurred for the first time.

Appendix: Answers

Chapter 1: Discussion *Answers*

The answers provide some outline points for discussion.

A1 Accounting is important because it is the language of business and provides a means of effective and understandable business communication. The general terminology of business is thus accounting-driven. Concepts such as profit and cash flow are accounting terms. In addition, accounting provides the backbone of a business's information system. It provides figures for performance measurement, for monitoring, planning and control and gives an infrastructure for decision-making. It enables businesses to answer key questions about past business performance and future business policy.

A3 There are many differences. The six listed below will do for starters!

(a) Financial accounting is designed to provide information on a business's recent financial performance and is targeted at external users such as shareholders. However, the information is also often used by managers. By contrast, management accounting is much more internally focused and is used solely by managers.

(b) Financial accounting operates within a regulatory framework set out by accounting standards and the Companies Acts. There is no such framework for management accounting.

(c) The main work of financial accounting is preparing financial statements such as the balance sheet and profit and loss account. By contrast, management accounting uses a wider range of techniques for planning, control and decision-making.

(d) Financial accounting is based upon double-entry bookkeeping, while management accounting is not.

(e) Financial accounting looks backwards, while management accounting is forward-looking.

(f) The end product of financial accounting is a standardised set of financial statements. By contrast, management accounting is very varied. Its output depends on the needs of its users.

Chapter 2: Discussion *Answers*

The answers provide some outline points for discussion.

A1 Financial accounting is essentially the provision of financial information to users for decision making. More formally:

> 'The objective of financial statements is to provide information about the financial position, performance and changes in financial position of an enterprise that is useful to a wide range of users in making economic decisions.'
>
> International Accounting Standards Board (2000), *Framework for the Preparation and Presentation of Financial Statements.*

In other words, financial accounting provides financial information (such as assets, liabilities, capital, expenses and income) to users (such as management and shareholders). This is useful because they can assess how well the managers run the company. On the basis of their assessment of the stewardship of management, they can make business decisions, for example, shareholders can decide whether or not to keep or sell their shares.

Financial accounting is central to any understanding of business. It provides the basic language for assessing a business's performance. Unless we understand financial accounting, it is difficult to see how we can truly understand business. It would be like trying to drive a car without taking driving lessons. For non-specialists, a knowledge of financial accounting will help them to operate effectively in a business world.

A5 True or false?

(a) *True.* Assets show what a business owns, while liabilities show what a business owes.

(b) *False.* The profit and loss account does show income earned and expenses incurred. However, net assets are the assets less liabilities which are shown in the balance sheet. Income less expenses equals profit.

(c) *False.* Stewardship used to be the main objective up until about the 1960s. However, now decision-making is generally recognised as the main objective.

(d) *True.* This is because of the entity concept where the business is separate from the owner. Therefore, business assets, liabilities, income and expenses must be separated from private ones.

(e) *True.* This is because the matching concept seeks to match income and expenses to the accounting period in which they arise. There is thus accounting symmetry. By contrast, prudence dictates that income should be matched to the year in which it is earned; any liabilities should be taken as soon as they are recognised. This means that if it is known that a liability would be incurred, say, in three years' time it would be included in the current accounting period. There is thus accounting asymmetry. The two principles thus clash.

Chapter 2: Numerical *Answers*

A1 Sharon Taylor **Profit and Loss Account**

	£	£
Sales		8,000
Less *Expenses*		
General expenses	4,000	
Trading expenses	3,000	7,000
Net Profit		1,000

Balance Sheet

	£
Assets	15,000
Liabilities	(3,000)
Net assets	12,000

	£
Opening capital employed	11,000
Add Profit	1,000
Closing capital employed	12,000

Cash Flow Statement

	£
Cash inflows	10,000
Cash outflows	(12,000)
Net cash outflow	(2,000)

Chapter 3: Discussion *Answers*

The answers provide some outline points for discussion.

A1 Double-entry bookkeeping is the essential underpinning of accounting. It provides an efficient mechanism by which organisations can record their financial transactions. For instance, large companies, such as Tesco or British Petroleum, may have millions of transactions per year. Double-entry bookkeeping provides a useful way of consolidating these. In a sense, therefore, the double-entry process permits organisations to make order from chaos. Double-entry bookkeeping enables the preparation of a trial balance. This, in turn, permits the construction of a profit and loss account and a balance sheet.

A6 True or false?

(a) *True.*

(b) *False.* We credit the sales account with sales, but debit the purchases account with purchases.

(c) *True.*

(d) *True.*

(e) *False.* Sales and capital are credits, but rent paid is a debit.

Chapter 3: Numerical *Answers*

A1

(i)	Assets = Liabilities	£25,000 = £25,000
(ii)	Assets = Liabilities + Capital	£25,000 = £15,000 + £10,000
(iii)	Assets = Liabilities + Capital + Profit	£40,000 = £15,000 + £10,000 + £15,000
(iv)	Assets = Liabilities + Capital + (Income – Expenses)	£40,000 = £15,000 + £10,000 + (£60,000 – £45,000)
(v)	Assets + Expenses = Liabilities + Capital + Income	£40,000 + £45,000 = £15,000 + £10,000 + £60,000

(vi)

'T' Account		'T' Account	
Assets + Expenses	Liabilities + Capital + Income	£40,000 + £45,000 = £85,000	£15,000 + £10,000 + £60,000 = £85,000

A2

	Account	*Debit*	*Account*	*Credit*
(a)	Wages	Increases an expense	Bank	Decreases an asset
(b)	Bank	Increases an asset	Capital	Increases capital
(c)	Hotel	Increases an asset	Bank	Decreases an asset
(d)	Electricity	Increases an expense	Bank	Decreases an asset
(e)	Bank	Increases an asset	Sales	Increases income
(f)	Purchases	Increases an expense	A. Taylor (creditor)	Increases a liability

A3 A. Bird

(i) Ledger accounts

Sales

	£		£
		6 June Thrush	4,000
		6 June Raven	7,000
7 June Bal. c/f	*17,000*	6 June Starling	6,000
	17,000		*17,000*
		7 June Bal. b/f	*17,000*

Purchases

	£		£
1 June Robin	8,000		
1 June Falcon	6,000		
1 June Sparrow	5,000	*7 June Bal. c/f*	*19,000*
	19,000		*19,000*
7 June Bal. b/f	*19,000*		

A3 A. Bird (continued)

Sales returns

	£		£
7 June Starling	1,000	*7 June Bal. c/f*	*1,000*
	1,000		*1,000*
7 June Bal. b/f	*1,000*		

Purchases returns

	£		£
		4 June Robin	1,000
7 June Bal. c/f	*3,000*	4 June Falcon	2,000
	3,000		*3,000*
		7 June Bal. b/f	*3,000*

Thrush (debtor)

	£		£
6 June Sales	4,000	*7 June Bal. c/f*	*4,000*
	4,000		*4,000*
8 June Bal. b/f	4,000		

Raven (debtor)

	£		£
6 June Sales	7,000	*7 June Bal. c/f*	*7,000*
	7,000		*7,000*
8 June Bal. b/f	*7,000*		

Starling (debtor)

	£		£
6 June Sales	6,000	7 June Sales Rets.	1,000
		7 June Bal. c/f	*5,000*
	6,000		*6,000*
8 June Bal. b/f	*5,000*		

Robin (debtor)

	£		£
4 June Purchases Rets.	1,000	1 June Purchases	8,000
7 June Bal. c/f	*7,000*		
	8,000		*8,000*
		8 June Bal. b/f	*7,000*

Falcon (creditor)

	£		£
4 June Purchases Rets.	2,000	1 June Purchases	6,000
7 June Bal. c/f	*4,000*		
	6,000		*6,000*
		8 June Bal. b/f	*4,000*

Sparrow (creditor)

	£		£
7 June Bal. c/f	*5,000*	1 June Purchases	5,000
	5,000		*5,000*
		8 June Bal. b/f	*5,000*

Note. To aid understanding the balancing off process is italicised.

A6 Katherine Jones: *(i) Ledger accounts*

Capital

	£		£
7 July Bal. c/f	*195,000*	1 July Bank	195,000
	195,000		*195,000*
		8 July Bal. b/f	*195,000*

Bank

	£		£
1 July Capital	195,000	2 July Premises	75,000
7 July Edwards	5,000	2 July Office equipment	9,000
7 July Smith	4,500	2 July Purchases	7,000
7 July Patel	3,500	5 July Wages	4,000
		5 July Electricity	2,000
		5 July Telephone	1,000
		7 July Johnston	1,250
		7 July Singh	500
		7 July Bal. c/f	*108,250*
	208,000		208,000
8 July Bal. b/f	*108,250*		

A6 Katherine Jones (continued)

Sales

	£		£
		3 July Edwards	10,000
		3 July Smith	9,000
7 July Bal. c/f	*26,000*	3 July Patel	7,000
	26,000		*26,000*
		7 July Bal. b/f	*26,000*

Purchases

	£		£
2 July Bank	7,000		
2 July Johnston	3,000		
2 July Singh	1,000	*7 July Bal. c/f*	*11,000*
	11,000		*11,000*
7 July Bal. b/f	*11,000*		

Telephone

	£		£
5 July Bank	1,000	*7 July Bal. c/f*	*1,000*
	1,000		*1,000*
7 July Bal. b/f	*1,000*		

Purchases returns

	£		£
7 July Bal. c/f	*500*	4 July Johnston	500
	500		500
		7 July Bal. b/f	*500*

Electricity

	£		£
5 July Bank	2,000	*7 July Bal. c/f*	*2,000*
	2,000		*2,000*
7 July Bal. b/f	*2,000*		

Wages

	£		£
5 July Bank	4,000	*7 July Bal. c/f*	*4,000*
	4,000		*4,000*
7 July Bal. b/f	*4,000*		

Premises

	£		£
2 July Bank	75,000	*7 July Bal. c/f*	*75,000*
	75,000		*75,000*
8 July Bal. b/f	*75,000*		

Office equipment

	£		£
2 July Bank	9,000	*7 July Bal. c/f*	*9,000*
	9,000		*9,000*
8 July Bal. b/f	*9,000*		

Edwards (debtor)

	£		£
3 July Sales	10,000	7 July Bank	5,000
		7 July Bal. c/f	*5,000*
	10,000		*10,000*
8 July Bal. b/f	*5,000*		

Smith (debtor)

	£		£
3 July Sales	9,000	7 July Bank	4,500
		7 July Bal. c/f	*4,500*
	9,000		*9,000*
8 July Bal. b/f	*4,500*		

Patel (debtor)

	£		£
3 July Sales	7,000	7 July Bank	3,500
		7 July Bal. c/f	*3,500*
	7,000		*7,000*
8 July Bal. b/f	*3,500*		

Johnston (creditor)

	£		£
4 July Purchases Rets.	500	2 July Purchases	3,000
7 July Bank	1,250		
7 July Bal. c/f	*1,250*		
	3,000		*3,000*
		8 July Bal. b/f	*1,250*

Singh (creditor)

	£		£
7 July Bank	500	2 July Purchases	1,000
7 July Bal. c/f	*500*		
	1,000		*1,000*
		8 July Bal. b/f	*500*

(ii)

Katherine Jones
Trial Balance as at 7 July

	£	£
Capital		195,000
Bank	108,250	
Sales		26,000
Purchases	11,000	
Purchases returns		500
Telephone	1,000	
Electricity	2,000	
Wages	4,000	
Premises	75,000	
Office equipment	9,000	
Edwards (debtor)	5,000	
Smith (debtor)	4,500	
Patel (debtor)	3,500	
Johnston (creditor)		1,250
Singh (creditor)		500
	223,250	223,250

A8 Jay Shah

Jay Shah
Trial Balance as at 31 December

	Debit £	Credit £	Type
Capital		45,300	Capital
Motor car	3,000		Asset
Building	70,000		Asset
Office furniture	400		Asset
A. Smith (debtor)	250		Asset
J. Andrews (creditor)		350	Liability
T. Williams (creditor)		550	Liability
G. Woolley (debtor)	150		Asset
Purchases returns		500	Income*
Bank	3,600		Asset
Electricity	1,400		Expense
Business rates	1,800		Expense
Rent	1,600		Expense
Wages	3,500		Expense
Long-term loan		9,000	Liability
Sales		100,000	Income
Purchases	70,000		Expense
	155,700	155,700	

*An income because it reduces the expense of purchases.

A10 Rajiv Sharma

Rajiv Sharma
Trial Balance as at 31 December

	Debit	Credit
	£	£
Shop	55,000	
Machinery	45,000	
Car	10,000	
Sales		135,000
Purchases	80,000	
Opening stock	15,000	
Debtors	12,000	
Creditors		8,000
Long-term loan		16,000
General expenses	300	
Telephone	400	
Light and heat	300	
Repairs	400	
Capital		59,400
	218,400	218,400

Chapter 4: Discussion *Answers*

The answers provide some outline points for discussion.

A1 The profit and loss account is used by a variety of users for a variety of purposes. Sole traders and partnerships use it to determine how well they are doing. They will want to know if the business is making a profit. This will enable them to make decisions such as how much they should take by way of salary (commonly called drawings). Similarly, the tax authorities will use the profit and loss account as a starting point to calculate the tax that these businesses owe the government.

For the shareholders of a limited company, profit enables them to assess the performance of the company's management. Shareholders can then make decisions about their investments. Company directors, on the other hand, may use profits to work out the dividends payable to shareholders, or to calculate their own profit-related bonuses. In short, profit has multiple uses.

A5 True or false?

(a) *False.* Profit is income earned less expenses incurred.
(b) *True.*
(c) *False.* Sales returns are returns by customers.
(d) *True.*
(e) *True.*

Chapter 4: Numerical *Answers*

A1 Joan Smith

Joan Smith
Trading and Profit and Loss Account Year Ended 31 December 2001

	£	£
Sales		100,000
Less *Cost of Sales*		
Opening stock	10,000	
Add Purchases	60,000	
	70,000	
Less Closing stock	5,000	65,000
Gross Profit		35,000
Less *Expenses*		
General expenses	10,000	
Other expenses	8,000	18,000
Net Profit		17,000

A2 Dale Reynolds

Dale Reynolds
Trading Account Year Ended 31 December 2001

	£	£	£
Sales			50,000
Less Sales returns			1,000
			49,000
Less *Cost of Sales*			
Opening stock		5,000	
Add Purchases	25,000		
Less Purchases returns	2,000		
	23,000		
Add Carriage inwards	1,000	24,000	
		29,000	
Less Closing stock		8,000	21,000
Gross Profit			28,000

Chapter 5: Discussion *Answers*

The answers provide some outline points for discussion.

A1 The balance sheet and profit and loss are indeed complementary. Both are prepared from a trial balance. The balance sheet takes the assets, liabilities and capital, and arranges them into a position statement. By contrast, the profit and loss account takes the income and expenses and arranges them into a performance statement. The balance sheet represents a snapshot of the business at a certain point in time. The profit and loss account represents a period, usually a month or a year. The balance sheet deals with liquidity, while the profit and loss account deals with performance. Both together, therefore, provide a complementary picture of an organisation both at a particular point in time and over a period.

A5 True or false?

(a) *True.*

(b) *False.* Stock and bank are indeed current assets, but creditors are current liabilities.

(c) *False.* Total net assets is fixed assets and current assets less current liabilities, but also less long-term creditors.

(d) *True.*

(e) *False.* An accrual is an expense owing, for example, an unpaid telephone bill. An amount prepaid (for example, rent paid in advance) is a prepayment.

Chapter 5: Numerical *Answers*

A1 Jane Bricker

Jane Bricker
Capital Employed as at 31 December 2001

	£
Opening capital	5,000
Add Profit	12,000
	17,000
Less Drawings	7,000
Closing capital	10,000

A2 Alpa Shah

Alpa Shah
Total Net Assets as at 30 June 2002

	£	£
Fixed Assets		100,000
Current Assets	50,000	
Current Liabilities	(30,000)	
Net current assets		20,000
Total assets less current liabilities		120,000
Long-term Creditors		(20,000)
Total net assets		100,000

Chapter 6: Discussion *Answer*

The answer provides some outline points for discussion.

A1 A sole trader is where only one person owns the business. For example, a retailer, such as a baker, might be a sole trader. It is important for sole traders to prepare accounts for several reasons. First, as a basis for assessing their own financial performance. This indicates whether they can take out more wages (known as drawings), whether they can expand or pay their workers more. Second, the tax authorities need to be assured that the profit figure, which is the basis for assessing tax, has been properly prepared. And, third, if any loans have been borrowed, for example, from the bank, then the bankers will be interested in assessing performance.

Chapter 6: Numerical *Answers*

A1 M. Anet

M. Anet
Trading and Profit and Loss Account for Year Ended 31 December 2001

	£	£
Sales		25,000
Less *Cost of Sales*		
Purchases		15,000
Gross Profit		10,000
Less *Expenses*		
Electricity	1,500	
Wages	2,500	4,000
Net Profit		6,000

M. Anet
Balance Sheet as at 31 December 2001

	£	£	£
Fixed Assets			
Hotel			40,000
Van			10,000
			50,000
Current Assets			
Bank	8,000		
A. Brush (debtor)	400	8,400	
Current Liabilities			
A. Painter (creditor)	(500)	(500)	

A1 M. Anet (continued)

	£
Net current assets	7,900
Total assets less current liabilities	57,900
Long-term Creditors	–
Total net assets	57,900
Capital Employed	£
Opening capital	51,900
Add Net Profit	6,000
Closing capital	57,900

A2 P. Icasso

P. Icasso
Trading and Profit and Loss Account for Year Ended 31 March 2002

	£	£
Sales		35,000
Less Sales returns		3,000
		32,000
Less *Cost of Sales*		
Purchases	25,000	
Less Purchases returns	4,000	21,000
Gross Profit		11,000
Less *Expenses*		
Electricity	1,000	
Advertising	800	1,800
Net Profit		9,200

P. Icasso
Balance Sheet as at 31 March 2002

	£	£	£
Fixed Assets			
Hotel			50,000
Van			8,000
			58,000
Current Assets			
Bank	9,000		
Shah (debtor)	1,250		
Chan (debtor)	2,250	12,500	
Current Liabilities			
Jones (creditor)	(1,250)	(1,250)	

A2 P. Icasso (continued)

	£
Net current assets	11,250
Total assets less current liabilities	69,250
Long-term Creditors	–
Total net assets	69,250
Capital Employed	£
Opening capital	60,050
Add Net Profit	9,200
Closing capital	69,250

A3 R. Ubens

R. Ubens
Trading and Profit and Loss Account for Year Ended 31 December 2001

	£	£	£
Sales			88,000
Less Sales returns			800
			87,200
Less *Cost of Sales*			
Opening stock		3,600	
Add Purchases	66,000		
Less Purchases returns	1,200	64,800	
		68,400	
Less Closing stock		4,000	64,400
Gross Profit			22,800
Less *Expenses*			
Electricity		1,500	
Advertising		300	
Printing and stationery		50	
Telephone		650	
Rent and rates		1,200	
Postage		150	3,850
Net Profit			18,950

R. Ubens
Balance Sheet as at 31 December 2001

	£	£	£
Fixed Assets			
Building			20,400
Motor van			3,500
			23,900

A3 R. Ubens (continued)

	£	£	£
Current Assets			
Stock	4,000		
Bank	4,400		
Debtors	2,600	11,000	
Current Liabilities			
Creditors	(3,800)	(3,800)	
Net current assets			7,200
Total assets less current liabilities			31,100
Long-Term Creditors			–
Total net assets			31,100
Capital Employed			£
Opening capital			19,950
Add Net Profit			18,950
			38,900
Less Drawings			7,800
Closing capital			31,100

A4 C. Onstable

Trading and Profit and Loss Account (extracts)

Expenses

Rent	£2,880 (i.e., 12 × £240)
Insurance	£360 (£480 – £120)

Balance Sheet (extracts)

Current liabilities

Prepayments £840*

*(Rent 3 × £240 = £720; Insurance £120)

A5 V. Gogh

V. Gogh
Trading and Profit and Loss Account Year Ended 31 December 2001

	£	£	£
Sales			40,000
Less Sales returns			500
			39,500
Less *Cost of Sales*			
Opening stock		5,500	
Purchases	25,000		
Less Purchases returns	450	24,550	
		30,050	
Less Closing stock		9,000	21,050
Gross Profit			18,450

A5 V. Gogh (continued)

	£	£
Less *Expenses*		
Business rates	1,000	
Rent	400	
Telephone	450	
Insurance	750	
General expenses	150	
Electricity	700	
Wages	10,500	13,950
Net Profit		4,500

V. Gogh
Balance Sheet as at 31 December 2001

	£	£	£
Fixed Assets			
Shop			9,000
Motor car			8,500
			17,500
Current Assets			
Stock	9,000		
Bank	1,300		
Debtors	3,500		
Prepayments	200	14,000	
Current Liabilities			
Creditors	(1,500)		
Accruals	(350)	(1,850)	
Net current assets			12,150
Total assets less current liabilities			29,650
Long-term Creditors			(3,700)
Total net assets			25,950
Capital Employed			£
Opening capital			34,350
Add Net profit			4,500
			38,850
Less Drawings			12,900
Closing capital			25,950

A6 L. Da Vinci

L. Da Vinci

Trading and Profit and Loss Account Year Ended 30 September 2001

	£	£	£
Sales			105,000
Less Sales returns			8,000
			97,000
Less *Cost of Sales*			
Opening stock		6,500	
Add Purchases	70,000		
Less Purchases returns	1,800		
	68,200		
Add Carriage inwards	250	68,450	
		74,950	
Less Closing stock		7,000	67,950
Gross Profit			29,050
Less *Expenses*			
Discounts allowed		300	
Electricity		2,025	
Telephone		500	
Wages		32,500	
Insurance		200	
Rent		1,000	
Business rates		1,000	37,525
Net loss			(8,475)

L. Da Vinci

Balance Sheet as at 30 September 2001

Fixed Assets	£	£	£
Business premises			18,000
Motor van			7,500
Computer			1,500
			27,000
Current Assets			
Stock	7,000		
Debtors	12,000		
Bank	1,800		
Prepayments	275	21,075	
Current Liabilities			
Creditors	(13,000)		
Accruals	(1,375)	(14,375)	

A6 L. Da Vinci (continued)

	£
Net current assets	6,700
Total assets less current liabilities	33,700
Long-term Creditors	(6,600)
Total net assets	27,100
Capital Employed	£
Opening capital	44,075
Less Net loss	8,475
	35,600
Less Drawings	8,500
Closing capital	27,100

A7 H. Ogarth

H. Ogarth
Trading and Profit and Loss Account year ended 31 December 2002 (extracts)

Expenses	£
Depreciation on buildings	10,000
Depreciation on machinery	3,000
Depreciation on motor van	2,000
	15,000

	£	£	£
Balance Sheet (extracts)	*Cost*	*Accumulated*	*Net book*
Fixed assets		*depreciation*	*value*
Buildings	100,000	(10,000)	90,000
Machine	50,000	(3,000)	47,000
Motor van	20,000	(2,000)	18,000
	170,000	(15,000)	155,000

Chapter 7: Discussion *Answers*

The answers provide some outline points for discussion.

A1 These three forms of business enterprise fit various niches. The sole trader form is good for very small businesses, such as a window cleaner, carpenter or small shopkeeper. There is a limited amount of capital needed and the individual can do most of the work. The accounting records needed for this type of business are not extensive. Partnerships are useful where the business is a little more complicated. They are suitable for situations where more than one person work together. There is then a need to sort out each partner's share of capital and profits. Companies are useful where a lot of capital is needed. Thus, they are particularly suitable for medium-sized and large businesses. They are particularly appropriate when raising

money externally because of the concept of limited liability. As shareholders are only liable for the amount of their initial investments, they will be keener to invest as their potential losses will be limited.

5 True or false?

(a) *False.* Drawings are withdrawal of capital by the partners, they are found in the partners' current accounts.
(b) *True.*
(c) *False.* Nominal value is the face value of the shares, normally the amount the shares were originally issued at. Market value is their stock-market value.
(d) *False.* Unsecured loans are secured on the general assets of the business. It is secured loans which are attached to specific assets.
(e) *False.* Reserves are accumulated profits and cannot directly be spent. Only cash can be spent.

Chapter 7: Numerical *Answers*

A1 Tom and Thumb

Tom and Thumb

Trading, Profit and Loss and Appropriation Account for Year Ended 31 December 2001

		£	£
Net Profit before Appropriation			100,000
Less Salaries:			
Tom		10,000	
Thumb		30,000	40,000
			60,000
Profits:			
Tom	3	45,000	
Thumb	1	15,000	60,000

Tom and Thumb

Balance Sheet as at 31 December 2001

Capital Employed	£	£	£
	Tom	Thumb	
Capital Accounts	8,000	6,000	14,000
Current Accounts			
Opening balances	3,000	(1,000)	

A1 Tom and Thumb (continued)

Add:			
Salaries	10,000	30,000	
Profit share	45,000	15,000	
	58,000	44,000	
Less Drawings	25,000	30,000	
Closing balances	33,000	14,000	47,000
Total partners' funds			61,000

A2 J. Waite and P. Watcher

J. Waite and P. Watcher

Trading, Profit and Loss and Appropriation Account for Year Ended 30 November 2001

		£	£
Sales			350,000
Less Cost of Sales			
Opening stock		9,000	
Add Purchases		245,000	
		254,000	
Less Closing stock		15,000	239,000
Gross Profit			111,000
Less *Expenses*			
Depreciation:			
Land and buildings		2,000	
Motor vehicles		3,000	
Electricity		3,406	
Wages		14,870	
Rent and business rates		6,960	
Telephone		1,350	
Interest on loan		2,800	
Other expenses		5,500	39,886
Net Profit before Appropriation			71,114
Less Salaries:			
Waite		18,000	
Watcher		16,000	34,000
			37,114
Profits:			
Waite	3	22,268	
Watcher	2	14,846	37,114

A2 J. Waite and P. Watcher (continued)

J. Waite and P. Watcher
Balance Sheet as at 30 November 2001

	£ *Cost*	£ *Accumulated depreciation*	£ *Net book value*
Fixed Assets			
Land and buildings	166,313	(2,000)	164,313
Motor vehicles	65,000	(3,000)	62,000
	231,313	(5,000)	226,313
Current Assets			
Stock	15,000		
Debtors	12,000		
Bank	6,501	33,501	
Current Liabilities			
Creditors	(18,500)		
Accruals	(300)	(18,800)	
Net current assets			14,701
Total assets less current liabilities			241,014
Long-term Creditors			(28,000)
Total net assets			213,014

Capital Employed	Waite £	Watcher £	£
Capital Accounts	88,000	64,000	152,000
Current Accounts			
Opening balances	(2,500)	12,000	
Add:			
Salaries	18,000	16,000	
Profit share	22,268	14,846	
	37,768	42,846	
Less Drawings	13,300	6,300	
Closing balances	24,468	36,546	61,014
Total partners' funds			213,014

A5 Red Devils Ltd

Red Devils Ltd
Profit and Loss Account for the Year Ended 30 November 2001 (unpublished)

	£	£
Sales	*[As per Accounts of Sole*	
Less Cost of Sales	*Trader or Partnership*]	
Gross Profit		150,000
Less *Expenses*		
Debenture interest	14,000	
General expenses	22,100	
Directors' fees	19,200	
Auditors' fees	7,500	62,800
Profit before Taxation		87,200
Taxation		(17,440)
Profit after Taxation		69,760
Proposed dividends on ordinary shares	(25,000)	
Proposed dividends on preference shares	(9,000)	
Transfer to general reserve	(3,500)	(37,500)
Retained Profit		32,260

Red Devils Ltd
Balance Sheet as at 30 November 2001

	£	£	£
Fixed Assets			680,900
Current Assets			
Stock	105,000		
Debtors	4,700		
Bank	5,300	115,000	
Current Liabilities			
Creditors	(12,200)		
Taxation	(17,440)		
Dividends (ordinary £25,000, preference £9,000)	(34,000)		
Auditors' fees	(7,500)		
Debenture interest	(14,000)	(85,140)	
Net current assets			29,860
Total assets less current liabilities			710,760
Long-term Creditors			(200,000)
Total net assets			510,760

A5 Red Devils Ltd (continued)

	£	£	£
Share Capital and Reserves			
Share Capital		Authorised	Issued
Ordinary share capital (£1 each)		400,000	250,000
6% preference shares		150,000	150,000
		550,000	400,000
Reserves			
Capital reserves			
Share premium account			55,000
Other reserves			
Opening general reserve	11,000		
Transfer for year	3,500		
Closing general reserve		14,500	
Opening profit and loss account	9,000		
Retained profit for year	32,260		
Closing profit and loss account		41,260	55,760
Total shareholders' funds			510,760

A6 Superprofit Ltd

Superprofit Ltd

Trading, Profit and Loss and Appropriation Account for the Year Ended 31 December 2001

	£000	£000	£000
Sales			351
Less *Cost of Sales*			
Opening stock		23	
Add Purchases		182	
		205	
Less Closing stock		26	179
Gross Profit			172
Less *Expenses*			
Depreciation:			
Land and buildings		18	
Motor vehicles		7	
Auditors' fees		2	
Loan interest		4	
Electricity		12	
Insurance		3	
Wages		24	
Light and heat		8	
Telephone		5	
Other expenses		26	109

A6 Superprofit Ltd (continued)

	£000	£000
Profit before Taxation		63
Taxation		(13)
Net Profit after Taxation		50
Ordinary dividends	(9)	
Preference dividends	(3)	(12)
Retained Profit		38

Superprofit Ltd
Balance Sheet as at 31 December 2001

	£000 *Cost*	£000 *Accumulated depreciation*	£000 *Net book value*
Fixed Assets			
Intangible Assets			
Patents			12
Tangible Assets			
Land and buildings	378	(18)	360
Motor vehicles	47	(7)	40
	425	(25)	400
Total fixed assets			412
Current Assets			
Stock	26		
Debtors	18		
Bank	31	75	
Current Liabilities			
Creditors	(33)		
Taxation payable	(13)		
Dividends payable	(12)		
Other accruals (see note below)	(6)	(64)	
Net current assets			11
Total assets less current liabilities			423
Long-term Creditors			(32)
Total net assets			391

A6 Superprofit Ltd (continued)

	£000	£000	£000
Share Capital and Reserves			
Share Capital		Authorised	Issued
Ordinary share capital		250	210
Preference share capital		50	25
		300	235
Reserves			
Capital reserves			
Share premium account		40	
Other reserves			
Revaluation reserve		35	
General reserve		15	
Opening profit and loss account	28		
Retained profit for year	38		
Closing profit and loss account		66	156
Total shareholders' funds			391

Note: Loan interest (£4) and auditors' fees (£2).

Chapter 8: Discussion *Answers*

The answers provide some outline points for discussion.

A1 Cash is king because it is central to the operations of a business. Unless you generate cash you cannot pay employees, suppliers or expenses, or buy new fixed assets. The end result of a lack of cash is the closure of a business. Cash is also objective. There is very little subjectivity involved in estimating cash. Either you have cash or your don't! With profits, however, there is much more subjectivity. Often one can alter the accounting policies of a business, for example, use a different rate of depreciation and thus alter the amount of profit. It is not as easy to manipulate cash.

A5 True or false?

(a) *False.* It is true that both items are non-cash flow items and that depreciation is added back to operating profit. However, profit from sale of fixed assets is deducted from operating profit.

(b) *False.* Stock and debtors are items of working capital. However, fixed assets are not.

(c) *True.*

(d) *True.*

(e) *False.* It is much more commonly used than the direct method.

Chapter 8: Numerical *Answers*

A1 Bingo

Included in profit and loss account	Included in cash flow statement
(a) Yes as part of charge for year	No
(b) No	Yes, financing
(c) Yes	No
(d) Yes	Yes, net cash outflow from operating activities
(e) No	Yes, capital expenditure and financial investment
(f) No	Yes, capital expenditure and financial investment
(g) No	Yes, returns on investments and servicing of finance
(h) Yes, part of charge for year	No
(i) No	Yes, financing
(j) Yes	Yes, returns on investments and servicing of finance.

2 Peter Piper

Peter Piper
Cash Flow Statement for Year Ended 31 December 2001

	£	£
Net Cash Inflow From Operating Activities		
Receipts from customers	250,000	
Payments to suppliers	(175,000)	
Payments to employees	(55,000)	
Expenses	(10,000)	10,000
Returns on Investments and Servicing of Finance		
Interest received	1,150	
Interest paid	(350)	800
Capital Expenditure and Financial Investment		
Sale of property	25,000	
Purchase of office equipment	(15,000)	10,000
Financing		
Loan repaid	(25,000)	(25,000)
Decrease in Cash		(4,200)

A4 D. Rink

D. Rink

Reconciliation of Operating Profit to Operating Cash Flow Year Ended 31 December 2001

	£	£
Operating Profit		95,000
Add:		
Decrease in stock	3,000	
Decrease in prepayments	1,500	
Increase in creditors	300	
Increase in accruals	250	
Depreciation	8,000	13,050
Deduct:		
Increase in debtors	(1,150)	
Profit on sale of fixed assets	(3,500)	(4,650)
Net Cash Inflow from Operating Activities		103,400

6 Grow Hire Ltd

Grow Hire Ltd

Reconciliation of Operating Profit to Operating Cash Flow Year Ended 31 December 2001

	£	£
Operating Profit (Note)		105,500
Add:		
Decrease in stock	2,000	
Decrease in debtors	7,000	
Increase in accruals	1,000	
Depreciation (£44,000 – £28,000)	16,000	26,000
Deduct:		
Decrease in creditors	25,000	25,000
Net Cash Inflow from Operating Activities		106,500

Note:

	£
Net Profit before Taxation	112,000
Add: Interest paid	6,500
Deduct: Interest received	(13,000)
Operating Profit	105,500

6 Grow Hire Ltd (continued)

Grow Hire Ltd
Cash Flow Statement for Year Ended 31 December 2001

	£	£
Net Cash Inflow from Operating Activities		106,500
Return on Investments and Servicing of Finance		
Interest received	13,000	
Interest paid	(6,500)	6,500
Taxation		
Taxation paid	(33,600)	(33,600)
Capital Expenditure and Financial Investment		
Patents purchased	(34,200)	
Land and buildings purchased	(20,000)	(54,200)
Equity Dividends Paid	(35,800)	(35,800)
Financing		
Increase in long-term creditors	12,000	
Increase in share capital	1,600	13,600
Increase in Cash		3,000

	£
Opening Cash	7,000
Increase in cash	3,000
Closing Cash	10,000

Chapter 9: Discussion *Answers*

The answers provide some outline points for discussion.

A1 Ratio analysis is simply the distillation of the figures in the accounts into certain key ratios so that a user can more easily interpret a company's performance. Ratio analysis is also known as financial statement analysis or the interpretation of accounts. There are traditionally thought to be six main types of ratios:

(a) *Profitability ratios*: Generally derived from the profit and loss account, they seek to determine how profitable the business has been. Main ratios: return on capital employed, gross profit, net profit.

(b) *Efficiency ratios*: Compare the profit and loss account to key balance sheet figures. Try to work out how efficiently the company is utilising its assets and liabilities. Main

ratios: debtor collection period, creditors collection period, stock turnover and asset turnover ratio.

(c) *Liquidity*: Assesses the short-term cash position of the company. They are derived from the balance sheet. Main ratios: current ratio and quick ratio.

(d) *Gearing*: Looks at the relationship between the owners' capital and the borrowed capital. This ratio is derived from the balance sheet.

(e) *Cash flow*: This ratio seeks to measure how the company's cash inflows and cash outflows compare. The cash flow ratio, unlike the other ratios, is derived from the cash flow statement.

(f) *Investment ratios*: These ratios are used by investors to determine how well their shares are performing. They generally compare share price with dividend or earnings information. The five main ratios are: dividend yield, dividend cover, earnings per share, price/earnings ratio and interest cover.

A5 True or false?

(a) *True.*

(b) *False.* More usually: $\text{Net profit} = \dfrac{\text{Net profit before taxation}}{\text{Sales}}$

(c) *False.* $\text{Current assets ratio} = \dfrac{\text{Current assets}}{\text{Current liabilities}}$. The ratio given was the quick ratio.

(d) *True.*

(e) *False.* This is actually the fixed assets turnover ratio.

$$\text{Asset turnover ratio} = \frac{\text{Sales}}{\text{Total assets}}$$

(f) *False.* $\text{Dividend yield} = \dfrac{\text{Dividend per ordinary share}}{\text{Share price}}$

(g) *True.*

Chapter 9: Numerical *Answers*

A1 John Parry

The ratios below are calculated in £s.

(a) $$\text{Return on capital employed} = \frac{\text{Net profit*}}{\text{Capital employed**}} = \frac{50{,}000}{(300{,}000 + 500{,}000) \div 2} = 12.5\%$$

*For sole traders, tax is not an issue.

**We take average for year.

A1 John Parry (continued)

(b) Gross profit ratio $= \dfrac{\text{Gross profit}}{\text{Sales}} = \dfrac{80{,}000}{150{,}000} = 53.3\%$

(c) Net profit ratio $= \dfrac{\text{Net profit*}}{\text{Sales}} = \dfrac{50{,}000}{150{,}000} = 33.3\%$

*For sole traders, tax is not an issue

(d) Debtors collection period $= \dfrac{\text{Average debtors}}{\text{Credit sales per day}}$

$$= \frac{(18{,}000 + 19{,}000) \div 2}{150{,}000 \div 365} = 45 \text{ days}$$

(e) Creditors collection period $= \dfrac{\text{Average creditors}}{\text{Credit purchases per day}}$

$$= \frac{(9{,}000 + 10{,}000) \div 2}{75{,}000 \div 365} = 46 \text{ days}$$

(f) Stock turnover ratio $= \dfrac{\text{Cost of sales}}{\text{Average stock}}$

$$= \frac{70{,}000}{(25{,}000 + 30{,}000) \div 2} = 2.5 \text{ times}$$

(g) Asset turnover ratio $= \dfrac{\text{Sales}}{\text{Average total assets}}$

$$= \frac{150{,}000}{(50{,}000 + 60{,}000) \div 2} = 2.7 \text{ times}$$

A2 Henry Mellet

£

(a) Current ratio $= \dfrac{\text{Current assets}}{\text{Current liabilities}} = \dfrac{31{,}903}{14{,}836} = 2.2 \text{ times}$

(b) Quick ratio $= \dfrac{\text{Current assets − stock}}{\text{Current liabilities}} = \dfrac{31{,}903 - 18{,}213}{14{,}836} = 0.9 \text{ times}$

(c) Gearing ratio $= \dfrac{\text{Long-term borrowings}}{\text{Total long-term capital*}} = \dfrac{30{,}000}{150{,}000 + 30{,}000} = 16.7\%$

*Remember, total net assets is equivalent to shareholders' funds.

A3 Jane Edwards

	Cash inflows	Cash outflows
	£	£
Customers	125,000	
Issue of shares	29,000	
Sale of fixed assets	35,000	
Employees		18,300
Suppliers		9,250
Buy back loan		8,000
Dividends		8,000
Taxation		16,000
Purchase of fixed assets		80,000
	189,000	139,550

$$\text{Cash flow ratio} = \frac{\text{Total cash inflows}}{\text{Total cash outflows}} = \frac{189{,}000}{139{,}550} = 1.35$$

A4 Clatworthy Plc

£000*

(a) $$\text{Dividend yield} = \frac{\text{Dividend per ordinary share}}{\text{Share price (in £s)}} = \frac{40 \div 500}{1.25} = 6.4\%$$

(b) $$\text{Dividend cover} = \frac{\text{Profit after tax and preference dividends}}{\text{Ordinary dividends}} = \frac{580}{40} = 14.5 \text{ times}$$

(c) $$\text{Earnings per share} = \frac{\text{Profit after tax and preference dividends}}{\text{Number of ordinary shares}} = \frac{580}{500} = £1.16$$

(d) $$\text{Price/earnings ratio} = \frac{\text{Share price}}{\text{Earnings per share}} = \frac{1.25}{1.16} = 1.08$$

* Except for price/earnings ratio.

(e) $$\text{Interest cover} = \frac{\text{Profit before tax and loan interest}}{\text{Loan interest}} = \frac{790}{40} = 19.8 \text{ times}$$

A7 Anteater plc

The ratios below are calculated from the accounts. There is insufficient information to average return on capital employed and the efficiency ratios. The closing year figure is, therefore, taken from the balance sheet. This is indicated by a single asterisk. Except where indicated, calculations are in £000s.

(a) Profitability ratios	Ratio	*Calculations (£000s)*		
Return on capital employed	Net profit before tax and loan interest/Average capital employed*	(100 + 3)/500* * = 300 + 20 + 10 + 70 + 100	=	20.6%
Gross profit ratio	Gross profit/Sales	250/1,000	=	25%
Net profit ratio	Net profit before tax/Sales	100/1,000	=	10%
(b) Efficiency ratios				
Debtors collection period	Average debtors*/Credit sales per day	50/(1,000 ÷ 365)	=	18.3 days
Creditors collection period	Average creditors*/Credit purchases per day (cost of sales taken)	40/(750 ÷ 365)	=	19.5 days
Stock turnover ratio	Cost of sales/Average stock*	750/40	=	18.8 times
Asset turnover ratio	Sales/Average total assets*	1,000/(420 + 120)	=	1.8 times
(c) Liquidity ratios				
Current ratio	Current assets/Current liabilities	120/40	=	3.0
Quick ratio	Current assets – stock/Current liabilities	(120 – 40)/40	=	2.0
(d) Gearing ratio				
Gearing	Long-term borrowings/Total long-term capital	(100 + 20)/(400 + 100)	=	24%
(e) Investment ratios				
Dividend yield	Dividend per ordinary share/Share price (in £s)	(40 ÷ 300)/2.00	=	6.7%
Dividend cover	Profit after tax and preference dividends/Ordinary dividends	70/40	=	1.8
Earnings per share	Profit after tax and preference dividends/Number of ordinary shares	70/300	=	23.3p
Price/earnings ratio +Calculation in pence.	Share price/Earnings per share+	200/23.3	=	8.6
Interest cover	Profit before tax and loan interest/Loan interest	103/3	=	34.3

Chapter 10: Discussion *Answers*

The answers provide some outline points for discussion.

A1 The role of the three is complementary.

(a) The directors run the business and prepare the accounts. They invest their labour and are rewarded, for example, by salaries and bonuses.

(b) The shareholders own the business and make decisions partly on the basis of the accounts they receive. They invest their capital and are rewarded, hopefully, by dividends and an increase in share price.

(c) The auditors check that the directors have prepared accounts that provide a 'true and fair' view of the company's performance over the year. They invest their labour and are rewarded by the auditors' fees. They are independent of the directors and, in theory, are responsible to the shareholders.

A2 The decision-making model is concerned with communicating economic information to users so that they can make decisions. Or more formally:

> 'The objective of financial statements is to provide information about the financial position, performance and changes in financial position of an enterprise that is useful to a wide range of users in making decisions' (*Framework for the Preparation and Presentation of Financial Statements*, International Accounting Standards Board, 1999, para. 12).

There are a variety of users. The International Accounting Standards Board, for example, identified shareholders, suppliers and other trade creditors, employees, customers, government and its agencies, and the general public. We can also add academics, management, analysts and advisers, and pressure groups. These users have a variety of different information needs, such as profitability (the shareholders) or future employment prospects (the employees). They also will want to make different decisions. For example, the shareholder will want to buy and sell shares whereas the employee may be wondering whether or not to change jobs. The financial information provided (i.e., income, expenses, assets, capital and liabilities) is normally provided in the form of the profit and loss account, balance sheet or cash flow statement.

This appears reasonable as far as it goes. However, critics argue that the decision-making model has several flaws. First, it assumes that one set of financial statements is appropriate for all users. Second, what is appropriate for shareholders is assumed to be appropriate for all users. Third, the focus on decision making neglects other important aspects such as stewardship. And, finally, and most radically, the decision-making model focuses on financial information, thus ignoring non-financial aspects such as the environment.

Chapter 11: Discussion *Answers*

The answers provide some outline points for discussion.

A1 An accounting measurement system is a method of determining the monetary amounts in the accounts. Different measurement systems will result in different valuations for net assets. This in turn will cause profit to be different. The measurement of assets and the determination of profit is key to the preparation of accounts. However, different accounting measurement systems can cause wide variation in both net assets and net profit. Thus, it is probably not unreasonable to call accounting measurement systems the skeleton that underpins the accounting body. The accounting measurement systems will determine the basic parameters of the accounting results.

A2 Historical cost is still widely used internationally. Indeed, it is certainly much more popular than the alternative measurement bases. However, historical cost has been widely criticised. In particular, there is concern that it fails to reflect changing asset values resulting from inflation or technological change. There has, therefore, been extensive debate over alternative measurement systems, most obviously by the Financial Accounting Standards Board in the US, the Accounting Standards Board in the UK and the International Accounting Standards Board. All these bodies have wrestled with alternative measurement systems, seeing them as essential to framing a successful conceptual theory for accounting.

So far, there has been little agreement on an alternative system. The first reason is that historical cost is widely used, easy to understand and objective. Users are, therefore, reluctant to abandon it. This is particularly true when the alternatives are often not easy to understand or use and are often subjective. In addition, the main incentive to change was the high level of inflation in the Western world, particularly in the UK and the US in the 1970s. However, more recently inflation rates have fallen and with them interest in alternatives to historical cost.

Chapter 12: Discussion *Answers*

The answers provide some outline points for discussion.

A1 The annual report plays a central role in corporate governance. Essentially, it is a key mechanism by which the directors report to the shareholders and other users on their stewardship of the company. The directors prepare the financial statements, which provide information on income and expenses, assets, liabilities and capital, so that shareholders can monitor the activities of the directors. This monitoring allows the shareholders to see that the managers are not abusing their position, for example, by paying themselves great salaries at the expense of the shareholders. In order to ensure that the accounts are true and fair, the auditors audit the financial statements. These financial statements will be included in an annual report along with other financial and non-financial information. The annual report itself is often prepared by design consultants.

A2 These three roles have the following functions

(i) *Stewardship and accountability*
The directors provide information to the shareholders so that they can monitor the directors' activities. This has grown out of ideas of accountability (i.e., the control and safeguarding of corporate assets). The idea has been extended to corporate governance, in particular, the monitoring of directors' remuneration.

(ii) *Decision making*
The decision-making aim is 'to communicate economic measures of, and information about, the resources and performance of the reporting entity useful to those having reasonable rights to such information' (*The Corporate Report*, 1975). The aim, therefore, in essence, is to provide users, such as shareholders, with information so that they can make decisions, such as buying or selling shares.

(iii) *Public relations*
The public relations objective of the annual report is simply recognising the incentives that management has to use the annual report to show the results in a good light. It is often associated with the idea of the annual report as a marketing tool.

The clash between these three objectives is that while the first two (accountability and decision making) are based on the idea of a true and fair view, public relations is not about truth and fairness. Public relations attempts to depart from truth and fairness to show the company in the best possible light.

Chapter 13: Discussion *Answers*

The answers provide some outline points for discussion.

A1 Creative accounting is using the flexibility within accounting to manage the measurement and presentation of the accounts so that they serve the interests of the managers. These interests may be to smooth profits, to increase profits or reduce gearing. The essential point of creative accounting is that it serves the interests of the preparers not the users. In particular, creative accounting may clash with the basic requirement that the financial statements should give a true and fair view of the company's financial position and financial performance. This is because creative accounting puts the interests of preparers first. The regulatory framework specifies that accounting should be unbiased and neutral and should faithfully represent the economic reality of the company. Accounting cannot do this if it is managed creatively for the benefit of the preparers.

A2 Creative accounting can exist, and indeed thrive, particularly in countries such as the UK where accounting is valued for its flexibility. The basic idea is that accounting professionals should be free to choose accounting policies which give a true and fair view. In other words, accounting should not be governed by a set of inflexible and constraining rules, as are found in some other countries, such as Germany.

However, the flexibility that allows accounting to present a 'true and fair view' is also the same flexibility that enables accounting to be used creatively. Indeed, it is all a matter of degree and intention. Creative accounting is when accounting policies are chosen to serve the managers' interests rather than to give a 'true and fair' view. However, in reality, it is extremely difficult for outsiders to distinguish between the two. This is especially true as there is generally more than one possible 'true and fair' view.

Chapter 14: Discussion *Answers*

The answers provide some outline points for discussion. However, they should not be taken as exhaustive or prescriptive. International accounting is a topic that is very fluid and open to interpretation.

A2 There are potentially eight divergent forces: objectives, users, sources of finance, regulation, taxation, the accounting profession, spheres of influence and culture. In their own way, they are all important. They are also interrelated. It is, therefore, very difficult to sort out the most important. Different individuals will legitimately have different views.

In my opinion, the most important divergent forces are the sources of finance, regulation, taxation and the accounting profession. This is because:

- Sources of finance provide the basic funding for industry (either debt or equity finance); they thus orientate the accounting system towards a balance sheet or a profit and loss account focus.
- Regulation determines the backbone of the accounting system, setting out the detailed rules and regulations.
- Taxation is one of the main drivers of financial accounting in some countries, such as Germany and France. Companies comply with taxation requirements which then dictate financial accounting practices.
- The accounting profession because qualified accountants provide the judgement and interpretation in countries such as the UK and US, which enable the achievement of a 'true and fair' view.

However, it must be stressed that this is just my opinion. Other viewpoints are equally justifiable.

A3 These two systems are distinguishable in many ways. In essence, the Anglo-American system is followed by countries such as the US and the UK, and is often termed a 'micro' system. By contrast the continental accounting system is followed by countries such as France and Germany and is often termed a 'macro' system. The main differences are as follows:

- **Objectives**. The Anglo-American system favours the true and fair/present fairly objective of financial reporting designed to show economic reality. The continental system is more about planning and control.
- **Users**. In the Anglo-American model, these are primarily shareholders; in the continental model, they are the government, the banks, and the tax authorities.

- **Sources of Finance**. In the Anglo-American model, shareholders provide risk capital; whereas in the continental model, banks provide loan capital.
- **Regulation**. In the Anglo-American model, standards are very important. In addition, in the UK, the Companies Acts provide detailed regulation, while in the US listed companies supply a detailed 10-K to the Securities Exchange Commission (an independent regulatory commission). In the continental model, detailed regulation comes from the government.
- **Taxation**. In the Anglo-American model, taxation is not an important driver of the financial accounts. In the continental model, taxation drives financial accounting.
- **Accounting Profession**. This is strong and influential in the Anglo-American model, comparatively weak and uninfluential in the continental model.

Chapter 15: Discussion *Answers*

The answers provide some outline points for discussion.

A1

Branches	*Functions*
(1) Cost Accounting	
(i) Costing	To recover costs as a basis for pricing and for stock valuation.
(ii) Planning and Control	
(a) Budgeting	To plan and control future costs through budgeting.
(b) Standard costing	To plan and control future costs through standard costing.
(2) Decision Making	
(i) Short-term Decisions	To make short-term decisions (such as whether to make a product) using techniques such as contribution analysis and break-even analysis. In addition, to minimise the cost of working capital.
(ii) Long-term Decisions	
(a) Strategic management	To make decisions of a strategic nature such as whether or not to diversify the business.
(b) Capital budgeting	To make decisions about whether or not to invest in long-term projects such as a new product or service.
(c) Sources of finance	To make decisions about whether to raise new finance via share or loan capital.

A6 True or False?

(a) *True.*

(b) *False.* Total absorption costing is where all the costs incurred by a company are totalled so that they can be recovered into the product or service's final selling price.

(c) *True.*

(d) *False.* Strategic management accounting is concerned with the long-term strategic direction of a firm.

(e) *True.*

Chapter 16: Discussion *Answers*

The answers provide some outline points for discussion.

A1 Costing is the process of recording, classifying, allocating and then absorbing costs into individual products and services. Costing is particularly important for two main reasons: stock valuation and pricing. In stock valuation, the aim is to recover the costs incurred in producing a good. The costs of stock valuation thus include direct materials, direct labour, direct expenses and appropriate production overheads. In absorption costing, we include all production overheads, both fixed and variable. In marginal costing, we include only variable production overheads. In effect, financial accounting drives the stock valuation process. This is because external reporting regulations allow attributable production overheads (i.e., allow absorption costing) to be included in stock. For pricing, all the overheads (i.e., total absorption costing) are included in the cost before a percentage is added for profit. Thus the total cost may include direct materials, direct labour, direct expenses, production and non-production overheads. Nowadays, with the decline of manufacturing industry, there has been a decline in the importance of direct costs and an increase in the importance of indirect costs or overheads.

A5 True or False?

(a) *False.* It is true that a cost can be an actual past expenditure. However, it is important to appreciate that a cost can also be an estimated, future expenditure.

(b) *True.*

(c) *False.* We identify activity cost drivers in activity-based costing.

(d) *True.*

(e) *False.* No, in financial reporting we use absorption costing.

Chapter 16: Numerical *Answers*

A1 *Sorter*

(i) Direct materials	*c* (purchase of raw materials)
(ii) Direct labour	*a* (machine workers' wages)
(iii) Production overheads	*b* (cost clerk's wages)*, *d* (machine repairs), *m* (depreciation on machinery), *o* (electricity for machines)

*Assumes cost clerks are based in factory.

(iv) Administrative expenses	*e* (finance director's salary), *f* (office cleaners), *h* (managing director's car expenses), *i* (depreciation on office furniture), *j* (computer running expenses for office), *k* (loan interest), *l* (auditors' fees), *p* (bank charges)
(v) Selling and distribution costs	*g* (delivery van staff's wages), *n* (advertising costs)

A2 Costa

Costa
Costing Statement for Costa

	£	£	£
Direct materials			320,000
Direct labour			200,000
Royalties			3,600
(i) Prime Cost			523,600
Production Overheads			
Factory supervisors' wages		120,000	
Depreciation (£8,000 + £5,000)		13,000	
Computer overheads		6,000	
Other overheads		70,000	209,000
(ii) Production Cost			732,600
Other Costs			
Administrative Expenses			
Administrative salaries	90,800		
Depreciation (£4,200 + £2,500)	6,700		
Computer overheads	3,000		
Interest on loans	3,000	103,500	
Selling and Distribution Costs			
Wages	18,300		
Marketing salaries	25,000		
Commission	1,200		
Depreciation (£3,500 + £2,500)	6,000	50,500	154,000
(iii) Total Cost			886,600

A3 Makemore

Makemore
Overhead Allocation Statement

	Total	Ratio Split	A	B	C
	£		£	£	£
Supervisors' salaries	25,000	1,000:2,000:500	7,143	14,286	3,571
Computer advisory	18,000	1,000:2,000:500	5,143	10,286	2,571
Rent and business rates	20,000	10,000:6,000:4,000	10,000	6,000	4,000
Depreciation	21,000	30,000:15,000	14,000	7,000	–
Repairs	4,000		2,800	1,100	100
Allocated	88,000		39,086	38,672	10,242
Reallocation of service support department C's costs			60%	40%	(100%)
			6,145	4,097	(10,242)
Total Allocation	88,000		45,231	42,769	–
			80,000	40,000	
(i) Labour hours Rate per hour			£0.57	£1.07	
(ii) Machine hours			100,000	200,000	
Rate per hour			£0.45	£0.21	

A8 Serveco

Calculate activity-cost driver rates

Activity	*Spare Parts Installed*	*Technical Support*	*Service Documentation*
Cost	£100,000	£125,000	£300,000
Cost driver	150,000 parts	175,000 minutes	125,000 units
Cost per unit of Cost driver	£0.667	£0.7143	£2.4

Costs absorbed to services				*Total cost*
Basic	50,000 × £0.667 = £33,333	75,000 × £0.7143 = £53,572	100,000 × £2.4 = £240,000	£326,905
Enhanced	100,000 × £0.667 = £66,667	100,000 × £0.7143 = £71,428	25,000 × £2.4 = £60,000	£198,095
Total costs	£100,000	£125,000	£300,000	£525,000

Total overhead costs	*Total overheads*	*Call outs*	*Overheads per call out*
Basic	£326,905	50,000	£ 6.54
Enhanced	£198,095	10,000	£19.81

Call Out Cost/Charge

	Basic		***Enhanced***	
	£		£	
Labour hours (including travelling time)	20.00	$\frac{(25{,}000+25{,}000) \times £20}{50{,}000 \text{ call outs}}$	106.25	$\frac{(37{,}500+5{,}000) \times £25}{10{,}000 \text{ call outs}}$
Overheads	6.54		19.81	
Total Cost	26.54		126.06	
Profit: 25% mark-up on cost	6.63		31.51	
Call out Charge	33.17		157.57	

Note: Some of the calculations have been rounded in this answer.

Chapter 17: Discussion *Answers*

The answers provide some outline points for discussion.

A1 A major advantage of budgeting is that it requires you to predict the future in economic terms. You can plan ahead and buy in extra resources and schedule workloads. In addition, budgeting enables you to predict your future performance. You can then tell how well or how badly you are actually doing against your predicted budget. Finally, budgeting has an important responsibility and control function. You can put somebody in charge of the budget, make them responsible and so control future activities. Variances from budget can thus be investigated.

The major disadvantages of budgets are that they may be constraining and dysfunctional to the organisation. The constraint is caused by the budget perhaps setting a straitjacket which inhibits innovation and the adoption of flexible business policies. The dysfunctional nature of budgeting is that the objectives of the individual may not necessarily be the same as that of the organisation. Individuals may, therefore, seek to 'pad' their budgets or manage them to their own personal advantage.

4 True or False?

(a) *True.*

(b) *False.* Although sometimes production is a limiting factor, the commonest limiting factor is sales.

(c) *True.*

(d) *False.* Depreciation is never found in a cash budget as it does not represent a cash flow.

(e) *True.*

Chapter 17: Numerical *Answers*

A1 Jill Lee

Jill Lee
Cash Budget for Six Months Ending June

	Jan.	Feb.	March	April	May	June	Total
	£	£	£	£	£	£	£
Opening cash	15,000	28,200	21,100	8,280	(5,942)	1,389	15,000
Add *Receipts*							
Sales	25,200	27,100	21,200	20,250	48,300	37,500	179,550
	25,200	27,100	21,200	20,250	48,300	37,500	179,550
Less Payments							
Goods	–	21,000	19,500	18,500	23,400	25,900	108,300
Expenses	12,000	13,200	14,520	15,972	17,569	19,326	92,587
	12,000	34,200	34,020	34,472	40,969	45,226	200,887
Cash flow	*13,200*	*(7,100)*	*(12,820)*	*(14,222)*	*7,331*	*(7,726)*	*(21,337)*
Closing cash	28,200	21,100	8,280	(5,942)	1,389	(6,337)	(6,337)

A3 Fly-by-Night

Fly-by-Night
Sales Budget for Six Months Ending June

	Jan.	Feb.	March	April	May	June	Total
	£	£	£	£	£	£	£
Moon (1)	20,000	21,000	22,000	28,750	30,000	31,250	153,000
Star (2)	20,000	21,000	22,000	23,000	24,000	25,000	135,000
	40,000	42,000	44,000	51,750	54,000	56,250	288,000
(1) Moon (units)	1,000	1,050	1,100	1,150	1,200	1,250	6,750
(2) Star (units)	2,000	2,100	2,200	2,300	2,400	2,500	13,500

Helpnote: Multiply the units by the price per unit.

A4 D. Ingo

D. Ingo
Debtors Budget for Six Months Ending June

	Jan.	Feb.	March	April	May	June	Total
	£	£	£	£	£	£	£
Opening debtors	2,400	2,000	2,100	2,310	2,541	2,795	2,400
Credit sales	1,000	1,100	1,210	1,331	1,464	1,610	7,715
	3,400	3,100	3,310	3,641	4,005	4,405	10,115
Cash received	(1,400)	(1,000)	(1,000)	(1,100)	(1,210)	(1,331)	(7,041)
Closing debtors	2,000	2,100	2,310	2,541	2,795	3,074	3,074

A6 B. Ear

B. Ear
Production Cost Budget Six Months Ending June

	Jan.	Feb.	March	April	May	June	Total
	£	£	£	£	£	£	£
Raw materials	3,500	3,750	4,000	4,250	4,500	4,750	24,750
Direct labour	3,850	4,125	4,400	4,675	4,950	5,225	27,225
Variable overheads	1,400	1,500	1,600	1,700	1,800	1,900	9,900
	8,750	9,375	10,000	10,625	11,250	11,875	61,875
Units	700	750	800	850	900	950	4,950

A7 R. Abbit

R. Abbit
Raw Materials Budget Six Months Ending June

	Jan.	Feb.	March	April	May	June	Total
	£	£	£	£	£	£	£
Opening stock	1,000	940	1,080	1,420	1,670	2,120	1,000
Purchases	900	1,100	1,300	1,500	1,700	1,900	8,400
	1,900	2,040	2,380	2,920	3,370	4,020	9,400
Used in production	(960)	(960)	(960)	(1,250)	(1,250)	(1,250)	(6,630)
Closing stock	940	1,080	1,420	1,670	2,120	2,770	2,770

Chapter 18: Discussion Answers

The answers provide some outline points for discussion.

A1 Setting the standards involves first of all making a decision about whether one is aiming for ideal, attainable or normal standards. Ideal standards are those that can be reached if everything goes perfectly. Attainable standards are those that can be reached with a little effort. Normal standards are those that are based on past experience and that the business normally meets. Attainable standards are probably the best because they include a motivational element.

The two main elements of a standard are quantity (i.e., hours, materials) and price (i.e., hourly rate, price per kilo). It follows, therefore, that we need to consider the factors that influence quantity and price. These might be based on past experience, prevailing market conditions and future expectations. For instance, if a firm were setting labour standards it might base its labour quantity standard on the hours taken in the past less an improvement element. The labour price standard might be based on the prevailing wage rate with an amount built in for any wage increases.

A5 True or False?

(a) *True.*

(b) *False.* Flexing the budget means adjusting the budget to take into account the actual quantity produced.

(c) *True.*

(d) *True.*

(e) *False.* The direct labour quantity variance is: (standard *quantity* of labour hours for actual production – actual *quantity* of labour hours used) × standard labour price per hour.

Chapter 18: Numerical *Answers*

A1 Stuffed
(i–iii)

	Budget	*Flexed Budget*	*Actual*	*Sales Price and Overall Cost Variances**	
Number of customers	10,000	12,000	12,000		
	£	£	£	£	
Sales	100,000	120,000	127,200	7,200	Fav.
Food cost (i.e., materials cost)	(30,000)	(36,000)	(37,200)	(1,200)	Unfav.
Labour cost	(35,000)	(42,000)	(36,000)	6,000	Fav.
Variable overheads	(5,000)	(6,000)	(6,000)	–	
Contribution	30,000	36,000	48,000	12,000	Fav.
Fixed overheads	(3,000)	(3,000)**	(3,100)	(100)	Unfav.
Profit	27,000	33,000	44,900	11,900	Fav.

Helpnotes

*The sales price variance is £7,200 Fav. The overall cost variances are food cost (i.e., direct materials) variance £1,200 Unfav., labour cost variance £6,000 Fav., zero overall variable overheads cost variance, and £100 Unfav. fixed overhead variance.

**Remember, fixed costs remain unchanged whatever the level of activity. We do not, therefore, flex these.

(iv) *Sales Quantity Variance* = (standard quantity of meals sold – actual quantity of meals sold) × standard contribution per unit = 10,000 – 12,000 × (£30,000 ÷ 10,000) = £6,000 Fav. Note this is simply the flexed profit (£33,000) less the original budget (£27,000) = £6,000 Fav.

(v) (a) *Sales Variances*. Both are favourable. 2,000 more customers visited the restaurant than anticipated. They paid £0.60 more per meal than expected.

(b) *Material Cost*. We paid £1,200 more than expected. This may be due to increased prices or increased quantity used. We need more information on this.

(c) *Labour Cost*. We paid £6,000 less than expected. Either we paid less per hour or we used fewer hours than expected. We need more information.

(d) *Variable Overheads*. These were as budgeted.

(e) *Fixed Overheads*. These were slightly more (£100 more) than expected.

A3 Birch Manufacturing

	£	
(i) Overall Direct Materials Variance		
Standard cost of materials for actual production		
(10 metres of wood* at 50p × 11,000)	55,000	
Actual cost of materials used in production		
(120,000 metres × 0.49 pence)	58,800	
	(3,800)	Unfav.

*Each bookcase is estimated to take 10 metres of wood (100,000 metres ÷ 10,000 bookcases)

	£	
(ii) Direct Materials Price Variance		
(standard price per unit of material – actual price per unit of material) × actual quantity of materials used		
= (50p – 49p) × 120,000 metres	1,200	Fav.
(iii) Direct Materials Quantity Variance		
(standard quantity of materials for actual production – actual quantity of materials used) × standard material price per unit		
= (11,000 × 10 metres – 120,000 metres) × 0.50p	(5,000)	Unfav.
	(3,800)	Unfav.

A4 Sweatshop

	£	
(i) Overall Direct Labour Variance		
Standard cost of labour for actual production		
(500 sweatshirts at 2 hours) × £3.50	3,500	
Actual cost of labour used in production	3,780	
	(280)	Unfav.

	£	
(ii) Direct Labour Price Variance		
(standard labour price per hour – actual price per hour) × actual quantity of labour used		
= (£3.50 – £3.60*) × 1,050 hours	(105)	Unfav.
*£3,780 labour cost divided by 1,050 hours		
(iii) Direct Labour Quantity Variance		
(Standard quantity of labour hours for actual production – actual quantity of labour hours used) × standard labour price per hour		
= (2 hours × 500 sweatshirts – 1,050 hours) × £3.50 per hour	(175)	Unfav.
	(280)	Unfav.

A6 Wonderworld

(i) Overall Variable Overheads Cost Variance		
Standard cost of variable overheads for actual production	£	
(110,000 teleporters at 2 labour hours at £2.50)	550,000	
Actual cost of variable overheads for production	517,500	
	32,500	Fav.

(ii) Variable Overheads Price Variance

(standard variable overheads price per hour – actual variable overheads price per hour) × actual quantity of labour hours used	£	
= (£2.50 – £2.25)* × 230,000	57,500	Fav.

*variable overheads £517,500 ÷ 230,000 actual labour hours

(iii) Variable Overheads Quantity Variance

(standard quantity of labour hours for actual production – actual quantity of labour hours used) × standard variable overheads price per hour		
= (110,000 × 2 hours – 230,000 hours) × £2.50	(25,000)	Unfav.
	32,500	Fav.

(iv) Fixed Overheads Variance

Standard fixed overheads less actual fixed overheads	£	
= £10,000 – £9,800	200	Fav.

A8 Peter Peacock plc

August's results

(i) Flexed budget	*Budget (i.e., standard)*	*Flexed Budget (i.e. standard quantity of actual production)*	*Actual*	*Sales Price and Overall Cost Variances*	
Volume	200,000	220,000	220,000		
	£	£	£	£	
Sales	560,000	616,000	611,600	(4,400)	Unfav.
Direct materials	(125,000)	(137,500)	(150,000)	(12,500)	Unfav.
Direct labour	(290,000)	(319,000)	(317,550)	1,450	Fav.
Variable overheads	(40,000)	(44,000)	(42,500)	1,500	Fav.
Contribution	105,000	115,500	101,550	(13,950)	Unfav.
Fixed overheads	(68,000)	(68,000)	(67,000)	1,000	Fav.
Profit	37,000	47,500	34,550	(12,950)	Unfav.

(ii) Individual Variances: Sales

Sales Quantity Variance

(Standard quantity of units sold – actual quantity of units sold) × standard contribution

£

(200,000 – 220,000) × 0.525* = 10,500 Fav.**

$$\frac{\text{*Contribution}}{\text{Budgeted sales volume}} = \frac{105{,}000}{200{,}000}$$

**Represents budgeted profit £37,000 – flexed budget profit £47,500

Sales Price Variance

This can be taken direct from the flexed budget, or it can be calculated as follows:
(standard selling price – actual selling price per unit) × actual quantity of units sold
(£2.80 – £2.78) × 220,000 = (£4,400) Unfav.

Costs

	Price	Quantity
	(standard price – actual price per unit) × actual quantity	*(standard quantity of actual production – actual quantity) × standard price*
Direct Material Variances	(£1.25 per sheet – £1.20 per sheet) × 125,000 sheets= £6,250 Fav.	(220,000 subcomponents × 0.50 sheets per subcomponent* gives 110,000 sheets – 125,000 sheets) × £1.25 = (£18,750) Unfav. *100,000 sheets ÷ 200,000 subcomponents. This gives the amount of material per subcomponent.
Direct Labour Variances	(£7.25 per hour – £7.30 per hour) × 43,500 hours = (£2,175) Unfav.	(220,000 subcomponents × 0.20 hours* gives 44,000 hours – 43,500 hours) × £7.25 = £3,625 Fav. *40,000 hours ÷ 200,000 subcomponents. This gives the amount of the labour per subcomponent
Variable Overhead Variances	(£1.00 per hour – £0.977 per hour)* × 43,500 hours = £1,000 Fav. *£42,500 variable overheads ÷ 43,500 hours	(220,000 subcomponents × 0.20 hours* gives 44,000 hours – 43,500 hours) × £1.00 = £500 Fav. *40,000 hours ÷ 200,000 subcomponents. This gives the amount of labour per subcomponent and we are recovering our variable overheads on the labour hours.
Fixed Overheads		£68,000 standard fixed overheads – £67,000 actual fixed overheads = £1,000 Fav.

(iii) **Peter Peacock plc**

Standard Cost Reconciliation Statement for August

	£	£	£	
Budgeted Profit			37,000	
Sales quantity variance			10,500	Fav.
Budgeted profit at actual sales			47,500	
Variances	*Fav.*	*Unfav.*		
Sales price		4,400		
Direct materials price	6,250			
Direct materials quantity		18,750		
Direct labour price		2,175		
Direct labour quantity	3,625			
Variable overheads price	1,000			
Variable overheads quantity	500			
Fixed overheads variance	1,000			
	12,375	25,325	(12,950)	
Actual Profit			34,550	

(iv) The actual profit for Peacock is £2,450 less than budgeted (£37,000 – £34,550). Peacock has actually sold 20,000 more units than anticipated, creating a favourable sales quantity variance of £10,500. However, it has done this by reducing the price slightly so there is an unfavourable sales price variance.

On the cost variances, there is a favourable direct materials price variance (£6,250) as the sheets are cheaper than anticipated. However, more sheets were used than anticipated, possibly because they were poorer quality. There is thus an unfavourable quantity variance of £18,750. The labour price variance of £2,175 is unfavourable since Peacock paid £7.30 per hour rather than the budgeted £7.25. However, perhaps because a better quality of labour was used, less hours were used creating a favourable quantity variance of £3,625. For variable overheads, less overheads than anticipated were incurred creating a favourable price variance of £1,000. Also because of the fewer hours used, less overheads were recovered into the product causing a favourable quantity variance. Finally, fixed overheads were less than anticipated.

Chapter 19: Discussion *Answers*

The answers provide some outline points for discussion.

A1 Fixed costs are those costs, like depreciation or insurance, which do not vary with production or sales. They remain fixed whatever the level of production or sales. This is not universally true as at a certain point, such as acquiring a new machine, fixed costs will vary. However, it is a reasonable working assumption.

Variable costs, by contrast, are those costs that do vary with production or sales. If we make more products or provide more services, then our variable costs will increase. Conversely, if we make fewer products or provide fewer services, then our variable costs will decrease.

Fixed costs are irrelevant for decision making because they will be incurred whatever the decision. They are fixed within the relevant range of activity. For example, insurance and depreciation will not vary whether we choose to produce more of product A or more of product B. We should, therefore, ignore these costs when making decisions.

A4 True or False?

(a) *False.* Wrong time horizon. Fixed costs do *not* vary with short-term changes in the level of sales or production.

(b) *True.*

(c) *False.* Wrong numerator. Break-even point is $\frac{\text{Fixed costs}}{\text{Contribution per unit}}$

(d) *False.* Contribution/sales ratio is $\frac{\text{Contribution}}{\text{Sales}}$

(e) *False.* Non-financial items do not feature directly in the calculations, but they are extremely important.

Chapter 19: Numerical *Answers*

A1 Jungle Animals

(i), (ii)

	Selling Price	*Variable Costs*	*Contribution*	*Contribution/ Sales Ratio*
	£	£	£	%
Alligators	1.00	1.05	(0.05)	(5.0)
Bears	1.20	1.00	0.20	16.7
Cougars	1.10	1.14	(0.04)	(3.6)
Donkeys	1.15	1.08	0.07	6.1
Eagles	1.20	0.96	0.24	20.0
Foxes	0.90	0.85	0.05	5.6
Giraffes	1.05	0.85	0.20	19.0
Hyenas	1.25	0.94	0.31	24.8
Iguanas	0.95	0.72	0.23	24.2
Jackals	0.80	0.73	0.07	8.8

(iii) The two toys with the highest contribution are Hyenas (£0.31 contribution) and Eagles (£0.24 contribution).

(iv) The three toys with the highest contribution/sales ratio are Hyenas (24.8%), Iguanas (24.2%) and Eagles (20%).

(v) Alligators and Cougars have a negative contribution so we would not make them.

A3 Scrooge

Internal bid:	£
Clerical labour	50,000
Supervisory labour	60,000
Direct materials	25,000
Variable overheads	30,000
	165,000
External bid	(160,000)
Thus saving by buying in	5,000

So on the straight accounting calculation Scrooge would outsource. The chief assumption is that all the labour is indeed variable and can be laid off or redeployed easily. Other factors are the impact upon industrial relations, long-term implications and confidentiality. The outside bid is marginally superior. However, when these other factors are taken into account it may be better to go with the status quo.

A6 Freya

		£	£
(i)	Contribution per hammer		
	Sales		10
	Less: *Variable Costs*		
	Direct materials	4	
	Variable expenses	3	7
	Contribution		3

Break-even point: $\frac{\text{Fixed costs}}{\text{Contribution per unit}} = \frac{£30,000}{£3}$ = 10,000 hammers

(ii) (a) If 4,000 sold	Contribution (4,000 × £3) =	£12,000
	Fixed costs	(£30,000)
	Loss	(£18,000)
(b) If 14,000 sold	Contribution (14,000 × £3) =	£42,000
	Fixed costs	(£30,000)
	Profit	£12,000

A6 Freya (continued)

(iii) Current margin of safety

(a) Units (i.e. Hammers): $\dfrac{\text{Actual hammers sold} - \text{hammers at break-even}}{\text{Hammers at break-even}}$

$$= \frac{20{,}000 - 10{,}000}{10{,}000} = 100\%$$

(b) £s: $\dfrac{\text{Actual sales} - \text{sales at break-even}}{\text{Sales at break-even}}$

$$= \frac{200{,}000 - 100{,}000}{100{,}000} = 100\%$$

(iv) Break-even chart

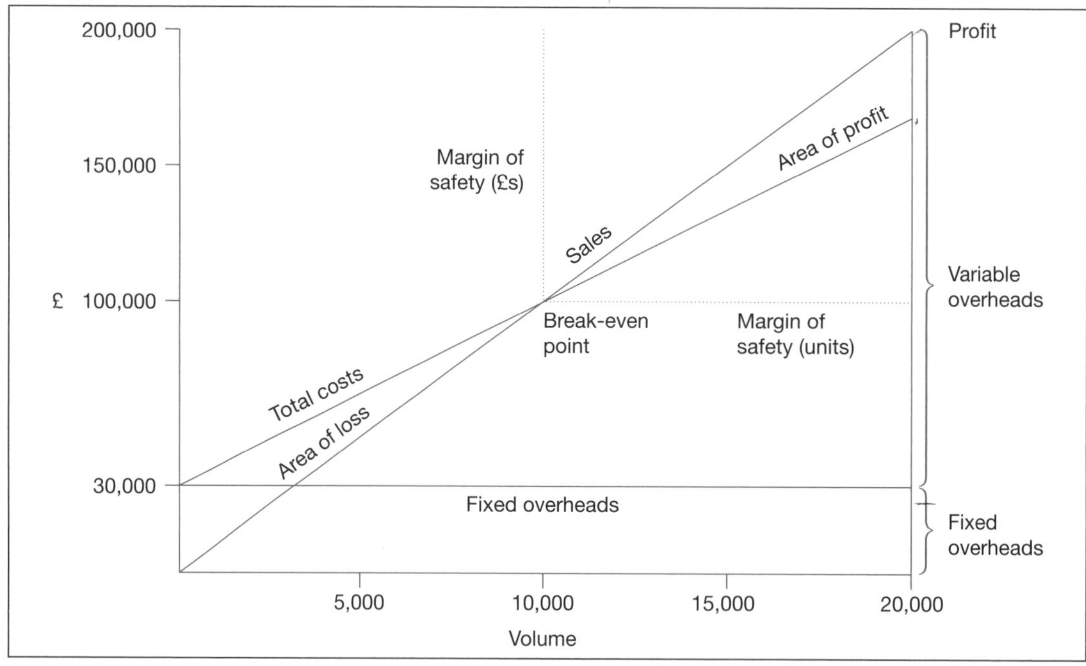

A8 Modem

(i)

Branch	Sales	Variable Costs	Contribution	Contribution/ Sales Ratio	Ranking
	£	£	£	%	
Cardiff	200,000	165,000	35,000	17.5 (£35,000/£200,000)	2
Edinburgh	300,000	230,000	70,000	23.3 (£70,000/£300,000)	1
London	1,000,000	870,000	130,000	13.0 (£130,000/£1,000,000)	3
	1,500,000	1,265,000	235,000		
Fixed costs			(150,000)		
Net Profit			85,000		

A8 Modem (continued)

(ii) Cumulative profit table in contribution/sales ratio ranking:

	Cumulative Sales	*Cumulative Contribution*	*Cumulative Profit/Loss*
	£	£	£
Fixed costs			(150,000)
Edinburgh	300,000	70,000	(80,000)
Cardiff	500,000	105,000	(45,000)
London	1,500,000	235,000	85,000

Modem's Contribution/Sales Graph

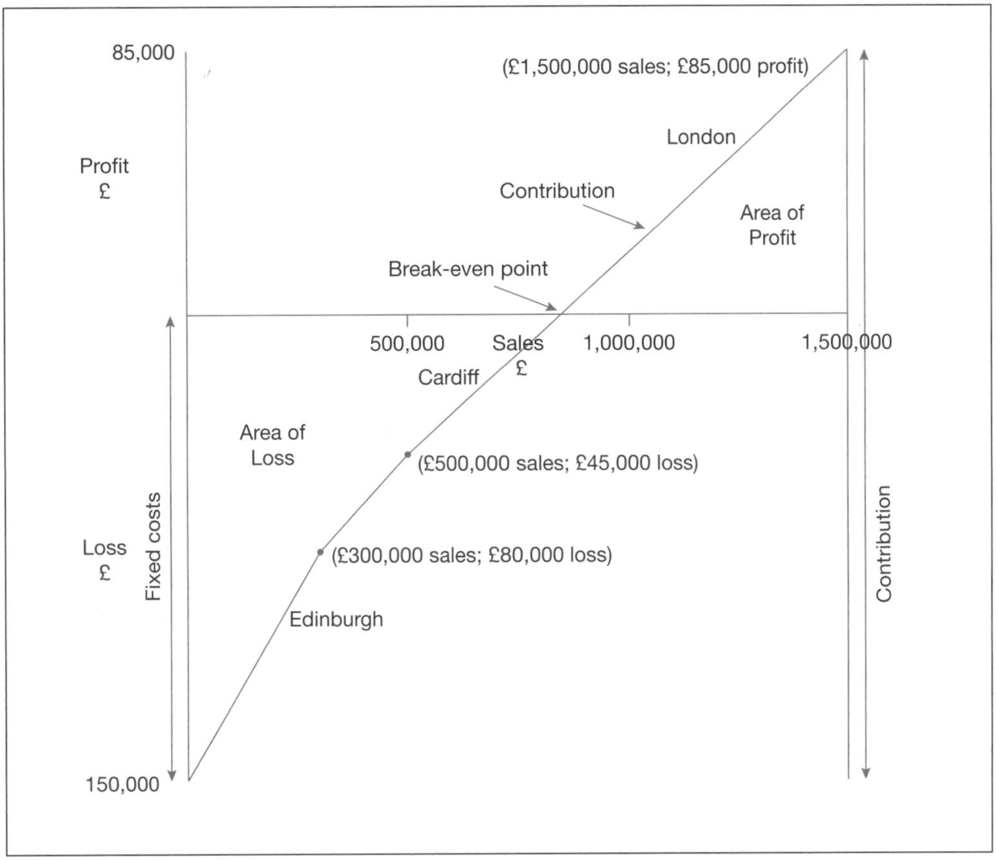

85,000
(£1,500,000 sales; £85,000 profit)
Profit
£
London
Contribution
Area of
Profit
Break-even point
500,000
Sales
£
1,000,000
1,500,000
Cardiff
Area of
Loss
(£500,000 sales; £45,000 loss)
(£300,000 sales; £80,000 loss)
Loss
£
Fixed costs
Contribution
Edinburgh
150,000

Chapter 20: Discussion *Answers*

The answers provide some outline points for discussion.

A1 Strategic management accounting is a relatively new branch of management accounting. It is concerned with a business's strategic future, long-term direction. A working definition of strategic management accounting is a form of management accounting which considers both an organisation's internal and external environments. Strategic management accounting comprises three main stages. First, an assessment of the current position of a business, which seeks to establish the internal and external elements of the business's environment. Second, an appraisal of the current position of the business. This establishes the current strengths and weaknesses. Finally, strategic management accounting looks at the possible strategic choices of the business, such as product diversification or mergers/acquisitions.

Strategic management accounting grows out of a concern with the traditional role of the management accountant. Traditionally, management accounting has been criticised as inward-looking and failing to respond to the external environment. Strategic management accounting attempts to lift management accountants out of this, traditionally fairly narrow, functional specialism. There is an emphasis on long-term strategic thinking embracing the whole business.

A5 True or False?

(a) *True.*
(b) *True.*
(c) *False.* No, the four components are stars, question marks, cash cows and dogs.
(d) *False.* The balanced scorecard uses a mixture of financial and non-financial performance indicators.
(e) *False.* Almost right. SWOT analysis stands for strengths, weaknesses, *opportunities* and threats.

Chapter 20: Numerical *Answers*

A1 Maxi, Mini, Midi

In year 3	Maxi – high sales, high profitability, thus	**Maturity**
	Mini – sales declining, profits declining thus	**Decline**
	Midi – sales growing fast, profitability growing, thus	**Growth**

A2 Apple, Orange, Pear and Banana

Apple:	Star (high market share; high market growth)
Orange:	Question mark (low market share; high market growth)
Pear:	Cash cow (high market share; low market growth)
Banana:	Dog (low market share; low market growth)

The pears are cash cows and likely to be profitable, whereas the bananas are dogs and probably unprofitable. The situation with apples and oranges is more complex. The apples are stars and it depends on the current capital expenditure. If low they would be profitable, but if high they might not be. The oranges are question marks and are probably unprofitable at the moment. They leave the company with the difficult question of whether to invest more resources or whether to pull out of the market.

Chapter 21: Discussion *Answers*

The answers provide some outline points for discussion.

A1 (a) Capital investment is necessary for the development of a company's infrastructure. Companies also expand or reorientate their strategic direction. Therefore, they need to invest in infrastructure assets, such as factories or plant and machinery. Essentially, in a competitive business world companies will either grow or stagnate. If they stagnate they will be likely to be taken over.

(b) (i) Ships or equipment for making ships such as dry docks, cranes, heavy equipment etc.
(ii) New hotels or refurbishment of old ones.
(iii) New factories, plant and machinery, computer technology

A5 True or False?

(a) *True.*
(b) *False.* Net present value and the internal rate of return use discounted cash flows. The payback period generally does not use discounted cash flows. It is, however, possible to incorporate them into payback models .
(c) *True.*
(d) *False.* The discount rate normally used for discounting cash flows is the company's weighted average cost of capital.
(e) *True.*

Chapter 21: Numerical *Answers*

A1 (i) 0.6209 (iv) 0.6355
(ii) 0.5983 (v) 0.6750
(iii) 0.3762 (vi) 0.1615

A2 Fairground

	Rocket	*Carousel*	*Dipper*
(i) **Payback**			
	£16,000 + (£2,000/£8,000) = 2.25 years	£10,000 + (£8,000/£14,000) = 2.57 years	£16,000 + (£2,000/£5,000) = 2.4 years
(ii) **Accounting rate of return**			
Annual average profit/ Initial investment	(£24,000 ÷ 3)/£18,000 = 44.4%	(£24,000 ÷ 3)/£18,000 = 44.4%	(£21,000 ÷ 3)/£18,000 = 38.9%

A2 Fairground (continued)

(iii) **Net present value**

Year	Rocket £	Carousel £	Dipper £	Discount Rate 8%	Rocket £	Carousel £	Dipper £
0	(18,000)	(18,000)	(18,000)	1	(18,000)	(18,000)	(18,000)
1	8,000	6,000	10,000	0.9259	7,407	5,555	9,259
2	8,000	4,000	6,000	0.8573	6,858	3,429	5,144
3	8,000	14,000	5,000	0.7938	6,350	11,113	3,969
Net Present Value (NPV)					2,615	2,097	372

So on (i)–(iii) we would choose the Rocket on all criteria as our project.

(iv) **Internal rate of return (IRR)**: choose 18% to achieve a negative NPV

Net present value Year	*Rocket* £	*Carousel* £	*Dipper* £	*Discount Rate 18%*	*Rocket* £	*Carousel* £	*Dipper* £
0	(18,000)	(18,000)	(18,000)	1	(18,000)	(18,000)	(18,000)
1	8,000	6,000	10,000	0.8475	6,780	5,085	8,475
2	8,000	4,000	6,000	0.7182	5,746	2,873	4,309
3	8,000	14,000	5,000	0.6086	4,869	8,520	3,043
Net Present Value (NPV)					(605)	(1,522)	(2,173)

Therefore, calculate IRR using formula:

$$\text{IRR} = \text{Lowest discount rate} + \text{difference in discount rates} \times \frac{\text{lowest discount rate NPV}}{\text{difference in NPVs}}$$

$$\text{Rocket} = 8\% + \left(10\% \times \frac{£2{,}615}{£2{,}615 + £605}\right) = 16.1\%$$

$$\text{Carousel} = 8\% + \left(10\% \times \frac{£2{,}097}{£2{,}097 + £1{,}522}\right) = 13.8\%$$

$$\text{Dipper} = 8\% + \left(10\% \times \frac{£372}{£372 + £2{,}173}\right) = 9.5\%$$

Therefore, as our cost of capital (8%) is less than the IRR, we could potentially undertake all the projects. We would choose to invest in Rocket because it has the highest IRR. Rocket is the preferred project under all four methods.

A3 Wetday

	Storm	*Cloud*	*Downpour*
(i) **Payback**	£15,000 + (£3,000/£7,000) = 3.43 years	£10,000 + (£2,000/£2,500) = 3.8 years	£12,000 + (£1,000/£4,000) = 3.25 years

(ii) **Accounting rate of return**

Average annual profit/ Initial investment	£6,000/£18,000 = 33.3%	£3,100/£12,000 = 25.8%	£4,000/£13,000 = 30.8%

(iii) **Net present value**

Year	*Storm* £	*Cloud* £	*Downpour* £	*Discount Rate* 12%	*Storm* £	*Cloud* £	*Downpour* £
0	(18,000)	(12,000)	(13,000)	1	(18,000)	(12,000)	(13,000)
1	4,000	5,000	4,000	0.8929	3,572	4,464	3,572
2	5,000	2,000	4,000	0.7972	3,986	1,594	3,189
3	6,000	3,000	4,000	0.7118	4,271	2,135	2,847
4	7,000	2,500	4,000	0.6355	4,449	1,589	2,542
5	8,000	3,000	4,000	0.5674	4,539	1,702	2,270
Net Present Value (NPV)					2,817	(516)	1,420

Therefore, depending upon the criteria, we will choose a different project. Downpour has the quickest payback; Storm the highest accounting rate of return and NPV. Probably, therefore, we will choose Storm. We will definitely not choose Cloud, because of its negative NPV.

(iv) **Internal rate of return (IRR)**. Choose 20% for Storm and Downpour to get a negative NPV. However, choose 8% for Cloud to get a positive NPV.

		Storm £	*Cloud* £	*Downpour* £	*Discount Rate* 20% £	*Discount Rate* 8% £	*Storm (20%)* £	*Cloud (8%)* £	*Downpour (20%)* £
Year	0	(18,000)	(12,000)	(13,000)	1	1	(18,000)	(12,000)	(13,000)
	1	4,000	5,000	4,000	0.8333	0.9259	3,333	4,629	3,333
	2	5,000	2,000	4,000	0.6944	0.8573	3,472	1,715	2,778
	3	6,000	3,000	4,000	0.5787	0.7938	3,472	2,381	2,315
	4	7,000	2,500	4,000	0.4823	0.7350	3,376	1,837	1,929
	5	8,000	3,000	4,000	0.4019	0.6806	3,215	2,042	1,608
Net Present Value (NPV)							(1,132)	604	(1,037)

Calculate IRR, using formula

$$\text{IRR} = \text{Lowest discount rate} + \text{difference in discount rates} \times \frac{\text{lowest discount rate NPV}}{\text{difference in NPVs}}$$

$$\text{Storm} = 12\% + \left(8\% \times \frac{£2,817}{£2,817 + £1,132}\right) = 17.7\%$$

A3 Wetday (continued)

$$\text{Cloud} = 8\% + \left(4\% \times \frac{£604}{£604 + £516}\right) = 10.2\%$$

$$\text{Downpour} = 12\% + \left(8\% \times \frac{£1{,}420}{£1{,}420 + £1{,}037}\right) = 16.6\%$$

As our cost of capital is 12%, we could potentially undertake Storm or Downpour. Storm has the highest IRR and we would choose this project if funds were limited. As Storm has the highest accounting rate of return, NPV and IRR, this is our preferred project.

Chapter 22: Discussion *Answers*

The answers provide some outline points for discussion.

A1 Firms are in some ways like living things. They need energy to survive and grow. In the case of living things, the energy is provided by sunlight. For firms, the energy is supplied by sources of finance. These may be short-term, like a bank overdraft, or long-term, like a long-term loan. These sources of finance enable a firm to buy stock, carry out day-to-day operations and expand by the purchase of new fixed assets. In essence, short-term finance should be used to sustain the firm's working capital, while long-term finance should fund the company's infrastructure. Some of the major sources of finance and the activities financed are outlined below:

Source of finance	*Activities financed*
Bank overdraft	Working capital, day-to-day operations
Debt factoring	Debtors
Invoice discounting	Debtors
Sale and buy back of stock	Stock
Leasing	Leased assets
Retained profits	Infrastructure assets, e.g., fixed assets
Share capital	Infrastructure assets, e.g., fixed assets
Long-term loan	Infrastructure assets, e.g., fixed assets

A5 **True or False?**

(a) *False.* They are normally used to finance long-term infrastructure assets.
(b) *True.*
(c) *True.*
(d) *False.* Retained profits is an internal, not an external, source of long-term finance.
(e) *True.*

Chapter 22: Numerical *Answers*

A1 Lathe

(i) (a) **Graphical Solution**

Order Quantity Q	*Number of Orders per annum*	*Order Cost* £	*Total Order Cost* £	*Average Quantity in Stock* £	*Carrying Cost* £	*Total Carrying Cost* £	*Total Cost* £
250	80	25	2,000	125	0.90	112.50	2,112.50
500	40	25	1,000	250	0.90	225.00	1,225.00
1,000	20	25	500	500	0.90	450.00	950.00
1,500	13.33	25	333	750	0.90	675.00	1,008.00
2,000	10	25	250	1,000	0.90	900.00	1,150.00

Graphical solution: Optimal order quantity about 1,000

A1 Lathe (continued)

(i) (b) Algebraic solution:

Economic order quantity formula

$$Q = \sqrt{\frac{2AC}{i}}$$

where:

Q = Economic order quantity
A = Average annual usage
C = Cost of each order being placed
i = Carrying cost per unit per annum

$$Q = \sqrt{\frac{2 \times 20{,}000 \times 25}{0.90}} = 1{,}054 \text{ Tweaks}$$

(ii) Total costs per annum

		£
Costs of purchase	£1.50 × 1,054	1,581
Order costs	$\frac{20{,}000}{1{,}054} \times £25$	474
Holding costs	$\frac{1{,}054}{2} \times 0.90$	474
Total costs		2,529

A4 Albatross

Source of Finance	*Current Market Value*		*Present Cost of Capital*	*Weighted Average Cost of Capital*
	£million	%	%	%
Ordinary shares	4	50.0	12	6.00
Preference shares	1	12.5	10	1.25
Long-term loan	3	37.5	8	3.00
	8	100.0		10.25

Index

Notes

Notes

Notes

Notes

Notes

Notes

Notes